AF598281

PROGRESS IN CLINICAL AND BIOLOGICAL RESEARCH

RECENT TITLES

Vol 248: **Advances in Cancer Control: The War on Cancer—15 Years of Progress,** Paul F. Engstrom, Lee E. Mortenson, Paul N. Anderson, *Editors*

Vol 249: **Mechanisms of Signal Transduction by Hormones and Growth Factors,** Myles C. Cabot, Wallace L. McKeehan, *Editors*

Vol 250: **Kawasaki Disease,** Stanford T. Shulman, *Editor*

Vol 251: **Developmental Control of Globin Gene Expression,** George Stamatoyannopoulos, Arthur W. Nienhuis, *Editors*

Vol 252: **Cellular Calcium and Phosphate Transport in Health and Disease,** Felix Bronner, Meinrad Peterlik, *Editors*

Vol 253: **Model Systems in Neurotoxicology: Alternative Approaches to Animal Testing,** Abraham Shahar, Alan M. Goldberg, *Editors*

Vol 254: **Genetics and Epithelial Cell Dysfunction in Cystic Fibrosis,** John R. Riordan, Manuel Buchwald, *Editors*

Vol 255: **Recent Aspects of Diagnosis and Treatment of Lipoprotein Disorders: Impact on Prevention of Atherosclerotic Diseases,** Kurt Widhalm, Herbert K. Naito, *Editors*

Vol 256: **Advances in Pigment Cell Research,** Joseph T. Bagnara, *Editor*

Vol 257: **Electromagnetic Fields and Neurobehavioral Function,** Mary Ellen O'Connor, Richard H. Lovely, *Editors*

Vol 258: **Membrane Biophysics III: Biological Transport,** Mumtaz A. Dinno, William McD. Armstrong, *Editors*

Vol 259: **Nutrition, Growth, and Cancer,** George P. Tryfiates, Kedar N. Prasad, *Editors*

Vol 260: **EORTC Genitourinary Group Monograph 4: Management of Advanced Cancer of Prostate and Bladder,** Philip H. Smith, Michele Pavone-Macaluso, *Editors*

Vol 261: **Nicotine Replacement: A Critical Evaluation,** Ovide F. Pomerleau, Cynthia S. Pomerleau, *Editors*

Vol 262: **Hormones, Cell Biology, and Cancer: Perspectives and Potentials,** W. David Hankins, David Puett, *Editors*

Vol 263: **Mechanisms in Asthma: Pharmacology, Physiology, and Management,** Carol L. Armour, Judith L. Black, *Editors*

Vol 264: **Perspectives in Shock Research,** Robert F. Bond, *Editor*

Vol 265: **Pathogenesis and New Approaches to the Study of Noninsulin-Dependent Diabetes Mellitus,** Albert Y. Chang, Arthur R. Diani, *Editors*

Vol 266: **Growth Factors and Other Aspects of Wound Healing: Biological and Clinical Implications,** Adrian Barbul, Eli Pines, Michael Caldwell, Thomas K. Hunt, *Editors*

Vol 267: **Meiotic Inhibition: Molecular Control of Meiosis,** Florence P. Haseltine, Neal L. First, *Editors*

Vol 268: **The Na^+,K^+-Pump,** Jens C. Skou, Jens G. Nørby, Arvid B. Maunsbach, Mikael Esmann, *Editors*. Published in two volumes: Part A: *Molecular Aspects*. Part B: *Cellular Aspects*.

Vol 269: **EORTC Genitourinary Group Monograph 5: Progress and Controversies in Oncological Urology II,** Fritz H. Schröder, Jan G.M. Klijn, Karl H. Kurth, Herbert M. Pinedo, Ted A.W. Splinter, Herman J. de Voogt, *Editors*

Vol 270: **Cell-Free Analysis of Membrane Traffic,** D. James Morré, Kathryn E. Howell, Geoffrey M.W. Cook, W. Howard Evans, *Editors*

Vol 271: **Advances in Neuroblastoma Research 2,** Audrey E. Evans, Giulio J. D'Angio, Alfred G. Knudson, Robert C. Seeger, *Editors*

Vol 272: **Bacterial Endotoxins: Pathophysiological Effects, Clinical Significance, and Pharmacological Control,** Jack Levin, Harry R. Büller, Jan W. ten Cate, Sander J.H. van Deventer, Augueste Sturk, *Editors*

Vol 273: **The Ion Pumps: Structure, Function, and Regulation,** Wilfred D. Stein, *Editor*

Vol 274: **Oxidases and Related Redox Systems,** Tsoo E. King, Howard S. Mason, Martin Morrison, *Editors*

Vol 275: **Electrophysiology of the Sinoatrial and Atrioventricular Nodes,** Todor N. Mazgalev, Leonard S. Dreifus, Eric L. Michelson, *Editors*

Vol 276: **Prediction of Response to Cancer Therapy,** Thomas C. Hall, *Editor*

Vol 277: **Advances in Urologic Oncology,** Nasser Javadpour, Gerald P. Murphy, *Editors*

Vol 278: **Advances in Cancer Control: Cancer Control Research and the Emergence of the Oncology Product Line,** Paul F. Engstrom, Paul N. Anderson, Lee E. Mortenson, *Editors*

Vol 279: **Basic and Clinical Perspectives of Colorectal Polyps and Cancer,** Glenn Steele, Jr., Randall W. Burt, Sidney J. Winawer, James P. Karr, *Editors*

Vol 280: **Plant Flavonoids in Biology and Medicine II: Biochemical, Cellular, and Medicinal Properties,** Vivian Cody, Elliott Middleton, Jr., Jeffrey B. Harborne, Alain Beretz, *Editors*

Vol 281: **Transplacental Effects on Fetal Health,** Dante G. Scarpelli, George Migaki, *Editors*

Vol 282: **Biological Membranes: Aberrations in Membrane Structure and Function,** Manfred L. Karnovsky, Alexander Leaf, Liana C. Bolis, *Editors*

Vol 283: **Platelet Membrane Receptors: Molecular Biology, Immunology, Biochemistry, and Pathology,** G.A. Jamieson, *Editor*

Vol 284: **Cellular Factors in Development and Differentiation: Embryos, Teratocarcinomas, and Differentiated Tissues,** Stephen E. Harris, Per-Erik Mansson, *Editors*

Vol 285: **Non-Radiometric Assays: Technology and Application in Polypeptide and Steroid Hormone Detection,** Barry D. Albertson, Florence P. Haseltine, *Editors*

Vol 286: **Molecular and Cellular Mechanisms of Septic Shock,** Bryan L. Roth, Thor B. Nielsen, Adam E. McKee, *Editors*

Vol 287: **Dietary Restriction and Aging,** David L. Snyder, *Editor*

Vol 288: **Immunity to Cancer. II,** Malcolm S. Mitchell, *Editor*

Vol 289: **Computer-Assisted Modeling of Receptor–Ligand Interactions: Theoretical Aspects and Applications to Drug Design,** Robert Rein, Amram Golombek, *Editors*

Vol 290: **Enzymology and Molecular Biology of Carbonyl Metabolism 2: Aldehyde Dehydrogenase, Alcohol Dehydrogenase, and Aldo-Keto Reductase,** Henry Weiner, T. Geoffrey Flynn, *Editors*

Vol 291: **QSAR: Quantitative Structure-Activity Relationships in Drug Design,** J.L. Fauchère, *Editor*

Vol 292: **Biological and Synthetic Membranes,** D. Allan Butterfield, *Editor*

Vol 293: **Advances in Cancer Control: Innovations and Research,** Paul N.Anderson, Paul F. Engstrom, Lee E. Mortenson, *Editors*

Vol 294: **Development of Preimplantation Embryos and Their Environment,** Koji Yoshinaga, Takahide Mori, *Editors*

Vol 295: **Cells and Tissues: A Three-Dimensional Approach by Modern Techniques in Microscopy,** Pietro M. Motta, *Editor*

Vol 296: **Developments in Ultrastructure of Reproduction,** Pietro M. Motta, *Editor*

Vol 297: **Biochemistry of the Acute Allergic Reactions: Fifth International Symposium,** Bruce U. Wintroub, Alfred I. Tauber, Arlene Stolper Simon, *Editors*

Vol 298: **Skin Carcinogenesis: Mechanisms and Human Relevance,** Thomas J. Slaga, Andre J.P. Klein-Szanto, R.K. Boutwell, Donald E. Stevenson, Hugh L. Spitzer, Bob D'Motto, *Editors*

Vol 299: **Perspectives in Shock Research: Metabolism, Immunology, Mediators, and Models,** John C. Passmore, *Editor*, Sherwood M. Reichard, David G. Reynolds, Daniel L. Traber, *Co-Editors*

Please contact the publisher for information about previous titles in this series.

CELLS AND TISSUES: A Three-Dimensional Approach by Modern Techniques in Microscopy

Portrait of the young M. Malpighi (1628–1694) in the Collection of the Royal Society of London, to which it was sent in 1680.

CELLS AND TISSUES: A Three-Dimensional Approach by Modern Techniques in Microscopy

A Celebrative Symposium: The "Opera Omnia" of Marcello Malpighi
Proceedings of the VIIIth International Symposium on Morphological Sciences, Held in Rome, Italy, July 10–15, 1988

Editor

Pietro M. Motta, MD, PhD
Chairman and Professor
Department of Anatomy
Faculty of Medicine
University "La Sapienza," Rome, Italy

ALAN R. LISS, INC. • NEW YORK

Address all Inquiries to the Publisher
Alan R. Liss, Inc., 41 East 11th Street, New York, NY 10003

Printed in the United States of America

Library of Congress Cataloging-in-Publication Data

International Symposium on Morphological Sciences (8th : 1988 : Rome, Italy)
Cells and tissues : a three-dimensional approach by modern techniques in microscopy : a celebrative symposium—the Opera omnia of Marcello Malpighi : proceedings of the VIIIth International Symposium on Morphological Sciences, Rome, Italy 10–15 July, 1988 / editor Pietro M. Motta.
p. cm.—(Progress in clinical and biological research : v. 295)
Includes index.
ISBN 0-8451-5145-2
1. Cytology—Congresses. 2. Histology—Congresses. 3. Electron microscopy—Congresses. I. Motta, Pietro M. II. Malpighi, Marcello, 1628–1694. III. Title. IV. Series.
[DNLM: 1. Cells—ultrastructure—congresses. 2. Microscopy, Electron, Scanning—congresses. W1 PR668E v. 295 / QH 212.S3 I597c 1988]
QH573.I54 1988
574.87—dc19
DNLM/DLC 89-2324
for Library of Congress CIP

Contents

Contributors xiii

Symposium Board xxi

Preface
Pietro M. Motta xxiii

Acknowledgments xxvii

HISTORY

Introduction. Marcello Malpighi and the Foundation of Microscopic Anatomy
Pietro M. Motta 3

Brief Survey of Malpighi's Life
Liberato J.A. DiDio 7

THE CELL

The Cytoplasm and Its Matrix
Keith R. Porter 15

Ultra-High Resolution Scanning Electron Microscopy of Biological Materials
Keiichi Tanaka, Akira Mitsushima, Yuzuru Kashima, Takashi Nakadera, and
Hatsuko Fukudome 21

Freeze-Fracture of 3T3 and HeLa Cells for High-Resolution SEM and Deep-Etch Replicas: Structure of the Interphase Nucleus
G.H. Haggis 31

A New Approach for the TEM Visualization of the 3-D Organization of Filamentous Structures in Cells and Tissues
José F. David-Ferreira and António J. Cidadão 39

Fracture-Flip: Nanoanatomy and Topochemistry of Cell Surfaces
Pedro Pinto da Silva, Catarina Anderson Forsman, and Kazushi Fujimoto 49

Chromosomal Structure Observed by Optical, Color Laser and Electron Microscopes
Akihiro Iino, Tomonori Naguro, Kenji Funaki, and Sumire Inaga 57

Mitochondrial Structure Revealed by Scanning Electron Microscopy (SEM)
Peter J. Lea and Martin J. Hollenberg 63

Development of the Quick-Freeze, Deep-Etch, Rotary-Replication Technique of Sample Preparation for 3-D Electron Microscopy
John E. Heuser 71

Golgi Apparatus Subcompartments as Visualized by Means of Lectincytochemistry
Margit Pavelka and Adolf Ellinger 85

TISSUES AND ORGANS — CONNECTIVE AND MUSCLE TISSUES

Freeze-Etching as a 3D Approach to the Collagen Fibril Structure
Alessandro Ruggeri, Maurizio Marchini, Vittoria Ottani, and Mario Raspanti 95

Extracellular Matrix: Functional Significance of Oxytalan, Elaunin and Elastic Fibers
G. Cotta-Pereira and M.L. Iruela-Arispe 101

Electron Microscope Studies of the Early Stage of the Calcification Process: Role of Matrix Vesicles
Ermanno Bonucci 109

Quantitative Assessment of Proteoglycans by Proton Microprobe
T. Cichocki, S. Divoux, B. Gonsior, L. Jarczyk, E. Rokita, A. Strzalkowski, and M. Sych 115

Age- and Radiation-Related Alterations of Bone Tissue: A Comparative Study
Antoine Dhem, Catherine Nyssen-Behets, and René Dambrain 121

Scanning and Transmission Electron Microscopy on the Synovial Membrane in the Rabbit
Wei Jun Chen, Gui Lan He, and Chang Jiang Sun 127

Isometric Contraction, Isotonic Contraction and Passive Contraction in Smooth Muscles
Giorgio Gabella 133

Techniques for Determining Morphological Characteristics of Smooth Muscle Cells Using Computer Assisted Reconstructions
Mary E. Todd, F.S.F. Chu, and B. Gowen 139

RESPIRATORY AND URINARY SYSTEMS

Design of Microvasculature and Gas Exchange
Ewald R. Weibel 147

Shape Changes in Kidney Glomerular Podocytes: Mechanisms and Possible Functional Significance
Peter M. Andrews 157

The Fine Structure of the *Laminae rarae* of the Glomerular Basement Membrane in the Rat
E. Reale, L. Luciano, and K. Kühn 167

Human Developing Metanephric Nephrons. Morphological, Immunohistochemical and Histochemical Studies
Cesare De Martino, Pier Giorgio Natali, Paola Nistico', and Lidia Accinni 173

Three-Dimensional Vascular Architecture of the Malpighi's Glomerular Capillary Beds as Studied by Vascular Corrosion Casting–SEM Method
Akio Kikuta and Takuro Murakami 181

Three Dimensional Organization of the Collecting Tubule of the Rabbit Kidney
Andrew P. Evan, Bret A. Connors, and James A. McAteer 189

The Three-Dimensional Structure of Renal Tubule Cells
Hiromi Takahashi-Iwanaga 203

Analysis of Structural and Functional Properties of the Urinary Bladder: The Impact of the SEM and Ancillary Approaches
Gisèle M. Hodges and Charles Rowlatt 213

GLANDS AND DEVELOPMENT

Three Dimensional Aspect on the Functional Morphology of the Thyroid Gland
Hisao Fujita and Masato Imada 227

Surface Microanatomy of Human Major Salivary Glands
A. Riva, L. Valentino, M.S. Lantini, E. Cotti, and F. Testa Riva 235

Comparative Studies of the Striated Ducts of Mammalian Salivary Glands
Bernard Tandler, Carleton J. Phillips, Kuniaki Toyoshima, and Toshikazu Nagato 243

The Golgi Apparatus of the Pancreatic Acinar Cells Observed by Scanning Electron Microscopy
Tomonori Naguro and Akihiro Iino 249

Three-Dimensional Structure of the Sinusoidal Wall in the Liver: A Golgi Study
Kenjiro Wake 257

The Morphology of Hepatic Glycogen Synthesis
Robert R. Cardell and Emma Lou Cardell 263

Morphological Aspects of Cholesterol Storage in the Human Gallbladder
Liliana Luciano 269

Human Fetal Liver Cells as Seen by Scanning Electron Microscopy
Sayoko Makabe, Guido Macchiarelli, and Pietro M. Motta 277

Development of Adipose Tissue in the Human Fetus
S. Labbe, J.P. Barbet, H. Copin, M. Maillet, and B. Schramm 281

Ultrastructural Evidence of Receptor-Mediated Endocytosis During Early Mouse Limb Morphogenesis: A Three Dimensional Analysis Using Deep Etching and Scanning Electron Microscopy
Robert O. Kelley 287

Three-Dimensional Localization of Contractile Proteins in Cultured Cardiac Myocytes by Immunogold Staining and Deep-Etching Replica Electron Microscopy
Yuji Isobe, Guan R. Hou, Dino A. Messina, and Larry F. Lemanski 295

DIGESTIVE TRACT

Comparative Studies on the Stereo Architecture of the Connective Tissue Papillae in Some Mammalian Tongues
Kan Kobayashi and Shin-ichi Iwasaki 303

Epithelial Cell Proliferation and Differentiation in the Gastric Mucosa: Comparisons Between Histogenetic and Cell Renewal Processes
Katsuko Kataoka, Akiko Kantani-Matsumoto, and Yasuko Takeoka 309

Corrosion Casts in Liver and Stomach Microcirculation
Osamu Ohtani 317

Gastric Mucosae as Studied by Scanning Electron Microscopy in 3,242 Human Biopsies
B. Foliguet, J.C. Guedenet, F. Vicari, L. Marchal, G. Jeanvoine, and G. Grignon 327

Structural and Ultrastructural Observations on Rat Gastric Mucosa Treated With Cytoprotective Agents Prior to Ethanol Exposure
F. Carpino, G. Prino, E. Gaudio, V. Petrozza, M. Mantovani, D. Bosco, P. Alberico, and M. Melis 333

Three-Dimensional Visualization of the Ganglionated Enteric Nerve Plexuses in the Small Intestine of the Pig
Dietrich W. Scheuermann, Werner Stach, and Jean-Pierre Timmermans 343

NEUROENDOCRINE, CARDIOCIRCULATORY AND IMMUNE SYSTEMS

Pinealocytes – Neuroendocrine Units Related to the Vascular and Cerebrospinal Fluid Compartments of the Brain
A. Oksche 351

Mammalian Tanycytes: Transmission and Scanning Electron Microscopy
J. Flament-Durand **363**

A Technique for Studying the Perikaryal Projections of Spinal Ganglion Neurons by Scanning Electron Microscopy
E. Pannese, M. Ledda, V. Conte, E. Marini, and S. Matsuda **371**

3-D Changes in Neuroblastoma × Glioma Hybrid (NG108-15) Cell Differentiation as Studied by SEM and TEM
A. Iavarone, M.L. Eboli, M. Osti, A. Redler, M. Pocchiari, and M.A. Russo **377**

Perivascular Glia–Endothelium Relationships in the Developing Cerebral Vessels. A Three-Dimensional Computer Aided Study
Luisa Roncali, Mirella Bertossi, Domenico Ribatti, Beatrice Nico, Daniela Virgintino, and Lucia Mancini **383**

SEM of Subdural Space in Mammals
Eduard Klika and Milan Richter **389**

Ultrastructural Morphology of the Hypophyseal Cleft in Some Mammals
Silvia Correr, Serena Petrillo, Marco Laureti, and Fabrizio Barberini **395**

The Endocrine Heart
Wolf-Georg Forssmann **401**

Surface Morphology of Myocardium, Purkinje Fibers, and Transitional Cells
Tatsuo Shimada and Tsuyoshi Noguchi **419**

Morphometry of Corrosion Casts
A. Lametschwandtner, T. Weiger, and G. Bernroider **427**

Microvascular Aspects of Myocardial Lymphatic Vessels in the Dog by Scanning Electron Microscopy
J.A. Esperança-Pina **435**

Corrosion Casts in the Microcirculation of Skeletal Muscle
E. Gaudio, L. Pannarale, and G. Marinozzi **443**

Morphological Aspects of Nasal Blood Vessels
Gerhard Grevers, Ulrich Heinzmann, and Joana Grevers-Colorian **451**

Statistical, Morphometric and Stereological Basics in the Morphological Sciences and Their Possible Applications in Corrosion Cast Studies
Nigel T. James **457**

Scanning Electron Microscopy of Microvessels and Perivascular Cells in Different Organs After KOH Digestion
Andrea Maggioni, Alberto Caggiati, and Guido Macchiarelli **469**

Scanning and Transmission Electron Microscopic Studies on the Vascular System of Xenotransplanted Human Tumors on Nude Mice
Moritz A. Konerding and Fritz Steinberg **475**

Cadmium Toxicity on the Aortic Media of Pregnant Rats: Transmission and Scanning Electron Microscopic Study
Mitsuaki Yoshizuka, Takeshi Maruyama, Naoki Mori, Hiroshi Ueda, and Sunao Fujimoto **481**

Morphological Characters of the Absorbing Peripheral Lymphatic Vessel by TEM, SEM and Three-Dimensional Models
Giacomo Azzali, Guido Orlandini, and Giovanna Bucci **487**

SEM of Immunohematopoietic Tissues
Tsuneo Fujita **493**

EYE AND EAR

Scanning Electron Microscopy in Ophthalmology
P. Versura . . . 503

Sclerococorneal Trabecula and Glaucoma. A Scanning Electron Microscopic Study
Domingo Ruano-Gil and Jesus Costa-Vila . . . 511

Scanning Electron Microscopic Study of the Microcirculation of the Rabbit's Anterior Uvea
J. Goyri O'Neill and J. Esperança-Pina . . . 515

SEM-Studies of the Nictitating Membrane of the Pigeon
Wolfgang Kühnel and Uda Schramm . . . 521

Application of the Corrosion Casts Method to the Anuran Amphibian Eye and Mammalian Cochlea
Adam J. Miodoński, Jan Kuś, and Thomas Bar . . . 527

Morphological Correlations of Hearing in the Phylogenetic Scale: *Mauremys caspica*
J. Morales, V. García-Martínez, D. Sánchez-Quintana, and A. Ambel . . . 537

The Endolymphatic Sac. A Scanning Electron Microscopic Study in Different Animal Species
Maurizio Barbara, Masaya Takumida, and Roberto Filipo . . . 543

PRESENT AND FUTURE OF MICROSCOPY

Tandem Scanning Microscope - A New Tool for Three-Dimensional Microanatomy
Mojmír Petráň and Milan Hadravský . . . 551

Application of Optical Diffractometry in the Analysis of Cytological and Histological Images
Kazimierz Ostrowski . . . 559

Osmium Impregnation and Micromanipulation. Their Association in Studies Using Secondary or Backscattered Electrons
Gebhard Reiss and Enrico Reale . . . 563

Microdissection by Ultrasonication for Scanning Electron Microscopy
Frank N. Low . . . 571

Micromorphology by Cryo-HVEM
Mircea Fotino . . . 581

Review of Scanning Tunneling Microscopy — New Biological Frontier?
Branislav Vidić . . . 589

Scanning Electron Microscopy in Biomedicine
Karlheinz A. Rosenbauer . . . 597

Scanning Electron Microscopy in Clinics
R. Laschi, G. Pasquinelli, P. Versura, and F. Bonvicini . . . 605

Three Dimensional Microarchitecture of Organs Reconstructed on the Basis of Modern Histological Observation Methods
R.V. Krstić . . . 623

Colored Scanning Electron Microscopic Pictures in Teaching Anatomy
Pietro M. Motta and Stefania A. Nottola . . . 629

Index . . . 635

Contributors

Lidia Accinni, Department of Electron Microscopy, Institute of Experimental Medicine-C.N.R., 00156 Rome, Italy **[173]**

P. Alberico, Crinos Research Department, Como, Italy **[333]**

A. Ambel, Departamentos de Otorrinolaringologia, Universidad de Extremadura, 06071 Badajoz, Spain **[537]**

Catarina Anderson Forsman, Membrane Biology Section, Laboratory of Mathematical Biology, National Cancer Institute, Frederick Cancer Research Facility, Frederick, MD 21701; present address: Department of Anatomy, Karolinska Institutet, S-10401 Stockholm, Sweden **[49]**

Peter M. Andrews, Department of Anatomy and Cell Biology, Georgetown University School of Medicine, Washington, D.C. 20007 **[157]**

Giacomo Azzali, Institute of Anatomy, University of Parma, 43100 Parma, Italy **[487]**

Thomas Bar, Max-Planck-Institut für Systemphysiologie, 4600 Dortmund, Federal Republic of Germany **[527]**

Maurizio Barbara, Department of Otolaryngology, University of Rome "La Sapienza," 00185 Rome, Italy **[543]**

Fabrizio Barberini, Department of Anatomy, Faculty of Medicine, University of Rome "La Sapienza," 00161 Rome, Italy **[395]**

J.P. Barbet, Department of Pathology, Hôpital Saint Vincent de Paul, 75014 Paris, France **[281]**

G. Bernroider, University of Salzburg, Institute of Zoology, Department of Experimental Zoology, A-5020 Salzburg, Austria **[427]**

Mirella Bertossi, Institute of Human Anatomy, Histology and General Embryology, University of Bari Medical School, 70124 Bari, Italy **[383]**

Ermanno Bonucci, Department of Human Biopathology, Section of Pathological Anatomy, University of Rome "La Sapienza," 00161 Rome, Italy **[109]**

F. Bonvicini, Institute of Clinical Electron Microscopy, University of Bologna, 40138 Bologna, Italy **[605]**

D. Bosco, Department of Human Biopathology, University of Rome "La Sapienza," 00161 Rome, Italy **[333]**

Giovanna Bucci, Institute of Anatomy, University of Parma, 43100 Parma, Italy **[487]**

Alberto Caggiati, Department of Anatomy, Faculty of Medicine, University of Rome "La Sapienza," 00161 Rome, Italy **[469]**

Emma Lou Cardell, Department of Anatomy and Cell Biology, University of Cincinnati College of Medicine, Cincinnati, OH 45267 **[263]**

Robert R. Cardell, Department of Anatomy and Cell Biology, University of Cincinnati College of Medicine, Cincinnati, OH 45267 **[263]**

F. Carpino, Department of Human Biopathology, University of Rome "La Sapienza," 00164 Rome, Italy **[333]**

The number in brackets is the opening page number of the contributor's article.

Wei Jun Chen, Department of Orthopedics, Dalian Medical College, Dalian, People's Republic of China **[127]**

F.S.F. Chu, Department of Anatomy, University of British Columbia, Vancouver, British Columbia V6T 1W5, Canada **[139]**

T. Cichocki, Department of Histology, Academy of Medicine, PL-31034 Krakow, Poland **[115]**

António J. Cidadão, Department of Cell Biology, Gulbenkian Institute of Science, Oeiras 2781, Portugal **[39]**

Bret A. Connors, Department of Anatomy, Indiana University School of Medicine, Indianapolis, IN 46223 **[189]**

V. Conte, Institute of Histology, Embryology and Neurocytology, University of Milan, I-20133 Milano, Italy **[371]**

H. Copin, Department of Histology-Embryology, and Biology of Reproduction, Hôpital Fernand Widal, 75010 Paris, France **[281]**

Silvia Correr, Department of Anatomy, Faculty of Medicine, University of Rome "La Sapienza," 00161 Rome, Italy **[395]**

Jesus Costa-Vila, Department of Human Anatomy, Faculty of Medicine, University of Barcelona, 08028 Barcelona, Spain **[511]**

G. Cotta-Pereira, Department of Histology and Embryology, State University of Rio de Janeiro, 20551 Rio de Janeiro, Brasil **[101]**

E. Cotti, Dipartimento di Citomorfologia, Universitá di Cagliari, 09124 Cagliari, Italy **[235]**

René Dambrain, Human Anatomy Research Unit, Université Catholique de Louvain, 1200 Brussels, Belgium **[121]**

José F. David-Ferreira, Department of Cell Biology, Gulbenkian Institute of Science, Oeiras 2781, Portugal **[39]**

Cesare De Martino, Department of Electron Microscopy, Regina Elena Institute for Cancer Research, 00158 Rome, Italy **[173]**

Antoine Dhem, Human Anatomy Research Unit, Université Catholique de Louvain, 1200 Brussels, Belgium **[121]**

Liberato J.A. DiDio, Department of Anatomy, Medical College of Ohio, Toledo, OH 43699 **[7]**

S. Divoux, Institute of Experimentalphysis III, Ruhr- Universität Bochum, D-4630 Bochum, Federal Republic of Germany **[115]**

M.L. Eboli, Istituto di Patologia Generale, Universita Cattolica, 00100 Roma, Italy **[377]**

Adolf Ellinger, Institute of Micromorphology and Electron Microscopy, University of Vienna, A-1090 Vienna, Austria **[85]**

J.A. Esperança-Pina, Department of Anatomy, Faculty of Medical Sciences, New University of Lisbon, 1198 Lisbon, Portugal **[435,515]**

Andrew P. Evan, Department of Anatomy, Indiana University, School of Medicine, Indianapolis, IN 46223 **[189]**

Roberto Filipo, Department of Otolaryngology, University of Rome "La Sapienza," 00185 Rome, Italy **[543]**

J. Flament-Durand, Department of Pathology, Hôpital Universitaire Erasme, Université Libre de Bruxelles, 1070 Brussels, Belgium **[363]**

B. Foliguet, Laboratoire de Microscopie Electronique, Faculté de Médecine de Nancy, 54505 Vandoeuvre, Nancy, France **[327]**

Wolf-Georg Forssmann, Department of Anatomy and Cell Biology, University of Heidelberg, D-6900 Heidelberg, Federal Republic of Germany **[401]**

Mircea Fotino, Department of Molecular, Cellular and Developmental Biology, University of Colorado, Boulder, CO 80309 **[581]**

Kazushi Fujimoto, Membrane Biology Section, Laboratory of Mathematical Biology, National Cancer Institute, Frederick Cancer Research Facility, Frederick, MD 21701 **[49]**

Sunao Fujimoto, Department of Anatomy, University of Occupational and Environmental Health, School of Medicine, Kitakyushu 807, Japan **[481]**

Hisao Fujita, Department of Anatomy, Osaka University Medical School, Osaka 530, Japan **[227]**

Tsuneo Fujita, Department of Anatomy, Niigata University School of Medicine, Niigata 951, Japan **[493]**

Hatsuko Fukudome, Department of Anatomy, Tottori University School of Medicine, Yonago 683, Japan **[21]**

Kenji Funaki, Department of Anatomy, Tottori University School of Medicine, Yonago 683, Japan **[57]**

Giorgio Gabella, Department of Anatomy, University College of London, London WC1E 6BT, United Kingdom **[133]**

V. García-Martínez, Departamentos de Ciencias Morfológicas, Universidad de Extremadura, 06071 Badajoz, Spain **[537]**

E. Gaudio, Department of Anatomy, University of Rome "La Sapienza," 00161 Rome, Italy; Department of Anatomy, Universitá degli Studi dell' Aquila, 67100 L'Aquila, Italy **[333,443]**

B. Gonsior, Institute of Experimentalphysis III, Ruhr-Universität Bochum, D-4630 Bochum, Federal Republic of Germany **[115]**

B. Gowen, Department of Anatomy, University of British Columbia, Vancouver, British Columbia V6T 1W5, Canada **[139]**

Gerhard Grevers, Department of Otorhinolaryngology, University of Munich, Klinikum Grosshadern, 8 München 70, Federal Republic of Germany **[451]**

Joana Grevers-Colorian, Department of Otorhinolaryngology, University of Munich, Klinikum Grosshadern, 8 München 70, Federal Republic of Germany **[451]**

G. Grignon, Laboratoire de Microscopie Electronique, Faculté de Médecine de Nancy, 54505 Vandoeuvre, Nancy, France **[327]**

J.C. Guedenet, Laboratoire de Microscopie Electronique, Faculté de Medécine de Nancy, 54505 Vandoeuvre, Nancy, France **[327]**

Milan Hadravský, Institute of Biophysics, Charles University Faculty of Medicine, 301 66 Plzeň, Czechoslovakia **[551]**

G.H. Haggis, Research Branch, Canada Department of Agriculture, Ottawa K1A 0C6, Canada **[31]**

Gui Lan He, Dalian Animal and Plant Quarantine Service, Dalian, People's Republic of China **[127]**

Ulrich Heinzmann, Institute of Pathology, University of Munich, Klinikum Grosshadern, 8000 München 70, Federal Republic of Germany **[451]**

John E. Heuser, Department of Cell Biology and Physiology, Washington University School of Medicine, St. Louis, MO 63110 **[71]**

Gisèle M. Hodges, Tissue Interaction Laboratory, Imperial Cancer Research Fund, London WC2A 3PX, United Kingdom **[213]**

Martin J. Hollenberg, Department of Anatomy, Faculty of Medicine, University of Toronto, Toronto, Ontario M5S 1A8, Canada **[63]**

Guan R. Hou, Department of Anatomy and Cell Biology, SUNY-Health Science Center at Syracuse, Syracuse, NY 13210 **[295]**

A. Iavarone, Istituto di Patologia Generale, Universitá Cattolica, 00100 Roma, Italy **[377]**

Akihiro Iino, Department of Anatomy, Tottori University, School of Medicine, Yonago 683, Japan **[57,249]**

Masato Imada, Department of Anatomy, Osaka University Medical School, Osaka 530, Japan **[227]**

Sumire Inaga, Department of Anatomy, Tottori University School of Medicine, Yonago 683, Japan **[57]**

M.L. Iruela-Arispe, Department of Histology and Embryology, State University of Rio de Janeiro, 20551 Rio de Janeiro, Brasil **[101]**

Yuji Isobe, Department of Anatomy and Cell Biology, SUNY-Health Science Center at Syracuse, Syracuse, NY 13210 **[295]**

Shin-ichi Iwasaki, Department of Anatomy, School of Dentistry at Niigata, The Nippon Dental University, Niigata 951, Japan **[303]**

Nigel T. James, Department of Biomedical Science, University of Sheffield, Sheffield S10 2TN, United Kingdom **[457]**

L. Jarczyk, Department of Nuclear Physics, Institute of Physics, Jagellonian University, PL-30059 Krakow, Poland **[115]**

G. Jeanvoine, Laboratoire de Microscopie Electronique, Faculté de Médecine de Nancy, 54505 Vandoeuvre, Nancy, France **[327]**

Akiko Kantani-Matsumoto, Department of Anatomy, Hiroshima University School of Medicine, Hiroshima 734, Japan **[309]**

Yuzuru Kashima, Department of Anatomy, Tottori University School of Medicine, Yonago 683, Japan **[21]**

Katsuko Kataoka, Department of Anatomy, Hiroshima University School of Medicine, Hiroshima 734, Japan **[309]**

Robert O. Kelley, Department of Anatomy, University of New Mexico, School of Medicine, Albuquerque, NM 87131 **[287]**

Akio Kikuta, Department of Anatomy, Okayama University Medical School, Okayama 700, Japan **[181]**

Eduard Klika, Department of Histology, Faculty of General Medicine, Charles University in Prague, 12800 Prague 2, Czechoslovakia **[389]**

Kan Kobayashi, Department of Anatomy, School of Dentistry at Niigata, The Nippon Dental University, Niigata 951, Japan **[303]**

Moritz A. Konerding, Institute of Anatomy, University Essen, D-4300 Essen, Federal Republic of Germany **[475]**

R.V. Krstić, Institute of Histology and Embryology, University of Lausanne, CH-1005 Lausanne, Switzerland **[623]**

K. Kühn, Division of Nephrology, School of Medicine, 3000 Hannover 61, Federal Republic of Germany **[167]**

Wolfgang Kühnel, Institut für Anatomie, Medizinische Universität zu Lübeck, 2400 Lübeck 1, Federal Republic of Germany **[521]**

Jan Kuś, SEM Laboratory of ENT Department of N. Copernicus Academy of Medicine, Jagellonian University, 31-501 Krakow, Poland **[527]**

S. Labbe, Department of Histology-Embryology, and Biology of Reproduction, Hôpital Fernand Widal, 75010 Paris, France **[281]**

A. Lametschwandtner, University of Salzburg, Institute of Zoology, Department of Experimental Zoology, A-5020 Salzburg, Austria **[427]**

M.S. Lantini, Dipartimento di Citomorfologia, Universitá di Cagliari, 09124 Cagliari, Italy **[235]**

R. Laschi, Institute of Clinical Electron Microscopy, University of Bologna, 40138 Bologna, Italy **[605]**

Marco Laureti, Department of Anatomy, Faculty of Medicine, University of Rome "La Sapienza," 00161 Rome, Italy **[395]**

Peter J. Lea, Department of Anatomy, Faculty of Medicine, University of Toronto, Toronto, Ontario M5S 1A8, Canada **[63]**

M. Ledda, Institute of Histology, Embryology and Neurocytology, University of Milan, I-20133 Milano, Italy **[371]**

Larry F. Lemanski, Department of Anatomy and Cell Biology, SUNY-Health Science Center at Syracuse, Syracuse, NY 13210 **[295]**

Frank N. Low, Department of Anatomy, Louisiana State University Medical Center, New Orleans, LA 70112 **[571]**

Liliana Luciano, Laboratory of Cell Biology and Electron Microscopy, School of Medicine, 3000 Hannover 61, Federal Republic of Germany **[167,269]**

Guido Macchiarelli, Department of Anatomy, University of Rome "La Sapienza," 00161 Rome, Italy **[277,469]**

Andrea Maggioni, Department of Anatomy, Faculty of Medicine, University of Rome "La Sapienza," 00161 Rome, Italy **[469]**

M. Maillet, Department of Histology-Embryology, and Biology of Reproduction, Hôpital Fernand Widal, 75010 Paris, France **[281]**

Sayoko Makabe, Department of Obstetrics and Gynecology, Toho University, Tokyo, Japan **[277]**

Lucia Mancini, Institute of Human Anatomy, Histology and General Embryology, University of Bari Medical School, 70124 Bari, Italy **[383]**

M. Mantovani, Crinos Research Department, Como, Italy **[333]**

L. Marchal, Laboratoire de Microscopie Electronique, Faculté de Médecine de Nancy, 54505 Vandoeuvre, Nancy, France **[327]**

Maurizio Marchini, Istituto di Anatomia Umana Normale, University of Bologna, 40126 Bologna, Italy **[95]**

E. Marini, Institute of Histology, Embryology and Neurocytology, University of Milan, I-20133 Milano, Italy **[371]**

G. Marinozzi, Department of Anatomy, Faculty of Medicine, University of Rome "La Sapienza," 00161 Rome, Italy **[443]**

Takeshi Maruyama, Department of Anatomy, University of Occupational and Environmental Health, School of Medicine, Kitakyushu 807, Japan **[481]**

S. Matsuda, Institute of Histology, Embryology and Neurocytology, University of Milan, I-20133 Milano, Italy **[371]**

James A. McAteer, Department of Anatomy, Indiana University School of Medicine, Indianapolis, IN 46223 **[189]**

M. Melis, Department of Human Biopathology, University of Rome "La Sapienza," 00161 Rome, Italy **[333]**

Dino A. Messina, Department of Anatomy and Cell Biology, SUNY-Health Science Center at Syracuse, Syracuse, NY 13210 **[295]**

Adam J. Miodoński, SEM Laboratory of ENT Department of N. Copernicus Academy of Medicine, Jagellonian University, 31-501 Krakow, Poland **[527]**

Akira Mitsushima, Department of Anatomy, Tottori University School of Medicine, Yonago 683, Japan **[21]**

J. Morales, Departamentos de Otorrinolaringologia, Universidad de Extremadura, 06071 Badajoz, Spain **[537]**

Naoki Mori, Department of Anatomy, University of Occupational and Environmental Health, School of Medicine, Kitakyushu 807, Japan **[481]**

Pietro M. Motta, Department of Anatomy, Faculty of Medicine, University of Rome "La Sapienza," 00161 Rome, Italy **[xxiii,xxvii,3,277,629]**

Takuro Murakami, Department of Anatomy, Okayama University Medical School, Okayama 700, Japan **[181]**

Toshikazu Nagato, Department of Oral Biology, School of Dentistry, Case Western Reserve University, Cleveland, OH 44106 **[243]**

Tomonori Naguro, Department of Anatomy, Tottori University School of Medicine, Yonago 683, Japan **[57,249]**

Takashi Nakadera, Department of Anatomy, Tottori University School of Medicine, Yonago 683, Japan **[21]**

Pier Giorgio Natali, Department of Immunology, Regina Elena Institute for Cancer Research, 00158 Rome, Italy **[173]**

Beatrice Nico, Institute of Human Anatomy, Histology and General Embryology, University of Bari Medical School, 70214 Bari, Italy **[383]**

Paola Nistico', Department of Immunology, Regina Elena Institute for Cancer Research, 00158 Rome, Italy **[173]**

Tsuyoshi Noguchi, Department of Anatomy, Medical College of Oita, 879-56 Oita, Japan **[419]**

Stefania A. Nottola, Department of Anatomy, Faculty of Medicine, University of Rome "La Sapienza," 00161 Rome, Italy **[629]**

Catherine Nyssen-Behets, Human Anatomy Research Unit, Université Catholique de Louvain, 1200 Brussels, Belgium **[121]**

Osamu Ohtani, Department of Anatomy, Okayama University School of Medicine, Okayama 700, Japan **[317]**

A. Oksche, Department of Anatomy and Cytobiology, Justus Liebig University of Giessen, D-6300 Giessen, Federal Republic of Germany **[351]**

J. Goyri O'Neill, Department of Anatomy, Faculty of Medical Sciences, New University of Lisbon, 1198 Lisbon, Portugal **[515]**

Guido Orlandini, Institute of Anatomy, University of Parma, 43100 Parma, Italy **[487]**

M. Osti, Dip. di Medicina Sperimentale, Universita "La Sapienza," 00161 Roma, Italy **[377]**

Kazimierz Ostrowski, Department of Histology, Institute of Biostructure, Medical Academy, 02-004 Warsaw, Poland **[559]**

Vittoria Ottani, Istituto di Anatomia Umana Normale, University of Bologna, 40126 Bologna, Italy **[95]**

L. Pannarale, Department of Anatomy, Faculty of Medicine, University of Rome "La Sapienza," 00161 Rome, Italy **[443]**

E. Pannese, Institute of Histology, Embryology and Neurocytology, University of Milan, I-20133 Milano, Italy **[371]**

G. Pasquinelli, Institute of Clinical Electron Microscopy, University of Bologna, 40138 Bologna, Italy **[605]**

Margit Pavelka, Institute of Micromorphology and Electron Microscopy, University of Vienna, A-1090 Vienna, Austria **[85]**

Mojmír Petráň, Institute of Biophysics, Charles University Faculty of Medicine, 301 66 Plzeň, Czechoslovakia **[551]**

Serena Petrillo, Department of Anatomy, Faculty of Medicine, University of Rome "La Sapienza," 00161 Rome, Italy **[395]**

V. Petrozza, Department of Human Biopathology, University of Rome "La Sapienza," 00161 Rome, Italy **[333]**

Carleton J. Phillips, Department of Biology, Hofstra University, Hempstead, NY 11550 **[243]**

Pedro Pinto da Silva, Membrane Biology Section, Laboratory of Mathematical Biology, National Cancer Institute, Frederick Cancer Research Facility, Frederick, MD 21701 **[49]**

M. Pocchiari, Istituto di Patologia Generale, Universitá Cattolica, 00100 Roma, Italy **[377]**

Keith R. Porter, Department of Biology, University of Pennsylvania, Philadelphia, PA 19104-6018 **[15]**

G. Prino, Crinos Research Department, Como, Italy **[333]**

Mario Raspanti, Istituto di Anatomia Umana Normale, University of Bologna, 40126 Bologna, Italy **[95]**

Enrico Reale, Laboratory of Cell Biology and Electron Microscopy, School of Medicine, D-3000 Hannover 61, Federal Republic of Germany **[167,563]**

A. Redler, Dip. di Medicina Sperimentale, Universitá "La Sapienza" di Roma, 00161 Roma, Italy **[377]**

Gebhard Reiss, Laboratory of Cell Biology and Electron Microscopy, School of Medicine, D-3000 Hannover 61, Federal Republic of Germany **[563]**

Domenico Ribatti, Institute of Human Anatomy, Histology and General Embryology, University of Bari Medical School, 70124 Bari, Italy **[383]**

Milan Richter, Department of Histology, Faculty of General Medicine, Charles University in Prague, 12800 Prague 2, Czechoslovakia **[389]**

A. Riva, Dipartimento di Citomorfologia, Universitá di Cagliari, 09124 Cagliari, Italy **[235]**

E. Rokita, Department of Nuclear Physics, Institute of Physics, Jagellonian University, PL-30059 Krakow, Poland **[115]**

Luisa Roncali, Institute of Human Anatomy, Histology and General Embryology, University of Bari Medical School, 70124 Bari, Italy **[383]**

Karlheinz A. Rosenbauer, Department of Anatomy, University of Düsseldorf, Faculty of Medicine, D-4000 Düsseldorf 1, Federal Republic of Germany **[597]**

Charles Rowlatt, Histopathology Laboratory, Imperial Cancer Research Fund, London WC2A 3PX, United Kingdom **[213]**

Domingo Ruano-Gil, Department of Human Anatomy, Faculty of Medicine, University of Barcelona, 08028 Barcelona, Spain **[511]**

Alessandro Ruggeri, Istituto di Anatomia Umana Normale, University of Bologna, 40126 Bologna, Italy **[95]**

M.A. Russo, Dip. di Medicina Sperimentale, Universitá "La Sapienza" di Roma, 00161 Roma, Italy **[377]**

D. Sánchez-Quintana, Departamento de Ciencias Morfológicas, Universidad de Extremadura, 06071 Badajoz, Spain **[537]**

Dietrich W. Scheuermann, Institute of Histology and Microscopic Anatomy, University of Antwerp, B-2020 Antwerp, Belgium **[343]**

B. Schramm, Department of Pathology, Hôpital Saint Vincent de Paul, 75014 Paris, France **[281]**

Uda Schramm, Institut für Anatomie, Medizinische Universität zu Lübeck, 2400 Lübeck 1, Federal Republic of Germany **[521]**

Tatsuo Shimada, Department of Anatomy, Medical College of Oita, 879-56 Oita, Japan **[419]**

Werner Stach, Institute of Anatomy, Wilhelm-Pieck-Universität, DDR-2500 Rostock, German Democratic Republic **[343]**

Fritz Steinberg, Institute of Med. Radiation Biology, University Essen, D-4300 Essen, Federal Republic of Germany **[475]**

A. Strzalkowski, Department of Nuclear Physics, Institute of Physics, Jagellonian University, PL-30059 Krakow, Poland **[115]**

Chang Jiang Sun, Dalian Animal and Plant Quarantine Service, Dalian, People's Republic of China **[127]**

M. Sych, Department of Histology, Academy of Medicine, PL-31034 Krakow, Poland **[115]**

Hiromi Takahashi-Iwanaga, Department of Anatomy, Niigata University School of Medicine, 951 Niigata, Japan **[203]**

Yasuko Takeoka, Department of Anatomy, Hiroshima University School of Medicine, Hiroshima 734, Japan; present address: Yodogawa Christian Hospital, Osaka 533, Japan **[309]**

Masaya Takumida, Karolinska Hospital, Stockholm, Sweden; present address: Department of Otolaryngology, University School of Medicine, Hiroshima 734, Japan **[543]**

Keiichi Tanaka, Department of Anatomy, Tottori University School of Medicine, Yonago 683, Japan **[21]**

Bernard Tandler, Department of Oral Biology, School of Dentistry, Case Western Reserve University, Cleveland, OH 44106 **[243]**

F. Testa Riva, Dipartimento di Citomorfologia, Universitá di Cagliari, 09124 Cagliari, Italy **[235]**

Jean-Pierre Timmermans, Institute of Histology and Microscopic Anatomy, University of Antwerp, B-2020 Antwerp, Belgium **[343]**

Mary E. Todd, Department of Anatomy, University of British Columbia, Vancouver, British Columbia V6T 1W5, Canada **[139]**

Kuniaki Toyoshima, Department of Oral Biology, School of Dentistry, Case Western Reserve University, Cleveland, OH 44106; present address: Department of Oral Anatomy II, Kyushu Dental College, Kitakyushu 803, Japan **[243]**

Hiroshi Ueda, Department of Anatomy, University of Occupational and Environmental Health, School of Medicine, Kitakyushu 807, Japan **[481]**

L. Valentino, Dipartimento di Citomorfologia, Universitá di Cagliari, 09124 Cagliari, Italy **[235]**

P. Versura, Institute of Clinical Electron Microscopy, University of Bologna, 40138 Bologna, Italy **[503,605]**

F. Vicari, Laboratoire de Microscopie Electronique, Faculté de Médecine de Nancy, 54505 Vandoeuvre, Nancy, France **[327]**

Branislav Vidić, Department of Anatomy and Cell Biology, Georgetown University School of Medicine, Washington, DC 20007 **[589]**

Daniela Virgintino, Institute of Human Anatomy, Histology and General Embryology, University of Bari Medical School, 70124 Bari, Italy **[383]**

Kenjiro Wake, Department of Anatomy, Faculty of Medicine, Tokyo Medical and Dental University, Tokyo 113, Japan **[257]**

Ewald R. Weibel, Department of Anatomy, University of Berne, CH-3000 Berne 9, Switzerland **[147]**

T. Weiger, University of Salzburg, Institute of Zoology, Department of Experimental Zoology, A-5020 Salzburg, Austria **[427]**

Mitsuaki Yoshizuka, Department of Anatomy, University of Occupational and Environmental Health, School of Medicine, Kitakyushu 807, Japan **[481]**

VIII INTERNATIONAL SYMPOSIUM ON MORPHOLOGICAL SCIENCES

Rome, July 10–15, 1988

President	P.M. Motta,	Italy
Vice Presidents	A. Dhem,	Belgium
	G. Grignon,	France
Honorary Presidents	L.J.A. DiDio,	U.S.A.
	T. Fujita,	Japan
	K.R. Porter,	U.S.A.

INTERNATIONAL COMMITTEE on MORPHOLOGICAL SCIENCES

T. Cichocki, Poland
A. Dhem, Belgium
L.J.A. DiDio, U.S.A.-Chairman
J.A. Esperança-Pina, Portugal-Secretary
C. Foroglu-Kerameos, Greece
T. Fujita, Japan
S. Gomez-Alvarez, Mexico
G. Grignon, France
E. Klika, Czechoslovakia
W. Kuhnel, West Germany
K. Moore, Canada
M. Moscovici, Brazil
P.M. Motta, Italy
O. Nilsson, Sweden
P.E. Olivares, Argentina
D. Ruano-Gil, Spain

A. Delmas, France, Honorary Member

Preface

The VIIIth International Symposium on Morphological Sciences, under the high patronage of Senator Professor F. Cossiga, President of the Republic of Italy, under the auspices of the International Federation of Associations of Anatomists, sponsored by the International Committee on Morphological Sciences, supported by the "Consiglio Nazionale delle Ricerche" and organized by the morphologists of the Department of Anatomy at the University of Rome "La Sapienza," was held in Rome, Italy, from July 10 to July 15, 1988.

The symposium, a periodical meeting of the world leaders in the morphological sciences, put forth an excellent overview of the many interesting current trends of our field, but also served to emphasize the most significant courses recently charted by the various branches of our discipline. Mindful of the prevailing role played by the microscope in our sector, both in research and in teaching, we designated several internationally famous colleagues to present contributions on general topics of interest to all the participants. The majority of these presentations on the applications of modern technology and methodology were selected to exemplify the fascinating microscopic world of the human body. The unique and hidden beauty of this microcosm began to unravel in the middle of the 17th century, thanks to the early microscopists, including Marcello Malpighi (1628–1694), the father of microscopic anatomy. While celebrating the 300th anniversary of the publication of Malpighi's Opera Omnia (1686–87), it was thus fitting that the VIIIth Symposium should emphasize his achievements. Holding the symposium in Rome was another tribute to Malpighi, as it was in the Eternal City that he culminated his career—as private physician to Pope Innoncent XII—and where he died, in the Quirinale Palace, the former papal residence.

Previous symposia were held in several countries and organized by distinguished members of our community of morphologists. These symposia were, are, and will be intended to create a forum of leaders for the dissemination of up-to-date morphological data and innovative trends, both in research and in education, to benefit investigators, teachers, and students.

The Ist Symposium was held in Mexico City, in 1971, under the presidency of Prof. S. Gomez-Alvarez. Envisioning the potential of the symposia, the officers of the world's main associations lent their support to the founders and participated actively in all aspects of the event. The following institutions and their representatives joined Prof. Gomez-Alvarez in the organization of the symposium: Profs. M. Niizima and A. Delmas, president and secretary general, respectively, of the International Federation of Associations of Anatomists; Prof. L.J.A. DiDio, president of the Pan American Association of Anatomists; Dr. A.A. Del Pozo, secretary of the Pan American Universities Association; Prof. D. Bodian, president of the American Association of Anatomists; Prof. V. Simic, president of the Anatomical Society of Yugoslavia; and Dr. M. Vasquez-Vigo, regional director of the World Health Organization.

The IInd Symposium took place in Cordoba, Argentina, in 1973, under the presidency of Prof. P.E. Olivares, and was opened with a special lecture delivered by Nobel Laureate L. Leloir.

The IIIrd Symposium was organized and presided over by the late Prof. H. Nathan, in 1977, in Tel Aviv, Israel. The opening lecture was given by S. Goren, the Chief Rabbi of Jerusalem.

The IVth Symposium was held in Toledo, Ohio, U.S.A., in 1979, under the presidency of Prof. L.J.A. DiDio. Special lectures included two Nobel laureates, Prof. G.E. Palade and Prof. C. Gadjuzek, and Prof. K.R. Porter, among others, who updated several anatomical topics of general interest.

The Vth Symposium took place in Rio de Janeiro, Brazil, in 1982, organized and presided over by Prof. M. Moscovici. Several courses ran simultaneously with the scientific sessions, and Prof. Don W. Fawcett was one of the special lecturers.

The VIth Symposium was held in Lisbon, Portugal, in 1984, and was organized and presided over by Prof. J.A. Esperança-Pina, rector of the New University of Lisbon. The symposium was opened by Dr. J. Ramalho-Eanes, president of the Republic of Portugal, who was accompanied by the prime minister and ministers of the government, and by the Cardinal Patriarch of Lisbon.

The VIIth Symposium was organized and presided over by Prof. A. Dhem, in 1986, at the Catholic University of Louvain, in Brussels, Belgium. During the administrative session, the permanent International Organizing Committee of Symposia on Morphology selected Rome, Italy as the site to host the VIIIth symposium, under the presidency of Prof. Pietro M. Motta.

The VIIIth Symposium was thus held at the Hotel Villa Pamphili from July 10 to July 15, 1988. At this symposium a permanent logo was proposed and adopted for subsequent meetings. The emblem represented a universal symbol, originally created by Leonardo da Vinci, to illustrate the perfection of proportions of the human body. This symbol was included in a flag and in a medal documenting the central importance and unity of anatomy through the creation of man. The medal, especially coined, carries the dates and sites of past symposia and will contain those of future symposia. The medal was donated by the President to the international institution, to be passed, at the end of each symposium, from the president to the president-elect, who will organize the subsequent event. It will thus emphasize the continuity of interest and solidarity among morphologists.

At this symposium, dedicated to Marcello Malpighi, a series of Malpighi Lectures was included: the first, which opened the symposium, was given by Prof. Liberato J.A. DiDio (U.S.A.) on "Biographic Aspects of Marcello Malpighi. Form and Beauty, Technology and Science." The second was delivered by Prof. Keith R. Porter (U.S.A.) on "The Cytoplasm and Its Matrix," and the third by Prof. Robert G. Edwards (U.K.) on "Early Human Embryos and the Fundamentals of Embryology."

To these three keynote speakers, to Prof. Tsuneo Fujita, and to the President of the VIIIth Symposium, the Director of the Department of Microscopy of the Carl Zeiss Foundation, Dr. M. Hiller gave for "outstanding scientific merit" a special award, represented by Ernst Abbe's version of the microscope.

Other Malpighi lectures were subsequently delivered by Prof. J.A. Esperança-Pina (Portugal), W.C. Forssmann (F.R. Germany), H. Fujita (Japan), T. Fujita (Japan),

P.M. Motta (Italy), A. Oksche (F.R. Germany), M. Petráň (Czechoslovakia), K. Tanaka (Japan), B. Vidić (U.S.A.), E.R. Weibel (Switzerland), and B. Zanobio (Italy).

These lectures, as well as other leading addresses relating to Malpighi's original discoveries, essentially served to mark the commemorative tone of the entire Rome Symposium, the proceedings of which are published here.

The president announced that a Malapighian Academy will be organized at the University "La Sapienza" of Rome to honor the great founder of microscopic anatomy.

Furthermore, during the closing ceremony, special plaques were awarded to colleagues who have added and continue to add significant contributions to the progress of the morphological disciplines and who have had a long-standing scientific activity, by means of publications and editorial work, to maintain high academic standards in our field.

The following morphologists were honored with plaques: Prof. G. Burnstock, L. Edvisson, V. Ersparmer, J.A. Esperança-Pina, G. Fazio, C. Fieschi, W.C. Forssmann, H. Fujita, W. Lierse, A.Oksche,I.H. Schumacher and K. Tanaka. To Mrs. and Prof. Z. Fumagalli, Professor Emeritus of Anatomy at the University of Rome, whom I had the privilege to succeed, were awarded special medals.

Closing the ceremony, the President passed the medal and flag of the International Morphological Symposia to his successor Prof. Georges Grignon, who will organize the IXth Symposium in Nancy, France, in 1990.

Selected contributions of participants of the VIIIth International Symposium on Morphological Sciences are published in two volumes. The first, corresponding to "Malpighi and the Foundations of Microscopic Anatomy," essentially contains the "Malpighi lectures" as well other pertinent results presented in the main celebrative symposium originally entitled "Microanatomy of Cells and Tissues by Scanning Electron Microscopy and Other Techniques: A Three-Dimensional Approach," which I organized with the kind help of Profs. T. Fujita and Prof. K. Tanaka from Japan.

The second celebrative volume commemorates "Malpighi as Pioneer in Embryology and Reproduction" and it also reports the main results presented in a parallel main symposium entitled "Ultrastructure of Reproduction," which I designed with the cordial suggestions of Prof. M. Dvorak (Czechoslovakia) and S. Suzuki (Japan).

A few relevant presentations related to the above topics, belonging to other sessions or minisymposia of the meeting have been also selected for inclusion in these books.

In any case, all 437 abstracts of various sessions of the meeting were published in the Supplementum of "Quaderni of Anatomia Pratica," Vol. 44: 1446–1988.

I am happy and proud to point out that, after having organized our scientific program and presided over the symposium, I was delighted to see that 500 colleagues from 37 nations attended, far exceeding our original predictions. Moreover, the 500 papers submitted by the participants, most of which were included in the scientific program, were more than expected. As a result, our organizational work was almost doubled. I have to recognize, however, that our efforts were rewarded by the warm and enthusiastic participation of all our colleagues who, with their excellent presentations and outstanding discussions, have given light to many exciting scientific sessions, including plenary and Malpighi Lectures, eight parallel symposia, two free communications sessions, two poster sessions, and a highly appreciated artistic exhibit on Art and Anatomy.

I believe that, owing to this array of remarkable scientific contributions from our many colleagues, the symposium of Rome has been not only an occasion to present new results, but also to have exchanged information in a cordial atmosphere of camaraderie and friendship. This is, in a sense, the way in which we Italians experience life that is called "Italian flavor."

I hope that these books will prove to be valid scientific resources to those involved in the morphological sciences, and to all those scientists in the world who are fascinated by the new dimension of our discipline. Do not forget that a continuity links the past achievements to the new frontiers of the structure of life.

Pietro M. Motta

Acknowledgments

Many individuals contributed to the success of the VIIIth International symposium on Morphological Sciences. The scientific committee included the following Professors: F. Barberini, F. Caramia, F. Carpino, M. Casini, C. Cavallotti, S. Correr, L. D'Este, G. Familiari, L. Fumagalli, E. Gaudio, M. Melis, T. Renda, M. Ripani, and A. Riva. Their cordial help was much appreciated, as was that of Drs. G. Macchiarelli and S.A. Nottola (scientific program coordinators) and Dr. D.W. Finn and Mr. G.F. Franchitto.

The specific organization of various symposia and sessions of the meeting was mainly dependent on their fine cooperation, as well as on that of other colleagues who acted as guest organizers, namely: Profs. T. Fujita, K. Tanaka, and S. Suzuki (Japan); L.J.A. DiDio and W.C. Hamlett (USA); M. Dvorak (Czechoslovakia); V. Kaitsas (Greece); C. Donati, L. Fonzi, L. Pannarale, G.F. Patrizi, A. Ruggeri, and B. Tota (Italy).

The invaluable assistance of all the above individuals, in addition to that of Dr. G. Girini and his staff at Barberini Tours (mainly Mr. G. Sibaud and Miss F. Garofolo), our students in the Department of Anatomy at the University of Rome, Mrs. E. Vergnano, and Mr. A. Familiari were of great benefit to the success of the meeting of Rome.

The C. Zeiss Foundation deserves a special mention in that its contributions aided much in subsidizing the main commemorative Malpighian microscopical events of the entire symposium and the work related to the publication of these proceedings.

Furthermore, I express my appreciation to Mr. G. Campbell, Ms. M. Walsh, Ms. L. Bass, Mr. T. Battle, and Ms. T. Cromwell at Alan R. Liss, Inc. (New York), for their kind assistance in completing these volumes.

Sincere thanks are also due to my colleagues for their enthusiastic participation and for the meticulous preparation of their articles.

Finally, my deep gratitude goes to my wife, Silvia, and to my daughter, Cecilia, without whose lovely presence and continuous tolerance during these last 9 months, I would never have been able to discover in myself the courage, energy, and enthusiasm necessary to organize the symposium of Rome and to condense its scientific results in these books.

P.M. Motta

History

Cells and Tissues: A Three-Dimensional
Approach by Modern Techniques in Microscopy,
pages 3–6

INTRODUCTION. MARCELLO MALPIGHI AND THE FOUNDATION OF MICROSCOPIC ANATOMY

Pietro M. Motta

Department of Anatomy
University "La Sapienza" of Rome
Via Alfonso Borelli 50, 00161 Rome, Italy

In keeping with the commemorative tone of this event, a main symposium will re-examine in greater detail - thanks to modern microscopy and the state-of-the-art technology available to us today - those areas of the human body first probed by Malpighi some 300 years ago. Thus, in a certain way, the classical portraits of cells, tissues and organs, many of which were provided by Malpighi himself, after careful observation through his rudimentary microscope, will breathe new life as the discoveries of this century, especially the extraordinary progresses of the past three decades, are described and correlated to the exquisitely complex but perfect body machinery referred to by Vesalius in his DE HUMANI CORPORIS FABRICA (1543).

Malpighi had ingeniously intuited almost immediately the enormous potential of the microscope. He stressed the importance of this new instrument throughout much of his work, even while defending himself from criticism. In fact, some among his peers, unable to recognize the dawn of a new era in scientific exploration, believed that most of Malpighi's results contained a great deal of artifacts. Indeed, "per microscopia incertum in anatomia judicium" was the opinion of T. Kerckring, (1670) another celebrated anatomist of Malpighi's time.

Yet, Malpighi persevered and subscribed, along with others, to the opposing view, writing, in 1662: "on understanding the world of nature it becomes easier when one's perception is honed by Art, through the instruments invented during this century, which, with two glances through the eyeglass of the immortal Galileo, saw more discovered than all the past millennia have speculated upon; and with the microscope turned towards the minutest parts of the living body, saw, in the smallest of animals, marvelous mechanics, oddities and freaks of nature - - in other, perfect organisms, it has revealed the structure of many machines: therefore, Anatomy can expect great progress from the increased use of this instrument".

The symposium will underscore the close relationship between structure and function and will offer a solid foundation for comprehending and correlating morphodynamics with biochemical and biophysical events. Such an approach might lead to a clearer picture of physiopathologic phenomena, earlier diagnosis and prevention of disease, so as to expedite an appropriate therapy. Here, again, Malpighi was not only an anatomist in the strictest sense, curious about the microscopic world of the human body, but also a physician. In fact, as a clinician he intended to use the new technology of the microscope to discover the functional basis of disease and the alterations which occur at the level of the most minute parts of the body. This approach was carried out further by Valsalva, who was not only Malpighi's student, but, in turn, was the mentor of Morgagni, the founder of pathologic anatomy, bringing to us the current basis of modern medicine.

The publication of Malpighi's DE PULMONIBUS OBSERVATIONES ANATOMICAE (1661) (Fig. 1) represents a milestone in the history of medicine, in that, with these observations, both the microanatomy of the frog lung and the capillary circulation were accurately described and interpreted, elucidating the connection between arteries and veins (Fig. 2). Malpighi thus confirmed what had been postulated by Harvey in his EXERCITATIO ANATOMICA DE MOTU CORDIS ET SANGUINIS IN ANIMALIBUS (1628).

Such important discoveries, made by means of the microscope, represent the very foundation of microscopic anatomy. Through the correct description of the lung structure, which had heretofore been considered, along with that of other viscera, to be made up of an amorphous "parenchyma", Malpighi put to rest the then-current view that "all organs were similarly flesh-like and arose from the blood".

Through his tireless efforts and continuous observations, Malpighi revealed the microscopic structure of many other organs, such as the kidney, liver, spleen, tongue, skin, and brain, in a series of papers published in 1666 in Bologna, entitled DE VISCERUM STRUCTURA. Several of Malpighi's discoveries were named after him, e.g., the follicles of the spleen, the renal corpuscles, and the tongue receptors. The most important common thread throughout his papers is represented by the use of the microscope, which Malpighi applied to every living object that came within his reach. His method was based on the unity of living matter, which can be fully appreciated through the microscope. This instrument allowed him to recognize basic similarities among otherwise apparently dissimilar things, in the animal and plant kingdoms, under both normal and pathological conditions. Malpighi considered microscopic anatomy essential to the study of physiology and pathology and as a strategic approach to the progress of medicine.

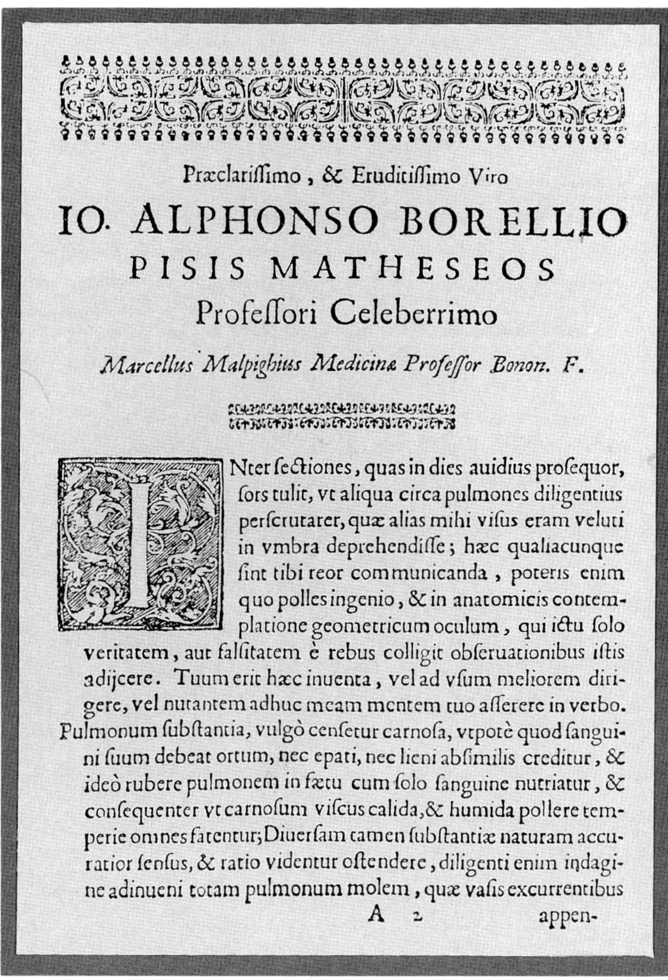

Præclariſſimo, & Eruditiſſimo Viro

IO. ALPHONSO BORELLIO

PISIS MATHESEOS

Profeſſori Celeberrimo

Marcellus Malpighius Medicinæ Profeſſor Bonon. F.

INter ſectiones, quas in dies auidius proſequor, ſors tulit, vt aliqua circa pulmones diligentius perſcrutarer, quæ alias mihi viſus eram veluti in vmbra deprehendiſſe; hæc qualiacunque ſint tibi reor communicanda, poteris enim quo polles ingenio, & in anatomicis contemplatione geometricum oculum, qui ictu ſolo veritatem, aut falſitatem è rebus colligit obſeruationibus iſtis adijcere. Tuum erit hæc inuenta, vel ad vſum meliorem dirigere, vel nutantem adhuc meam mentem tuo aſſerere in verbo. Pulmonum ſubſtantia, vulgò cenſetur carnoſa, vtpotè quod ſanguini ſuum debeat ortum, nec epati, nec lieni abſimilis creditur, & ideò rubere pulmonem in fætu cum ſolo ſanguine nutriatur, & conſequenter vt carnoſum viſcus calida, & humida pollere temperie omnes fatentur; Diuerſam tamen ſubſtantiæ naturam accuratior ſenſus, & ratio videntur oſtendere, diligenti enim indagine adinueni totam pulmonum molem, quæ vaſis excurrentibus

A 2 appen-

Fig. 1 Frontispiece of Malpighi's work on the structure of the lungs early addressed as a letter in Pisa to his friend and mentor, the mathematician Alfonso Borelli.

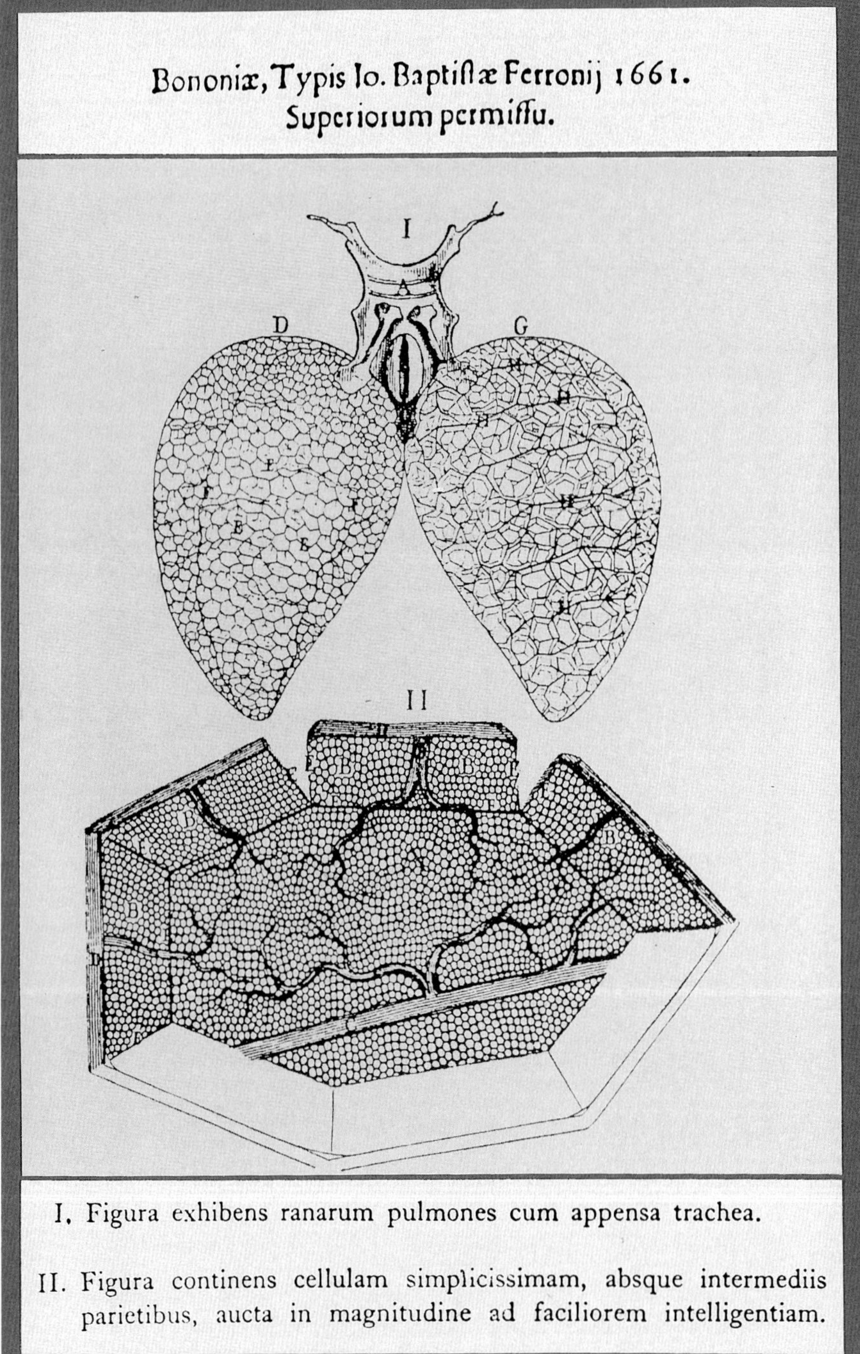
Bononiæ, Typis Io. Baptistæ Ferronij 1661.
Superiorum permissu.

I. Figura exhibens ranarum pulmones cum appensa trachea.

II. Figura continens cellulam simplicissimam, absque intermediis parietibus, aucta in magnitudine ad faciliorem intelligentiam.

Fig. 2 Original pictures from Malpighi's work, "De Pulmonibus Observationes Anatomicae"

Cells and Tissues: A Three-Dimensional
Approach by Modern Techniques in Microscopy,
pages 7–11

BRIEF SURVEY OF MALPIGHI'S LIFE*

Prof. Dr. Liberato J. A. DiDio

President, International Federation of Associations of Anatomists, Department of Anatomy, Medical College of Ohio, C. S. 10008, Toledo, Ohio 43699, USA

Marcello Malpighi was born in Crevalcore, near Bologna, on March 6, 1628, the eldest of eight children of Marcantonio Malpighi and Maria Cremonini. The year of Malpighi's birth coincided with the publication of William Harvey's Exercitatio Anatomica de Motu Cordis et Sanguinis in Animalibus, to which the Italian scientist would contribute in 1661, by demonstrating the presence of capillaries.

Malpighi, as a young boy, moved from Crevalcore to Bologna, where he attended the "Scuole Pie". In this institution, Malpighi learned reading, writing, singing, the use of the abacus, studied Christianity and acquired "beautiful virtues". Then, he was taught logic, philosophy, arts, and sciences, in preparation for admission to an institution of higher education.

In the beginning of 1646, Malpighi was admitted as a freshman at the University of Bologna. Unfortunately both his parents died in 1649, forcing him to interrupt his studies at the University, in order to provide guidance and assistance to his brothers and sisters.

One year later, Malpighi married Francesca Massari, a much older lady than his age and sister of his preceptor and professor of anatomy, Bartolommeo Massari.

---

*The entire First Malpighian Lecture, presented on July 11, 1988, at the opening ceremony of the VIII International Symposium on Morphological Sciences, held in Rome, Italy, will be published in Quaderni di Anatomia Pratica (1988).

Malpighi was accepted as one of the nine members of the Chorus Anatomicus (Aldemann, 1966), an academy founded by Massari to perform dissections and experiments and discuss a variety of controversial and scientific subjects.

After graduation, Malpighi practiced medicine and began to teach in the Archiginnasio, later the Studio of the University of Bologna. In spite of his quiet personality, Malpighi had several enemies, who were highly critical of his discoveries because these were breaking old traditions but mainly because of jealousy. To the antagonistic attitudes, taken by a few of his colleagues and even by some of his so-called disciples, Malpighi responded always as a gentleman and never showed any animosity toward his opponents.

In 1656, Malpighi moved from Bologna to Pisa, where he was attracted by Giovanni Alfonso Borelli and became the chairman of Theoretical Medicine at the University. Borelli was a follower of Galileo's ideas and exerted a strong influence on the new pupil, just arrived from the University of Bologna.

During a triennium, in Pisa, Malpighi continued projects which he had started in Bologna, completed some, developed new techniques and studied the minute structures of animals and plants under the compound microscope.

In 1659, Malpighi returned to his alma mater, the University of Bologna, where he resumed the observation of the alveolar structure of the frog's lung.

After three years, the University of Messina, in Sicily, was able to convince Malpighi to become the successor of Pietro Castelli, whose death had vacated a chairmanship at the institution (Sotgiu, 1965). The environment in Messina provided Malpighi with abundant and diverse marine material, which he used for his biological explorations. Malpighi made, then, many contributions to comparative morphology and to microbotany.

Malpighi has been recognized as the father of histology (Field and Harrison, 1957), the founder of microscopic anatomy of both animals and plants, and one of the pioneers of micro-biology.

In 1666, after a productive period of four years in Messina, Malpighi returned, again, to Bologna to become professor of medicine at the University.

In 1667, the Royal Society of London, after admitting Malpighi as a member, extended to him an invitation to send periodical reports of his observations for publication. In 1669, the Royal Society of London bestowed upon him an honorary membership as a special award to a foreign scientist.

The most known disciples of Malpighi and the closest friends of their magister were Giorgio Baglivi and Antonio Maria Valsalva. The latter had in turn, as a disciple, the great Giovanni Battista Morgagni, considered the founder of Pathology.

Many outstanding members of the aristocracy in the community of Bologna sought Malpighi to take advantage of his clinical skills and ability and most were persuaded to allow or request the performance of autopsy, thus overcoming the fear, superstition or prejudice, which at that time involved necroscopy.

Before eponyms were deleted from the anatomical terminology, though still found in medical nomenclature, the name of Malpighi was attached to several structures, which he had discovered, studied and/or described. Among the most known of these structures, the following can be mentioned (de Terra, 1913):

1. Malpighi's renal acinus or renal corpuscle
2. Malpighi's splenic or lienal corpuscle
3. Malpighi's renal glomerulus or corpuscle or glandular body
4. Malpighi's nervous body (papillary) of the skin or corium papillary body
5. Malpighi's papillary layer of the skin
6. Malpighi's splenic or lienal corpuscles or lymph nodes or glands
7. Malpighi's renal pyramids
8. Malpighi's lienal stigmata, that is, the openings of venules in larger veins of the lienal or splenic pulp
9. Malpighi's rete or network or germinative layer of the epiderm
10. Malpighi's air vesicles or membranous vesicles or pulmonary alveoli
11. Malpighi's lingual receptors.

Anatomical expressions were also designated under Malpighi's name in association with the name or names of other authors, such as Gärtner (e.g., duct of Malpighi and Gärtner).

Ironically, the description of the spiral bundles of the myocardium, that Borelli claimed to had been the first to discover and that Malpighi later pointed out to be his own, belonged to neither. In fact, Stenon (Stensen) had preceded both, according to Belloni (1967).

In 1689, owing to his frail health, Malpighi was under clinical treatment instituted by the physician Salamone. Apparently the latter was successful at least in the treatment of Malpighi's symptoms, as he received an extremely grateful letter from his already famous patient.

Two years later, Malpighi was invited to become the private physician of Pope Innocent XII. He was very well acquainted with the Pope because the latter had been the Pontifical Legate in Bologna. Malpighi accepted the invitation and took residence in Rome, more precisely in the Palazzo Quirinale.

Malpighi was unable, however, to be of great help as a physician to the Pope, because his own health had deteriorated. On his deathbed, Malpighi requested to his nephew, Dr. Antonio M. Fabri, to give his scientific papers, through Dr. S. Bonfiglioli, to the Royal Society of London (Belloni, 1967), the institution that published them as Malpighi's Opera Posthuma. Malpighi requested also a necroscopy of his body 30 hours after his death, because he was concerned with "resuscitation". The necroscopy was performed by one of his disciples, Dr. Giorgio Baglivi, and by Dr. Gian Maria Lancisi.

Malpighi had an "apoplectic attack" followed by a recurrency, and died on November 29, 1694 (Busacchi, 1966), six years before his most important patient, Pope Innocent XII. Funeral services were held in Rome and approximately one year later in Bologna, followed by his burial in the church of Saint Gregorio.

Malpighi's death was mourned worldwide, especially by the scientific world, and periodically his life and achievements have been commemorated.

Malpighi's plaza was named in Bologna in 1896 and his monument was dedicated one year later in Crevalcore. His portrait was included among worldwide renown scientists in the Royal Society of London next to that of William Harvey, and his profile served as the symbol of the Federative International Congress of Anatomy held in Milano, in 1936. Another monument was erected in his honor, in 1965, by the University of Bologna in the church of Saints Gregorio and Siro.

The VIII International Symposium on the Morphological Sciences honored Malpighi in its logo, next to the Palazzo Quirinale, the symbol of Rome, where he last resided and died. Malpighi fully deserves to be considered an honorary Roman anatomist.

At the same Symposium, the University "La Sapienza" of Rome instituted the Malpighian Lectures, to be delivered periodically in honor of the founder of microscopic anatomy and biology.

As we approach 1994, the year of the third centennial of Marcello Malpighi's death, all scientists expect that in addition to Bologna, Pisa, Messina and Rome, every science center of the world will join in the commemoration. This should be the celebration of the pioneer of the modern era of natural sciences.

REFERENCES

ALDEMANN, Howard B. -1966- Marcello Malpighi and the Evolution of Embryology. - Ithaca, New York, Cornell University press.

BELLONI, Luigi -1967- Opere Scelte di Marcello Malpighi. - Torino, Unione Tipografico-Editr. Torinese.

BUSACCHI, Vincenzo -1966- Celebrazioni Malpighiane. Discorsi e Scritti. - Bologna, Azzoguidi Soc. Tip. Editoriale.

de TERRA, Paul -1913- Vademecum anatomicum. Kritisch-etymologisches Wörterbuch der systematischen Anatomie. - Jena, Verlag von G. Fisher.

FIELD, E. J. and HARRISON, R. J. -1957- Anatomical Terms: Their Origin and Derivation. - Cambridge, W. Heffer and Sons Ltd., 2nd ed.

SOTGIU, Giulio -1965- Marcello Malpighi e l'inizio dell' era scientifica. - Bollettino della Biblioteca Comunale di Bologna, 60:10-30.

The Cell

Cells and Tissues: A Three-Dimensional
Approach by Modern Techniques in Microscopy,
pages 15–20

THE CYTOPLASM AND ITS MATRIX

Keith R. Porter

Department of Biology, University of Pennsylvania,
Philadelphia, PA 19104-6018

INTRODUCTION

The microscopic anatomy of the cytoplasm is proving to be an area of increasing fascination for the cell biologist. It includes microtubules which support anisotropic cell forms and provide guidance for intracellular translocations. Then also there are actin-rich stress fibers which have the capacity to change position and orientation as they contribute to cell motion. And finally the cytoplasmic anatomy includes in some instances remarkable quantities of intermediate (100 A) filaments which appear in a diversity of forms and composition in different cell types and which seem consistent in providing structural strength to an otherwise gelatinous cytoplasm. These three filamentous forms are non-randomly distributed in the cytoplasm and appear, in fact, to repeat in their plan of distribution in cells of the same kind. Just how this is achieved is not precisely known, but certain observations relative to the question are emerging. There is, for example, evidence that the filaments exist in a structured matrix which may control their distribution and orientation. This matrix comprises innumerable fine filaments (microtrabeculae) which, together with the water-rich phase are thought to give to the cytoplasm its well-known gel consistency. This view of the matrix structure is also supported by diffusion studies (Luby-Phelps, Taylor, and Lanni, 1986) using particles of known size and shape. Pertinent also are the observations that cell fragments, created by microsurgery, have a demonstrable ability to support the assembly of microtubules as well as their polarity, orientation and distribution (McNiven and Porter, 1986). This is in line with the belief that there exists in the matrix organization information, in some instances, in the form of dense bodies in the matrix and in other instances, devoid of any morphological entity (McNiven and Porter, 1988).

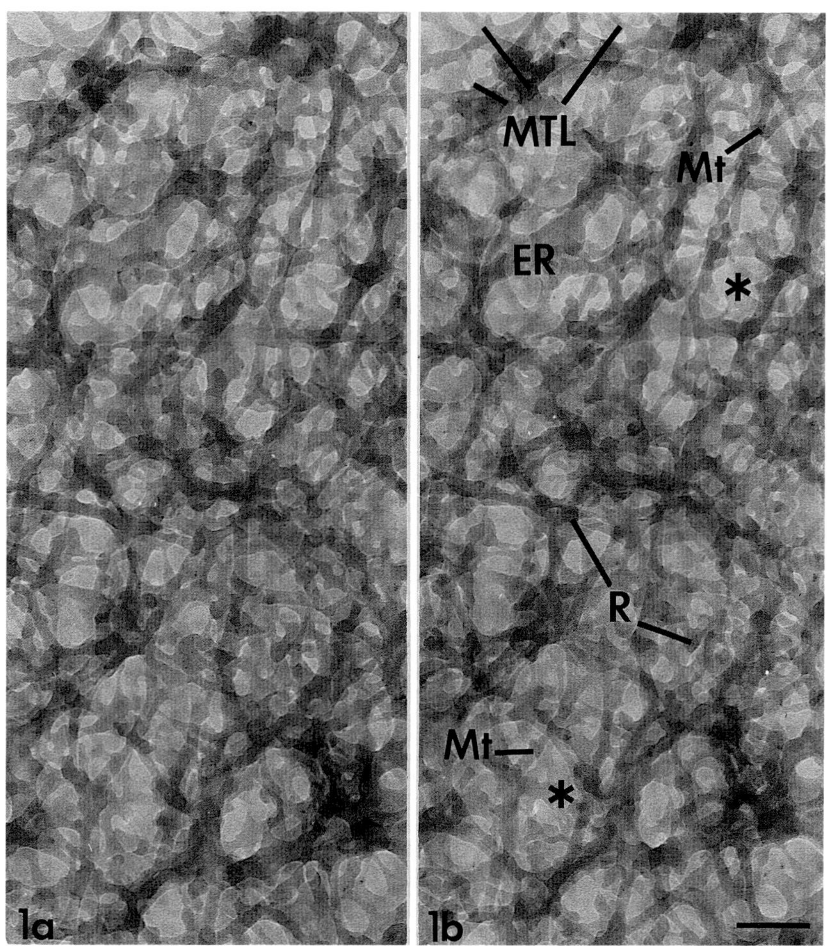

Fig. 1a, b - Magnification x 108.000; Bar = 0,1 um (for figure legend see text).

RESULTS

These observations on the structure of the cytoplasmic matrix have been made largely on cultured cells grown directly on grids (electron microscope) coated with formvar films and in turn coated with evaporated carbon. In most of the early studies the cells were fixed in glutaraldehyde and thence dried by the critical-point method (Wolosewick and Porter, 1979). Such whole cell preparations required high voltage microscopy (1 MeV) in order to obtain electron penetration and useful image formation. The question of artifact production, especially by the procedures of fixation and drying, was constantly in mind and was investigated by varying the procedures. Of the several tried, that which involved freezing the cells at liquid nitrogen temperatures and thence drying them while maintained in the frozen state (-95°C) proved to be the most valuable (see figs. 1a and 1b). Control experiments in which the cells were fixed first with glutaraldehyde and then frozen-dried yielded results that were not strikingly different from those in which glutaraldehyde and critical point drying were avoided (Porter and Anderson, 1982).

From numerous micrographs of different kinds of cells, prepared for observation in various ways, it has been possible to reach a consensus as to the general structural features of the cytoplasmic matrix. These are included in the schematic shown in Fig. 2. Perhaps the finding of greatest significance is that which reveals the matrix to be a two phase system, one protein-rich and the other water-rich. This is essentially the structure of a gel and one that is anticipated in the older literature. What is original in this view of the matrix is the fact that ribosomes (polysomes) are suspended in the protein-rich phase and by virtue of this arrangement are subject to matrix control over their non-random distribution. Presumably by this arrangement, the product of synthesis is contained in the protein-rich phase and controlled as to distribution in the cytoplasm (see figure legends for other details).

There is much food for thought in these new images of the cytoplasmic matrix and the phenomena revealed. One may presume that further investigations of its variations in cells of different types will better define its several roles. Some of these are considered in recent publications. (See Porter, 1988.)

FIGURE LEGENDS

Figs. 1a and 1b. These represent a pair of stereo electron micrographs of a small area of the cytoplasm of a PtK2 cell prepared for high voltage microscopy by the freeze dry procedure. The cell was cultured on a carbon-coated gold grid, frozen instantaneously by immersion in liquid nitrogen, and dried in a vacuum (10-7) while maintained at a temperature of -95°C. Certain well known features of the cytoplasm are easily identified.

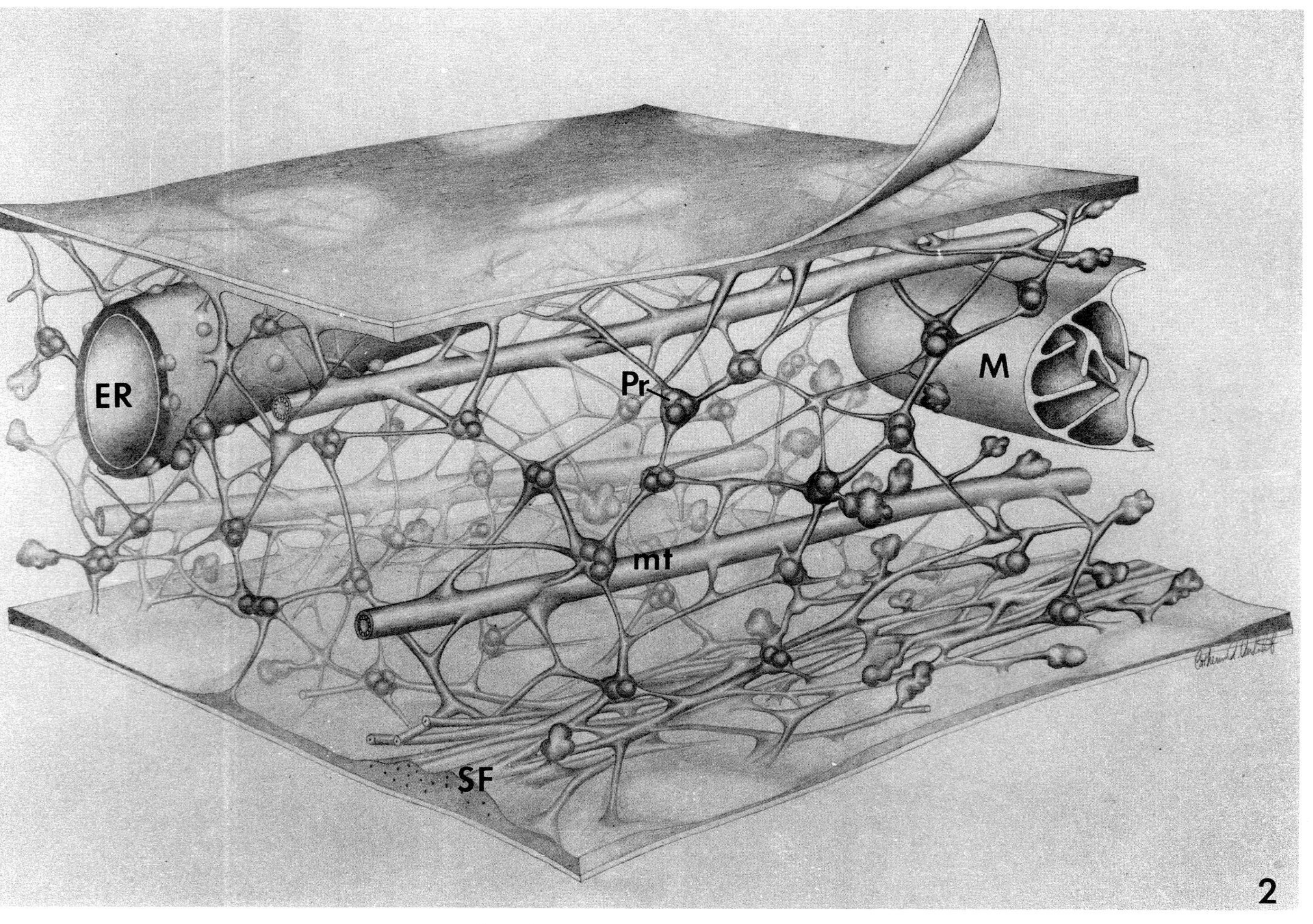

Fig. 2 - For legende see text.

A few cisterna of the endoplasmic reticulum are identified at ER, microtubules at Mt, ribosomes and polysomes at R and microtrabeculae at MTL. A few water-rich spaces are indicated by asterisks. Special note should be taken of places (unmarked) where microtrabeculae attach to ER vesicles or to microtubules or to other MTL's. It is evident that many attach to the surfaces of microtubules. Presumably, if a microtubule disassembles, the tubulin molecules retain a linear position within the microtrabecular lattice. Ribosomes also appear to be contained within the microtrabecular lattice and are assumed to be non-randomly positioned within the cytoplasm by this connection. It is further attractive to assume that the protein synthesized by polysomes remains within the MTL where synthesized. This obviously assigns to the MTL a role in controlling the organization of the cytoplasm.

The water-rich phase we identify with the intertrabecular spaces relatively free of electron scattering matter. This phase occupies about 70% of the cytoplasmic volume and can be involved in the diffusion of small molecule metabolites.

Fig. 2. An interpretive schema representing the structure of the cytoplasmic matrix as observed in stereo electron micrographs of whole cells. The intertrabecular spaces (the water-rich phase) are substantially exaggerated in size for greater visability. A cisterna of the endoplasmic reticulum is represented at ER; a mitochondrion at M; a microtubule among several at mt and a polyribosome at pr. The upper plasma membrane is pealed back to emphasize the existence of a cytoplasmic cortex; the lower membrane is flat on the substrate. A stress fiber (SF) is depicted as intimately associated with the lower cortex. This structure is rich in actin filaments which are shown as coated by the same material that makes up the system of trabeculae (MTL for microtrabecular lattice). In both instances, actin filaments and microtubules, the available evidence suggests that they can disassemble (depolymerize) and reassemble in much the same position. The drawing is made to emphasize our belief that the MTL is a continuous phase which surrounds (or contains) the formed filamentous structures of the cytoplasm.

REFERENCES

Luby-Phelps K, Taylor DL, Lanni F (1986). Probing the structure of cytoplasm. J Cell Biology 102:2015-2022.

McNiven MA, Porter KR (1986). Microtubule polarity confers direction to pigment transport in chromatophores. J Cell Biology 103:1547-1555.

McNiven MA, Porter KR (1988). Organization of microtubules in centrosome-free cytoplasm. J Cell Biology 106:1593-1605.

Porter KR (1988). Cytomatrix. In Arias IM, Jakoby WB, Popper H, Schachter D, Shafritz DA (eds): "The Liver: Biology and Pathology. Second Edition," New York: Raven Press Ltd, pp 29-45.

Porter KR, Anderson KL (1982). The structure of the cytoplasmic matrix preserved by freeze-drying and freeze-substitution. European J Cell Biology 29:83-96.

Wolosewick JJ, Porter KR (1979). Preparation of cultured cells for electron microscopy. In "Practical Tissue Culture Applications," Academic Press Inc, pp 59-85.

Cells and Tissues: A Three-Dimensional Approach by Modern Techniques in Microscopy, pages 21–30

ULTRA-HIGH RESOLUTION SCANNING ELECTRON MICROSCOPY OF BIOLOGICAL MATERIALS

Keiichi Tanaka, Akira Mitsushima, Yuzuru Kashima, Takashi Nakadera and Hatsuko Fukudome

Department of Anatomy, Tottori Univerisity School of medicine, Yonago 683 Japan

INTRODUCTION

Three-dimensional architecture of cell components has been usually studied by reconstruction method from serial thin sections, high voltage electron microscopy and replica method. Though scannig electron microscope (SEM) is also a powerful tool for this study, it has been used mostly for researches at low magnifications such as surfaces of cells and tissues with little attention given to high resolution field as like as intracellular components. The reason depended on lack of a good specimen preparation technique for revealing intracelular structures and inferiority of the resolving power of SEM.

About the specimen preparation technique, the osmium-DMSO-osmium method was developed in recent years (Tanaka and Naguro, 1981; Tanaka and Mitsushima, 1984). By this method, three-dimensional structure of membranous intracellular organelles such as mitochondria, endoplasmic reticulum, Golgi complex were successively studied (Tanaka, 1987). On the other hand, instrumental resolution of SEM was also improved rapidly. In 1985 we developed together with Hitachi Ltd. an ultra-high resolution SEM, named UHS-T1 (Tanaka et al., 1985; Tanaka et al., 1986), which was equipped with a field emission electron source and an objective lens with a very short focal length. It showed a resolution of 0.5nm at a biological specimen coated with platinum.

Due to the both advances, intracellular structures became to be easily studied by scanning electron microscopy, and not only intracellular structures but also viruses, bacteriophages and biological macromolecules were also clearly observed (Tanaka et al., 1988). In this paper outline of the ultra-high resolution SEM and findings of some biological materials observed by the SEM are reported.

ULTRA-HIGH RESOLUTION SEM (UHS-T1)

This SEM is equipped with a field emission source and an objective lens with a very short focal length. The diameter of the electron probe was calculated to be 0.45 nm at 30 kV, and a probe diameter of 0.5 nm was confirmed by observation in STEM mode. On observation of a biological material coated with platinum the SEM showed 0.5nm resolution when dark space criterion was used.

The main production design specifications are as follows;

Electron Optics
- Electron gun: field emission electron source
- Accelerating voltages: 1-30 kV (1kV/step)
- Magnification range: 150-1,000,000x
- Lens system: two stage electromagnetic lens system
- Objective lens: f 3.6mm, Cs 1.6mm, Cc 2.0mm

Specimen stage
- Specimen stage: side-entry specimen stage (eucentric)
- Specimen size: 2mm dia. x 2mm high

Vacuum system
- Gun chamber: Ion pump (60 l/sec) x 1
- First intermediate chamber: Ion pump (20 l/sec) x 1
- Second intermediate chamber: Ion pump (20 l/sec) x 1
- Specimen chamber: Turbo molecular pump (340 l/sec) x 1, Turbo molecular pump (60 l/sec) x 1 and Rotary pump (174 l/min) x 1
- Specimen exchange chamber: Turbo molecular pump (60 l/min) x 1 and rotary pump x 1

Ultimate vacuum
- Electron gun Chamber: 1×10^{-8} Pa
- First intermediate chamber: 3×10^{-8} Pa
- Second intermediate chamber: 3×10^{-7} Pa
- Specimen chamber: 6×10^{-6} Pa

Anti-contamination devices
- cold finger located close to the specimen stage x 1
- cold trap located over the turbo molecular pump x 1

CRT: 9" x 3 (photo-CRT: high resolution type)

Photo-apparatus
- lens: Rodagon (Rodenstock) 135mm (f:5.6)
- film: 4"x5" sheet film

MATERIALS AND METHODS

Cell organelles:

Tissue samples were prepared mainly by the aldehyde-prefixed osmium-DMSO-osmium method (A-ODO method, Tanaka and Mitsushima, 1984). The procedures were as follows; 1)After removal of blood, animals were perfused with the fixative; a mixture of 0.5% glutaraldehyde and 0.5% formaldehyde in M/15 phosphate buffer solution at pH 7.4. 2)Small blocks of tissues (about 1x1x3mm) were removed from the animals. The blocks were rinsed in the buffer and post-fixed with 1% osmium tetroxide solution in the same buffer for 1-2 hrs. 3)After having been well rinsed with the buffer they were immersed in 25 and 50% dimethyl sulfoxide (DMSO) aqueous solution for 30 min each. 4)The specimens were frozen on a metal plate chilled with liquid nitrogen, and split with a razor blade and a hammer. We have used a freeze-cracking apparatus (TF-2, EIKO Engineering Co. Ltd., Japan) for this purpose. The split pieces were immediately placed in 50% DMSO and thawed at room temperature. Then they were rinsed in the buffer until DMSO had been completely removed. 5)The specimens were treated with 1% osmium tetroxide for 1 hr and then transferred into 0.1% osmium tetroxide solution buffered at pH 7.4 with M/15 phosphate buffer and left standing for three days or more at 20°C. 6)The specimens were fixed again for 1 hr in 1% osmium tetroxide. They were then treated with 1% tannic acid solution for 2 hrs and then 1% osmium tetroxide for 1hr. 7)The specimens were dehydrated through a graded alcohol series. After treatment with iso-amyl acetate, they were dried in a critical point dryer (HCP-2, Hitachi Koki Co. Ltd., Japan). 8)Dried specimens were coated lightly (about 1-2nm) with platinum in an ion-beam sputter coater, equipped with a quartz crystal thickness monitor (VA-10S, Hitachi Koki Co. Ltd.) or an ion sputter coater with a rotating stage (VX-10R, EIKO Engineering Co. Ltd.). 10)The metal-coated specimens were observed at accelerating voltages of 15-30 kV.

Clathrin coated vesicles:

Cells of a fibroblast strain (3Y1) which had been cultured on plastic cover slips were used as materials. They were treated with a saponin solution (50 μg/ml, PHEM buffer: Schliwa and van Blerkom, 1981) for 5 min at room temperature, then fixed with 0.5% glutaraldehyde in the same buffer for 30 min, followed by 0.5% osmium tetroxide for 5 min. After being thoroughly rinsed, the samples were conductively stained in 0.2% tannic acid solution for 30 min, 0.5% osmium tetroxide for 5 min and then saturated uranyl acetate solution for 5 min. The specimens were then dehydrated through an ascending alcohol series, and critical point dried. Adhesive ("Scotch") tape was lightly touched to the dried cells on the coverslip and then taken off. By this procedure (Mitsushima et al., 1986), the upper membranes of cells were removed exposing intracellular structures in the cells on the cover slip. They were observed by the SEM with or without metal coating, at an accelerating voltage of 15 kV.

Bacteriophage:

A suspension of Escherichia coli which had been infected by T_2 phage (in a phosphate buffered saline solution at pH 7.2: PBS) were used as material. For fixation 2.5% glutaraldehyde solution was added to the suspension to a final concentration of 1%. After centrifugation, sediments were resuspended with a small amount of the PBS. Droplets of this suspension were placed on polished carbon plates which had been previously treated with ultraviolet light to render them hydrophilic, and left standing for a few minutes. Excess fluid was removed with filter paper, and 1% osmium tetroxide in PBS was dropped on the carbon plates. This post-fixation was performed for 20 min. Then the specimens were treated with 0.5% tannic acid for 30 min and 1% osmium tetroxide for 20 min. After being rinsed they were lead-stained for 2-3 min (Sato, 1968) and again treated with 1% osmium tetroxide for 10 min. The specimens were then dehydrated through an alcohol series and critical point dryed. The specimens were coated with platinum under 1 nm with an ion-beam sputter coater (VA-10S) and observed at an accelerating voltage of 30 kV.

Vaccinia virus:

Droplets of a dilute vaccinia virus suspension were placed on polished carbon plates and left 5 min. After being rinsed the specimen were fixed with 1% glutaraldehyde in PBS for 10 min and then conductively stained with tannic acid and osmium solution. After dehydration through an ascending alcohol series the specimens were dried in a critical point dryer. The dried specimens were coated with platinum thinny and observed with the SEM.

Ferritin:

Horse liver ferritin (commercial preparation; Polyscience, Inc., U.S.A.) was used. Droplets of the dilute ferritin suspension (100ug/ml) were placed on polished carbon plates and left for 5 min. After having been rinsed in PBS, the specimens were fixed with 0.2% glutaraldehyde in PBS for 3-5 min. They were rinsed in distilled water, dehydrated through a graded alcohol series and freeze-dried with t-butyl alcohol (Inoue and Osatake, 1988). The dried specimens were observed without metal coating at an accelerating voltage of 30 kV.

RESULTS

Three dimensional architecture of cell organelles has been studied by ordinary SEMs using specimens prepared by the ODO method and consequently many reports have been published. Although it might be unneccessary to use an ultra-high resolution SEM for such kind of research, the images of cell organelles observed with the SEM were markedly shaper than those observed by ordinary SEMs and hence we could easily obtain the SEM pictures of good quality. Figure 1 shows intracellular structures of a rat liver cell. Various membraneous structures, such as endoplasmic reticulum, Golgi complex, mitochondria are clearly seen. By the ultra-high resolution SEM, on the other hand, coated platinum particles were plainly discerned as rounded "pebbles" at magnification over 100,000x, when an ordinary ion-sputter coater had been used, although they were not resolved with ordinary SEMs. On observations over 100,000x therefore it is better to use an ion-beam sputter coater as like as VA-10s (Hitachi Koki, Ltd., Japan), because its coated platinum particles are much smaller than those by ordinary ion sputter coaters.

By the ODO method, however, not all intracellular structures were revealed, though membraneous cell organelles were

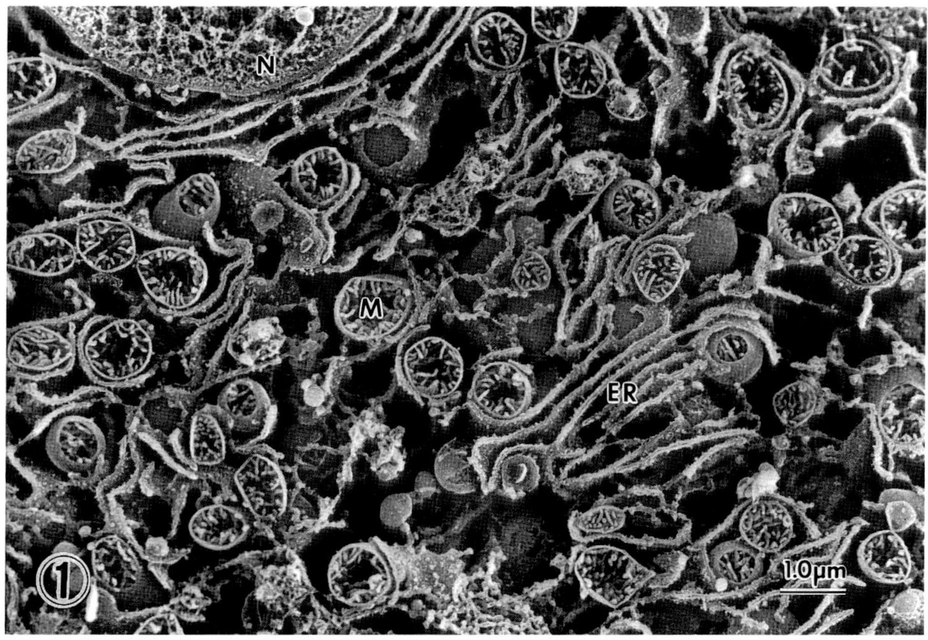

Figure 1. Intracellular structures of a rat liver observed by the ultra-high resolution SEM. N:nucleus, ER:endoplasmic reticulum, M:mitochondria.

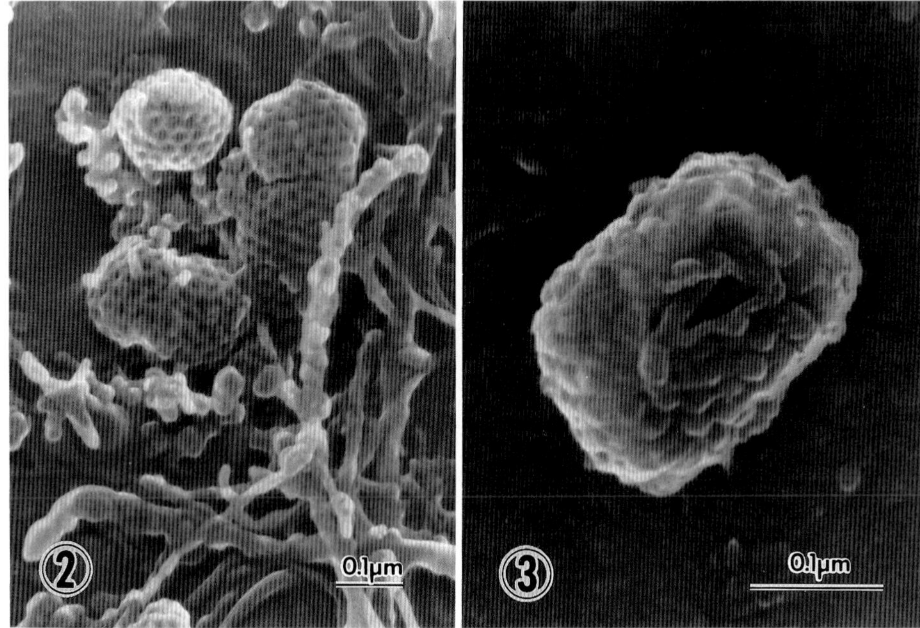

Figure 2. Reverse side of the cell membrane of a cultured fibroblast . Coated vesicles are seen.
Figure 3. Vaccinia virus.

well disclosed. Filamentous structures, such as microtubules, 10-nm filaments and actin filaments were all removed during the osmium maceration procedure. Clathrin coats of coated vesicles were also melted away. As a method for observing these structures by SEM, the scotch tape method (See Material and Methods) was of use. By this method we could see the reverse side of cell membrane of cultured fibroblasts and observed endocytotic coated vesicles. They were found sporadically among many filamentous structures which might correspond to microtubules, 10-nm filaments or actin filaments (Fig.2). The clathrin coated vesicles showed a basket-like structure and their diameter was about 140nm.

Viruses and bacteriophages were very suitable materials for researches by the ultra-high resolution SEM. We could easily observed their fine structures. Figure 3 shows Vaccinia viruses. They had blick-like form and unevenness on their surfaces. In Figure 4 we presented T_2 bacteriophages attached on an E. coli. In this picture even their tail fibers are seen. For making samples of viruses or bacteriophages, we recommend that the samples are coated with platinum very lightly (under 1 nm) with an ion-beam sputter coater after a conductive staining.

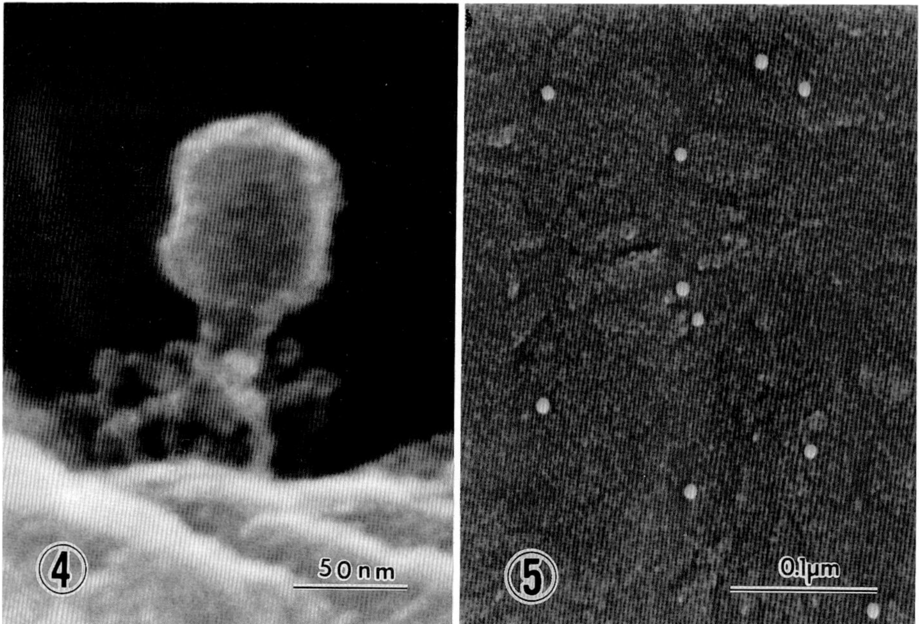

Figure 4. T_2 bacteriophage.
Figure 5. Ferritin particles fixed with glutaraldehyde.

Then you can obtain good pictures of them.

Figure 5 shows ferritin particles of horse liver. The ferritin particles appeared as rounded bodies of almost same size without any suggestion of the polyhedral substructure which was obtained on the surface of negatively stained apoferritin molecules (Esterbrook, 1970). The diameter of the ferritin particles fixed with glutarladehyde solution showed a size of 9±1nm in diameter, but it became larger up to 11±1nm after a conductive staining.

DISCUSSION

Until recent years, it was generally accepted that the resolution of SEM was unable to surpass 2nm because of the escape depth of the secondary electrons from a solid surface. However we suggested that it is possible to achieve a resolution under 1nm, because we could verify a resolution of 1.5nm at 30kV in a biological specimen using "zero working distance method" (Nagatani and Okura, 1977) on a conventional field emission SEM (Tanaka, 1981). Kuroda and Komoda (1985) experimented with a field emission TEM with an SEM attachment and showed a resolution under 1nm on the surface of a tungusten tip.

On the other hand, the demand for ultra-high resolution SEM was increasing in biomedical field year after year, because the resolution of ordinary SEM was markedly inferior to that of the TEM and insufficient for cytological researches. In 1985 we developed together with Hitachi Ltd. an ultra-high resolution SEM, named UHS-T1. It showed a resolution of 0.5nm at a biological specimens coated with platinum. Thus we could pass the limit of the resolution of SEM, 2nm, by this instrument. Following this instrument, some commercialized types of ultra-high resolution SEMs have successively developed, i.e. Hitachi S-900 (Nagatani et al., 1987), JEOL JSM-890, Akashi beam Technology DS-130F.

When intracellular structures prepared by the osmium-DMSO-osmium method were observed by our new SEM, the images were seen markedly shaper than those by ordinary field esmission SEM. Further, not only intracellular structures but also viruses, bacteriophages and even biological macromolecules were clearly observed.

With improvement of the instrumental resolution, some problems came to the fore. One of the problems is specimen contamination by the electron beam. At the observation by the new SEM, the current density increased remarkably, because the observations are usually performed at very high magnifications. Consequently the contamination instantly covers the tiny samples on substratum. To avoid such contamination, vacuum of the SEM should be made clean as much as possible. In UHS-T1, the diffusion oil pump were changed to two turbo molecular pumps connected in series. At the specimen exchange chamber, we also equipped a turbo molecular pump between the chamber and a rotary pump. By this improvement, the ultimate vacuum degree of the specimen chamber raised up to $6x10^{-6}$ Pa and the amount of the contamination was remarkably decreased. On the other hand, the contamination varies with the sort of specimens and substratum. It is necessary therefore to make an effort to diminish contamination coming from specimen area hereafter.

The next problem is metal-coating of specimens. SEM specimens are usually coated with metal such as platinum for obtaining electric conductivity. By the new SEM the platinum particles are plainly seen as rounded "pebbles" at magnifications over 100,000x. Even if we use an ion-beam sputter coater whose metal particles are much smaller than those of ordinary ion-sputter coaters, the platinum particles are clearly seen at magnifications over 250,000x. At high magnifications therefore specimens should be observed without metal coating. For this aim, conductively stained specimens (for example, by the tannin-osmium method) can be used. Using such specimens, ribosomes and clathrin coated vesicles were studied at very high magnifications (Tanaka et al., 1988). By this technique, unfortunately, the surface of specimens are sometimes decorated by the tannin-osmium complex produced during the conductive staining. Then the best way for ultra-high resolution scanning electron microscopy of very small samples is the observation without metal-coating and conductive staining. Fortunately, very samll samples, i.e. biological macromolecules, can be successfully observed without charging, if they are directly placed on polished carbon plates, fixed with a aldehyde-fixative and dried in a critical point dryer. By this technique we could observe ferritin and immunoglobulin G in good contrast (Tanaka et al., 1988). In this case, almost all incident electrons may penetrate through the samples and do not accumulate in them. In short, we should use specimens without metal coating at observations of very high magnifications, though metal-coated specimens are of use at magnifications under 250,000x.

Although some problems remain unsettled as above described, we believe that the ultra-high resolution scanning electron microscopy surely will become an effective means for observing biological fine structures in the near future.

REFERENCES

Esterbrook K B (1970). The arrangement of subunits in the shell of ferritin, Electron microscopic observations of molecules negatively stainined with uranyl acetate. J Ultrastruct Res 33:442-450.

Inoue T, Osatake H (1988). A new drying method of biological specimens for scanning electron microscopy: The t-butyl alcohol freeze-drying method. Arch Histol Cytol 51:53-59

Mitsushima A, Katsumoto T, Tanaka K (1986). Scanning electron microscopic observation of cytoskeletal networks lying under the cell membrane. Biomed SEM 15:74-77

Nagatani T, Okura A (1977). Enhanced secondary electron detection at small working distance in the field emission SEM. In Johari O (ed) "Scanning Electron Microscopy," O'Hare: SEM Inc., vol.1, pp 695-702

Nagatani T, Saito S, Sato M, Yamada M (1987). Development of an ultra high resolution scanning electron microscpe by means of a field emission source and In-lens system. Scannig Microsc 1:901-909.

Sato T (1968) A modified method for lead staining of thin sections. J Electron Microsc 17:158-159

Schiliwa M, van Blerkom J (1981). Structural interaction of cytoskeletal components. J Cell Biol 90:222-235

Tanaka K (1981). Demonstration of intracellular structures by high resolution scanning electron microscopy. In Johari O (ed): "Scanning Electron Microscopy," O'Hare: SEM Inc., vol.2, pp 1-8.

Tanaka K (1987). Eukaryotes: Scanning electron microscopy of intracellular structures. Int Rev Cytol Suppl 17:89-120.

Tanaka K, Naguro, T (1981). High resolution scanning electron microscopy of cell organelles by a new specimen preparation method. Biomed Res 2 (Suppl):63-70.

Tanaka K, Mitsushima, A (1981). A preparation method for observing intracellular structures by scanning electron microscopy. J Microsc (Oxford) 133:213-222.

Tanaka K, Matsui I, Kuroda K, Mitsushima A (1985). A new ultra-high resolution scanning electron microscope (UHS-T1). Bio-medical SEM 14:23-25.

Tanaka K, Mitsushima A, Kashima Y, Nakadera T, Osatake H (1988). Application of an ultra-high resolution scanning electron microscope (UHS-T1) to biological specimens. J. Electron Microsc Tech, in press.

Tanaka K, Mitsushima A, Kashima Y, Osatake H (1986). A new high resolution scanning electron microscpe and its application to biological materials. In Proc XIth Int Congr EM, Kyoto, vol. III:2097-2100.

Cells and Tissues: A Three-Dimensional
Approach by Modern Techniques in Microscopy,
pages 31–37

FREEZE-FRACTURE OF 3T3 AND HeLa CELLS FOR HIGH-RESOLUTION SEM AND DEEP-ETCH REPLICAS: STRUCTURE OF THE INTERPHASE NUCLEUS

G.H. Haggis

Canada Department of Agriculture, Ottawa, Canada

In developing new techniques for the study of internal cell structure by high-resolution scanning electron microscopy (SEM) we have found it useful to make direct comparison between micrographs of the same structure, prepared on the one hand for viewing by SEM or alternatively, for viewing by the rapid-freeze, deep-etch technique of Heuser and Hirokawa (see later papers in this volume). The preparative methods we have developed prove particularly useful in the three-dimensional study of the structure of interphase chromatin, which forms my contribution to this symposium.

3T3 and HeLa cell cultures are maintained in the laboratory of Dr. Natalie Chaly (Biology Department, Carleton University, Ottawa) and we have grown these cells on fibrin films 20 μm thick (a thickness chosen for reasons given below). The preparations are sandwiched between thin sheets of copper, separated by a 30 μm spacer and rapidly frozen with jets of propane at -180°C. The cells are thus frozen in a 10 μm space between the fibrin and the copper sheet on one side. After freezing, the copper sheet adjacent to the cells and the spacer are removed, under liquid nitrogen. The preparation is then placed in a Balzers freeze-etch unit, and the cooled microtome knife of this unit is brought down to the sample surface and then, in four 2 μm steps, brought to a level at which it passes through and fractures the cells. (The cells are cultured on 20 μm fibrin films because it is difficult to set the knife exactly parallel to the specimen block and with thinner films the knife tends to catch on the specimen holder). After fracture, the sample can be removed and critical-point dried for viewing by SEM or a replica can be taken from the fracture face for viewing by transmission electron microscopy. A 3T3 cell fractured in this way is shown, by SEM, in Fig. 1. In the preparation of Fig. 1 the cell has been fixed before freezing and little detail is seen

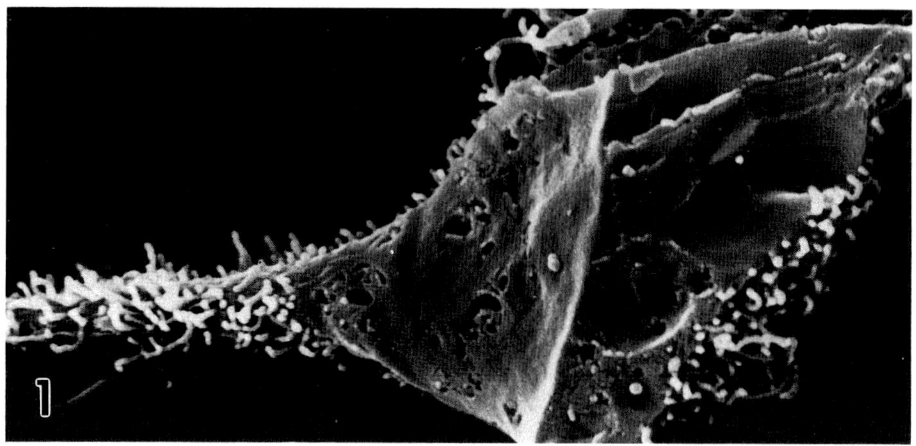

Figure 1. A 3T3 cell fixed before freezing, fractured at -170°C by the method described in the text, critical point dried and viewed by SEM. Magnification X 6300. Further experimental details for Figs. 1 and 2 will be found in Haggis (1987).

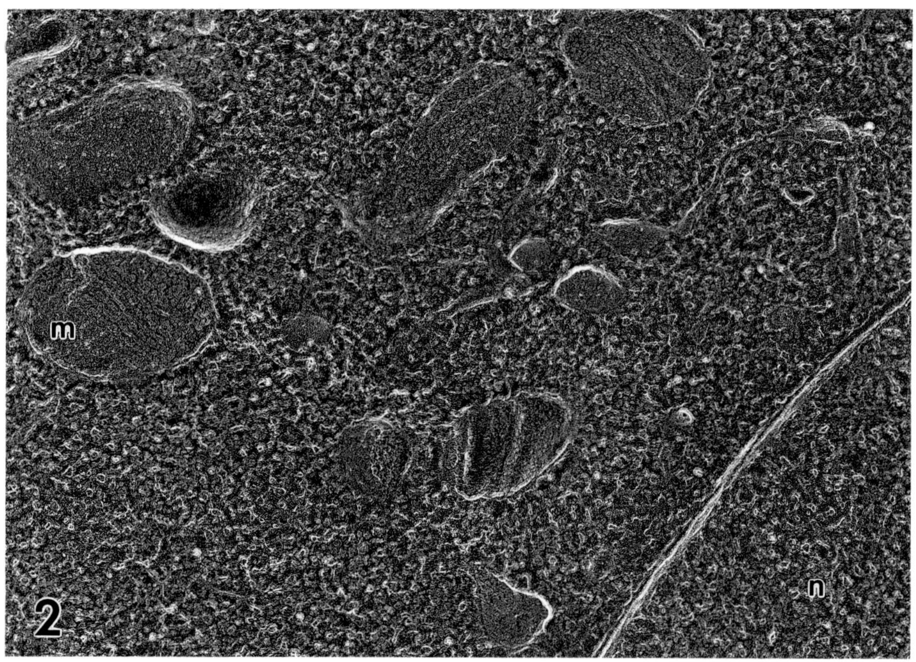

Figure 2. A platinum-carbon replica micrograph of a 3T3 cell fresh-frozen, fractured and etched for 2 mins at -100°C m=mitochondrion, n=nucleus. Magnification X 40,000.

within the cell. The reason for this becomes clear in Fig. 2, which shows a replica from a fractured 3T3 cell fresh-frozen, without any pretreatment, etched after fracture for 2 mins at -100°C. This 'etching' is an *in vacuo* sublimation of ice from the fracture face. At -100°C ice sublimes, but salts, solutes and soluble proteins remain, preventing any deep view of structure such as cytoskeletal filaments or nuclear chromatin. The cross linking of soluble and structural components during fixation is the reason that no internal structure is seen in Fig. 1.

One way to reveal internal structure in preparations of this kind is to introduce a detergent extraction step prior to rapid freezing. Fig. 3 shows a 3T3 cell which has been extracted with 0.5%. Triton X100 for 15 mins in a buffer of 50mM imidazole, 5mM KCl, 0.5 mM $MgCl_2$, 1mM EGTA, 0.1mM EDTA, 1mM mercaptoethanol, 4M glycerol, pH 6.7 (Bershadsky et al. 1978) then fixed with 1.5% glutaraldehyde in this buffer. The preparation was washed in distilled water after fixation, dehydrated to 70% ethanol, frozen, fractured at -170°C, thawed back into 70% ethanol, then taken to 100% ethanol and critical-point dried.

In this experiment a 15 min. extraction time was chosen to allow as extensive an extraction as possible for the nucleus. The chromatin seen in Fig. 4 has thus been exposed for this time to a buffer which mimics but does not totally maintain *in vivo* conditions, before fixation, allowing the possibility of ultrastructural change during this time.

Some years ago we introduced a thaw-fix technique (Haggis & Bond 1979) for preparation of this kind in which cells were rapidly frozen unfixed, fractured, then thawed into fixative. In this initial work we were concerned about structural damage due to ice crystal formation, during the freezing and thawing steps, and infiltrated the cells with glycerol or dimethylsulphoxide to 25-30% before freezing. This infiltration took 15-60 mins. so that, again, we were exposing cells to unnatural conditions for a significant time prior to fixation.

Recently we have achieved freezing and thawing rates sufficiently rapid to avoid ice crystal damage, starting with fresh material. Fig. 5 and 6 show replicas of HeLa cells which have been fresh-frozen then thawed into 0.3% glutaraldehyde in a buffer of 80 mM PIPES, 5 mM EGTA, 1 mM $MgCl_2$, pH 6.8 (Drubin and Kirschner, 1986). After 10 min. fixation they have been washed in distilled water, infiltrated with 15% methanol (in distilled water) rapidly frozen again, fractured, etched and replicated. The cytoplasm shows no evidence of ice damage. Although ice forms more readily in the nucleus than in the cytoplasm, if freezing and thawing are

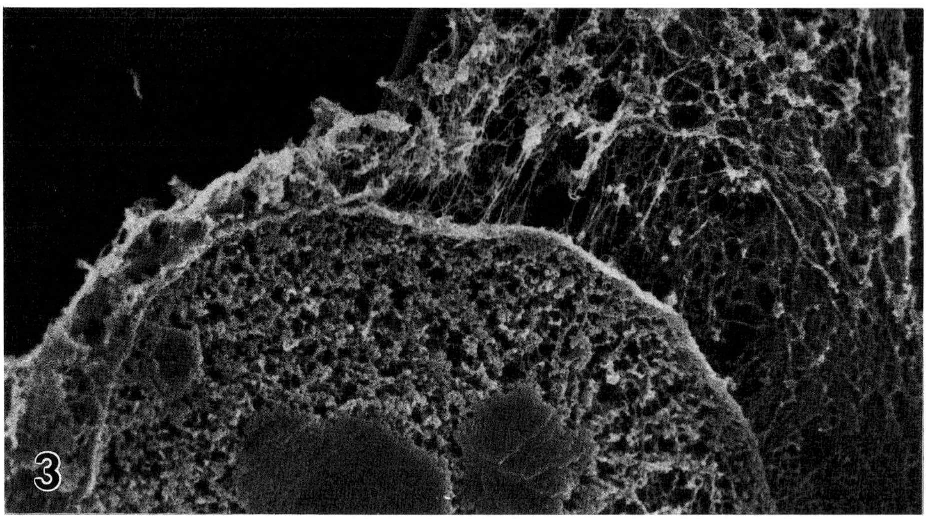

Figure 3. A 3T3 cell extracted for 15 mins with 0.5% Triton, fixed, frozen in 70% ethanol, fractured at -170°C, thawed back into 70% ethanol and critical-point dried for SEM viewing. Magnification X 12,000. Further experimental detail for Figs. 3 and 4 will be found in Haggis & Pawley (1988).

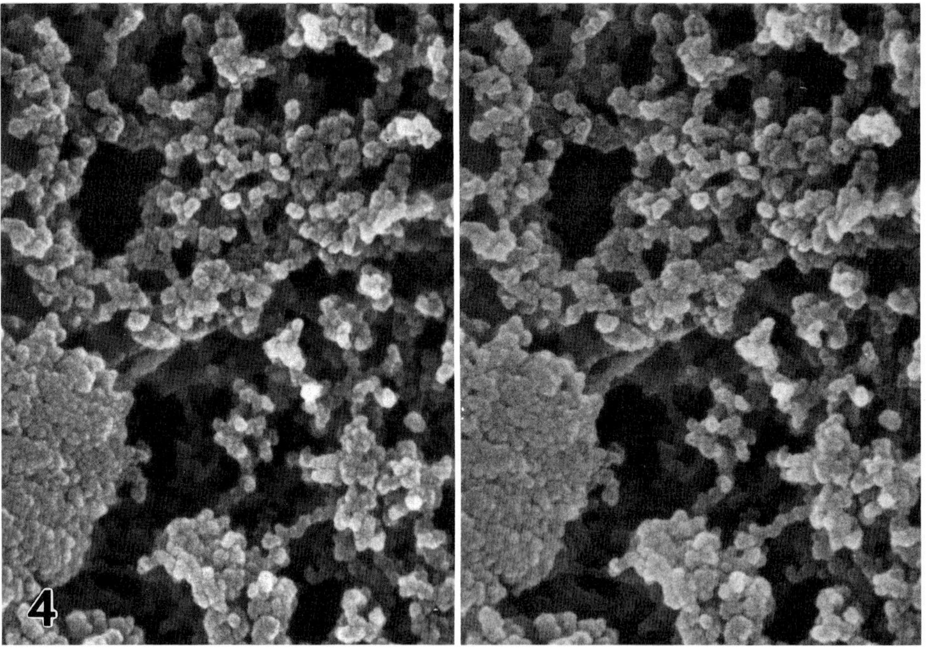

Figure 4. Stereoview of chromatin in a fractured nucleus, prepared as for Fig. 3, examined in a Hitachi S900 field-emission SEM. Magnification X 60,000.

not rapid enough, we do not believe there has been any extensive ice formation in the nuclei of these cells because of the close similarity between the structure seen in these micrographs and in earlier preparations after Triton extraction (Fig. 4) (with freezing in 70% ethanol) and after glycerol infiltration to 30% (Haggis et al. 1983) where, in each case, under the rapid freezing and thawing conditions used and the high level of cryoprotectant (ethanol, glycerol) we can reasonably eliminate the possibility of ice-crystal artifact. Preliminary results have also been obtained both by SEM and deep-etched replication for cells fresh-frozen, fractured and thawed into fixative. In this case the preparation of the replica is achieved by refreezing in 15% methanol (with no further fracture) with a long etch time, to bring the level of the frozen 15% methanol down to the level of the previously fractured cells. The composition of the fixation buffer, the concentration of glutaraldehyde (0.1 to 1.5%) the time of fixation (1 to 10 min.) and the effect of delaying fixation (by thawing into buffer, with later transfer to fixative) are being studied. Results will be reported in a later paper (Haggis 1988).

One of the advantages of the thaw-fix technique is that initial fixation can be light. We currently fix with 0.3% glutaraldehyde for 10 min. This fixation leaves a wide variety of nuclear and cytoplasmic antigenic sites still capable of binding antibodies (D.L. Brown, private communication) and the fracture allows direct access of antibody to all compartments of the cell. After antibody labelling, samples can be post-fixed to strengthen structure against critical-point drying. Freezing and thawing take place within milliseconds and initial fixation is rapid, since the glutaraldehyde has only to penetrate 2-5 μm into the fractured cell, without any membrane barrier to diffusion. One can hope, therefore, that the chromatin structure is preserved close to the in vivo state.

It will be interesting, in future work , to achieve colloidal-gold antibody labelling of selected proteins within this structure, to explore new methods of high-resolution SEM, as for examples working at low voltage, so that the detail-obscuring effect of platinum coating can be eliminated (Osumi et al., 1988) also to try to reduce Mg^{++} in the buffer and time of fixation to reveal more molecular detail than is seen in Fig. 4. It should be possible also to follow the condensation of chromatin to form chromosomes in dividing cells (Fig. 6 & 7 of Haggis and Pawley 1988).

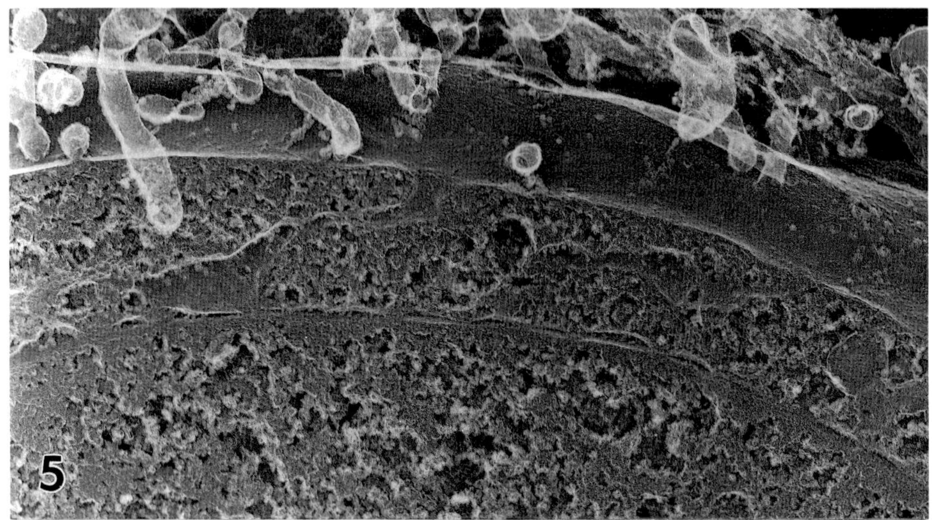

Figure 5. HeLa cell fresh frozen, thawed into fixative. After fixation, refrozen in 15% methanol (in distilled water) fractured at -95°C and deep-etched. Replica micrograph. Magnification X 24,000. Further experimental details for Figs. 5 and 6 will be found in Haggis (1988).

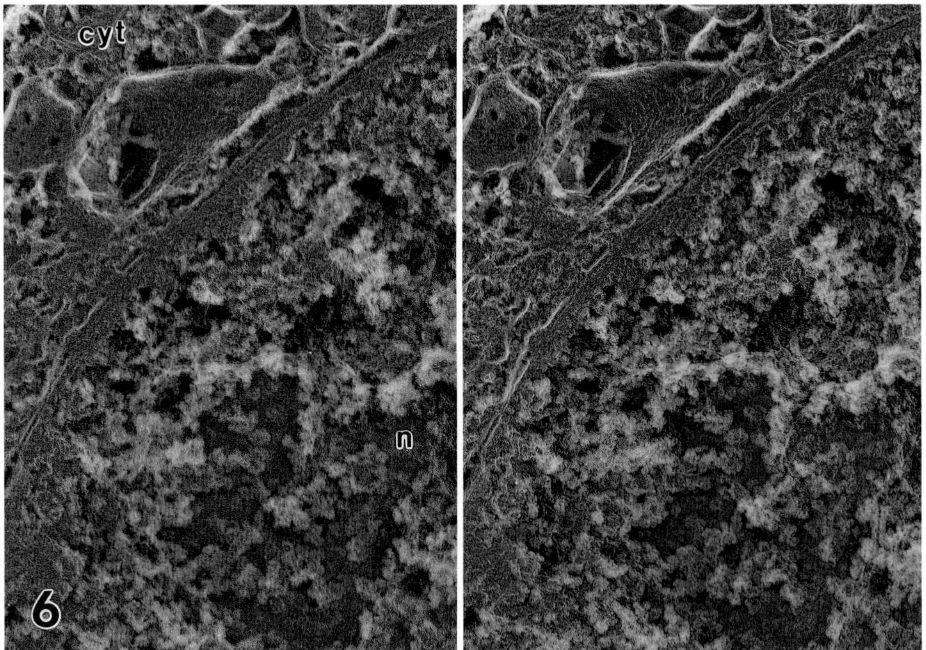

Figure 6. Stereoview of chromatin in a fractured nucleus prepared as for Fig. 5. n=nucleus, cyt=cytoplasm. Replica micrographs. Magnification X 32,000.

REFERENCES

Bershadsky AD, Gefland VI, Svitkina TM, Tint IS (1978). Microtubules in mouse fibroblasts extracted with Triton X100. Cell Biol. Internat. Rep. 2:425-432.

Drubin DG, Kirschner MW (1986). Tau protein function in living cells. J. Cell. Biol. 103:2739-2746

Haggis GH, Bond EF (1979). Three-dimensional view of the chromatin in freeze-fractured chicken erythrocyte nuclei. J. Microsc. 115:225-234.

Haggis GH, Schweitzer I, Hall R, Bladon T (1983). Freeze-fracture through the cytoskeleton, nucleus and nuclear matrix of lymphyocyte nuclei studied by scanning electron microscopy J. Microsc. 132:185-194.

Haggis GH (1987). Freeze-fracture of cell nuclei for high-resolution SEM and deep-etch TEM. Proc. 45th Ann. Meeting of Electron Microsc. Soc. of America. Bailey GW (ed) San Francisco Press, pp. 560-563.

Haggis GH, Pawley JB (1988). Freeze fracture of 3T3 cells for high-resolution scanning elecron microscopy. J. Microsc. 150: June issue.

Haggis GH (1988). Three-dimensional viewing of internal cell structure, in Science of Biological Specimen Preparation for Microscopy and Microanalysis. Proceedings of the 7th Pfefferkorn Conference, Pub. SEM Inc. Chicago.

Osumi M, Baba M, Naito N, Taki A, Yamada N, Nagatani T.(1988). High-resolution, low voltage scanning electron microscopy of uncoated yeast cells fixed by the freeze-substitution method. J. Electron Microsc. 37:17-30.

Cells and Tissues: A Three-Dimensional
Approach by Modern Techniques in Microscopy,
pages 39–48

A NEW APPROACH FOR THE TEM VISUALIZATION OF THE 3-D ORGANIZATION OF FILAMENTOUS STRUCTURES IN CELLS AND TISSUES

José F. David-Ferreira and António J. Cidadão

Department of Cell Biology - Gulbenkian Institute of Science, Oeiras, Portugal

The organization and relationships of cell and tissue filamentous networks are difficult to analyze by conventional TEM of ultrathin sections. In fact, a thin section offers only a 2-D view of 3-D structures, and the embedding medium masks by electron scattering many filamentous structures present on cellular and extracellular matrices (ECMs).

Those limitations of thin sectioning, which are particularly evident in the study of complex filamentous networks, can be partially surpassed by the use of whole mount preparations. Valuable information about the organization of cytoplasmic filamentous systems and their relationships with cell organelles can be obtained by conventional or high voltage electron microscopy in whole mounts of cell preparations.The usefulness of this method is, however, limited to cells which can be isolated and grown on formvar-coated grids. Whole mounts cannot be used for the observation of tissue cells *in situ* and their relationships with ECM components.

The introduction of extractable embedding media like polyethylene glycol (PEG) to electron microscopy (Wolosewick, 1980) overcame the masking effect of dense epoxy resins, allowing the use of thicker sections so useful for 3-D reconstruction.This embedding technique provides a good ultrastructural preservation (Guatelli *et al.*, 1982; Nagele *et al.*, 1984) and allows the use of immunocytochemical methods (Wolosewick *et al.*, 1983). Unfortunately, structures observed in embedment-free sections lack surface details, and their overlapping puts some difficulties in the interpretation of their spatial organization and relationships.

For the above reasons we developed a method (Cidadão and David-Ferreira, 1986) to obtain high resolution 3-D images. It consists of TEM observation of rotary-shadowed Pt-C replicas obtained from critical-point dried resinless PEG sections. We combined and adapted the technique developed by Wolosewick (1980) for the preparation of resinless sections with the method used by Heuser for the observation of quick-freeze deep-etched surfaces (Heuser, 1981).

PROCEDURE

Only a brief description of the main steps of the method will be made (Fig. 1), but for further details see Cidadão and David-Ferreira (1986). Specimens are fixed according to the aim of the study, but a conventional EM fixation (glutaraldehyde-osmium) is, in most cases, well suited for a first morphological approach of the ECM organization. Protocols employing cationic dyes or detergent extractions can also be used.Extraction procedures are, in our opinion, necessary for the study of cells. After fixation, tissue fragments are dehydrated, embedded in polyethylene glycol (PEG), and sectioned with glass knives in thickness ranging from ~200nm-2μm. Sections are attached to formvar-carbon coated gold grids, or alternatively to glass coverslips. Both grids and coverslips are usually covered with 0.1% poly-L-lysine to prevent section detachment. However, poly-L-lysine is not always necessary, particularly when the thinnest sections are employed. Sections are de-embedded by immersing grids (or coverslips) in

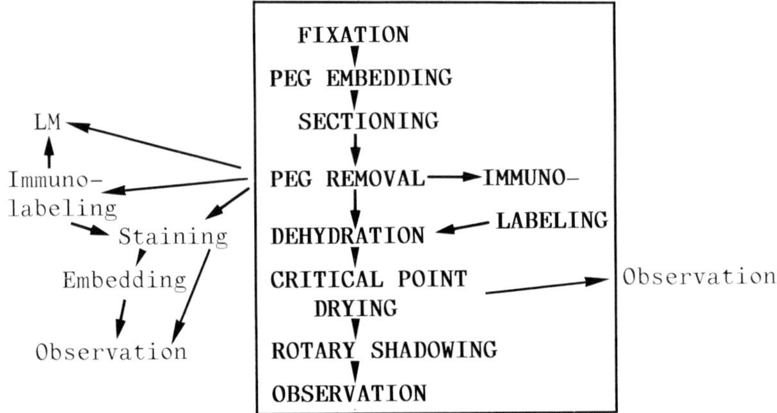

Fig.1- The main steps of our replica method are enclosed within the rectangle (see text for details). However, PEG is a very versatile embedding medium, allowing various alternative techniques to be performed using the same tissue block. After PEG removal, and optional immunolabeling, sections can be observed by LM or EM. For EM, preparations can be a) positively stained with uranyl acetate and embedded in methyl cellulose, or b) negatively stained. Sections can also be observed after critical-point drying without metal shadowing.

distilled water, dehydrated in crescent ethanol concentrations, and critical-point dried with CO_2. A special care must be taken in order to prevent accidental air drying of the preparations before critical-point drying, otherwise an artifactual collapse of the structures will occur. Replicas are rapidly transferred to a metal-shadowing device equipped with a rotating table(in our case a Balzers BAF 300 freeze-fracture apparatus) and replicated at room temperature. Shadowing is performed when the vacuum is better than 10^{-5} Torr using a specimen table rotation speed of 60rpm, Pt evaporation from 25° during 8sec. (2000V; 80mA), and carbon reinforcement from 80° during 10sec. (1200V; 200mA). After shadowing, gold grids are immersed in bleach for replica cleaning, washed several times in distilled water and dried with filter paper. Alternatively, replicas are detached from coverslips with hydrofluoric acid, floated in distilled water, transferred with a clean coverslip to bleach in order to remove the biological material and washed several times distilled water. Replicas are recovered by putting an EM grid (coated side down) over the floating replica, destroying the surrounding unwanted replica with a forceps and drying the grid with filter paper. Preparations are observed with a TEM operating at 80 or 100kV.

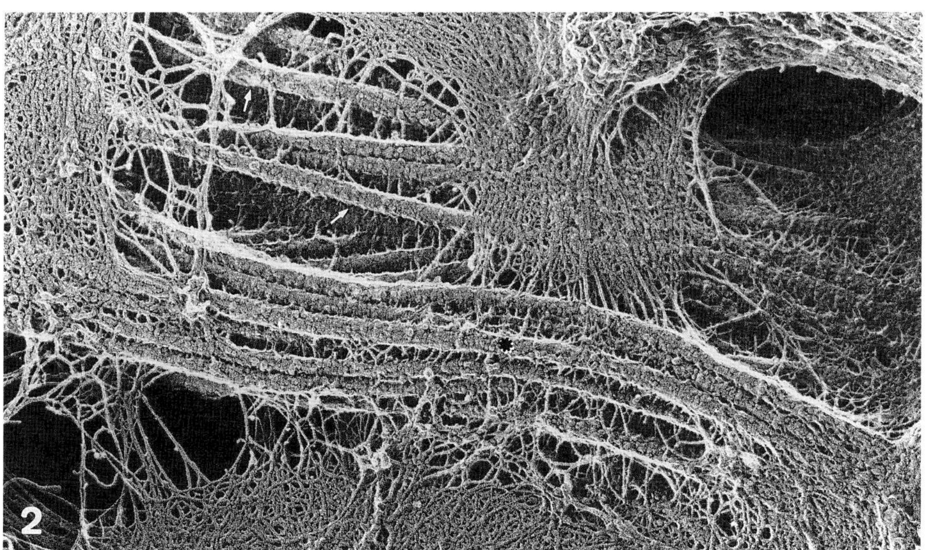

Fig.2- Low magnification of the ECM from the rooster comb. Glutaraldehyde-osmium fixation. Collagen fibrils, either isolated (arrows) or forming bundles (✱), are seen surrounded by a proteoglycan fibrillar network. Such network is remarkably apparent even without special fixation schedules (e.g. cationic dyes). The image provides a large amount of data about the 3-D organization/relationships of ECM components that could not be obtained by the analysis of conventional ultrathin sections. Photographic inversion of the negative. (x54,000).

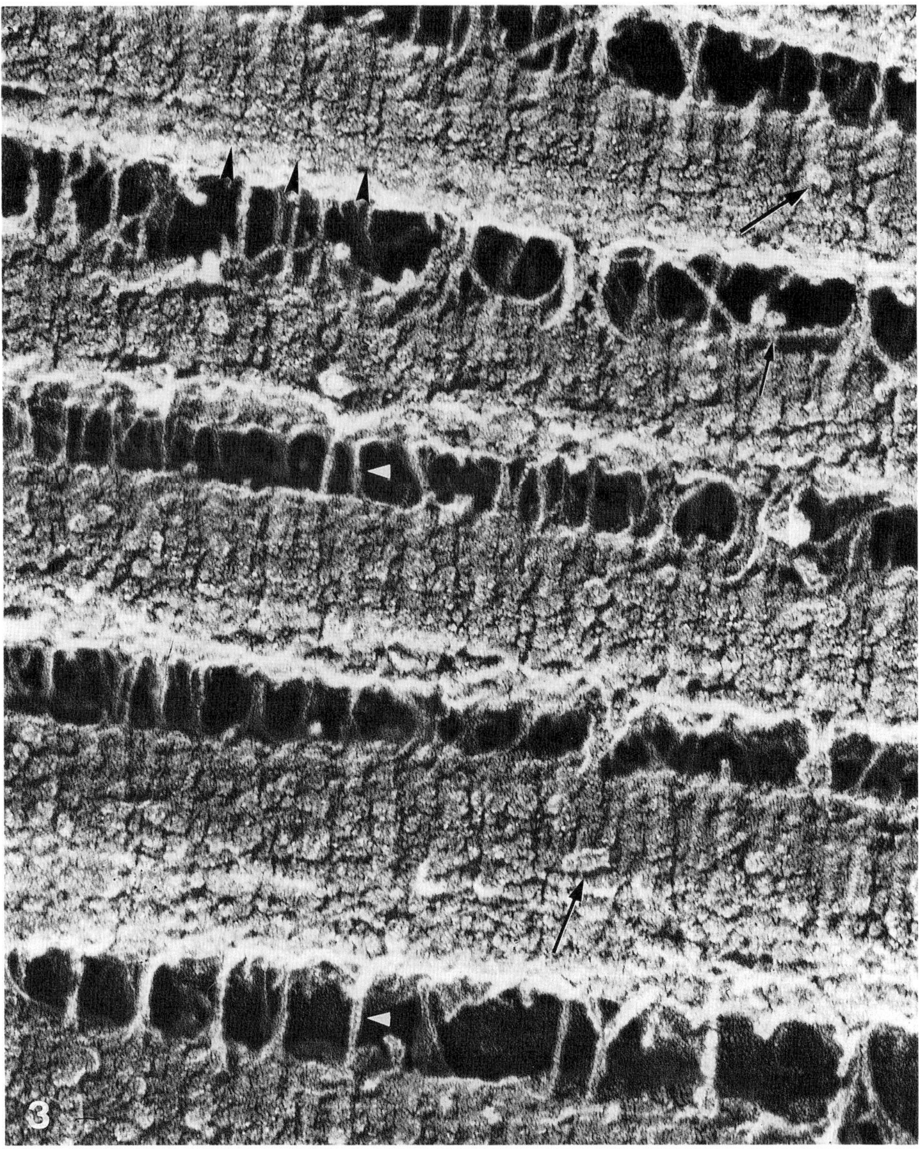

Fig.3- Bundle of thick collagen fibrils from rat tail tendon. Glutaraldehyde fixation. Numerous surface features, namely transversal cross-striation (large arrowheads), granules (large arrows) and longitudinal filaments (small arrow), are seen over fibrils. Although cationic dyes were not added to the fixative to improve glycosaminoglycan preservation/staining, numerous filamentous bridges are seen linking adjacent collagen fibrils (small arrowheads). Such bridges are not apparent in conventional ultrathin sections without cationic dye fixation. Photographic inversion of the negative. (x200,000).

RESULTS

ECM components like collagen fibrils (Figs. 2, 3, 5 and 9), elastic fibers (Fig. 6), connective tissue microfibrils (Figs.6 and 7), and proteoglycan networks (Figs. 2 and 4) are clearly identified in replicas. A 3-D appearance is obvious, even without the use of stereo-pairs. This is particularly true when photographic inversion of the negatives is performed prior to printing. However, one must remind that there may be some loss of information during inversion due to the increase of image contrast that normally occurs. A great amount of information can be obtained from replicas of tissues fixed without cationic dyes. This is particularly evident for the proteoglycan fibrillo-granular network (Fig. 2) and collagen-associated filaments and granules (Fig. 3). In contrast, most studies on the ECM organization using conventional ultrathin sections rely on cationic dyes to preserve/stain labile ECM components (e.g. proteoglycans, glycosaminoglycans). Therefore, it appears that

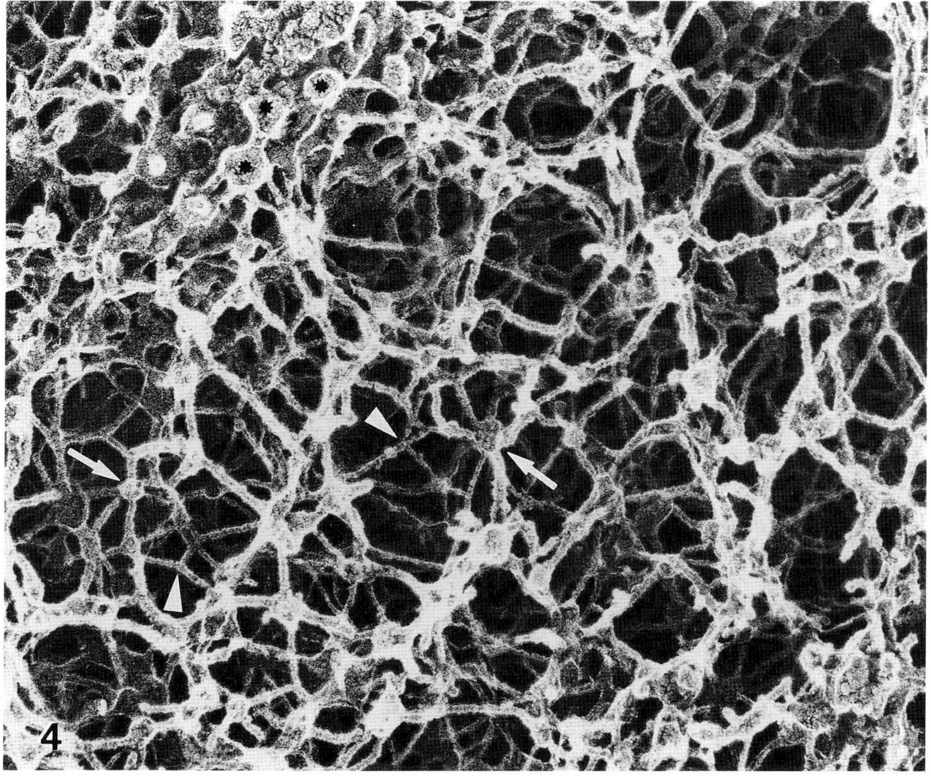

Fig.4- Rooster comb, fixed in the presence of ruthenium hexammine trichloride. A hyaluronan-proteoglycan network formed by branching filaments (arrowheads) and small granules (arrows) is well preserved, occupying most of the space around collagen fibrils (*). Photographic inversion of the negative. (x120,000).

replicas are able to demonstrate structures that, although present, would require additional electron density (e.g. cationic dyes) to become visible in ultrathin sections. This may be potentially important since cationic dyes often prevent the use of immunocytochemistry.

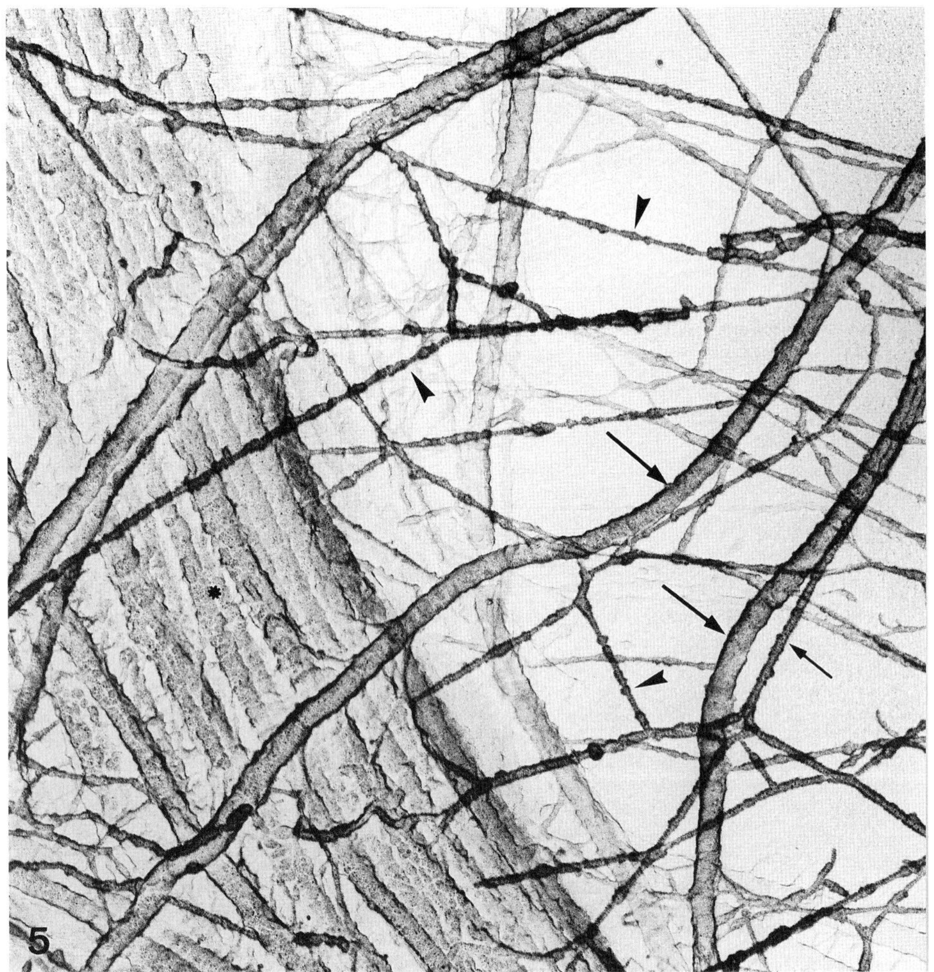

Fig.5- Triton-extracted and glutaraldehyde-osmium fixed rooster comb. Following the extraction procedure labile extracellular matrix components (proteoglycans) are extracted, allowing a good visualization of the remaining elements. These include collagen fibrils, either isolated (large arrows) or forming bundles (*), a network of thin filaments showing a 100nm periodicity (most probably collagen type VI; arrowheads), and some uniform or beaded thin filaments (thin collagen fibrils or microfibrils; small arrow). Compare with Figs. 2 and 4. (x70,000).

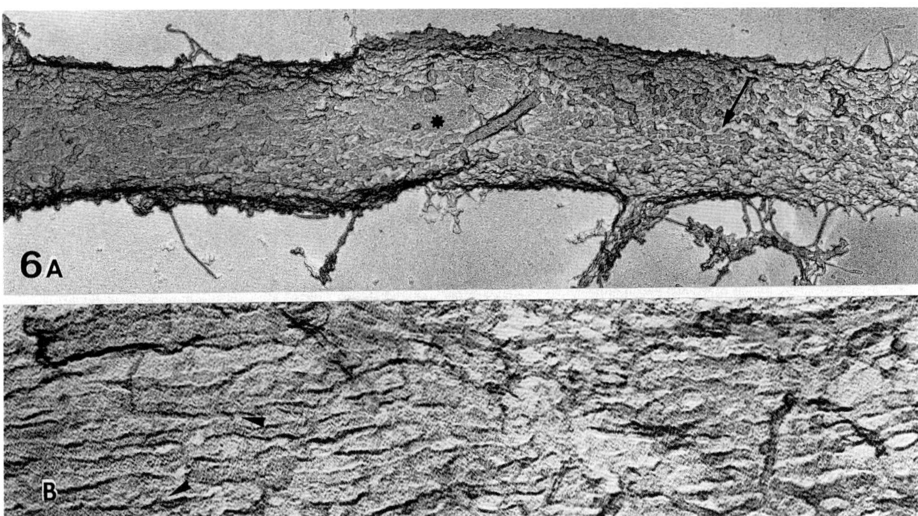

Fig.6- Elastic fibers from the rooster comb, fixed with glutaraldehyde-osmium (A) or Triton-extracted and glutaraldehyde-osmium fixed (B). Both the amorphous (✱) and fibrillar (arrow) components are seen. Elastic fiber microfibrils (B) have a beaded appearance and are linked by thin filaments (arrowheads). (A-x32,000; B-x120,000).

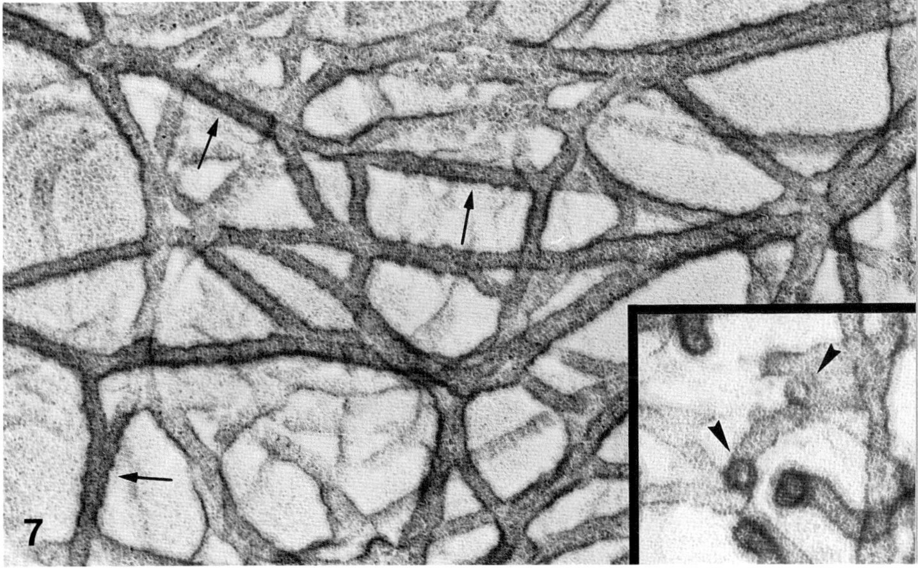

Fig.7- Connective tissue microfibrils from the rat foot pad. Paraformaldehyde fixation. Microfibrils form complex meshworks, have a striking beaded appearance (arrows) and are seen as hollow structures in cross-sections (arrowheads). Compare with Fig. 6B. (x200,000; inset-x280,000).

Although problems may arise in the study of cells, namely poor membrane preservation and compact appearance of both nucleus and cytoplasm in conventionally fixed tissues, the method can be used if applied in combination with extraction procedures (Fig. 7). It is particularly useful for the analysis of the interactions between ECM components and filamentous structures of cells. However, when strictly an analysis of cell components is required alternative methods may prove advantageous (Heuser, 1981; Fey et al., 1984).

A great advantage of the method is that it can be used in combination with post-embedding EM immunocytochemical techniques (Fig. 9). It is necessary to choose a fixation protocol that maintains tissue antigenicity (e.g. 5% paraformaldehyde),but all other steps are unchanged (Fig. 1). Incubation with adequate primary and secondary antibodies is performed after sectioning and PEG removal, and is followed by dehydration,critical-point drying and metal shadowing. When labeling intensity is not satisfactory, it may be useful to fix the labeled sections prior to the critical-point drying procedure (e.g. 1% glutaraldehyde in distilled water for 1 min.). On the other hand, when secondary antibodies are tagged to small diameter gold particles, it is advisable to evaporate a thinner metal layer otherwise the labeling will be masked by the electron-density of platinum. In our experience (using a Balzers BAF 300 freeze-fracture device) a 5 sec. evaporation time at 2,000V and 80mA produces light replicas that allow a clear visualization of 5nm gold particles.Of course, some surface detail of the tissue components will be lost.

CONCLUSIONS

A morphological approach of cell and ECM organization must rely on the use of multiple and complementary techniques. The present method, although useful to investigate the 3-D organization of both extracellular and cellular filamentous structures, is particularly well suited for the study of the ECM and cell-ECM interactions. The technical procedure is simple, rapid, with high rates of sample recovery, and can be combined with EM immunocytochemical methods to identify tissue components.

Fig.8- Triton-extracted and glutaraldehyde-osmium fixed fibroblasts from the rooster comb. The relationships between ECM components and cytoplasmic filaments at the cell cortex are demonstrated when the section plane runs above the cell surface (A). In sectioned cells (B and C),cytoplasmic filament meshworks, filament bundles (✱), and clathrin baskets (arrowheads) are well visualized. C- photographic inversion of the negative.(A-x54,000; B-x108,000; C-x100,000).

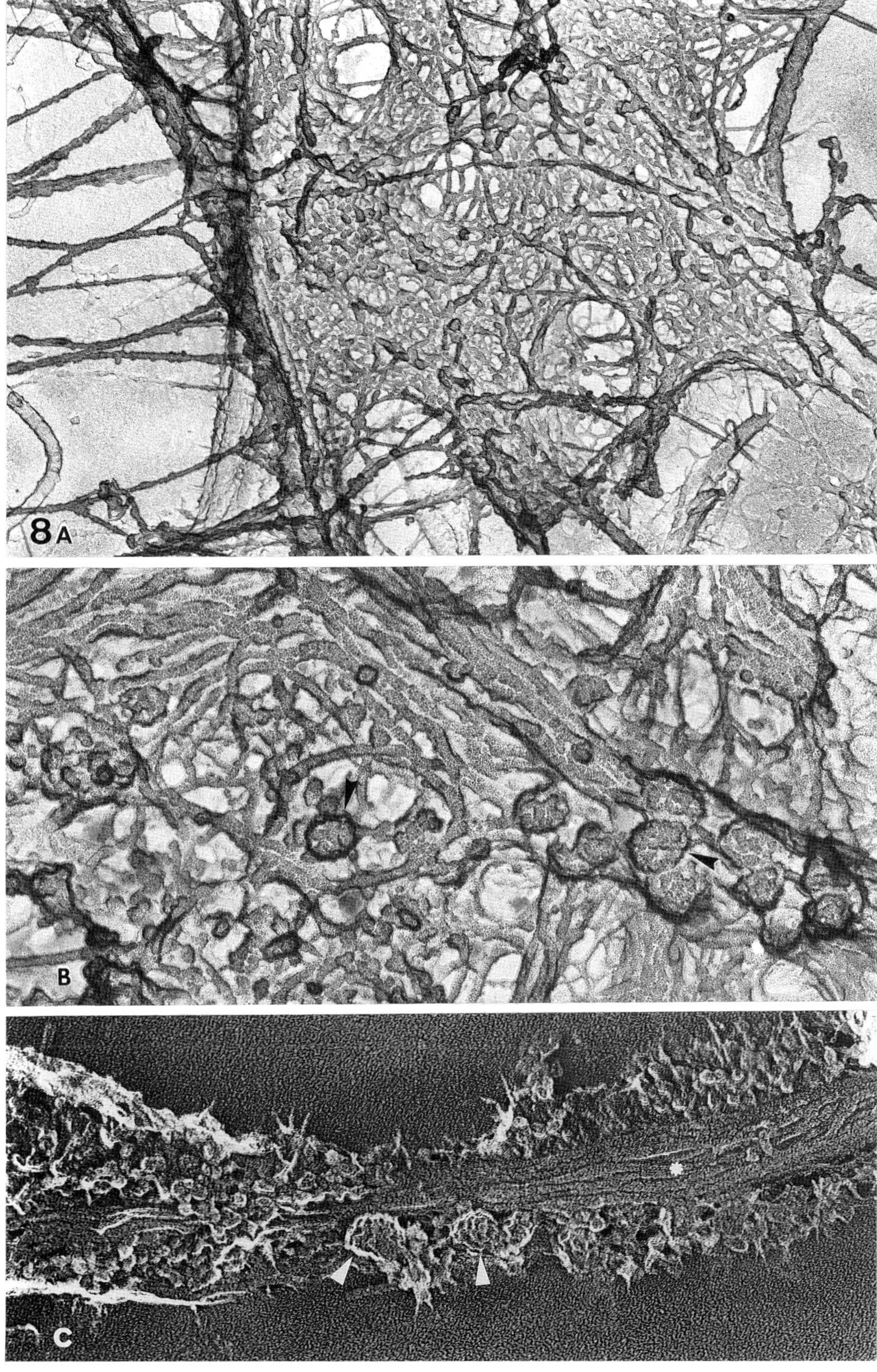
8A
B
C

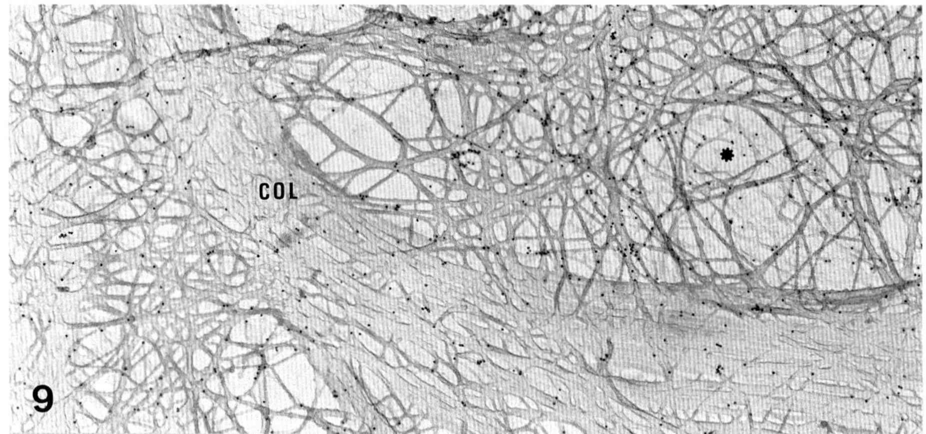

Fig.9- Paraformaldehyde-fixed paratendinous tissue from the rat tail, labeled for fibronectin by a post-embedding immuno-gold method. Fibronectin is scarcely distributed over bundled collagen fibrils (COL), but is frequently found at the complex network formed by isolated collagen fibrils and microfibrils (✱). (x27,000).

REFERENCES

Cidadão AJ, David-Ferreira JF (1986). A method for TEM visualization of the extracellular matrix three-dimensional organization in tissues. J Microsc (Oxf) 142:49-62.

Fey EG, Wan LM, Penman S (1984).Epithelial cytoskeletal framework and nuclear matrix-intermediate filament scaffold: three dimensional organization and protein composition. J Cell Biol 98:1973-1984.

Guatelli JC, Porter KR, Anderson KL, Boggs DP (1982). Ultrastructure of the cytoplasmic and nuclear matrice of human lymphocytes observed using high voltage electron microscopy of embedment-free sections. Biol Cell 43:69-80.

Heuser JE (1981). Preparing biological samples for stereomicroscopy by the quick-freeze, deep-etch, rotary-replication technique. Three-dimensional Ultrastructure in Biology. Methods in Cell Biology, Vol.22 (ed.JN Turner), pp.97-122. Academic Press, London.

Nagele RG, Doane KJ, Lee H, Wilson FJ, Roisen FJ (1984). A method for exposing the internal anatomy of small and delicate tissues for correlated SEM/TEM studies using polyethylene glycol embedding. J Microsc (Oxf) 133:177-183.

Wolosewick JJ (1980). The application of polyethylene glycol(PEG) to electron microscopy. J Cell Biol 86:675-681.

Wolosewick JJ, De Mey J, Meininger V (1983). Ultrastructural localization of tubulin and actin in polyethylene glycol-embedded rat seminiferous epithelium by immunogold staining. Biol Cell 49:219-226.

Cells and Tissues: A Three-Dimensional
Approach by Modern Techniques in Microscopy,
pages 49–56

Fracture-Flip: Nanoanatomy and Topochemistry of Cell Surfaces

Pedro Pinto da Silva, Catarina Anderson Forsman and Kazushi Fujimoto
Membrane Biology Section, Laboratory of Mathematical Biology, NCI-Frederick Cancer Research Facility, Frederick, MD, 21701, USA and Department of Anatomy, Karolinska Institutet, Stockholm, Sweden

We describe here "fracture-flip," a method to study the nanoanatomy and topochemistry of cell surfaces. Fracture-flip requires only routine equipment and quickly produces macromolecular resolution images of the ultrastructure of cell surfaces. We propose the word "nanoanatomy" to refer to this new realm. Fracture-flip can be combined with immunocytochemical methods to deliver topochemical maps of the cell surface with a resolution better than 10 nm. Fracture-flip is, therefore, one of those rare instances where the development of a new method requires no novel technologies. It is simply the product of a new idea, one that remained available, ready to be picked, for over 20 years.

Ontologically, fracture-flip (Andersson Forsman and Pinto da Silva, 1988) is the last of a series of methods that we have developed to study the topology, structure and dynamics of plasma and of intracellular membranes (for reviews see Pinto da Silva, 1987a,b):

Freeze-etching (1970) was developed as the first cytochemical approach to characterize the freeze-fracture image. It produced high resolution, contiguous images of a fracture-face and an actual membrane surface. It was used first to prove splitting of biomembranes and to show that intramembrane particles corresponded to transmembrane proteins (Pinto da Silva and Branton, 1970; Pinto da Silva et al., 1971; Pinto da Silva and Nicolson, 1974).

Fracture-label (1981), the cytochemical and topological characterization of membrane halves split by freeze-fracture led to the identification, planar distribution and mode of partition of transmembrane (glyco)proteins (Pinto da Silva et al., 1981a,b), glycolipids (Barbosa and Pinto da Silva, 1983), allowed the study of intracellular traffic of

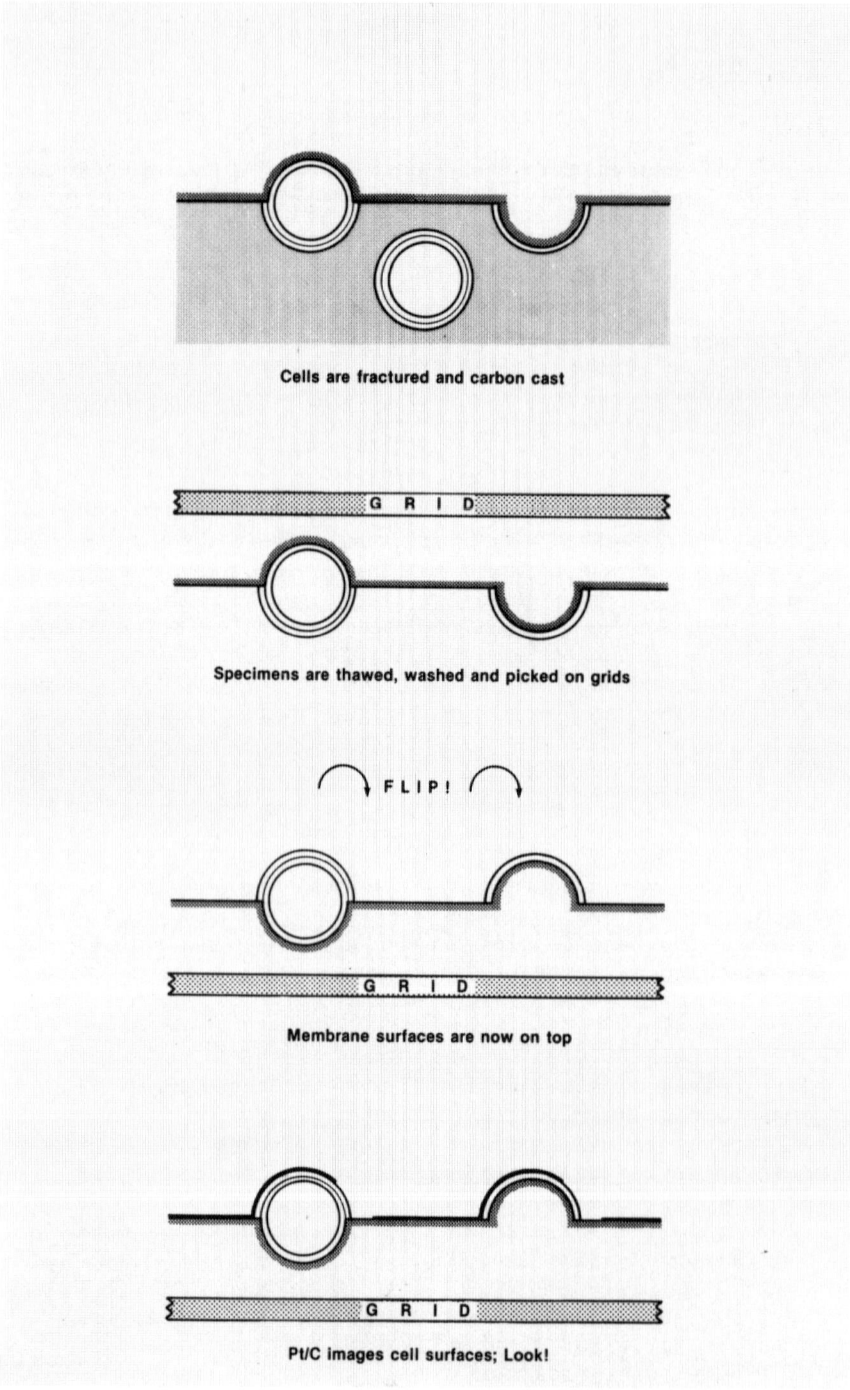

Fig. 1. How to do a fracture-flip.

membrane glycoconjugates (Torrisi and Pinto da Silva, 1984), and led to the identification of glycoproteins in non-membranous components such as the nuclear matrix (Kan and Pinto da Silva, 1986).

Label-fracture (1984), was the first method to successfully combine high resolution immunogold labeling of cell surfaces with conventional freeze-fracture. With label-fracture, we can easily observe co-incident, superimposed images of the exoplasmic fracture-face of membranes and the distribution of surface receptors and antigens. High resolution topochemical maps of cell surfaces became possible (Pinto da Silva and Kan, 1984; Kan and Pinto da Silva, 1987).

Fracture-permeation (1986) was developed to assess the compactness of cytomatrices. With fracture-permeation the entrance of macromolecular tracers into cross-fractured cytoplasm and nucleoplasm demonstrated that the compactness of cytoplasmic matrices depends on the physiologic state and can be far more compact than that predicted by micro-trabecular theories of cytoplasm (Barbosa and Pinto da Silva, 1986).

Fracture-flip derives directly from label-fracture as it is based on the stabilization of the exoplasmic half of a freeze-fractured membrane by the Pt/C replica. The steps in fracture-flip are illustrated in Fig. 1. Isolated cells, cell monolayers and isolated epithelia (but not tissues) are chemically fixed, impregnated in a cryoprotectant (glycerol) and frozen in a liquid/solid nitrogen slush (or in partially frozen Freon 22). [Unfixed cells can also be used but their freezing must be adequately fast.] The specimens in their carriers are then transferred to a freeze-fracture machine and fractured. The next step consists of "carbon fixation" of the fractured membranes; this is achieved by evaporation of pure carbon at an angle of 90 degrees. The fractured, carbon-fixed specimens are then removed from the freeze-fracture machine, thawed and the replicas floated and extensively washed in distilled water. The replicas are picked from above with formvar coated grids and dried. This step, which we call "flip," reverses the relative position of carbon cast fractured membranes and leads to exposure of cell surfaces (see Fig. 1). Accurate casts of the cell surface can now be easily obtained by re-introducing the replicas into a freeze-fracture machine (or in any other platinum evaporator) to produce high-resolution, high contrast casts by unidirectional or rotary deposition of platinum/carbon (85%/15%) from a platinum gun or electrode. These casts are observed with the transmission electron microscope.

Examination of fracture-flip replicas immediately shows that cell surfaces are studded by particulate structures whose size, density, texture and distribution are characteristic of each cell.

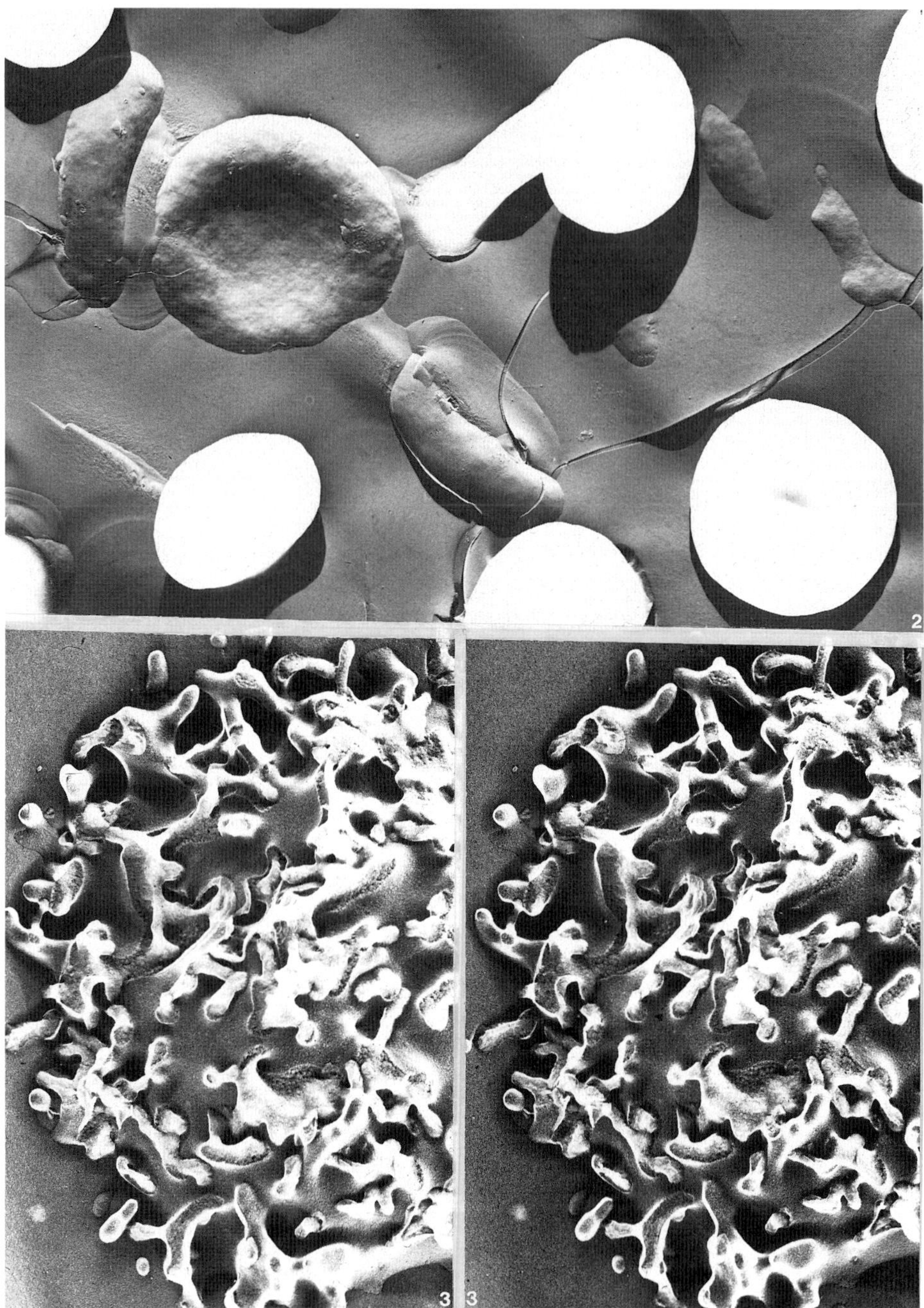

Fig. 2. Fracture-flipped erythrocytes (rev. print). Cell surfaces are found amidst cross-fractured or cells attached by the protoplasmic half of the membrane (x5500). Fig. 3. The human lymphocyte shows a high density of surface particles. Cross-fractured cell processes remain attached to the cast and are seen as whitish remnants (x20,000).

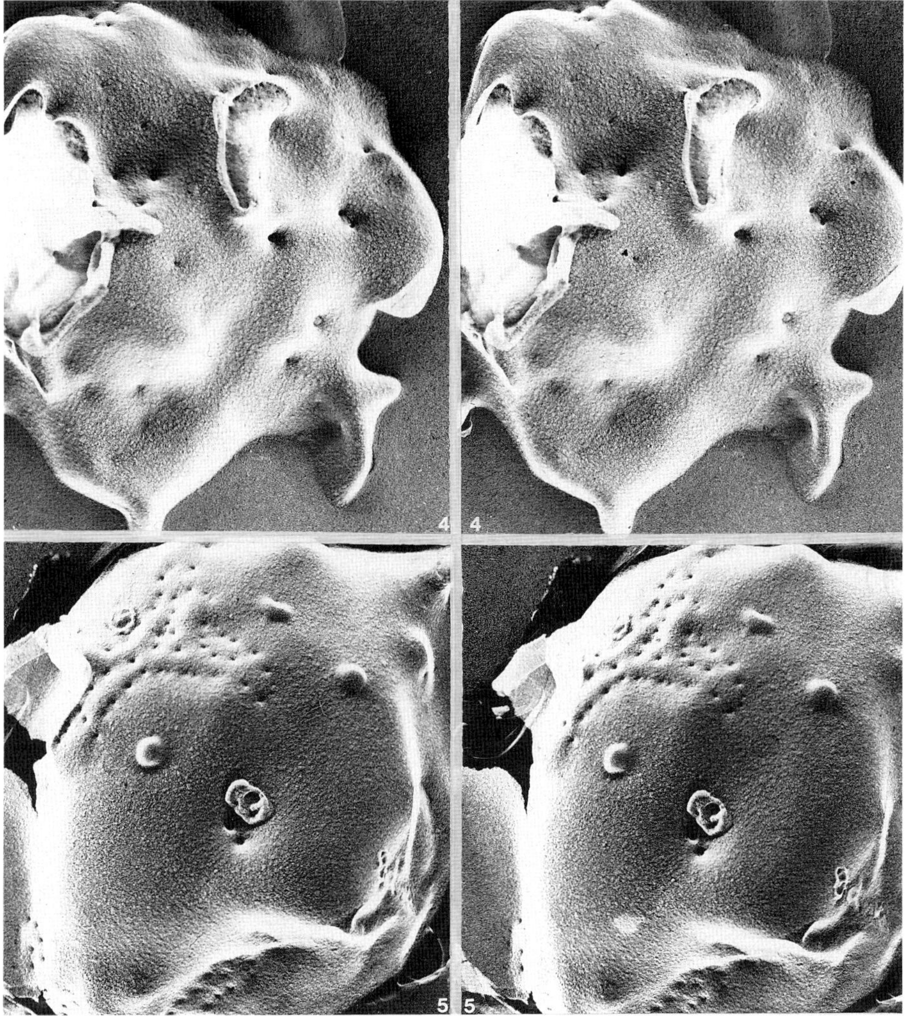

Figs. 4 and 5. Human platelets display a flocculent texture. The indentations and openings correspond to the open canalicular system (OCS) ([a]x25,000 and [b]x28,000).

Some examples are illustrated in Figs. 2 to 8. We encourage you to examine and compare these micrographs among each other. You will see that their high resolution, while similar to that of freeze-etched membranes surfaces (note, however, that etching can only reveal small areas of the membrane and that prolonged etching can cause tearing and distortion), is clearly higher than that provided by any conventional scanning electron microscope. The resolution obtained by Tanaka and his colleges (see chapter this volume) with the high resolution scanning electron microscope and specimen coater is, in some instances, higher. Tanaka's approach,

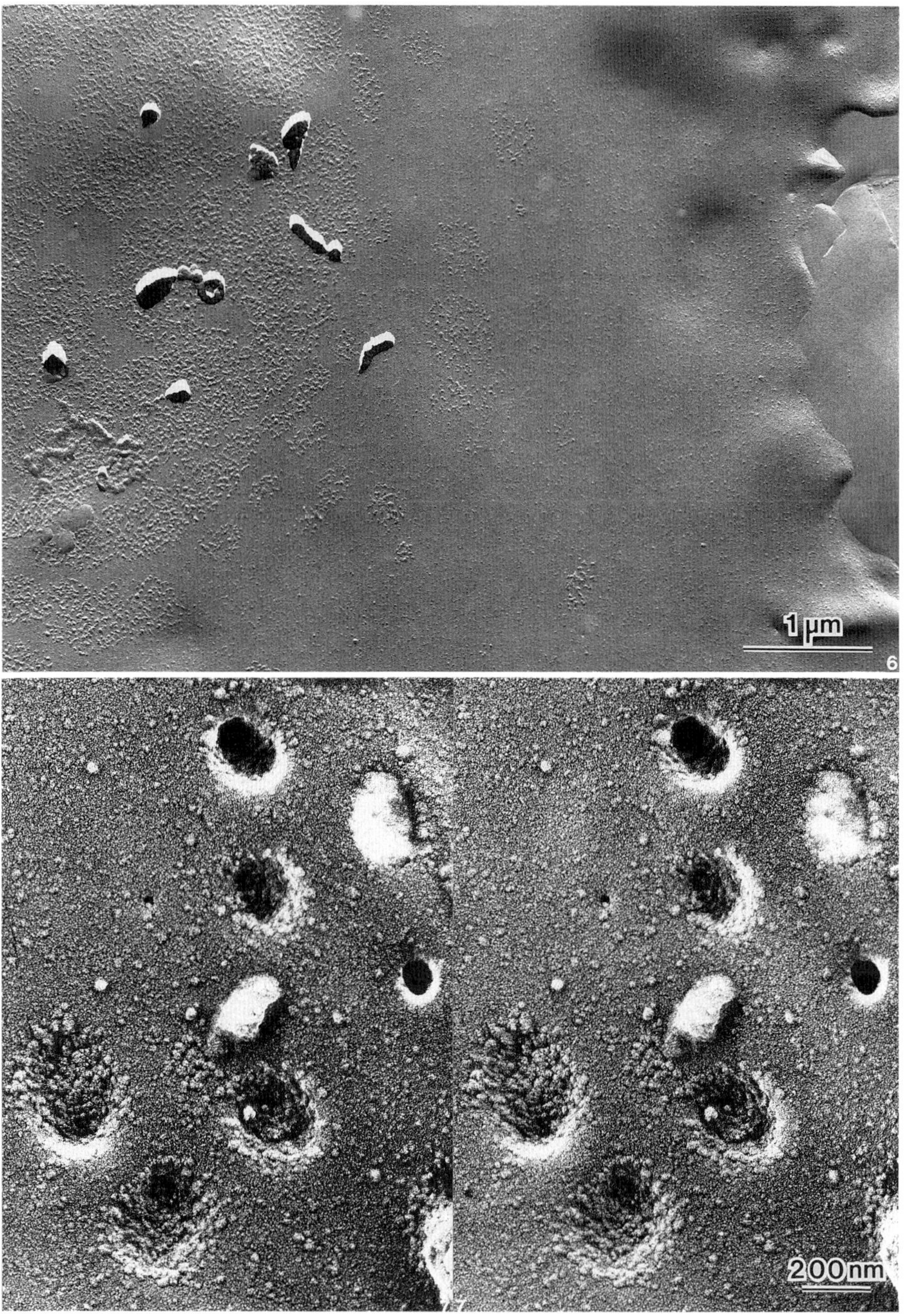

Fig. 6. Adherent surface of a macrophage warmed up to 37°C for 15 min. Membrane particles (15-25 nm) are clustered into patches (area 0.5-4 μm^2). Fig. 7. Stereo pair micrographs of rotary-shadowed replica of a macrophage incubated at 37°C for 30 min. The particle clusters are deeply invaginated.

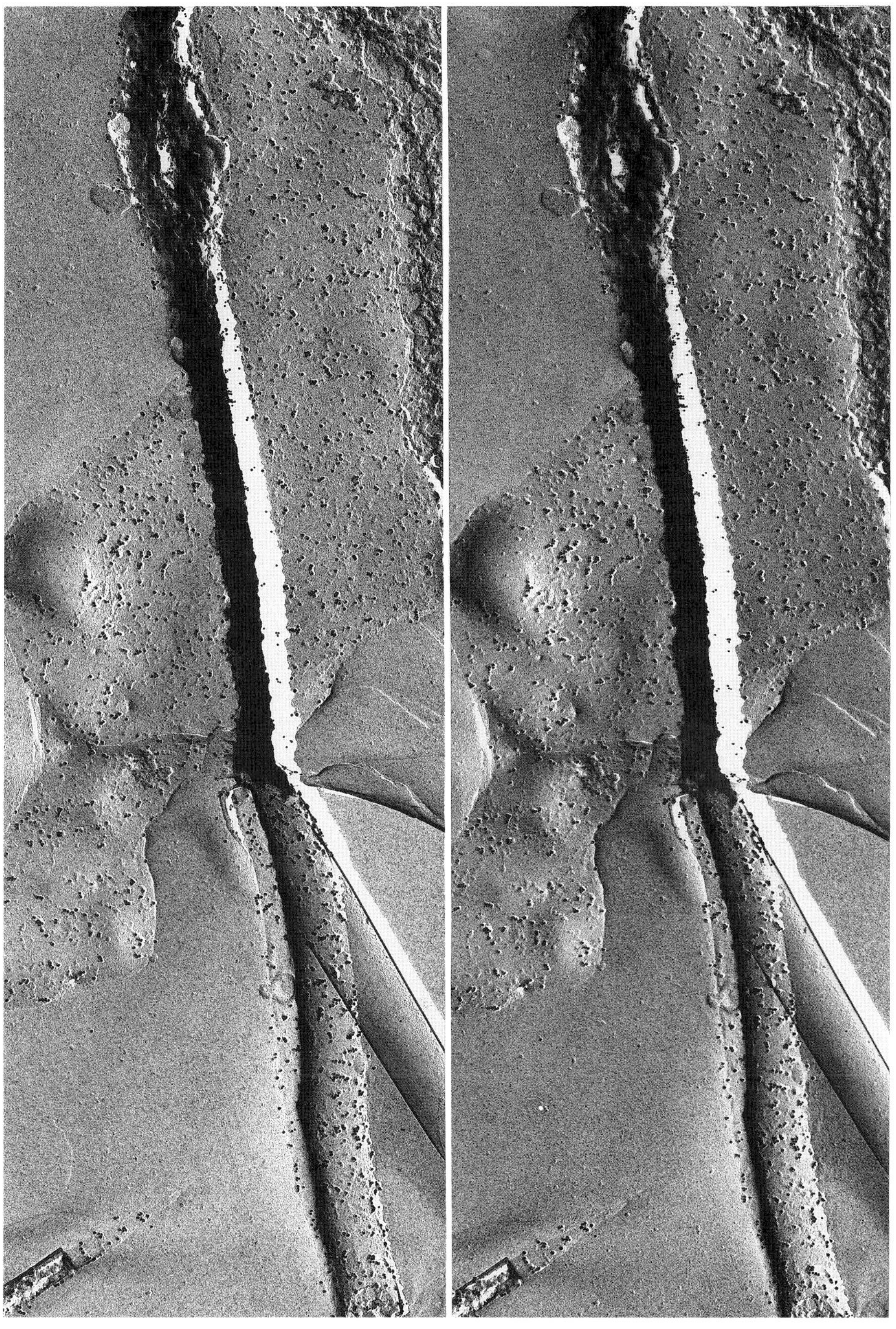

Fig. 8. Primary culture of rat cerebella granule cells. These cells have been labeled with conA-colloidal gold. The label is found both over the perikarya and the neurites (x32000).

however, is limited to the anatomical exploration of cellular structures and is, at present, technically difficult to combine with cytochemistry.

Fracture-flip is an easy method that requires only readily available equipment. It can be applied to any isolated cell or membrane, as well as to cell monolayers or to isolated epithelia. Fracture-flip can be combined with immunogold methods to provide high resolution topochemical maps of the surface of cells. An example is shown in Fig. 8.

Fracture-flip is not a trick, rather, the product of thought. It leads to the unexpected paradox that freeze fracture transmission electron-microscopy, a method that revealed images of the apolar interior of membranes, originates the method for routine study of the nanoanatomy and topochemistry of cell surfaces. We encourage you to try it.

REFERENCES

Andersson Forsman C, Pinto da Silva P (1988). J Cell Sci 90:531-541.

Barbosa MLF, Pinto da Silva P (1983). Cell 33:959-966.

Barbosa MLF, Pinto da Silva P (1986). J Electron Microsc Tech 4:385-397.

Kan FWK, Pinto da Silva P (1986). J Cell Biol 102:576-686.

Kan FWK, Pinto da Silva P (1987). J Histochem Cytochem 35:1069-1078.

Pinto da Silva P (1987a). Molecular cytochemistry of freeze-fractured cells: Freeze-etching, fracture-label, fracture-permeation, and label fracture. Adv Cell Biol 1:157-190.

Pinto da Silva P (1987b). Toplogy, dynamics and molecular cytochemistry of integral membrane proteins: A freeze-fracture view. In Harris JR, Horne RW (eds): "Electron Microscopy of Proteins," Vol. 6 (Membranous Structures), London: Academic Presss, pp 1-38.

Pinto da Silva P, Branton D (1970). J Cell Biol 45:598-605.

Pinto da Silva P, Douglas SD, Branton D (1971). Nature 232:194-196.

Pinto da Silva P, Kan FWK (1984). J Cell Biol 99:1156-1161.

Pinto da Silva P, Nicolson GL (1974). Biochim Biophys Acta 363:161-181.

Pinto da Silva P, Parkison C, Dwyer N (1981a). Proc Natl Acad Sci USA 78:343-347.

Pinto da Silva P, Parkison C, Dwyer N (1981b). J Histochem Cytochem 29:917-928.

Torrisi MR, Pinto da Silva P (1984). J Cell Biol 98:29-34.

note: Figs. 6 and 7 are taken from Fujimoto K, Pinto da Silva P (1988) J Cell Sci., in press.

Cells and Tissues: A Three-Dimensional
Approach by Modern Techniques in Microscopy,
pages 57–62

CHROMOSOMAL STRUCTURE OBSERVED BY OPTICAL, COLOR LASER AND ELECTRON MICROSCOPES

Akihiro Iino, Tomonori Naguro, Kenji Funaki and Sumire Inaga

Department of Anatomy, Tottori Uneversity
School of Medicine, Yonago 683 Japan.

INTRODUCTION

The chromosomes are dense and intensely stainable small body, ,first named by Waldeyer exactly 100 years ago, 1888.

The chromosomes of higher organisms are studied most frequently at mitotic metaphase. This is the stage at which the chromosomes reach their greatest condensation. In general, metaphase chromosomes appear as small threads or sticks about several micron in length. That is to say that during mitotic metaphase the condensed chromosomes appear in identifiable shape characteristic of the karyotype of species beeing studied.

The color laser, scanning electron, transmission electron and ultra high voltage electron microscopes were used for this study in addition to the conventional optical microscope.

1. Conventional optical microscope

Cultured human peripheral lymphocytes were used for investigating the spiral structure of the chromosomes. Cells were harvested by centrifugation and treated with special hypotonic solution, mixture of 0.01M citric acid and 0.02M sodium secondary phosphate controlled at pH 5.6 for 20 m (Iino, 1975).

Several types of spiral structures were observed with the conventional optical microscope.

Although the real reason why the hypotonic solution controlled at pH 5.6 made coiled chromosomes fibers was uncertain,it is suggested that some cytoplasmic materials surrounding the chromosome were more or less released during the course of the treatment (Fig. 1).

2. Color Laser Microscope

The color laser microscope (2LM 11, Lasertec, Yokohama, Japan) using a new color laser imaging system has been especially developed for the visual inspection of

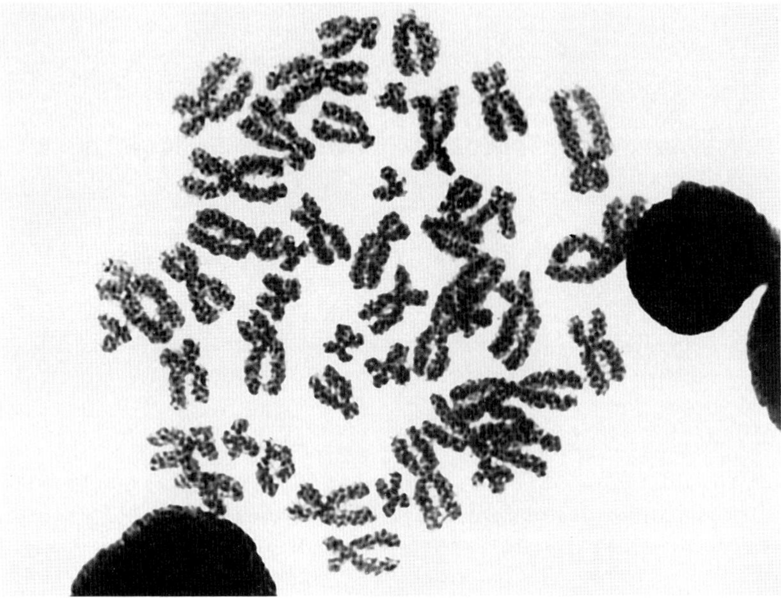

Fig. 1. Spiral structure of human somatic chromosomes treated with special hypotonic solution.

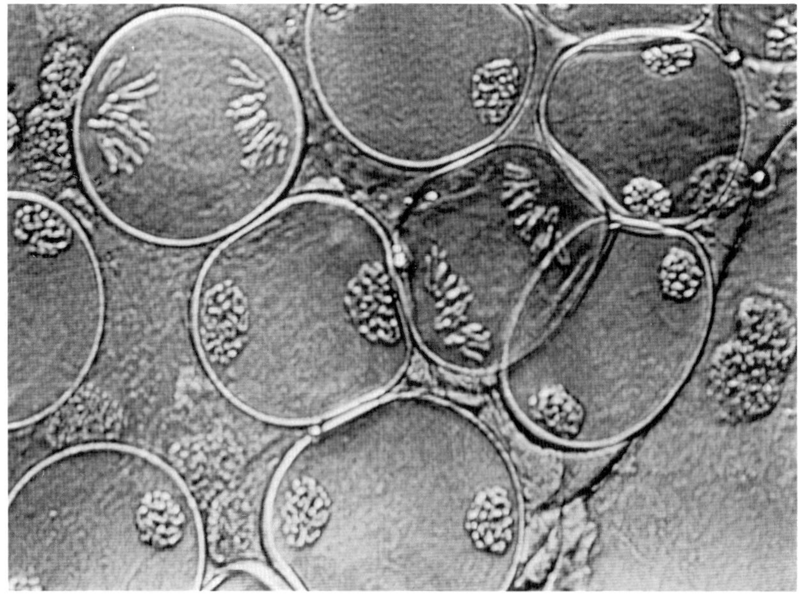

Fig. 2. Color laser micrograph of Lilium longiflorum pollen mother cells (original picture is in color).

semiconductors (Awamura et al., 1987). The light source is produced by three lasers (Red ; He-Ne 633 nm, Green ; Ar 515 nm, Blue ; Ar 488 nm).

The laser beam is focused in a small spot which is scanned over the sample at high speed. The microscope uses a reflective confocal optic system which produces a deeper focus than that of the conventional microscope.

Fig. 2 is a picture of pollen mother cell of <u>Lilium longiflorum</u> taken by the color laser microscope. Original picture is in color.

Many kinds of cell division phases can be recognized in the picture.

3. Transmission electron microscope

In the present study, a vacuum evaporated silicon monoxide (SiO) film was used for supporting the chromosomes on the grids.

Chromosomes themselves were obtained from cultured human

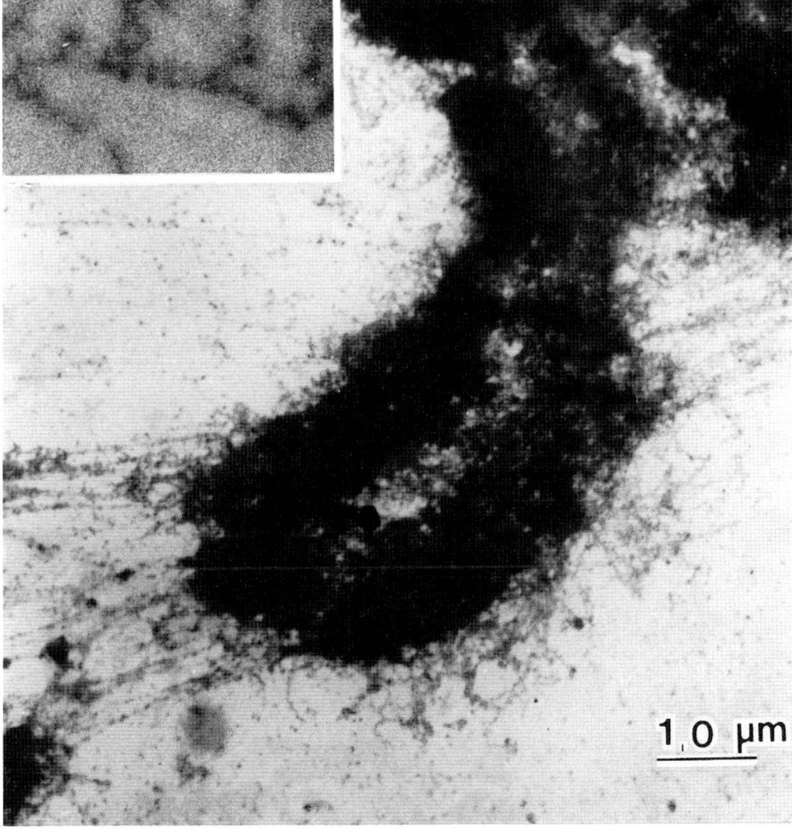

Fig.3. Transmission electron microscopic picture of human somatic chromosome on SiO thin film. Threads and beads structure can be observed.

lymphocytes of peripheral blood. Cells were harvested by centrifugation and were treated with same hypotonic solution for the conventional optical microscopy mentioned above controlled at pH 5.6.

Metaphase chromosomes were constructed of fibrous materials which usually made loops around chromosome body. Some of the inter chromosomal fibers show beaded or spiral structure as shown in Fig. 3.

An ultrahigh voltage electron microscope was more useful to observe thick whole mount materials. The direct comparison was made between Fig. 4 and Fig. 5. Fig. 4 was observed by conventional transmission electron microscope and 100 KV electron beam was too weak to penetrate into the thick chromosome body. On the other hand a 1000 KV ultrahigh voltage microscope crosses the chromosome body to observe inside thin fibers as shown in Fig. 5.

For both transmission electron microscopes the vacuum evaporated SiO films were applied for supporting whole mount chromosomes. The SiO film was very tough and strong against the electron beam. It is not much difficult to make the uniformly thin film from a small lump of pure SiO. In addition to the toughness the film is very convenient for

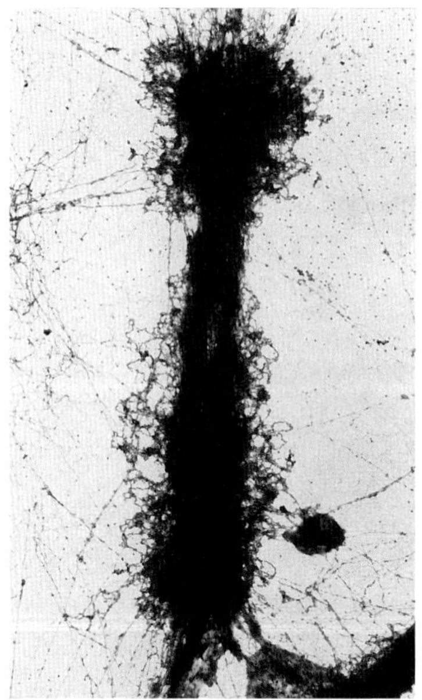

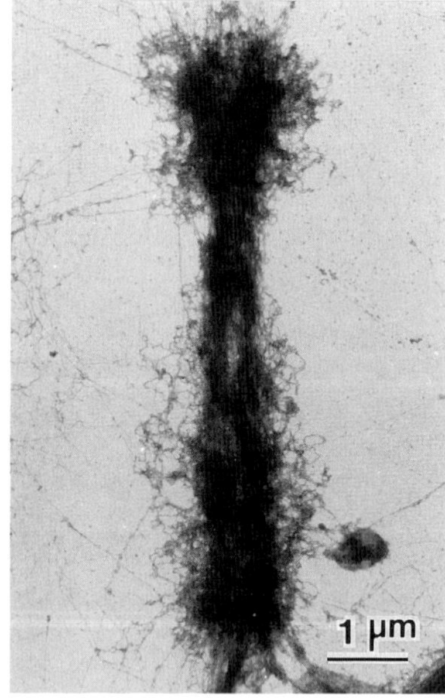

Fig.4. (left) Conventional transmission electron micrograph (100 KV) of one of C group human chromosome. Chromosome body is too thick to be penetrated with electron beam.

Fig.5. (right) Same chromosome observed by ultrahigh transmission electron microscope (1000 KV). The fibers of chromosome body become clearer than Fig.4.

chromosome spreading, because the surface of the film is similar to that of the slide glass.

4. Scanning electron microscope

Human peripheral lymphocytes were also employed for this observation. After cultivation the cells were collected by centrifugation and were treated with the same hypotonic solution as used in optical and transmission electron microscopies.

A drop of the mixture of cells and solution was put onto a cover slip and covered with another small cover slip. Then the cover slips were heated gently over a flame until the cytoplasmic matrices adhering to the chromosomes were removed completely.

After heating, the preparation was left in a Petridish for 24 h with a filter paper moistened with 45% acetic acid solution, and then immersed in equal parts of glacial acetic acid and absolute ethanol for 20 m and finally transferred to

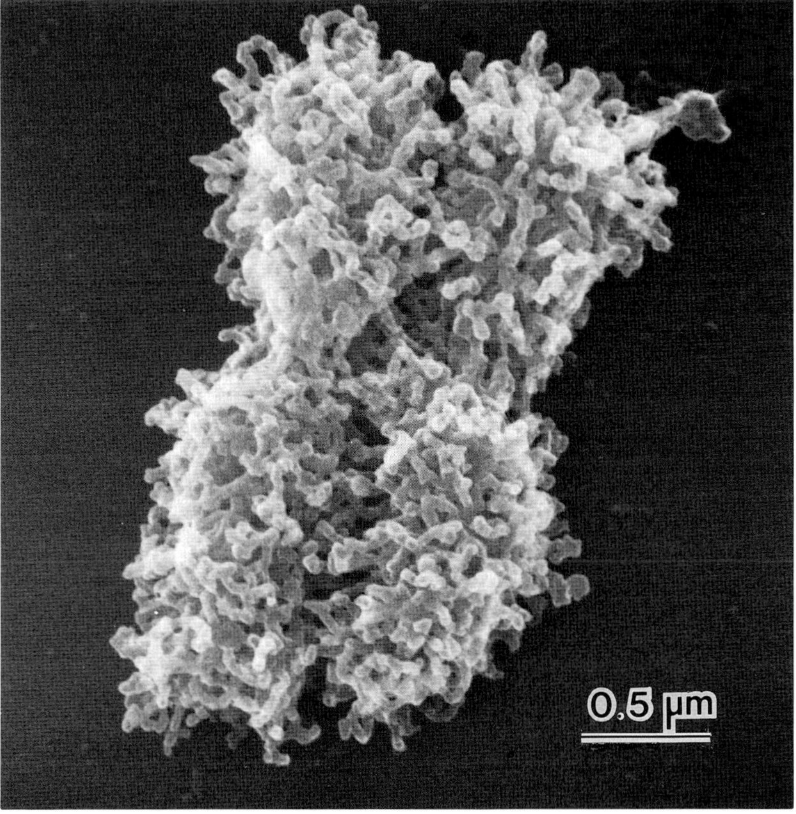

Fig.6. Scanning electron micrograph of one of human chromosomes which was observed to be made up of fibrous components.

absolute ethanol for 30 m. The bigger cover slip was then inverted in the final absolute ethanol to remove the smaller cover slip. The cover slip mounting chromosomes were critical point dried using dry ice (Tanaka & Iino, 1974). After immersing in isoamyl acetate, the required area was cut off and coated with platinum and observed with the scanning electron miroscope.

A chromosome was ovserved to be made up of fibrous components as shown in Fig. 6. The fibers were found to be roughly 30 nm in diameter. The fibrous bridges connecting each sister chromatid were present. The kinetochore region was seen as a constriction between the tortuous fiber masses of chromosome, but any particular structure was not found in our investigation.

CONCLUSION

Human somatic chromosomes essentially appear spiral or meshlike structures when observed eather with optical or electron microscope. By the surface spreading technique, metaphase chromosomes proved to be composed of fibers about 30nm in diameter. The fibers are nodular, twisted and looping especially observed with scanning electron microscope. The diameter of these fibers coincides well with that of the solenoidal model.

We hope the improvement of the color laser microscope enables the structure of chromosome clearer and better specimen preparation techniques for electron microscopesmakes chromosome details more obvious.

REFERENCES

Awamura D, Ode T, Yonezawa M (1987). Color laser microscope. SPIE 765 : 53-60.

Iino A (1975). Human somatic chromosomes observed by scanning electron microscope. Cytobios 14 : 39-48.

Tanaka K, Iino A (1974). Critical point drying method using dry ice. Stain technol 49 : 203-206.

Cells and Tissues: A Three-Dimensional
Approach by Modern Techniques in Microscopy,
pages 63–70

MITOCHONDRIAL STRUCTURE REVEALED BY SCANNING ELECTRON MICROSCOPY (SEM).

Peter J. Lea and Martin J. Hollenberg
Department of Anatomy, Faculty of Medicine, University of Toronto, Ontario, Canada M5S 1A8

INTRODUCTION

Our current understanding of mitochondrial structure has been mainly derived from the study of thin sections by transmission electron microscopy (TEM). To develop a three dimensional (3D) perspective, artist's impressions (Krstic, 1979) and computer based reconstructions from serial sections (Lea and Pawlowski, 1986) have been used. However, the resolution of 3D reconstructions in the third dimension is limited by section thickness and misinterpretation can result.

Direct and simultaneous observation of both internal and external ultrastructure of mitochondria (Lea and Hollenberg, 1987; Hollenberg and Lea, 1988) is now possible due to the improved resolution of the modern scanning electron microscope coupled with specimen preparation techniques which expose intracellular structures in 3D. As a result, our micrographs provide a new perspective on the microanatomy of mitochondria in rat hepatocytes, retinal pigment epithelial cells, renal proximal tubular epithelial cells and corneal endothelal cells. Each high resolution scanning electron micrograph presented herein, due to the depth of field inherent in the technique, corresponds to a series of conventional, perfectly aligned, serial, TEM sections.

Our results indicate that mitochondria, prepared by our technique and viewed by high resolution electron microscopy (HRSEM), for the most part, display cristae that are tubular in shape. The one exeption identified so far is in the corneal endothelial cell. We hypothesize, that, most mature mitochondria have tubular, not shelf-like cristae and that, in mitochondria with tubular cristae, the walls of the tubes are continuous with the inner mitochondrial membrane and the lumen of each tube is in continuity with the intermembranous space of the mitochondrial wall.

MATERIALS AND METHODS

Specimens fixed by immersion perfusion in 0.5% glutaraldehyde were washed in phosphate buffer, placed in 1% osmium tetroxide in phosphate buffer for 1.5 hrs at 20 degrees Centigrade, washed again and stored in buffer at 4 degrees overnight. The blocks were placed in 25% dimethylsulfoxide (DMSO) in distilled water for 0.5 hr, then 50% DMSO for 0.5 hr, frozen by immersion in liquid freon 22, transferred to liquid nitrogen and cleaved with a razor blade in a brass well while immersed in liquid nitrogen. Thawing took place in vials containing 50% DMSO at 20 degrees. The DMSO was removed by washing in phosphate buffer. The specimens were postfixed in 1% osmium tetroxide in phosphate buffer for 1 hr and extraction of the cytosol was accomplished by immersion of the samples in 0.1% osmium tetroxide in phosphate buffer for up to 6 days at room temperature. Samples were then washed in buffer, placed in 1% tannic acid in distilled water for 1 hr., washed, placed in 1% osmium teroxide in buffer for 1 hr and returned to buffer. Dehydration was accomplished in a graded ethanol series and the specimens dried using a critical point method employing liquid carbon dioxide (Lea and Ramjohn, 1980).

The specimens were sputter coated with gold or gold/palladium alloy and viewed using a Hitachi, Model S-570 scanning electron microscope. The specimen preparation techniques used in this study allowed us to make direct comparisons at similar magnifications between transmission electron micrographs of mitochondria and those in our HRSEM micrographs.

RESULTS

Hepatocyte mitochondria, examined at low (Fig.1), and higher magnification (Fig.2), varied both in size and shape and contained numerous tubular cristae. Shelf-like cristae were not found. Many tubular cristae were continuous with the inner mitochondrial membrane at one end and were fractured at the other (Figs.1,2). Some tubular cristae protruded above the freeze cleavage plane. However, many tubular cristae remained intact and some of these were found to span the matrix (Fig.2), with each end continuous with the inner mitochondrial membrane. The rough endoplasmic reticulum of the hepatocyte was found in close association with the mitochondria and was clearly visible in three dimensions (Fig.2).

Mitchondria of retinal pigment epithelial cells also contained only tubular cristae (Figs.3,4). In some cristae,

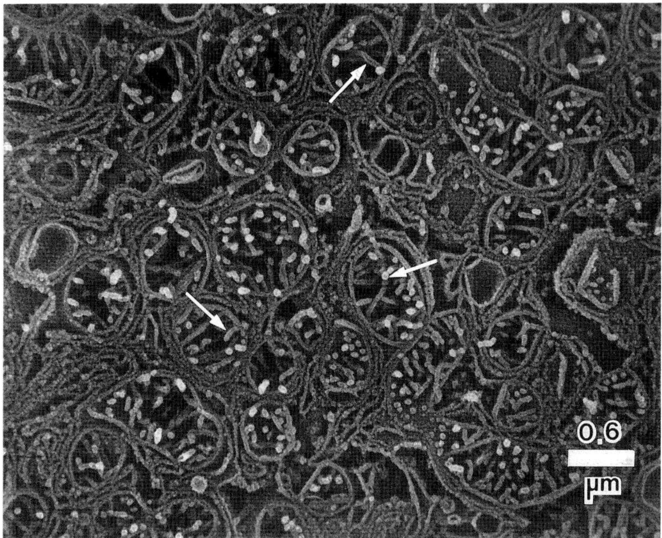

Figure 1. Rat hepatocyte mitochondria with tubular cristae. (arrows)

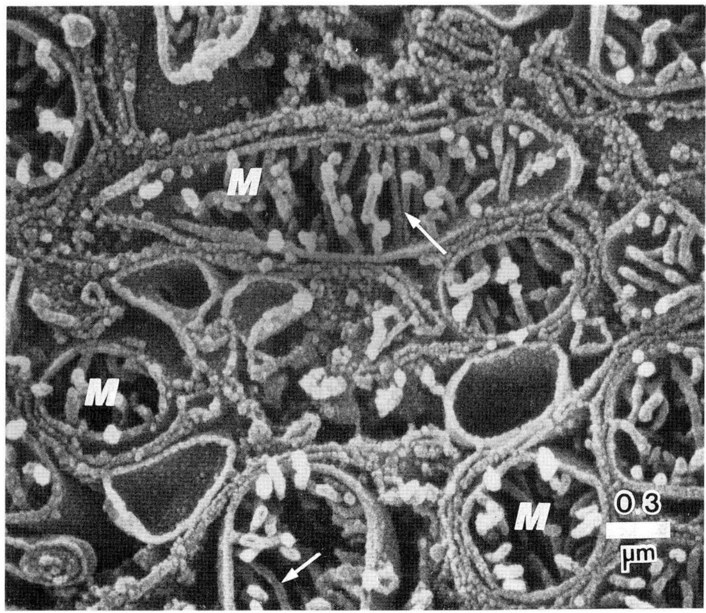

Figure 2. Rat hepatocyte mitochondria (M). Some of the cristae span the matrix (arrows).

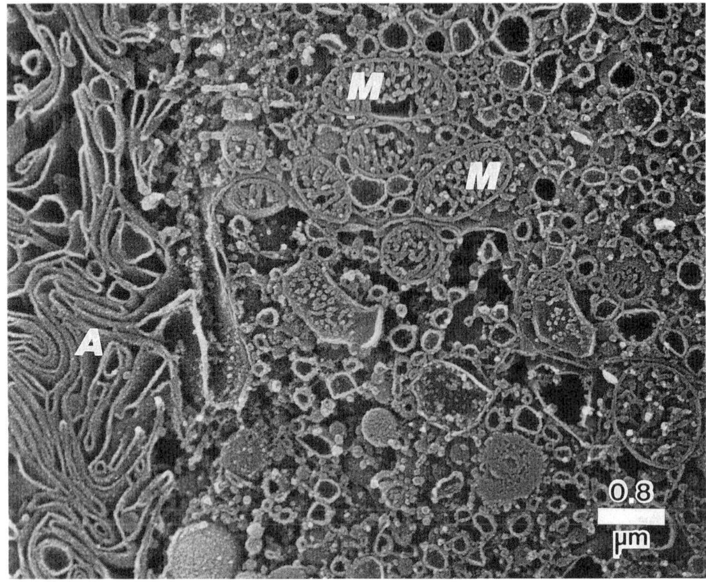

Figure 3. Mitochondria (M) in a rat retinal pigment epithelial cell (A) apical plasma membrane.

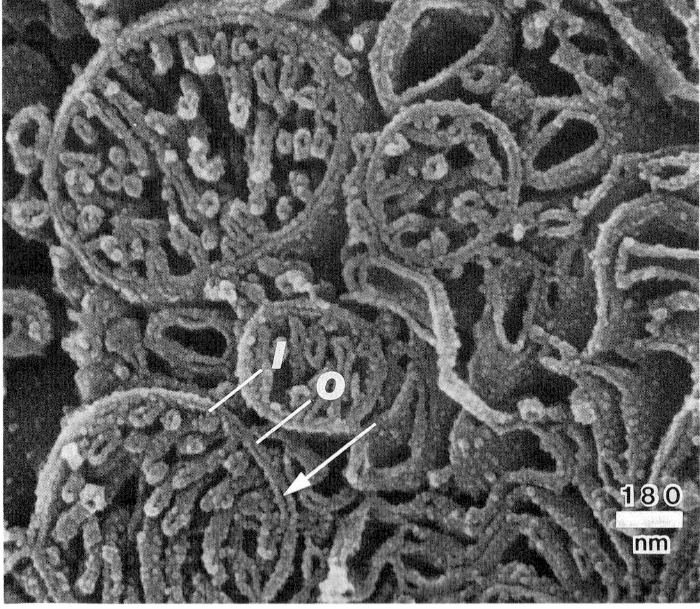

Figure 4. Inner (I) and outer (O) mitochondrial membrane of rpe separated by intermembranous space (arrow).

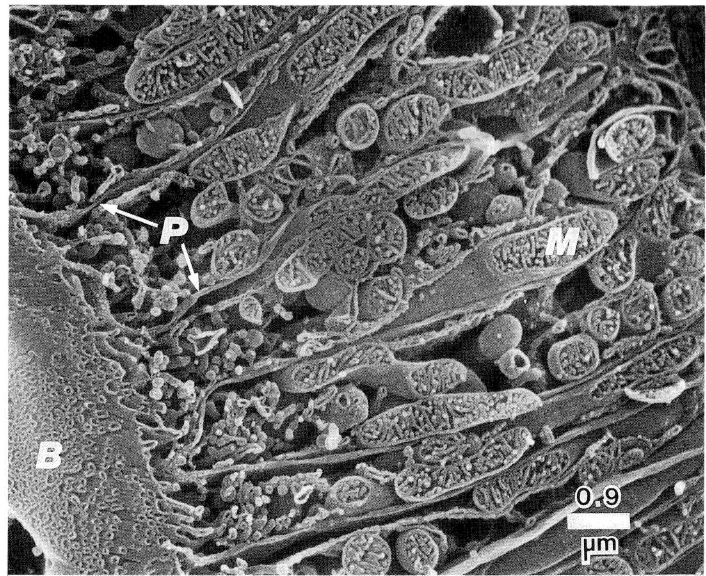

Figure 5. Rat renal proximal tubular cells with mitochondria (M). (B)= Brush border, (P)= plasma membrane.

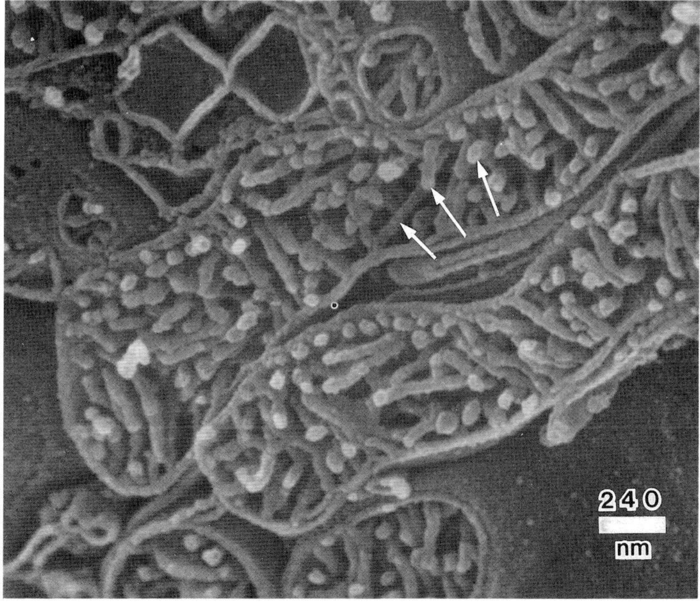

Figure 6. Mitochondria of rat proximal tubular cells with tubular cristae (arrows)

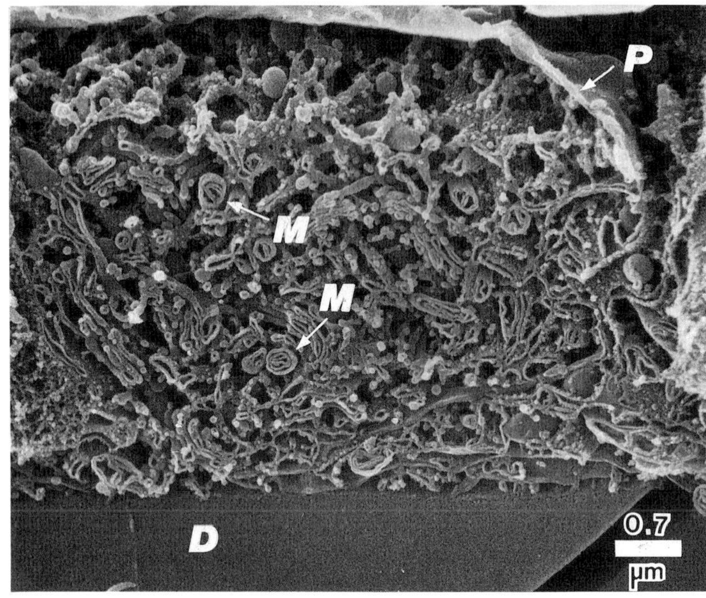

Figure 7. Rabbit corneal endothelial cell. (M)=mitochondria (D)=Descemet's membrane, (P)=plasma membrane

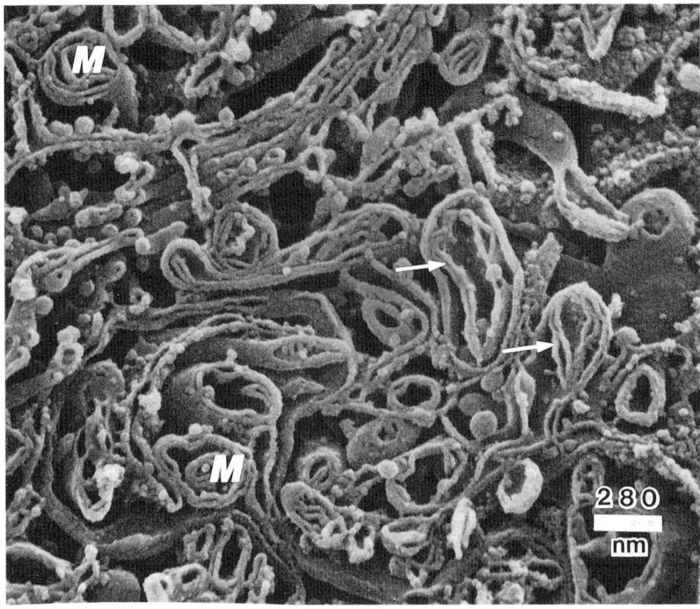

Figure 8. Mitochondria of a rabbit corneal endothelial cell. The mitochondria (M) display internal membranous sheets. (arrows).

cleaved longitudinally, lumens were be seen to be continuous with the space between the inner and outer mitochondrial membranes (Fig.4). Similarly, all of the many mitochondria examined in rat renal, proximal tubular cells (Figs.5,6) were found to contain only tubular cristae.

In marked contrast, mitochondria, observed at both low (Fig.7) and at higher magnification (Fig.8), found in rabbit corneal endothelial cells, had a different internal structure. Their cristae often appeared as layered, membranous sheets (Fig.8), while distinct tubular cristae were only rarely encountered.

DISCUSSION

There are a number of advantages to the HRSEM technique used in this study. Principal among these is the ability to examine intra and extracellular structures, prepared and viewed in a new and different way, at a resolution close to that of the conventional TEM. Direct comparisons between TEM and HRSEM micrographs of cellular structures at similar magnifications can now be made, thereby greatly strengthening our ability to interpret cellular ultrastructure. The design of the Hitachi S-570 SEM used to obtain our micrographs is well suited for high resolution work since it permits, (1) insertion of a specimen approximately 2 cubic mm in size directly into the bore of the objective lens and (2) secondary collection in the column above the objective lens. Specimens can readily be viewed at 50 diameters to gain an overall appreciation of tissue organization and then the components of an individual cell examined in the same sample, clearly and easily, at over 50,000 diameters, using stereo pair micrographs if desired. This flexibility in tissue examination is ideal for the study of pathological specimens, clearly a major advantage of the technique which has not been generally appreciated as yet. Further, embedding and sectioning is not required and the cytosol removal technique, pioneered by Tanaka (e.g. Tanaka and Mitsushima, 1984) allows examination of organelles, such as mitochondria and Golgi in 3D, both internally and externally. At the same time an appreciation of inter-organelle relationships can be gained (Hollenberg et al., 1988)

Although mitochondria with tubular cristae have been described (Wheatly, 1968), the current view is that most mitochondrial cristae are shaped as shelves or plates which stretch part or all of the way across the organelle (Palade, 1952; Andersson-Cedergren, 1959; Alberts et al., 1983). In contrast, in this study, mitochondria viewed by HRSEM contained tubular, not plate-like cristae with the exception of those in the corneal endothelial cell.

We have studied, by TEM, the structure of the cristae in the hepatocyte of the adult rat and have found numerous circular profiles corresponding in size to the tubular cristae seen in our HRSEM preparations cut in cross section.

REFERENCES

Alberts B, Bray D, Lewis J, Raff M, Roberts K, Watson JD (1983). The molecular biology of the cell. pp 487. Garland Publ Inc, New York.

Andersson-Cedergren E (1959). Ultrastructure of motor end plates and sarcoplasmic components of mouse skeletal muscle fiber as revealed by three-dimensional reconstruction from serial sections. J Ultrastruct Res Supl 1:1-191.

Krstic RV, (1979). Ultrastructure of the mammalian cell, An Atlas. Springer Verlag.

Hollenberg MJ, Cormack DH, Lea PJ (1988) Stereo Atlas of the Cell. B.C. Decker, Toronto-New York.

Hollenberg MJ, Lea PJ (1988) High resolution scanning electron microscopy of the retinal pigment epithelium and Bruch's layer. J Ophthalmol Vis Scie (In Press, Sept issue).

Lea PJ, Hollenberg MJ (1987). High resolution 3-dimensional scanning electron microscopy of rat hepatocytes. Proc Micros Soc Canada 14:24-25.

Lea PJ, Pawlowski A, (1986). Human melanocytic nevi.I. Electron microscopy and three dimensional computer reconstruction of nevi and basement membrane zone from ultrathin serial sections. Acta Derm Venereol Supl 127:5-15.

Lea PJ, Ramjohn SA (1980). Investigating the substitution of ethanol with liquid carbon dioxide during critical point drying. Micros Acta 83:291-296.

Palade G (1952). The fine structure of mitochondria. Anat Rec 114:427-451.

Tanaka K, Mitsushima A (1984). A preparation method for observing intracellular structures by scanning electron microscopy. J Micros 133:213-222.

Wheatly DN (1968). Mitochondrial tubules in the rat adrenal cortex. J Anat 103:151-154.

Acknowledgements: RP Eye Research Foundation of Canada, Medical Research Council of Canada, Technical Assistance; R. Temkin

Cells and Tissues: A Three-Dimensional
Approach by Modern Techniques in Microscopy,
pages 71–83

Development of the Quick-Freeze, Deep-Etch, Rotary-Replication Technique of Sample Preparation for 3-D Electron Microscopy

John E. Heuser, M.D.

Department of Cell Biology & Physiology
Washington University School of Medicine
St. Louis, Missouri 63110

The goal of every electron microscopic preparative technique is to preserve as closely as possible the original, living structure of biological samples. With the aim of avoiding the distorting effects of chemical fixation, we developed several years ago* a method for quick freezing cells in preparation for freeze fracture electron microscopy (Heuser et al., 1979). This was a natural wedding of two techniques that had already been described: freezing against a cryogenically cooled metal block (van Harreveld & Crowell, 1964)) and freeze fracture with metal replication (Moor & Muhlethaler, 1963). Initially, our goal was to capture fleeting cellular processes for visualization in the electron microscope. Processes like nerve transmission and muscle contraction are interrupted and distorted by chemical fixation, and occur on a time scale (msec) that can only be captured by quick-freezing.

---

*ACKNOWLEDGEMENTS: The developmental work described here was done some years ago in close collaboration with Dr. Thomas S. Reese. His support and inspiration is gratefully acknowledged. Current work is supported by grants from the USPHS (#GM26947) and the Muscular Dystrophy Association of America.

Subsequently, we found that quick-frozen tissues of all sorts were ideal starting materials for freeze-drying or freeze-etching ("deep-etching" being a limited form of freeze-drying; Heuser & Salpeter, 1979). This brought into view all sorts of natural surfaces that could not be seen by freeze fracture or thin sectioning, including the surfaces of biological crystals and, most importantly, membrane surfaces. Replicating these surfaces by rotary deposition of platinum, rather than unidirectional shadow-casting as was usually done for freeze fracture, yielded images that looked like unusually high resolution scanning electron micrographs, provided that photographic reversal was interposed to make the platinum look white in the final image (Heuser, 1980a).

Finally, this approach was extended to the level of molecular imaging by developing a procedure to adsorb individual proteins to tiny flakes of natural mineral mica and then to prepare the mica flakes like cells in the above freeze-etch procedures (Heuser, 1983a). This turned out to yield informative 3-D images of macromolecular assemblies like viruses and certain cellular organelles as well.

Examples of the application of these preparative procedures to various electron microscopic samples are presented in the following brief overview. All of the electron micrographs presented here are printed at the same high magnification (175,000x or 1cm = 55nm) in order to facilitate direct comparison. Also, all the micrographs are presented in stereo, to illustrate how much 3-D detail is visible in platinum replicas of quick-frozen and deep-etched samples.

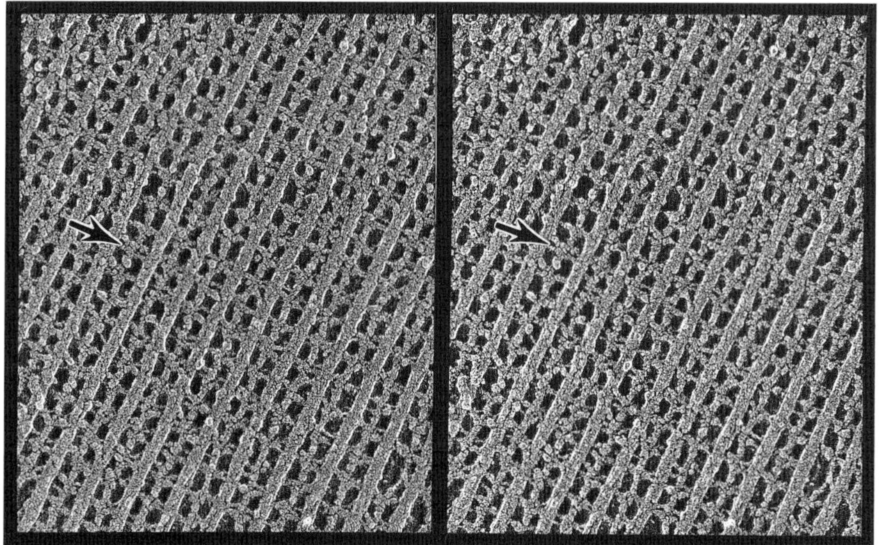

Figure 1: Interior of a muscle fiber showing barb-like crossbridges between interdigitating thick and thin filaments (arrow).

One of our primary reasons for developing quick-freezing in the first place was to capture the structural changes that occur during muscle contraction. These have long been thought to involve the movements of crossbridges from myosin thick filaments upon interposed actin thin filaments. However, the movements occur on a millisecond time scale and the crossbridges are too small to see clearly in traditionally fixed and thin-sectioned muscles. The sample seen here represents the classical insect flight muscle of the water bug Lethocerus, quick frozen after glycerination to remove intervening soluble proteins and after ATP-depletion to create a state of "rigor" or complete crossbridge attachment. The unusual clarity of the myriads of delicate crossbridges thus revealed (arrow) has permitted a detailed analysis of intricate geometrical patterns of actin-myosin interaction during muscle contraction (Heuser, 1983b; 1987).

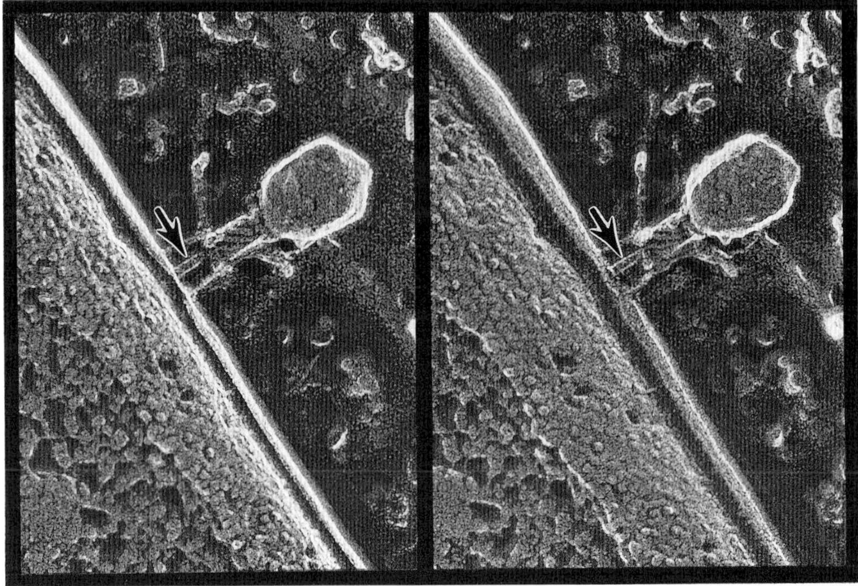

Figure 2: T4 bacteriophage on the surface of an E. coli bacterium, injecting its DNA into the bacterium (arrow).

An unexpected outcome of the quick-freezing technique was that it reduced ice crystal formation to a minimum, thus avoiding the major source of distortion that plagued earlier attempts to freeze-dry biological samples for electron microscopy. Quick-frozen samples turned out to yield remarkably lifelike and undistorted images upon deep-etching (Heuser & Salpeter, 1979). Particularly impressive is the degree of structural preservation of unusually delicate structures like the hollow polyhedral "head" of the T4 phage shown here. Never before could 3-D images be obtained of structures as fine as the "tail fibers", also seen here holding the virus to the outer envelope of the bacterium. In this particular example, the actual injection of DNA appears to have been captured (arrow).

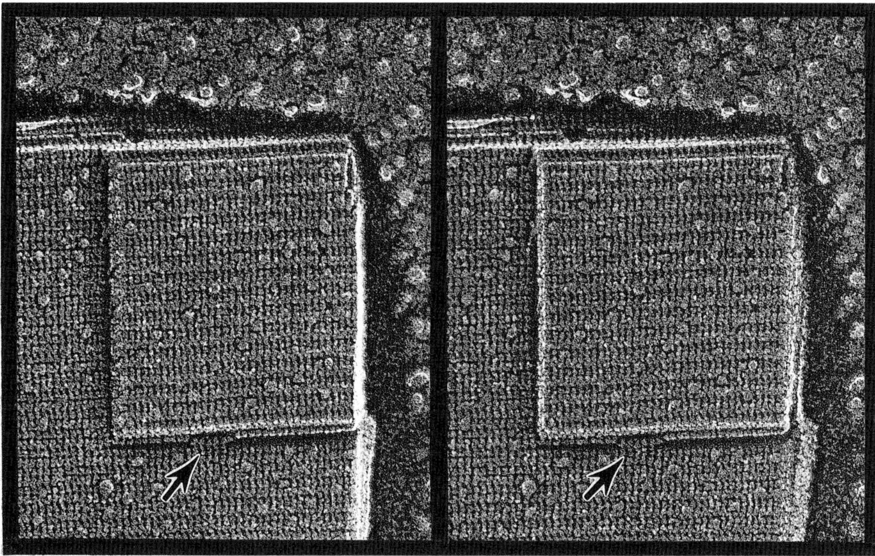

Figure 3: Natural surface of a crystal of catalase molecules, showing crystal dislocation beneath the uppermost orthogonal plate (arrow).

The opportunity to view natural surfaces by quick-freezing and freeze-etching permits renewed analysis of protein crystals of various degrees of complexity. The commonplace beef liver catalyze crystal shown here is often used as a calibration standard for electron microscopy, primarily because the horizontal 16nm stripes show up well after negative staining. More apparent after freeze etching, however, are the vertical 6nm cracks that separate adjacent molecules in the crystal lattice. 3-D analysis of such images has turned out to be a useful starting point for determining unit-cell dimensions in x-ray diffraction studies of new and unknown crystals. Liquid crystals and other unstable paracrystalline arrays and emulsions can also be imaged without distortion by the quick-freeze, deep-etch approach (Solans et al., 1988).

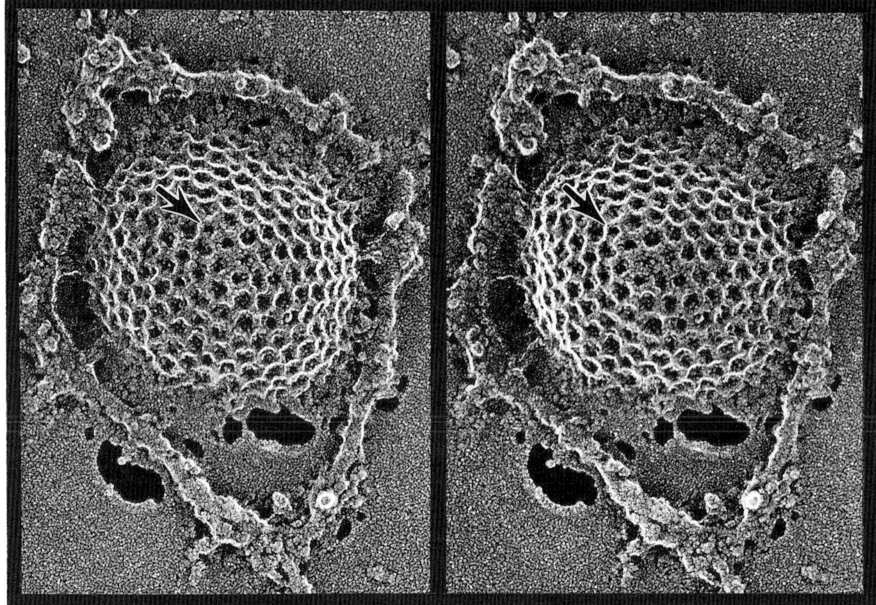

Figure 4: Geodetic lattice of clathrin forming a "coated pit" on the inside of the plasma membrane of a cultured cell engaging in receptor mediated endocytosis.

Such images are obtained by rupturing cultured cells with a jet of liquid, then quick-freezing and freeze-drying the glass coverslips upon which the cells were grown (Heuser, 1980b). The remaining fragments of cell membrane display many such geodetic domes in various stages of formation, illustrating the natural sequence of receptor-mediated endocytosis (Heuser, 1988). Clathrin lattices begin as flat meshworks composed exclusively of hexagonal facets, then subsequently undergo progressively greater inward curvature by converting certain hexagonal facets into pentagons (arrow). In the process, membrane receptors are sequestered and internalized by the cell. Surrounding this particular clathrin-coated pit is a necklace of endoplasmic reticulum also stuck to the inside of the plasma membrane; this is an unusual occurrence.

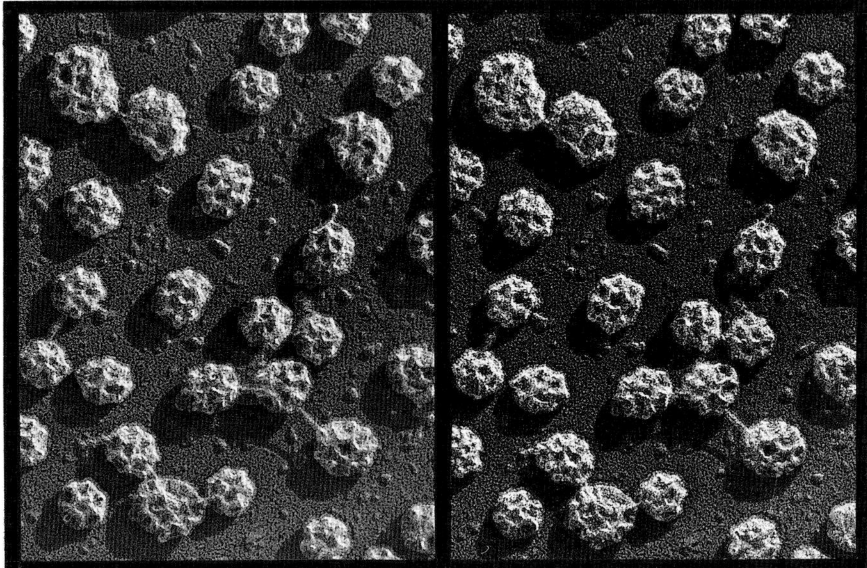

Figure 5: Clathrin coated vesicles extracted from a cow brain and freeze-dried after attachment to a flat piece of mica.

The polygonal construction of these small clathrin polymers is entirely analogous to that seen in the lattice of a coated pit shown in Fig. 4, but these are more richly endowed with pentagons because their radius of curvature is more extreme. Structures as small as this turn out to be best visualized after adsorption to mica, which supports them during freeze-drying and provides a flat background. During the initial development of this application of freeze-etching (Heuser, 1983a), we typically replicated such samples by unidirectional evaporation of platinum. This created unusually harsh highlights and shadows, as can be seen here. Subsequent Figures will illustrate that improved visibility is provided by low-angle rotary-replication with platinum instead of unidirectional shadow-casting. Though not apparent here, other studies have shown that many of the clathrin cages in brain preparations such as this enclose small spheres of membrane. Presumably this is membrane that has been removed from the plasma membrane in the process of recycling synaptic vesicles during transmitter discharge (Heuser & Reese, 1973).

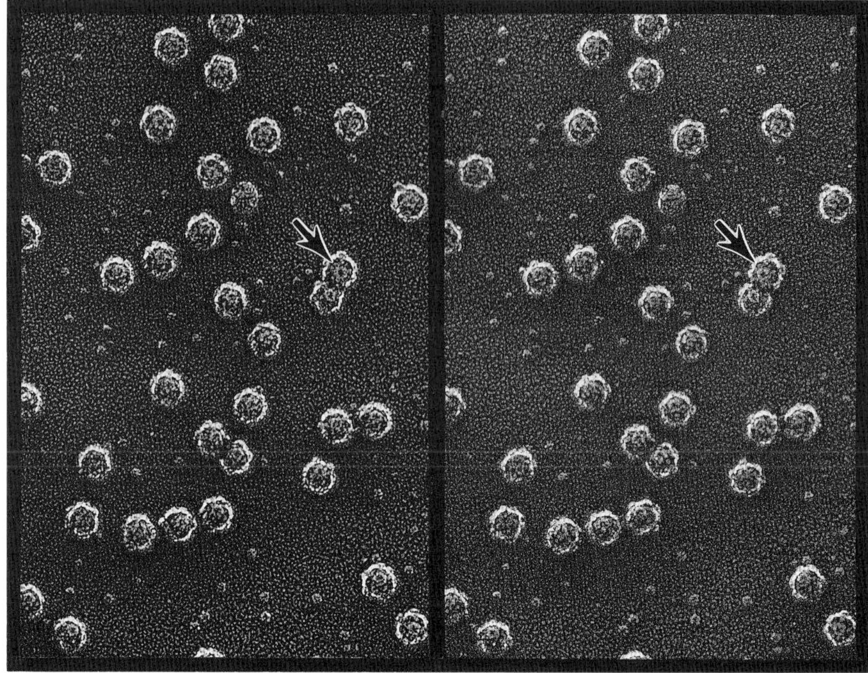

Figure 6: Tiny 25nm viruses (ϕX174 bacteriophages) adsorbed to a mica flake, quick-frozen, and freeze-dried.

Surface replication with platinum reveals new aspects of the icosahedral shell of viruses such as this, aspects that have not been visible by other techniques. These features include the five-sided ringlike conformation of each major surface projection and the tiny dot-like projections that occur between the major projections (arrow). Such details show up because rotary platinum replication provides strictly top-surface views and thus avoids the complicated problems of image-overlap that occur in other methods of imaging macromolecular assemblies. Furthermore, details are not obscured by the staining or drying artifacts that plague other methods. Since equal detail can also be obtained in platinum replicas of whole cells after freeze-drying, this technique also provides a unique opportunity to analyze the assembly and budding of viruses from the plasma membrane (Heuser et al., 1986).

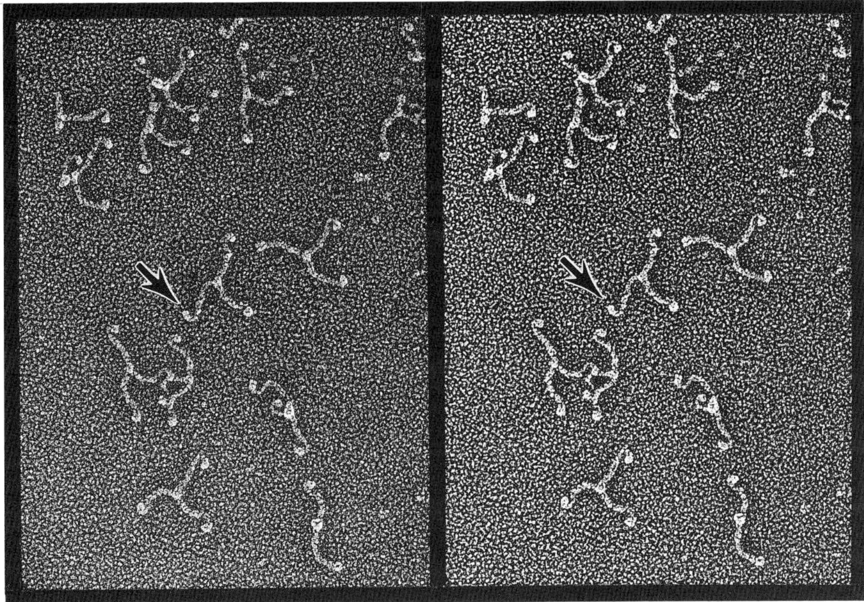

Figure 7: Clathrin "triskelia" generated by dissolving cages of the sort seen in Fig. 5 and then attaching their monomeric subunits onto mica before freeze-etching and platinum replication.

Alkaline 0.5M Tris solutions turn out to dissociate clathrin lattices into individual ~700kD molecules composed of three sets of heavy and light polypeptide chains. These are somehow held together in the form of 3-legged "triskelia" which, upon removal of the Tris buffer, will again interact with each other by intertwining their legs to reform the characteristic geodetic lattice. To generate the above image, dissociated "triskelia" were adsorbed to mica while still in the presence of 0.5M Tris, to keep them from polymerizing, and then the Tris was washed away so that it would not leave an obscuring film after freeze drying. Molecules prepared in this way turn out to display interesting structural details not visible by other techniques (Heuser & Kirchhausen, 1986), including distinct scroll-shaped hooks at the end of each leg (the points where clathrin is thought to attach to the membrane; cf. arrow). Various degrees of distortion occur during adsorption of such thin molecules to mica; but in this case, the legs themselves tend to bend clockwise while the terminal hooks always bend counterclockwise, thus making each leg S-shaped (arrow).

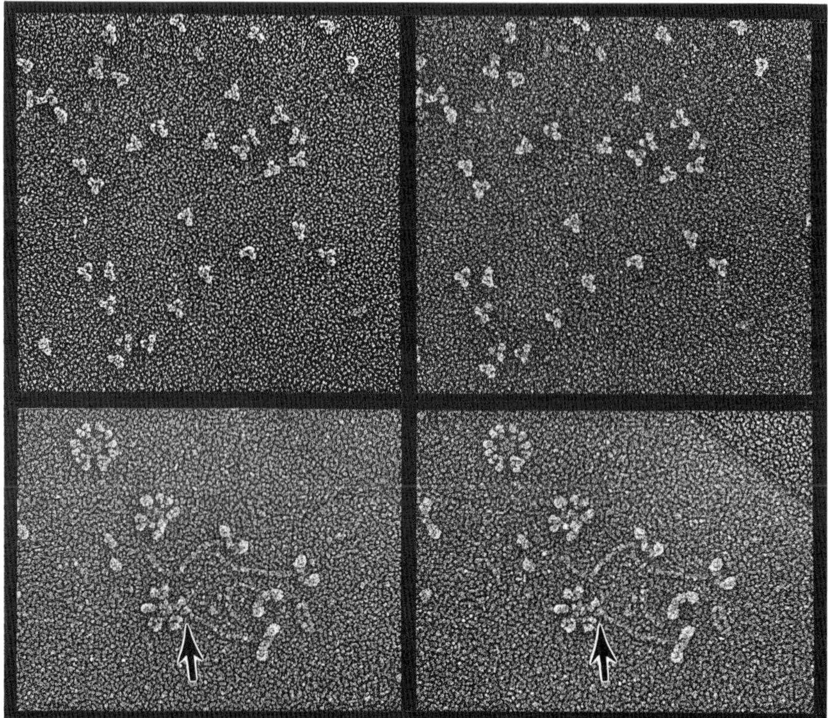

Figure 8: Monoclonal IgG molecules (top) and IgM molecules attached to the tails of myosin molecules (bottom).

Current techniques for generating specific monoclonal antibodies have created the need for an electron microscopic imaging technique that can reliably demonstrate individual molecular interactions. Freeze drying and rotary replication of antigen-antibody complexes adsorbed to mica turns out to be ideal for this purpose. Resolution is sufficient to see antibodies directly, thus indirect labeling with electron-dense markers such as gold or ferritin becomes unnecessary. In fact, substructure can be seen within the antibodies themselves. The IgG molecules in the upper field, for example, display distinct partitioning into three relatively globular subunits (their 2 Fab and single Fc portions) while the IgM molecules in the lower field display the expected 10 globular Fab's around a central ring. With this resolution, mapping the binding sites of monoclonal antibodies on extended molecules becomes straightforward; the IgM's shown here, for example, clearly bind to the ends of the tails of the myosin molecules also present in the lower field (arrow).

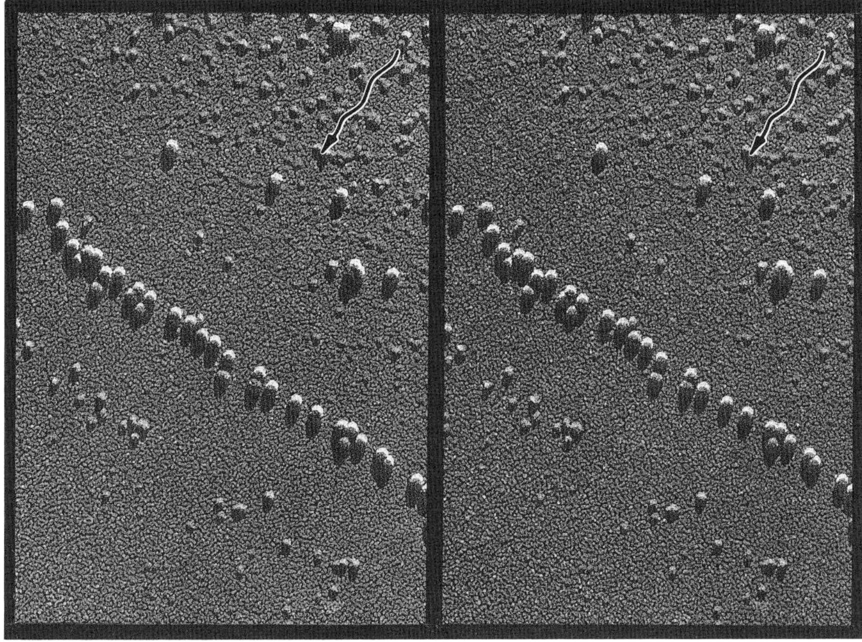

Figure 9: Separation of two differently-sized molecules by "permeation-chromatography" into a mica slurry.

Here a mixture of ferritin (the large 15nm balls) and albumin (the small 8nm particles) was mixed with mica flakes and allowed to percolate into the cracks between adjacent flakes. After quick-freezing, the mica flakes were separated during freeze fracture, in order to expose the advancing front of molecules (diffusing from upper right to lower left, cf. arrow). Since the crack between the mica flakes was initially wedge-shaped, the smaller albumin molecules penetrated further and became segregated from the larger ferritin molecules. The latter focused particularly well as a "front" perpendicular to the direction of diffusion. This unusual form of "permeation chromatography" proves to be useful for distinguishing the different molecules in complex mixtures such as human plasma. In the future, it should prove useful for characterizing complex mixtures of unknown proteins.

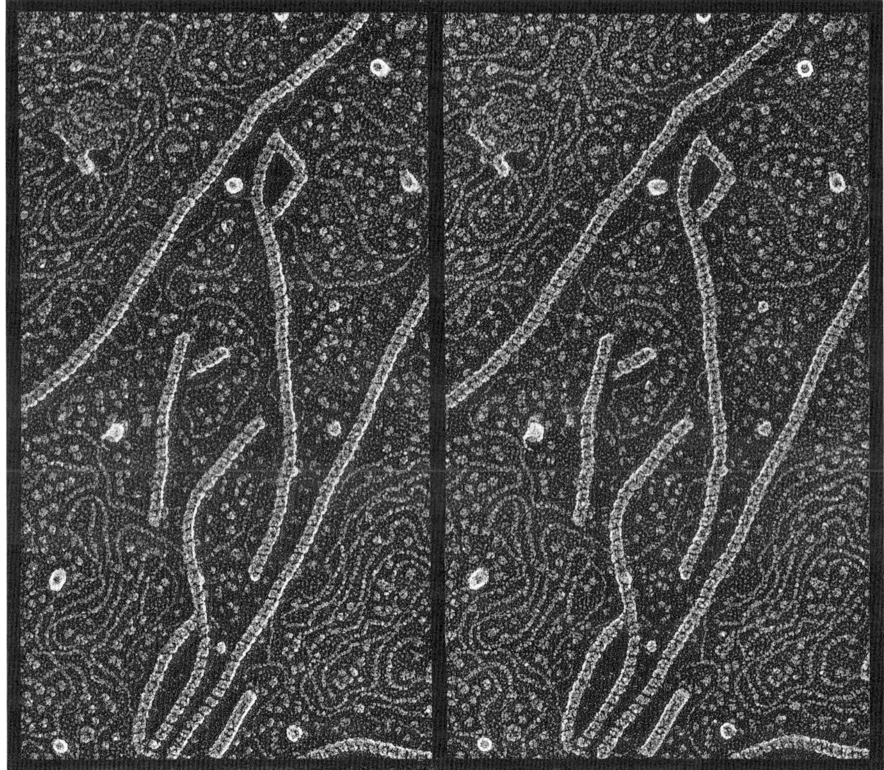

Figure 10: DNA molecules decorated with helices of bacterial Rec A protein that catalyzes gene recombination and repair.

Perhaps the major goal of current cell biological research is to characterize the DNA-protein interactions that underlie genetic regulation. One of the classical prototypes of such interactions is that of E. coli Rec A protein and DNA, pictured above. In the process of forming such beautiful helices around DNA, this protein uses ATP to drive a conformational change in the DNA. To generate the above image, the helices were stabilized in one conformation by including the non-hydrolyzable analog ATPγS in the reaction mixture during application to mica. In the background are undecorated DNA filaments and unreacted Rec A protein monomers. We find that when ATP is present in such a mixture instead of ATPγS, helices occur with several different degrees of twist. These conformational changes have suggested a mechanism by which Rec A might in fact catalyze genetic recombination (Heuser and Griffith, 1988).

SUMMARY: The images presented here, though covering a size-range from whole cells to individual molecules, have all be generated by the exact same technique, as was outlined in the introduction. All depend upon the high quality of the initial freezing of the sample, which is accomplished by the abrupt application of an ultra-pure copper block cooled with liquid helium. Copies of the freezing machine we have developed to accomplish this are currently manufactured and supplied to other laboratories at cost, by former university machinists who now operate under the name of Med-Vac, Inc., St. Louis, MO. Any laboratory wishing to generate images of the sort shown here are welcome to obtain one of these machines and to call us for further assistance and instruction.

REFERENCES:

Heuser, J.E. and T.S. Reese. 1973. Evidence for recycling of synaptic vesicle membrane during transmitter release at the frog neuromuscular junction. J. Cell Biol. 57: 315-344.

Heuser, J.E., T.S. Reese, L.Y. Jan, Y.N. Jan, M.J. Dennis and L. Evans. 1979. Synaptic vesicle exocytosis captured by quick-freezing and correlated with quantal transmitter release. J. Cell Biol. 81: 275-300.

Heuser, J.E. and S.R. Salpeter. 1979. Receptors in quick-frozen, deep-etched and rotary-replicated postsynaptic membranes. J. Cell Biol. 82: 150-175.

Heuser, J.E. 1980a. Three-dimensional visualization of coated vesicle formation in fibroblasts. J. Cell Biol. 84: 560-583.

Heuser, J.E. 1980b. The quick-freeze, deep-etch method of preparing samples for high resolution, 3-D electron microscopy. Trends in Biochem. Sci. 6: 64-68.

Heuser, J.E. 1983a. Procedure for freeze-drying molecules adsorbed to mica flakes. J. Mol. Biol. 169: 155-195.

Heuser, J.E. 1983b. Structure of the myosin crossbridge lattice in insect flight muscle. J. Mol. Biol. 169: 123-154.

Heuser, J.E. and T. Kirchhausen. 1986. Deep-etch views of clathrin assemblies. J. Ultrastruct. Res. 92: 1-27.

Heuser, J.E., M. DuBois-Dalcq, R.N. Hogan, C.A. Jordan and R.A. Lazzarini. 1986. Deep etch analysis of defects in budding of measles virus mutants. J. Cell Biol. 103: 113a.

Heuser, J.E. 1987. Deep etch visualization of crossbridge patterns in insect flight muscles exposed to various nucleotides and vanadate. J. Musc. Res. Cell Motil. 8: 303-321.

Heuser, J.E. 1988. Effects of cytoplasmic acidification on clathrin lattice formation. J. Cell Biol. (in press).

Heuser, J.E. & J. Griffith. 1988. Deep-etch demonstration of two different conformational states in Rec A protein helices. J. Mol. Biol. (submitted).

Moor, H. and K. Muhlethaler. 1963. Fine structure in frozen-etched yeast cells. J. Cell Biol. 17: 609-628.

Solans, C., J.G. Dominguez, J.L. Parra, J. Heuser and S.E. Friberg. 1988. Gelled emulsions with a high water content. Colloid Polym. Sci. 266: 570-574.

Cells and Tissues: A Three-Dimensional Approach by Modern Techniques in Microscopy, pages 85–92

GOLGI APPARATUS SUBCOMPARTMENTS AS VISUALIZED BY MEANS OF LECTINCYTOCHEMISTRY

Margit Pavelka and Adolf Ellinger

Institute of Micromorphology and Electron Microscopy
University of Vienna, Vienna, Austria

INTRODUCTION

The Golgi apparatus is a continuous membrane system composed of stacks of flat cisternae and interconnecting tubules that constitutes a central cellular crossroad (for review Farquhar and Palade, 1981; Tartakoff, 1987). Multiple newly synthesized as well as endocytic and recycling molecules visit the Golgi apparatus, where they are subject to a number of modifications; in the case of newly synthesized proteins, these include posttranslational modifications, such as terminal glycosylation, sulfation and proteolytic cleavage; endocytic and recycling molecules are subject to changes, e.g. resialylation, aiming at a restoration of the molecules (Snider and Rogers, 1985; Woods et al., 1986). Furthermore, it is from the Golgi apparatus that molecules are targeted to their final destinations.

Both cytochemical and biochemical findings indicate that different Golgi-associated processes, e.g. different glycosylation steps, occur at different sites within the complex system, thus pointing to the existence of functional subcompartments. Insight into the locations of Golgi apparatus subcompartments has been derived from immunocytochemical studies that have localized a number of endogenous Golgi apparatus molecules, including Golgi membrane receptor proteins (Brown and Farquhar, 1984; Geuze et al., 1984) and sugar-transferring enzymes (Dunphy et al., 1985; Roth et al., 1986).

The use of lectins capable of specifically binding to certain sugars and sugar sequences (Liener et al., 1986) provides another approach for elucidating the organization of Golgi apparatus subcompartments. Lectincytochemistry permits the identification of Golgi subcompartments, in which glycoconjugates bearing certain classes of glycans are present. The present paper shows the arrangement of Golgi apparatus subcompartments in various cell types, as visualized with the aid of lectins of diverse sugar specificities.

RESULTS

Two different techniques, a pre- and a postembedment affinitycytochemical technique, were employed. 1./ For preembedment incubations (Figs.1,3) 10 µm thick cryosections of prefixed tissue (4% paraformaldehyde, 0.5% glutaraldehyde, 30min, 4°C) were treated with lectin-horseradish peroxidase (HRP) conjugates (50-100µg/ml; 4h; room temperature); the peroxidase activity was visualized by means of diamine benzidine and H2O2 and the cryosections subsequently embedded in Epon (Pavelka and Ellinger, 1985). 2./ Thin sections of LRWhite resin-embedded tissue were used for postembedment lectin reactions (Figs.2,4,5) and incubated in solutions (30-100µg/ml phosphate-buffered saline) of lectin colloidal gold conjugates (Ellinger and Pavelka, 1985).

In the following table, lectins used and sugar specificities are summarized.

Table

Concanavalin A	Con A	αMan>αGlc>GlcNAc
Lens culinaris lectin	LCA*	αMan>αGlc>GlcNAc (+Fuc)
Pisum sativum lectin	PSA	αMan>αGlc=GlcNAc (+Fuc)
Ricinus communis I lectin	RCA I	βGal>αGal>>GalNAc
Erythrina cristagalli lectin	ECA	Galβ1,4GlcNAc>GalNAc>Gal
Helix pomatia lectin	HPA	αGalNAc>αGlcNAc>>Gal
Vicia villosa lectin	VVA	αGalNAc>αGal
Griffonia simplicifolia I-A$_4$	GS I-A$_4$	αGalNAc>αGal
Ulex europeus I	UEA I	αFuc

Man mannose, Glc glucose, GlcNAc N-acetyl-glucosamine

GalNAc N-acetyl-galactosamine, Gal galactose, Fuc fucose

*) "A" stays for agglutinin

Concanavalin A (ConA; Fig.1)

In most cell types studied, label for ConA, which preferably binds to high-mannose-type N-linked oligosaccharide chains, was intense in nuclear envelope and cisternae of the endoplasmic reticulum (ER); in the Golgi apparatus, ConA reactions were mostly concentrated in cisternae of the cis and medial subsections of the stacks; trans and transmost cisternae were frequently free of reaction products.

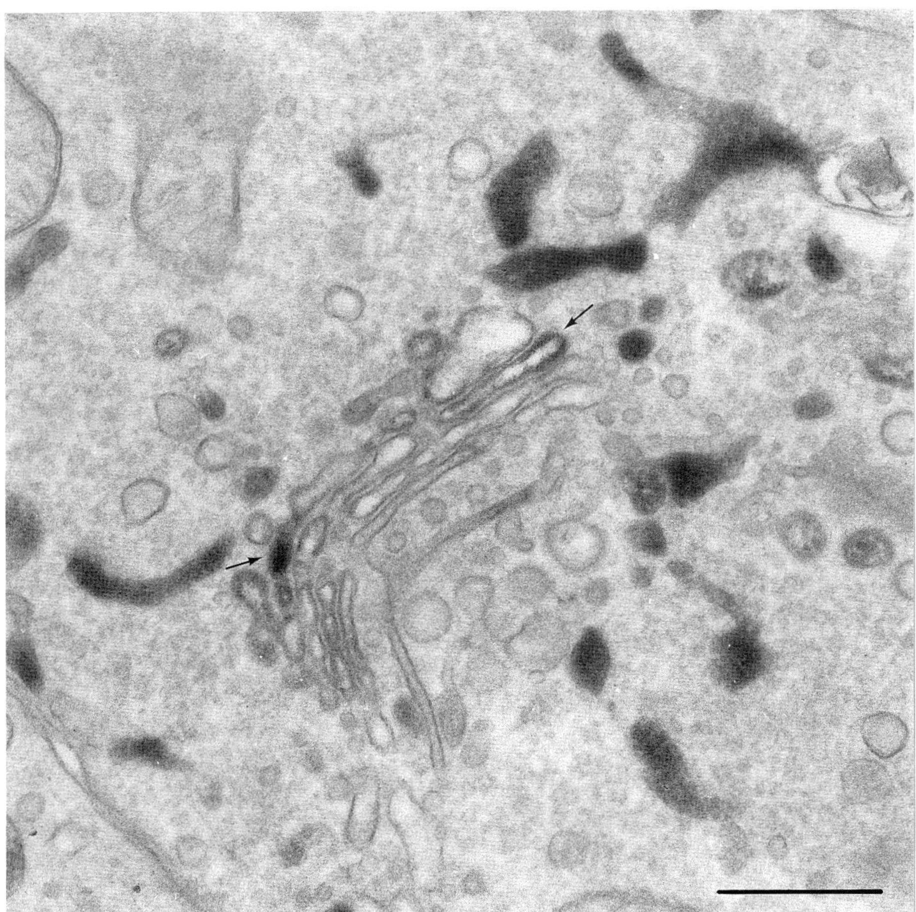

Figure 1. Embryonic pancreatic acinar cell of the rat at day 16 of gestation. ConA reactions are intense in the ER and in limited segments of cis and medial Golgi cisternae (arrow). Trans and transmost cisternae are weakly stained or devoid of reactions.

bar - 0.5μm

Ricinus Communis I Agglutinin (RCA I; Fig.2)

Label for RCA I, which particularly interacts with terminal ß-galactosyl residues, was, in the majority of cells, concentrated in medial, trans and transmost Golgi apparatus subsections (Fig.2); nevertheless, in embryonic pancreatic acinar cells, RCA I-binding molecules were found in all subsections of the Golgi apparatus stacks (Pavelka and Ellinger, 1986).

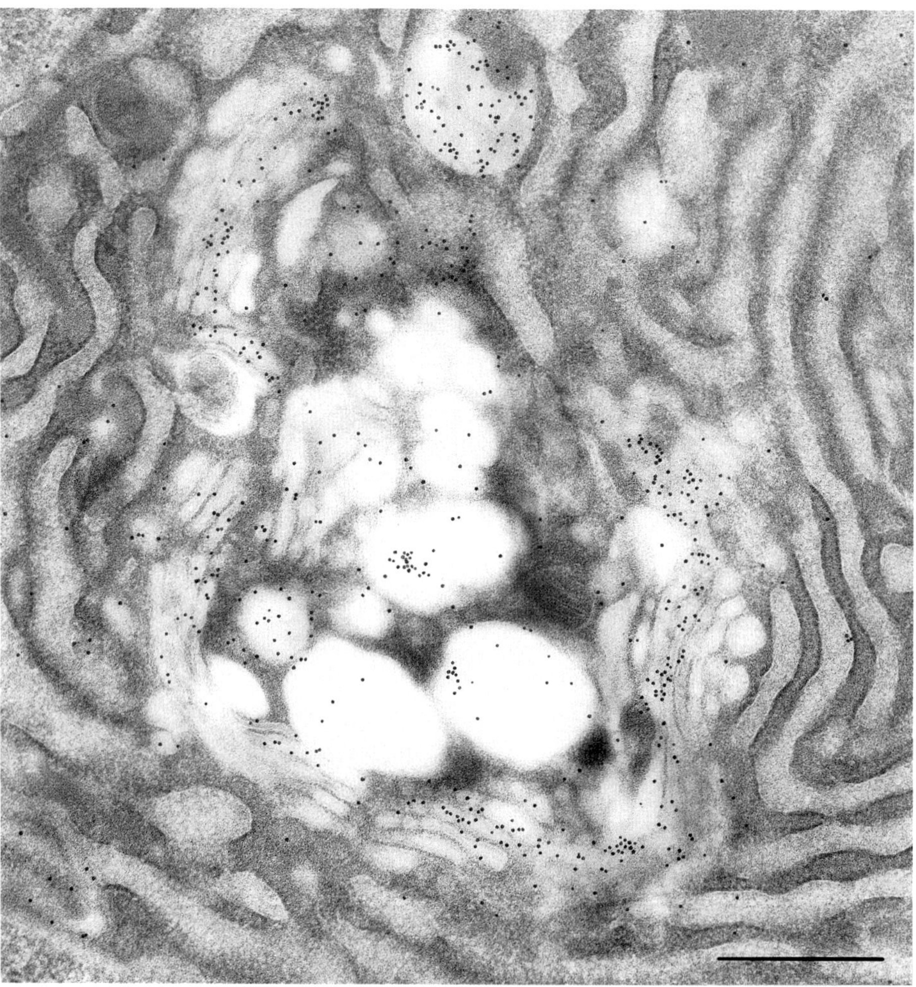

Figure 2. Goblet cell of the rat duodenum. Colloidal gold marker particles indicating RCA I-binding molecules are apparent in medial, trans and transmost elements of this Golgi apparatus stack.

bar - 0.5μm

Erythrina Cristagalli Lectin (ECA)

ECA, which binds to N-acetyl-lactosamine, induced weak reactions in medial, trans and transmost Golgi cisternae of fibroblasts and absorptive enterocytes; intense ECA label was apparent in tubular and vesicular structures in the transmost position of the Golgi stacks.

Pisum Sativum and Lens Culinaris Lectins (PSA; LCA; Fig.3)

Both lectins, which bind at high affinity to the fucosylated core-region of N-glycosidically linked glycans, showed rather scarce and punctated reactions of the ER. In the Golgi apparatus, PSA- and LCA-binding molecules predominated in in cis and medial cisternae of the stacks; frequently, the penultimate cis cisterna was the most intensely labeled one.

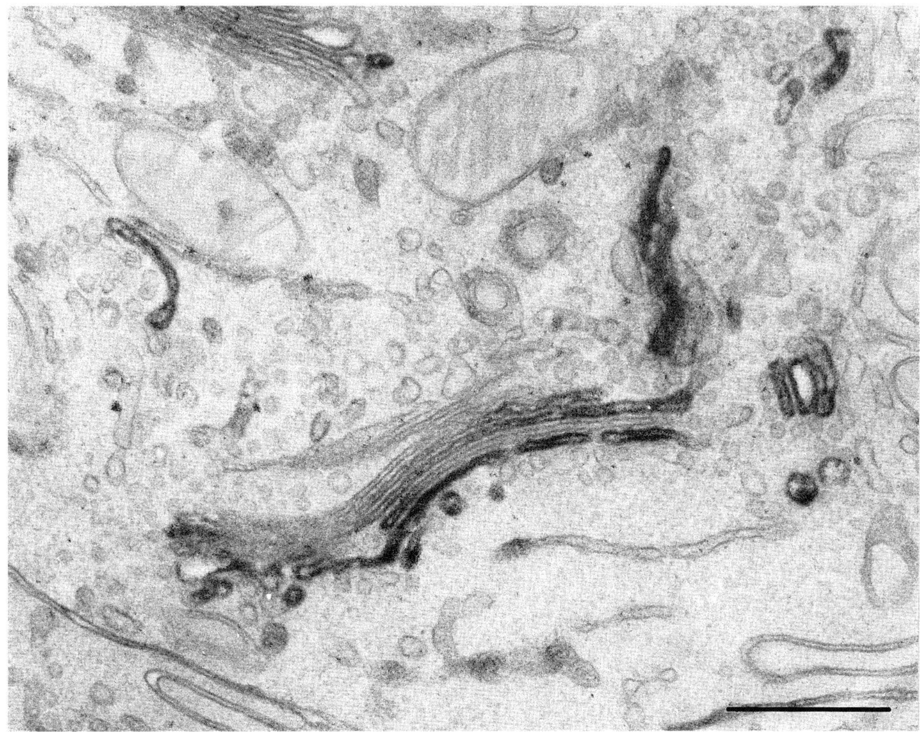

Figure 3. Small intestinal absorptive cell of the rat. Weak PSA reactions are apparent in the ER; reaction products for PSA are concentrated in one cisterna at the cis Golgi side and in limited regions of medial cisternae, as well as in some Golgi-associated vesicles.

bar - 0.5µm

Ulex Europaeus I Lectin (UEA I; Fig.4)

Binding reactions for the fucose-recognizing UEA I were weak in medial cisternae of the Golgi stacks. In small intestinal absorptive cells as well as goblet cells, UEA I label was most intense in cisternae of the trans/transmost subsections of the Golgi stacks.

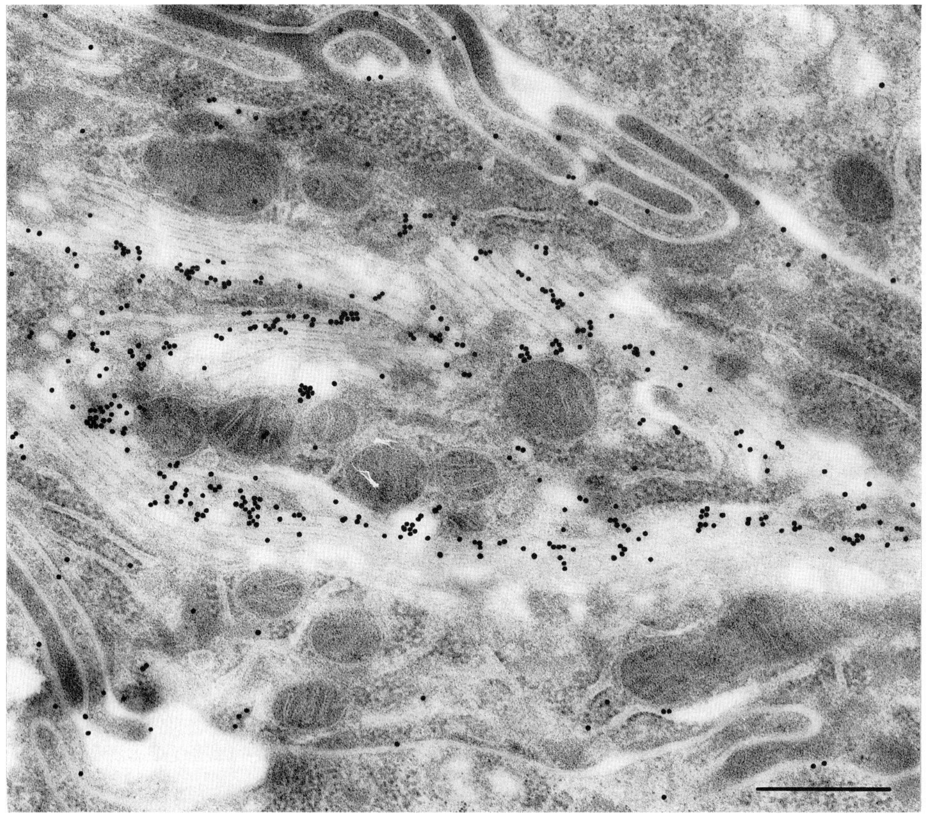

Figure 4. Small intestinal absorptive cell of the rat. UEA I-gold particles predominate in trans Golgi cisternae.
bar - 0.5μm

Helix Pomatia Lectin and Griffonia Simplicifolia Lectin (HPA; GS I-A4; Fig.5)

Reactions for HPA and GS I-A4, both recognizing N-acetylgalactosamine, were intense in cis cisternae. In several cell types, such as in duodenal goblet cells, intense HPA and GS I-A4-label was also found in the trans and transmost Golgi cisternae as well as in secretory granules.

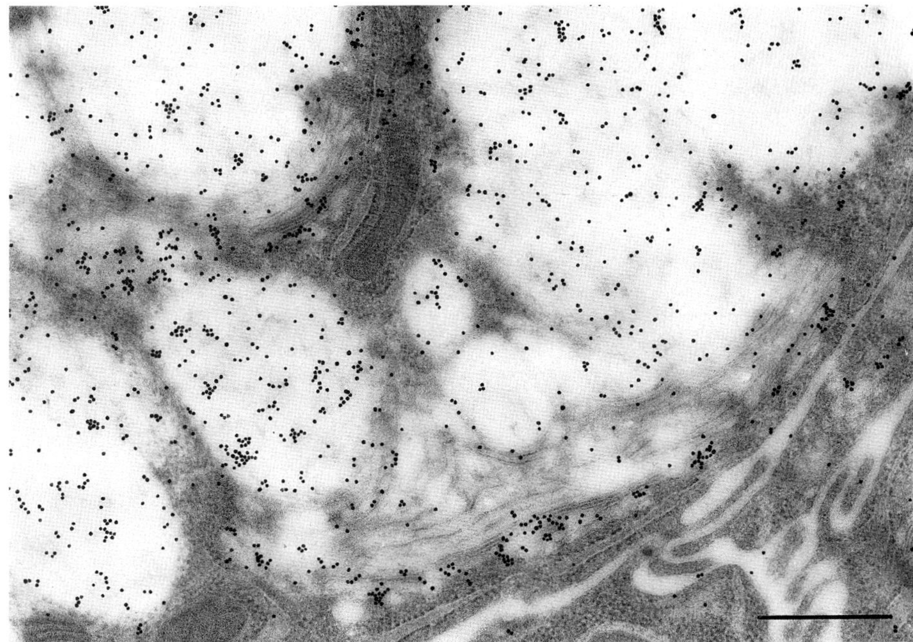

Figure 5. Duodenal goblet cell of the rat. GS I-A4 reactions are intense in a cis cisterna of this stack and are apparent in cisternae of the trans side and in secretory granules. bar - 0.5μm

Summarizing, the reactions obtained permit a differentiation between various subcompartments localized at either cis, medial or trans/transmost regions of the Golgi stacks. The findings show that Golgi apparatus cisternae, in which high-mannose type glycans are present, predominate at the cis side of the Golgi apparatus stacks; by constrast, glycoconjugates bearing complex-type N-linked glycans in most cell types are concentrated in trans/transmost Golgi elements. These patterns possibly mirror conversion from high-mannose- to complex-type glycan species of glycoconjugates occurring while the molecules traverse the Golgi stacks.

The cis and trans/transmost-dominant Golgi reactions obtained with HPA and GS I-A4 possibly correspond to initial and terminal O-glycosylation steps, respectively.

Furthermore, the subtle patterns obtained with some of the lectins (e.g. ConA/Fig.1; PSA/Fig.3) showing dense reactions near weakly stained and unreactive sites permit a differentiation between subregions of the individual Golgi cisternae.

REFERENCES

Brown WJ, Farquhar MG (1984). The mannose-6-phosphate receptor for lysosomal enzymes is concentrated in cis Golgi cisternae. Cell 36:295-307.

Dunphy WG, Brands R, Rothman JE (1985). Attachment of terminal N-acetyl-glucosamine to asparagine-linked oligosaccharides occurs in central cisternae of the Golgi stack. Cell 40: 463-472.

Ellinger A, Pavelka M (1985). Post-embedding localization of glycoconjugates by means of lectins on thin sections of tissues embedded in LR White. Histochem J 17:1321-1336.

Farquhar MG, Palade GE (1981). The Golgi apparatus (complex) - (1954-1981) - from artifact to center stage. J Cell Biol 91:77s-103s.

Geuze HJ, Slot JW, Strous GJAM, Hasilik A, von Figura K (1984). Ultrastructural localization of the mannose-6-phosphate receptor in rat liver. J Cell Biol 98:2047-2054.

Liener IE, Sharon N, Goldstein IJ (1986). "The lectins: Properties, functions, and applications in biology and medicine." Orlando: Academic Press.

Pavelka M, Ellinger A (1985). Localization of binding sites for concanavalin A, Ricinus communis I and Helix pomatia lectin in the Golgi apparatus of rat small intestinal absorptive cells. J Histochem Cytochem 33:905-914.

Pavelka M, Ellinger A (1986). RCA I binding patterns of the Golgi apparatus. Eur J Cell Biol 41:270-278.

Roth J, Taatjes DJ, Weinstein J, Paulson JC, Greenwell P, Watkins WM (1986). Differential subcompartmentation of terminal glycosylation in the Golgi apparatus of intestinal absorptive and goblet cells. J Biol Chem 261:14307-14312.

Snider MD, Rogers OC (1985) Intracellular movement of cell surface receptors after endocytosis: Resialylation of asialo-transferrin receptor in human erythroleukemia cells. J Cell Biol 100:826-834.

Tartakoff AM (1987) "The secretory and endocytic paths." New York: John Wiley and Sons.

Woods JW, Doriaux M, Farquhar MG (1986). Transferrin receptors recycle to cis and middle as well as trans Golgi cisternae in Ig-secreting myeloma cells. J Cell Biol 103: 277-286.

ACKNOWLEDGEMENTS

The authors gratefully acknowledge the excellent technical assistance of Mrs.Jutta Selbmann, Mrs.Elfriede Scherzer, Mrs.Gerlinde Hartl, Mr.Helmut Oslansky and Mr.Richard Reichhart.

This work was supported by the "Hochschuljubiläumsstiftung der Stadt Wien".

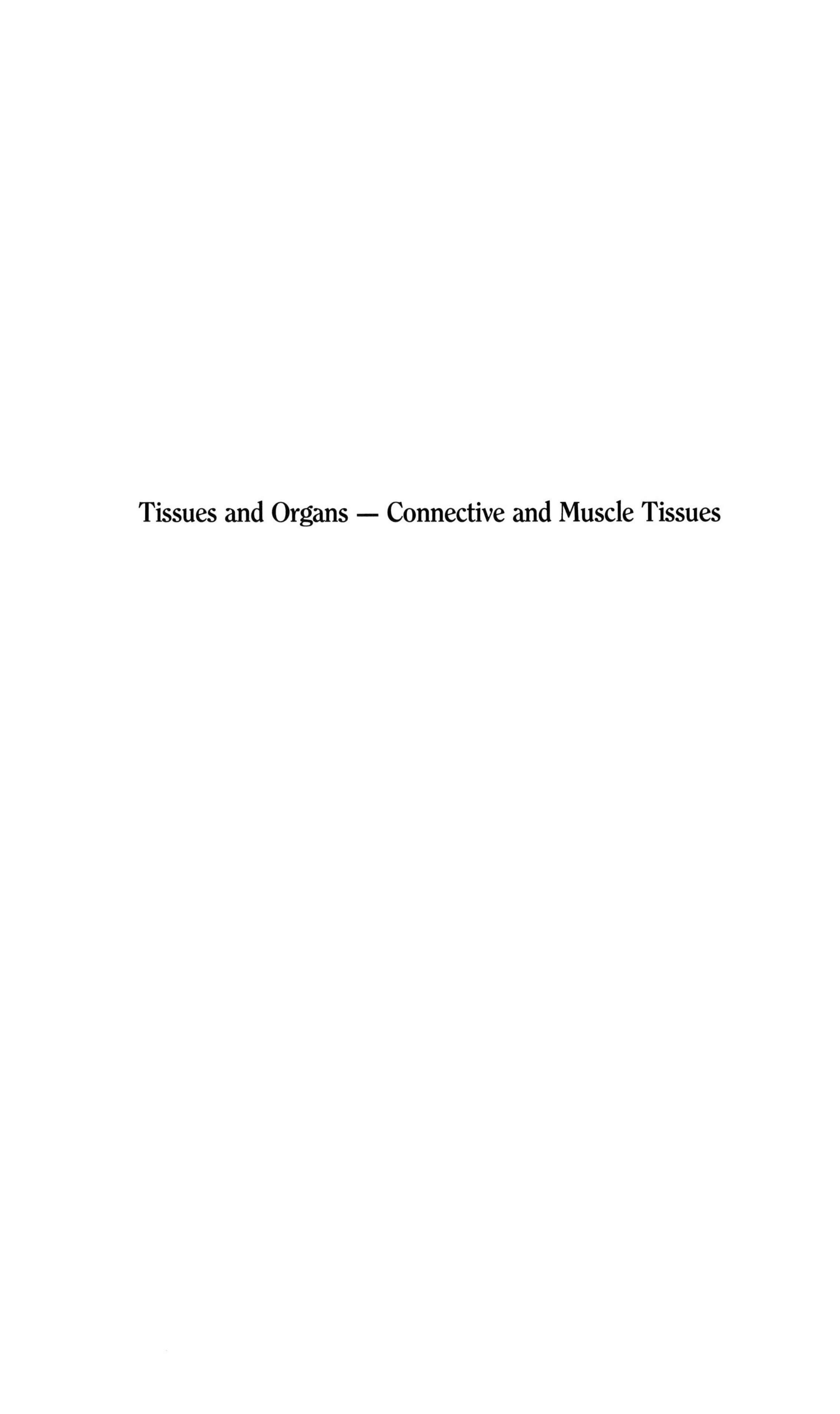

Tissues and Organs — Connective and Muscle Tissues

Cells and Tissues: A Three-Dimensional
Approach by Modern Techniques in Microscopy,
pages 95–100

FREEZE-ETCHING AS A 3D APPROACH TO THE COLLAGEN FIBRIL STRUCTURE

Alessandro Ruggeri, Maurizio Marchini, Vittoria Ottani and Mario Raspanti

Istituto di Anatomia Umana Normale, Bologna, Italy

INTRODUCTION

Collagen fibrils observed in replicas of freeze-fractured and replicated specimens may appear under two different aspects, showing a cross-banding pattern very similar to that observed after negative staining (Hashimoto, 1974; Marchini et al. 1983) or showing a parallel array of filaments running along the fibril (Ruggeri et al. 1979). A similar filamentous aspect is observable in thin sections when collagen fibrils are treated with extracting chemicals (Lillie et al. 1977).

THE CROSS-BANDING

Fig. 1 (below) shows two freeze-etched collagen fibrils from rat tail tendon and from bovine cornea. An evident banding pattern is determined by an alternated succession of elevated

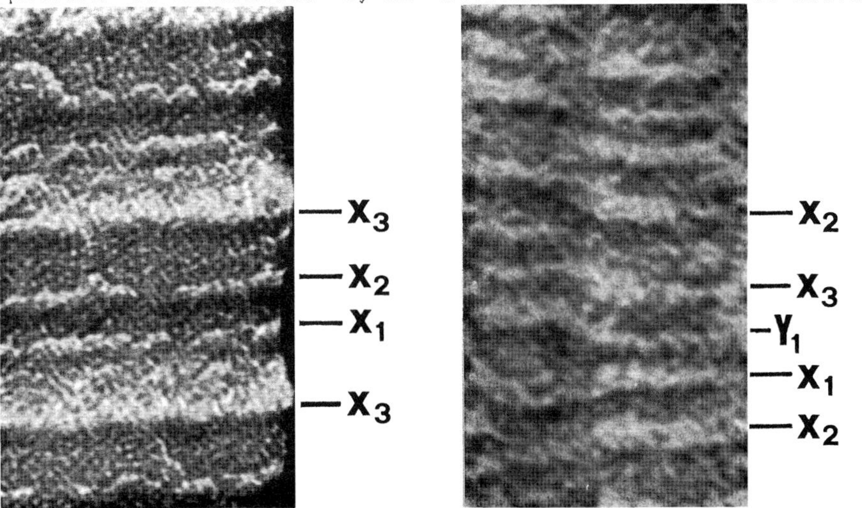

and depressed segments along the fibril and is usually observed after deep-etching when cryoprotection is avoided (Hashimoto, 1974). It is interpreted as the morphological evidence of the periodic distribution of proteic units along the collagen fibril. The axial period measured in tendons is 67 nm, a value tightly close to that determined by X-ray analysis. Also evident are two ridges (X_2 and X_3) located at the margins of the elevated segment and a third faint ridge (X_1) situated at an intermediate point of the depressed segment. These three ridges correspond to the widest light bands observable in negatively-stained fibrils (Marchini et al., 1983; Fig. 2, on the right) and represent sites of higher molecular density. In cornea (Marchini et al. 1986) as well as in skin and several other tissues (research in progress), a fourth ridge (Y_1) is also evident in the gap zone.

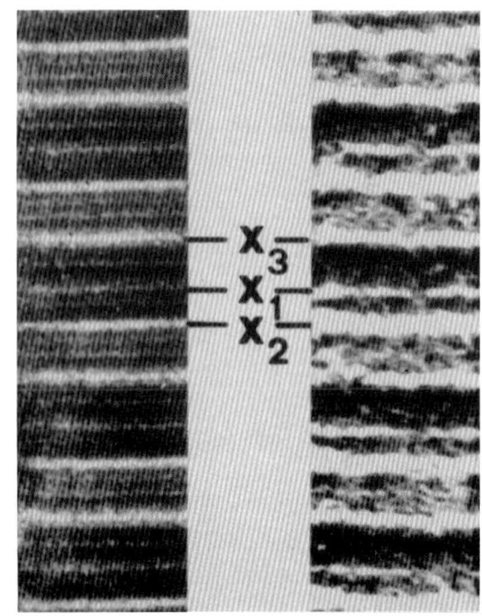

In order to obtain further data on the correspondence between the elevated and depressed segments as well as the ridge pattern observed on replicas and the distribution of density in the collagen fibril, a three-dimensional computer reconstruction of the collagen fibril has been developed (Raspanti et al., in press). An outline of type I collagen molecule was built with two $\alpha1(I)$ and one $\alpha2(I)$ chains and convoluted with a spacing of 67 nm. (234 residues). The telopeptides were unevenly contracted according to their non-helical conformation (Helseth et al. 1979; Capaldi & Chapman 1982). The distribution of volumes was computed by adding, line by line, the volume value of each residue (Zamyatnin 1972) and the volume of the bound mono- and diglycosides (Hulmes et al., 1980) according to the location and extent of glycosilation reported by Fietzek and Kuhn (1976) and Kivirikko and Myllyla (1979). The resulting sequence was then smoothed down and converted into a sequence of diameters and revolved into a 3D model. Fig. 3 (next page) shows the final model of a collagen fibril obtained on a CRAY X-MP 48 supercomputer and smooth-shaded from the same angle as the replicas. The sequence of elevated and depressed segments - corresponding to the overlap and gap zones- are easily seen, with the overlap zone spanning 106 residues and representing, as in the fibril, 45% of the axial period. The model also shows on the step zone two ridges corresponding to the X_2 and X_3 ridges of freeze-etched fibrils and identically spaced of 38% of the period (Fig. 4, next page). On the contrary, the gap zone presents only small ripples and neither the X_1 nor the Y_1 ridge are detectable.

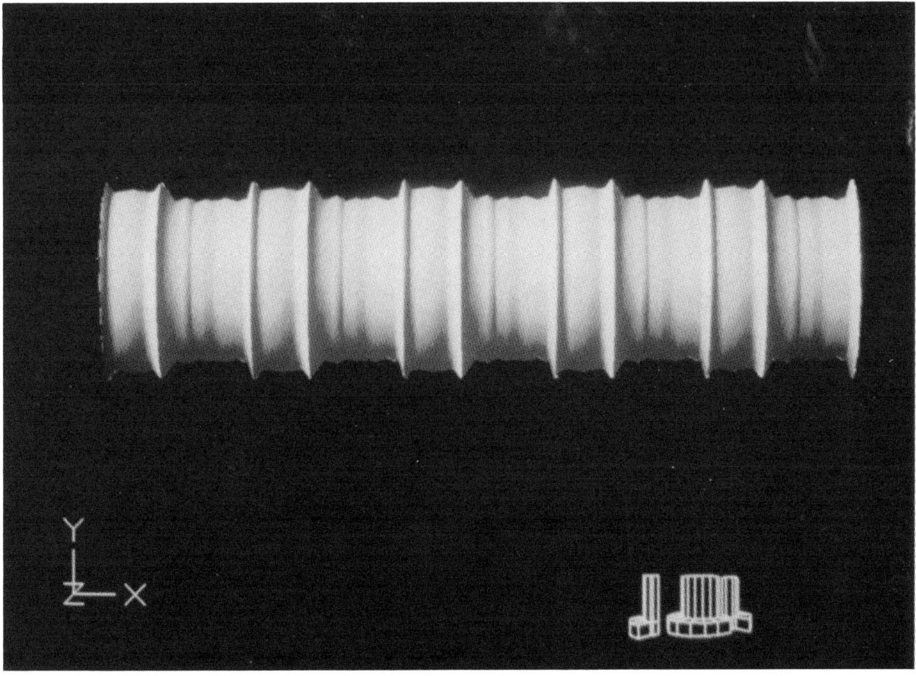

Fig. 3. Three-dimensional computer model of a collagen fibril.

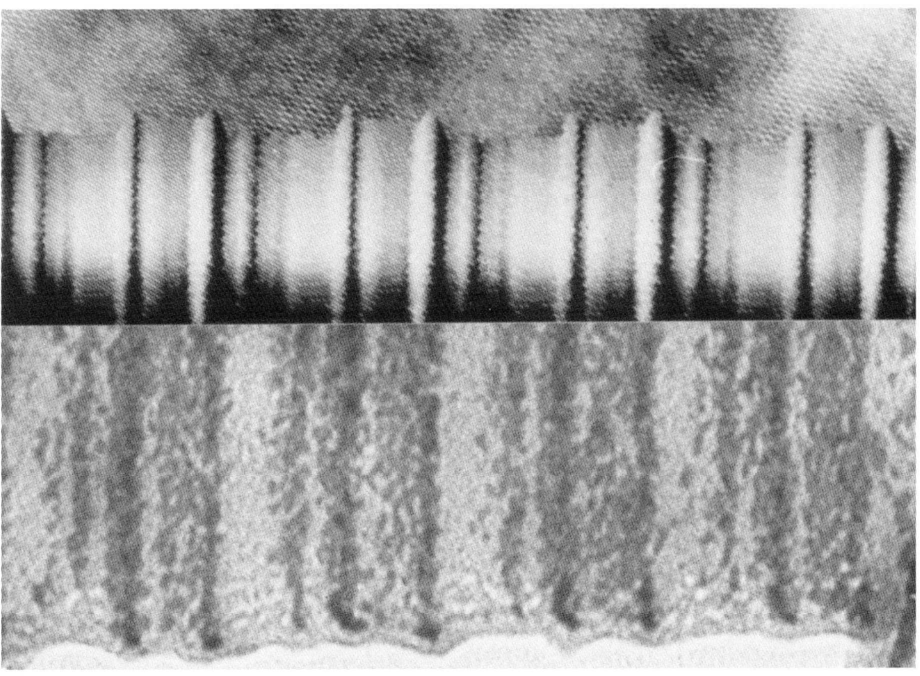

Fig. 4. Comparation of a freeze-etched fibril and a computer model. The computer model resolution has been lowered in order to match the resolution of the freeze-etching technique.

Since the model takes in careful account the amino acid sequence of collagen and its post-translational glycosilation, these ridges appear to be related to neither of these parameters but perhaps caused by other components of the extracellular matrix. In effect, observations carried out by other Authors have demonstrated that proteoglycans filaments are bound to the collagen fibril with a periodic relationship. Namely, in tendon, skin and cornea a DS-PG filament is located on the d-e bands while in cornea another PG, the KS-PG, is located on the a and c bands (Scott, 1986). Two of these positions (d-e and a bands) lie close to the location of the X_1 and Y_1 ridges observable in the gap zone of freeze-fractured fibrils and therefore it is possible that these ridges may represent proteoglycan particles linked to the fibril surface.

THE FIBRIL STRUCTURE

When glycerol solution is used as a cryoprotectant, the collagen fibril slighly swell so that the cross-banding is no longer detectable while longitudinal filaments become evident (Fig. 5, below). The average diameter of these filaments, once

subtracted the width of the replica, is close to 4 nm, a value consistent with the microfibril model proposed by Smith (1968). We have observed two mutually exclusive arrangements of the microfibrils within the collagen fibril: an helical arrangement, with a coiling angle of 18°, or a straight or slightly wavy arrangement. Differences in the microfibril arrangement involve also differences in the length of the axial period: our studies, carried out with freeze-etching and X-ray diffraction (Marchini et al. 1986) report in some tissues a periodicity shortened down to 64-65 nm, consistently with the values found by other Authors (Brodsky et al. 1980, Stinson & Sweeny 1980, Gathercole et al. 1986).

The arrangement is dependent on the anatomical source, the straight pattern being present in tendons and, in general, in fibrils with large and variable diameters (from 20 to 300nm and more). The helical pattern is mainly found in small fibrils, whose diameter is variable from site to site but remains fairly constant within the same anatomical region. Measurements carried out on many tissues indicate that the angle is independent from the fibril diameter (Ruggeri et al. 1979; Reale et al. 1981) as well as from the age of the specimen (unpublished results). In particular we studied skin samples of rat from newborn to old and found fibrils growing from 30 to 120 nm in diameter while maintaining the same 18° arrangement.

The two classes of fibrils, straight or helical, presumably play different functional roles. The spiral organization of the microfibrils gives the helical fibrils relative extensibility under tension, high flexibility and a capability to withstand off-axial loading without damage. The straight arrangement, on the contrary, makes the fibrils more unextensible under tension, which is a marked advantage in tissues transmitting mechanical loads, but on the other hand it causes the fibrils to be more prone to split longitudinally if loaded otherwise than along their axis. In effect fibrils with helical microfibrils appear gathered in vivo in randomly-oriented flexuous bundles or in three-dimensional networks and are commonly found, in association with elastic fibers, in highly deformable tissues, while straight fibrils are fastened in massive bundles of parallel fibrils bearing only unidirectional forces and are mainly found in unextensible tissues as tendons and ligaments.

REFERENCES

Brodsky B, Eikenberry EF, Cassidy K (1980). An unusual collagen periodicity in skin. Biophys.Biochim.Acta 621:162-166.

Capaldi MJ and Chapman JA (1982). The C-terminal extrahelical peptide of type I collagen and its role in fibrillogenesis in vitro. Biopolymers 21:2291-2313.

Fietzek PP and Kuhn KK (1976). The primary structure of collagen. Int.Rev.Connect.Tissue Res. 1-56.

Gathercole LJ, Shah JS and Nave C (1987). Skin-tendon differences in collagen D-period are not geometric or stretch-related artefacts. Int.J.Biol.Macromol. 9:181-183.

Hashimoto K (1974). Ultrastructure of freeze-cleaved dermal collagen. Acta Derm.Venerol. 54:241-248.

Helseth DL Jr, Lechner JH and Veis A (1979). Role of the amino-terminal extrahelical region of type I collagen in directing the 4D Overlap in fibrillogenesis. Biopolymers 18:3005-3014.

Hulmes DJS, Miller A, White SW, Timmins PA and Berthet-Colominas C. (1980). Interpretation of the low-angle meridional neutron diffraction patterns from collagen fibres in terms of the amino acid sequence. Int.J.Biol.Macromol. 2, 338-346.

Kivirikko KI and Myllyla R (1979). Collagen Glycosyltransferases. Int.Rev.Connect.Tissue Res. 23-72.

Lillie JH, MacCallum DK, Scaletta LJ and Occhino JC (1977). Collagen structure: evidence for a helical organization of the collagen fibril. J.Ultrastruct.Res. 58:134-143.

Marchini M, Morocutti M, Castellani PP, Leonardi L and Ruggeri A (1983). The banding pattern of rat tail tendon freeze-etched collagen fibrils. Connect.Tissue Res. 11:175-184

Marchini M, Morocutti M, Ruggeri A, Koch MHJ, Bigi A and Roveri N (1986). Differences in the fibril structure of corneal and tendon collagen. An electron microscopy and X-ray diffraction investigation. Connect.Tissue Res. 15:269-281.

Raspanti M, Marchini M, Ortolani F and Ruggeri A. Collagen fibril surface structures: freeze-etching data and computer modeling. Int.J.Biol.Macromol., in press.

Reale E, Benazzo F and Ruggeri A (1981). Differences in the microfibrillar arrangement of collagen fibrils. Distribution and possible significance. J.Submicr.Cytol. 13/2:135-143.

Ruggeri A, Benazzo F and Reale E (1979). Collagen fibrils with straight and helicoidal microfibrils: a freeze-fracture and thin section study. J.Ultrastruct.Res. 68:101-108.

Scott JE (1986). Proteoglycan-collagen interactions. CIBA Symposium 124:105-124.

Smith JW (1968). Molecular pattern in native collagen. Nature 219:157-158.

Stinson RH and Sweeny PR (1980). Skin collagen has an unusual d-spacing. Biophys.Biochim.Acta 621:158-161.

Zamyatnin AA (1972). Protein volume in solution. Progr.Biophys.Mol.Biol. 24:109-123.

Cells and Tissues: A Three-Dimensional
Approach by Modern Techniques in Microscopy,
pages 101–107

EXTRACELLULAR MATRIX: FUNCTIONAL SIGNIFICANCE OF OXYTALAN,ELAUNIN AND ELASTIC FIBERS

G.Cotta-Pereira and M.L.Iruela-Arispe

Department of Histology and Embryology, State University of Rio de Janeiro (U.E.R.J.), Rio de Janeiro,RJ 20551 , Brasil

INTRODUCTION

Much attention has been paid to the extracellular complex of macromolecules that includes collagens, elastin, proteoglycans and glycoproteins. They influence the cell behaviour in general during development and adult life and are sources of strength,resilience and cohesiveness to the tissues (Hay,1981).

Significant advances have contributed for elucidating many aspects of the nature of different constituents of extracellular matrices (Deyl and Adam,1981; Hukins,1984; Reddi,1985; Mecham,1986), but little is known about the composition and functional significance of the tissue structures that contain elastin.Elastic fibers are responsible for most of the elastic properties of many vertebrate tissues. These fibers have been described by Ross and Bornstein(1969) as consisting of an amorphous core of elastin surrounded by microfibrils of 10-12 nm in diameter. Other elastic-related fibers (oxytalan and elaunin fibers) have been visualized as formed by bundles of elastic microfibrils (oxytalan) or microfibrils intermingled with patches of amorphous elastin (elaunin) (Fullmer and Lillie,1958; Gawlik,1965; Cotta-Pereira et al.,1975, 1976a,1977).

The dermal elastic system fibers described by Cotta-Pereira and colleagues(1976a,1978) and confirmed by Schwartz and Fleischmajer(1986) as including oxytalan,elaunin and elastic fibers represent a useful model for studying the interaction between microfibrils and amorphous elastin during the maturation of elastic fibers.The work of Schwartz and Fleischmajer (1986)also suggests that oxytalan fibers may represent the initial step of elastogenesis. However, there are several other tissues where the study of elastic system fibers may help to clarify the nature and functional significance of those extracellular components.In this respect,the morphology of ciliary zonule and the annular ligament in rats were studied using light and electron microscopy.Also,aortae of chick embryos and adult unvertebrates(annelids,molluscs and crustaceans) were studied in the same way in order to show some aspects of ontogenesis and phylogenesis of elastic fibers.

MATERIAL AND METHODS

Fragments of eyeballs and temporal bones of young adult rats were fixed in 10% formaldehyde and embedded in paraffin.Other fragments were fixed in 3% glutaraldehyde dissolved in 0.1 Millonig buffer (pH=7.3) containing 0.25% of tannic acid,post-fixed in 1% osmium tetroxide and embedded in Epon (Cotta-Pereira,1976b).Temporal bones were previously decalcified before embedding.Also fragments of chick embryos from the 24th until 36th stage (Hamburger and Hamilton,1951) and adult earthworms (P.hawayana),bivalve mollusc(P.perna) and river shrimp(M.carcinus)were fixed both for light and electron microscopy.

The paraffin sections for light microscopic observations were stained with haematoxylin and eosin, iron haematoxylin(Verhoeff , 1908) and resorcin-fuchsin(Weigert,1898).While Verhoeff's iron haematoxylin demonstrates fully mature elastic fibers selectively,elaunin and oxytalan fibers are not stained by this method.Weigert's resorcin-fuchsin selectively stains both elastic and elaunin fibers,whereas oxytalan fibers remain unstained when they are not previously oxidized. Oxidation was performed using oxone as previously described (Fullmer et al.,1974).

RESULTS

Ciliary zonule. All fragments obtained from eyeballs of rats showed the ciliary zonule as formed by fibers that stained selectively for oxytalan fiber.At the electron microscope they disclosed typical pattern of oxytalan fibers (bundles of microfibrils of 10-12 nm in diameter,with a tubular appearance)(Figure 1).

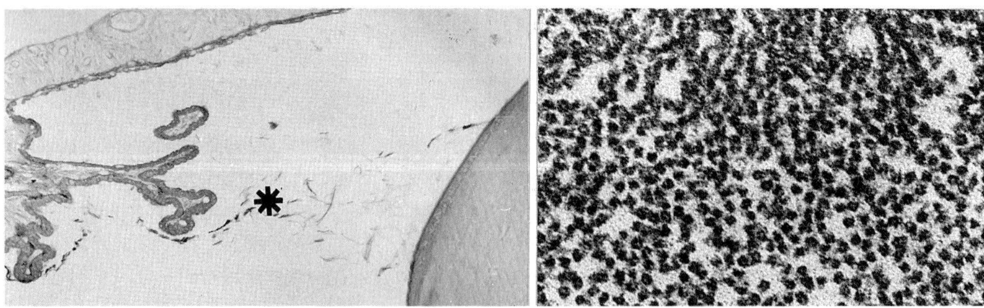

Figure 1.Left:Photomicrograph of ciliary zonule(asterisk) binding the lens to the ciliary body.Resorcin-fuchsin after oxidation,x200 Right:Electron micrograph of a zonular fiber showing cross sectioned microfibrils. Tannic acid fization, x80,000.

Annular ligament. The articulation between the base of the stapes and the margin of the oval window showed fibers with tinctorial and ultrastructural characteristics of elaunin fibers, anchoring into both articular surfaces as oxytalan fibers (Figure 2).

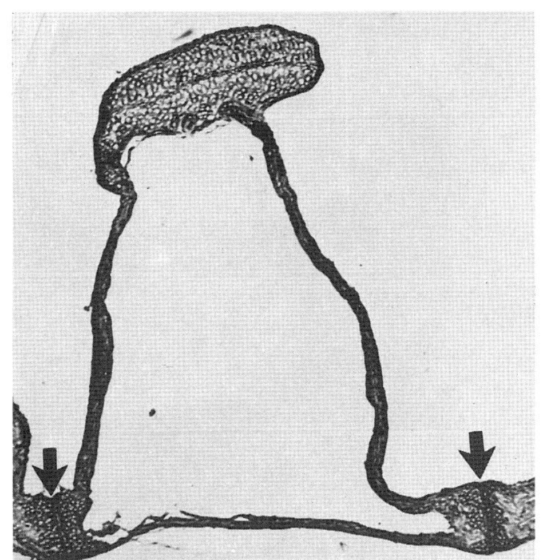
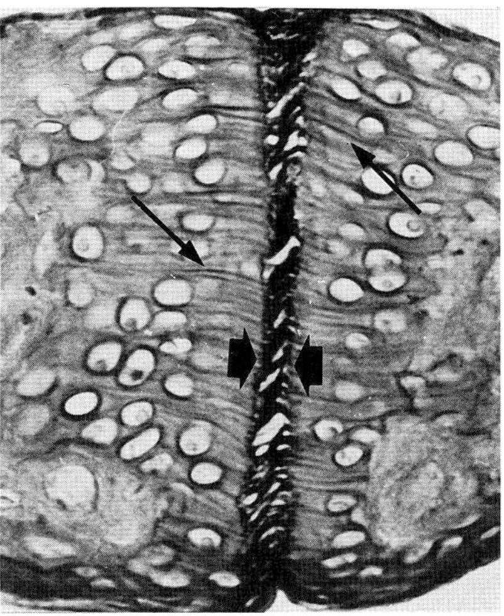

Figure 2. Left:Photomicrograph of the stapes and its articulation (arrows)with the oval window.H-E stain, x300. Right:Photomicrograph of the annular ligament showing elaunin fibers binding both articular surfaces(thick arrows)and oxytalan fibers(thin arrows)embedded into the cartilages.Resorcin-fuchsin after oxidation, x1,500.

Chick embryo aortae. Under the light microscope, the transverse sections of chick embryos have shown the presence of oxytalan fibers in the wall of developing aortae during the 29th stage, while elaunin fibers were visualized after 34th stage and elastic ones only after 36th stage. At the electron microscope patterns of oxytalan, elaunin and elastic fibers were coincident with the light microscopic observations (Figure 3).

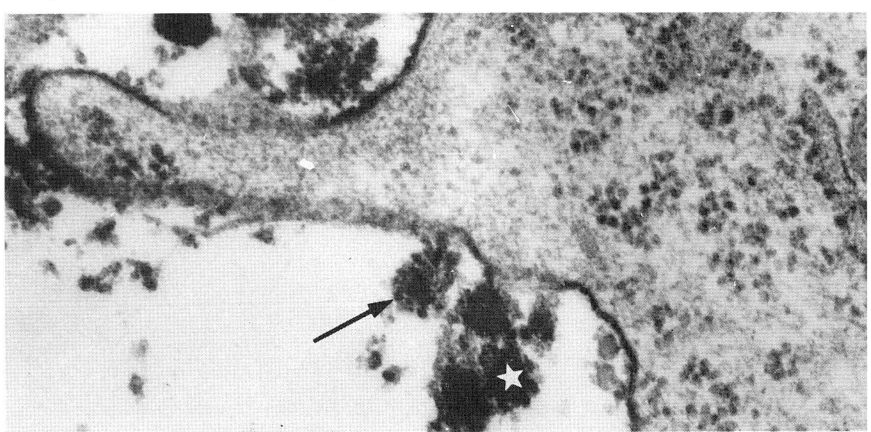

Figure 3.Electron micrograph of the wall of a dorsal aorta at the 34th stage (8th day of incubation)in the chick embryo. Observe microfibrils (arrow) intermingled with scarce dense amorphous material (star) in the extracellular matrix sinthesized by a smooth muscle cell. Tannic acid fixation, x 60,000.

Unvertebrate aortae. The fragments of the studied animals showed, in paraffin sections, positive reaction to oxytalan stains. At the electron microscope it was visualized the oxytalan pattern. (Figure 4).

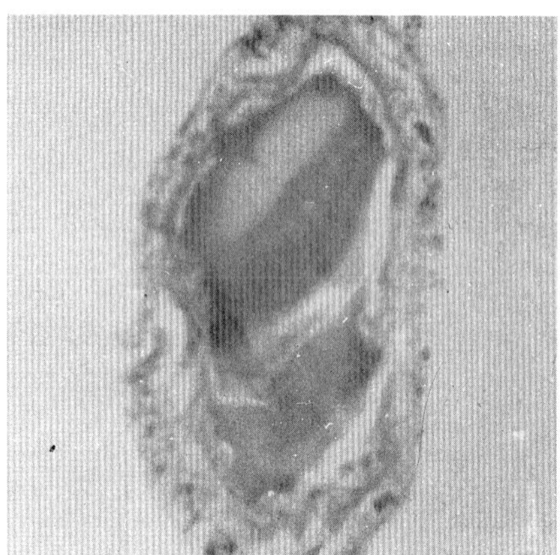

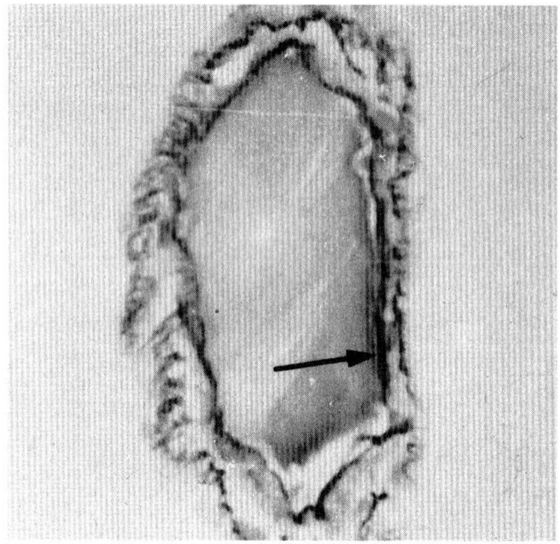

Figure 4. Left: Photomicrograph of dorsal artery of P.hawayana(an earthworm)stained with resorcin-fuchsin without previous oxidation. We don't observe any fiber . x 450. Right: Photomicrograph of the same artery stained with resorcin-fuchsin after oxidation. Observe oxytalan fibers in the artery wall(arrow), x 400.

DISCUSSION AND CONCLUSIONS

The present concept on the chemical composition of elastic fibers is basically that one appointed by Ross and Bornstein (1969): microfibrils and amorphous material containing elastin,which is responsible for the elastic properties of such fibers(Ross and Bornstein,1969; Robert et al.,1971; Cotta-Pereira et al.,1976a,1977; Robert and Robert,1980)

During the elastic fiber formation, the microfibrils appear first, followed by the amorphous component, which is gradually deposited between the m icrofibrils until the fiber is full matured (Fahrenbach et al.,1966).

However, in certain anatomical sites,fibers that are essentially formed by bundles of microfibrils remain without depositing amorphous elastin during the adult life. These fibers have been identified as oxytalan fibers and if they are devoid of the amorphous elastic material probably show a lesser tendency to elongate under mechanical stress (Cotta-Pereira et al.,1976a,1977).This hypothesis is consistent with the fact that oxytalan fibers have been found in locations where resistance to mechanical stress is required,such as periodontium (Fullmer and Lillie,1958),dermoepidermal junction (Cotta-Pereira et al.,1976a,1978),cartilage (Cotta-Pereira et al.,1984) and, as unequivocally now demonstrated,in the ciliary zonule. Our findings corroborate the observation of cross-reactivity between zonules and elastic tissue microfibrils by immunocitochemistry(Streeten,1982).It is possible that the zonular fibers are involved in a similar function of mechanical resistance; these fibers,departing from the ciliary body,reach the lens capsule and act by supporting the strength of the ciliary muscle upon the lens during the visual accommodation.

Elaunin fibers,containing microfibrils and a little amorphous material is expected to display elastic properties intermediate between those of elastic and oxytalan fibers (Cotta-Pereira et al., 1975,1976a,1977).As a matter of fact elaunin fibers may have an important role in the articulation between the base of the stapes and the margin of the oval window (fenestra vestibuli) because they predominate in this so-called annular ligament and are continuous with oxytalan fibers which are embedded in the matrix of the articular cartilage. Although the functional implications of these morphologic findings are a matter of speculation,the absence of elastic fibers and the presence of elaunin and oxytalan fibers instead suggest that a modulation of elasticity is required by this particular articulation.

Finally, it appear expedient to study the mechanism of elastogenesis both ontogenetically and phylogenetically.

During the formation of the chick aorta,mesenchymal cells of the wall of such developing arteries synthesize bundles of microfibrils which stain as oxytalan fibers.These fibers change the stainability to elaunin (34th stage) and elastic fibers (36th stage), when elastin is deposited between microfibrils. These findings , obtained by light and electron microscopic techniques corroborate a former observation of Gawlik(1965) who suggested the sequence: oxytalan-elaunin-elastic fibers after studying human fetuses. Recently, Schwartz and Fleischmajer (1986) also demonstrated,after studying the dermal elastic system fibers, by immunocitochemistry, that oxytalan,elaunin and elastic fibers constitute a sequence during the elastogenetic process.

One question, that remains to be answered , refers to which factors would be responsible for the interruption of elastogenesis remaining the fiber as oxytalan or elaunin during the adult life, in certain tissues and disclosing relevant functions of mechanical resistance in ciliary zonule,modulate elasticity in the stapes-oval window articulation or, when mature,improving full elasticity to the artery wall?

Probably lesser complex organisms, like unvertebrates,could led us to any speculation. The artery wall of the unvertebrates studied in the present work, is formed only by oxytalan fibers,identified by both light and electron microscopy. It is reasonable to assume that the simpleness of these animals doesn't require mechanisms of elasticity in their circulatory system. As a matter of fact unvertebrates doesn't possess cross-links as detected by Sage (1982) only in vertebrate aortae after using biochemical methods. Such findings corroborate our observations on the presence of oxytalan fibers and absence of elastic fibers in unvertebrate arteries

If in lower animals only the initial steps of elastogenesis are observed, it can be speculated that this is another example of "ontogenesis recapitulating phylogenesis" (Haeckel,1874).

ACKNOWLEDGEMENTS

This work was supported by CNPq,CAPES,FAPERJ,FAPPUERJ,Instituto Estadual de Hematologia Artur de Siqueira Cavalcante.

REFERENCES

Cotta-Pereira G, Del-Caro L, Montes GS (1984). Distribution of elastic system fibers in hyaline and fibrous cartilages of the rat. Acta Anat 119:80-85

Cotta-Pereira G, Rodrigo FG, Bittencourt-Sampaio S (1975). Ultrastructural study of elaunin fibers in the secretory coil of human eccrine sweat glands. Br J Derm 93:623-629.

Cotta-Pereira G, Rodrigo FG, Bittencourt-Sampaio S (1976a). Oxytalan,elaunin and elastic fibers in the human skin. J Invest Derm 66:146-148.

Cotta-Pereira G, Rodrigo FG, David-Ferreira JF (1976b). The use of tannic acid-glutaraldehyde in the study of elastic and elastic-related fibers. Stain Technol 51:7-11.

Cotta-Pereira G, Rodrigo FG, David-Ferreira JF (1977). The elastic system fibers. In Sandberg LB, Gray WR, Franzblau C (eds):"Elastin and Elastic Tissue," New York: Plenum Press, pp 19-30.

Cotta-Pereira G, Rodrigo FG, David-Ferreira JF (1978). Comparative study between the elastic system fibers in human thin and thick skin. Biol Cell 31:297-302.

Deyl Z, Adam M (1981) "Connective Tissue Reserach:Chemistry,Biology,and Physiology." New York: Alan R.Liss, pp 1-251.

Fahrenbach WH, Sandberg LB, Cleary EG (1966). Ultrastructural studies on early elastogenesis. Anat Rec 155-563-576.

Fullmer HM, Lillie RD (1958). The oxytalan fiber: a previously undescribed connective tissue fiber. J Histochem Cytochem 6:425-430.

Fullmer HM, Sheetz JH, Narkates AJ (1974). Oxytalan connective tissue fibers:a review. J Oral Pathol 3:291-316.

Gawlik Z (1965). Morphological and morphochemical properties of the motor organ of man. Folia Histochem Cytochem 3:233-251.

Haeckel E (1874). "Histoire de la création des êtres organisées d' aprés les lois naturelles." Paris: C Reinwald et Cie, p 274.
Hamburger V, Hamilton HL (1951). A series of normal stages in the development of the chick embryo. J Morphol 88:49-92.
Hay ED (1981). " Cell Biology of Extracellular Matrix." New York: Plenum Press, pp 1-4.
Hukins DW (1984). " Connective Tissue Matrix." Weinheim: Verlag Chemie pp 1-245.
Mecham RP (1986). " Regulation of Matrix Accumulation." Orlando: Academic Press pp 1-461.
Reddi AH (1985). " Extracellular Matrix: Structure and Function." New York: Alan R.Liss pp 1-436.
Robert L, Robert AM (1980). Elastin,elastase and arteriosclerosis. In Robert L,Robert AM (eds): "Biology and Pathology of Elastic Tissues," Basel: Karger, pp 130-173.
Robert B, Szigeti M, Deroette JC, Robert L, Bouissou H, Fabre MT (1971). Studies on the "microfibrillar" component of elastic fibers. Eur J Biochem 21:507-516.
Ross R, Bornstein P (1969). The elastic fiber.I. The separation and partial characterization of its macromolecular components.J Cell Biol 40:366-381.
Sage H (1982). Structure-function relationships in the evolution of elastin. J Invest Derm 79:146s-153s.
Schwartz E, Fleischmajer R (1986). Association of elastin with oxytalan fibers of the dermis and with extracellular microfibrils of cultured skin fibroblasts. J Histochem Cytochem 34:1063-1068.
Streeten BW (1982). Zonular apparatus. In Jakobiec FS (ed):"Ocular Anatomy, Embryology and Teratology," Philadelphia:Harper & Row, Publishers, pp 331-353.
Verhoeff FH (1908). Some new staining methods of wide applicability Including a rapid differential stain for elastic tissues. J Am Med Ass 50: 876-877.
Weigert C (1898). Über eine Methode zur Färbung elastischer Fasern Zentbl allg Path 9:189-292.

Cells and Tissues: A Three-Dimensional Approach by Modern Techniques in Microscopy, pages 109–114

ELECTRON MICROSCOPE STUDIES OF THE EARLY STAGE OF THE CALCIFICATION PROCESS: ROLE OF MATRIX VESICLES

Ermanno Bonucci

Department of Human Biopathology, Section of Pathological Anatomy, La Sapienza University, Rome, Italy

The mechanism of calcium phosphate deposition in organic matrix is not yet completely known, in spite of the innumerable investigations carried out on bone, epiphyseal cartilage, dentine, enamel, and many other calcifying tissues. Electron microscopy has furnished many data on the structure and organization of different calcified tissues which, althougth not resolutive, have proved useful for understanding the mechanism of the process. This paper concisely reviews the most important ultrastructural data on the so-called matrix vesicles.

The theory that calcification occurs in the holes of collagen fibrils by heterogenous nucleation, proposed by Glimcher (1959) for bone matrix, cannot be applied to other calcifying matrices which have other types of organization. This is the case of epiphyseal cartilage, because its matrix consists of abundant proteoglycan components and relatively few and thin collagen fibrils, probably lacking the hole zones present in bone collagen. Actually, in calcifying epiphyseal cartilage the aggregates of inorganic crystallites are irregularly scattered through the calcifying matrix without any relationship with the collagen fibrils or their periodic banding (Bonucci, 1984 a). Moreover, in types of bone characterized by a loose structure, such as the medullary bone of birds, the early aggregates of inorganic crystals are placed between, and not within the collagen fibrils (Bonucci and Gherardi, 1975). These findings suggest that in this type of tissues the early phase of the calcification process is not in relationship with the collagen fibrils, but with other structures of the matrix.

Looking for these structures, the areas of early calcification of epiphyseal cartilage, bone and other tissues have been studied under the electron microscope, with particular consideration for the so-called maturative and hypertrophyc zones of the cartilage, where the earliest crystal aggregates are formed. These studies have shown that the pericellular matrix of these zones contains small electron dense bodies irregularly distributed between thin collagen fibrils (Bonucci, 1967). At high enlargement (Fig. 1a), these bodies appear as roundish structures consisting of a homogeneous, osmiophilic matrix surrounded by a dense border which, in equatorial sections, appears as a trilaminar membrane (Anderson, 1969; Bonucci, 1970). They have been called matrix vesicles (Anderson, 1969).

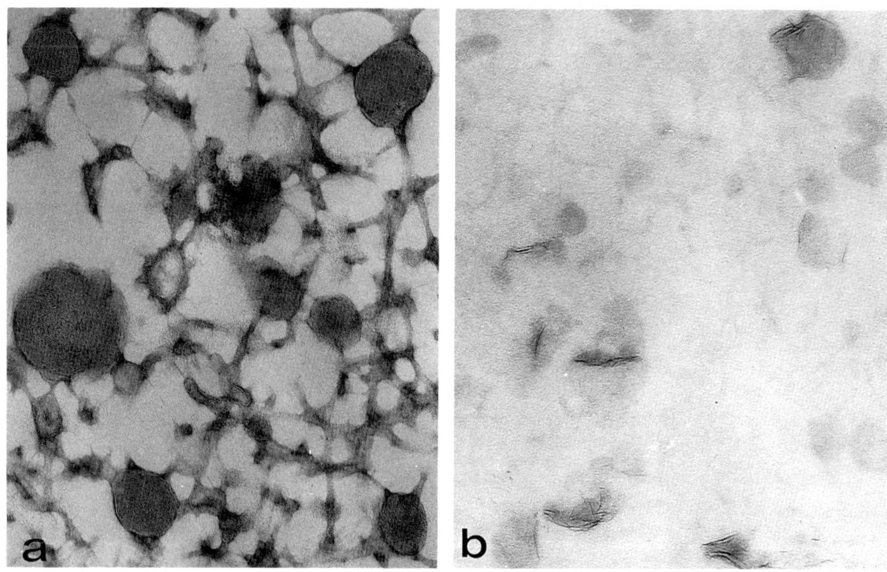

Fig. 1 - Calcifying epiphyseal cartilage: (a) staining with uranium and lead shows the homogeneous matrix, peripheral membrane, and relationships with collagen fibrils of matrix vesicles; (b) relationships of inorganic crystals with matrix vesicles are evident in unstained sections. x 60,000.

Because of their peripheral membrane, matrix vesicles are considered to be of cellular origin. However, they have no direct connection with the chondrocytes or their processes. This has been shown in serial sections of epiphyseal cartilage (Bonucci, 1970, 1978) and confirmed by the fact that matrix vesicles can be isolated from cartilage matrix by enzymatic digestion and differential centrifugation (Ali et al., 1970).

The cellular origin of matrix vesicles is generally accepted; however, the mechanism of their formation is not completely known. Most of them are formed by swelling of the tip of cell processes, followed by detachment of the swollen part. This forms a matrix vesicle which remains in the cartilage matrix, whereas the cell process retracts (Hale and Wuthier, 1987). In some cases, whole cytoplasmic processes swell and break into several fragments which become matrix vesicles. In other cases, these are formed by degeneration and disruption of chondrocytes, a process known since 1930 as "Verdammerung der Zellen", that is, disappearance of cells. The fragmentation of the cytoplasm of these cells forms group of matrix vesicles which remain in the lacuna of the degenerate chondrocyte. In other cases, matrix vesicles are formed by a process of exocytosis and direct extrusion from the chondrocyte. What causes the formation of matrix vesicles is not known. It is possible that swelling, degeneration and fragmentation of cytoplasm and cytoplasmic processes may be due to changes of oxygen tension and pH values.

The interest on matrix vesicles is mainly due to the fact that they represent areas of initial calcification. This is shown in unstained sections (Fig. 1), where isolated crystals and small aggregates of crystals can be found both within the vesicles (data confirmed in serial sections; Bonucci, 1976) and on their peripheral membrane (Bonucci, 1984 b) when the collagen fibrils of the cartilage matrix are still uncalcified (Gay et al., 1978). The shape of the calcifying matrix vesicles is preserved until they are full of crystals. Successively, as the number of crystals increases, the roundish shape is lost, the vesicle peripheral membrane disappears and the crystals spread into the surrounding matrix forming a calcification nodule. Calcification of collagen fibrils seems to begin at this stage.

The mechanism of matrix vesicle disruption is not known. It has been suggested that the peripheral membrane of the vesicles might be broken by the pressure of the needle-like crystals growing within them (Anderson, 1976). However, because remnants of matrix vesicles are not present in the calcified areas, another possibility is that they are completely dissolved, perhaps by an enzymatic mechanism.

Matrix vesicles similar to those found in cartilage are present in bone, mantle dentine, and other calcifying tissues (Anderson, 1976; Bonucci, 1984 a,b). However, their frequency in these tissues is variable and usually smaller than in cartilage.

To understand in which way matrix vesicles are calcified it is necessary to know their composition. The PAS method for light microscopy and the periodic acid - silver nitrate method for electron microscopy show that matrix vesicles contain glycoproteins (Bonucci, 1967, 1970). The colloidal iron method at pH 2.5 shows that their membrane is coated by acid proteoglycans in the same way as the chondrocyte membrane (Bonucci, 1970). Moreover, biochemical analysis carried out on isolated vesicles shows that they contain phospholipids (Wuthier, 1975).

The most interesting finding is probably that concerning the high content of alkaline phosphatase in matrix vesicles. This has been shown both by histochemical (Matsuzawa and Anderson, 1971; Bernard, 1978; Göthlin and Ericsson, 1973; Khan et al., 1978; Väänänen and Korhonen, 1980), biochemical (Ali et al., 1970; Fortuna et al., 1978; Majeska and Wuthier, 1975) and immunohistochemical (De Bernard et al., 1986; Morris et al., 1986) methods. The enzyme, which is a glycoprotein with Ca-binding properties synthesized by chondrocytes and exported into the extracellular matrix by matrix vesicles, is active in the zone of maturing and hypertrophic chondrocytes and is then incorporated in calcified matrix and inactivated (De Bernard et al., 1986). This and other findings suggest that a direct association exists between alkaline phosphatase and the developing mineral (McLean et al., 1987).

Many data are still lacking about the structure, composition and function of matrix vesicles. However, the available results show that, although the presence of matrix vesicles, and consequently their role in calcification, is variable, to reach a minimum in compact bone, they represent the areas of initial calcification in cartilage, bone, dentine, and other normal and pathological calcifying tissues. Calcification of collagen fibrils seems to begin when matrix vesicles are completely calcified and the crystals spread from them into the matrix. The presence of phospholipids in matrix vesicles suggests that calcium ions are accumulated in the vesicles by reacting with their acidic groups. Accumulation of phosphate groups could be due to the activity of the alkaline phosphatase. Although further investigation is needed, it is possible to speculate that these two mechanisms, by accumulating calcium and phosphate ions into the vesicles, might be responsible of the induction of the calcification process.

REFERENCES

Ali SY, Sajdera SW, Anderson HC (1970). Isolation and characterization of calcifying matrix vesicles from epiphyseal cartilage. Proc Natl Acad Sci 67: 1513-1520.

Anderson HC (1969). Vesicles associated with calcification in the matrix of epiphyseal cartilage. J Cell Biol 41: 59-72.

Anderson HC (1976). Matrix vesicles of cartilage and bone. In Bourne GH (ed): "The biochemistry and physiology of bone", New York, Academic Press, v. 4, pp 135-157.

Bernard GW (1978). Ultrastructural localization of alkaline phosphatase in initial intramembranous osteogenesis. Clin Orthop 135: 218-225.

Bonucci E (1967). Fine structure of early cartilage calcification.calcification. J Ultrastruct Res 20: 33-50.

Bonucci E (1970). Fine structure and histochemistry of "calcifying globules" in epiphyseal cartilage. Z Zellforsch 103: 192-217.

Bonucci E (1971). The locus of initial calcification in cartilage and bone. Clin Orthop 78: 108-139.

Bonucci E (1978). Matrix vesicles formation in cartilage of scorbutic guinea pigs: electron microscope study of serial sections. Metab Bone Dis Rel Res 1: 205-212.

Bonucci E (1984 a). The structural basis of calcification. In Ruggeri A, Motta PM (eds):"Ultrastructure of the connective tissue matrix", Boston, M Nijhoff, pp 165-191.

Bonucci E (1984 b). Matrix vesicles: their role in calcification. In Linde A (ed): "Dentin and dentinogenesis", Boca Raton, CRC Press, v. 1, pp 135-154.

Bonucci E, Dearden LC (1976). Matrix vesicles in aging cartilage. Fed Proc 35: 163-168.

Bonucci E, Gherardi G (1975). Histochemical and electron microscope investigations on medullary bone. Cell Tissue Res 163: 81-97.

De Bernard B, Bianco P, Bonucci E, Costantini M, Lunazzi GC, Martinuzzi P, Modricky C, Moro L, Panfili E, Pollesello P, Stagni N, Vittur F (1986). Biochemical and immunohistochemical evidence that in cartilage an alkaline phosphatase is a Ca-binding glycoprotein. J Cell Biol 103: 1615-1623.

Fortuna R, Anderson HC, Carty RP, Sajdera SW (1978). The purification and molecular characterization of alkaline phosphatases from chondrocytes and matrix vesicles of bovine fetal epiphyseal cartilage. Metab Bone Dis Rel Res 1: 161-168.

Gay CV, Schraer H, Hargest TEJr (1978). Ultrastructure of matrix vesicles and mineral in unfixed embryonic bone. Metab Bone Dis Rel Res 1: 105-108.

Glimcher MJ (1959). Molecular biology of mineralized tissues with particular reference to bone. Rev Modern Phys 31: 359-393.

Göthlin G, Ericsson JLE (1973). Fine structural localization of alkaline phosphatase in the fracture callus of the rat. Histochemie 36: 225-235.

Hale JE, Wuthier RE (1987). The mechanism of matrix vesicle formation. Studies on the composition of chondrocyte microvilli and on the effects of microfilament-perturbing agents on cellular vesiculation. J Biol Chem 262: 1916-1925.

Kahn SE, Jafri AM, Lewis NJ, Arsenis C (1978). Purification of alkaline phosphatase from extracellular vesicles of fracture callus cartilage. Calcif Tiss Res 25: 85-92.

Majeska RJ, Wuthier RE (1975). Studies on matrix vesicles isolated from chick epiphyseal cartilage. Association of pyrophosphatase and ATPase activities with alkaline phosphatase. Biochim Biophys Acta 391: 51-60

Matsuzawa T, Anderson HC (1971). Phosphatase of epiphyseal cartilage studied by electron microscopic cytochemical methods. J Histochem Cytochem 19: 801-808.

McLean FM, Keller PJ, Genge BR, Walters SA, Wuthier RE (1987). Disposition of preformed mineral in matrix vesicles. Internal localization and association with alkaline phosphatase. J Biol Chem 262: 10481-10488.

Morris DC, Väänänen HK, Munoz P, Anderson HC (1986). Light and electron microscopic immunolocalization of alkaline phosphatase in bovine growth plate cartilage. In Ali SY (ed): "Cell mediated calcification and matrix vesicles", Amsterdam, Excerpta Medica, pp 21-26.

Väänänen HK, Korhonen LK (1980). Purification of matrix-vesicle alkaline phosphatase from chicken epiphyseal cartilage and experiments on its ATP-hydrolyzing properties. Clin Orthop 148: 291-296.

Wuthier RE (1975). Lipid composition of isolated epiphyseal cartilage cells, membranes and matrix vesicles. Biochim Biophys Acta 409: 128-143.

Cells and Tissues: A Three-Dimensional Approach by Modern Techniques in Microscopy, pages 115–119

QUANTITATIVE ASSESSMENT OF PROTEOGLYCANS BY PROTON MICROPROBE

T. Cichocki, S. Divoux*, B. Gonsior*, L. Jarczyk**, E. Rokita**, A. Strzalkowski, M. Sych.

Academy of Medicine, Kopernika 7,PL-31034 Krakow,Poland.
*Institute of Experimentalphysis III,Ruhr-Universitat, P.O.B.102148,D-4630 Bochum,Fed.Rep.Germany
**Institute of Physic,Jagellonian University,Reymonta 4,PL-30059 Krakow,Poland

INTRODUCTION

Colloidal iron and alcian blue methods have been used for determination of glycosaminoglycans within tissue sections for more than 40 years. The original methods have undergone many modifications (Pearse, 1985) but their principles remained unchanged: if reactions are carried out at low pH, only strongly acidic groups will be dissociated and thus reactive. Under such conditons the reaction will identify sulphate groups of keratin and chondroitin sulphates and not carboxyl groups of uronic acids. Since the results of a routine histochemical staining are qualitatively only, the specificity of reactions is still a subject of much controversy.

The aim of this paper is to quantitate the colloidal iron and alcian blue reactions by measuring the amount of Fe and Cu bound mainly by sulphate groups of mucopolysaccharides. Since many factors may also influence Fe and Cu binding, the content of S was measured parellelly at the same sites to assess the amount of sulphates more directly. In order to compare both reactions and to find out how they are interrelated, the Fe/S and Cu/S ratios were calculated.

MATERIALS AND METHODS

Cartilage from the femur of 18 days old embryos and from ribs of mature CBA strain mice was investigated. The animals and embryos were killed by decapitation. Hind limbs and skull of embryos as well as costal parts of the thoracic wall of mature mice were quickly dissected and snap frozen on metal holders. 8 µm cryomicrotome sections cut at -25°C were air-dried. Adjacent sections were placed on a milipore backing (thickness 5 µm) for micro-PIXE measurements, and on glass slides for routine histochemical procedures. The former sections were photographed and then incubated for 30 min. floating in the colloidal iron medium acc. to Mueller (Pearse, 1985) or in 0.1% alcian blue medium at pH 1.0 and pH 2.5 respectively.

The micro-PIXE measurements were made using the proton microprobe at the Dynamitron Tandem Accelerator of the Ruhr Universitat Bochum. Details of the method are given elsewhere (Cichocki et al. 1988a, 1988b). For each sample, the mean content of each element in the selected parts of the measured region was calculated from element contents in the irradiated points.

For light microscopic histochemistry, adjacent sections were processed in the identical way, but with additional colour-developing treatment with $K_4Fe(CN)_6$, in order to visualize the absorbed iron by Prussian Blue deposits.

RESULTS AND DISCUSSION

An example of micro-PIXE results is presented in Fig.1. The

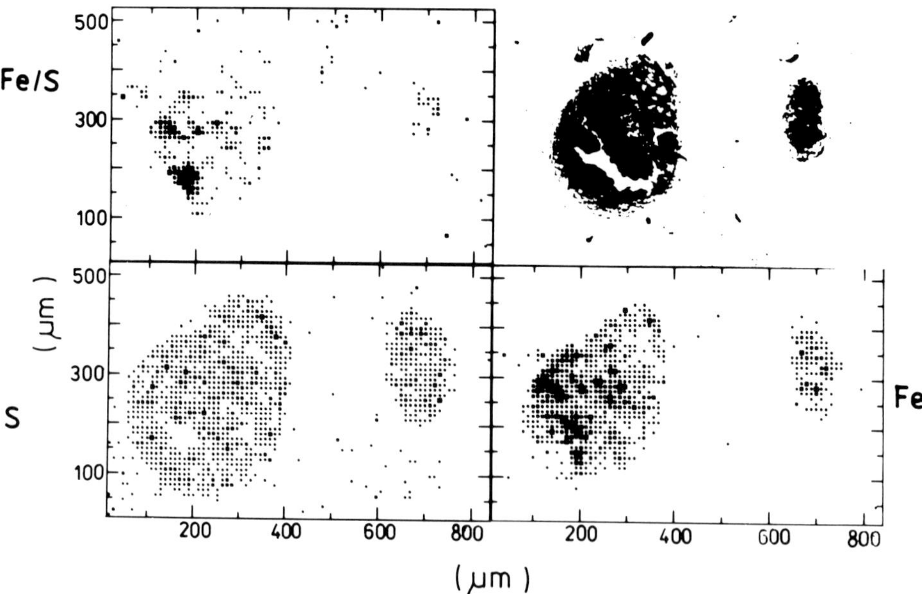

Fig.1 Costal cartilage of mature mouse. Raster scan plots for Fe, S and Fe/S respectively made in the area shown in the upper right corner micrograph with exact matching in size and point localization. The picture in the upper right corner shows colloidal iron-potassium ferrocyanide stainig of the adjacent section. Shading of each point is correlated with element content. Ten-grade linear scale was used.

TABLE 1
Contents of elements in the investigated samples (ng/cm) and rations of contents. Abbreviations: C - cartilage, T - another types of tissue.

Sample		S		Cu		Cu/S	
		C	T	C	T	C	T
1/87	pH=1.0	4060	1250	1485	31	0.37	0.02
1/87	pH=2.5	2910	812	1080	28	0.37	0.03
2/87	pH=1.0	6030	1200	2370	100	0.39	0.08
2/87	pH=2.5	7100	128	3830	50	0.54	0.05
		S		Fe		Fe/S	
		C	T	C	T	C	T
2/87	pH=1.8	1450	458	4220	699	2.91	1.53
3/87	pH=1.8	1970	859	4760	926	2.42	1.08
4/87	pH=1.8	3910	1650	4640	952	1.19	0.58

irradiated region consists of two mouse ribs. The costal hyaline cartilage stains strongly with Prussian Blue with the Hale method (Fig.1 -upper right corner picture), the intensity of staining shows local differences. The micro-PIXE raster scan measurements (Fig.1) show a preferential localization of colloidal iron in the rib territory, however, the distribution of iron concentration is not homogeneous within the rib area. The sulphur content is also much higher than elsewhere, its distribution, however, is more even. Mean S and Fe contents calculated for rib territory as well as for other types of tissues are presented in Table 1 (sample 4/87). Generally, the average Fe content exceeds the mean S content. In the rib cartilage alone, the level of iron is slightly above the level of sulphur, therefore the Fe/S ratio amounts to 1.19. Other tissues contain much less iron (by factor 5 if compared to cartilage) and less sulphur, thus their ratio is around 2 times lower than in the cartilage. It should be pointed out that the amount of S found in nonchondronal tissues of mature animals (muscle and dense, collagen-rich, connective tissue) indicates the presence of protein sulphur.

For embryonal cartilage (Table 1 - sample 3/87) the level of sulphur in the cartilage plates reflects the presence of glycosaminoglycan sulphate groups. This was confirmed by a strong binding of colloidal iron; the absolute amount of Fe is the same as for the rib cartilage, although the mount of S is 2 times lower. The high level of iron and high Fe/S proportions (above 2) show a marked domination of sulphates, as in the embryonal cartilage. In the rib cartilage, the high level of iron is accompanied by very high concetration of sulphur, so that the Fe/S proportions are relatively low. This indicates that aging cartilage becomes fibrous, and though it still contains substantial amounts of sulphate groups its protein conent is much higher than that of the embryonal cartilage (Table 1 sample 4/87 and 3/87 respectively).

For newly formed cartilage from the skull (samples 1/87 and 2/87 the amount of the bound copper is only slightly lower than that of iron. Moreover, we have found that more copper binds at the sites of high sulphur content (cartilage) in strictly proportional manner but the amount of bound iron was independent from S level within the cartilage territory. This observation results from simultaneous, higly sensitive measurements of two elements (S and Fe or S and Cu) in the same points. Unfortunately, no straighforward conclusions may be drawn from such observation. The suphur comes from two main sources: sulphate-containing carbohydrates and sulphur-containing proteins; the mutual proportion of the two will influence the Fe/S and Cu/S ratios.

The influence of pH on alcian blue binding by reactive anionic groups needs further exploration.

With this information it would be difficult to construct a model of the binding process: in the cartilage itself there is no constant Cu/S or Fe/S proportion; we found for istance points with relatively low content of sulphur and very strong Fe binding. This indicates that other factors - not only the amount of dissociated anionic groups - influence the binding of iron. Such factors be stereological in nature. The distance between SO_3 groups and their relation to the size of iron and alcian blue particles might additionally influence the binding. Therefore, since the alcian blue molecule is characterized by a smaller diameter than the colloidal iron particle, there are more sulphate groups available in a section what in turn diminishes or even cancels the stereological inhibition of the dye molecule binding. The observation that cartilage in certain regions of the growth plate binds more iron per sulphur-containing molecule may be therefore explained by changes in the spatial arrangement of chemical groups rather than by changes in the chemical composition.

CONCLUSION

The results presented both as raster plot and as calculated mean values show that colloidal iron and alcian blue are bound mostly by the cartilage and that their level may be about 10 times higher than the mean bound levels in other tissues. This provides quantitative data for the crucial step of the method, and confirms the known fact of cartilage staining with cationic dyes. However, it is still impossible to answer the question whether carboxyl groups may contribute to colloidal iron and alcian blue staining. The lack of stereological limitations makes the binding of alcian blue molecule to sulphate groups of glycosaminoglycans much more specific than that of colloidal iron.

PIXE spectroscopy in combination with the proton microprobe has the advantage of a high power of detection and easy matching of organ structure with analytical data. It enables to detect both endogenous elements as well as those introduced by histochemical procedures, which makes the technique a useful tool for quantitative histochemistry.

This work was partially supported by the Polish Central Research Program CPBP - 01.09 no III.1.2.12.

REFERENCES

Cichocki T, Gonsior B, Hofert M, Jarczyk L, Rokita E, Strzalkowski A, Sych M. (1988a), Measurement of colloidal iron binding at low pH in cartilage using the proton microprobe. Histochem. J. 20, 201-6.

Cichocki T, Gonsior B, Hofert M, Jarczyk L, Rait B, Rokita E, Strzalkowski A, Sych M. (1988b), Measurements of mineralization process in the femour growth plate and rib cartilage of the mouse using pixe in combination with a proton microprobe. Histochemistry 89, 99-104.

Pearse AGE. (1985) Histochemistry. Theoretical and Applied 4th edn. Edinburgh, London, New York.

Cells and Tissues: A Three-Dimensional Approach by Modern Techniques in Microscopy, pages 121–125

AGE- AND RADIATION-RELATED ALTERATIONS OF BONE TISSUE: A COMPARATIVE STUDY

Antoine Dhem, Catherine Nyssen-Behets and René Dambrain

Human Anatomy Research Unit, Université Catholique de Louvain, 52 Avenue E. Mounier (5240), 1200 Brussels, Belgium

INTRODUCTION

Radiation damage has been regarded for a long time as the result of an accelerated ageing process because it corresponds to a decrease of living cells either by direct effect or through vascular alteration (Ewing, 1926). However, bone tissue not only contains specialized cells but also a tremendous amount of calcified matrix which has to be considered too. The aim of this study was to compare, by microradiographic analysis, the respective alterations of bone tissue after irradiation and in ageing. This latter process was also studied by scanning electron microscopy.

MATERIAL AND METHODS

Fifty-five large mandibular fragments were resected in patients suffering from severe osteoradionecrosis. The respective clinical data and treatments were published earlier (Dambrain et al., 1979). The bone pieces were embedded in methyl-methacrylate and the 80-micron-thick undecalcified sections obtained from these samples were microradiographed according to our usual and previously described method (Dhem, 1967).

The same technique was also applied to femoral and tibial diaphyses of 36 subjects aged 18 to 93 years. Furthermore, 3 femoral and 7 tibial samples were prepared for scanning electron microscopy (SEM) examination by fixation in neutral formaldehyde, cutting into 1-mm-thick longitudinal sections, immersion in 5.25% NaOCl, dehydration, and coating with a thin layer of gold (Boyde, 1972).

RESULTS

Microradiographic analysis of undecalcified sections from mandibular fragments resected in all the subjects suffering from osteoradionecrosis always shows the same peculiar aspect of bone destruction. The foci of bone resorption present an extremely striking polymorphism which is easily distinguishable from normal osteoclastic resorption cavities. Unlike these latter, they do not form tunnels with a well-definite direction, but very irregular anfractuosities (fig. 1) which are visible in any section plane.

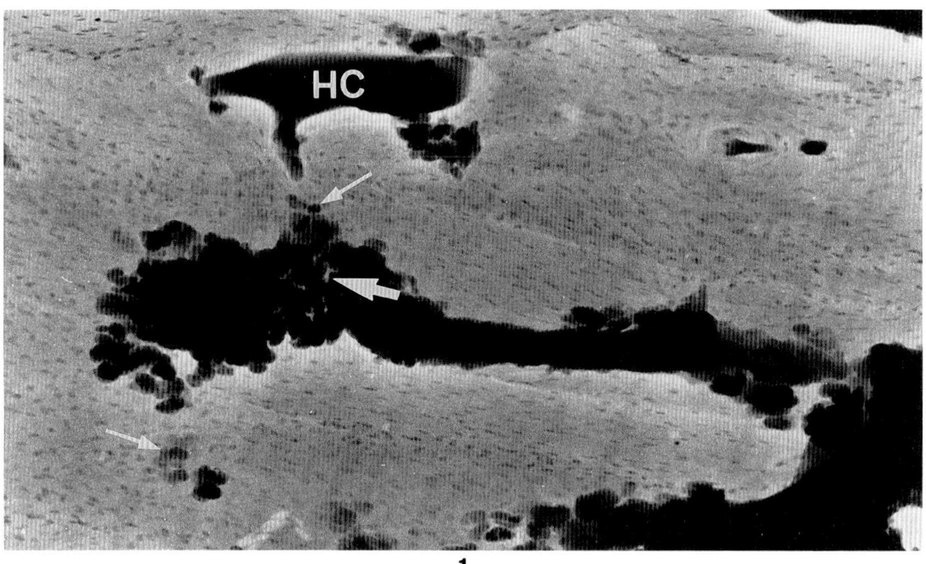

Figure 1. Microradiograph of a section through the mandibular fragment of a 55-year-old man (x 93). HC: Haversian canal. Thick arrow: persistence of bone spicules. Thin arrows: smallest bone destruction cavities.

Numerous small fragments of bone tissue (thick arrow) often persist giving rise to a multilocular configuration of the anfractuosities. This phenomenon always develops around pre-existent Haversian canals (HC) but it does not necessary involve all of them. The smallest cavities (thin arrows) appear spherical with a diameter strikingly similar to the largest dimension of an osteocytic lacuna. These cavities do not contain osteoclasts but only small mononucleated cells. In most cases, this bone destruction process is so tremendous that it affects the anatomical shape of the bone piece.

Compact bone ageing studied on microradiographs of femoral and tibial sections is characterized by a progressive structural change in the osteonic aspects. The process starts by occurrence of a hypercalcified ring concentric to the Haversian canal. Hypercalcified osteocytic lacunae inside the ring attest the cellular death (Dhem, 1970). Afterwards, the hypermineralization of the inner lamellae is observed as well as the disjoining of these latter, giving rise to a wide and irregular Haversian canal. All these phenomena can be present simultaneously in the same osteon, as summarized in the microradiograph of a longitudinal section (fig. 2). They have been interpreted as the result of a specific bone destruction process called "delitescence" by analogy

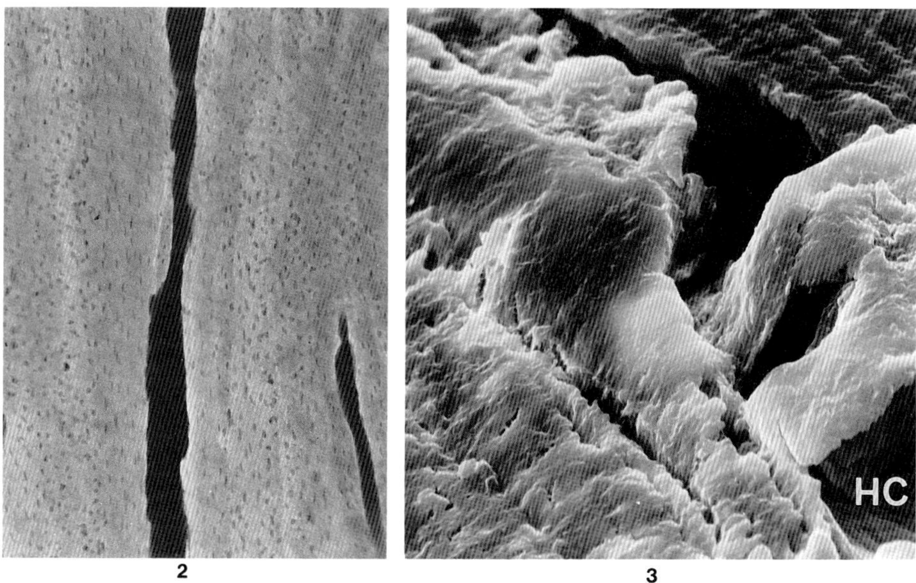

Figure 2. Microradiograph of a longitudinal section through the tibial diaphysis of a 86-year-old woman (x 85).

Figure 3. SEM of a longitudinal section through the femoral diaphysis of a 76-year-old man (x 1000). HC: Haversian canal.

with what occurs during crystal dehydration (Dhem, 1970, 1973, 1980; Dhem and Robert, 1986). It should be noted that the Haversian canal (fig. 2) has an inconstant diameter.

SEM examination of longitudinal sections in femoral and tibial diaphyses confirms the microradiographic observations. In some Haversian canals, the walls are fissured and fragments are detaching. Others have irregular outlines. It is also possible to observe the intact inner lamella which then hides partly the

Haversian canal (HC), as in fig. 3. However, fissures and detaching fragments are most often present in this type of osteon.

Analysis of the corresponding stained sections has given evidence of the osteoclasts absence in this process (Jaworski et al., 1972; Dhem and Piret, 1975).

DISCUSSION

Microradiographic analysis of undecalcified sections clearly shows that irradiation, whatever the type of ray used, is responsible for bone alterations (fig. 1) which are quite different from compact bone ageing (fig. 2). The former corresponds to very polymorphic foci of bone destruction, different from osteoclastic resorption cavities (Dhem, 1967), and are probably of osteocytic origin, as explained in experimental osteoradionecrosis of the cat mandible (Dambrain et al., 1986).

Bone loss observed in compact bone ageing is associated with hypercalcification of some lamellae and subsequent disjoining of these latter (Dhem, 1970, 1973, 1980; Dhem and Robert, 1986). A good correlation exists between the SEM and the microradiographic aspect of ageing osteons. Indeed, it is impossible to detect the hypercalcified ring by ultramicroscopic examination but the other modifications can de illustrated with this technique. Then detaching inner lamellae, as in fig. 3, and Haversian canals with irregular diameter can be observed. Comparison of our observations with classical description of Haversian canals and Howship's lacunae (Boyde, 1972) leads to the conclusion that delitescence is not the result of osteoclastic resorption, as previously reported by Jaworski et al. (1972). Pathogenic factors responsible for bone delitescence are still unknown.

Therefore it may be concluded that irradiation and ageing result in two distinct non-osteoclastic bone destruction processes. The former affects the macroscopic aspect of the skeletal piece, contrary to the latter. That osteoradionecrosis could be explained by an accelerated ageing phenomenon is excluded by our observations.

REFERENCES

Boyde A (1972). Scanning electron microscope studies of bone. In Bourne GH (ed): "The biochemistry and physiology of bone," 2nd ed, Vol I, New York, London: Academic Press, pp 259-310.

Dambrain R, Barrelier P, Billet J, Lecacheux B (1979). L'ostéoradionécrose mandibulaire. I. Etude clinique.Acta Stomatol Belg 76: 227-242.

Dambrain R, Dhem A, Winant M, Wambersie A (1986). Microradiographical aspects of maxillo-dental lesions in the irradiated

cat. J Eur Radiother 7: 111-118.
Dhem A (1967). Le remaniement de l'os adulte. Thèse Univ Louvain, Bruxelles, Paris: Arscia, Maloine.
Dhem A (1970). Microradiographic study of ageing of compact bone in man. In Jelliffe AM, Strickland B (eds): "Symposium ossium," London, Edinburgh: Livingstone pp 289-291.
Dhem A (1973). Les mécanismes de destruction du tissu osseux. Acta Orthop Belg 39: 423-443.
Dhem A (1980). Etude histologique et microradiographique des manifestations biologiques propres au tissu osseux compact. Bull Acad Med Belg 135: 368-381.
Dhem A, Piret N (1975). Remarques à propos de la résorption ostéoclastique. Bull Assoc Anat (Nancy) 59: 157-162.
Dhem A, Robert V (1986). Morphology of bone tissue aging. In Uhthoff HK (ed): "Current concepts of bone fragility," Berlin, Heidelberg: Springer-Verlag, pp 363-370.
Ewing J (1926). Radiation osteitis. Acta Radiol 6: 399-412.
Jaworski ZF, Meunier P, Frost HM (1972). Observations on two types of resorption cavities in human lamellar cortical bone. Clin Orthop 83: 279-285.

Cells and Tissues: A Three-Dimensional
Approach by Modern Techniques in Microscopy,
pages 127–131

SCANNING AND TRANSMISSION ELECTRON MICROSCOPY ON THE SYNOVIAL MEMBRANE IN THE RABBIT

Wei Jun Chen, Gui Lan He and Chang Jiang Sun

Department of Orthopedics,Dalian Medical College(W.J.C)
Dalian Animal and Plant Quarantine Service. China
(G.L.H., C.J.S),Dalian. People's Republic of China

INTRODUCTION

The synovial membrane as one of the basic component of synovial joint has an important function. It was often involved in the articular trauma, infection, degeneration and tumor. The synovial membrane consists of two parts, namely the synovial intima adjacent to the joint space and a supportive layer called the subsynovial tissue.

As early as 1928, the synovial membranes were classified into areolar, adipose and fibrous types (Key, 1928). Then, some authors reported the microscopic structures of normal and pathological synovial membranes. The ultrastructures of normal synovial membranes were studied in some mammals (Watanabe , 1974; Krey,1973; Linck,1978; Okada, 1981;Ghadially,1983) . The topographic structures of normal and abnormal synovial membranes were investigated by scanning electron microscopy (Cryfe et al., 1969; Wysocki,1972). Ultrastructural pathological changes of abnormal synovial membranes were also observed (Roy, Ghadially and Grane,1966; Bhan and Roy,1971) .

In the study of pathological synovium with SEM and TEM , it is important to identify the histological type of synovial membrane or synovial intima affected and to determine the extent to which normal surface features have been changed. In the present paper, to provide the morphological basis for investigating ultrastructural pathological changes of synovial membranes, the synovial membranes of the rabbit knee joints were studied with SEM and TEM .

MATERIALS AND METHODS

Seven adult healthy rabbits weighing 2 to 3 kg were used in the present study. The synovial membrane specimens were immediately obtained after killing rabbit, and consisted of the anterior aspect of the rabbit knee joint, and then were gently washed with normal saline solution. The tissue blocks were removed from areas representing areolar, adipose and fibrous synovium, then were rinsed in phosphate buffer, and were placed immediately into a buffered 2.5 % glutaraldehyde fixative (PH 7.4) for 2 to 4 hours. The specimens were prepared with the method described by Wysocki and were examined with SEM and TEM.

RESULTS

Scanning Electron Microscopic Appearances

As mentioned above, the synovial membranes of the rabbit knee joint can be divided into three kinds : areolar, adipose and fibrous synovium with SEM. The areolar synovium possesses an undulating surface structure with the folds arranged in a parallel fashion (fig.1). The synovial intimal cells cover the surface .

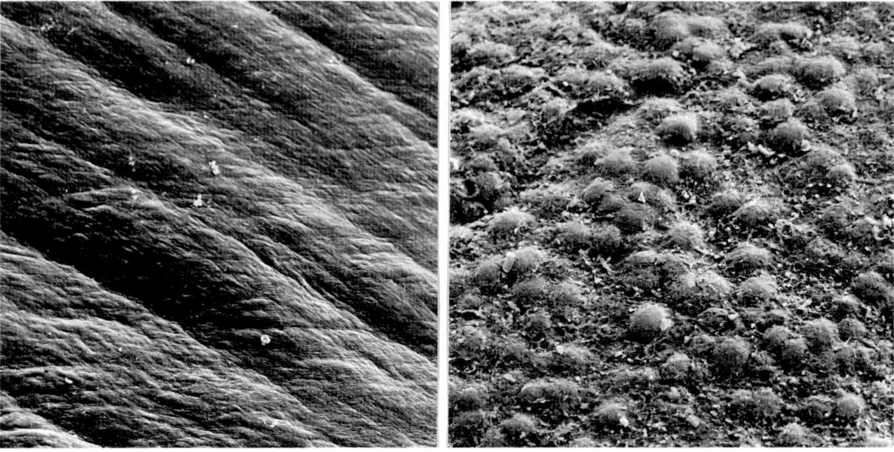

Figure 1. Scanning electron micrograph showing the surface structure of the areolar synovium of the rabbit knee joint (reduced from x100).

Figure 2. Scanning electron micrograph showing the surface structure of the adipose synovium of the rabbit knee joint (reduced from x100) .

The adipose synovium possesses a characteristic " cobblestone " surface pattern caused by clustering of underlying fat cells which are covered by a thin layer of synovial intimal cells (fig.2). At higher magnification, the " cobblestone " surface appearance is quite obvious, and synovial intimal cells with their cytoplasmic prominences can be identified. The fibrous synovium shows a smoother surface in the SEM (fig. 3). At higher magnification, the slight undulating topographic pattern and superficial synovial intimal cells with their cytoplasmic processes can be identified.

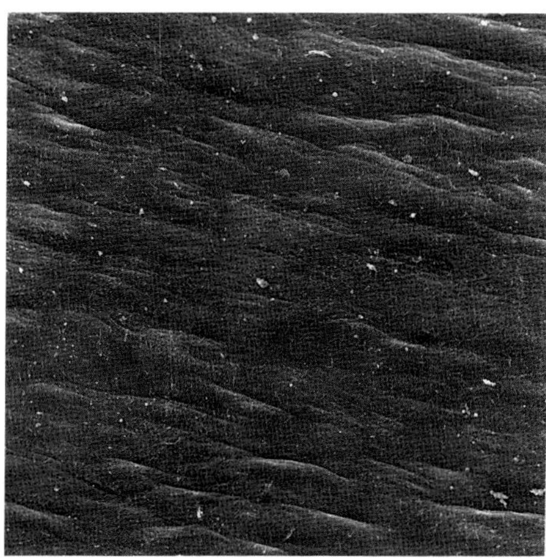

figure 3.Scanning electron micrograph showing topographic structure of the fibrous synovium of the rabbit knee joint (reduced from x100).

Transmission Electron Microscopic Appearances

The synovial intima of rabbit knee joint comprises numerous cells set in matrix in the TEM. In the synovial intima, type A, type B and type AB synovial cells were found by the TEM. Type A synovial cell is characterized by numerous cytoplasmic processes and containing many vesicles and vacuoles, but little rough endoplasmic reticulum (fig. 4), while type B synovial cell contains a lot of rough endoplasmic reticulum, but Golgi complex and vesicles and vacuoles are scanty (fig. 5). The Golgi complex is relatively prominent in type A cells than in type B cells. Type AB synovial cell is characterized by well developed Golgi complex and rough endoplasmic reticulum. The present results showed that type A synovial cells are more numerous in number.

Differences of opinion exist regarding the nature and function of type A and type B cells. We considered that their morphological differences suggested possibly that they are during different functional states.

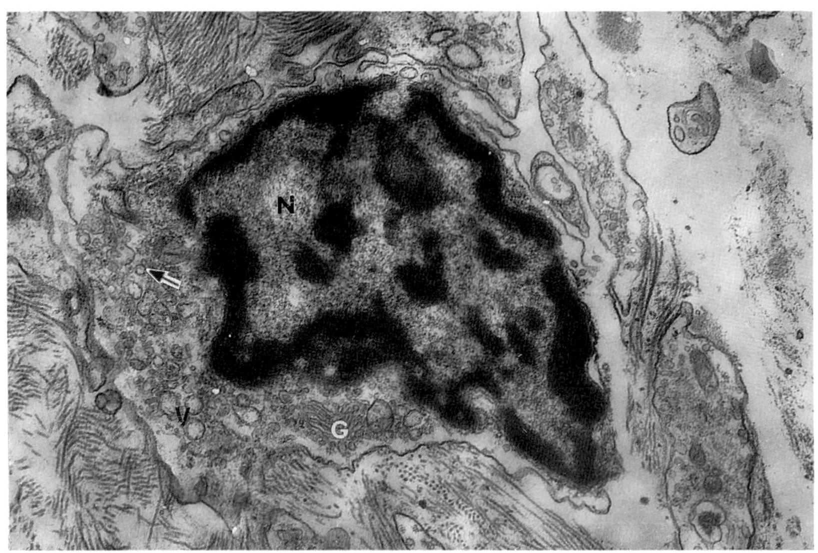

Figure 4. Synovial membrane of normal rabbit knee joint. Type A cell showing its numerous cytoplasmic processes, vacuoles (V), numerous vesicles (arrow), Golgi complex (G) and nucleus (N) . x10000

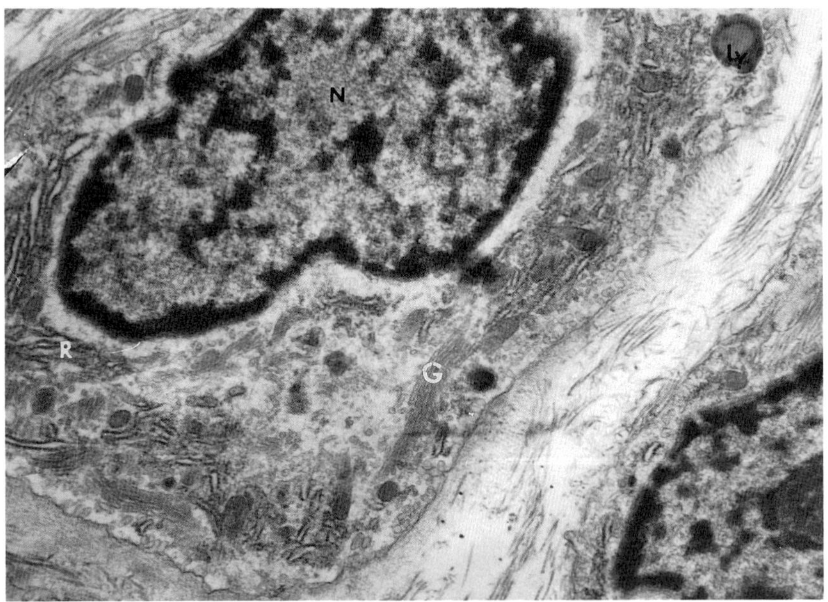

Figure 5. Synovial membrane of normal rabbit knee joint. Type B cell with its numerous rough endoplasmic reticulum (R), Golgi complex (G), lysosome (Ly) and nucleus (N). x13000

REFERENCES

Bhan AK, Roy S (1971). Synovial giant cells in rheumatoid arthritis and other joint diseases. Ann Rheum Dis 30: 294

Wysocki GP , Brinkhous KM (1972). Scanning electron microscopy of synovial membranes. Arch Path 93: 172.

Ghadially FN (1983). Fine structure of synovial joints. London: Butterworths .PP. 4-28 .

Gryfe A, Gardner DL, Woodward DH (1969). Scanning electron microscopy of normal and inflamed synovial tissue from a rheumatoid patient. Lancet 2: 156 .

Johansson HE, Rejnö S (1976). Light and electron microscopic investigation of equine synovial membrane. Acta Vet Scand 17:153 .

Key JA (1928). The synovial membrane of joints and bursae , in Cowdry EV(ed) : Special cytology, New York: Paul B Hoeber inc, Vol 1, PP 735-766.

Krey PR,Cohen AS (1973). Fine structural analysis of rabbit synovial cells. 1. The normal synovium and changes in organ culture. Arthritis Rheum 14: 319.

Linck G, Porte A (1978a). B-cell of the synovial membrane. 1.A comparative ultrastructural study in some mammals. Cell Tiss Res 187:251.

Okada Y, Nakanishi L, Kajikawa K (1981b). Ultrastructure of the mouse synovial membrane. Arthritis Rheum 24:835.

Roy S, Chadially FN, Grane WAJ (1966). Synovial membrane in traumatic effusion . Ultrastructure and autoradiography with triated leucine. Ann Rheum Dis 25:259.

Watanabe H, Spycher MA, Rüttner JR (1974). Ultrastructural study of the normal rabbit synovium. Pathol Microbiol 41:283 .

Cells and Tissues: A Three-Dimensional
Approach by Modern Techniques in Microscopy,
pages 133–138

ISOMETRIC CONTRACTION, ISOTONIC CONTRACTION AND PASSIVE CONTRACTION IN SMOOTH MUSCLES

Giorgio Gabella

Department of Anatomy, University College London, Gower Street, London, WC1E 6BT

Two contrasting modes of contraction in smooth muscle, as in all muscles, are the isometric contraction (development of tension and hardening without change in length) and the isotonic contraction (shortening and bulging out). In the former mode the muscle is able to generate an amount of force that is comparable to that produced by skeletal muscles which have a much higher myosin content (hence a larger number of potential cross-bridges) (Murphy et al., 1977), whereas in the latter mode the muscle can shorten, in certain conditions, to as little as one fifth its resting length. There are also passive changes in shape (imposed, for example by other muscles), including passive contraction (reviewed in Gabella, 1988).

Isometric contraction

When contracting isometrically a strip of smooth muscle (for example a taenia of the guinea-pig caecum, as in all the illustrations of this paper) practically does not change in length; within the muscle, however, there are structural changes, including small changes in length to take up any slack that might have been present either in the contractile elements of in the cytoskeleton of the muscle cells or in the stroma between cells. Therefore, isometric contraction does not imply absence of longitudinal displacenment of the muscle components, although this displacement is almost negligible by comparison with that occurring in isotonic contraction.

A distinctive structural feature accompanying isometric contraction is a chequered appearance of the

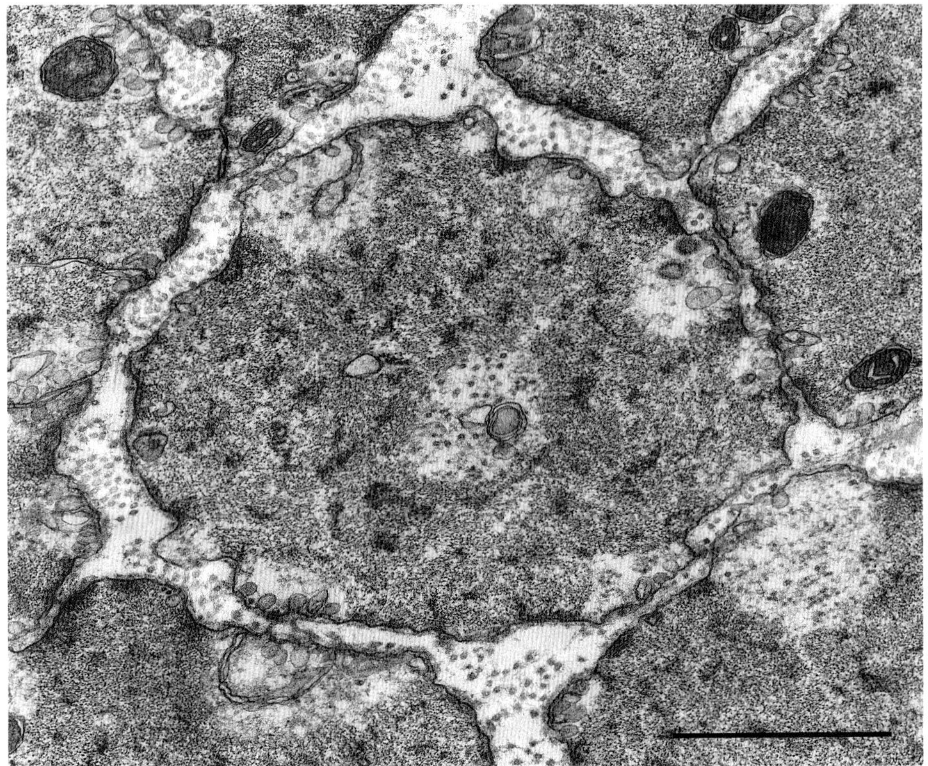

Figure 1. Guinea-pig taenia coli isometrically contracted in vitro. Transverse section. Note in each cell profile several sharply-outlined domains of low electron density, devoid of myofilaments. Collagen fibrils lie in the intercellular space. Bar: 1 µm

---

muscle cell profiles due to the presence of several distinct areas of the cytoplasm that are devoid of myofilaments and are therefore of lighter appearance (Fig. 1). These domains are mostly distributed beneath the plasma membrane, sometimes in a regular array around the perimeter of the cell, and often causing small protrusions of the cell profile. Light areas of similar appearance form a cone tens of micrometres in length at either side of the nucleus. These large conical domains of non-filamentous cytoplasm are centrally placed near the nucleus (as the nucleus itself is), but often become eccentric along their course and reach the cell membrane. The non-filamentous domains are sharply outlined and contain mitochondria, ribosomes, glycogen-like granules, cisternae of smooth and rough endoplasmic reticulum, Golgi apparatus, a few

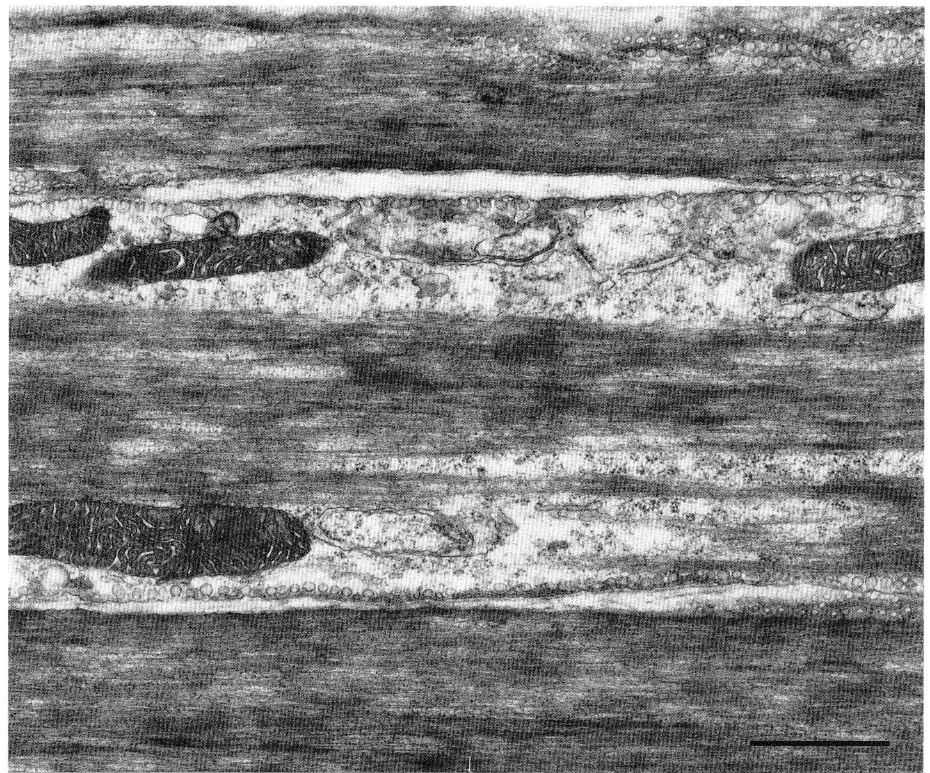

Figure 2. Guinea-pig taenia coli isometrically contracted in vitro. Longitudinal section. Note the cytoplasmic domains of low electron density occupied by several organelles but no myofilaments. Bar: 1 μm

filaments and many microtubules. Myofilaments have a higher packing density than in the cell at rest and they appear to be perfectly longitudinally arranged (Fig. 2). Gillis et al. (1988) observed in the rat anococcygeus muscle that the packing density of myofilaments is increased by a factor of 2 during isometric contraction, and in this preparations the contracted muscle cells are reported to be of smaller volume than when at rest.

Isotonic contraction

The extent of shortening in an isotonically contracted smooth muscle can be very large and it is limited by structural constraints in the wall of the organ (presence of adjacent layers of tissue, or presence of incompressible spaces, for example the collapsed lumen of a contracted urinary bladder); in vitro, isolated muscle strips can contract more extensively then in situ, but still fully reversibly, and isolated muscle cells can contract up to a point when they are almost spherical in shape (these contractions, however, are irreversible).

The shortening of the muscle is accompanied by a lateral expansion (in one or in both axes, depending mainly on the arrangement of the ´skeletal´ elements of the wall of a viscus) and the increase in cross sectional area is about equal to the decrease in length. The size of muscle cell profiles is much increased; because of this, in the taenia coli there are about 18000 cell profiles per square millimetre in a contracted strip as opposed to over 90000 in a strip at rest. The long axis of the contracted muscle cells is no longer parallel to the length of the muscle, but it deviates appreciably, a change more apparent in muscles containing many intramuscular septa. Moreover, the muscle cells are no longer straight and virtually parallel to one another (as they are in the muscle at rest) but somewhat contorted; a torsion of their long axis gives them the shape of irregular and long-pitched helices. These changes in shape are in part imposed by the stroma and by adjacent muscle cells; however, Warshaw et al. (1987) have observed a corkscrew-like shortening in single isolated muscle cells, and have discussed this observation in the light of a possible helical arrangement of the cytoskeleton and of the contractile apparatus.

The surface of the isotonically contracted muscle cells is thrown into myriad projections that are mainly laminar and run at an angle to the cell length; the projections bear mainly caveolae, whereas the dense bands are mainly located in the regions of the membrane that have expanded least and appear therefore invaginating (Fig. 3). Other evaginations of the cell membrane in contracted cells, more irregular and less numerous, are bulbous and connected to the bulk of the cell through a short peduncle; these projections develop rapidly at the onset of contraction and disappear slowly during relaxation. Similar projections are seen on a greater scale and covering the entire cell surface in isolated muscle cells contracting in a fluid medium (Fay & Delise, 1973).

While in the muscle at rest the collagen fibrils surrounding the muscle cells run mainly longitudinally or at a small angle with the axis of the muscle cells, in the shortened muscle the fibrils are wound around

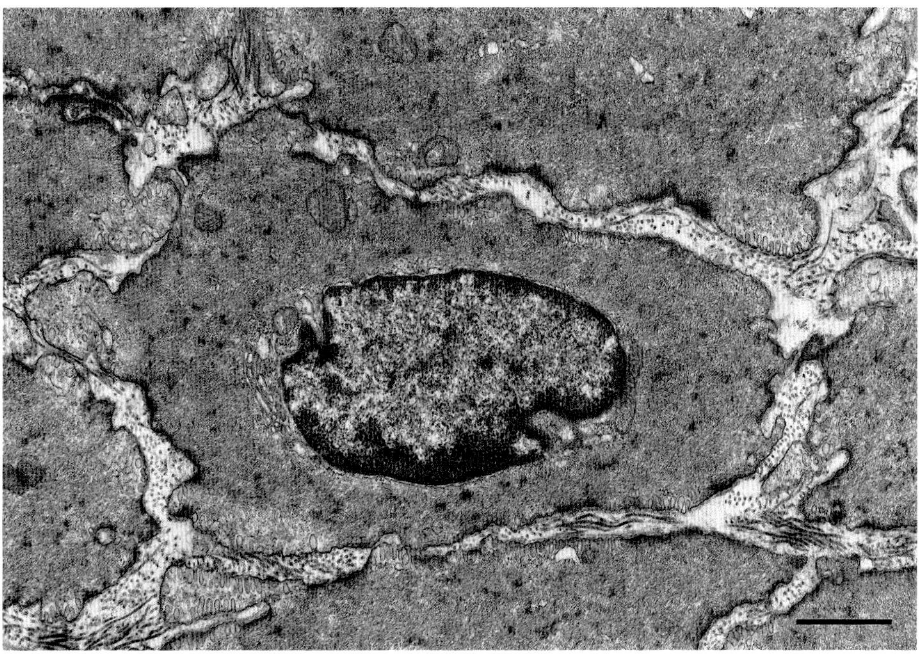

Figure 3. Guinea-pig taenia coli isotonically contracted in vitro. Transverse section. Note the large size of cell profiles, the crenations of the nucleus and the numerous invaginations of the cell membranes. Some of the collagen fibrils run very obliquely to the long axis of the muscle. Bar: 1 μm

---

the cells and run obliquely or almost transversely to the cell long axis. The experimental observations in visceral and vascular muscles support the theoretical model (Mullins & Guntheroth, 1965) according to which collagen provides a mechanism for binding cells together and for transmitting the tension generated by the myofilaments in directions up to 90 away from the direction of the myofilaments.

Similarly in the intramuscular septa the collagen fibrils run obliquely to the length of the muscle; when the muscle shortens the septa also shorten and their depth increases and the collagen fibrils increase the angle they form with the long axis of the muscle. In smooth muscles, because of the arrangement of collagen (and of cell junctions) a mechanical advantage is obtained by both the cell shortening and the cell fattening; indeed, given the tightly-knit 3-dimensional mesh of the muscle stroma, any change in muscle cell shape probably contributes to the mechanical performance of the tissue.

Passive contraction

Within a strip of muscle, especially when it is set up and stimulated in vitro, there are often some cells which fail to be activated and are shortened passively by the action of the adjacent contracting cells. The cells appear regularly coiled and their surface is devoid of evaginations. Cells have been found that are passively contracted over a portion of their length and actively contracted over the remaining part.

There are also changes in shape of the muscle cells that occur by the effect of contraction in adjacent muscles. A contraction of the longitudinal muscle of the intestine shortens the gut and compresses sideways the muscle cells of the circular layer which thus acquire a highly elliptical profile; when the circular muscle contracts, the shortening and fattening of the muscle cells produce a lateral expansion of the muscle so that the gut elongates and the longitudinal muscle cells grow thinner and longer.

The profile of muscle cells can change markedly depending on forces imposed by nearby tissues; the profiles can be approximately circular or very flat. The arrangement of the cytoskeleton and of the contractile material is such that this changes in shape are possible, are fully reversible and do not seem to affect the ability of the cells to contract.

References

Fay FS, Delise CM (1973) Contraction of isolated smooth-muscle cells - structural changes. Proc Nat Acad Sci USA 70:741

Gabella G (1988) Structure of the intestinal musculature. In: Handbook of Physiology, vol IV: Motility and Circulation. Ed. by JD Wood. Bethesda

Gillis JM, Cao ML, Godfraind-De Becker A (1988). Density of myosin filaments in the rat anococcygeus muscle, at rest and in contraction. II. J Muscle Res Cell Motil 9:18-28.

Mullins GL, Guntheroth WG (1965) A collagen net hypothesis for force transference of smooth msucle. Nature 206:592-594.

Murphy RA, Driska SP, Cohen DM (1977). Variations in actin to myosin ratios and cellular force generation in vertebrate smooth muscles. In: Excitation-Contrac tion Coupling in Smooth Muscle, ed. by R Casteels, p. 417-424. Amsterdam: Elsevier/North Holland.

Warshaw DM, McBride WJ, Work SS (1987). Corkscrew-like shortening in single smooth muscle cells. Science 236:1457-1459.

Cells and Tissues: A Three-Dimensional Approach by Modern Techniques in Microscopy, pages 139–144

TECHNIQUES FOR DETERMINING MORPHOLOGICAL CHARACTERISTICS OF SMOOTH MUSCLE CELLS USING COMPUTER ASSISTED RECONSTRUCTIONS

Mary E. Todd, F.S.F. Chu and B. Gowen

Department of Anatomy, The University of British Columbia, Vancouver, B.C., Canada, V6T 1W5

INTRODUCTION

A critical aspect of morphological studies is the ability to measure accurately all functionally significant structural features of cells, including volume, surface area, shape and orientation. Quantitative measurements are essential in deriving practical values such as surface area to volume ratios in different tissues. Some techniques can be so time consuming and labour intensive that they become impractical. If cells match a standard geometric shape like a sphere or cone, quantitative morphological measurements using computer assisted reconstructions are readily obtained. Problems arise with vascular smooth muscle cells because of their irregularity and heterogeneity. These cells are flattened and elongate, they change in shape, size and orientation in different types of blood vessels, for example elastic versus muscular artery, and artery versus vein (Todd et al., 1983). Even the same type of artery (for example, small versus large muscular) can have cells with quite different dimensions (Todd, 1985). Further complications arise when conditions of disease, vessel development, or experimental manipulations like denervation, are investigated.

Standard 0.5µm plastic sections can be utilized for computer assisted reconstructions under many conditions (Todd et al., 1983), but there are limitations such as for surface area calculations because the irregularity of the plasma membrane cannot be accurately visualized at the light microscopic level. Ultrastructural serial reconstructions solve all visual difficulties but the number of sections required for each smooth muscle sample (at least 800-1000) presents difficulties of technique and labour. In this report we describe the results obtained for vascular smooth muscle cells and the light and electron microscopic techniques developed to overcome the difficulties of quantitative morphological measurements, and achieve the greatest efficiency.

METHODS

Samples of perfusion fixed blood vessels, oriented to give a cross section, were embedded in Epon/Araldite (Todd et al., 1983). The block was trimmed to form a mesa and the bottom of the block (first in contact with the knife) was coated with dilute contact cement (Fahrenbach, 1984). One hundred to 150 serial sections (0.5µm thick) were cut with a diamond knife and the cement ensured that a ribbon formed on the water in the boat, to keep the sections in order. After 15 section ribbons were cut, they were stretched using ethylene dichloride. With eyelash probes and a pickup loop, individual sections were separated and placed (15 per glass slide), each on a drop of Millipore filtered ethylene glycol (after Burnett, 1975). Heating at 70°C ensures wrinkle free adherence to the slide.

When smooth muscle cells are closely apposed, as in developing blood vessels where paracellular matrix is still sparse, an etching technique (Thurley and Mouel, 1974), aids in defining the cell profiles. The slides were placed in a saturated solution of sodium hydroxide in absolute ethanol in a covered coplin jar overnight. They were then rinsed in 3 changes of absolute ethanol, and then in 3 changes of distilled water. The dried sections were stained with a 1% Toluidene Blue/1% Azure Blue II mixture for 5 minutes at 70°C and then permanently mounted. Each section was photographed at a low magnification and profiles of cells occurring within the series were located and rephotographed at a higher magnification. To test the effect of the etching on the tissue, whole sections and individual cell profiles of 4 complete cells were photographed before and after etching, in which case the pre-etched sections were temporarily mounted using immersion oil.

The same blocks were longitudinally sectioned at 70nm and sections were placed on numbered formvar coated slot grids. The same region of the vessel wall was photographed in 30 consecutive sections over a range of ultrastructural magnifications.

The photographs of cell profiles at light microscopic level were oriented within the vessel wall and in relation to the preceding and succeeding profiles using the corners of the mesa and/or landmarks in the vessel wall. The orientation of profiles at the ultrastructural level is critical since minor errors in register distort the surface area values.

Cytoplasmic organelles were traced along with the cell profile to ensure "stacking" of the slices as in the cell originally. Each cell profile, in correct orientation, was delineated by recording x and y coordinates by means of a digitizer (Todd et al., 1983; Todd, 1985). The plotted images provided the information on size, shape, and orientation. Cell volume was obtained from a program which summates individual profile areas and section thickness. Surface area values were estimated by perimeter times section thickness (slab method) and by finding the optimal surface (Fuchs, 1984), where a series of small triangles connecting coordinates of adjacent sections are summated (triangle method). Results are given as mean $\pm$ SEM, and statistical analysis was carried out using Student's _t_-test.

RESULTS AND CONCLUSIONS

The etching technique was tested on tissue in which the cell profiles were the most difficult to define, that is cells in developing (3 week old) denervated saphenous artery from the Spontaneously Hypertensive Rat strain (SHR). Measurements were taken of the lengths and widths of cell profiles, and of complete cross sections of the vessel, to ensure that the technique was not altering either the individual cell profiles or any component of the vessel wall. The photos of etched vessel wall and its components were also found to superimpose precisely on the pre-etched photos for all sections at all magnifications. Photographs were taken with a 100X objective (phase: oil immersion, final magnification 3000X) and four complete smooth muscle cells were reconstructed. The area of each individual profile making up the cell was traced before and after etching. When all profile areas from all 4 cells were paired and compared, there was a significant difference following etching (mean values: 110 ± 11 versus $99 \pm 10 \mu m^2$, $p < 0.01$).

The next stage involved reconstructing the same 4 cells using a lower magnification to avoid having to assemble montages as occurred when profiles were photographed using the 100X objective. A 63X bright field objective (final magnification 1800X) was used. The same profiles from the same cells in the etched sections were rephotographed. There were no significant differences when mean area of the profiles, total cell volume, and length of the cell (measured from the computer plots) were compared (Fig. 1).

Once all the profiles of a cell have been traced and the x and y coordinates recorded, the images can be rotated, viewed and plotted from any axis, and the plots give information about the shape of the cell, position of the nucleus, and orientation within the vessel wall.

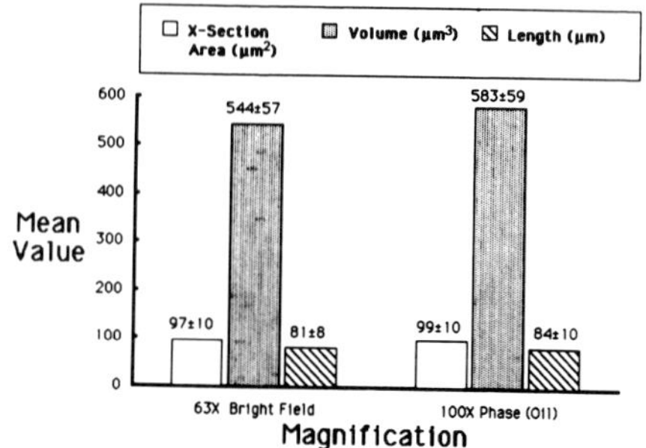

Figure 1. A comparison of profiles of four complete cells photographed using either 63X or 100X objectives. There were no significant differences in the areas of the profiles, or in total cell volume or cell length.

The plasma membrane irregularities cannot be accurately traced in any of these light microscopic images. Therefore, ultrastructural sections must be analyzed. The information from light microscopy is used in reorienting for ultrathin sectioning. Since very thin slices (60-70nm) were involved, our original assumption was that perimeter times section thickness would provide a reasonable assessment of the surface area (Todd, 1985). When this slab method was compared with the triangle method, the latter resulted in a significant increase in the estimated surface area in portions of three adult rat tail artery cells (mean values for the three cells combined were 52 ± 8 versus $122 \pm 22 \mu m^2$, $p<0.01$). These cells are much more irregular than smooth muscle cells in a developing muscular artery. The more irregular the profiles and the greater the variation in shape along the length of the cell, the greater the difference between the two methods. For example, if the shape of the cell were comparable to a cylinder, there would be no difference in surface areas with the two methods. To ensure that the surface area was not being overestimated with the triangle method, a geometric shape, a cone, was digitized. The two methods were used to calculate surface areas from the digitized profiles. The slab method markedly underestimated, and the triangle method was within 1% of the mathematically calculated values for a cone with those dimensions.

The saphenous artery tissue that was analyzed using light microscopy had an ultrathin serial set of portions of 5 cells analyzed at a range of magnifications to determine the lowest magnification that would give accurate surface area measurements. For this tissue, the negative magnification was 3300X with a final magnification of 8750X. The lower the magnification, the greater the number of profiles that can be included on one negative. The orientation within the vessel wall, determined from computer plots (light microscopy), allows reorientation so that slices cut across the cells. Although sections along the long axis are advantageous for reconstructing whole cells because fewer sections are required, cutting across the cells gives more profiles for ultrastructural reconstructions. At very low magnifications

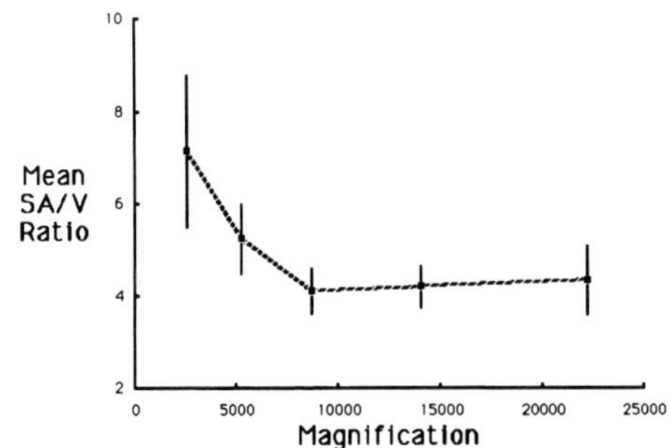

Figure 2. Surface area to volume (SA/V) ratios for serial sets of 30 of portions of five cells. Bars indicate ± SE. Higher ratios at lower magnifications were due to difficulties in aligning adjacent profiles.

with light microscopy, perimeter measurement is underestimated because the irregularities cannot be defined (Weibel, 1979). We found this also with the vascular smooth muscle cells. However, at the magnifications that span the overlap between light and electron microscopy, the surface areas are overestimated because of the difficulty in precisely aligning adjacent profiles. This overestimation of the surface area is reflected in high surface area to volume ratios (Fig. 2). Volumes can be accurately estimated at much lower magnifications, including the light microscopic range (Weibel, 1979) and we confirmed this in vascular smooth muscle.

As an example from another blood vessel, preliminary reconstructions and analyses from ultrathin serial sections were carried out on smooth muscle cells from the portal vein, and samples from 16 week old SHR and their Wistar-Kyoto controls were compared. Plots of reconstructed cell profiles illustrate the change in cell shape and size in hypertensive animals (Fig. 3), resulting in the significantly different surface area to volume ratios (Table 1).

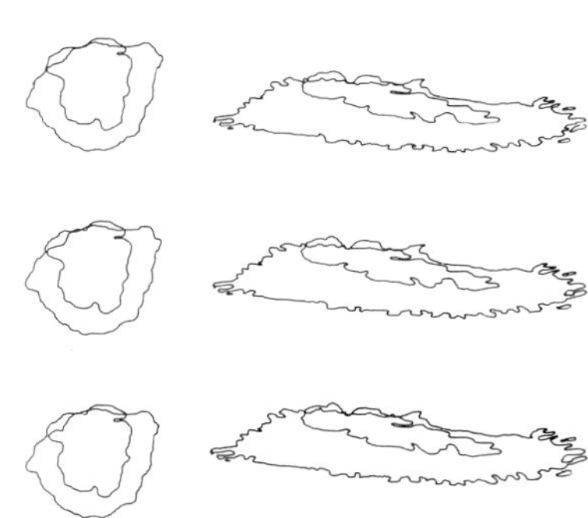

Figure 3. Computer plots of three adjacent profiles of portal vein smooth muscle from SHR (right) and controls (left) to illustrate the increased irregularity of the plasma membrane in cells from hypertensive animals.

TABLE 1. Portal vein smooth muscle cells. Results from four cells (sets of 10 consecutive sections) for each type of rat (16 weeks old), using 64nm thick sections and electron microscopy.

	Surface Area† (μm^2)	Volume (μm^3)	Surface Area/ Volume Ratio
WKY	12 ± 2	5 ± 1	2.6 ± 0.5
SHR	128 ± 54	22 ± 6*	5.4 ± 0.8*

† Area of the convex surface only
* Significant, $P < 0.05$ between WKY and SHR.

We conclude that a combination of light and electron microscopic computer assisted reconstructions and analyses provide the best solution for determining the dimensional characteristics of vascular smooth muscle with all its heterogeneity.

REFERENCES

Burnett BR (1975). A new method for serially mounting resin sections (Spurr) for light microscopy. Stain Technol 50:288-290.

Fahrenbach WH (1984). Continuous serial thin sectioning for electron microscopy. J Electron Microsc Tech 1:387-398.

Fuchs H, Kedem ZM, Uselton SP (1984). Optimal surface reconstruction from planar contours. Comm ACM 20:693-702.

Thurley KW, Mouel WC (1974). The etching of thick araldite-embedded sections for scanning electron microscopy. J Microsc 101:215-218.

Todd ME (1985). Vascular smooth muscle cell parameters from computer assisted reconstructions. In Bevan JA, Godfraind T, Maxwell RA, Stoclet JC, Worcel M (eds): "Vascular Neuroeffector Mechanisms", Netherlands: Elsevier, pp 265-269.

Todd ME, Laye CG, Osborne DN (1983). The dimensional characteristics of smooth muscle in rat blood vessels. A computer assisted analysis. Circ Res 53:319-331.

Weibel ER (1979). "Stereological Methods". New York: Academic Press, pp 153-157.

Respiratory and Urinary Systems

Cells and Tissues: A Three-Dimensional
Approach by Modern Techniques in Microscopy,
pages 147–155

DESIGN OF MICROVASCULATURE AND GAS EXCHANGE

Ewald R. Weibel, M.D.
Department of Anatomy, University of Berne
Bühlstrasse 26, CH-3000 Berne 9

MARCELLO MALPIGHI AND LUNG CAPILLARIES

Marcello Malpighi's Opera Omnia contains two letters to G.A. Borelli in Pisa in which he describes his observations on the lung, done around 1660. He had first discovered that the lungs were made of a large number of small alveoli and had seen the blood enter this structure through branches of the arteries; it appeared to vanish "into empty spaces" but then was collected again - "full of air" - in the veins. But then, choosing the simpler frog lung as a model, and using "better lenses" (perhaps even one of Galileo's compound microscopes) he observed networks confined to the membranous wall through which he saw the blood stream along predetermined paths. He concluded that this rete "anastomosed" arteries and veins. The drawing he produced of his observation (Fig. 1) was remarkably correct. With even much better lenses - even with the scanning electron microscope - we can simply confirm his fundamental discovery of alveolar capillaries and of their connection to arteries and veins (Fig. 2), even to the point where he said that the arteries were in the center of the unit and the veins at its periphery.

Together with his discovery of capillaries in the omentum, and hence on the systemic part of circulation, this description of pulmonary capillaries completed the scientific revolution started by William Harvey in 1628: the blood was now seen to circulate in a closed vascular compartment; it did not disappear into "porosities" or "empty spaces", as was first presumed. Malpighi had also discovered the (red) blood corpuscules, but it took over a century until their functional significance as carriers for O_2 became known.

I would here like to examine, how the microvasculature in the lung and in muscle is designed to serve the transport of O_2 from air to mitochondria, a process evidently influenced by the presence of red blood cells.

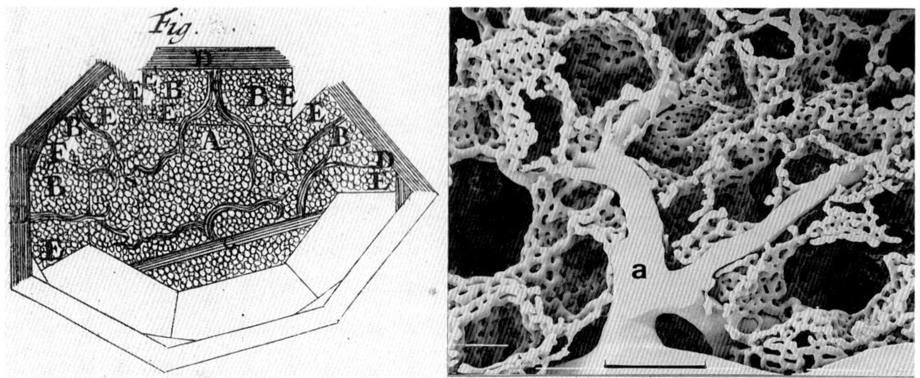

Figure 1 (left): Marcello Malpighi's original drawing of the capillary network in the frog lung, connected to an arterial branch (C) in the center and to veins (D) at the periphery of the unit. [From Malpighi (1687)]

Figure 2 (right): Scanning electron micrograph of capillary network with terminal branches of an artery (a) in the rat lung. Scale marker: 100μm

PULMONARY CAPILLARIES

The pulmonary capillaries are disposed as a single very dense network (Fig. 2) which is incorporated into the walls of alveoli that number 300 million in the human lung and establish an air-blood contact surface about the size of a tennis court (Weibel, 1984). The capillary network is continuous in three dimensions and is tapped by arterial and venous end branches at more or less regular distances (Fig. 2). Microvascular units are therefore ill-defined.

The tissue framework of the alveolar walls is reduced to thin sheets that ensheath the capillaries and form a barrier for the diffusion of gases between air and blood (Fig. 3a). But this sheet is highly organized into three basic layers (Weibel,1984): an epithelium that lines the alveolar surface, an endothelium that lines the capillaries, and a very slim interstitium that binds the two cell layers together. This structural design favours gas exchange, but it leads to the question, how such a minimal tissue framework can support the capillaries that are essentially suspended in air in a sort of foam. Two tricks are noted:

a) The support fibers form a network within the alveolar walls that is interlaced with the capillaries so that one side of the capillary surface remains free of fibers (Fig. 3b). These fibers are anchored in coarse fiber systems that, on the one hand, follow the airways into the alveolar ducts (axial fibers), and, on the other hand, penetrate from the pleura

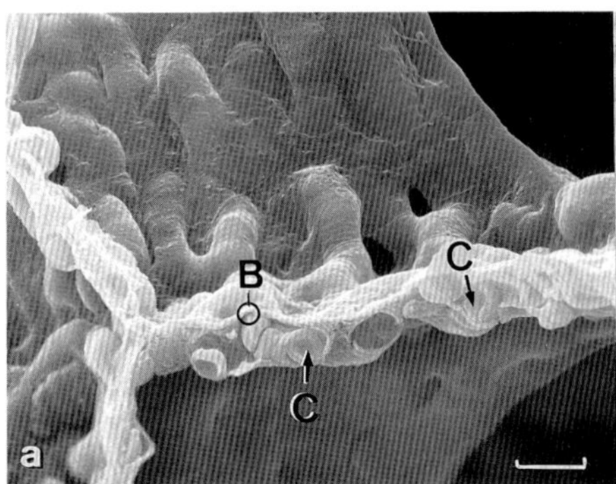

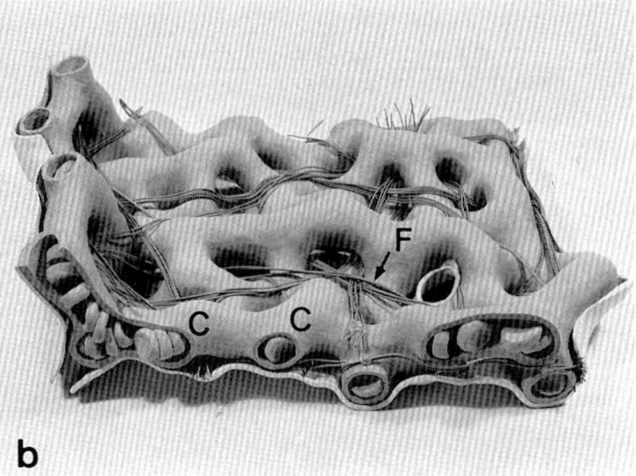

Figure 3: In the alveolar wall, shown in (a) in a scanning electron micrograph from a human lung, the capillary blood (C) is separated from the air by a very thin tissue barrier (B). The model (b) shows the capillary network (C) to be interwoven with the meshwork of septal fibers (F). Scale marker: 10μm. [From Weibel (1984)]

and interlobular septa into the lung parenchyma (peripheral fibers). This way, the fibers of the alveolar wall can be very delicate and still form an adequate mechanical support for the capillary network (Weibel and Gil, 1977).

b) The potentially high forces that act on the alveolar wall due to surface tension are made very small by the presence of a surface lining layer with surfactant. This allows the capillaries to be wide, and the alveolar walls to remain expanded (Gil *et al.*, 1979).

How well is the lung's microvasculature designed as a gas exchanger? To examine this we must attempt to estimate the lung's gas exchange capacity from morphometric data obtained on histological specimens. Fig. 4 shows the model by which the diffusing capacity D_{LO_2} of the lung can be calculated. We need reasonable estimates of some physical coefficients, and morphometric measurements. As shown in table 1 we estimate D_{LO_2} for the human lung to be about 200 ml $O_2 \cdot min^{-1} \cdot mmHg^{-1}$! This means that one needs to have a driving force in form of an alveolar-capillary P_{O_2} difference of 10 mmHg in order to take up 2 liters of O_2 per minute, the amount an average person uses when doing heavy work. It can be shown that the average normal individual never really exploits the gas exchange capacity of his lungs, but that well-trained people doing heavy exercise at a moderate altitude, where ambient P_{O_2} is low, may need their diffusing capacity completely. The design of the pulmonary gas exchanger with its microvasculature therefore contains a certain reserve capacity that allows the lung to function well also under unfavorable conditions (Weibel *et al.*, 1987).

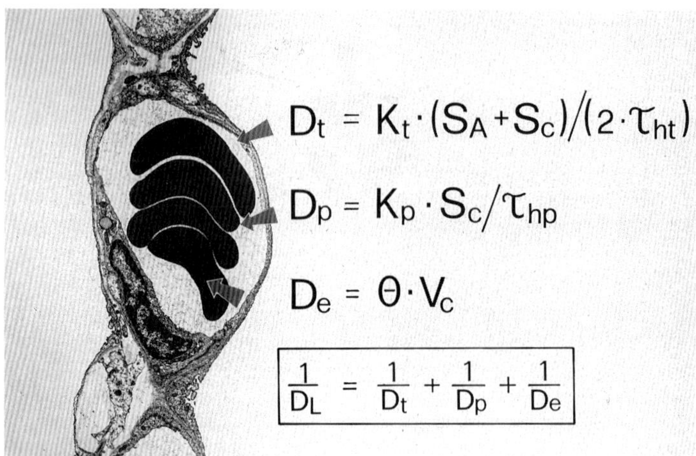

Figure 4: Model for estimating pulmonary diffusing capacity (D_L) from morphometric information on the design of three partial conductances: D_t for the tissue barrier, D_p for plasma, and D_e for erythrocytes. K_t, K_p and θ are physical coefficients, S_A, S_c are alveolar and capillary surface area, V_c is capillary volume, and τ_{ht}, τ_{hp} are the harmonic mean thicknesses of tissue and plasma barrier, respectively. See Weibel (1984) for details.

Table 1: Morphometry of human lung: estimation of pulmonary diffusing capacity D_{LO_2}

Body mass	74	± 4	kg
Alveolar surface	130	±12	m^2
Capillary surface	115	±12	m^2
Capillary volume	194	±30	ml
Tissue barrier thickness	0.62	± 0.04	µm
Plasma barrier thickness	0.15	± 0.01	µm
Diffusing capacity D_{LO_2}	200 - 270	$mlO_2 \cdot min^{-1} \cdot mmHg^{-1}$	

MUSCLE CAPILLARIES

When looking at peripheral microvasculature in terms of O_2 supply to the tissues, muscle is the most interesting because in heavy work over 90 % of the O_2 taken up in the lung is consumed by muscle cells.

The design of muscle microvasculature differs markedly from that in the lung: we find networks of long capillary tubes arranged more or less parallel to the muscle fibers with interconnections that make this network continuous over large parts of a muscle (Fig. 5). These networks are tapped by arteries and veins, but cannot be separated into distinct microvascular units.

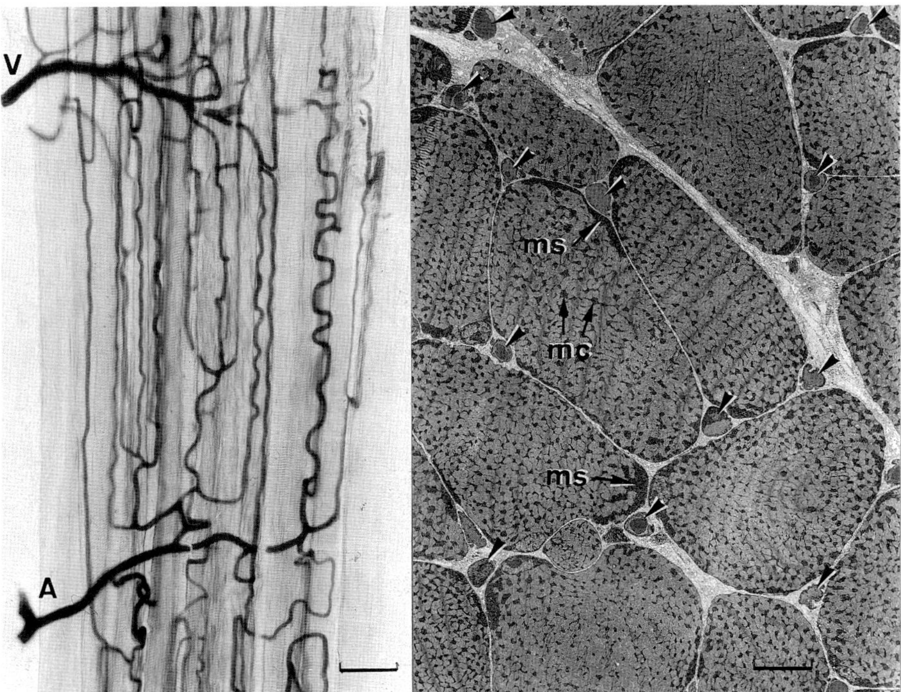

Figure 5 (left): The microvasculature of skeletal muscle, extending from arteriole (A) to venule (V), is revealed by perfusion of stained gelatin. Note that capillaries course predominantly parallel to muscle fibers, and that the capillary density increases somewhat toward the venule. Scale marker: 100μm. [From Weibel (1984)]

Figure 6 (right): Electron micrograph of a cross-section of muscle fibers shows central (mc) and subsarcolemmal (ms) mitochondria, and capillaries (arrow heads). Scale marker: 10μm. [From Hoppeler *et al.* (1981)]

Is the design of muscle capillaries quantitatively related to O_2 consumption? We must first note two features (Hoppeler and Lindstedt, 1985): (1) O_2 consumption by muscle increases linearly with the work performed and reaches a limit, $V_{O_2}max$, which is set by the capacity of mitochondria to perform oxidative phosphorylation; (2) $V_{O_2}max$ seems determined by the amount of mitochondria in the muscle cells, such that 1 ml of mitochondria can consume 4 - 5 ml O_2 per minute. The question therefore is whether the muscle capillaries are matched to local O_2 demand as set by the quantity of mitochondria in the muscle cells supplied.

The usual way of approaching this question is to cut cross-sections of muscle (Fig. 6). By using stereological methods (Weibel, 1979) one can then estimate the volume density of mitochondria, $V_V(mi)$, by simple point counting, and the capillary supply by counting capillary profiles per unit area of cross-section, $N_A(c)$. In most instances, and following the pioneering work of A. Krogh (1922), one uses $N_A(c)$ as the direct measure of "capillarity". However, this leaves a considerable degree of uncertainty (Fig. 7). The true measure of capillary supply is, in fact, their length density in the muscle volume, $J_V(c)$ (Hoppeler *et al.*, 1981); but its relation to $N_A(c)$ depends on the degree of preferential orientation of capillaries to the longitudinal axis of muscle fibers (Mathieu *et al.*, 1983). If the capillaries are perfectly straight parallel tubes

$$J_V(c)=N_A(c),$$

but if they form a random network in space

$$J_V(c)=2 \cdot N_A(c).$$

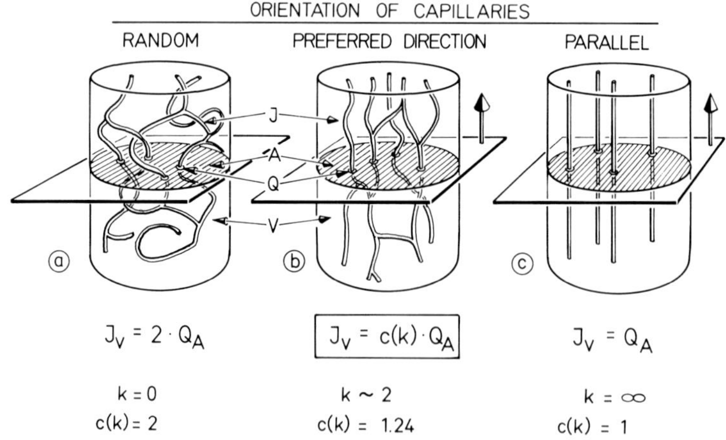

Figure 7: Estimating capillary length density $J_V(c)$ depends on the degree of orientation (k).

Using $N_A(c)$ as a measure of capillarity, therefore, leaves an uncertainty by a factor of up to 2. By estimating the degree of preferred orientation we have found that

$$J_V(c) = 1.24 \cdot N_A(c)$$

is a reasonable estimate of capillary length density in skeletal muscle (Conley *et al.*, 1987). Since the capillary volume density can be calculated from the length density by estimating mean capillary cross-sectional area we can now approach the question whether the capillary supply of muscles is matched to the quantity of mitochondria in the muscle cells.

If we first compare, in the same animal, two types of muscle, white and red, which differ markedly in their capacity for oxidative metabolism, we find that capillary length and mitochondrial volume are matched: in both cases we find for each ml of blood 3 ml of mitochondria. When looking at many different animals, from the Etruscan shrew weighing 2 g to the 400 kg cow, and considering skeletal as well as heart muscle we find a good

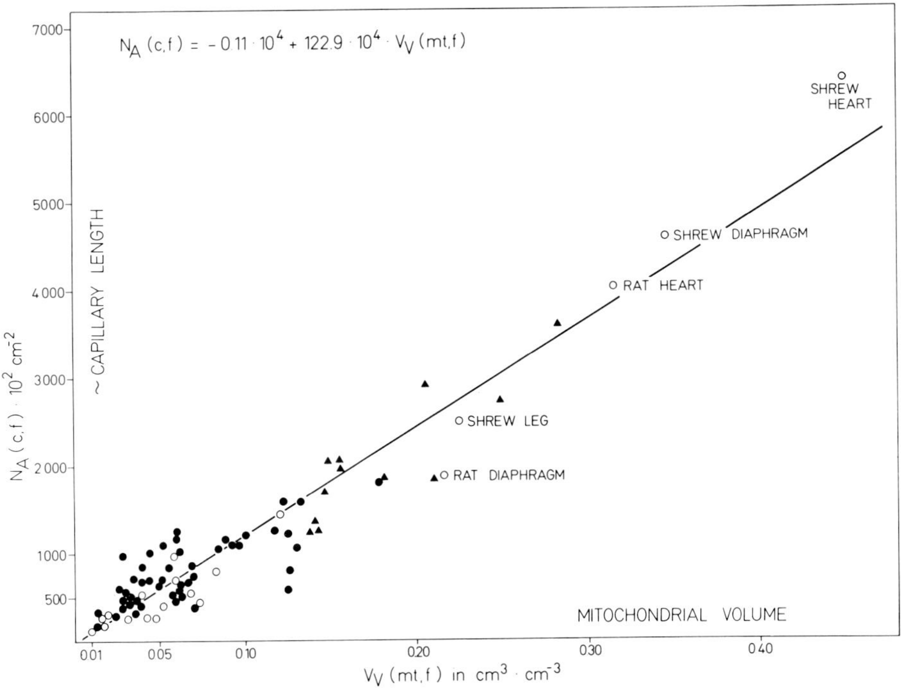

Figure 8: Relating capillary length to mitochondrial volume for large range of animals from the 2 g shrew to cows, and for skeletal and heart muscle. [From Hoppeler *et al.* (1981)]

correlation between capillary length and mitochondrial volume (Fig. 8), such that each ml of mitochondria is associated with 12 km of capillaries.

However, this simple correlation is not always borne out. When studying dogs and goats (Conley *et al.*, 1987) we found that dogs consumed 2.4 times as much O_2 at $V_{O_2}max$ than goats, and that their mitochondrial volume was enlarged in proportion to $V_{O_2}max$ (table 2). But the capillary supply of muscle was only 1.8 times larger in the dogs than in the goats. The solution was found when we looked at the hematocrit, or at the hemoglobin concentration of blood, which were 1.7 times higher in the dogs. The same was found when comparing ponies with calves. The product of hemoglobin concentration with capillary volume then was found to be proportional to the mitochondrial volume. It thus appears that the capacity of the microvascular blood to carry and supply O_2 is matched to the capacity of the cell's mitochondria to consume O_2 in oxidative phosphorylation.

Table 2: O_2 delivery by microcirculation to mitochondria

	$V_{O_2}max$	V(mi)	V(c)	M(Hb)	M(Hb)/V(mi)
Dog	2.28	40.6	8.2	18.8	0.038
Goat	0.90	13.8	4.5	10.7	0.035
Dog/Goat	**2.4**	**2.9**	**1.8**	**1.7**	**1.1**
Pony	1.48	19.5	5.1	17.0	0.045
Calf	0.61	9.1	3.2	10.3	0.037
Pony/Calf	**2.4**	**2.1**	**1.6**	**1.7**	**1.2**
	$ml \cdot sec^{-1}kg^{-1}$	$ml \cdot kg^{-1}$	$ml \cdot kg^{-1}$	g/100ml	g/ml

CONCLUSION

The design of the "minute vessels" of Marcello Malpighi proves to be related or even matched to functional demand, at least in those microvascular systems that predominantly serve gas exchange. The design of pulmonary microvessels is such that a high diffusion conductance is achieved, allowing the blood flowing through to be equilibrated with air in less than 1/2 second. In muscle the size of microvasculature appears matched to the O_2 needs established by the capacity for oxidative phosphorylation in mitochondria. But in both systems the third discovery of Malpighi, the red blood corpuscules, play an important role in determining the capacity for O_2 uptake and delivery.

REFERENCES

Conley KE, Kayar SR, Rösler K, Hoppeler H, Weibel ER, Taylor CR (1987). Adaptive variation in the mammalian respiratory system in relation to energy demand: IV. Capillaries and their relation to oxidative capacity. Respir Physiol 69:47-64.
Gil J, Bachofen H, Gehr P, Weibel ER (1979). Alveolar volume-surface area relation in air- and saline-filled lungs fixed by vascular perfusion. J Appl Physiol 47:990-1001.
Hoppeler H, Lindstedt SL (1985). Malleability of skeletal muscle in overcoming limitations: structural elements. J Exp Biol 115:355-364.
Hoppeler H, Mathieu O, Weibel ER, Krauer R, Lindstedt SL, Taylor CR (1981). Design of the mammalian respiratory system. VIII. Capillaries in skeletal muscles. Respir Physiol 44:129-150.
Malpiphi M (1687). "Opera Omnia". London: P. Vander, pp 320-379.
Mathieu O, Cruz-Orive LM, Hoppeler H, Weibel ER (1983). Estimating length density and quantifying anisotropy in skeletal muscle capillaries. J Microsc 131:131-146.
Weibel ER (1979). "Stereological Methods. Volume 1". London: Academic Press.
Weibel ER (1984). "The Pathway for Oxygen". Cambridge Mass.: Harvard University Press.
Weibel ER, Gil J (1977). Structure-function relationships at the alveolar level. In West JB (ed.) "Bioengineering Aspects of the Lung". New York: Marcel Dekker, pp 1-81.
Weibel ER, Marques LB, Constantinopol M, Doffey F, Gehr P, Taylor CR (1987). Adaptive variation in the mammalian respiratory system in relation to energy demand: VI: The pulmonary gas exchanger. Respir Physiol 69:81-100.

ACKNOWLEDGEMENTS

Supported by grants from the Swiss National Science Foundation.

Cells and Tissues: A Three-Dimensional
Approach by Modern Techniques in Microscopy,
pages 157–166

SHAPE CHANGES IN KIDNEY GLOMERULAR PODOCYTES: MECHANISMS AND POSSIBLE FUNCTIONAL SIGNIFICANCE

Peter M. Andrews

Department of Anatomy and Cell Biology, Georgetown University School of Medicine, Washington, D.C. 20007

INTRODUCTION

The glomerular (visceral) epithelial layer of Bowman's capsule is comprised of cells termed "podocytes" which surround the underlying glomerular capillaries and are designed to facilitate movement of a large quantity of solute across the glomerular wall. Both pathological and experimental findings indicate that glomerular podocytes possess the ability to undergo significant morphological changes which may play an important role in the pathophysiology of some kidney diseases. In this manuscript, I will summerize the normal morphological characteristics of the glomerular epithelium, discuss the morphological changes that this epithelium undergoes in response to the nephrotic syndrome and different forms of acute renal failure, then describe the morphological alterations of the glomerular epithelium in response to in vitro and in vivo experimental procedures.

MORPHOLOGY OF GLOMERULAR PODOCYTES

The glomerular podocytes consist of a central nucleated cell body from which thick major processes arise and loop around underlying glomerular capillary loops. The cell bodies of the podocytes are typically displaced away from the underlying glomerular basement membrane (GBM), thereby permitting podocyte processes to make contact with the GBM and maintain the porosity of this epithelium over the capillary loops. Each podocyte cell body contains a prominent and often deeply infolded nucleus, numerous free ribosomes, prominent Golgi, scattered profiles of granular (rough) and agranular (smooth) endoplasmic reticulum, slender mitochondria, pinosomes, lysosomes, characteristic amorphous bodies, and numerous microtubules and microfilaments. Podocyte major processes arise from the cell body not unlike arms from an octopus and may themselves branch to form secondary and tertiary processes. The major processes are variable in length, width, and diameter, appear rounded or flattened in cross-sectional profile, and contain mitochondria, pinosomes, and a prominent population of microtubules and microfilaments which run parallel to the long axis of these processes. The major processes can frequently be found crossing under cell bodies and over and under the major process of other podocytes. Arising at nearly right angles from the major processes in a fern leaf pattern are numerous smaller and more uniformly shaped processes termed pedicels or foot processes. These smaller processes may be of slightly different lengths and some may be considerably thicker than others. Typically the bases of foot processes ,which appear embedded in the lamina rara externa of the GBM, are expanded giving each foot process a characteristic bell shape when viewed in cross sectional profile. Foot processes contain a dense matrix of filamentous material which has been identified as filamentous actin (i.e. F-actin) (Andrews and Bates, 1984; Trenchev et al., 1976). Also, peroxidase-labelled antibodies at the electron microscopic level have demonstrated the presence of heavy meromysin in the bases of foot processes (Trenchev et al., 1976). These myosin rich sites appear associated with electron dense regions adjacent to the plasmalemma where the slit diaphragm inserts. As will be described lator, it is these contractile elements (i.e. actin and myosin) which are responsible for the morphological alterations that podocyte foot processes undergo in response to some disease states and experimental situations. An interesting feature of foot processes is that they appear to always interdigitate with foot processes from other podocytes rather than with foot processes arising from the same cell. Adjacent foot processes are separated by spaces of 20-40 nm ,termed the filtration slits, which provide the openings necessary to permit the efflux of glomerular filtrate across the glomerular wall. The filtration slits ,however, are not open but are spanned by thin membranous-like structures (4-6 nm thick) termed slit diaphragms. Using tannic acid staining to enhance morphological features, Rodewald and Karnovsky (1974) observed that the slit diaphragm has a porous substructure consisting of two parallel rows of rectangular pores separated by a central bar. These investigators observed that these rectangular pores measure 4 by 14 nm and are therefore small enough to impede the movement of albumin across the glomerular wall. A morphometric analysis of the filtration surface area provided by the porous substructure of the filtration slit diaphragms indicates that this filtration space plays a major role in determining rate of solute efflux across the glomerular wall (Shea and Morrison, 1975). Although most of the length of each foot process is separated from adjacent foot processes by the filtration slit space, there are also focal regions of contact between adjacent foot processes (Rodewald and Karnovsky, 1974).

Even in the normal kidney a certain percentage (e.g. 20%) of the foot processes will appear flattened and expanded to varying degrees, a characteristic of foot process loss (Andrews, 1977a; Pinto and Douglas, 1974). As in response to pathological states, filtration slits between these distorted foot processes are often replaced with tight junctional complexes, and the slit diaphragms may be folded and displaced away from the GBM. These facts should therefore be taken into consideration whenever attempting to evaluate podocyte foot process loss in response to pathological and experimental situations.

The free surface of the glomerular epithelium normally exhibits a sparse population of finger or bleb-shaped microprojections. These microprojections are found on the cell bodies, major processes, and even occasionally on the small foot processes. Because they are relatively few in number, podocyte microprojections are often not detected in the thin sections studied by transmission electron microscopy and are best seen in the panoramic three dimensional views provided by scanning electron microscopy. There is evidence to suggest that in certain species, podocyte microprojections may vary significantly in number in response to differentiation (Hay and Evan, 1979), dietary protein regime, and aging (Johnson and Barrows, 1980). In addition, in response to certain disease states (e.g. the nephrotic syndrome,) and experimental conditions, podocyte microprojections may undergo significant changes in size, shape, and number (Andrews, 1977a).

In addition to microprojections, glomerular podocytes have been seen to occasionally exhibit "rudimentary" or "primary cilia" (Andrews, 1975). The latter finding is not surprising in that both the parietal epithelial cells and proximal tubule cells with which podocytes are continuous with at the vascular and urinary poles respectively, typically exhibit one or two rudimentary cilia per cell. Such cilia exhibit an irregular intracellular pattern of microtubule singlets and doublets, and there is evidence that similar solitary cilia on other tissues (e.g. rabbit oviductal epithelium) may exhibit a vortical or funnel-type motility (Odor and Blandau, 1985).

Podocytes exhibit a thick negatively charged free surface glycocalyx which has been demonstrated using a variety of techniques including colloidal iron (Mohos and Skoza, 1969), alcian blue (Michael et al., 1970), ruthenium red (Latta et al., 1975), and polycationized ferritin (Andrews, 1980). This surface coat is especially rich in a sialoprotein which has been termed podocalyxin (Kerjaschki et al., 1984). However, the podocyte glycocalyx may not be homogenous in composition in that the glycocalyx associated with the bases of foot processes appears to be different in composition than the glycocalyx associated with the free surfaces of these processes (Roth et al., 1983). The podocyte glycocalyx also extends over the filtration slit spaces between foot processes and has proposed to possibly represent another barrier (e.g. physical and/or electrostatic) to the movement of molecules across the glomerular wall (Latta et al., 1975).

MORPHOLOGICAL ALTERATIONS OF THE GLOMERULAR EPITHELIUM IN RESPONSE TO THE NEPHROTIC SYNDROME

In response to the nephrotic condition, glomerular podocytes undergo a series of morphological alterations which result in the loss of their many foot processes and intervening filtration slits. Puromycin aminonucleoside-induced nephrosis (PAN) is an experimental kidney disease which simulates human lipoid nephrosis (i.e. childhood nephrosis) and which has been used extensively to elucidate ultrastructural changes which characterize the nephrotic condition. Loss of foot processes appears to begin as a flattening of the tips of these processes, which then progresses down the rest of the length of foot processes. Occurring coincident with this flattening is a gradual broadening of the processes. Numerous small knob-like protrusions which are evident along the sides of these flattened processes appear to meet to form what have been termed interpedicular microbridges (Andrews, 1979). It has been proposed that these microbridges may represent the initial sites for the eventual formation of broader tight junctional complexes between adjacent podocytes. An analysis of thin sections and freeze-fractured material indicate that the junctions which eventually form between podocytes in PAN are typical of "leaky" tight junctions (Caulfield et al., 1976; Ryan et al., 1975a). The flattening, broadening and formation of tight junctions between podocyte processes results in a smudged or smeared appearance when viewed by scanning electron microscopy. As these morphological events continue, most if not all of the foot processes and filtration slits are eventually lost and the glomerular capillaries appear covered by a continuous sheet of glomerular epithelium. The many slit diaphragms which previously spanned the filtration slits adjacent to the GBM now appear displaced away from the GBM and stacked above junctional complexes which have formed between remaining podocyte processes. Ryan et al. (1975b), have provided ultrastructural evidence to indicate that the apparent stacking of the slit diaphragms actually represents an extensive folding of these structures which occurs coincident with the loss of foot processes. However, despite this folding and displacement, slit diaphragms still retain their characteristic porous substructure (Ryan et al., 1975b).

In addition to the loss of foot processes and filtration slits, with the onset of the nephrotic syndrome glomerular podocytes become swollen due to an accumulation of cytoplasmic vacuoles and protein absorption droplets. It has been reported that in some cases the vacuoles appear to open to the GBM on one side and be continuous with the urinary space on the other side (Venkatachalam et al., 1969). Such channels compromise the integrity of the glomerular epithelial layer over the glomerular capillaries and may represent sites of plasma protein leakage across the glomerular wall. Also, a certain percentage of the glomeruli (e.g. 30% at the onset of PAN) exhibit what have been termed "blow-out" sites, in which there is a loss of glomerular epithelium over the capillary loops (Ryan and Karnovsky, 1975). Ryan and Karnovsky (1975) noted that the percentage of glomeruli exhibiting blow-out sites corresponds to the number of proximal tubules reported to have very high amounts of protein in their tubules (Oken and Flamenbaum, 1971), and proposed that these focal regions of glomerular epithelial loss may represent the main sites of plasma albumin loss in the nephrotic state. Similar blow-out sites have been reported in other experimental models of the nephrotic syndrome (Grisham and Churg, 1975) and in biopsies taken from nephrotic patients (Carroll et al., 1974).

MORPHOLOGICAL ALTERATIONS OF THE GLOMERULAR EPITHELIUM IN RESPONSE TO ACUTE RENAL FAILURE

A number of investigators have reported changes in the glomerular epithelium in response to renal ischemia. Cox et al. (1974) reported an extensive loss of foot processes 48 hours following norepinephrine induced renal ischemia in dogs. Bulger et al. (1980) ,however, reported only focal loss of podocyte foot processes when they evaluated the same norepinephrine model of ischemia. Thirty minutes following one hour or total ischemia, Barnes et al. (1981) noted spreading of podocyte foot processes associated with the loss of these processes, loss and displacement of the slit diaphragms, and an increase in the number of microvillous projections. Solez et al. (1981) reported that in response to renal ischemia induced by renal pedicle clamping in rabbits, podocyte cell bodies and major processes appear flattened. This flattening increased as the ischemic time increased and did not occur if the kidneys were protected from ischemic damage by administration of clonidine or mannitol (Solez et al., 1981; Racusen and Solez, 1984). These same researchers reported similar flattening of human podocytes in biopsies taken from the kidneys of patients who had suffered from renal ischemia. Recently, we evaluated rat kidney glomeruli which were fixed by vascular perfusion immediately following one hour of total renal ischemia. In response to this ischemic insult many foot processes appeared swollen, and focal sites of swelling were evident over the podocyte free surface (Andrews, unpublished observations). We believe that this apparent swelling of the podocytes is similar in nature to that which characterizes the uriniferous tubules in response to ischemia (Andrews and Coffey, 1982), and is therefore probably due to an inability of the podocytes to regulate cell volume during the ischemic insult (Leaf, 1958).

There have also been some observations regarding the effects of nephrotoxic forms of acute renal failure on the glomerular epithelium. Stein et al. (1975) reported that 48 hours following administration of the nephrotoxin uranyl nitrate to dogs, there is an extensive loss of foot processes. In support of these findings, Flamenbaum et al. (1976) noted similar changes 24 hours following uranyl nitrate administration to rats. However, we recently evaluated the glomerular epithelium of rats six days following the induction of uranyl nitrate induced acute renal failure, and failed to detect significant alterations of the glomerular epithelium including loss of foot processes (Andrews, unpublished observations). Also, Baehler et al. (1977) reported that following administration of the nephrotoxin $HgCl_2$ to dogs, podocyte cell bodies and processes appear swollen but that no loss of foot processes is evident.

MORPHOLOGICAL ALTERATIONS OF THE GLOMERULAR EPITHELIUM IN RESPONSE TO IN VITRO AND IN VIVO EXPERIMENTAL PROCEDURES

In the early seventies, Michael, Blau and their co-workers (Michael et al, 1970; Blau and Haas, 1973) reported that coincident with the onset of PAN and human glomerular disease characterized by proteinuria, there is a significant reduction in the sialic acid component of the glomerular surface glycocalyx. These investigators suggested that a reduction in cell surface polyanionic charge resulting from this loss of sialic acid may lead to reduced electrostatic repulsion between foot processes and the resultant loss of the filtration slits and podocyte processes seen in proteinuric states. Support for this theory came in 1975, when Seiler and his co-workers reported that in response to neutralization of the glomerular polyanionic surface charge by in situ vascular perfusion of cationic molecules (e.g. protamine sulfate, poly-l-lysine), the glomerular epithelium exhibits a series of changes which mimic those seen in the nephrotic state. These changes include flattening and retraction of foot processes, narrowing of filtration slits, and the formation of tight junctions between adjacent podocyte processes. These researchers also reported that these dramatic morphological changes could be reversed by subsequent perfusion of the cation-treated kidneys with anionic heparin to reestablish the glomerular polyanion.

In 1978, both Kerjaschki and myself reported that similar cation-induced alterations of the glomerular epithelium can be induced in vitro. In addition, I noted that prolonged in vitro treatment of podocytes with high concentrations of polycations (i.e. 1000 ug/ml of protamine sulfate for one hour), can transform these once elaborate cells into simple polygonal cells (Andrews, 1978). The intercellular borders between such morphologically transformed podocytes become filled with stacks of membranes which are similar in structure to slit diaphragms. Additional in vitro studies were undertaken to determine the effects of removal of the sialic acid component of the glomerular polyanion with purified neuraminidase (Andrews, 1979). This latter procedure resulted not only in the flattening and loss of foot processes, but the appearance of irregular knob-like free surface microprojections which is another morphological characteristic of the nephrotic condition (Andrews, 1979). Together, the above studies have provided strong support for the concept that loss of foot processes in the nephrotic state results from a reduction in the polyanionic sialic acid component of the glomerular surface coat.

As noted earlier, both cytoplasmic microtubules and actin-like microfilaments are abundant within glomerular podocyte cell bodies and their major processes, while filamentous actin and myosin are especially abundant within podocyte foot processes. These cytoplasmic organelles are known to play a number of important roles in cells, including the maintenance and alteration of cell shape (Spooner, 1975). In a series of investigations, we and others have evaluated the roles of these cytoplasmic elements in maintaining and altering the shape of glomerular podocytes

The cytochalasins are fungal metabolites which are able to inhibit the retractile capacity of actin-like cytoplasmic microfilaments (Miranda, 1974). When we added small quantities of either cytochalasin D or B to the culture medium of glomeruli being incubated in vitro, podocytes were prevented from undergoing the loss of foot processes and filtration slits which otherwise occur in response to long term in vitro incubation (Andrews, 1981b). We also found that the cytochalasins prevented the loss of foot processes which otherwise occurs in response to neutralization of the glomerular polyanion with cationic molecules, and removal of the sialic acid component of the glomerular glycocalyx by neuraminidase treatment (Andrews, 1981b). Studying isolated glomeruli in vitro, Kerjaschki (1978) demonstrated that calcium was necessary for polycation-induced foot loss to take place, and found that cold temperatures also inhibited foot process loss. Because in vitro and in situ foot process loss appear to mimic the loss of podocyte processes seen in the nephrotic state, it is likely that calcium dependent retraction of filamentous actin also mediate the morphological changes which these cells undergo in response to the nephrotic condition. It is interesting to note that, in addition to its inhibition of foot process loss, in vitro treatment with cytochalasin prevents the formation of free surface microprojections on the glomerular epithelium (Andrews, 1981b). This latter observation is not unexpected in that free surface microprojections are supported by a cytoskeletal core of microfilaments.

In addition to the role of contractile elements in foot process loss, studies have shown that exposure to cytochalasin induced inhibition of contractile elements causes foot processes to change in shape from short processes with more or less flattened bases to taller processes with narrow bases (Andrews, 1981b). By affecting this narrowing of the foot process bases, adjacent foot processes move apart to fully open up previously narrowed filtration slits. As a result, the number of fully patent filtration slits appear to increase significantly. These latter morphological changes represent alterations that are opposite to the flattening of foot processes and subsequent narrowing and loss of filtration slits seen in neuraminidase-treated, polycation-treated, and the diseased kidney. It appears therefore that the broadened bases which are normally exhibited by many foot processes may in fact represent an initial stage in actin initiated flattening of these processes. Since the filtration area defined by the filtration slits is a major determinent of solute efflux across the glomerular wall (Shea and Morrison, 1971), it is tempting to speculate that by regulating the shapes of foot processes (i.e. by expanding and contracting the bases of the foot processes), glomerular podocytes are able to precisely control solute efflux across any given region of the glomerular wall. This potential of the glomerular epithelium to regulate glomerular filtration may account for some of the reported alterations of the ultrafiltration coefficient (i.e. the product of the porosity of the glomerular wall and the surface area available for ultrafiltration) which appear to occur in response to a variety of physiological stimuli (Ichikawa and Brenner, 1977; Baylis et al., 1977). In support of this theory, it has been reported that angiotensin, which is known to reduce the ultrafiltration coefficient, results in a flattening of the foot processes both in vivo (Hornych et al. 1972) and in vitro (Racusen and Solez, 1984). However, changes that alter the ultrafiltration coefficient at the level of the filtration slits may be subtle and/or focal in nature and may require serial sectioning and a meticulous morphometric analysis of glomeruli in order to detect.

Both in vivo and in vitro treatment with compounds which depolymerize cytoplasmic microtubules (e.g. vinblastine) have been informative in revealing roles that microtubules may or may not play in morphological alterations of kidney podocytes. In response to in vivo or in vitro treatment with vinblastine, podocyte cell bodies swell and the major processes become extremely attenuated in some regions along their lengths and abnormally swollen in other regions (Andrews, 1977; Andrews, 1981b). Podocyte foot processes ,however, appear to be unaffected by treatment with any of a wide variety of compounds which depolymeriz cytoplasmic microtubules (e.g. vinblastine, colchicine, colcemid, and podophyllotoxin). Also, treatment with any of the foregoing agents failed to inhibit the loss of foot processes which otherwise occurs in response to neutralization of the glomerular polyanion with cationic molecules or long term in vitro incubation

(Andrews, 1981b). Loss of microtubules did, however, result in the formation of unusually long microvillous projections on the free surfaces of podocytes during in vitro incubation (Andrews, 1981b). In view of these observations it would appear that cytoplasmic microtubules play an important role in maintaining the structural integrity of podocyte cell bodies and their major processes, but do not play a significant role in the affecting the characteristic loss of foot processes.

REFERENCES

Andrews, PM (1975). Scanning electron microscopy of human and rhesus monkey kidneys. Lab Invest 32:610-618.

Andrews, PM (1977a). A scanning and transmission electron microscopic comparison of puromycin aminonucleoside-induced nephrosis to hyperalbuminemia-induced proteinuria with emphasis on kidney podocyte pedicel loss. Lab Invest 36:183-197.

Andrews, PM (1977b). The effect of vinblastine-induced microtubule loss on kidney podocyte morphology. Am J Anat 150:53-62.

Andrews, PM (1978). Scanning electron microscopy of the kidney glomerular epithelium after in situ and in vitro treatment with polycations. Amer J Anat 153:291-304.

Andrews, PM (1979). Glomerular epithelial alterations resulting from sialic acid surface coat removal. Kidney Int 15:376-385.

Andrews, PM (1980). Cationized ferritin binding to anionic surfaces in normal and aminonucleoside nephrotic kidneys. Amer J Anat 162:89-106.

Andrews, PM (1981a). Characterization of free surface microprojections on the kidney glomerular epithelium. In Vidrio EA, Galina MA (eds): "Progress in Clinical and Biological Research" New York: Allan Liss, pp.21-35.

Andrews, PM (1981b). Investigations of cytoplasmic contractile and cytoskeletal elements in the kidney glomerulus. Kidney Int 20:549-562.

Andrews, PM and Bates, SB (1984). Localization of F-actin in the rat kidney. Anat Rec 210:1-9.

Andrews, PM and Bates, SB (1985). Dose dependent movement of cationic molecules across the glomerular wall. Anat Re, 212:223-231.

Andrews, PM and Coffey, AK (1980). In vitro studies of kidney glomerular cells. Scanning Electron Microscopy 2:179-191.

Andrews, PM and Coffey, AK (1982). Factors that improve the preservation of nephron morphology during cold storage. Lab Invest 46:100-120.

Andrews, PM and Ccffey, AK (1983). Cytoplasmic contractile elements in glomerular cells. Federation Proc 42:3046-3052.

Baehler, RW, Kotchen, TA, Burke, JA, Galla, JH, and Bhathena, D (1977). Considerations of the pathophysiology of mercuric chloride-induced acute renal failure. J Lab Clin Med 90:330-340.

Barnes, JL, Osgood, RW, Reineck, HJ, and Stein, JH (1981). Glomerular alterations in an ischemic model of acute renal failure. Lab Invest, 45:378-386.

Baylis, C, Ichikawa, I, Willis, WT, Wilson, CB, and BM Brenner 1977. Dynamics of glomerular ultrafiltration. IX. Effects of plasma protein concentration. Am J Physiol, 232:F58-F71.

Blau, EB and Haas, JE (1973). Glomerular sialic acid and proteinuria in human renal disease. Lab Invest, 28:477-481.

Bulger, RE, Cronin, RE, and Dobyan,DC (1980). Glomerular architectural changes after a two-hour infusion of norepinephrine. Am J Anat, 159:379-384.

Carroll, N, Crock, GW, Funder, CC, Green, CR, Ham, KN, and Tange, JD (1974). Glomerular epithelial cell lesions induced by N,N'-diacetylbenzidine. Lab Invest, 31:239-245.

Caulfield, JP, Reid, JJ and Farquhar, MJ (1976). Alterations of the glomerular epithelium in acute aminonucleoside nephrosis: evidence for formation of occluding junctions and epithelial cell detachment. Lab Invest, 34:43-59.

Cox, JW, Baehler, RW, Sharma, H, O'Dorisio, T, Osgood, RW, Stein, JH, and Ferris, TF (1974). Studies on the mechanisms of oliguria in a model of unilateral acute renal failure. J Clin Invest, 53:1546-1558.

Cronin, RE, DeTorrente, A, Miller, PD, Bulger, RE, Burke, TJ, and Schrier, RW (1978). Pathogenic mechanisms in early norepinephrine-induced acute renal failure: Functional and histological correlates of protection. Kidney Int, 14:115-125.

Flamenbaum, E, Hamburger, RJ, Huddleston, ML, Daufman, J, McNeil, JS, Schwartz, JS, and Nagle, RB (1976). The initiation phase of experimental acute renal failure: An evaluation of uranyl nitrate-induced acute renal failure in the rat. Kidney Int, 10:S115-122.

Grisham, E, and Churg, J (1975). Focal glomerular sclerosis in nephrotic patients: An electron microscopic study of glomerular podocytes. Kidney Int, 7:111-122.

Hay, DA, and Evan, AP (1979). Maturation of the glomerular visceral epithelial and capillary endothelium in the puppy kidney. Anat Rec, 193:1-22.

Hornych, H, Beaufils, M, and Richet, G (1972). The effect of exogenous angiotensin on superficial and deep glomeruli in the rat kidney. Kidney Int, 2:336-343.

Ichikawa, I and Brenner, BM (1977). Evidence for glomerular actions of ADH and dibutyryl AMP in the rat. Am J Physiol, 233:F102-F117.

Johnson, JE, and Barrows, CH (1980). Effects of age and dietary restriction on the kidney glomeruli of mice: Observations by scanning electron microscopy. Anat Rec, 196:145-151.

Kerjaschki, D (1978). The effects of low temperature, divalent cations, colchicine and cytochalasin B. Lab Invest, 39:430-440.

Kerjaschki, D, Sharkey, DJ, and Farquhar, MG (1984). Identification and characterization of podocalyxin-the major sialoprotein of the renal glomerular epithelial cell. J Cell Biol, 98:1591-1596.

Latta, H Johnston, W, and Stanley, T (1975). Sialoglycoproteins and filtration barriers in the glomerular capillary wall. J Ultrastr Res, 51:354-376.

Leaf, A (1959). Maintenance of concentration gradients and regulation of cell volume. Ann NY Acad Sci, 72:396-404.

Michael, AF, Blau, E, and Vernier, RL (1970). Glomerular polyanion: Alteration in aminonucleoside nephrosis. Lab Invest, 23:649-657.

Miranda, AR, Godman, GC, Deitch, AD, and Tanenbaum, SW (1974). Action of cytochalasin D on cells of established lines: Early events. J Cell Biol, 61481-500.

Mohos, SC and Skoza, L (1969). Glomerular sialoprotein. Science, 164:1519-1521.

Odor, DL, and Blandau, RJ (1985). Observations on the solitary cilium of rabbit oviductal epithelium: its motility and ultrastructure. Am J Anat, 174:437-453.

Oken, DE, and Flamenbaum, W (1971). Micropuncture studies of proximal tubule albumin concentrations in normal and nephrotic rats. J Clin Invest, 50:1498-1505.

Pinto, JA, and Douglas, BB (1974). Glomerular morphometry. I. Combined light and electron microscope studies in normal rats. Lab Invest, 30:657-663.

Racusen, LC and Solez, K (1984). Podocyte changes in postischemic acute renal failure. In Solez K, Whelton A (eds): "Correlations between Morphology and Function" New York: Marcel Dekker, Inc., pp 135-145.

Rodewald, R and Karnovsky, MJ (1974). Porous structure of the glomerular slit diaphragm in the rat and mouse. J Cell Biol, 60:423-433.

Roth, J, Brown, D, and Orci, L (1983). Regional distribution of N-acetyl-D-galactosamine residues in the glycocalyx of glomerular podocytes. J Cell Biol, 96:1189-1196.

Ryan, GB, and Karnovsky, MJ (1975). An ultrastructural study of the mechanisms of proteinuria in aminonucleoside nephrosis. Kidney Int, 8:219-232.

Ryan, GG, Leventhal, M, and Karnovsky, MJ (1975a). A freeze-fracture study of junctions between glomerular epithelial cells in Aminonucleoside nephrosis. Lab Invest, 32:397-403.

Ryan, GB, Rodewald, P, and Karnovsky, MJ (1975b). An ultrastructural study of the glomerular slit diaphragm in aminonucleoside nephrosis. Lab. Invest, 33:461-468.

Seiler, MR, Rennke, HG, Venkatachalam, MG, and Cotran, RS (1977). Pathogenesis of polycation-induced alteration (fusion) of glomerular epithelium. Lab Invest, 36:48-61.

Seiler, MW, Venkatachalam, MA, and Cotran, RS (1975). Glomerular epithelium: Structural alterations induced by polycations. Science, 189:390-393.

Shea, SM, and Morrison, AB (1975). A stereological study of the glomerular filter in the rat. Morphometry of the slit diaphragm and basement membrane. J Cell Biol, 67:436-443.

Solez, K, Racusen, LC, and Whelton, A (1981). Glomerular epithelial cell changes in early postischemic acute renal failure in rabbits and man. Am J Pathol, 103:163-173.

Spooner, BS (1975). Microfilaments, microtubules, and extracellular materials in morphogenesis. Bioscience, 25:440-451.

Stein, JH, Gottschall, RW, Osgood, and Ferris, TF (1975). Pathophysiology of a nephrotoxic model of acute renal failure. Kidney Int,8 :27-41.

Trenchev, P, Dorling, J, Webb, J, and Holborrow, EJ (1976). Localization of smooth muscle-like contractile proteins in kidney by immunoelectron microscopy. J Anat, 121:85-95.

Venkatachalam, MA, Karnovsky, MJ, and Cotran, RS (1969). Glomerular permeability: Ultrastructural studies in experimental nephrosis using horseradish peroxidase as a tracer. J Exp Med, 130:381-399.

Cells and Tissues: A Three-Dimensional Approach by Modern Techniques in Microscopy, pages 167–172

THE FINE STRUCTURE OF THE LAMINAE RARAE OF THE GLOMERULAR BASEMENT MEMBRANE IN THE RAT 1)

E. Reale 2), L. Luciano 2) and K. Kühn 3)

2) Laboratory of Cell Biology and Electron Microscopy; 3)Division of Nephrology, School of Medicine, Konstanty-Gutschow-Str. 8, 3000 Hannover 61, FRG

INTRODUCTION

The extracellular matrix of the renal glomerulus is essentially represented by the glomerular basement membrane (GBM) and by the mesangial matrix. Usually, the GBM is described as a three-layered structure composed of a lamina densa (LD) with a fine fibrillar structure and two laminae rarae, the interna (LRI, on the endothelial side) and the externa (LRE, on the epithelial side). The LD contains collagen IV; the LRI can display "tubular fibrils" (Farquhar, 1978) which, as those of the mesangial matrix, contain fibronectin (Courtoy et al., 1980). Other glycoproteins (GP) of the glomerular extracellular matrix (with debated localization in one or both the laminae rarae) are laminin, entactin, nidogen. Also glycosaminoglycans (GAG) occur in the GBM and in the mesangial matrix. These are hyaluronic acid and, in the form of proteoglycans (PG), heparan sulfate and chondroitin sulfate (rev. by Timpl and Dziadek, 1986).

Using different cationic dyes (Alcian blue, Safranine O, Acridine orange or Ruthenium red) we demonstrated that the aspect of the complexes formed by PG within the rat GBM varied according to the different charge of the dye. Polycations generate complexes or precipitates like particles; monocations appear as filaments occasionally showing lateral projections (Reale et al., 1983).

Glycoconjugates can be histochemically revealed also by the ferrocyanide-reduced OsO_4 (Karnovsky, 1971). With this method we have recently demonstrated the glycoconjugates of the articular cartilage as a tight network of threads which interconnected all other cartilage components (Reale, 1986). The network was comparable to that revealed in the same

1) Supported by the DFG (SFB 146; EN 65/15-1)

tissue which was quick frozen (under high pressure), cryosubstituted and embedded by low temperature in Lowicryl (Hunzicker and Schenk, 1983).

Recently, Goldberg and Escaig-Haye (1986) studied the basement membrane in unfixed, quick frozen (at liquid helium temperature) and cryosubstituted tissues of the oral cavity of the rat. These authors have demonstrated that the subdivision of the basement membrane into lamina rara, lamina densa and lamina fibroreticularis was missing and a single homogeneous structure was seen. This structure, in thin sections stained by Alcian blue at acidic pH (this reaction reveals sulfated GAG), was homogeneously electron dense.

The results summarized above demonstrate that the PG can be histochemically revealed with very different aspects, from particles to a homogeneous structure. In the present study we tried to demonstrate the reason for these morphological differences, applying methods we have used previously and cryosubstitution.

MATERIALS AND METHODS

Rat kidneys were fixed by vascular perfusion with a solution containing either (*a*) glutaraldehyde 2% in 0.1 M phosphate buffer, or (*b*) glutaraldehyde, Alcian blue 8 GX and 0.3 M $MgCl_2$ (Reale *et al.*, 1983). After 2 h fixation small samples of the cortex prefixed in solution *a* were transferred into the ferrocyanide-reduced OsO_4 solution (Karnovsky, 1971) and then embedded in Epon. Samples prefixed in solution *b* were cryoprotected (25% glycerol in Ringer) for 30', frozen in liquid Freon 22 (about -150C) and transferred into a cryosubstitution apparatus (CS auto-device, Reichert-Jung, Heidelberg). The ice was substituted first by absolute acetone containing about 3% OsO_4 at -80C for 34 h; then the samples were allowed to reach room temperature over 24 h and finally embedded in Epon.

In addition samples of the kidneys perfused with solutions *a* and *b* were postfixed in phosphate or cacodylate buffered 1% OsO_4 and conventionally embedded in Epon.

Thin sections were collected on formvar-carbon membranes, stained with uranyl acetate and lead citrate and examined in a Siemens Elmiskop 101 electron microscope.

RESULTS

Fixative solution *a* followed by reduced-OsO_4 revealed the endothelial and epithelial surface coat as a structure composed of particles interconnected by fine threads in a polygonal network (Fig. 1). The particles had a diameter of about 10 nm, the polygonal meshes about 20 nm.

Fixative solution *b* followed by conventional embedding procedure stained the surface coat of the endothelial and

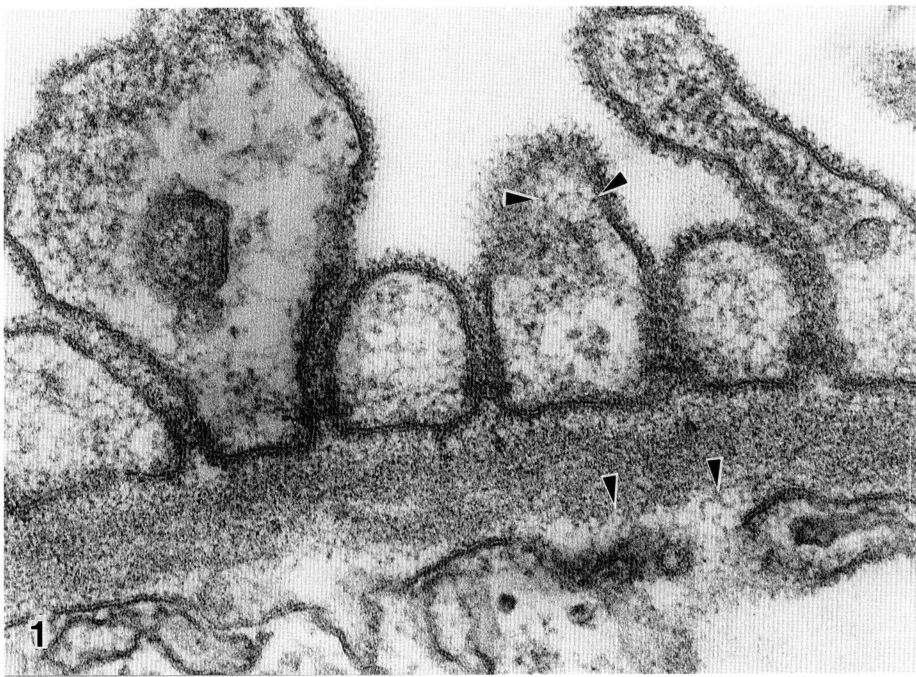

Fig. 1. Postfixation in the reduced-OsO4 solution. Surface coats and GBM show particles interconnected by fine thread (arrowheads). x 108.000.

epithelial cells, as well as PG located in the laminae rarae as previously described (Reale et al., 1983).

After cryosubstitution, the glomerular capillary wall presented two different aspects with transitions from the first to the second. In the first aspect (Fig. 2) the surface coat of epithelial and endothelial cells was composed of electron dense particles ordered in a distinct layer more evident on the outer surface of the cells (towards the urinary space and the blood, respectively) than on the inner aspect (towards the LD). The LD of the GBM was bordered, as usual, by the laminae rarae. Across the LD and in the laminae rarae a network was seen, formed by electron dense threads. In the laminae rarae the borders with the particles of the surface coat were frequently indistinct. The LRE was evident, especially between LD and slit diaphragms. In its second aspect (Fig. 3), the glomerular capillary wall showed on the outer surface of endothelial and epithelial cells (urinary space, blood) a homogeneous, uninterrupted, deeply electron dense surface coat. On the opposite side (LD side) the surface coat was less stained and merged without sharp contours into the LD. The GBM extended from the surface coat of the pedicles and from the interposed slit diaphragms to the surface coat of the endothelial cells (and to the endothelial

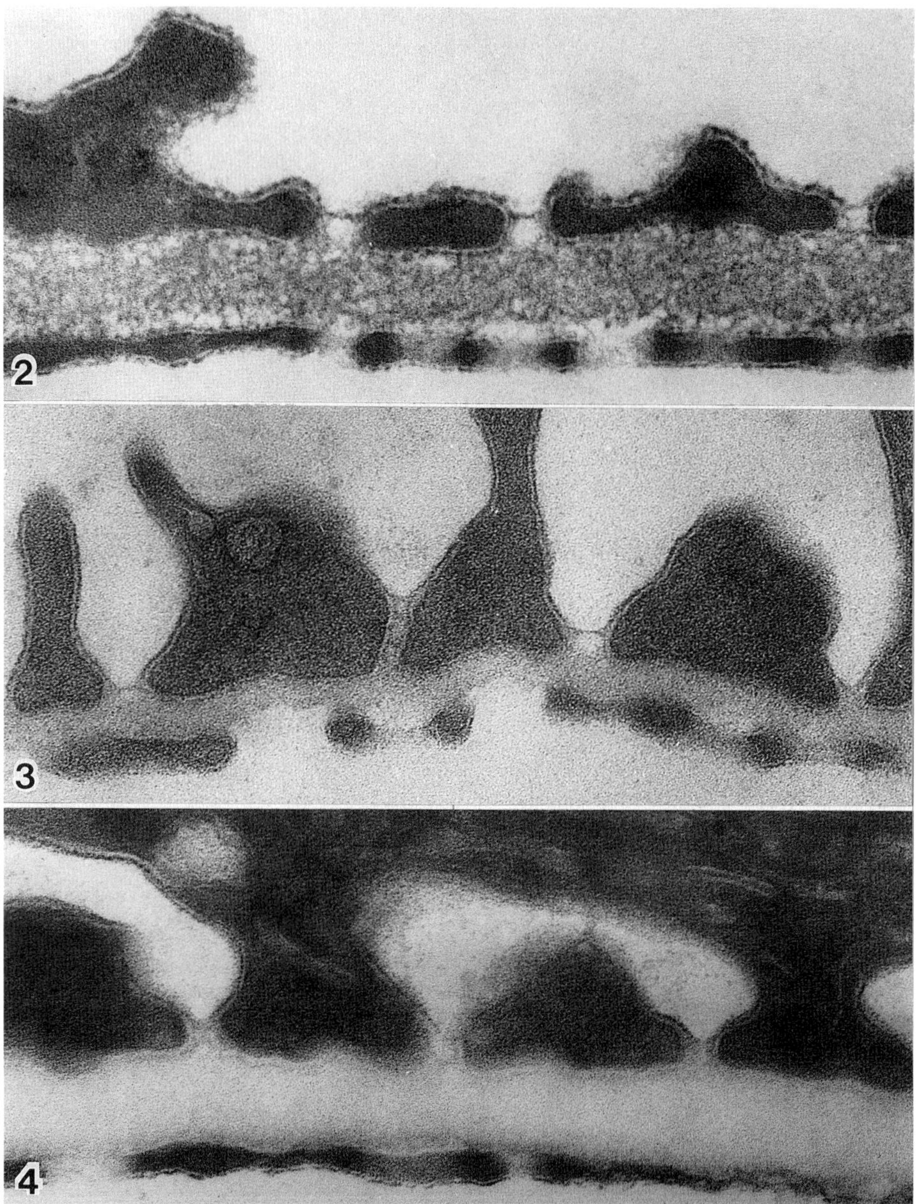

Figs. 2, 3, 4. Different aspects of the glomerular capillary wall after cryosubstitution. Surface coat composed of particles in Fig. 2; uninterrupted in Figs. 3 and 4. GBM with distinct LD, LRI and LRE as well as with network-like substructure in Fig. 2; LD and laminae rarae not visible in Figs. 3 and 4. Figs. 2 - 4, x 108.000.

pores) as a homogeneous structure. Instead of laminae rarae a layer as homogeneous and electron dense as the LD was visible. Close examination of the GBM shows here and there a very delicate, indistinct network of electron dense meshes through the whole GBM (Fig. 4). This aspect can be compared to the previous one shown in Fig. 2. However, the spaces among the meshes were not devoid of structures but homogeneous and lightly electron dense.

In conclusion, endothelial and epithelial surface coats as well as PG of the GBM of the rat renal glomerulus can display different morphological aspects. They can appear: (1) as particles if polycations are used as markers. In this case, the particles should be considered as completely collapsed glycoconjugate molecules as suggested by G.K. Hascall (1980) for the PG. (2) as filaments with some lateral projections. These forms could represent partially extended PG-molecules (Reale et al., 1983; 1985). (3) as a tight network of particles and threads, as demonstrated by the reduced-OsO_4 procedure (this study). A comparable result has been obtained in quick-frozen cryosubstituted cartilage specimens (Hunzicker and Schenk, 1983; Arsenault, 1985) and in the glomerular capillary wall by using the deep-etching replica method (Kubosawa and Kondo, 1985). In our experiments with cryosubstituted specimens a network was occasionally seen as well. However, most of our specimens showed (4) a completely uninterrupted surface coat surrounding endothelial and epithelial cells and, in addition, a homogeneous structure of the GBM. This result agrees with the observations of Goldberg and Escaig-Haye (1986) showing the absence of a lamina rara in other basement membranes.

From the different aspects described, it can be suggested that particles, filaments and networks represent glycoconjugate molecules in different stages of collapse. These molecules, in vivo or after application of procedures preserving their full hydrated form, do not have a substructure and form homogeneous areas.

REFERENCES

Arsenault AL (1985). Fine structure and elemental maps of the calcifying epiphysis preserved by slam freezing and freeze substitution. In Butler WT (ed): " The Chemistry and Biology of Mineralized Tissues", Birmingham, Ala.: Ebsco Media, pp 364-367.

Courtoy PJ, Kanwar YS, Hynes RO, Farquhar MG (1980). Fibronectin localization in the rat glomerulus. J Cell Biol 87:691-696.

Farquhar MG (1978). Structure and function in glomerular capillaries. In Kefalides NA (ed): "Biology and Chemistry of Basement Membranes", New York: Academic Press, pp 43-80.

Goldberg M, Escaig-Haye F (1986). Is the lamina lucida of the basement membrane a fixation artefact? Europ J Cell Biol 42:365-368.

Hascall GK (1980). Cartilage proteoglycans: comparison of sectioned and spread whole molecules. J Ultrastruct Res 70:369-375.

Hunziker EB, Schenk RK (1984). Cartilage ultrastructure after high pressure freezing, freeze substitution, and low temperature embedding. II. Intercellular matrix ultrastructure - preservation of proteoglycans in their native state. J Cell Biol 98:277-282.

Karnovsky MJ (1971). Use of ferrocyanide - reduced osmium tetroxide in electron microscopy. Am Soc Cell Biol, 11th Meeting, p 146.

Kubosawa H, Kondo Y (1985). Ultrastructural organization of the glomerular basement membrane as revealed by a deep-etch replica method. Cell Tissue Res 242:33-39.

Reale E (1986). The ground substance of the human articular cartilage after postfixation in ferrocyanide-reduced osmium tetroxide. J Clin Chem Clin Biochem 24:918-920.

Reale E, Luciano L, Kühn K (1983). Ultrastructural aspects of proteoglycans in the glomerular basement membrane. A cytochemical approach. J Histochem Cytochem 31:662-668.

Reale E, Luciano L, Kühn K (1985). Cationic dyes reveal proteoglycans on the surface of epithelial and endothelial kidney cells. Histochem 82:513-518.

Timpl R, Dziadek M (1986). Structure, development, and molecular pathology of basement membranes. Int Rev Exp Pathol 29:1-112.

Cells and Tissues: A Three-Dimensional Approach by Modern Techniques in Microscopy, pages 173–179

HUMAN DEVELOPING METANEPHRIC NEPHRONS. MORPHOLOGICAL, IMMUNOHISTOCHEMICAL AND HISTOCHEMICAL STUDIES

Cesare De Martino, Pier Giorgio Natali, Paola Nistico' and Lidia Accinni

Regina Elena Institute for Cancer Research, Rome, Italy; Institute of Experimental Medicine - C.N.R., Rome, Italy

INTRODUCTION

The metanephros originates from two different systems: the metanephric duct and the metanephrogenic blastema, which derives from the caudal portion of the intermediate mesoderm. The ureter, calyces, and collecting tubules develop from the first system, whereas the proximal and distal tubules, Henle's loops, and glomeruli derive from the second. In the present report we describe the main steps of metanephric nephron development through light and electron microscopic analysis, histochemical assays and through the use of immunohistochemical methods employing a panel of polyclonal (anti smooth muscle actin and myosin, anti glomerular basement membrane) and monoclonal (Mab) antibodies recognizing structural as well as differentiation antigens (URO2,URO3, URO4,J5,MBR1,345-134S). The source and specificity of these antibodies have been reported in previous papers (Canevari et al., 1983; Imai et al.,1982; Metzgar et al.,1981; Natali et al.,1984; Ueda et al.,1981).

RESULTS AND DISCUSSION

Although the differentiation of metanephric nephron is a continuously evolving process, for descriptive purposes it may be divided into the following main stages (De Martino and Accinni 1985; Zamboni and De Martino,1968): 1) Stage of "metanephric cap" during which the cells of the metanephrogenic blastema condense around the tip of the expanded extremities of collecting tubules or ampullae. In this stage a basal lamina is present along the ampulla. 2) Stage of "metanephric vesicle" in which each solid cell nest, originating from the extremities of the cap, develops a central lumen and rapidly establishes a communication with the collecting ampulla. A discontinuous basal lamina appears along the metanephric vesicle. 3) Stage of "S-shaped body" which originates from the elongated metanephric vesicle for the appearance of two indentations. The progressive deepening of both indentations transforms the vesicle in a S-shaped structure composed of three limbs, namely "the upper" from which the distal tubule develops, "the middle" which gives rise to the macula densa and proximal tubule, and "the lower" from which the glomerulus originates. A very thin, continuous basal lamina surrounds the S-body. 4) Stage of intracleftal growing of glomerular vessels, characterized initially by the invagination of a capillary from the surrounding mesenchyme into the lower cleft. 5) Stage of "double spiral body" in which the architecture of the different segments of early nephron is clearly recognizable. This stage is characterized by the growth and differentiation of capillary loops and an increase of the convolutions of proximal and distal tubules. 6) This last stage collects several continuous evolving development processes of metanephric nephrons in which the glomeruli reach a complete differentiation with the final maturation of their cellular and acellular components. At this stage an immature Henle's loop appears (Figs. 1 and 2).

As revealed by the immunohistochemical study, these morphological modifications are accompained by the appearance of some structural as well as differentiation antigens. In fact, basal lamina antigens are detected at very early stages of development in the structures originating from the metanephrogenic blastema while in those developing from the metanephric duct the antigens follow the budding of ureter, calyces, and collecting tubules. Actin and myosin are recognizable in the cells of metanephrogenic blastema before this structure undergoes organization into the various nephron segments where they then localize in the apical portion of the cells and precisely on the plasma membrane facing the lower cleft of S-shaped body and the lumina of different tubular segments. In mature glomeruli actin is mainly localized in the mesangial cells, even if the linear distribution along the glomerular basement membranes is suggestive of an extracellular presence as well, similarly to that previously observed in the glomeruli of adult kidney (Accinni et al.,1975). On the contrary, in mature glomeruli myosin is strictly localized in the mesangial area (Figs. 3-8). The immunohistochemical analysis performed with Mab showed that the differentiation antigens appear only in mature nephrons and are mainly detected in the proximal tubular cells, even if some of them are also present in the epithelial cells of glomeruli (URO2, URO4, J5) and of Henle's loops (URO4), suggesting that these antigens undergo a programmed stage and cell related expression. Whether this phenomenon is related to distinct renal functional activities remains to be established (Figs. 9-12).

The experiments performed with PASM and PAS reactions showed that the basement membrane of developing glomeruli is negative to these histochemical reactions until the capillary loops

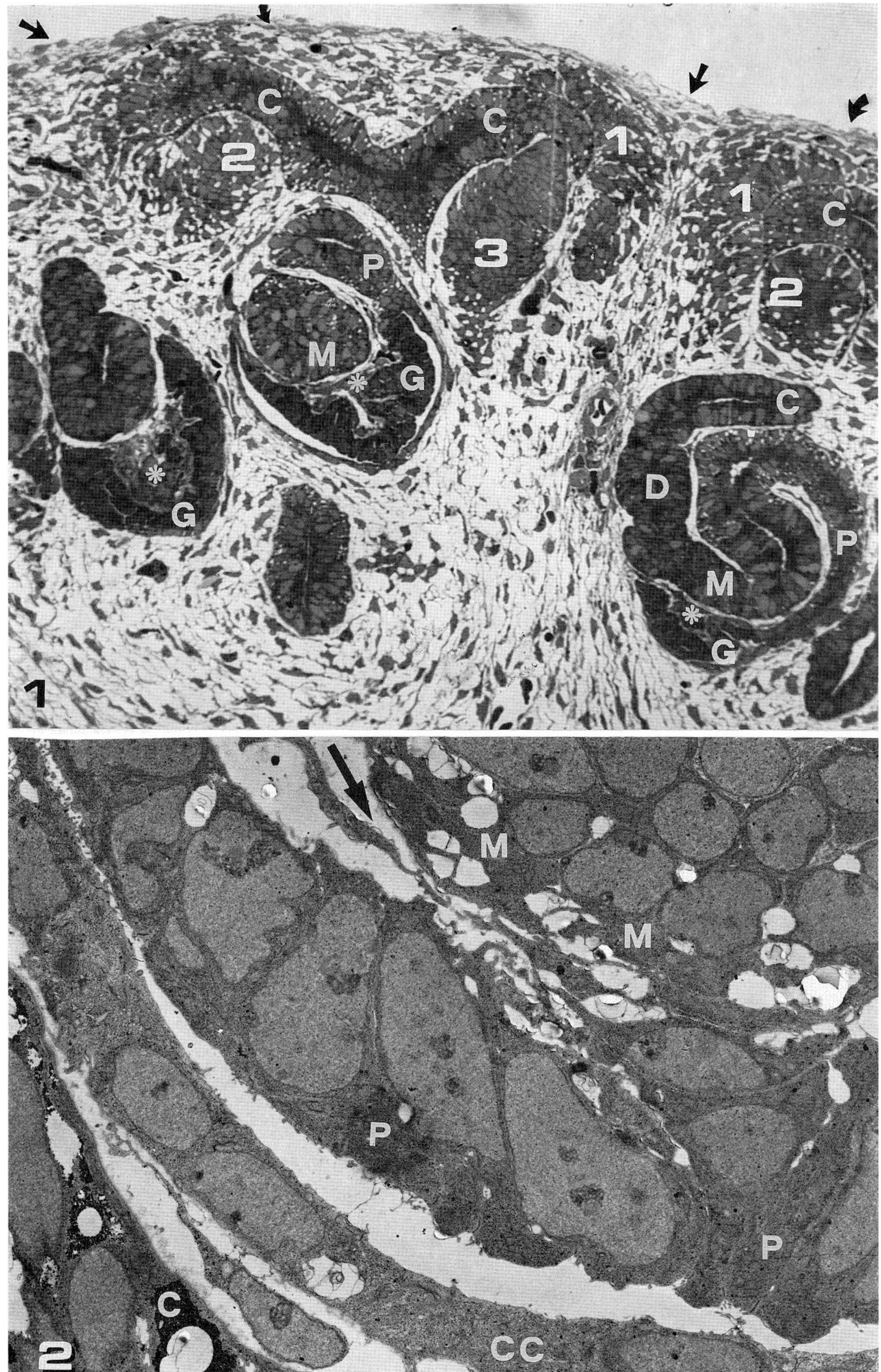
C
C
2
1
1
C
3
P
2
M
G
C
G
D
P
M
G
1
M
M
P
P
C
2
CC

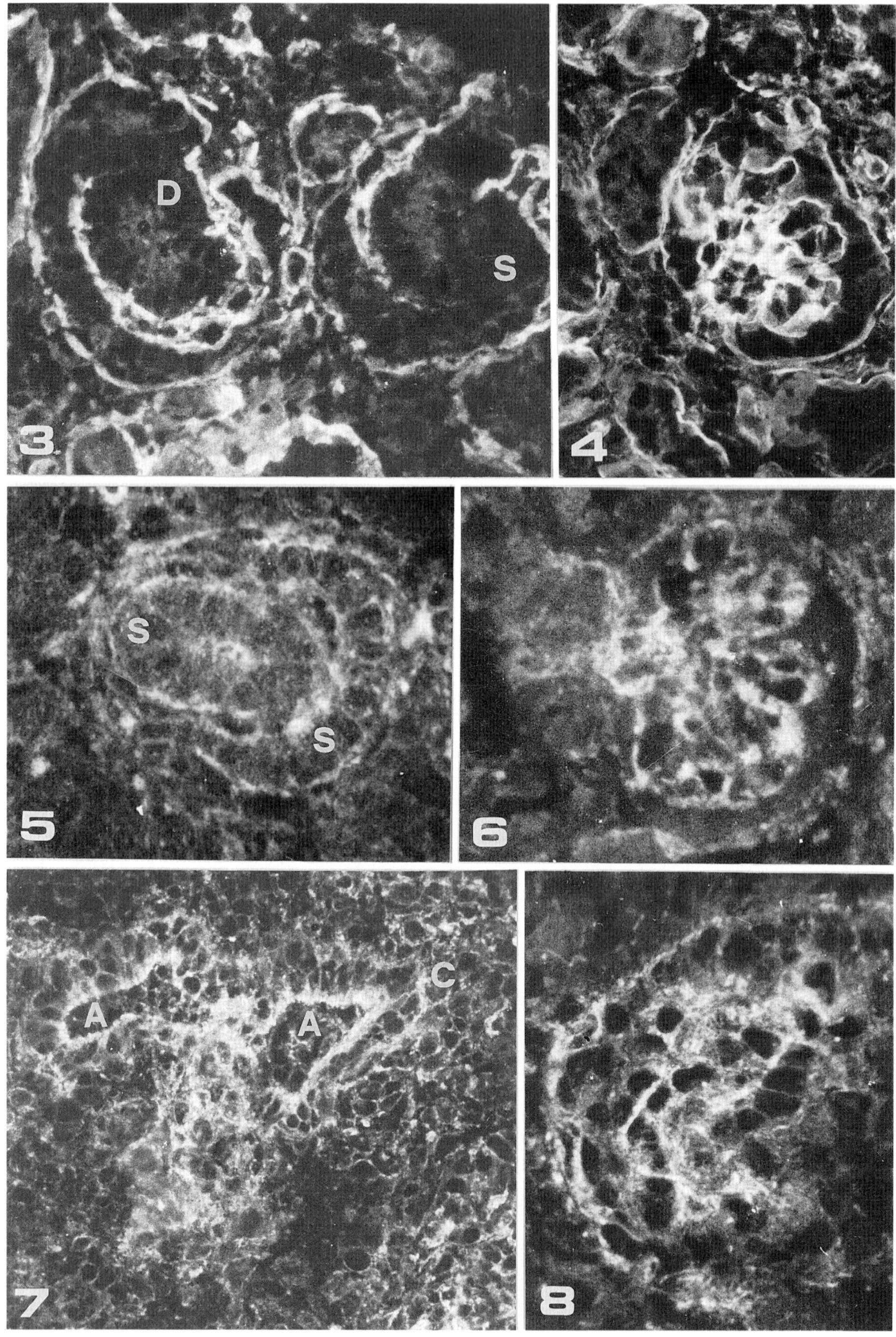
D
S
3
4
S
S
5
6
A
A
C
7
8

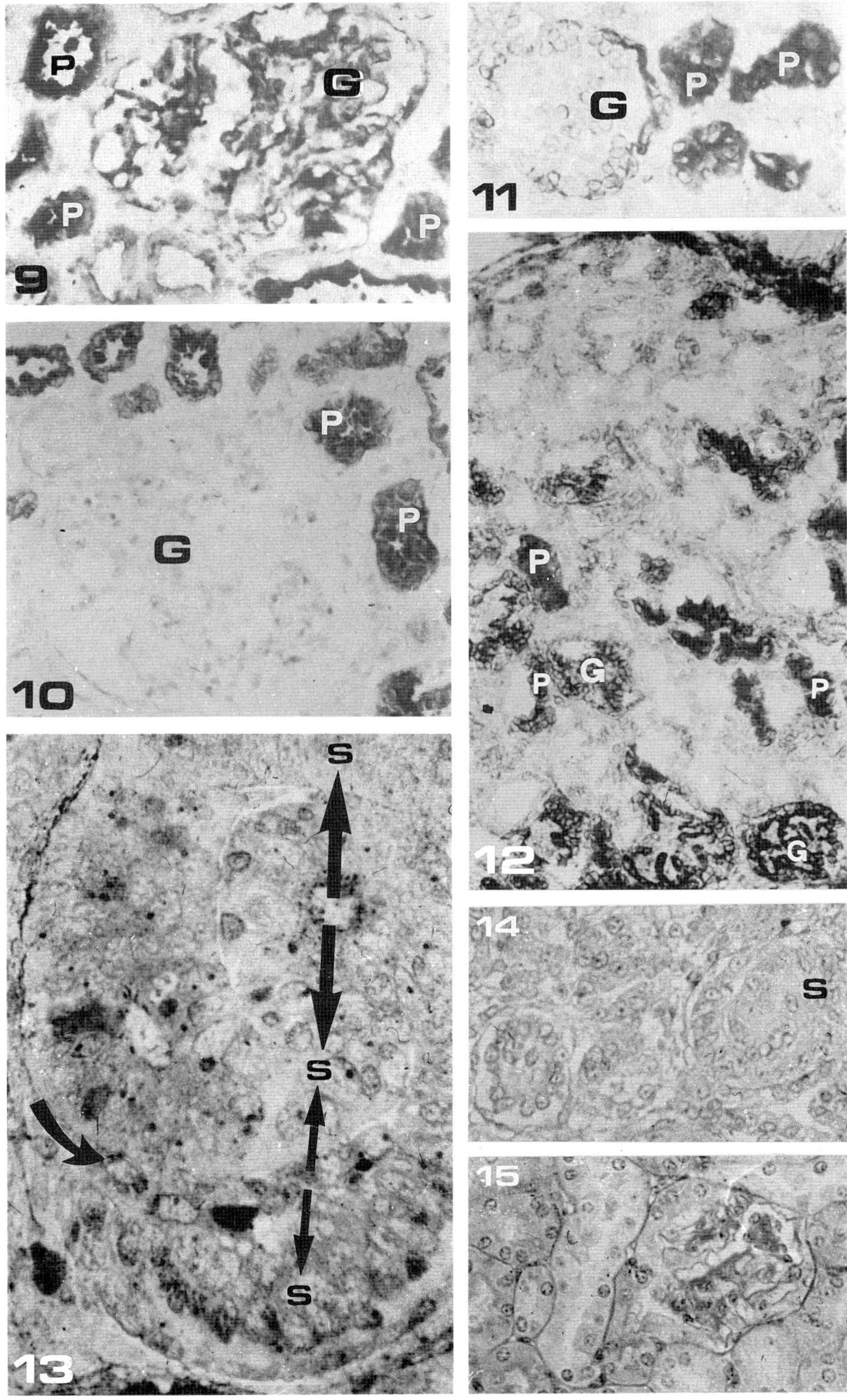
9
P
G
P
P
P
10
P
P
G
11
G
P
P
12
P
G
P
P
G
13
S
S
S
14
S
15

reach the complete differentiation. Moreover, the basal lamina of proximal and distal tubules and Bowman's capsule shows a positivity to PASM and PAS earlier than the basement membrane of glomerular capillaries. The discrepancy between the results obtained by the immunohistochemical analysis and those obtained by the histochemical reactions is likely to be due to a low amount of sugars in the immature glomerular basement membrane, which precludes the histochemical reactions and the visualization of this structure. On the other hand, the positivity to the specific antibodies suggests that the antigenic determinants are localized mainly in the proteic components of the glomerular basement membrane. An alternative hypothesis (De Martino et al., 1976) is that during the early stages of glomerular differentiation glycoproteins are present in amount sufficient to react with specific antibodies, whereas the glycidic moiety, normally revealed by PASM and PAS, may be sterically hindered and thus inaccessible to periodic acid oxidation (Figs. 13-18).

We thank Dr. C. Cordon Cardo, Memorial Sloan-Kettering Cancer Center, New York, for providing URO2,3,4, and Dr. M.I. Colnaghi, Istituto Nazionale per lo Studio e la Cura dei Tumori, Milano, for the gift of MBR1.

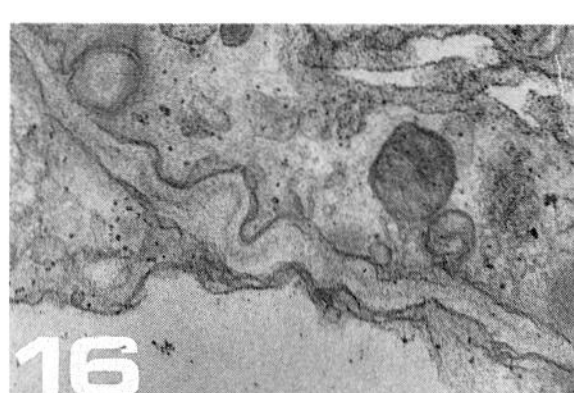

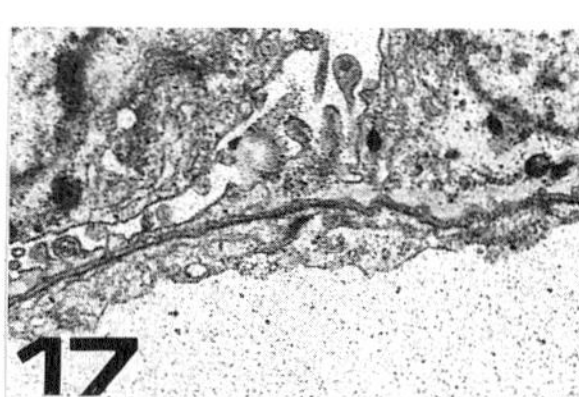

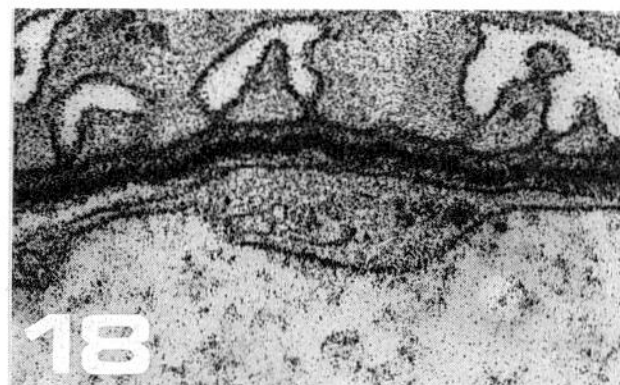

REFERENCES

Accinni L, Natali PG, Vassallo L, Hsu K, De Martino C (1975). Immunoelectron microscopic evidence of contractile proteins in the cellular and acellular components of mouse kidney glomeruli. Cell Tiss Res 162:297-312.

Canevari S, Fossati G, Balsari A, Sonnino S, Colnaghi MI (1983). Immunochemical analysis of the deteminant recognized by a monoclonal antibody (MBR1) which specifically binds to human mammary epithelial cells. Cancer Res 43: 1301-1305.

De Martino C, Accinni L, Nicotra MR, Natali PG (1976). Embryogenesis of mammalian glomerulus. Immunologic and histochemical studies. J. Submicr Cytol 8:249-250.

De Martino C, Accinni L (1985). Embryogenesis of the kidney. In DiDio LJA, Motta PM (eds): "Basic, Clinical and Surgical Nephrology", Boston: Martinus Nijhoff Publishing, pp 25-51.

Imai K, Natali PG, Kay NF, Wilson BS, Ferrone S (1982). Tissue distribution and molecular profile of a differentiation antigen detected by monoclonal antibody (345-134S) produced against human melanoma cells. Cancer Immunol Immunother 12:159-166.

Metzgar RS, Borowitz MJ, Jones NH, Dowell BL (1981). Distribution of common acute lymphoblastic leukemia antigen in nonhematopoietic tissues. J. Exp Med 154:1249-1254.

Natali PG, Giacomini P, Bigotti G, Nicotra MR, Bellocci M, De Martino C (1984). Heterogeneous distribution of actin, myosin, fibronectin and basement membrane antigens in primary and metastatic human breast cancer. Virchows Arch (Pathol Anat) 405: 69-83.

Ueda R, Ogata SI, Morrissey DM, Finstad CL, Szkudlarek J, Whitmore WF, Oettgen HF, Lloyd KO, Old LJ (1981). Cell surface antigens of human renal cancer defined by mouse monoclonal antibodies: identification of tissue-specific kidney glycoproteins. Proc Natl Acad Sci USA 78:5122-5126.

Zamboni L, De Martino C (1968). Embryogenesis of the human glomerulus. I. A histological study. Arch Path 86:279-291.

LEGENDS

Fig. 1 11 week old human embryo. Light microscopy. Below the renal capsule (arrows) collecting tubules (C) progressively organize the cells of metanephrogenic blastema in caps (1) and nests (2) which give rise to metanephric vesicles (3). The latter differentiating in S- and double spiral bodies form the primitive nephrons composed by an immature glomerulus (G), a short proximal (P) and distal (D) tubule with its macula densa (M) in close relationship with the primitive vascular pole (asterisks).

Fig. 2 11 week old embryo. Electron microscopy. Lower indentation of S-body. CC: capsular cells, P: primitive podocytes, M: macula densa. The arrow points to the space where the primitive glomerular blood vessel penetrates. C: collecting tubules.

Figs. 3-8 11 week old embryo. Immunofluorescence. Glomerular basement membrane (GBM) antigens are localized in a S (S) and double spiral (D) body (Fig. 3) and in a mature glomerulus (Fig. 4). Figs. 5 and 6 illustrate the localization of actin in a tangentially sectioned S-body (S) and in a mature glomerulus, respectively. Fig. 7: myosin is visible in the cells of ampulla (A) and metanephric cap (C); Fig. 8: mesangial localization of myosin in a mature glomerulus.

Figs. 9-12 Immunoperoxidase. Localization of URO2 in adult kidney (Fig.9) and of URO3, URO4, J5 in 22 week old embryo (Figs. 10, 11, 12, respectively). G: glomerulus; P: proximal tubule.

Figs. 13-18 The pictures show PASM and PAS negative reactions in immature nephrons (Figs. 13 and 14) and PAS positive reaction in mature nephrons (Fig. 15). The electron microscopy (Figs. 16-18) confirms the previous results. Fig. 16: PASM-negative immature GBM; Fig. 17: PASM-slightly positive maturating GBM; Fig. 18: PASM-positive mature GBM. In Fig. 13 curved arrow indicates the vascular low indentation of S-body (S) whose extension is pointed by straight arrows. Figs. 13-18: cortical and medullary nephrons of 11 week old embryo. Fig. 13: light microscopy on plastic section;Figs.14-15: paraffin sections.

Cells and Tissues: A Three-Dimensional Approach by Modern Techniques in Microscopy, pages 181–188

THREE-DIMENSIONAL VASCULAR ARCHITECTURE OF THE MALPIGHI'S GLOMERULAR CAPILLARY BEDS AS STUDIED BY VASCULAR CORROSION CASTING-SEM METHOD

Akio Kikuta, Takuro Murakami

Department of Anatomy, Okayama University Medical School, Okayama 700, Japan

INTRODUCTION

Mammalian glomerular vascular arrangement had been intensely investigated by light microscopy since the Malpighi's first description of the renal corpuscle (Malpighi's corpuscle). However, because of the conglomerated vascular structure of the glomerulus and the limitation in resolution and depth of focus, several contradictory glomerular vascular models were presented.

A 'capillary loop' model first proposed by Bowman (1842) was supported by other investigators (Vimtrup, 1928; Wilmer, 1941) and was the most popular glomerular vascular model. In this model, glomerulus consisted of several tortuous capillary loops which directly connected the afferent and efferent arterioles and only exceptionally anastomosed with each other. This model is still often referred in medical illustrations of the glomerulus. Johnston (1899), based on the reconstruction from the serial sections of glomeruli, proposed a 'capillary plexus' model in which glomerular capillaries divided and anastomosed to form a capillary plexus. After long ignorance, Hall (1955), Boyer (1956) and Lewis (1958) confirmed the presence of numerous anastomoses among the glomerular capillaries. Hall (1955) microdissected glomerular vascular casts and observed them by light microscopy and described independent lobulation of glomerular capillary plexus. Trabucco and Marquez (1952) proposed another 'blind sacculation' model.

This contradiction in views on glomerular vascular arrangement was solved by a new method for investigating the vascular arrangement of organs, vascular corrosion casting - scanning electron microscopy (SEM) method introduced by one of the present authors, Murakami (1971). The present paper presents several scanning electron micrographs of rat glomerular vascular casts and reviews mammalian glomerular vascular architecture as observed by this method.

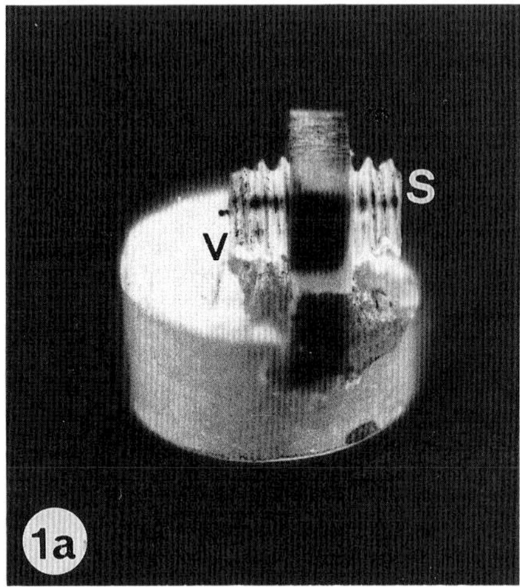

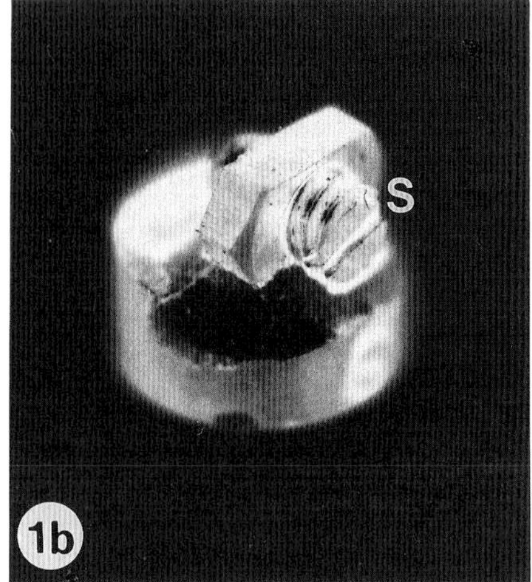

Figure 1. A new specimen holder for rotating a specimen about its horizontal axis. A specimen is glued to the vertical plane (V) of the holder screw (S), which can be rotated freely. x 4. (a) Frontal view. (b) Slight lateral, overhead view.

VASCULAR CORROSION CASTING AND SEM OBSERVATIONS

Vascular corrosion casts were prepared by a modified method of the original Murakami's one (Murakami, 1971). Fresh kidneys of the rats (Wistar strain) were thoroughly perfused with Ringer's solution to remove blood, and then perfused and fixed with a fixative (2.5% glutaraldehyde in 0.1 M phosphate buffer). Semi-polymerized casting medium Mercox (CL-2R, Japan Vilane Co., Tokyo) was injected entirely or partially. The resin-injected organs were immersed for 2-3 hr in a warm water bath about 60°C, then were corroded overnight in a warm 10-20% NaOH solution, and washed in running tap water. The blood vascular corrosion casts thus prepared were frozen in water, cut into several slices, and air-dried. Glomerular casts were separately microdissected out from the renal cortical three layers, subcapsular, middle, and juxtamedullary layers. Some of the glomerular casts were heated up to about 200°C and the softened casts were pressed or stretched with a forceps on a heated metal block. Then the casts were quickly cooled by a gentle blowing in order to harden in the extended form. Each glomerular cast was mounted on a special specimen holder on which specimen could be rotated freely around its horizontal axis (Fig. 1). The glomerular cast was fixed by gluing the efferent vessels to the vertical edge plane of the specimen holder screw in such a position as the afferent arteriole projecting at right angle against the vertical plane (Fig. 1). The glomerular casts were coated with gold from both sides, vascular and urinary pole sides and were observed in a scanning electron microscope (JSM-U3, JEOL, or HHS-2R, Hitachi) using an accelerating voltage of 5 kV, rotating the specimens around a line passing through their afferent-efferent arterioles.

VASCULAR ARCHITECTURE OF THE GLOMERULUS

Outer surface viewing of a same glomerular cast from various directions could be easily performed by using the present new specimen holder (Fig. 2). Each vessel running on the outer surface of the glomerulus could be traced. On the other hand, the renal glomerulus is a conglomerated structure of capillaries. Thus, it is essential to observe its inside view in

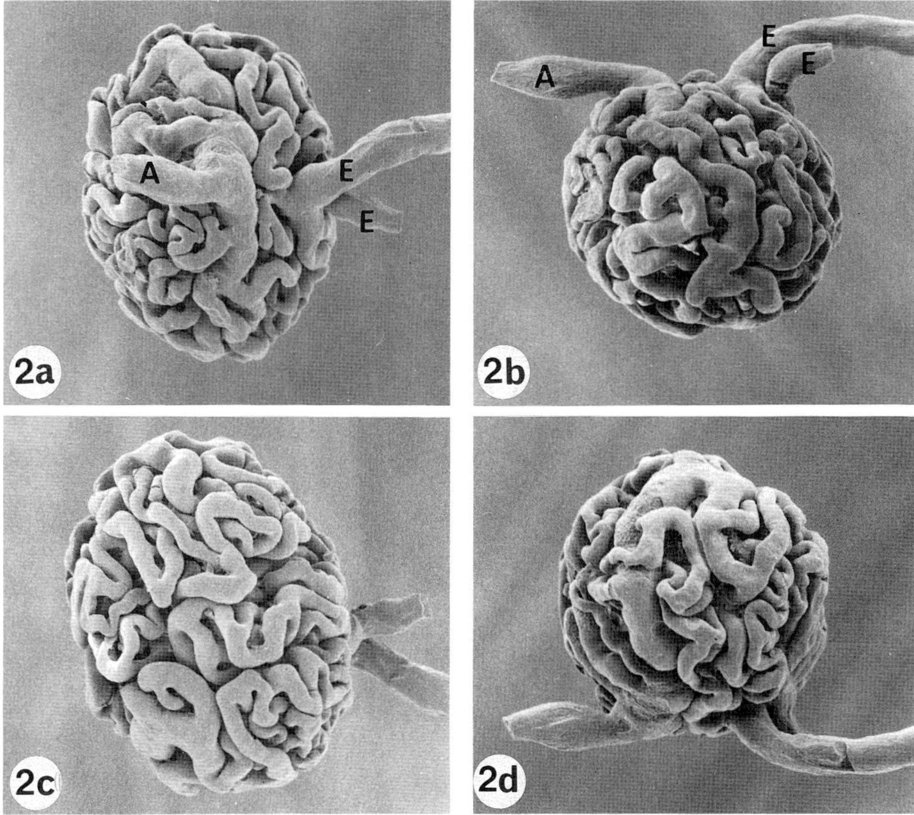

Figure 2. SEM views of a rat glomerular cast of the middle cortical layer with double efferent arterioles (E). x 200. (a) Vascular pole-side view. (b) Lateral side view. (c) Urinary pole-side view. (d) Opposite lateral side view of (b). A, afferent arteriole.

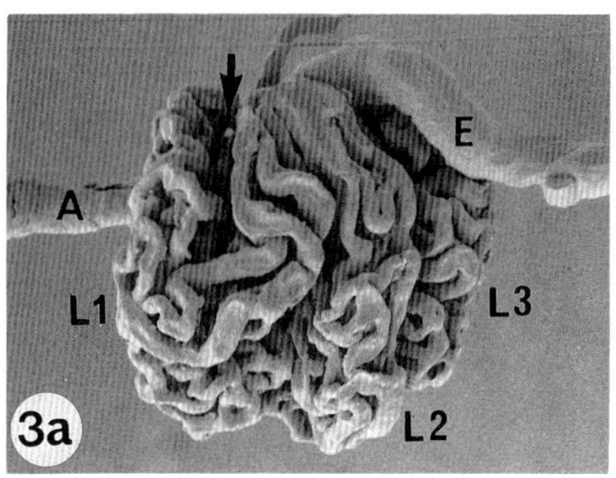

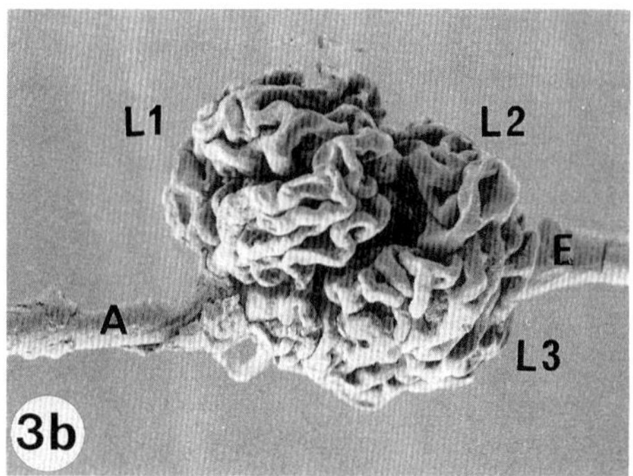

Figure 3. SEM views of a rat glomerular cast of the juxtamedullary layer. This cast is extended by gentle pressing showing lobulation of the glomerular capillary tuft. x 200. An arrow indicates a cleft between the afferent (A) and efferent (E) arterioles. L1, L2, L3, glomerular lobules (a) View from the lateral side. (b) View from the urinary pole.

order to study the glomerular vascular arrangement. Murakami (1972) treated vascular casts of glomeruli prepared of methyl methacrylate resin in a warm ethanol bath to soften the resin and pressed or pulled the casts to extend the tangled capillaries. On the other hand, the Mercox resin is one of the most popular injection medium and easier to handle than other injection media. However, this Mercox resin can not be softened by immersion in any hot organic solvents. In the present paper, we heated Mercox-casts up to a considerablely high temperature, about 200°C, and then pressed or stretched the softened capillary tufts. The extended casts were then quickly cooled by blowing, since the softened Mercox resin is prone to contract to its original form if the resin is cooled gradually. Thus, we could extend Mercox-prepared casts of glomerulus, although not so extensive (Figs. 3, 4) as methyl methacrylate-casts (Murakami et al., 1971; Murakami 1972). Figure 3 is the cast slightly pressed to show the glomerular capillary plexus lobulation and Figure 4 is a extended form of the glomerular cast the afferent and efferent arterioles of which were pulled in opposite directions.

A glomerular afferent arteriole arising from an interlobular artery enters a renal corpuscle through its vascular pole, where it divides into several thick segments. These segments further break up into capillaries to form a capillary plexus. These capillaries again diverge into an arteriole, efferent arteriole, at the vascular pole (Figs. 2-4).

Microdissection analysis combined with SEM observation

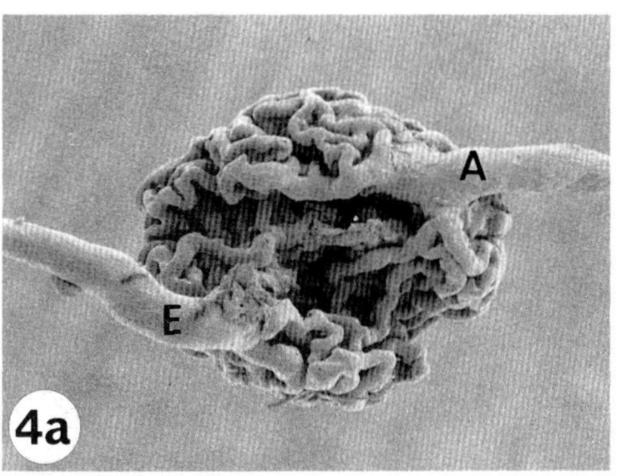

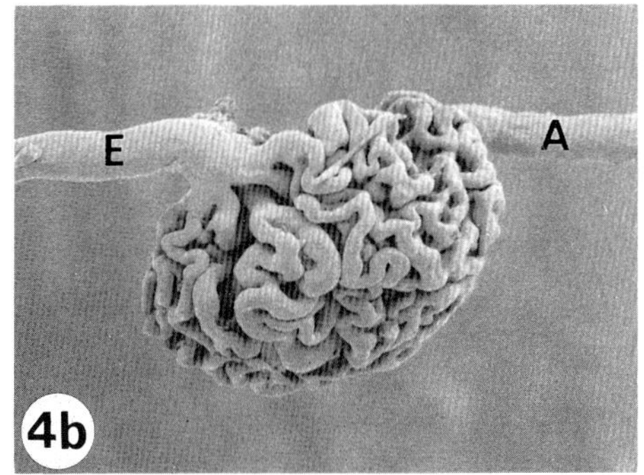

Figure 4. SEM views of a rat glomerular cast of the juxtamedullary layer. This cast was heated and stretched out by pulling the afferent (A) and efferent (E) arterioles in opposite directions. A deep cleft is exposed. Note few direct communication between the crura of the afferent and efferent arterioles and that the glomerular capillary plexus is folded in loop-like structure. x 200. (a) View from vascular pole. (b) View from lateral side.

unequivocally revealed the lobulation of the glomerular capillary plexus (Murakami, 1972), and confirmed the Hall's description by light microscopy (Hall, 1955), although this view was considered an artifact resulting from uncontrolled injection. When the softened glomerular casts were gently pressed, capillary tufts were easily separated into several blocks, indicating the presence of the lobulation within the glomerular capillary plexus (Fig. 3) (Murakami, 1972). Similar glomerular capillary plexus lobulation has been described by the corrosion casting-SEM method in dog (Spinelli et al., 1972; Spinelli, 1974; Anderson and Anderson, 1976), rhesus monkey (Unehira, 1981), and man (Murakami, 1978; Murakami et al., 1985). Recent reconstruction examination from resin-embedded serial sections of the rat glomerular capillary tufts also confirmed glomerular lobulation (Yang and Morrison, 1980).

Each glomerular lobule begins as a single short afferent lobular branch and terminates as a single short efferent rootlet (Hall. 1955; Murakami, 1972, 1978). A clear cleft has been described between the both crura of the afferent and efferent arterioles (Murakami, 1972). When the afferent and efferent arterioles of a softened glomerular casts are pulled in the opposite directions, the cast is stretched and this cleft is opened (Fig. 4) (Murakami, 1972; Unehira, 1981; Murakami et al., 1985). This shows that each lobule extends to the urinary pole of the corpuscle where it turns and then the ascending limb returns towards the vascular pole. Thus, the lobules can be divided into three segments, proximal descending limb, the middle turning segment, and distal ascending limb. Furthermore there are few directly connecting vessels between the proximal descending and distal ascending limbs (Fig. 3). These facts shows that each glomerular capillary plexus lobule, as a whole, is folded in a loop-like form within a renal corpuscle.

Hall (1955) described no interlobular connecting vessels and considered that the glomerular lobules were completely independent structures. SEM observations of the glomerular corrosion casts revealed that a few vessels usually bridge adjacent lobules in various mammals including man (Murakami, 1972, 1978; Spinelli et al, 1972; Spinelli, 1974; Anderson and Anderson, 1976; Unehira, 1981; Murakami et al., 1985). The lateral or large lobules sometimes showed sublobulations with many interlobular anastomoses (Murakami, 1972).

In the subcapsular and middle cortical layers, afferent arterioles are usually thicker than efferent arterioles, while vice versa in the juxtamedullary layer. The efferent arterioles of the glomeruli of the juxtamedullary layer course down deep into the medulla as vasa recta. Glomeruli of the juxtamedullary layer are larger and more complex in capillary plexus and lobulation configuration (Murakami, 1972; Murakami et al., 1985) However, vascular architecture of the glomerulus is fundamentally similar in the three layers.

Glomeruli with more than one efferent arterioles have been reported by light microscopy (Boenig, 1936; Shonyo and Mann, 1944; Okano et al., 1959; Moffat and Fourman, 1963) and by vascular corrosion casting-scanning electron microscopy (Murakami et al., 1971, 1972). Murakami (1976) found a rat glomerulus having two afferent arterioles. Casellas and Mimran (1979, 1981) considered that one of the double efferent arterioles functions as by-pass of the glomerular tuft. Microdissection analysis of the configuration and course of the the double efferent arterioles suggested that the circulatory course is essentially similar to that of normal efferent arterioles (Murakami et al., 1971).

It has been suggested that a few large vessels serve as preferential pathways for blood from afferent to efferent arterioles (Hall, 1955). Such direct connections have been reported in rat (Ljungqvist, 1975; Casellas and Mimran, 1979) and man (Ljungqvist, 1964). Most of the recent corrosion casting-SEM studies, however, have not confirmed any direct channels nor main routes in mammalian glomerulus so far examined (Murakami et al., 1971, 1985; Murakami, 1972).

Various models proposed on light microscopic observations thus reflected one aspect of the glomerular vascular architecture. The 'capillary loop' model revealed the loop-like structure of capillary plexus lobules of the glomeruli. The 'capillary plexus' model exaggerated the presence of the numerous anastomoses among the glomerular capillaries, but failed to account for the loop-like, lobular structure of the glomerular capillary plexus. Vascular corrosion casting-SEM method combined with microdissection clearly elucidated the complex three-dimensional vascular arrangement of the glomerulus, i.e., mammalian glomerulus is considered a capillary plexus consisting of several lobules which are fairly independent and folded in loop-like structure.

REFERENCES

Anderson BG, Anderson WD (1976). Renal vasculature of the trout demonstrated by scanning electron microscopy, compared with canine glomerular vessels. Am J Anat 145:443-458.

Boenig H (1936). Beitrage zur Kenntnis der Vasa efferentia in der menschlichen Niere. Z Mikrosk Anat Forsch 39:105-115.

Bowman W (1842). On the structure and use of the Malpighian bodies of the kidney, with observations on the circulation through that gland. Phil Trans Roy Soc London 1:57-80.

Boyer CC (1956). The vascular pattern of the renal glomerulus as revealed by plastic reconstruction from serial sections. Anat Rec 125:433-441.

Casellas D, Mimran A (1979). Aglomerular pathways in intrarenal microvasculature of aged rats. Am J Anat 156:293-298.

Casellas D, Mimran A (1981). Shunting in renal microvasculature of the rat: A scanning electron microscopic study of corrosion casts. Anat Rec 201: 237-248.

Hall BV (1955). Further studies of the normal structure of the renal glomerulus. In: "Proceedings of 6th Ann. Conf. Nephrotic Syndrome" New York: Nat. Nephrosis Found., pp 1-39.

Johnston WB (1899). A reconstruction of a glomerulus of the human kidney. Anat Anz 16:260-266.

Lewis OJ (1958). The vascular arrangement of the mammalian renal glomerulus as revealed by a study of its development. J Anat 92:433-440.

Ljungqvist A (1964). Structure of the arteriole-glomerular units in different zones of the kidney. Nephron 1: 329-337.

Ljungqvist A (1975). Ultrastructural demonstration of a connection between afferent and efferent juxtamedullary glomerular arterioles. Kidney International 8:239-244.

Moffat DB, Fourman J (1963). The vascular pattern of the rat kidney. J Anat 97:543-553.

Murakami T (1971). Application of the scanning electron microscope to the study of fine distribution of the blood vessels. Arch Histol Jpn 32:445-454.

Murakami T (1972). Vascular arrangement of the rat renal glomerulus. A scanning electron microscope study of corrosion casts. Arch Histol Jpn 34: 87-107.

Murakami T (1976). Double afferent arterioles of the rat

renal glomerulus as studied by the injection replica scanning electron microscope method. Arch Histol Jpn 39:327-332.
Murakami T (1978). Methyl methacrylate injection replica method. In Hayat AM (ed): "Principles and techniques of scanning electron microscopy." Vol. 6. New York: Van Nostrand Reinhold, pp 159-169.
Murakami T, Kikuta A, Ohtsuka A, Kaneshige T (1985). Renal vasculature as observed by SEM of vascular casts. Basic architecture, development and aging of glomerular capillary beds. In DiDio LJA, Motta PM (eds): "Basic Clinical, and Surgical Nephrology." Boston: Martinus Nijhoff Publishers, pp 83-98.
Murakami T, Miyoshi M, Fujita T (1971). Glomerular vessels of the rat kidney with special reference to double efferent arterioles. A scanning electron microscope study of corrosion casts. Arch Histol Jpn 33:179-198.
Okano H, Ohta Y, Fujimoto T, Ohtani Y (1959). Cubical anatomy of several ducts and vessels by injection method of acrylic resin. VI. On the vascular system of the kidney in Macacus cynomolgus. Okajimas Fol Anat Jpn 34:27-41.
Shonyo ES, Mann FC (1944). An experimental investigation of renal circulation. Arch Pathol 38: 287-296.
Spinelli F (1974). Structure and development of the renal glomerulus as revealed by scanning electron microscopy. In Bourne GH, Danielli JF (eds): "International Rev Cytol." vol. 39 New York: Academic Press, pp 345-381.
Spinelli F, Wirtz H, Brucher C, Pehling G (1972). Non-existence of shunts between afferent and efferent arterioles of juxtamedullary glomeruli in dog and rat kidneys. Nephron 9: 123-128.
Trabucco A, Marquez F (1952). Structure of the glomeruli tuft. J Urol 67:235-255.
Unehira M (1981). Blood vascular architecture of the rhesus monkey kidney glomerulus. A scanning electron microscope study of corrosion casts. Okayama Igakkai Zasshi 93:951-961 (in Japanese).
Vimtrup BJ (1928). On the number, shape, structure, and surface area of the glomeruli in the kidneys of man and mammals. Am J Anat 41:123-151.
Wilmer HA (1941). The arrangement of the capillary tuft of the human glomerulus. Anat Rec 80:507-518.
Yang GCH, Morrison AB (1980). Three large dissectable rat glomerular models reconstructed from wide-field electron micrographs. Anat Rec 196:431-440.

Cells and Tissues: A Three-Dimensional Approach by Modern Techniques in Microscopy, pages 189–202

THREE DIMENSIONAL ORGANIZATION OF THE COLLECTING TUBULE OF THE RABBIT KIDNEY

Andrew P. Evan, Bret A. Connors and James A. McAteer

Department of Anatomy, Indiana University, School of Medicine, Indianapolis, Indiana 46223

GENERAL

The collecting duct system of the mammalian kidney consists of two divisions, the connecting tubule and the collecting ducts (Morel et al., 1976; Imai, 1979; Kaissling and Kriz, 1979; Garg et al., 1981). The collecting ducts run from the kidney capsule to the tip of the renal papilla and can be divided further, based on location within the cortex and medulla into several subdivisions. These subdivisions are termed the cortical (outer and inner), outer medullary (outer stripe and inner stripe), and inner medullary collecting ducts. Each segment of the collecting duct system, from the connecting tubule to the inner medullary collecting duct, exhibits a distinctive cellular organization characterized by identifiable cell types. So it is that each segment is morphologically identifiable and exhibits specialized function.

CONNECTING TUBULE

The manner by which the distal tubule attaches to the collecting duct system varies according to the nephron type and species. In all species so far examined, each superficial nephron is linked to the collecting duct by a specialized connecting tubule (first segment of the collecting duct system) (Fig. 1). In man, both juxtamedullary and deep midcortical nephrons are directly joined to a collecting duct by way of a single connecting tubule (Oliver, 1968; Sperber, 1944). In the rabbit, however, the connecting tubules from several different juxtamedullary or deep midcortical nephrons join together to form an arcade. These arcades first ascend to the outermost layer of the renal cortex and then course toward a medullary ray.

In the rabbit, the boundaries of the connecting tubule are sharp and well defined (Kaissling and Kriz, 1979). This is not the case for all species (Jamison and Kriz, 1982) and has led to uncertainity regarding the embryonic origin of this tubule segment. It is unclear if connecting tubules are derived from the metanephric blastema or the ureteric bud.

Connecting tubules are found in the cortical labyrinth. Those associated with superficial nephrons may reach the kidney surface. The arcades formed by the union of connecting tubules of deep midcortical and juxtamedullary nephrons parallel the course of interlobular arteries. The tubules of such arcades and their associated arteries are usually separated from each other by only a thin margin of interstitial cells and connective tissue elements.

Two morphologically distinct cell types, the connecting tubule cell and the intercalated cell, line the connecting tubules of superficial nephrons and the arcades (Figs. 2, 3, 4). Both cell types are distinctly different from the distal convoluted tubule cell. The intercalated cell also lines the cortical and medullary segment of the collecting duct. In the connecting tubule the ratio of connecting tubule cells to intercalated cells is about 5:4. Figure 2 shows that connecting tubule cells are normally contiguous with one another while intercalated cells are always separated by adjacent connecting tubule cells.

The connecting tubule cell is characterized by a smooth apical surface (except for a central cilium and few stubby microvilli) and two types of grooves at the basal surface (Figs. 2,3,6). One type of groove appears as two parallel edges with interdigitating microvilli (Fig. 6). Such grooves form large polygonal patterns and represent the basal openings of the lateral intercellular channels. These parallel edges do not form the large interdigitating lateral processes that are typical of cells of the proximal tubule (Evan et al., 1978). The second type of basal groove forms irregular patterns of small, tortuous channels that do not

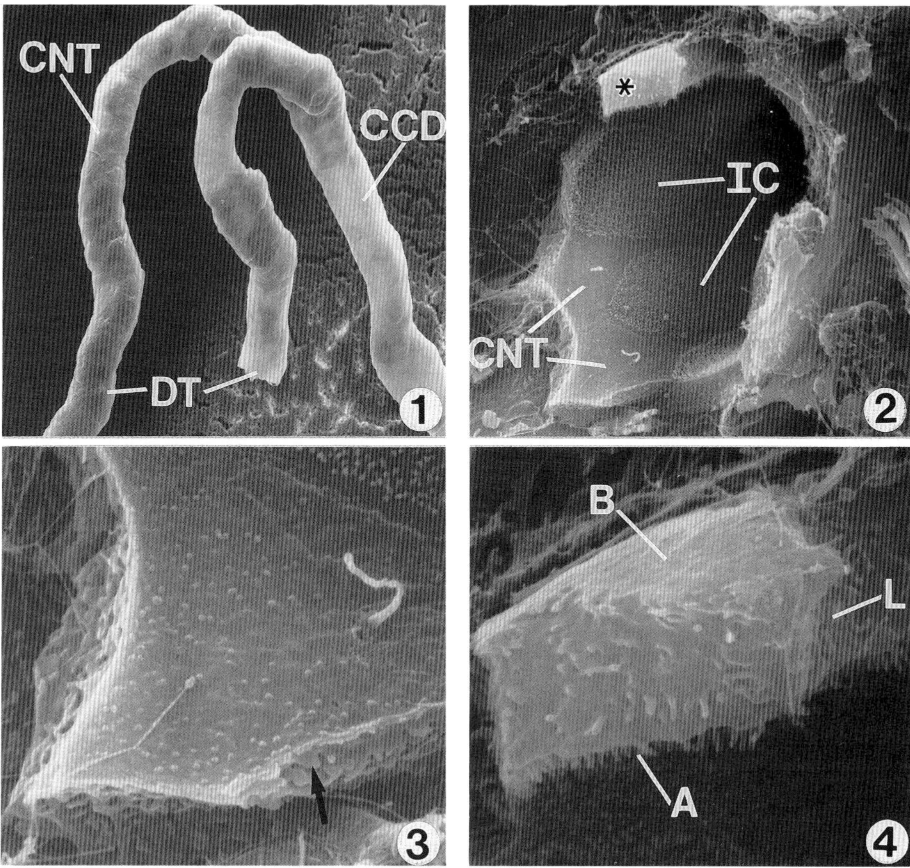

Figure 1. Scanning electron micrograph of an isolated and enzymatically treated distal nephron from the outer cortex of an adult rabbit showing the location of two separate connecting tubules (CNT). Each is positioned between a distal tubule (DT) and a cortical collecting tubule (CCD). X400

Figure 2. This scanning electron micrograph shows a connecting tubule that has been fractured, thereby revealing both the apical and lateral surfaces of connecting tubule cells (CNT) and intercalated cells (IC). One intercalated cell marked by an asterisk is also shown in figure 4. X2,000

Figure 3. Higher magnification micrograph showing the smooth apical surface of a connecting tubule cell. The lateral cell surface has numerous short, microvilli (arrow). X7,500

Figure 4. Higher magnification micrograph showing lateral (L), apical (A) and basal (B) surfaces of an intercalated cell from figure 2. The apical surface is characterized by numerous microvilli while the basal surface of this cell is smooth. X7,500

connect with the lateral intercellular spaces (Figs. 6,9,10). The basal membrane of these grooves forms numerous, true infoldings (Figs. 9,10) (Kaissling and Kriz, 1979; LeFurgey and Tisher, 1979; Welling et al., 1983). Furthermore, these infoldings extend nearly to the apical cell surface forming the "membranous labyrinth" (Fig. 10). The walls of these infoldings are smooth and parallel. As the basal membrane reaches the apical surface, it arches back to return to the cell base some distance from its origin. The lateral cell surface forms channels that extend radially from the cell apex to cell base. Although the lateral intercellular channels are parallel, they are somewhat irregular in form due to the presence of numerous laterally projecting ridges and short microvilli (Figs. 3,9,10). It is of interest that the amount of membrane associated with the basal infoldings increases when an animal is placed on a low-sodium/high-potassium diet or given mineralocorticoids together with a low-potassium diet (Kaissling and Le Hir, 1982).

Connecting tubule cells have a distinctive cytoplasmic ultrastructure. The nucleus is found in the center of the cell. Mitochondria are small and are positioned between the nucleus and the apical cell surface. These cells possess few other organelles and appear electron lucent. Thus they have also been termed "light" cells.

The intercalated cell is characterized by an elaborate apical surface of microplicae/microvilli and basal grooves primarily associated with lateral intercellular channels (Figs. 2,4,6,11,12) (LeFurgey and Tisher, 1979; Welling et al., 1983). A few simple basal channels are connected to basal infoldings which extend for only a short distance through the cell height. The lateral intercellular channels of this cell type are identical to those described for the connecting tubule cell. The intercalated cells are also called "dark" cells, due to their electron dense cytoplasm, numerous polysomes and a greater number of mitochondria compared to the connecting tubule cell. It is important to point out the Kaissling and Kriz (1979) have identified gradations in the staining properties of intercalated cells of the connecting tubule. They describe "black", "gray" and intermediate cell forms distinguished by the presence, in "gray" cells, of flattened or invaginated cytoplasmic vesicles.

In order to describe in a quantitative way the cells of the connecting tubule, Welling et al., (1981; 1983) combined two techniques, the visualization of cells by scanning electron microscopy and the precise measurement of cell dimensions by means of transmission electron microscopy and morphometric analysis. Based on the morphometric values collected, a model of the tubule was generated. Each cell was viewed as a truncated, equilateral hexagonal pyramid with a calculated luminal circumference of 49.2 microns (Fig. 7). The calculated cell volume for the two cell types was similar (connecting tubule cell = 1016 ± 6 um^3; intercalated cell = 1017 ± 38 um^3). Scanning and transmission electron microscopic observations suggest that the basal surface of the intercalated and connecting tubule cells are approximately equal in area while the apical surface of the intercalated cell often is considerably smaller than that of the connecting tubule cell. If one assigns this tubule the dimensions of 35 microns for the outer diameter and 25 microns for the inner diameter, eight cells are calculated to form the tubular circumference (Fig. 8). At these dimensions, approximately 450 cells occupy one millimeter of tubule length (Garg et al., 1981).

Welling et al., (1983) determined the relative areas of apical surface, basal surface and basement membrane for the connecting tubule cell and intercalated cell. First, the two cells had similar membrane areas (Figs. 9-12). For the intercalated cell, the mean area differences among the three surfaces were not statistically different from zero (Figs. 11,12). However, the connecting tubule cells possess a significantly smaller apical surface vs basal or basement membrane surface (Figs. 9,10). When these area measurements are adjusted according to the proper ratio of cell types per millimeter length of connecting tubule, it is noted that the connecting tubule cell contributes a greater amount of apical (55%), and basal (65%) surface area than the intercalated cell.

The surface concentrations of the lateral and infolded cell membranes were also calculated. In the connecting tubule cell approximately 78% of the total (lateral + infolded) cell membrane area was of the infolded type. However, in the intercalated cell the infolded membrane area accounted for only 26% of the total. Furthermore, the ratio of infolded cell membrane area to cell volume was 2.0 ± 0.1 in the connecting tubule cells and 0.31 ± 0.05 um^2/um^3 in the intercalated cell. This means that approximately 93% of all infolded membrane in the connecting tubule is contributed by the connecting tubule cells.

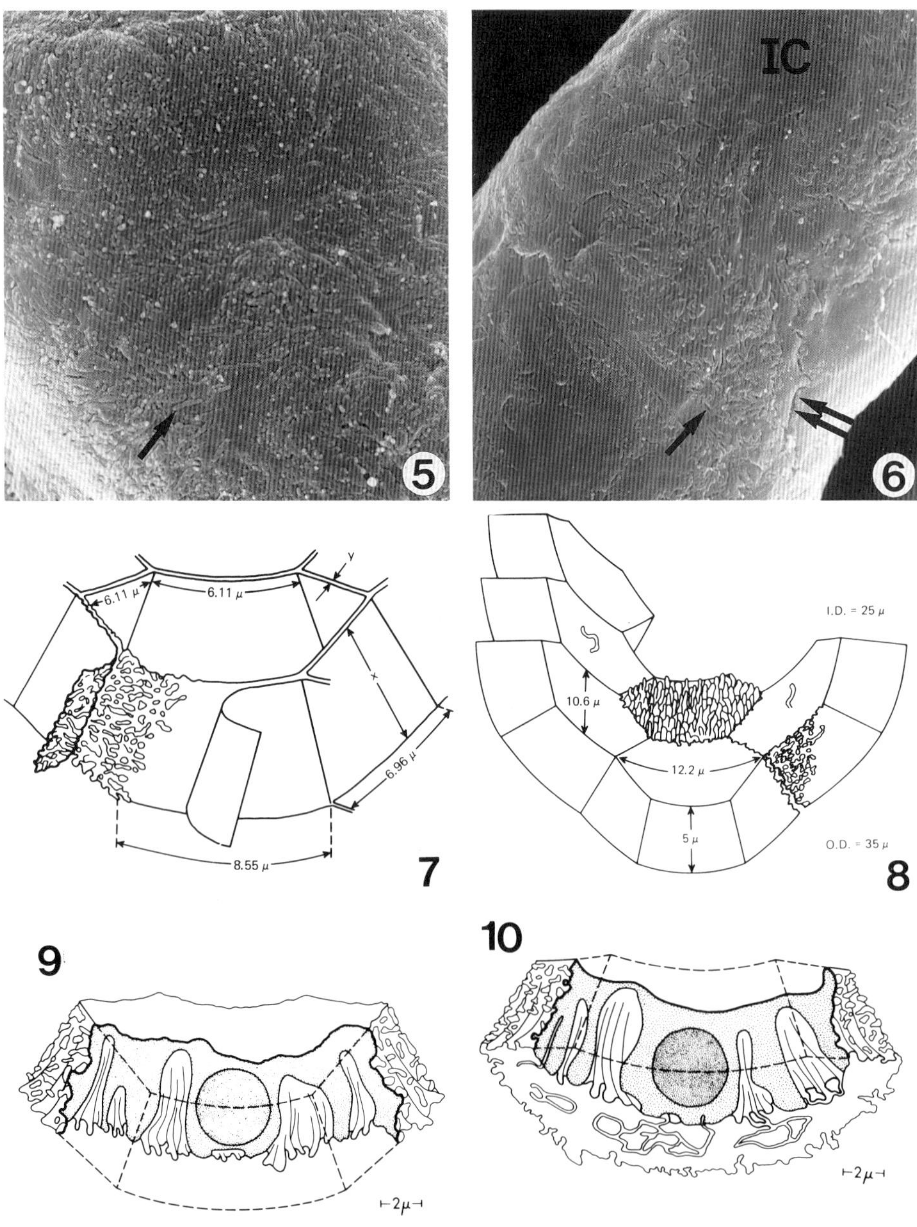

Figure 5. Scanning electron micrograph showing the basal surface of cells from a distal convoluted tubule. The basal surface of the tubule is homogeneous, indicative of the cellular homogeneity of the distal convoluted tubule. The cell surface is characterized by numerous basal microvilli and plates (arrow). X3,000

Figure 6. This scanning electron micrograph shows the basal surface of a connecting tubule. Principal cells possess grooves that are continuous with true basal infoldings (arrow) and basal slits (double arrows) which are associated with the lateral intercellular channels. Intercalated cells (IC) have a rather smooth basal surface and possess only one type of basal slit. X4,000

Figure 7. Drawing depicting the shape and dimensions of cell types (connecting tubule cell or intercalated cell) of the connecting tubule. Note that the lateral cell surface possesses numerous microvilli and ridges.

Figure 8. Model of a connecting tubule based on morphometric measurements. The two cell types of this tubule are positioned such that eight individual cells form the tubular circumference.

Figures 9 & 10. Both figures depict the shape and basal infolded channels of a typical connecting tubule cell. The cell possesses true basal infoldings that rise almost to the apical cell surface. These infoldings connect to the basal grooves seen in figure 6. (Figures 5,7,8,9,10 reprinted from Kidney International with permission).

COLLECTING DUCT

The collecting ducts begin in the outer cortex within medullary rays. In each medullary ray, collecting ducts occur in groups of four (Kaissling and Kriz, 1979). This is consistent and predictable pattern in the rabbit. As a collecting duct progresses from the outer to inner cortex, it is first joined by individual connecting tubules and then by a series of connecting tubule arcades. At the level of the outer medulla, collecting ducts no longer receive additional connecting tubules. As collecting ducts enter the inner medulla, they fuse to form a single collecting duct. This morphological feature is the result of diachotamus branching of the embryonic collecting duct. After about eight to ten such unions, large papillary ducts, termed ducts of Bellini, are formed. These eventually open into the renal pelvis.

As collecting ducts descend along their course through the kidney the tubular epithelium shows gradual changes in its cytological characteristics. The height of individual cells increases. At the level of the papillary duct cell height increases dramatically. Also, the outer diameter of the collecting duct gradually increases towards the inner medulla. Two distinct cells types, principal and intercalated, are easily identified in the cortex at a ratio of 2:1. This ratio approaches unity at the level of the outer medulla and few, if any, intercalated cells can be positively identified within inner medullary collecting ducts. There is considerable species variation in these ratios (Kaissling and Kriz, 1979; Myers et al., 1966).

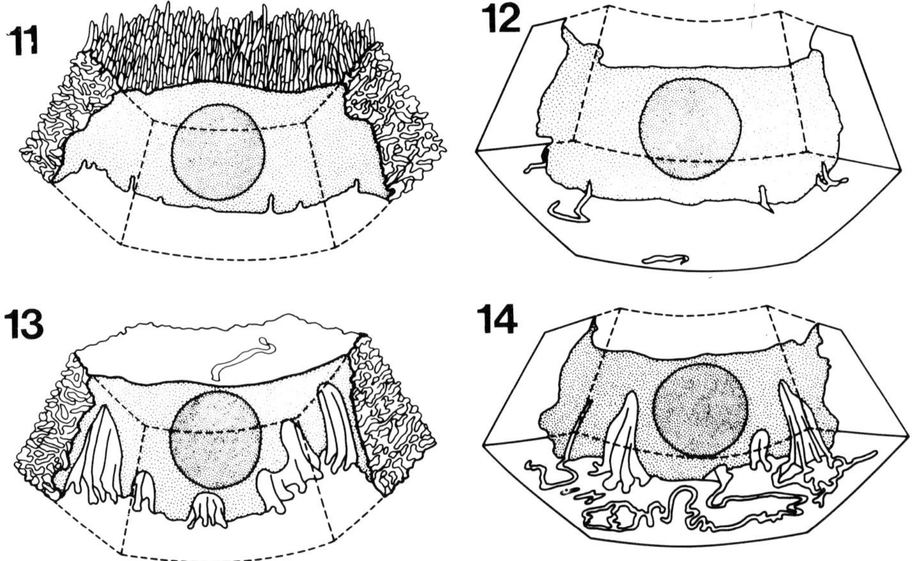

Figures 11 & 12. These drawings depict an intercalated cell typical of either the connecting tubule or collecting duct. Figure 11 shows the elaborate apical and lateral cell surfaces of the intercalated cell. Figure 12 shows the paucity of true infoldings this cell possesses.

Figures 13 & 14. These drawings depict a typical principal cell of the collecting duct. This cell has a smooth apical surface and numerous true basal infoldings positioned in the basal 40% of the cell height. (Figures 11,12,13, and 14 are reprinted from Kidney International with permission).

The principal cell of the collecting duct has many morphological characteristics in common with the connecting tubule cell. The principal cell has a simple apical cell surface featuring a cilium and a few blunt microvilli (Figs. 13,15,16,20,21,22). The basal surface presents the same two types of grooves described for the connecting tubule cell (Figs. 10,14,18). However, in the cortical collecting duct basal infoldings of the prinicpal cell are limited to the basal 40% of cell height and constitute 62% of all channel-associated membrane (Welling et al., 1981). The lateral cell surface of the principal cell is very similar to that described for the connecting tubule cell. Welling et al., (1981) determined that each cell possesses approximately 670 microvilli or about 110 microvilli per each of six lateral cell faces. This microvillar-type specialization (depicted in figure 7) is calculated to increase the lateral cell membrane by 1.8 fold. In the rabbit, this amplification of the basal infoldings gradually decreases toward the inner medulla. Furthermore, the amount of basolateral membrane has been found to be lower in animals with a very low endogenous aldosterone level (Kaissling and He Lir, 1982) and higher in rabbits with high mineralocorticoid levels (Kaissling and He Lir, 1982; Wade et al., 1979). In the rat, a high potassium intake results in an increase in the basolateral membrane area (Stanton et al., 1981).

Principal cells are also termed "light" cells, and may appear pale or electron lucent compared to most intercalated cells. This staining pattern is not consistent and the cytoplasm may actually appear more electron dense than adjacent intercalated cells. Principal cells possess a rather unremarkable complement of membranous organelles including cisternae of the rough endoplasmic reticulum and golgi apparatus, lysosomes and somewhat shortened mitochondria. The volume density of these organelles decreases as the collecting duct descends through the kidney except for the number of lysosomes which increase toward the medulla.

Intercalated cells of the collecting duct possess many morphological features in common with those of the connecting tubule. They are characterized by an elaborate apical cell surface and numerous microvilli projecting from the lateral cell surface and basal grooves (Figs. 15-22). The lateral surface of the intercalated cell is identical to that already described for the principal cell

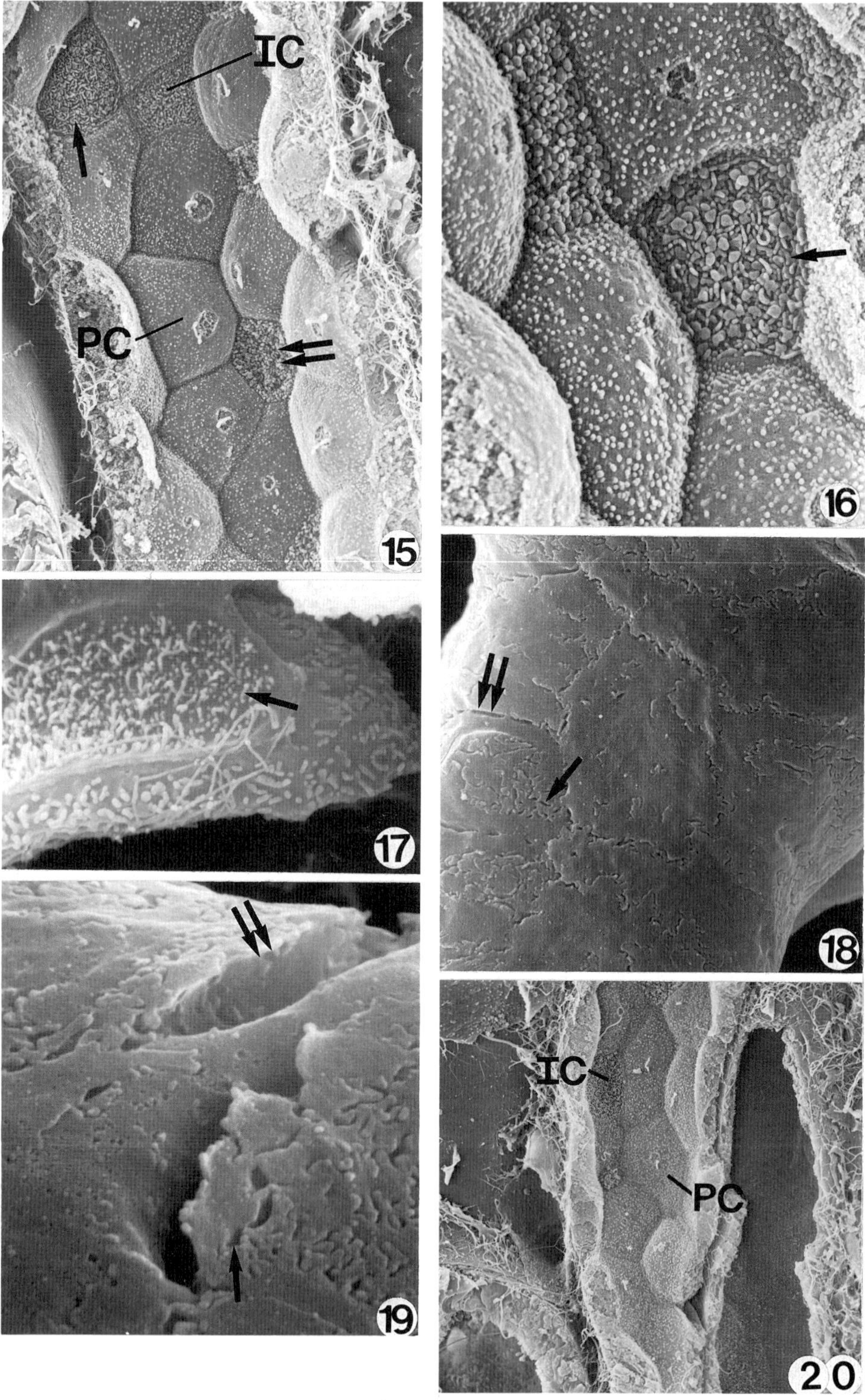
IC
PC
15
16
17
18
19
IC
PC
20

Figures 15 & 16. Scanning electron micrographs of outer cortical collecting ducts. Both principal (PC) and intercalated cells (IC) are noted. Two different intercalated cells can be identified, based on the characteristics of their apical surface. One cell possesses numerous apical microplicae (arrow) while the other presents many microvilli (double arrow). The principal cells (PC) possess a cilium and short, stubby microvilli. Fig. 15 X2,200; Fig. 16 X16,000
Figure 17. High magnification of an intercalated cell of the cortical collecting duct. This cell bears numerous apical (arrow) and lateral microvilli. X7,200
Figure 18. This scanning electron micrograph shows the basal surface of an isolated and enzymatically treated cortical collecting duct. Two types of grooves are seen at the basal cell surface. One groove is associated with the lateral cell boundary (double arrow) while another type (arrow) is connected to the basal infoldings. Only the principal cell possesses both types. X4,000
Figure 19. Higher magnification scanning electron micrograph showing the basal surface of two adjacent principal cells. The groove associated with the lateral intercellular channel is easily seen (double arrow). The grooves connected to the true basal infoldings show a pattern characterized by numerous microvilli (arrow). X10,000
Figure 20. Low magnification scanning electron micrograph showing the luminal view of an inner cortical collecting duct. Both principal (PC) and intercalated cells (IC) are noted. X1,500

(Figs. 11,13,17). When the tubular basement membrane is removed from the basal cell surface by collagenase treatment, two different groove-like infoldings or slits are noted (Figs. 12,18,19). One type clearly is associated with the lateral intercellular channels. The other form is within the lateral boundaries of each cell and connects with the basal infoldings. The intercalated cells of the rabbit possess few, if any, basal infoldings. When present, these infoldings are located in the basal 20% of the cell height. The precise functional correlate for these differences in the structure of the basal and lateral surface between cell types is yet to be determined. It is noteworthy, however, that the lateral cell surface of the intercalated cell does not respond to corticosteroid hormones, which is in contrast to that seen in principal cells (Wade et al., 1979).

Although the intercalated cell has many predictable features, there would appear to be a number of morphological and functional variations of this cell type. Based on staining characteristics, Kaissling and Kriz (1982) have identified a "black" and "grey" manifestation as well as several intermediate forms. Figures 15, 16, 17, 20, 21, and 22 demonstrate the different apical surface features of the intercalated cell. At the outer cortex, most of these cells possess numerous microvilli of varying lengths (Figs. 15,17). A second, less frequent pattern is characterized by numerous microplicae (Fig. 15). At the inner cortex, the ratio of these two apical appendages reverses such that apical microvilli now predominate (Fig. 20). All intercalated cells located along the medullary collecting duct of the outer stripe possess apical microplicae (Figs. 21,22). In the rat, it is interesting that potassium depletion induces large increases in the cell surface of intercalated cells of the outer medullary collecting duct (Stetson et al., 1980; Evan et al., 1980).

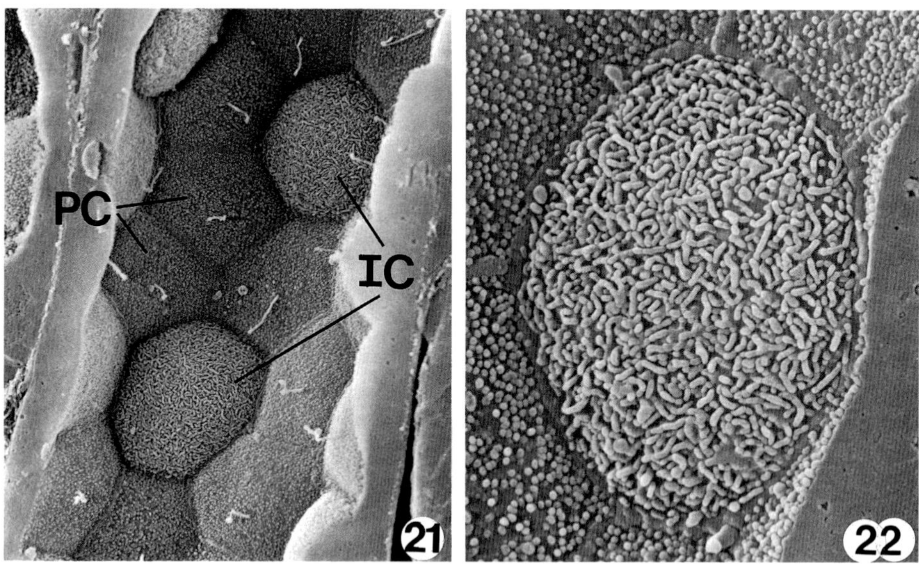

Figure 21. Luminal view of a collecting duct of the outer stripe of the outer medulla. Both principal (PC) and intercalated (IC) cells are seen. All the intercalated cells show microplicae at their apical surface. X2,000

Figure 22. Scanning electron micrograph of an intercalated cell of the outer stripe. This cell bears a dense array of microplicae at its apical surface. X6,000

Immunocytochemical studies have identified two types of intercalated cells. One cell type is peanut lectin positive and probably secretes HCO_3^-. The other cell type is band 3-positive and may reabsorb HCO_3^- (secrete H^+) (Schuster et al., 1986). The functional description of the intercalated cell may be even more complex, Schwartz et al., (1985) have presented evidence for functional plasticity of the intercalated cell. They observed that acid loading caused HCO_3^- secreting cells to alter their functional polarity to that of H^+ secreting cell.

In summary, the new data that has been published on the three dimensional organization of the connecting tubule and collecting ducts demonstrates the complexity of this cellular system. It will require the expertise of both the physiologist and morphologist to unravel the functional role of each cell.

ACKNOWLEDGEMENTS

We wish to thank Patty Summerlin for her expert technical assistance.

REFERENCES

Evan AP, Hay DA, Dail WG (1978). SEM of the proximal tubule of the adult rabbit kidney. Anat Rec 191:397-414.

Evan AP, Huser J, Bengele HH, Alexander EA (1980). The effect of alterations in dietary potassium on collecting system morphology in the rat. Lab Invest 42:668-675.

Garg LC, Knepper MA, Burg MB (1981). Mineralocortical effects on Na-K-ATPase in individual nephron segments. Am J Physiol 240:F536-F544.

Imai M (1979). The connecting tubule: A functional subdivision of the rabbit distal nephron segment. Kidney Int 15:346-356.

Jamison RL, Kriz W (1982). "Urinary concentrating mechanisms: Structure and function." New York: Oxford University Press, pp 189-219.

Kaissling B, Kriz W (1979). "Structural analysis of the rabbit kidney." New York: Springer-Verlag, pp 1-123.

Kaissling B, Le Hir M (1982). Distal tubular segments in the rabbit kidney after adaption to altered Na- and K- intake. I. Structural changes. Cell Tissue Res 224:469-492.

LeFurgey A, Tisher CC (1979). Morphology of rabbit collecting duct. Am J Anat 155:111-124.

Morel M, Charbardes D, Imbert M (1976). Functional segmentation of the rabbit distal tubule by microdetermination of hormone-dependent adenylate cyclase activity. Kidney Int 9:264-277.

Myers CE, Bulger RE, Tisher CC, Trump BF (1966). Human renal ultrastructure. IV. Collecting duct of healthy individuals. Lab Invest 15:1921-1950.

Oliver J (1968). "Nephrons and Kidneys." New York: Hoeber Medical Division-Harper and Row Publishers, pp 1-117.

Schuster VL, Bonsib SM, Jennings ML (1986). Two types of collecting duct mitochondria-rich (intercalated) cells: Lectin and band 3 cytochemistry. Am J Physiol 251:C347-C355.

Schwartz GJ, Barasch J, Al-Awqati Q (1985). Plasticity of functional epithelial polarity. Nature 318:368-371.

Sperber J (1944). Studies on the mammalian kidney. Zool Bidrag 22:249-431.

Stanton BA, Biemesderfer D, Wade JB, Giebisch G (1981). Structural and functional study of the rat distal nephron: Effects of potassium adaption and depletion. Kidney Int 19:36-48.

Stetson DL, Wade JB, Giebisch G (1980). Morphologic alterations in the rat medullary collecting duct following potassium depletion. Kidney Int 17:45-56.

Wade JB, O'Neil RG, Pryor JL, Boulpaep EL (1979). Modulation of cell membrane area in renal collecting tubules by corticosteroid hormones. J Cell Biol 81:439-445.

Welling LW, Evan AP, Welling DJ (1981). Shape of cells and extracellular channels in rabbit cortical collecting ducts. Kidney Int 20:211-222.

Welling LW, Evan AP, Welling DJ, Gattone VH (1983). Morphometric comparison of rabbit cortical connecting tubules and collecting ducts. Kidney Int 23:358-367.

Cells and Tissues: A Three-Dimensional Approach by Modern Techniques in Microscopy, pages 203–212

THE THREE-DIMENSIONAL STRUCTURE OF RENAL TUBULE CELLS

Hiromi Takahashi-Iwanaga

Department of Anatomy, Niigata University
School of Medicine, Asahimachi Niigata 951
Japan

INTRODUCTION

Light microscopists in the 19th and early 20th centuries realized that the renal tubule cells interdigitate with each other by numerous lateral processes (Zimmermann,1911). Heidenhain(1874) observed isolated proximal tubule cells. Zimmermann(1911) revealed undulated lateral surfaces of proximal and distal tubule cells by a silver impregnation method which occasionally stains epithelial cell boundaries. Zimmermann further depicted epithelial cells of various shapes in the thin limbs of Henle's loop.

Transmission electron microscopy (TEM) of proximal and distal tubule cells showed deep invaginations of basal plasma membranes extended perpendicularly to the basement membrane. Early investigators considered these structures as intracellular infoldings. Clark(1957) was probably the first to relate the TEM findings to the previous light microscopic studies, and to recognize the narrow gaps between the invaginating membranes as intercellular spaces. Extensive TEM studies were carried out by numerous investigators, confirming the idea of Clark(1957). TEM observations indicated,moreover, that epithelial cells lining different portions of the nephron showed different patters of interdigitation, and that in some portions they might possess true or intracellular infoldings in addition to the intercellular interdigitations (Rhodin,1958; Bulger,1965; Dieterich et al.,1975).

Scanning electron microscopy (SEM) was expected to be an useful weapon to visualize the complex three-dimensional shapes of urinary tubule cells. Evan et al.(1978) demonstrated by SEM the basal and lateral aspects of proximal tubule cells after removing the basement membrane by their HCl-collagenase method. Welling and his co-workers combined the SEM method of Evan with a morphometric analysis of TEM pictures, and revealed the three-dimensional structure of the proximal tubule cells and collecting tubule cells (Welling and Welling,1979; Welling et al.,1981). Their studies demonstrated, for the first time, numerous microvilli on the basal aspect of the proximal tubule cells and a complex system of extracellular channels formed by true basal infoldings in the collecting tubule cells.

With respect to other portions of the urinary tubules, however, SEM information about the epithelial cells is inadequate to be correlated with accumulating TEM findings. Although Jones (1985) reported SEM findings of the tubule cells exposed by an enzyme digestion method in most portions of the nephron, his observation was mainly concerned with their basal aspects.

The present study aims to re-examine by SEM the basolateral aspects of urinary tubule cells in various portions of the nephron, after treatment of rat renal tissue by the NaOH maceration method (Takahashi-Iwanaga and Fujita,1986), which removes basement membranes and facilitates detachment between the cells.

MATERIALS AND METHODS

Adult male Wistar rats weighing about 200 g were examined in this study. The rats were anesthetized with sodium pentobarbital and perfused through the ascending aorta with Locke's solution followed by 2.5 % glutaraldehyde in 0.1 M phosphate buffer, pH 7.3. The kidney was excised and cut into small columns which contained both cortex and medulla, and measured 1.5 x 1.5 x 5 mm in size. The tissue pieces were immersed in the same fixative for 3 hr at room temperature. Subsequently, the specimens were rinsed in phosphate buffer (pH 7.3) and placed in 6N NaOH for about 15 min at 60°C. After the NaOH maceration, the tissue was rinsed in the phosphate buffer and conductive-stained by the tannin-osmium method by Murakami(1974). The osmicated tissue blocks were dehydrated through a graded series of ethanol, transferred to isoamyl acetate and critical-point dried using liquid CO_2. The dried specimens were dissected with thin needles under a dissecting microscope, in order to expose urinary tubules. The specimens were evaporation-coated with gold-palladium and examined in a field emission SEM (Hitachi HFS-2) at an acceleration voltage of 10 kV.

RESULTS

The renal tubule cells revealed cytoplasmic processes of two different orders, i.e., large lateral processes and small microprojections. The former were ridge-like in shape and oriented perpendicularly to the basement membrane. They interdigitated with corresponding processes of adjacent cells. On the other hand, the latter were finger- or plate-like in shape, and showed no exact interdigitation between the cells.

On the basal aspect of the epithelial cells, some lateral processes were attached directly to the basement membrane, showing smooth basal surfaces which were belt-like in shape. Other lateral processes exhibited incisions or cleavages on their basal side and branched into several rows of basal microprojections of various shapes (Fig. 1b). The basal microprojections did not always interdigitate with each other between adjacent cells. In the center of cell base was found a stellate area, which were devoid of interdigitations. Intracellular basal infoldings were found in these areas in certain segments of the nephron.

1. The proximal tubules

The epithelial cells of the proximal convoluted tubules exhibited more complex shapes than those of the straight portions, because the former cells possessed more numerous lateral processes and basal microprojections than the latter. Detailed description of the convoluted portion will be given below.

The epithelial cells extended two types of lateral processes. One measured 1-5 µm in thickness and extended from the apical side to the basal side. Another was its secondary branch, measuring 0.3-0.4 µm in thickness, and restricted to the basal region. The lateral processes fanned out in the basal region, becomming plate-like structures which stood straight on the basement membrane. Microprojections were rarely found on the lateral surfaces of the epithelial cells (Fig. 1).

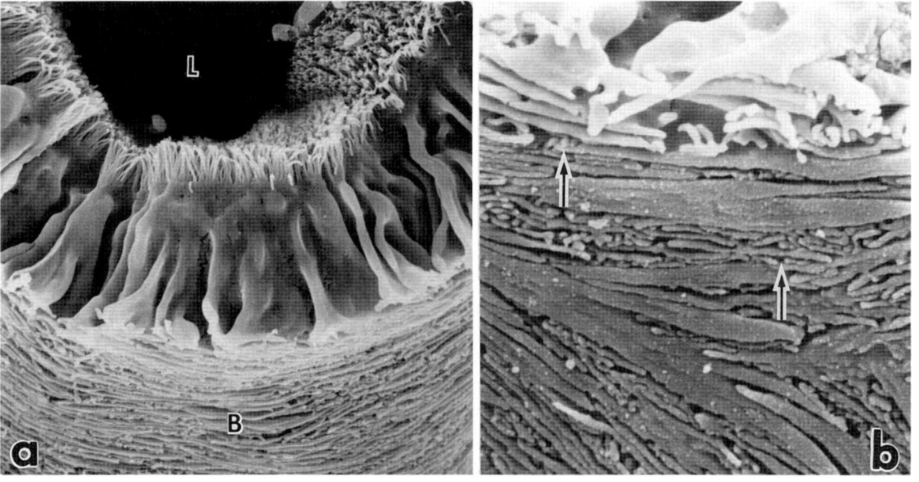

Figure 1. Proximal convoluted portion of rat urinary tubule.
a. Epithelial cells with numerous ridge-like side processes.
L.: Lumen. B.: Basal surface of the epithelium. x 3,300.
b. High magnification of the basal aspect. Basal microprojections are recognized as small rods or dots (arrows). x 11,000.

In contrast to the smooth appearance of the lateral surfaces, the basal surfaces of the proximal tubule cells were rough, because most lateral processes were divided into numerous microprojections near the cell base. The basal microprojections were plate- or finger-like in shape, measuring about 0.1 µm in thickness, and appeared on the basal surface in the shape of a short rod or a small dot. The lateral processes and their plate-like microprojections were oriented parallel to each other and transversely to the long axis of the tubule, showing lamellar arrangement on the basal aspect (Fig. 1). The epithelial cells interdigitated deeply, leaving only a small, smooth area devoid of interdigitations in the center of cell base. The area measured 3-4 µm in size.

2. The thin limbs of Henle's loop

The thin limbs of Henle's loop exposed their smooth bases with scanty microprojections. The thin limbs could be classified into three portions by the patterns of epithelial interdigitation, i.e., the thin limbs of short-looped nephrons, the descending thin limbs of long-looped nephrons, and the ascending thin limbs of long-looped nephrons.

In the thin limbs of short-looped nephrons, epithelial cells extended no lateral processes, showing simple polygonal patterns

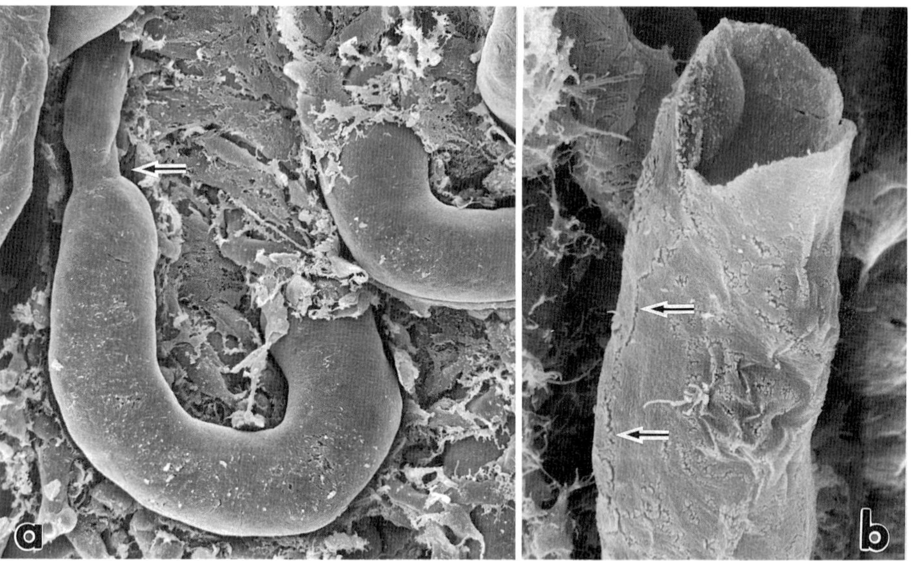

Figure 2. Thin limbs of short-looped nephrons in the outer medulla. a. The arrow indicates abrupt transition from a thin to a thick limb of Henle's loop. x 1.000. b. High magnification of the base of a limb. The arrows indicate cell boundaries. x 3,300.

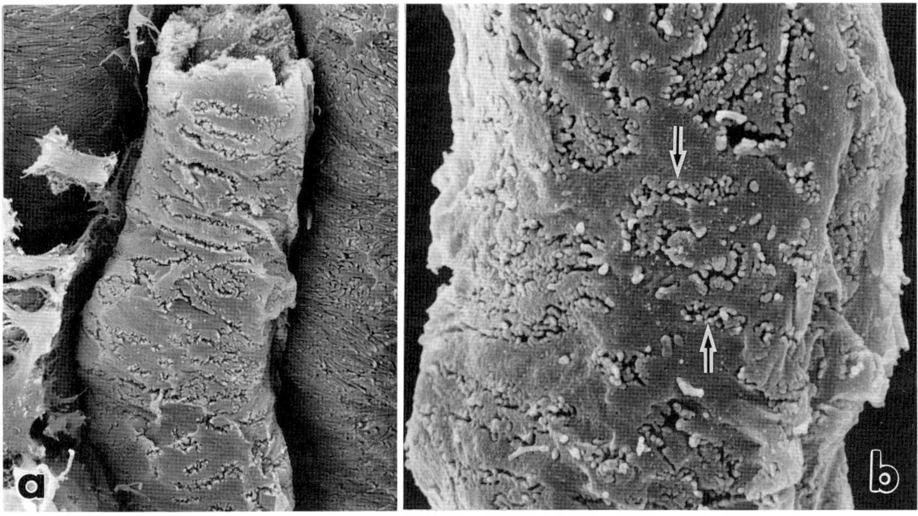

Figure 3. Descending thin limbs of long-looped nephrons in the outer medulla. The arrows indicate true basal infoldings. a. x 2,000. b. x 5,600.

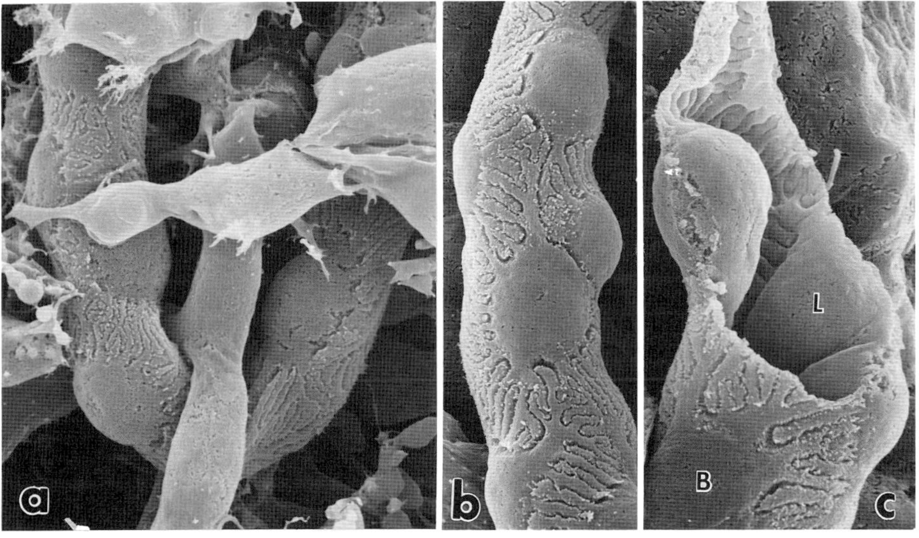

Figure 4. Ascending thin limbs of long-looped nephrons in the inner medulla. a. A hairpin portion. x 3,000. b. Epithelial cells literally interdigitating with each other. x 2,500. c. High magnification of the basal(B) and luminal(L) surfaces. x 3,500.

of cell boundaries on the basal aspect. Microplicae were sparsely found along the cell boundaries and on the cell base. This type of thin limbs was connected to the thick limb in its descending course in the outer medulla with abrupt changes in diameter and in basal structure (Fig. 2).

In the descending thin limbs of long-looped nephrons, epithelial cells radiated thick lateral processes which measured 1-2 μm in thickness. On the basal aspect, marginal regions of the lateral processes were covered with numerous microvilli or microplicae, while their central regions were smooth. Epithelial cells extended numerous microprojections of similar shapes into short wandering grooves on the basal surfaces of cell bodies. These grooves measured 1-5 μm in length and corresponded to intracellular basal infoldings (Fig.3). This type of thin limbs showed longer lateral processes with more numerous basal microprojections in the outer medulla than in the inner medulla. In the former portion, the lateral processes tended to be directed circumferentially,and the microprojections gathered in some areas of the basal surfaces, making the cell boundaries unclear.

In the ascending thick limbs of long-looped nephrons, epithelial cells radiated from the rounded cell bodies thin lateral processes, which measured about 0.7 μm in width and 3-4 μm in length. Adjacent cells geared with these uniform processes in a regular manner both on the luminal and the basal surfaces. Basal microprojections, which were stubby microvili, were scanty and restricted to the cell boundaries. Intracellular basal infoldings were not found in this segment (Fig. 4). This type of tubules formed hairpin and ascending portions of the long loopes and showed an abrupt transition to the thick ascending limbs at the boundary between the inner and the outer medulla.

3. The distal tubules

The distal tubule cells interdigitated deeply with each other in a manner similar to that of the proximal tubule cells. Their lateral processes were plate-like in shape, measuring 0.3-0.4 μm in thickness, and oriented circumferentially. The distal tubule cells showed microprojections on the lateral surfaces, which were more numerous than those of the proximal tubule cells. Intercellular spaces were slightly enlarged in some places of the epithelium to form canaliculi which started at the cell base and extended toward the lumen. The cells issued numerous microplicae into the intercellular canaliculi (Fig. 5).

The pars recta and the pars convoluta were different from each other in their basal aspects. In the former, the lateral processes were only partly divided into basal microprojections, though their attachment to the basement membrane was interrupted

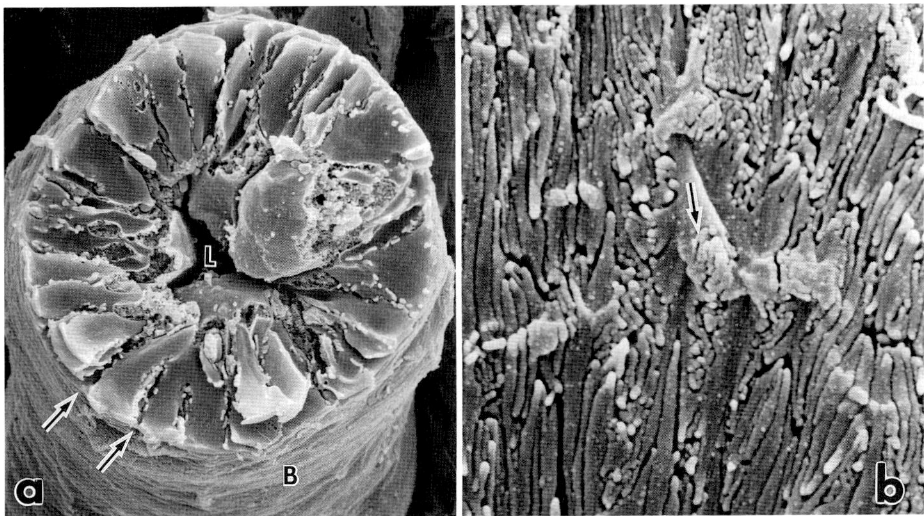

Figure 5. Pars recta of a distal tubule. a. Side aspect of epithelium showing numerous intercellular canaliculi (arrows). B.: Basal surface. C.: Luminal surface. x 3,500. b. High magnification of the basal aspect. The arrow indicates an intracellular basal infolding. x 11,000.

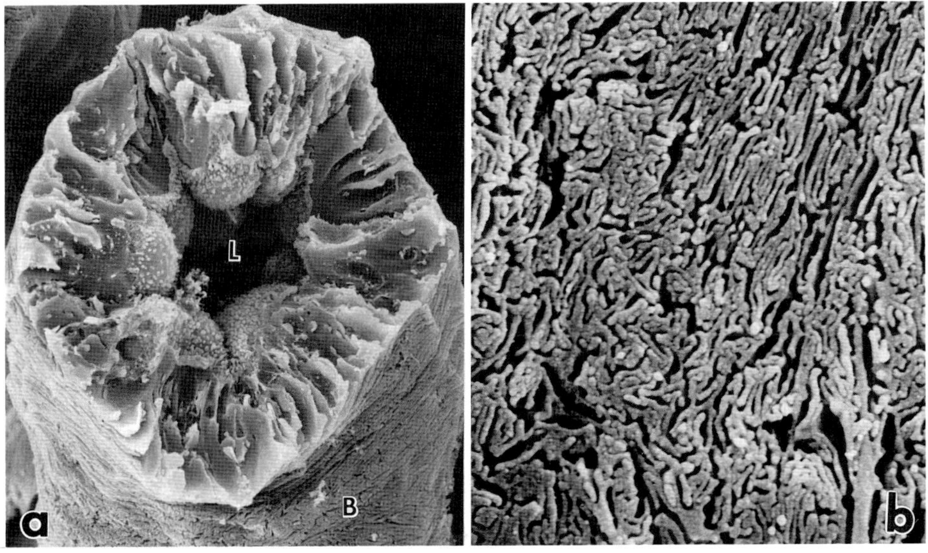

Figure 6. Pars convoluta of a distal tubule. a. lateral aspect of the tubular epithelium. L.: Lumen. B.: Basal surface. x 2,100. b. High magnification of the epithelial base. Microprojections showing complex branching patterns. x 8,000.

every 2 or 3 μm by deep incisions, which corresponded to the intercellular canaliculi (Fig. 5). In the latter, the lateral processes showed on their basal side numerous microprojections, which appeared long labyrinthine ridges on the basal aspect (Fig. 6). In both portions of the distal tubules, intracellular basal infoldings were found in the central areas of the cell base.

DISCUSSION

The present study corroborates previous light and electron microscopic studies (Zimmermann,1911; Clark,1957; Bulger,1965; Evan et al.,1978) which revealed intercellular interdigitation in the renal tubules, and demonstrates more ditailed characteristics of each segment of the tubule, with respect to the three-dimensional structure of intercellular intergiditation, the occurrences of intracellular basal infoldings, and the fine structures of the basolateral microprojections.

1. The proximal tubules

The proximal tubule cells are characterized by dense microprojections on the basal aspect, which form a striking contrast to the smooth lateral aspect. These microprojections have been described as "basal microvilli" by TEM (Waugh et al.,1967) and SEM (Evan et al.,1978; Jones,1985). In the present study, however, the basal microprojections are more various in shape, and more numerous in number, occupying a larger area of the basal surface than estimated in the previous SEM studies (Fig.1). Although the basal microprojections do not always interdigitate between adjacent cells, they serve to increase the intercellular spaces of the epithelium as well as to greatly increase the basal surface area. The spaces between the microprojections are communicated to each other and to those between the interdigitating lateral processes, forming a continuous intercellular labyrinth. Intracellular basal infoldings, which would increase only the basal surface area, are not found in this segment of nephrons.

2. The thin limbs of Henle's loop

Zimmermann(1911) suggested that the epithelia lining the thin limbs of Henle's loop might have morphological heterogeniety. Detailed TEM studies by Schwartz and Venkatachalam(1974), Dieterich et al.(1975) and other investigators have revealed that the thin limbs are classified into four portions, i.e., the thin limbs of short-looped nephrons, the descending thin limbs of long-looped nephrons in the outer medulla, those coursing in the inner medulla, and the ascending thin limbs of long-looped nephrons. Those authors described that the first and the third portions lacked intercellular interdigitations, while the second and the fourth portions showed deep ones. The present findings are in accord with this histological classification.

Schwartz and Venkatachalam(1974) depicted deep intercellular interdigitations both in the descending and in the ascending thin limbs of long-looped nephrons. Under the SEM, however, these two portions exhibited conspicuous differences in the pattern of their interdigitation, and in the amount of intracellular basal infoldings. The epithelium lining the descending thin limbs shows a complex intercellular labyringh which is formed by interdigitation of thick lateral processes plus numerous microprojections on them. In addition, the epithelial cells of this portion possess large amounts of intracellular basal infoldings (Fig. 3). These structures serve to increase the basolateral surface area of the cells. On the other hand, the cells of the ascending thin limbs interdigitate to each other by thin long processes with less numerous microprojections (Fig. 4). This type of interdigitation may presumably serve to increase the length of tight junctions between the cells per given length of urinary tubules, rather than to increase the basolateral surfaces. Intracellular infoldings are not foundiin this portion.

Physiological studies have revealed that the descending and the ascending thin limbs are different from each other with respect to their permeability to water and solutes. The former portions show high permeability to water and low permeability to solutes, while the latter possess low water permeability and high solute permeability. Although the mechanism of permeation in each portion has not been clarified yet, the present study suggests that the elaboration of basolateral cell membrane of the descending thin limb epithelium might increase permeability to water via an intracellular route, and that the long tight junctions in the ascending thin limbs might provide an intercellular or paracellular route for solutes transported across the epithelium.

3. The distal tubules

The present study is the first to show an intercellular canalicular system of the distal tubule epithelium. Small cytoplasmic compartments, which represent profiles of lateral microprojections, have been observed by TEM in the intercellular spaces of the tubule epithelium, forming small groups at various distances from the basement membrane (Bulger,1965; Allen and Tisher,1976). Such focal accumulations of microprojections probably correspond to the intercellular canaliculi sectioned transversely, though these structures have not draw attention of previous TEM investigators.

Although functional significance of this canalicular system is not clear, it might be involved in the active transport of electrolites. Physiological studies have shown that the distal tubule is a site of active NaCl reabsorption. Intercellular channels with numerous microprojections have been reported in other ion-transporting epithelia, e.g., the eccrine sweat glands (Bulger,1965). SEM will serve to elucidate the function-morphology relationship in various portions of the urinary tubules.

Acknowledgment. The author wish to thank Mr.K.Adachi in Niigata University and Miss U. Miura in Cinescience Co.Ltd. for their technical assistance and helpful discussion. This work was supported by a grand from The Ichiro Kanehara Foundation.

REFERENCES

Allen F, Tisher CC (1976). Morphology of the ascending thick limb of Henle. Kidney Int 9:8-22.
Bulger RE (1965). The shape of rat kidney tubular cells. Am J Anat 116:237-256.
Clark SL Jr (1957). Cellular differentiation in the kidneys of newborn mice studied with the electron microscope. J Biophys Biochem Cytol 3:349-362.
Dieterich HJ, Barrett JM, Kriz W, Bülhoff JP (1975). The ultrastructure of the thin loop limbs of the mouse kidney. Anat Embryol 147:1-18.
Evan AP, Hay DA, Dail WG (1978). SEM of the proximal tubule of the adult rabbit kidney. Anat Rec 191:397-414.
Heidenhain R (1874). Mikroskopische Beiträge zur Anatomie und Physiologie der Nieren. Arch Mikrosk Anat 10:1-50.
Jones DB (1985). Scanning electron microscopy of basolateral surfaces of rat renal tubules isolated by sequential digestion. Anat Rec 213:121-130.
Murakami T (1974). A revised tannin-osmium method for non-coated scanning electron microscope specimens.Arch Histol Jpn36:189-193.
Rhodin J (1958). Anatomy of kidney tubules.Int Rev Cytol 7:485-534.
Schwartz MM, Venkatachalam MA (1974). Structural differences in thin limbs of Henle:physiological implications. Kidney Int 6: 193-208.
Takahashi-Iwanaga H, Fujita T (1986). Application of an NaOH maceration method to a scanning electron microscopic observation of Ito cells in the rat liver. Arch Histol Jpn 49:349-357.
Waugh D, Prentice RSA, Yadav D (1967). The structure of the proximal tubule: a morphological study of basement membrane cristae and their relationships in the renal tubule of the rat. Am J Anat 121:775-786.
Welling DJ, Welling LW (1979). Cell shape as an indicator of volume reabsorption in proximal nephron. Fed Proc 38:121-127.
Welling LW, Evan AP, Welling DJ (1981). Shape of cells and extracellular channels in rabbit cortical collecting ducts. Kidney Int 20:211-222.
Zimmermann KW (1911). Zur Morphologie der Epithelzellen der Säugetierniere. Arch Mikrosk Anat 78:199-231.

Cells and Tissues: A Three-Dimensional
Approach by Modern Techniques in Microscopy,
pages 213–224

ANALYSIS OF STRUCTURAL AND FUNCTIONAL PROPERTIES OF THE URINARY BLADDER: THE IMPACT OF THE SEM AND ANCILLARY APPROACHES

Gisèle M. Hodges[1] and Charles Rowlatt[2]

Tissue Interaction[1] and Histopathology[2] Laboratories, Imperial Cancer Research Fund, London. WC2A 3PX, UK.

INTRODUCTION

As many studies have shown, SEM investigations of the surface structure of cells and tissues add information to the findings of LM and TEM, allow large areas of the specimen to be imaged tridimensionally at the same time and at high magnification, and provide further possibilities to study different parameters of cells and tissues.

With SEM it is possible to study not only superficial surface structures but also the cell surface of deeper layers of epithelia, subepithelial connective tissues, the epithelium-connective tissue interface, and various intracellular elements. Also, it is possible to establish structural changes associated with various pathological conditions; there is evidence that SEM analysis may be valid in the early diagnosis of premalignant lesions; and the association of morphometric methods with SEM using computerized image analysis systems may provide more information about pathological conditions. Furthermore, cell markers and labelling techniques which have been adapted for use with the SEM can provide new information on the distribution and dynamics both of specific membrane components on cell surfaces and of specific molecular units situated on intracellular or extracellular elements. This review addresses a selection of these topics and summarizes data highlighting the effectiveness of structural SEM and SEM immunocytochemistry in the field of urinary bladder research.

METHODOLOGY OF SEM STUDIES

The general methodological approaches and problems of preparing biological material for SEM have been the subject of a great many papers and discussed in various compendia (including Boyde and Wood, 1969; Nowell and Pawley 1980; Waterman 1980) while methodologies pertaining specifically to the bladder have been detailed by Hodges (1979) and Cano et al (1986).

The choice of method to be used in SEM studies depends on the specific question of interest. Combinations of methodologies and types of microscopy improve the reliability of morphological data, help in identification of artefacts, and provide a more complete analysis (structural, compositional, numerical): no single technique reveals the total picture. In the first instance it is important to consider the resolution and the information that can be obtained from the type of microscope used and from the signals available; also on the application of low or high voltage electron microscopy - this can have a tremendous effect on the information content of the final image. SEM, with its multiple signal detection facilities, has become a standard analytical tool in biological studies, each of the various signals obtained from the specimen during its interaction with the primary electron beam containing different information about the sample. In general, three types of emissions are used for most SEM studies in biology and medicine. Secondary electron imaging (SEI) is used to provide morphological information about surfaces using topographic contrast mechanisms; backscattered electron imaging (BEI) provides information on subsurface features of specimens and allows a qualitative analysis based on atomic number contrast (eg detection of medium to heavy atomic number elements in low atomic number biological matrices); and X-ray microanalysis detects and measures elements and is used to identify the elemental composition of specimens.

Analytical SEM instrumentation offers, therefore, surface characterization techniques which can conveniently be used to consider the surface morphology and composition of cells, tissues and organs. The morphological analysis gives information which can be related to description and measurement of the structural features (topography) and to quantitation of the interrelationships of these features (topology). The compositional analysis gives information which can be related to the elemental, molecular or macromolecular composition of the surface. The SEM characterization of biological structures may involve, therefore, more than just the topographic image and include coordinated indepth analysis of the molecular or macromolecular profile of the structures. To this end a variety of methods have been combined with SEM, including autoradiography (Cheng et al 1985), elemental chemical analysis (Baker et al 1985), histochemistry (Albrecht and Wetzel 1979) and immunocytochemistry (Hodges et al 1987). Recent developments, notably the use of colloidal gold (established as an excellent marker for SEM) and backscattered electron-imaging, which, combined with secondary electron-imaging provides for an optimal correlation between labelled site and structure, have firmly established the potential of SEM immunocytochemistry as a routine analytical procedure. Silver enhancement of colloidal gold-labelling patterns in SEM images, shows potential as an effective method for enhancing the localization of specific sites on cell surfaces (Goode and Maugel 1987).

A further analytical approach is in the use of metallic impregnation methods. These have been routinely used in LM histopathology for many years, more occasionally in TEM, and recently in SEM. The most common metal used is silver, usually in the form of silver nitrate. SEM can be used for simultaneous observation of the surface structure in SEI mode and of the internal structure of the same specimen in BEI mode after such heavy metal impregnation (Becker and Sogard 1979). This approach has been used to visualize cell nuclei, connective tissue fibres, basement membranes and mucopolysacchorides as well as nerve fibres.

There is interest also in using the SEM to study the internal structure of organs, tissues and cells. A variety of approaches have been used to reveal extracellular, cellular or cytoplasmic surfaces of interest and associated functional properties: several strategies are detailed in this volume and pertinent bibliographies may be found in other publications (Nowell and Pawley 1980; Waterman 1980; Borwein 1985).

THE BLADDER UROTHELIUM: NORMAL ASPECTS

The function of the mammalian urinary bladder as a reservoir devoted to the storage and periodic elimination of urine results in a highly specialized organ system capable both of accommodating large changes of volume, and of maintaining a relatively impermeable barrier between the hypertonic urine stored in the bladder and the adjacent tissues. This is possible because of the highly specialized nature of the urothelium lining the bladder lumen: this both allows for considerable expansion and contraction of the mucosal surface during the cycle of urine storage and release, and restricts free passage of water and small ions between tissue fluids and urine. The specialized function of the bladder urothelium is reflected in the somewhat unusual structural and compositional profiles of this lining epithelium. Investigative approaches to gain insight into the biological complexities of the urothelial tissue have included thin sectioning, negative staining, freeze-etching, freeze-fracture and frozen-surface replica techniques, and a range of bio-and cyto-chemical analyses: these are detailed in a series of compendia reviewing the development, histology, ultrastructure, turnover and function of normal bladder urothelium (Hicks 1975; Hicks et al 1978; Pauli et al 1983). SEM provides a significant contribution to this analysis of the morphology and composition of the bladder urothelium: this is detailed and illustrated in a number of reviews (Hodges 1979; Hodges and Kenemans 1982; Jacobs et al 1983) and outlined below.

The appearance of the normal postnatal mucosal surface depends on the state of contraction and distension of the urinary bladder. Both LM and SEM clearly identify the broad, uniform but irregularly distributed folds or rugae which mark the contracted mucosal surface and the attenuation of these folds in the distended bladder: but only SEM clearly resolves the early localized changes which may arise under pathological conditions (see below) (Figs. 1-3). Similarly, the fine complex surface patterning of microplicae which extend across groups of superficial cells in wave-like arrangements, while identifiable by LM, are best detailed by SEM. The superficial cells present a flat uniform aspect, are polygonal in shape, up to 100μm across and form a regular pavement-like arrangement with tight contact margins visible as well-defined ridges with a central groove.

Three distinct and superimposed cell zones mark the normal postnatal bladder urothelium: the basal and superficial zones consist of single layers of cells while the intermediate zone consists of a variable number of cell layers depending on species and distension state of the bladder (Fig. 4). Cells in all zones are polymorphic, undergo shape transitions, and show in the intermediate zone varying complex intercalations of plasma membranes with changes in distension. These features can be identified by SEM-SEI and SEM-BEI imaging both of histological sections and of planed surfaces of tissue blocks, and correlated to LM and TEM data and to SEM analysis of the lumenal surface of the same specimen. Further details of urothelial organization have been obtained by using silver impregnation and BEI imaging techniques: in particular, the organizational relationship of the superficial cells with the underlying intermediate cells suggestive of the proliferative unit organization seen in the epidermis (Fig. 5) (Thièbaut et al 1986), and the almost exclusively binucleate nature of the superficial layer of cells observed in the normal rat bladder (Williams AE, unpublished communication).

The lumenal membrane of the superficial cell presents a unique and characteristic angular scalloped appearance (Figs. 4,6) due to alternating patterns of plaque and interplaque regions: these have been established by TEM as two distinctive components. The concave plaques are an asymmetric unit membrane constituted by inner thin (~4nm) and outer thick (~8nm) electron-dense leaflets, which appear in freeze-fracture replicas as polygonal, rigid patches 0.2–0.5μm in diameter and consisting of a hexagonal array of plaque particles which in turn consist of 12 subunits (~3nm dia) arranged in a stellate configuration. The thickened membrane regions form from 5 to 75% of the lumenal membrane depending on age and species. Short segments of symmetrical interplaque membrane of ordinary unit membrane appearance (~8nm) interconnect the asymmetric unit membrane (AUM) plaques and form thinner more flexible "hinge" areas. These are often sharply folded particularly during physiological contraction

---

Figures 1, 3. Surface map of mucosal surface of bissected rat bladder (MNU treated) showing focal areas of modified tissue architecture: (a) thickened rugae; (b) 'doughnut' rugae configuration (see Fig. 3); (c) tumour. (Fig. 1 x175; Fig. 3 x190)

Figure 2. Ordered cellular organization as demonstrated on a few folds of normal bladder mucosal surface showing (a) pavement-like arrangement of the polygonal superficial cells; (b) surface patterning of microplicae extending over groups of superficial cells. (x300)

Figure 4. LM of normal rat urothelium showing (a) basal, intermediate and superficial cell zones; (b) scalloped angular aspect of the lumenal cell surface. PEG (polyethyleneglycol) histological section. (x160)

Figure 5. SEM-BEI of silver-stained normal rat bladder tissue showing organizational relationship between superficial (S) and intermediate (I) cells. (x550) (Courtsey of Dr. A.E. Williams).

Figures 6, SEM-SEI of normal rat bladder showing plaque-interplaque configuration (▼) of the lumenal plasma menbrane of terminally-differentiated superficial cells and tight intercellular contact margin marked by line of microvilli (▼). (x15600)

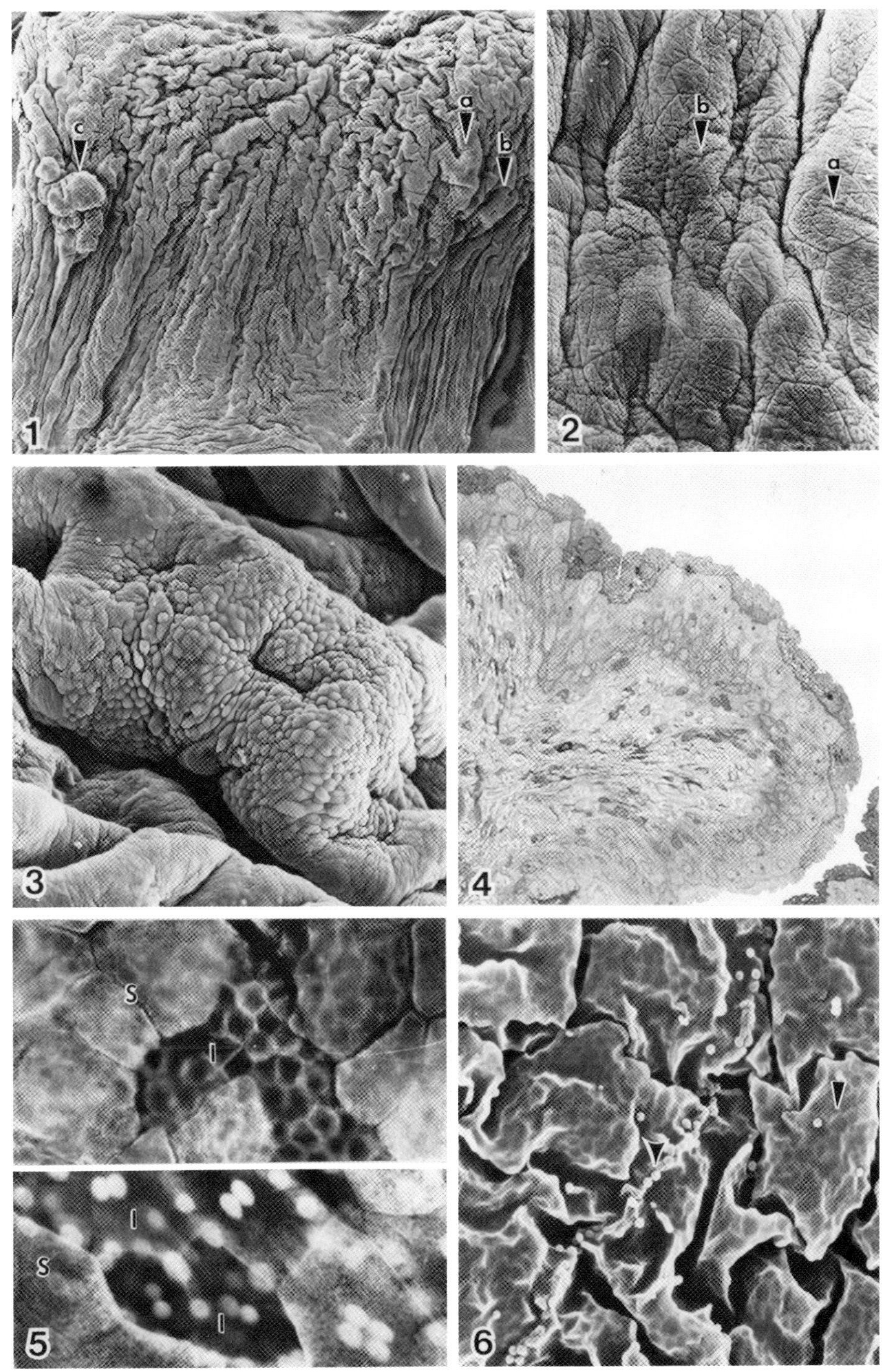
1
c
a
b
2
b
a
3
4
S
I
I
S
I
5
6

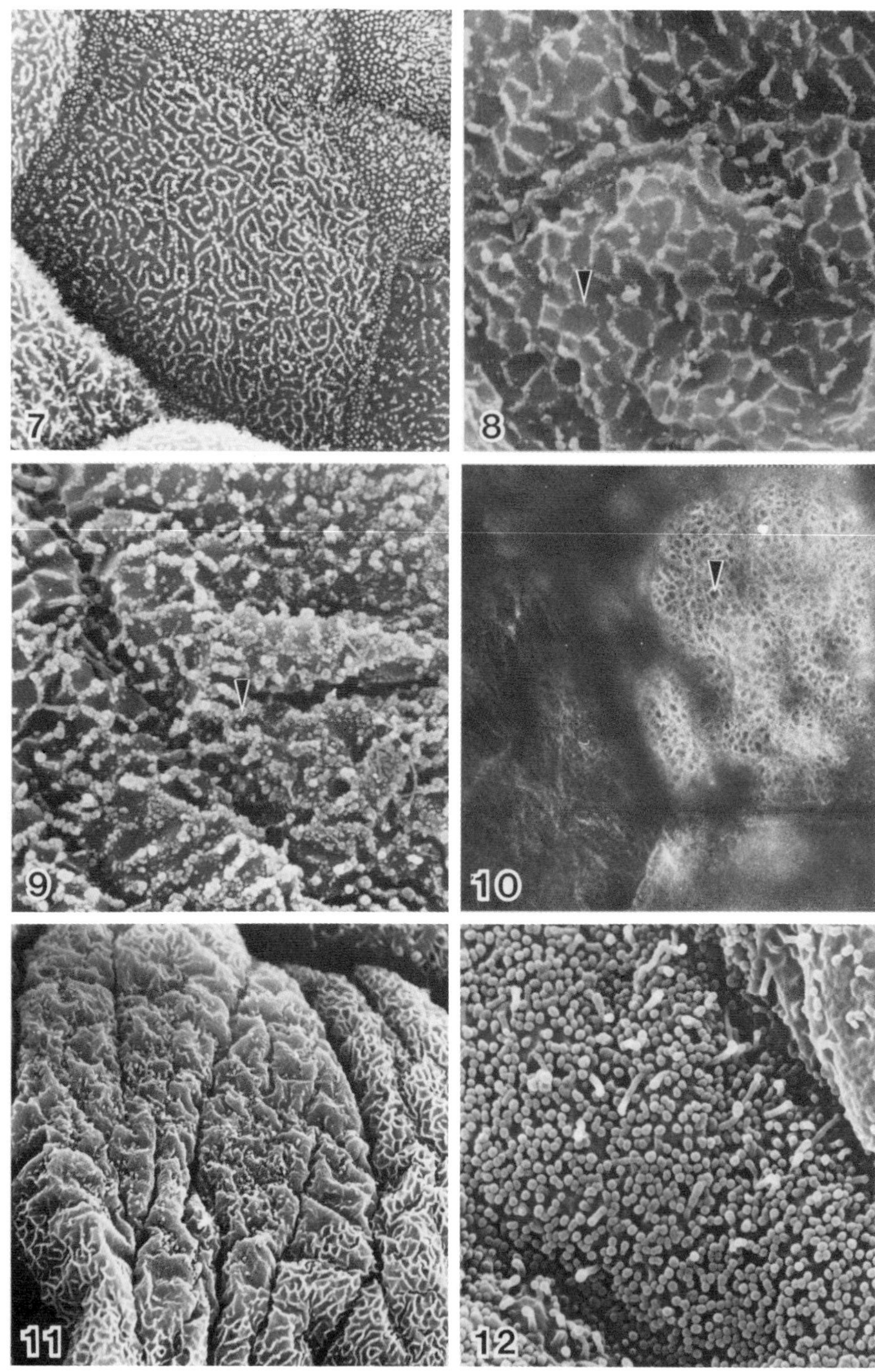
7
8
9
10
11
12

of the urinary bladder when the controlled infolding of the plasma membrane and reversible internalization of AUM plaques are seen as intraepithelial fusiform vesicles. The AUM structure, while unique to the urothelium and used as a marker of normal differentiation associated with terminal end-point features, may vary in specificity of expression. It is not present exclusively at the lumenal surface of the urothelium but can be found also at non-lumenal surfaces of superficial and intermediate cells. Furthermore, there is evidence that AUM plaques, although a normal component of mammalian bladder urothelium in many species, show a decrease in numbers with physiological aging in humans and primates.

In SEM the plaque-interplaque configuration of the lumenal plasma membrane is clearly profiled and the AUM plaques appear as distinctive polygonal domains (Fig. 6). SEM-SEI imaging also readily profiles the progressively more complex cell surface topographical patterns associated with normal urothelial cell maturation and differentiation (Fig. 7) (Hodges 1979). Further characterization of the lumenal membrane can be undertaken by localization and mapping studies of target molecules using SEM and colloidal gold immunocytochemistry as recently reviewed (Hodges et al 1987). The analytical potential of this approach to urinary bladder studies has been clearly underlined by the surface distribution studies of the urothelial membrane-associated differentiation antigen (UMA) shown by SEM to vary in expression and localization with stage of urothelial differentiation and maturation, and to be specifically localized to the interplaque region of the lumenal membrane of terminally differentiated cells (Hodges et al 1982). Glucose and mannose-rich carbohydrate moieties identified by Con A-binding studies of mouse urothelium (Hodges et al 1985) have been similarly localized to the interplaque region (Figs. 8, 9): while ABH antigens have been reported as expressed on both plaque and interplaque membrane regions of human urothelium (de Harven et al 1987). There is evidence that lumenal and intraepithelial AUM plaques may serve as membrane attachment sites for cytoplasmic microfilaments involved in

---

Figure 7. SEM-SEI of normal rat bladder showing surface topographical pattern typical of late differentiating intermediate (I) cell. (x4000)

Figures 8, 9. SEM-SEI of normal (Fig. 8) and modified (Fig. 9) plaque-interplaque distribution of Con A receptor sites over lumenal plasma membrane of normal rat bladder (Fig. 8) and irradiated mouse bladder (here showing variation across a single cell) (Fig. 9; Hodges et al 1985) and marked by indirect labelling with 30nm gold particles (▼). (Fig. 8 x16640; Fig. 9 x15400)

Figure 10. Plaque-interplaque configuration of lumenal plasma membrane in terminally-differentiated superficial cells of mouse bladder urothelium visualized by LAS86 immunfluorescence (Trejdosiewicz et al 1988) and marking localization of K19+ keratin filaments to the interplaque region (▼). (x2000)

Figures 11, 12. Cell surface topopgraphy of rat bladder urothelium (MNU treated) showing normal reticular patterning and areas of atypia marked here (Fig. 11) by short microvilli or including (Fig. 12) more pleomorphic microvilli. (Fig. 11 x4800, Fig. 12 x15400)

coordinated cell surface area modulation during the distension-contraction micturition cycle (Pauli et al 1983): immunofluorescence-derived data would indicate localization of K19+ filaments to the interplaque region (Fig. 10) and K18+ filaments to both plaque and interplaque regions. More detailed SEM analysis of intermediate filament localization patterns, of keratin epitope expression in bladder urothelium, and of the differential AUM integration of cytoskeletal elements should follow recent SEM-BEI cytoskeleton studies of immunogold-labelled cell cultures (Hodges et al 1987; Goode & Maugel 1987; Bell et al 1987).

LM and TEM studies from a variety of species suggest that the tissue architecture, cellular organization and surface topography of the mammalian bladder urothelium have a similar basic design with slight histological and ultrastructural differences (Hicks et al 1978), although few investigations have examined in detail different regions of the bladder. SEM mapping of whole bladder surfaces provides, however, a convenient approach (cf Fig. 1). First, to the detailing of possible regional topographical variations associated with species or developmental status (embryonic, neonatal or postnatal) (Nelson et al 1979; Feren and Reitan 1986). Second, to the establishment of normal baseline reference points such as may be required in studies assessing urothelial reactions to experimental treatments or evaluating bladder biopsies for pathophysiologic changes (Hodges and Kenemans 1982; Jacobs et al 1983; Feren and Reitan 1986).

THE BLADDER UROTHELIUM: PATHOLOGICAL ASPECTS IN NEOPLASIA

Attention has focussed on the possible relevance of SEM to clinical problems such as the differential diagnosis between reactive hyperplasia, dysplasia and neoplasia. Experimental urothelial neoplasia presents a most obvious opportunity to assess the contribution of the SEM to an earlier diagnosis in neoplastic disease and to the detection of minimal changes in tissue or cell properties. It has been established from comparative LM and TEM that the SEM is considerably more sensitive than LM in detecting early urothelial changes (Jacobs et al 1983). There is also evidence that SEM surface topography analysis of experimental animal or human bladder biopsy specimens may be capable of demonstrating field changes not apparent on LM (Newman and Hicks 1981; Jacobs et al 1983). Furthermore, correlation of SEM data with classical histopathological investigations is facilitated by the capability of reprocessing SEM specimens to LM- or TEM-embedded sections, of reprocessing paraffin wax blocks for SEM and correlating with matching histological sections, of SEM-SEI or SEM-BEI imaging of adjacent histological or semi-thin resin sections, and of relating internal tissue architecture to surface topography (Carr and Toner 1981; Laschi et al 1987) (Fig. 13).

Both in non-neoplastic and neoplastic disease, the bladder urothelium can express a series of changes from normal to varying grades of atypia which may be elucidated by SEM as changes in tissue architecture, cellular organization, or cell surface topography. The analytical impact of the SEM

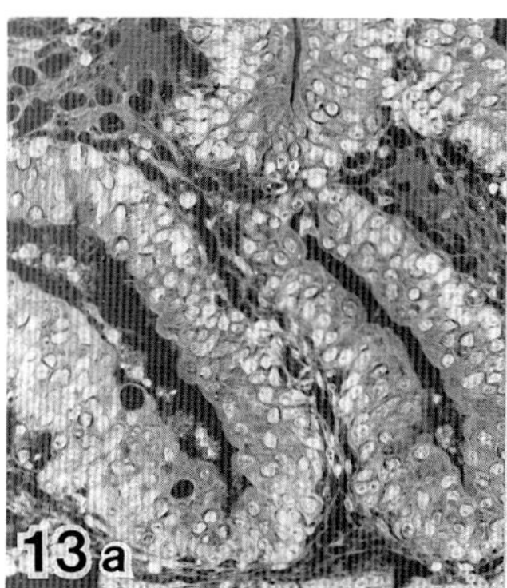

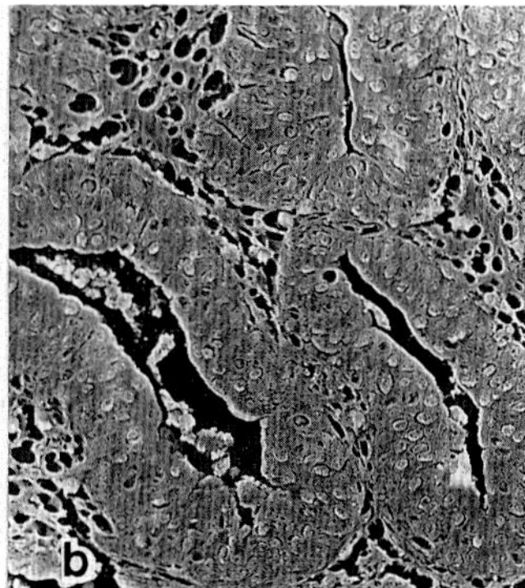

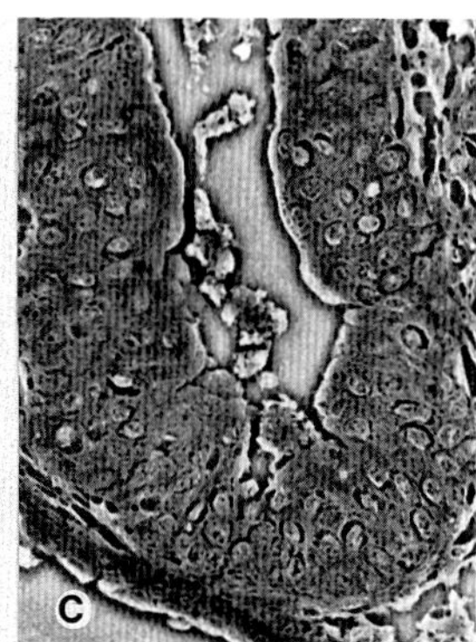

Figure 13. LM (a), SEM-SEI (b) and SEM-BEI (c) of a paraffin wax histological section from a MNU-induced rat bladder tumour. a) x360; b) x360; c) x480).

has been illustrated in a range of experimental bladder carcinogenesis studies and detailed by Hodges (1979), Hodges and Kenemans (1982) and Jacobs et al (1983). In the rat bladder N-methyl-N-nitrosurea (MNU) carcinogenesis model, focal areas of modified tissue architecture marked by contorted or 'doughnut' configurations of thickened rugae are indicative of tissue atypia (Figs. 1,3); altered cell organization marked by variable cobblestone arrangements of cells and associated variation in cell size and shape are indicators of some epithelial disorder (Fig. 3); while structural alterations of the lumenal plasma membrane marked by devolvement of the reticular microridge patternings into short uniform microvilli or gross expression of pleomorphic microvilli (PMV) serve as bioindicators of some physiopathological condition in individual cells (Figs. 11, 12).

Not surprisingly, attention has been drawn to the possible importance of PMV as a tumour marker in that microvilli of variable form, width, length and arrangement have been reported in carcinomas both during carcinogenesis in experimental animals and during oncological disease in humans (see Hodges and Kenemans 1982; Jacobs et al 1983). But the specificity of PMV as a marker of irreversible neoplastic transformation has been questioned as both PMV and short uniform microvilli may be expressed in non-neoplastic lesions and in regenerative or hyperplastic areas. However, quantitative SEM image analysis based on measurements of cell surface area, diameter, length, width, parameter, orientation and number of PMV per unit area would suggest a possible quantitiative distinction in PMV expression between patients with neoplastic and non-neoplastic disorders of the bladder (Cano et al 1985). What may be proposed at present is that PMV expression although not necessarily pathognomonic of malignancy provides a morphologic indicator which reflects some state of atypia at a given time, and may serve as a diagnostic indicator that the likelihood of tumour recurrence is greater in patients in whom the urothelium distant from the tumour itself bears PMV (Herd and Williams 1984).

A further dimension to establishing urothelial status in physiopathological studies is by SEM analysis of composition. To this end a significant contribution is the application of cell surface markers and labelling techniques adapted for use with the SEM: these can provide new information on the distribution and dynamics of specific membrane components on urothelial cell surfaces and identify urothelial atypia with ease.

CONCLUSION

SEM provides an excellent morphological technique for the study of mucosal surfaces: the range of magnification available permits survey studies of the mucosal architecture down to detailed ultrastructural analysis of the single epithelial cell. In consequence, SEM has played a fundamental role in deepening basic morphological knowledge of the urinary bladder with information about tissue architecture and cellular arrangement according to normal or physiopathological condition not easily provided by other techniques. Similarly, topographical, topological, or compositional alterations of the lumenal plasma membrane, as identified by SEM can serve as useful indicators of urothelial status, and combined with mapping techniques identify subpopulations of cells indistinguishable by conventional histological techniques. There is no doubt, as demonstrated in this review, that a significant analytical framework is provided by SEM and by various ancilliary approaches in investigations of structural and functional properties of the urinary bladder.

REFERENCES

Albrecht RM, Wetzel B (1979). Ancillary methods for biological scanning microscopy. Scanning Electron Microscopy/1979/III: 203-222

Baker D, Kupke KG, Ingram P, Roggli VL, Shelburne JD (1985). Microprobe analysis in human pathology. Scanning Electron Microscopy/1985/II: 659-680

Becker RP, Sogard M (1979). Visulization of subsurface structures in cells and tissues by backscattered electron imaging. Scanning Electron Microscopy/1979/II: 835-870.

Borwein B (1985). Scanning electron microscopy in retinal research. Scanning Electron Microscopy/1985/I: 279-301.

Boyde A, Wood C (1969). Preparation of animal tissues for surface-scanning electron microscopy. J. Microscopy 90: 221-249.

Cano M, Johansson SL. Wilson RB, Ellwein LB, Sakata T, Cohen SM (1986). Preparation methods for light microscopic and ultrastructural studies of fetal rat bladder. Scanning Electron Microscopy/1986/IV: 1357-1362.

Cano M, Wilson RB, Cohen SM (1985). Quantitation of scanning electron microscopic urinary cytology. Scanning Electron Microscopy/1985/III: 1273-1278.

Carr KE, Toner PG (1981). Scanning electron microscopy in biomedical research and routine pathology. J. Micros. 123: 147-159.

Cheng G, Hodges GM, Trejdosiewicz LK (1985). A methodological basis for SEM autoradiography: biosynthesis and radioligand binding. J. Microscopy 137: 9-16.

de Harven E, He S, Hanna W, Bootsma G, Connolly JG (1987). Phenotypically heterogeneous deletion of the ABH antigen from the transformed bladder urothelium. A scanning electron microscope study. J. Submicrosc. Cytol. 19: 639-649.

Feren K, Reitan JB (1986). Lack of regional surface differences in mouse bladder urothelium: a scanning electron microscopic study. Scanning Electron Microscopy/1986/II: 767-772.

Goode G, Maugel TK (1987). Backscattered electron imaging of immunogold-labelled and silver-enhanced microtubules in cultured mammalian cells. J. Electr. Micr. Tech. 5: 263-273.

Herd ME, Williams G (1984). Scanning electron microscopy in early human bladder neoplasia. Histopathology 8: 611-618.

Hicks RM (1975). The mammalian urinary bladder: an accommodating organ. Biol. Rev. Cambridge Phil. Soc. 50: 215-246.

Hicks RM, Chowaniec J (1978). Experimental induction, histology, and ultrastructure of hyperplasia and neoplasia of the urinary bladder epithelium. Int. Rev. Exptl. Pathol. 18: 199-280.

Hodges GM (1979). The urinary system-bladder. In Hodges GM, Hallowes RC (eds): "Biomedical Research Applications of Scanning Electron Microscopy. Vol 1. London: Academic Press, pp 307-353.

Hodges GM, Carr KE, Hume SP, Marigold JCL, Southgate J, Marshall JP (1985). Changes in surface structure and Concanavalin A binding capacity of urothelium in the mouse bladder after whole-body neutron irradiation. Scanning Electron Microscopy/1985/IV: 1603-1614.

Hodges GM, Kenemans P (1982). Diagnostic applications of SEM to oncology. In Hafez ESE, Kenemans P (eds): "Atlas of Human Reproduction by Scanning Electron Microscopy". Lancaster, Boston, The Hague: MTP Press Ltd, pp 325-337.

Hodges GM, Smolira MA, Trejdosiewicz LK (1982). Urothelium-specific antibody and lectin surface mapping of bladder urothelium. Histochem. J. 14: 755-766.

Hodges GM, Southgate J, Toulson EC (1987). Colloidal gold - a powerful tool in scanning electron microscope immunocytochemistry: an overview of bioapplications. Scanning Microscopy 1: 301-318.

Jacobs JB, Cohen SM, Friedell GH (1983). Scanning electron microscopy of the lower urinary tract. In Bryan GT, Cohen SM (eds): "The Pathology of Bladder Cancer". Vol II. Boca Raton, Florida: CRC Press Inc, pp 141-181.

Laschi R, Pasquinelli G, Versura P (1987). Scanning electron microscopy application in clinical research. Scanning Microscopy 1: 1771-1795.

Nelson CE, Croft WA, Nilsson T (1979). Characterization of different regions of the normal human urinary bladder by means of scanning electron microscopy. Scand. J. Urol. Nephol. 13: 23-30.

Newman J, Hicks RM (1981). Diffuse neoplastic change in urothelium from tumour-bearing human lower urinary tract. Scanning Electrion Microscopy/1981/III: 1-10.

Nowell JA, Pawley JB (1980). Preparation of experimental animal tissue for SEM. Scanning Electron Microscopy/1980/II: 1-19.

Pauli BU, Alroy J. Weinstein RS (1983). Ultrastructure and pathobiology of urinary bladder cancer. In Bryan GT, Cohen SM (eds): "The Pathology of Bladder Cancer". Vol II. Boca Raton, Florida: CRC Press Inc, pp 41-140.

Thiebaut F, Reitan JB, Feren K, Rigaut JP, Reith A (1986). An epidermal prolifeative unit-like structure in the epithelium of mouse bladder observed by backscattered electron imaging. Cell Tissue Res. 246: 1-6.

Trejdosiewicz LK, Southgate J. Hodges GM (1988). Unmasking of a novel lineage-specific keratin epitope during normal urothelial cytodifferentiation. In Rousset B(ed): "Structure and Functions of the Cytoskeleton". INSERM - John Libbey Eurotext Ltd. Montrouge.

Waterman RE (1980). Preparation of embryonic tissues for SEM. Scanning Electron Microscopy/1980/II: 21-44.

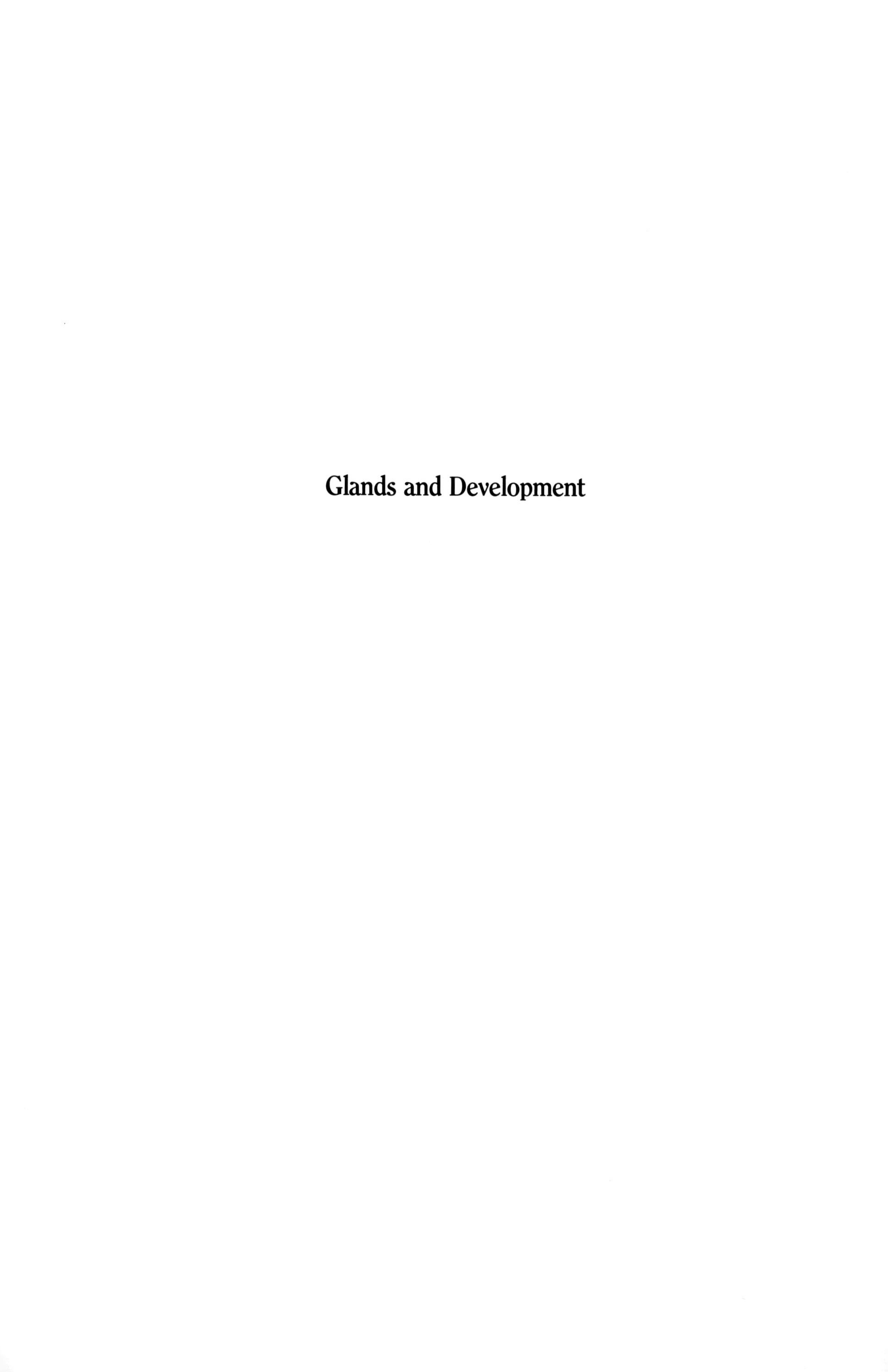

Glands and Development

Cells and Tissues: A Three-Dimensional Approach by Modern Techniques in Microscopy, pages 227–233

THREE DIMENSIONAL ASPECT ON THE FUNCTIONAL MORPHOLOGY OF THE THYROID GLAND.

Hisao Fujita and Masato Imada

Department of Anatomy
Osaka University Medical School
Osaka 530, Japan

INTRODUCTION

The thyroid hormones, thyroxine (T_4) and triiodothyronine (T_3), are secreted from the follicle epithelial cell under the control of the thyroid stimulating hormone (TSH). The follicle epithelial cell takes up raw materials necessary for the synthesis of thyroid hormones from the tissue fluid coming from the blood capillary and the hormones synthesized in this cell are released into the connective tissue and taken up into the blood capillary passing through the endothelial fenestrations. From these facts, it is easily speculated that the thyroid function may closely be related to the morphology of the blood capillary in this gland.

The aim of this presentation is to show vascular structures of normal, hyperstimulated, and hypofunctional thyroids mainly by scanning electron microscopy of the corrosion casts, and we wish to stress the structural flexibility and plasticity of the blood vessels of the thyroid according to functional states of the gland.

MATERIALS AND METHODS

1)Vascular corrosion casts of thyroids of male Wistar rats, 1, 2, 8 weeks, 6 months, and 1, 2 years in age, were observed by the scanning electron microscope.

2)Male Wistar rats, 8 weeks of age, were used for experimental studies. Sixty-five animals were divided into 5 groups. Ten animals were fed standard pellets (Oriental Yeast Co. Ltd.) and tap water. Ten animals were fed a low iodine diet (Oriental Yeast Co. Ltd.) and tap water for 4 weeks. Twenty animals were fed a low iodine diet and tap water for 6 months, and 12 months. Five animals were subcutaneously injected twice a day with 5 units/day of thyroid stimulating hormone (TSH, Sigma) for 7 days and fed the standard pellets and tap water. Ten animals were fed the standard pellets and tap water containing 0.02% propylthiouracil (PTU, Sigma) for 4 weeks. Ten animals were fed standard pellets and tap water containing 0.5% levothyroxine sodium (Thyradin-S, Teikoku-zoki Co, Ltd.) for 4 weeks. Thyroid glands of some animals in each group were examined by a transmission electron microscope, and the vascular corrosion casts of the thyroid of the other animals were examined by a scanning electron microscope.

(3) Four DDY mice, 8 weeks of age, fed with a low iodine diet for 1, 2, 3, and 4 weeks, and one normal animal, 8 weeks of age were injected intraperitoneally with 1 mCi of ^{3}H-thymidine respectively, and the autoradiographs of the thyroids were made in light as well as electron microscopic level.

RESULTS and DISCUSSION

Normal: Scanning electron microscopy of vascular corrosion casts of thyroids of normal rats, 8 weeks in age, shows a small number of deep, wide and long fissures corresponding to the interlobular connective tissue, where the interlobular arteries and veins reside, and many narrow grooves corresponding to the interfollicular connective tissue.

Each follicle is densely enclosed by a clearly defined basket-like capillary network, which is as a rule not confluent with adjacent networks, though a few anastomoses or common capillaries are sometimes seen (Figs. 1,2). About 60-70% of the capillary networks are completely independent of the neighbouring ones, and the other 30-40% are not entirely independent (Fig. 3a). The diameters of the capillaries (5-15 μm) do not vary greatly. The capillaries surrounding each follicle are anastomotic and the capillary bed occupies about 50% of the follicular surface area.

Hyperstimulated: The thyroid gland secretes thyroxine by stimulation of TSH. In long-term low iodine diet-treated, or long-term PTU-treated animals, the low concentration of thyroxine in blood stimulates the hypothalamus-pituitary system to secrete an excess dose of TSH by a negative feedback mechanism, and the follicle epithelial cell is hyperstimulated chronically by an excess dose of TSH, and the cell is considered to be in chronic hyperfunctional state cytologically. In TSH-treated animals, low iodine diet-treated animals, or PTU-treated animals three dimensional images of the blood vessels in the thyroid show almost similar pattern principally. By thin sectioning images, the follicle epithelial cells become markedly taller in height and the rough endoplasmic reticulum and Golgi apparatus are extremely well developed. The follicle lumen becomes smaller in size, especially in low iodine-treated-, or PTU-treated-animals.

By scanning electron microscopy, the blood capillaries in each basket-like network are markedly dilated in all kinds of the hyperstimulate animals (Figs. 3-7). The diameter of the capillary is the most heterogeneous in PTU-treated animals, and it is about 60 μm in the thickest part and 2 μm in the thinnest part. In low iodine diet-treated animal the thickest part is about 25 μm and the thinnest part is 5 μm in diameter. Many buds (sprouts), fusions and anastomoses of the capillaries are recognized in or between the capillaries throughout the basket by the scanning as well as transmission electron microscope (Figs. 3-7). The capillary endothelial cells of the thyroid show protein synthesizing signs such as well developed rough endoplasmic reticulum and the Golgi apparatus in low iodine diet-treated- or PTU-treated-animals. Using freeze-fracture method, we reported an increase of endothelial fenestrations in number in the long-term TSH-treated mice (Ishimura et al. 1976).

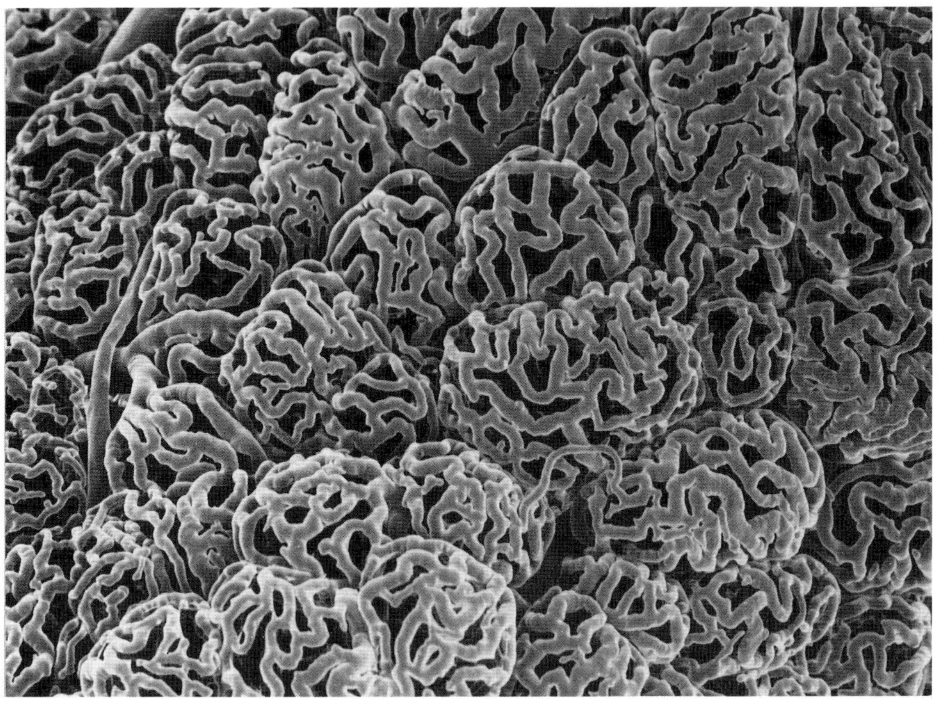

Fig 1. Normal rat (12 weeks old) thyroid. x360

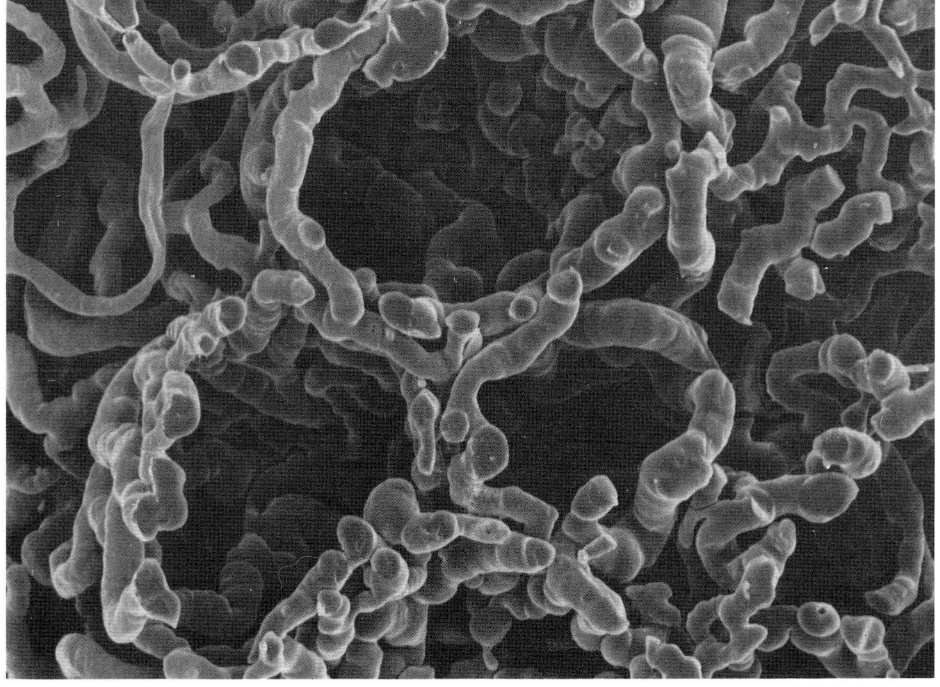

Fig 2. Normal rat (12 weeks old) thyroid. x930

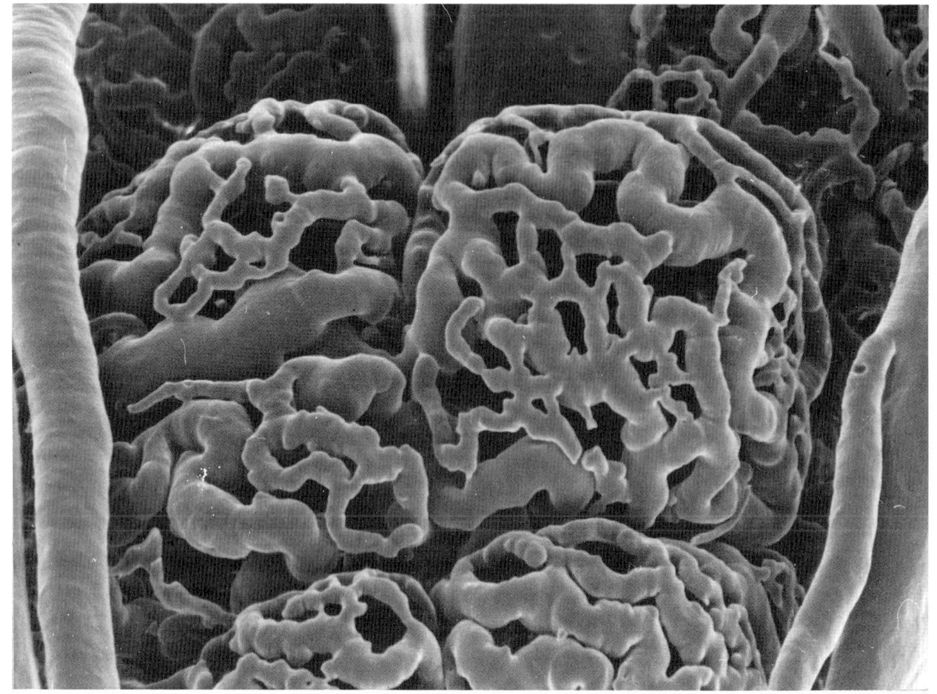

Fig 3. TSH-treated (for 7 days) rat thyroid. x540

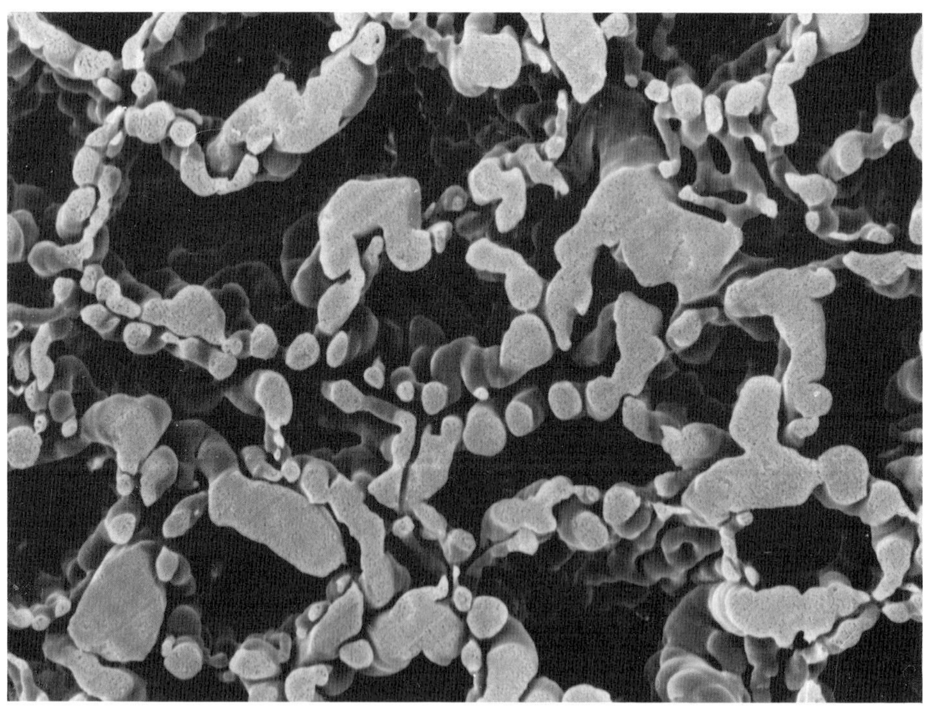

Fig 4. TSH-treated (for 7 days) rat thyroid. x600

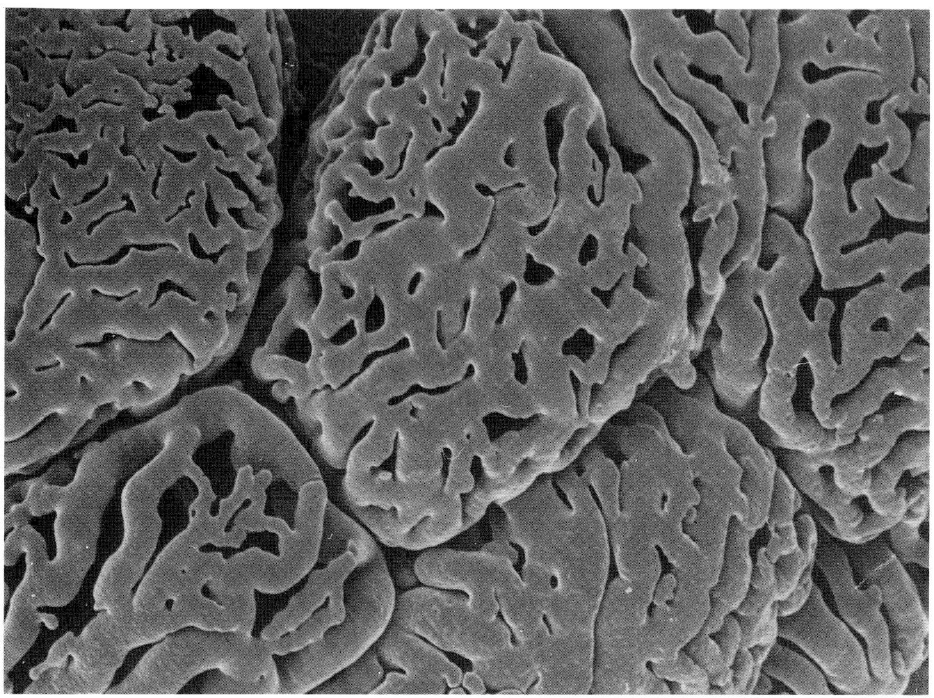

Fig 5. Low iodine-treated (for 4 weeks) rat thyroid. x740

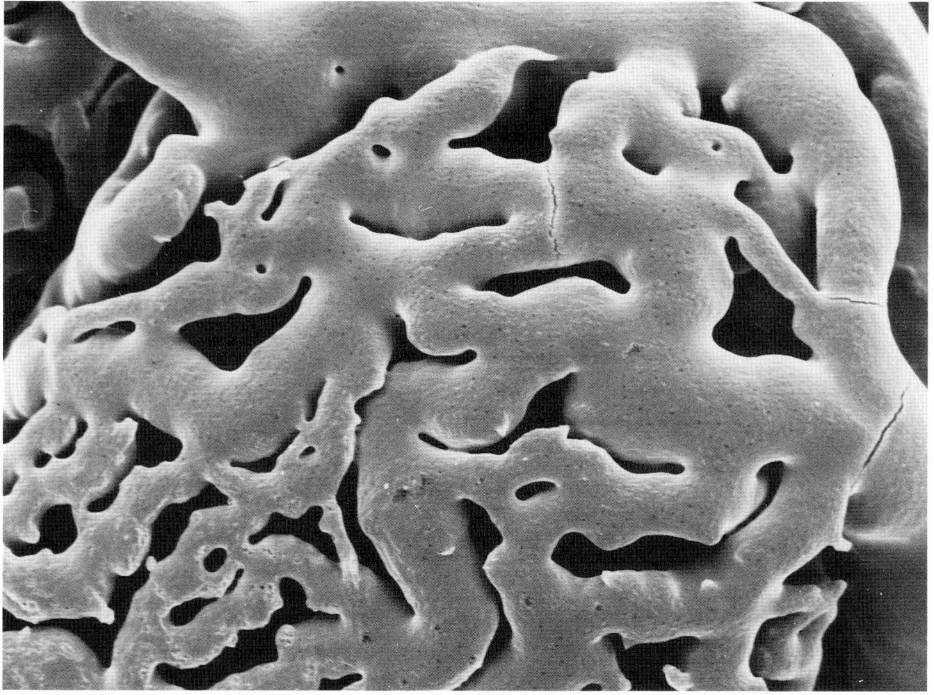

Fig 6. Low iodine-treated (for 4 weeks) rat thyroid. x1,700

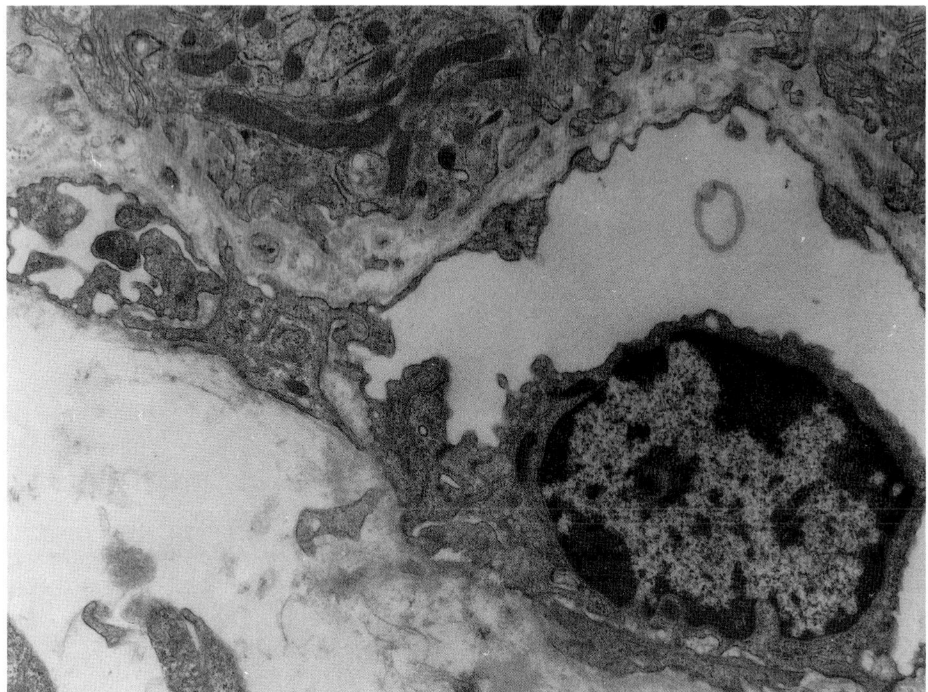

Fig 7. Budding of the blood capillary. Low iodine-treated rat thyroid. x15,000

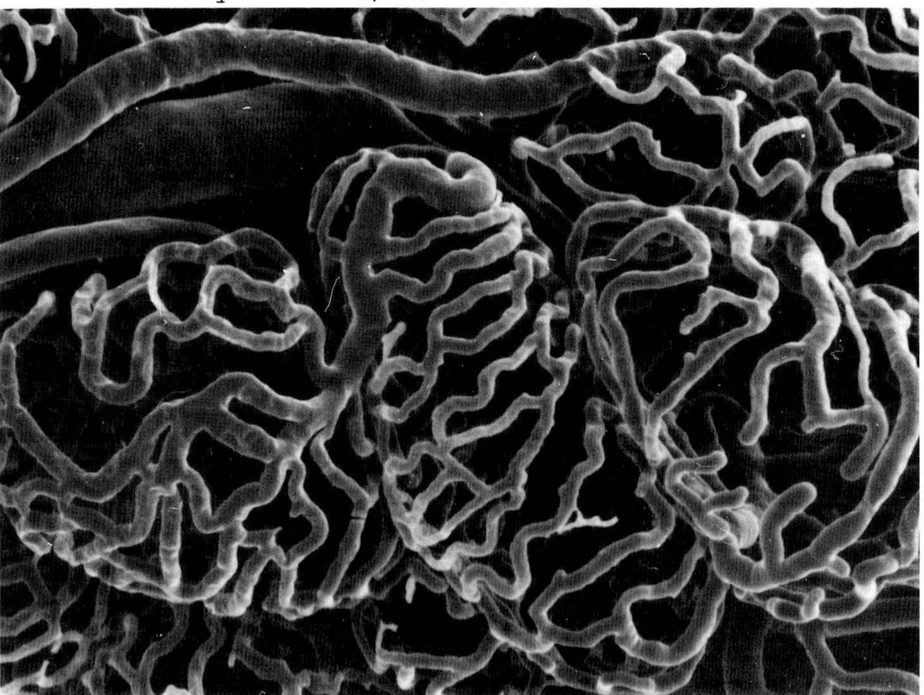

Fig 8. Levothyroxine sodium-treated rat thyroid. x790

Hypofunction: In long-term levothyroxine-treated animals, the thyroxine level in the blood is usually high, and the TSH secretion is suppressed by a negative feedback mechanism. The thyroid is considered to be in hypofunctional state in secretion. The thyroid glands of the long-term levothyroxine-treated animals are by far smaller in size than normal. The follicle epithelial cells become excessively lower in height, and the follicle lumina are intensively enlarged as compared with those of normal rats. The rough endoplasmic reticulum and the Golgi apparatus are reduced in size.

The basket-like capillary network surrounding each follicle becomes very poor in distribution and each capillary is very narrow in diameter (Fig. 8). The most thickest part is about 8 μm and the thinnest part is 3 μm in diameter. The capillary bed covers about 20-30% of the follicular surface area. Anastomoses of the capillaries within each basket-like network are markedly decreased in number, and the features of buddings and fusions are hardly to recognize.

The endothelial fenestrations are markedly decreased in number in long-term thyroxine-treated mice (Ishimura et al. 1976). These facts also mean the endothelial structures are changeable by the functional state of the gland.

CONCLUSIONS

We wish to emphasize that the microvascular system in the thyroid is flexible and changeable and shows a plasticity in three dimensional images, distribution patterns, and endothelial developments and morphologies, as a mirror of the functional state of the gland.

REFERENCES

Fujita H, Murakami T (1974). Scanning electron microscopy on the distribution of the minute blood vessel in the thyroid gland of the dog, rat and rhesus monkey. Arch histol jap 36:181-188.

Imada A, Kurosumi K, Fujita H (1986). Three-dimensional aspects of blood vessels in thyroids from normal, low iodine diet-treated, TSH-treated, and PTU-treated rats. Cell Tissue Res 245:291-296.

Imada A, Kurosumi K, Fujita H (1986). Three-dimensional imaging of blood vessels in thyroids from normal and levothyroxine sodium-treated rats. Arch histol jap 49:359-367.

Ishimura K, Okamoto H, Fujita H (1976). Freeze-etching studies on ultrastructural changes of endothelial cells in the thyroid of normal, TSH-treated and Thyradin-treated mice. Cell Tissue Res 175:313-317.

Cells and Tissues: A Three-Dimensional Approach by Modern Techniques in Microscopy, pages 235-241

SURFACE MICROANATOMY OF HUMAN MAJOR SALIVARY GLANDS

A. Riva, L. Valentino, M.S. Lantini, E. Cotti and F. Testa Riva
Dipartimento di Citomorfologia, Universita' di Cagliari, 09124 Cagliari, ITALY

TEM studies have shown that cells of human salivary glands possess a complex 3-d morphology. To investigate their cytoarchitecture we have employed the SEM using a variety of techniques that permit the visualization of isolated endpieces and of individual acinar and ductal cells.

For visualization of lateral cell surfaces, the samples, fixed for a minimum of 6 hrs. in a mixture of aldehydes, were cut into 100 um thick sections by a TC2 Sorvall tissue sectioner, postfixed with 1% OsO_4, dehydrated, and critical point dried. Some of the sections were subjected to sonication during the dehydration steps or, after C.P.D., further fractured using an Oxford Vibratome. To investigate gland organization, as well as the myoepithelium, we employed the methods of both Evan et al.(1976) and Nagato et al.(1980). In addition to the hydrolysis with 8N HCl used in both methods, we employed alternatively or in sequence a brief treatment with 6N NaOH or with 30% KOH. Following the treatment with HCl and collagenase, the specimens were, in some cases, briefly sonicated at 50 Hz. Observations were made with an ISI SS40 SEM. After removal of connective tissue the lobules of each gland resemble a bunch of grapes, a configuration similar to the models obtained by LM (Maziarski, 1900). The grapes correspond to clusters of secretory cells while the progressively anastomosing and enlarging stems represent the system of excurrent ducts.

Fractured secretory cells are characterized by the presence of round granules (Fig.1). Contrary to TEM findings (Riva et al., 1974), however, it is very

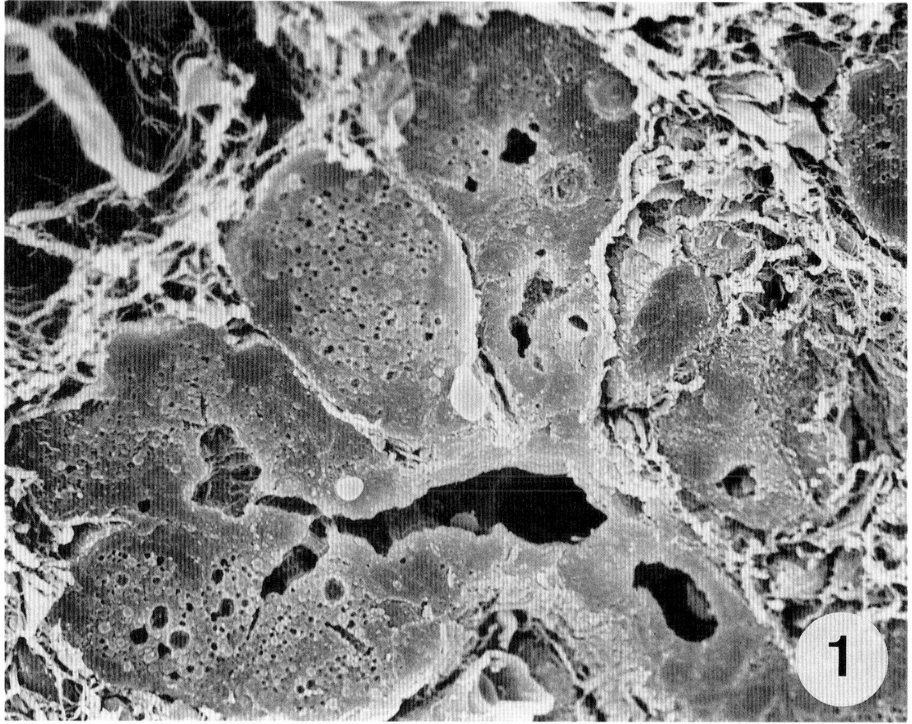

Fig. 1. Fractured serous endpieces of the submandibular gland. Several acini, recognizable by the presence of the secretory granules, drain into the intercalated duct. X 2,000.

difficult to distinguish by SEM the secretory products of the different salivary glands purely on morphological grounds. Secretory cells, particularly the serous cells, show a complex cytoarchitecture. In secretory cells whose basolateral surfaces have been exposed, the upper portion of the cell lateral surfaces are covered by frond like expansions of the plasmalemma oriented at random (Fig.2). Toward the cell base, these expansions become more numerous and consist of a series of parallel lamellae oriented at right angle to the basal lamina (Fig.3). The basal folds increase the cell basal surface by a factor of ten and are probably involved in the mechanism of saliva production (Tandler and Riva,1986).

Intercalated ducts are difficult to identify because their constituent cells lack basal and lateral specializations. In the few intercalated ducts that

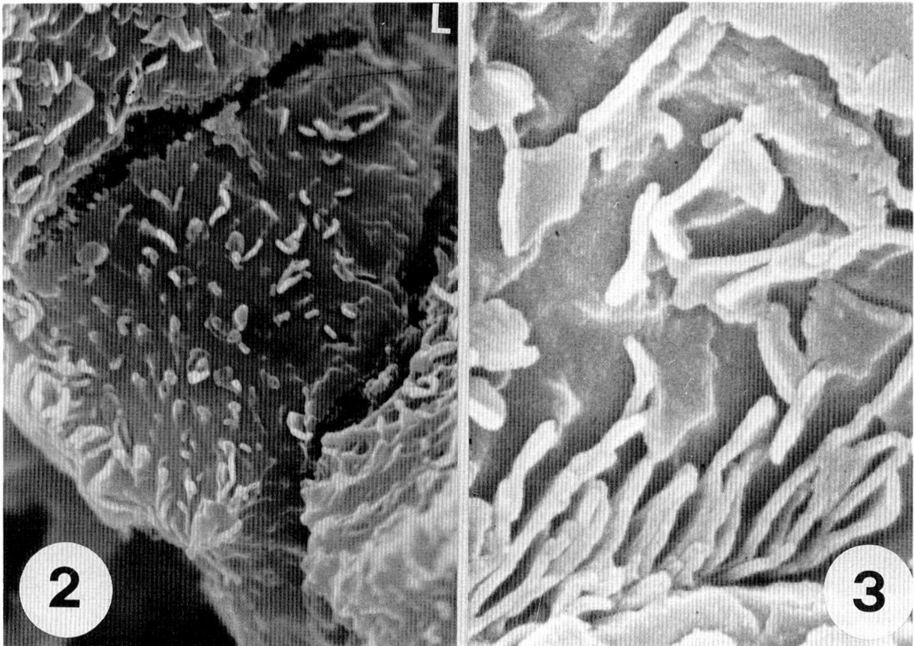

Fig. 2. Several surfaces of secretory cells exposed from the basal lamina to the lumen (L). Note the presence of an intercellular canaliculus extending toward the cell base. Submandibular. X 15,000.

Fig. 3. Higher magnification of the basolateral region of a secretory cell. The upper portion of the cell lateral surface is covered by frond-like folds while, basally, the processes of the plasmalemma are aligned vertically. Submandibular. X 32,000.

were unequivocally identified, the lateral surfaces of the duct cells appear rather smooth. In contrast, striated ducts are always clearly recognizable. Their exposed lateral portions vary according to the zone considered (Fig.4). Just beneath the luminal plasmalemma there is a smooth band that surrounds the cellular apex. This band probably corresponds to the junctional complexes. More basally the cell body splits into 4-5 large basal portions whose surface is covered by the typical processes responsible for the striations seen with the TEM. These processes appear as a series of long parallel laminae alternating with vertical grooves where processes of adjacent cells were inserted. After removal of the connective tissue

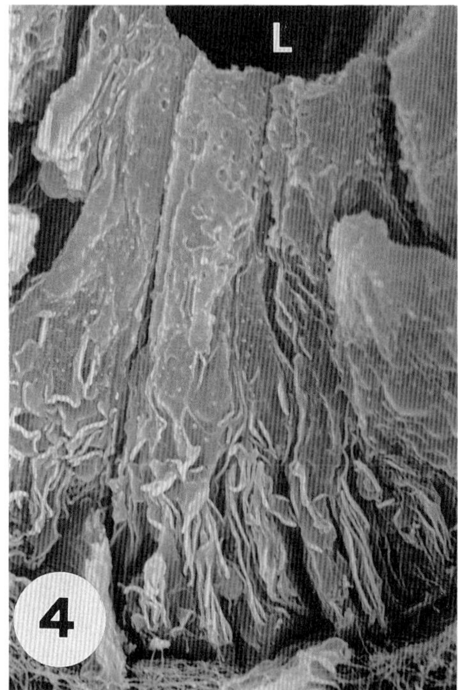

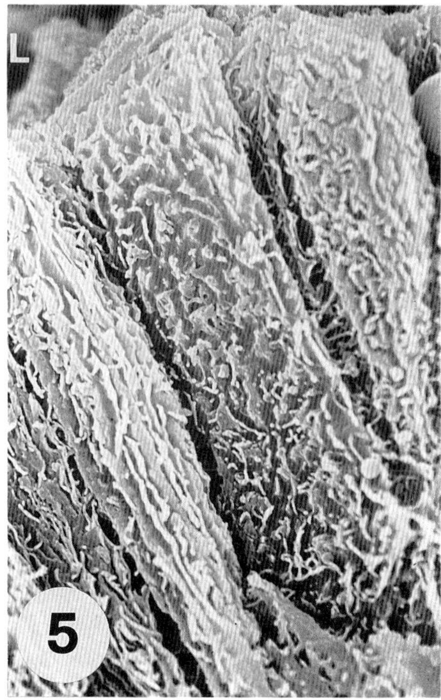

Fig. 4. Cells of a striated duct exposed from the basal lamina to the lumen (L). Submandibular. X 4,000.

Fig. 5. Cells from an excretory duct. The lateral surfaces are completely covered by short processes oriented at random. Submandibular. X 6,000.

and digestion of the basal lamina, the processes of striated cells seen from below, exhibit (Fig.6) a radial orientation with an extensive interlocking of laminae and grooves of neighboring cells that closely matches that observed in TEM (Fig.7). The cytoarchitecture of the ductal cells varies gradually as the duct approaches the interlobular septa wherein the extralobular or excretory ducts reside. Cells of excretory ducts (Fig.5) have a body that remains undivided and are completely covered by short processes oriented at random instead of vertically. Basal cells also are present.

Acini are capped by myoepithelial cells of different shapes (Fig.8). Their processes are tapering and their surface is generally smooth (Fig.9) but there are also myoepithelial cells endowed with a

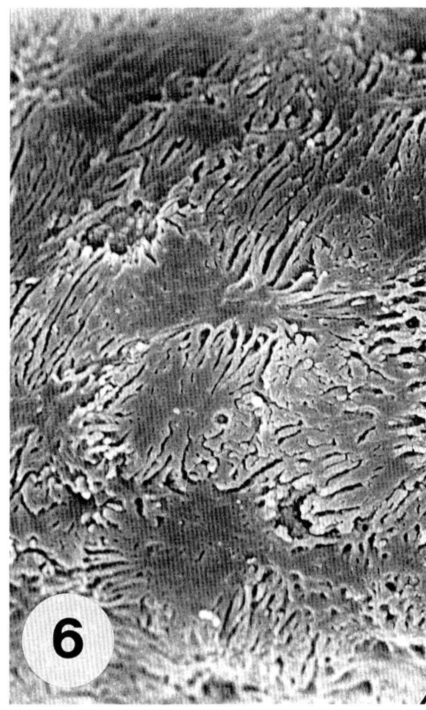

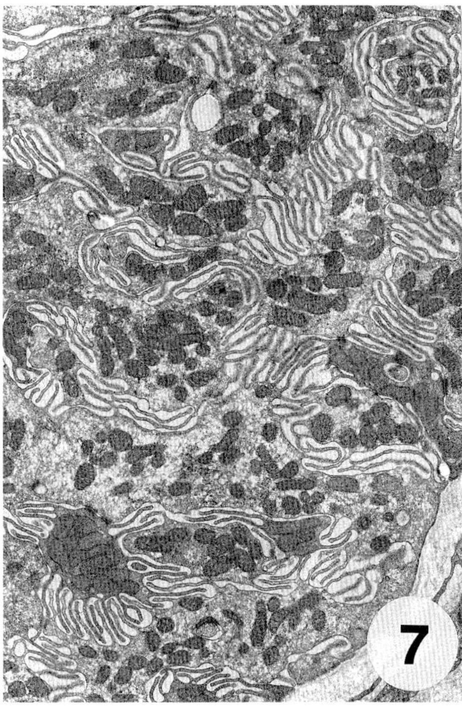

Fig. 6. Base of a striated duct seen from below after removal of connective tissue and digestion of the basal lamina according to Evan et al. (1976). Submandibular. X 4,400.

Fig. 7. TEM. Cross section of the basal portions of striated duct cells. Compare with figure 6. Submandibular. X 4,000.

ruffled surface that might represent contracted cells. Myoepithelial cells have been demonstrated also in the intercalated and, more rarely, in the intralobular striated ducts. If sonication is prolonged, myoepithelial cells are partially detached from underlying cells and it is possible to observe, on the acinus, the grooves formerly occupied by the myoepithelial processes. No differences in location and morphology have been detected among the myoepithelial cells in different human salivary glands; this finding is at variance with those obtained in rodents by Brocco and Tamarin (1979) and by Nagato et al. (1980).

Determination of the three-dimensional cytoarchitecture of human salivary glands by LM and TEM

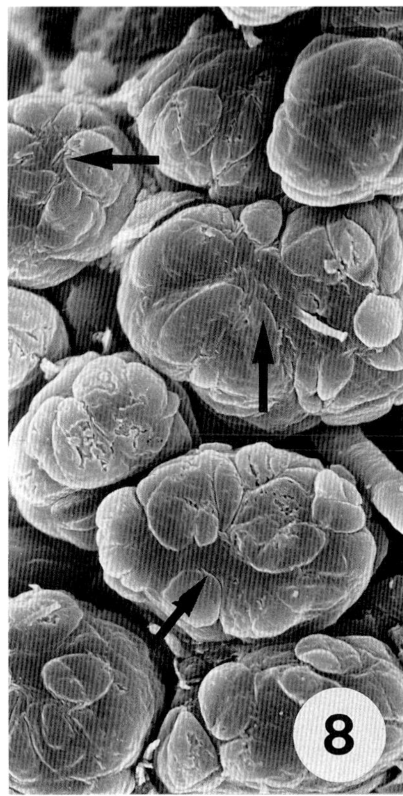

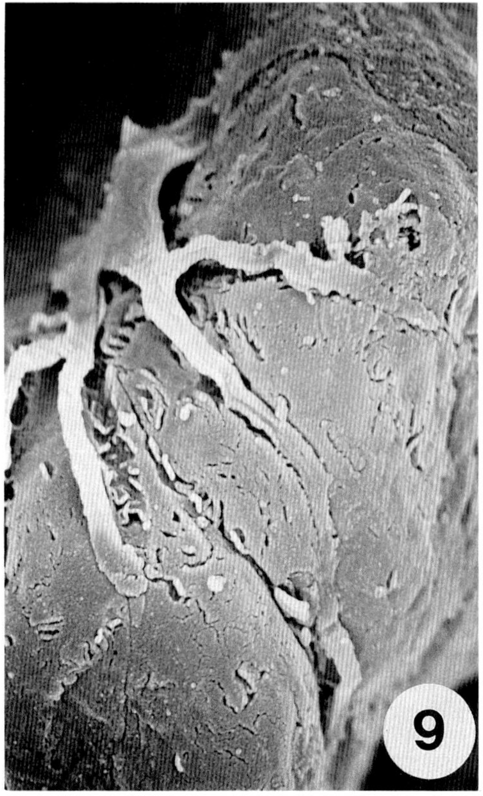

Fig. 8. Submandibular gland acini treated with the method of Evan et al. (1976). The arrows indicate the myoepithelial cells. X 2,000.

Fig. 9. Higher magnification of a myoepithelial cell. Parotid. X 6,000.

requires laborious and tedious reconstruction. The use of SEM allows direct visualization of the parenchymal components of these organs, and reveals the complexity of the surfaces of the constituent cells, as well as their interrelationships. Our work serves to confirm certain observations previously made by other workers (Tandler, 1962; Tamarin and Sreebny, 1965; Tandler, 1965), and, in certain cases, corrects their misinterpretations resulting from the limitations of their techniques. Further refinements in our techniques may reveal additional structural features of human salivary glands.

REFERENCES

Brocco SL, Tamarin A (1979). The topography of rat submandibular gland parenchyma as observed with SEM. Anat Rec 194:445-460.

Evan AP, Dail WG, Dammrose D, Palmer C (1976). Scanning electron microscopy of cell surfaces following removal of extracellular material. Anat Rec 185:433-447.

Maziarski S (1900). Über den Bau der Speichendrusen. Bull Acad Sci Cracovia 41, 22S.

Nagato T, Yoshida H, Yoshida A, Uehara Y (1980). A scanning electron microscope study of myoepithelial cells in exocrine glands. Cell Tissue Res 209:1-10.

Riva A, Motta G, Riva Testa F (1974). Ultrastructural diversity in secretory granules of human major salivary glands. Amer J Anat 139:293-298.

Tamarin A, Sreebny LM (1965). The rat submaxillary salivary gland. A correlative study by light and electron microscopy. J Morphol 117:295-352

Tandler B (1962). Ultrastructure of the human submaxillary gland. I. Architecture and histological relationships of the secretory cells. Amer J Anat 111:287-307.

Tandler B (1965). Ultrastructure of the human submaxillary gland. III. Myoepithelium. Z Zellforsch 68:852-863.

Tandler B, Riva A (1986). Salivary glands. In Mjör IA, Fejerskov O (eds): "Human Oral Embryology and Histology," Copenhagen: Munksgaard, pp. 243-284.

Cells and Tissues: A Three-Dimensional Approach by Modern Techniques in Microscopy, pages 243–248

COMPARATIVE STUDIES OF THE STRIATED DUCTS OF MAMMALIAN SALIVARY GLANDS

Bernard Tandler, Carleton J. Phillips, Kuniaki Toyoshima and Toshikazu Nagato

Department of Oral Biology (BT, KT, TN), School of Dentistry, Case Western Reserve University, Cleveland, Ohio 44106, and Department of Biology (CJP), Hofstra University, Hempstead, New York 11550

In the major salivary glands of mammals, striated ducts usually are the most prominent segments of the excurrent duct system. The basal striations that give these ducts their name are due to multiple folds, flutings, and digitiform processes at the bases of the tall cells that constitute the principal component of the ductal epithelium. These surface projections are complexly interlocked to form a series of basal compartments in which vertically-oriented, ostensibly rod-shaped mitochondria are lodged (Tandler, 1963). By analogy with the renal proximal and distal convoluted tubules, it is generally accepted that striated ducts are a major site of electrolyte resorption from initially isotonic saliva to produce the hypotonic saliva that is delivered to the mouth, but it should be noted that there is little direct evidence to bolster this supposition.

Striated duct cells show few departures from the basic structural design in those animal species that have an omnivorous diet. In human salivary glands, for example, the luminal surface of the duct cells bears only a few short, scattered microvilli. In contrast, these ducts may exhibit significant modifications, especially of their surfaces, in those species that favor a highly restricted diet. Vampire bats have striated ducts in both principal and accessory submandibular glands that have an array of microvilli matching the brush borders in kidney (Fig. 1). Because they consume only blood, these animals take on a huge quantity of sodium chloride with each feeding. Just as certain intestinal cells possessing a striated border are capable of secreting electrolytes, so too may the striated duct cells in the vampire salivary glands engage in secretion of sodium. Another possibility is that the striated ducts may be involved in selective resorption of an electrolyte that has been secreted into the saliva in tandem with sodium chloride.

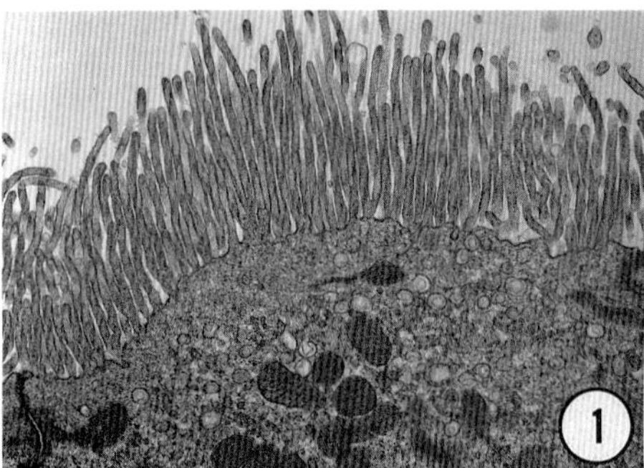

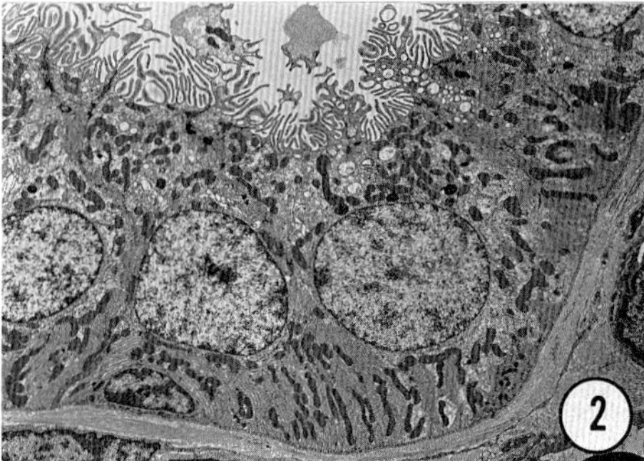

Fig. 1. The brush border of a striated duct in the principal submandibular gland of the vampire bat, Desmodus rotundus. X 8,500.

Fig. 2. Frondose processes in a striated duct in the submandibular gland of the Jamaican fruit bat, Artibeus jamacensis. X 2,200.

Fruit-eating bats are faced with the opposite problem. Since fruit is high in potassium content but extremely low in sodium, such bats have evolved mechanisms for retention of sodium. Instead of microvilli, fruit bats have extensive leaf-like or frondose processes on the luminal borders of their striated duct cells (Fig. 2). A series of repeating units is present on the cytoplasmic aspect of the limiting membrane of these processes. Similar units, termed portasomes, have been observed in a variety of invertebrates where they have been implicated in sodium transport (Harvey et al., 1981). It therefore is likely that in fruit bats, the portasome-like structures are involved in sodium conservation. In the North American beaver, an aquatic mammal, portasome-like units are present not at the cell apex, but underlying the folded plasma membranes at the cell base. Because these animals probably swallow a great deal of fresh water, sodium-retention mechanisms must be present to prevent excessive dilution and loss of plasma electrolytes.

In addition to electrolyte transport, striated ducts can modify the initial saliva by uptake of proteins from the luminal saliva. Using retrograde infusion of exogenous proteins such as ferritin, myoglobin, or horseradish peroxidase, and by immunogold localization of endogenous proteins, it has been shown that striated duct cells can take up these organic molecules and sequester them in small vesicles, multivesicular bodies, and lysosomes (Coleman and Hand, 1987; Hand et al., 1987). During degradation of such proteins, crystalloids may form within duct cell lysosomes (Tandler et al., 1979).

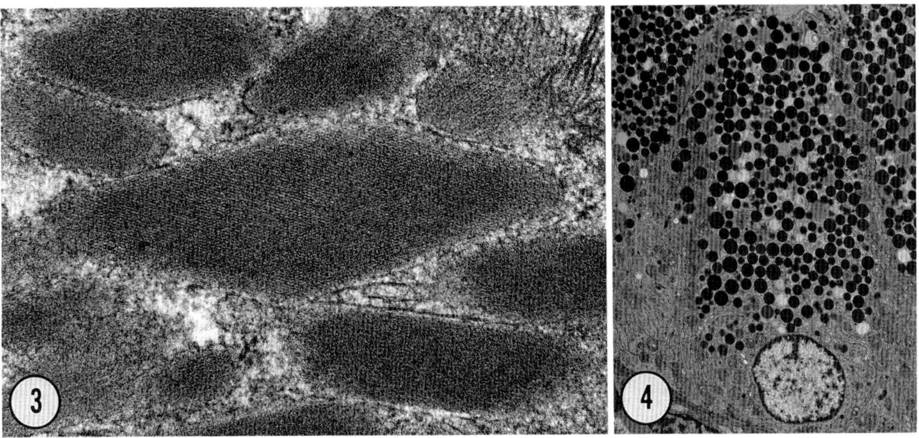

Fig. 3. A crystalloid in the striated duct of the parotid gland of the North American mink, _Mustela vison_. X 58,000.
Fig. 4. A striated duct in the parotid gland of the vampire bat. Secretory granules are extremely abundant. X 2,400.

In the submandibular gland of rodents, a segment of the duct system lying between the intercalated and striated ducts and known as the granular convoluted tubule is specialized for secretion of organic products. Cells in these ducts, which are especially prominent in males of many rodent species, have large secretory granules that contain a variety of growth factors and vasoactive substances (Barka, 1980). Until recently, it has been less clear whether or not striated duct cells have a synthetic function. Hand (1979) has demonstrated that in rats the striated ducts synthesize glycoproteins, some of which are stored in small apical granules, presumably secretory in nature; another portion consists of membrane glycoproteins which are incorporated into the plasma membrane of these cells.

Although striated duct secretory granules have now been observed in many different mammalian species, the precise nature of their contents is largely undetermined. The only mammals in which at least one granule component has been identified are cats and human beings; both have been shown by immunocytochemistry to have kallikrein activity at the apex of the duct cells (Schachter et al., 1980)-- presumably this activity resides in the small granules that are present in this zone (König and Kühnel, 1986; Tandler and Riva, 1986). Dogs lack such granules in their duct cells; instead they possess a series of tubule-like structures, some of which terminate in a bulbous expansion, that are perpendicularly appended to the inner surface of the apical membrane (Nagato and Tandler, 1986). Two other carnivores, the North American mink and the ferret, also lack duct granules; rather, the ducts contain an abundance of rhomboidal crystalloids that are present in equal numbers in the parotid and submandibular glands (Fig. 3) (Tandler, 1983; Jacobs and Poddar, 1987a, 1987b).

Striated duct granules generally range in size from 0.1 to 1.0 µm. An exception to this rule is found in the slow loris, where individual secretory granules may exceed the nucleus in size.

The number of secretory granules in striated duct cells varies from species to species. Striated duct cells in the parotid gland of the vampire bat seem to be the most highly specialized for secretion. The supranuclear region of the duct cells in this animal is crowded with small, dense granules; basal folds and their attendant mitochondria are absent and there is substantial rough endoplasmic reticulum in the infranuclear cytoplasm (Fig. 4). Little brown bats exhibit a band of small granules just below the lumen of the striated ducts; these granules have a crystalline repeat pattern (Tandler and Cohan, 1984). Similar crystalline granules are present in striated ducts of the mongoose (Schramm et al., 1979).

The abundant secretory granules in the striated ducts of the European hedgehog show an anomalous response to fixatives.

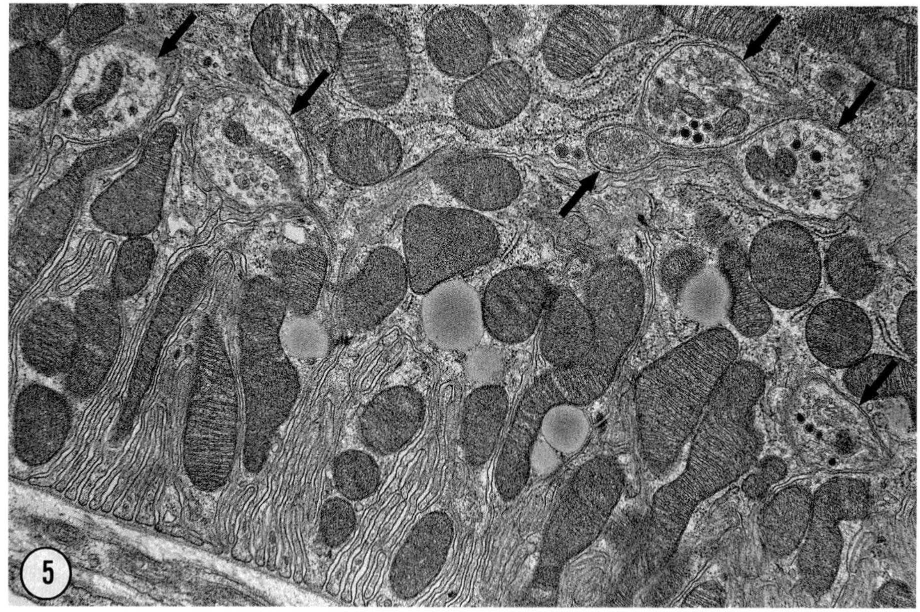

Fig. 5. The basal portion of a striated duct in the principal submandibular gland of the little brown bat, *Myotis lucifugus*. Nerve terminals are indicated by arrows. X 24,000.

In most species, initial fixation with osmium tetroxide yields "empty" appearing duct granules, whereas aldehyde fixation followed by osmium results in dense-appearing ones. In the hedgehog, the reverse is true: aldehyde fixation leads to empty granules, whereas initial fixation with osmium yields dense ones (Tandler and MacCallum, 1974). A possible explanation for these peculiar results is that the granules contain a water-soluble lipid.

Salivary gland function is largely under nervous control. Hypolemmal nerve terminals (naked axons that have penetrated the basal lamina to end between adjacent parenchymal cells) are present to a limited extent in acini, are considerably more abundant in intercalated ducts, but are very rare in striated ducts of most mammals (Cope, 1977). There are, however, a few species, notably in bats, where nerve terminals are extremely common in striated ducts. Fig. 5 shows a representative field at the base of a striated duct in the principal submandibular gland of the little brown bat-- nerve terminals are numerous. Such terminals generally contain a mixed population of synaptic vesicles. In the vampire bat, duct cells in the principal submandibular gland occasionally are linked by gap junctions, increasing the efficiency of nervous control.

In the accessory submandibular gland of the leaf-chinned bat, the duct that extends between intercalated and excretory ducts lacks the typical morphology of striated ducts. This highly vascularized duct segment is extremely convoluted, forming a structure that is somewhat reminiscent of a renal glomerulus. There are no basal striations. The luminal membrane of the duct cells is invaginated to form a capacious intracellular canaliculus rather in the manner of a gastric parietal cell. Unlike parietal cells, however, the duct cells lack the tubulovesicular array that has been implicated in secretion of hydrogen ions. Little is known of the feeding behavior of this bat, so the necessity for so unusual a duct architecture is unclear.

In conclusion, it is obvious that salivary striated ducts are more than mere conduits for saliva. They play an important role in electrolyte homeostasis and, in many species, modify the organic content of saliva either by selective uptake of molecules from the lumen or by the addition of secretory products to the saliva as this important biological fluid flows toward the mouth.

REFERENCES

Barka T (1980). Biologically active polypeptides in submandibular glands. J Histochem Cytochem 28: 836-859.

Coleman R, Hand AR (1987). Endocytosis of native and cationized ferritin by intralobular duct cells of the rat parotid gland. Cell Tissue Res 249: 577-586.

Cope CH (1977). Ultrastructure of the innervation of the ducts within the rabbit parotid gland. Arch Oral Biol 22: 715-719.

Hand AR (1979). Synthesis of secretory and plasma membrane glycoproteins by striated duct cells of rat salivary glands as visualized by radioautography after ^{3}H-fucose injection. Anat Rec 195: 317-340.

Hand AR, Coleman R, Mazariegos MR, Lustmann J, Lotti LV (1987). Endocytosis of proteins by salivary gland duct cells. J Dent Res 66: 412-419.

Harvey WR, Cioffi M, Wolfersberger MG (1981). Portasomes as coupling factors in active ion transport and oxidative phosphorylation. Am Zool 21: 775-791.

Jacob S, Poddar S (1987a). Ultrastructure of the ferret parotid gland. J Anat 152: 37-45.

Jacob S, Poddar S (1987b). Ultrastructure of the ferret submandibular gland. J Anat 154: 39-46.

König B Jr., Kühnel W (1986). Licht- und elektronenmikroskopische Untersuchungen an der Glandula parotis und der Glandula submandibularis der Hauskatze. Z Mikrosk-Anat Forsch 100: 469-483.

Nagato T, Tandler B (1986). Ultrastructure of dog parotid gland. J Submicrosc Cytol 18: 67-74.

Schachter M, Peret MW, Moriwaki C, Rodrigues JAA (1980). Localization of kallikrein in submandibular gland of cat, guinea pig, dog, and man by the immunoperoxidase method. J Histochem Cytochem 28: 1295-1300.

Schramm U, Flegler C, Ginsbach G (1979). Ultrastruktur der Glandula parotis und der Glandula submandibularis der Schleichkatze _Herpestes edwardsi_ (Viverridae). Anat Anzeiger 146: 113-132.

Tandler B (1963). Ultrastructure of the human submaxillary gland. II. The base of the striated duct cells. J. Ultrastruc Res 9: 65-75.

Tandler B (1983). Ultrastructure of the mink submandibular gland. J Submicrosc Cytol 15: 519-530.

Tandler B, Cohan RP (1984). Ultrastructure of the parotid gland in the little brown bat. Anat Rec 210: 491-502.

Tandler B, Lillie JH, MacCallum DK (1979). Cytoplasmic crystalloids in rat parotid glands. Arch Oral Biol 24: 327-335.

Tandler B, MacCallum DK (1974). Ultrastructure and histochemistry of the submandibular gland of the European hedgehog, _Erinaceous europaeus_ L. II. Intercalated ducts and granular striated ducts. J Anat 117: 117-131.

Tandler B, Riva A (1986). Salivary glands. In Mjör IA, Fejerskov O (eds): "Human Oral Embryology and Histology," Copenhagen: Munksgaard, pp. 243-284.

Cells and Tissues: A Three-Dimensional Approach by Modern Techniques in Microscopy, pages 249–256

THE GOLGI APPARATUS OF THE PANCREATIC ACINAR CELLS OBSERVED BY SCANNING ELECTRON MICROSCOPY

Tomonori Naguro and Akihiro Iino

Department of Anatomy, Tottori University School of Medicine, Yonago 683, Japan

INTRODUCTION

With the progress of specimen preparation methods and together with the improvement in the resolving power of the scanning electron microscope (SEM), the studies on the pancreas with SEM have been greatly advanced in recent years. The Osmium-DMSO-Osmium method (The ODO method) devised by Tanaka and Naguro (1981) and a revised ODO method (Tanaka and Mitsushima 1983) are widely used as the most suitable techniques for observing the intracellular structures with SEM. As previously reported we have applied the ODO method to observe the internal cell structures of the pancreatic cells and obtained good results (Naguro 1980, Naguro and Iino 1988). In this report the authors present detailed study on the intracellular structures of the pancreatic acinar cells with special attention to the three-dimensional structure of the Golgi apparatus.

Materials and Methods

The pancreata of the Wistar rats were used for this study. The practical procedure of the ODO method was descrived in the previous report (for detailes see Naguro and Iino 1988). In short, this method is characterized by the use of the maceration procedure with dilute osmium solution. In this procedure the soluble cytoplasmic matrices are removed from the cracked surface of the cells and mainly the membranous cell organelles such as endoplasmic reticulum, mitochondria and Golgi apparatus can be revealed three-dimensionally under the SEM.

RESULT

Identification of the Golgi apparatus

It is not difficult to identify the Golgi apparatus in the pancreatic acinar cells by its morphological features. The Golgi apparatus is recognized to be a complex which is composed of three elements: Golgi stack, Golgi vesicles and Golgi vacuoles (Fig.1). The Golgi stack, a main component of the Golgi aparatus, is easily distinguishable from other membranous cell organelles, since it shows a characteristic structure in that several flattened cisternae, made of smooth surfaced membranes, are piled up in parallel.

Position of the Golgi apparatus.

As one of the typical serous cells the pancreatic acinar cells are obviously polarized in the arrangement of the cell organelles(Fig.2). The Golgi apparatus in the cells is always localized at the position of the supranuclear region in contrast with that of some nerve cells where it is widely distributed all over the cytoplasm. It is interposed between the granular ER, which occupies the basal and perinuclear region, and the zymogen granules which accumulate in the apical region.

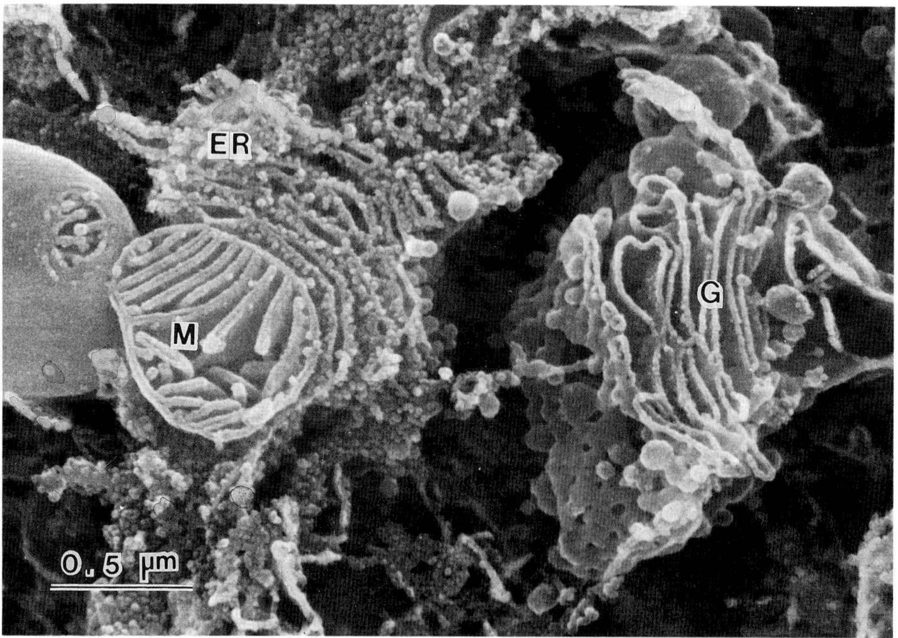

Figure 1. The Golgi apparatus (G) is easily distinguishable from other cell organelles. Granular endoplasmic reticulum (ER), Mitochondria (M).

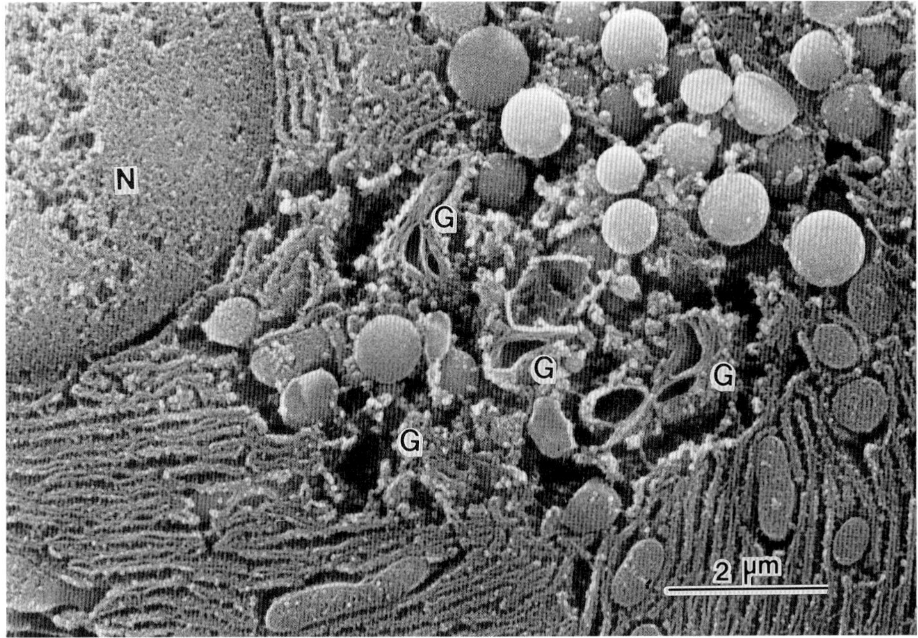

Figure 2. A pancreatic acinar cell. Golgi apparatus (G) is interposed between ER and zymogen granules. Nucleus (N).

Continuity of the Golgi apparatus

In the case of the thin sections used for TEM observation, more than one fragment of the Golgi apparatus appears to be isolated from one another in an acinar cell. Whether all of the fragments are virtually connected to each other or they exist one by one independently is a matter for argument and it is one of the most fundamental questions concerning the morphology of the Golgi apparatus. To answer the question by SEM the time cotrol for the maceration procedure is important. Zymogen granules, which accumulate at the apical cytoplasm and conc the most part of the Golgi complex from the view, tend to be extruted by the long period of the maceration procedure. The elimination of the zymogen granules from the cracked surface permits visualization of the three-dimensional structure of the Golgi apparatus(Fig. 3). From this Figure one can easily recognize the continuity of the Golgi apparatus. That is to say, on a plane surface many Golgi apparatus are found to be dispersed in a cell. However, the fact is that all Golgi complexes in a cell are three-dimensionally connected to each other to form a interconnected Golgi apparatus.

Structure of the Golgi stack

The Golgi stack in the pancreatic acinar cells consists of three to six flattened cisternae which are piled one upon the other in close parallel array. Though it may vary in its form in the different cell activity, generally it is in the form of a pile of ribbon-like cisternae, but sometimes it appears to be a pile of plate-like cisternae or a pile of slender tubules. In the case of the stack composed of the ribbon-like cisternae, it branches off into two to four at several portions and some of these branches join again to each other forming a complicated structure (Fig. 3). None of the generalized schemes about the Golgi stack as presented in the past is satisfactory for the explanation of this complicated structure. Therefore, Noda and Ogawa (1984) introduced the three-dimensional and topographical coordinates (X, Y and Z) on the Golgi stack in the study observed with the high voltage electron microscope. They termed that X axis is the elongating direction of ribbon-like Golgi cisternae, Y axis is the rectangular line to X axis on the plane formed by the cis-most cisterna of the Golgi stack and Z axis is the line perforating the Golgi stack from cis to trans vertically to both X and Y axes. We explain the feature of the Golgi stack observed by SEM in the present study in accordance with their proposition described above.

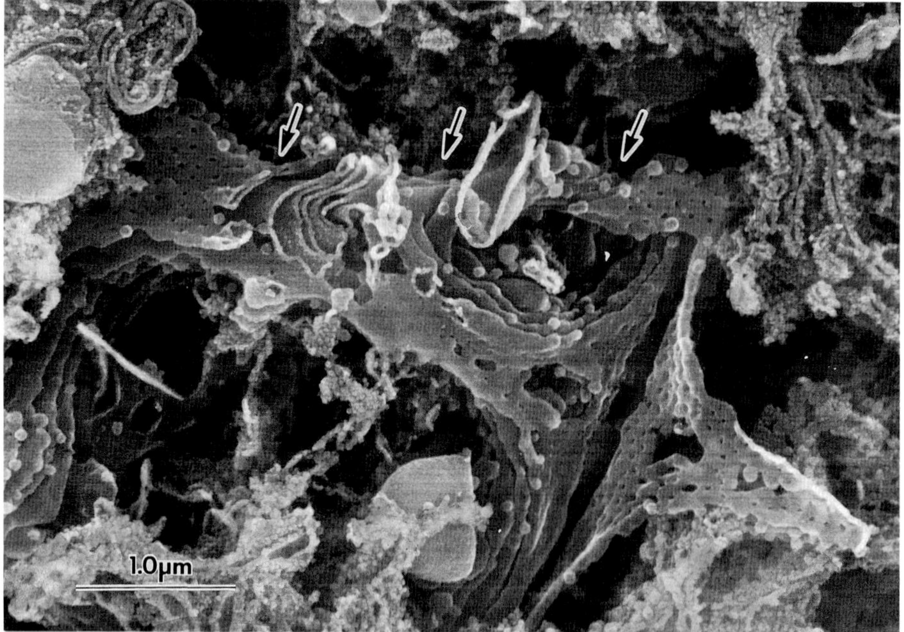

Figure 3. Ribbon-shaped Golgi stack. A stack twists and turns on its axis (arrows). Some portion of cis-most cisterna broked and turned over during specimen preparation.

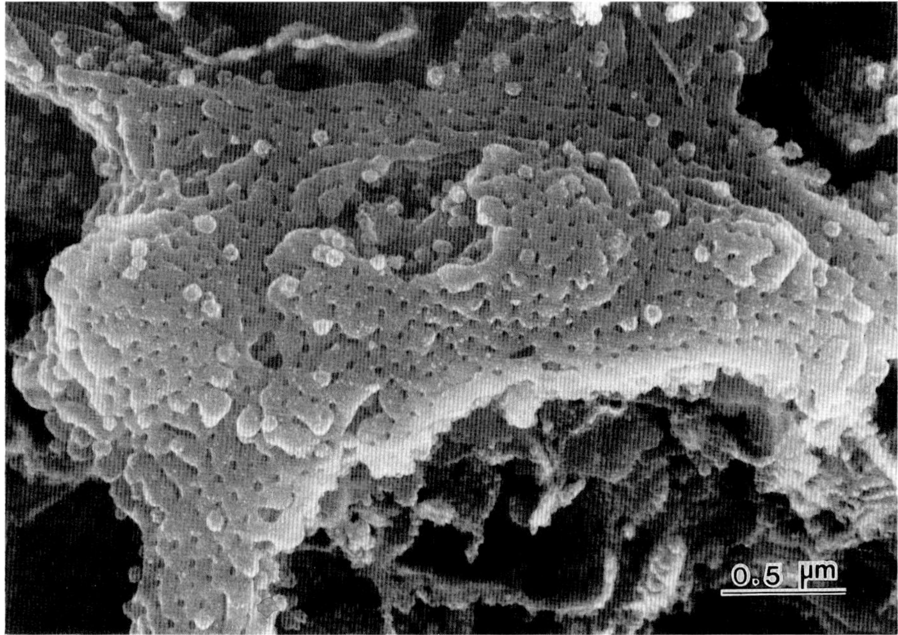

Figure 4. Cis face of the Golgi stack.

Width of the Golgi stack

The width of the ribbon-shaped Golgi stack, the distance along Y axis, usually ranges between 0.5 and 1um, but there are sometimes other portions in the form of plates of more than 2um or in the form of slender tubules of less than 0.2um. Although the stack varies in width in different portions, each individual cisterna piled along Z axis keeps its width in the fairly same size each other at any portion.

Polarity of the Golgi stack.

According to the TEM studies, the Golgi stack is said to be polarized and have two different faces: one facing to the endoplasmic reticulum is called the forming, immature, or cis face, which is usually convex and the other facing to zymogen granules is termed the mature or trans face, which is generally concave. Practically it is hard to identify the face merely by the morphological pattern of curved stack, since this pattern is often reversed.

The cis face of the Golgi stack shows a peculiar structure. Numerous small pores of about 50nm are evenly distributed on the entire surface of cis-most cisterna of the Golgi stack (Fig.4). The pores are rarely found on other cisternae. In addition to the presence of the pores, a few large openings of about 100 - 500nm are observed on the cis-most cisterna. They penetrate the whole Golgi stack along Z axis tapering off from cis to trans side.

The trans face of the Golgi stack shows relatively flat appearance as compared with the cis face, but sometimes it appears as a complicated plexus composed of anastomotic tubules (Fig.5). The continuity between these tubules and the trans-most cisterna of the Golgi stack can be found at several places. Golgi vacuoles of various size are also present at the trans face.

Rotation in the Golgi stack

In the SEM observation it is possible to identify the face of the Golgi stack by the morphological appearance, consequently, we can recognize the spatial relationships between the cis and trans sides of the Golgi stack. As shown in Figure 3, some portions of the Golgi stack twists and turns on X axis. In thin sections used for TEM studies, sometimes two adjacent Golgi stacks show their polarities in reverse direction. This phenomenon is easily understandable with the introduction of the three-dimensional interpretation that the two-dimensional image of the thin section was obtained from the portion where the Golgi stack showed rotation.

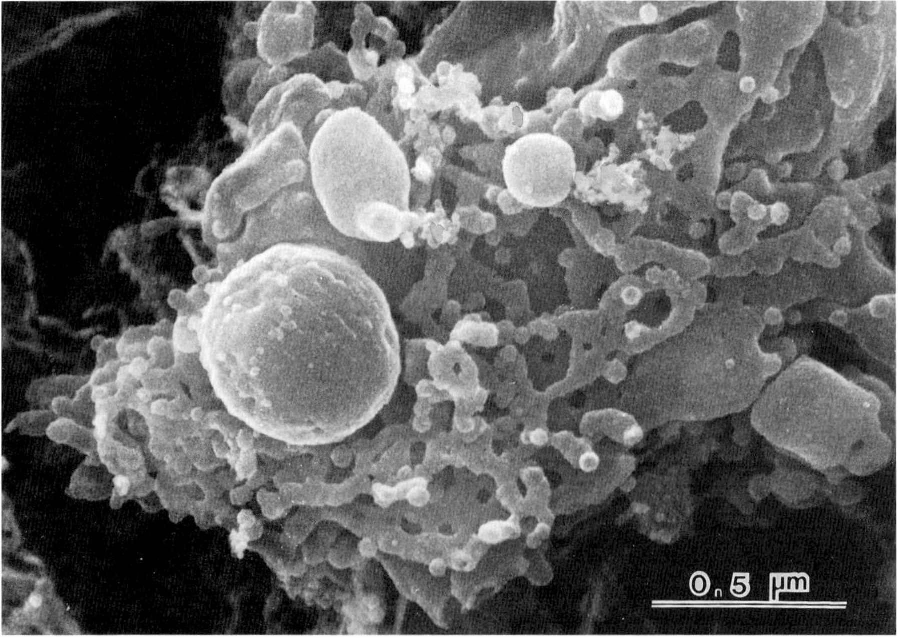

Figure 5. Trans face of the Golgi stack.

Golgi vesicles

Small vesicles of about 60 - 80nm in diameter are observed on and around the Golgi stack (Fig.3,5 & 6). They are called as Golgi vesicles, transitional vesicles or transport vesicles. Newly synthesized proteins are considered to be transported via these vesicles from ER to cis-most cisterna of the Golgi stack. Besides these vesicles, many other vesicles are supposed to take part in: transportation between the Golgi cisternae, reverse transportation from Golgi stack to ER, and formation of lysosomes. It is difficult to observe the vesicles suspended in the cytoplasm having no connection with membranous structure, since most of them disappear during specimen preparation.

CONCLUSION

Three-dimensional feature of the Golgi apparatus in the pancreatic acinar cells was reported in the present study. Tanaka et al (1986) reported that Golgi stack in the rat extralacrimal gland cell is made up of only one helically wound cisterna and in the rat motor nerve cells there are direct connections between Golgi stack and rough ER. The questions whether these findings are general or not in all kinds of cells and whether they are constant or temporary in different states of cell activity would be answered by detailed observation trough the SEM studeis.

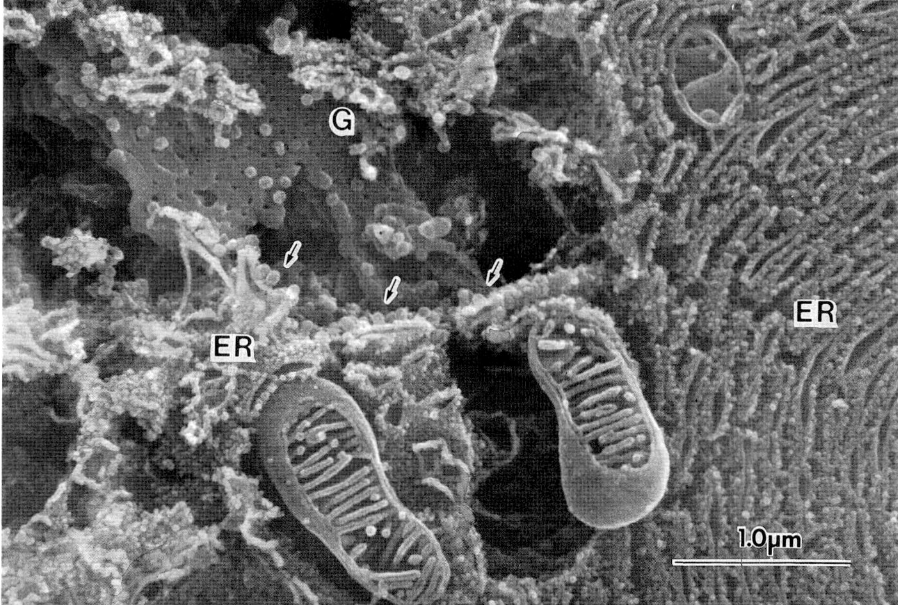

Figure 6. Small vesicles of about 60-80um are observed on the Golgi stack. Budding vesicles from ER (arrows).

REFERENCES

Tanaka T, Naguro T (1981). High resolution scanning electron microscopy of cell organelles by a new specimen preparation method. Biomed Res 2 (suppl.) 63-70.

Tanaka K, Mitsushima A (1983). A preparation method for observing intracellular structures by scanning electron microscopy. J Microscopy 133:213-222.

Tanaka K, Mitsushima A, Fukudome H, Kashima Y (1986). J Submicrosc Cytol 18:1-9.

Naguro T (1980). Three-dimensional view of the intracellular structures of pancreatic acinar cells. Yonago Acta Med 24: 81-92.

Naguro T, Iino A (1988). Three dimentional features of the pancreatic cells. In "Ultrastructure of the extraparietal glands of the alimentary canal" (A.Riva and P.M.Motta,eds) Electron Microscopy in Biology and Medicine. vol. VI, M.Nijhoff Med Pub, Boston (in press)

Noda T, Ogawa K (1984). Golgi apparatus is one continuous organalle in pancreatic exocrine cell of mouse. Acta Histochem Cytochem 17:436-451.

Cells and Tissues: A Three-Dimensional Approach by Modern Techniques in Microscopy, pages 257–262

THREE-DIMENSIONAL STRUCTURE OF THE SINUSOIDAL WALL IN THE LIVER: A GOLGI STUDY

Kenjiro Wake

Department of Anatomy, Faculty of Medicine, Tokyo Medical and Dental University, Yushima, Bunkyo-ku, Tokyo 113, Japan

INTRODUCTION

Besides the hepatocytes proper, the parenchyma of the liver contains a set of sinusoidal cells; endothelial cells, Kupffer cells, stellate cells, and pit cells (liver-associated NK cells). These cells lie in and around the hepatic sinusoid, constituting the sinusoidal residents. Some of their significant functions have been but recently elucidated.

As concerns the cellular and extracellular components of the sinusoids, the ultrastructural study has been striking in depth and enormous in amount, while the three-dimensional morphology remains an open question. The full extent of the wide-spreading endothelial cell and the vast-stretching stellate cell is not seen to advantage by conventional electron microscopy. In regard to this, the author has re-evaluated the traditional rapid Golgi method, studying with the combined use of modern methodologies, and has been able to demonstrate the tridimensional configuration, presented with minute surface details, of the constituent cells that form the sinusoidal wall.

MATERIALS and METHODS

Livers of several vertebrates were perfused via portal veins with 1.5 % cacodylate-buffered glutaraldehyde (pH 7.4). Blocks of tissue thus fixed were subjected to the rapid Golgi silver method (Wake et al. 1988), and prepared in 50 µm-thick sections for light microscopy. Quantitative measurements were done under an oil-immersion objective and stereo-pair micrographs of individual cells were recorded by a BH-2 Stereo-microscope (Olympus, Tokyo). Some of the sections were re-embedded in Poly-Bed 812 and cut at 0.5 - 5.0 µm thickness and were examined in TEM's at accelerating voltages of 200, 300, and 1000 kV. For observation in a SEM under backscattered mode, the Golgi-treated specimens were cracked in liquid nitrogen to create fractured surfaces. Routine ultrathin sections of the livers without the Golgi staining were produced parallelly and observed by TEM at 100kV.

RESULTS

The Golgi method unveils well distinguishably the endothelial cells and the stellate cells with remarkable details, each cell type being though stained at an unpredictable frequency. These cells are readily in their entirety by light microscopy, due to the viewable silver precipitates that occur equally in the cell body and the cytoplasmic processes throughout. These cell types are definitely different in appearance; membranous cytoplasmic processes of the endothelial cells are observed with thick and thin portions, while the dendritic processes of the stellate cells are found surrounding the sinusoids circumferentially. Despite the treatment with Golgi staining, fine structures of these cells are well preserved.

Endothelial cells

The whole profile of individual endothelial cells can be appreciated three-dimensionally from the stereo-pair micrographs (Fig. 1). Their membranous processes, which involve thick and

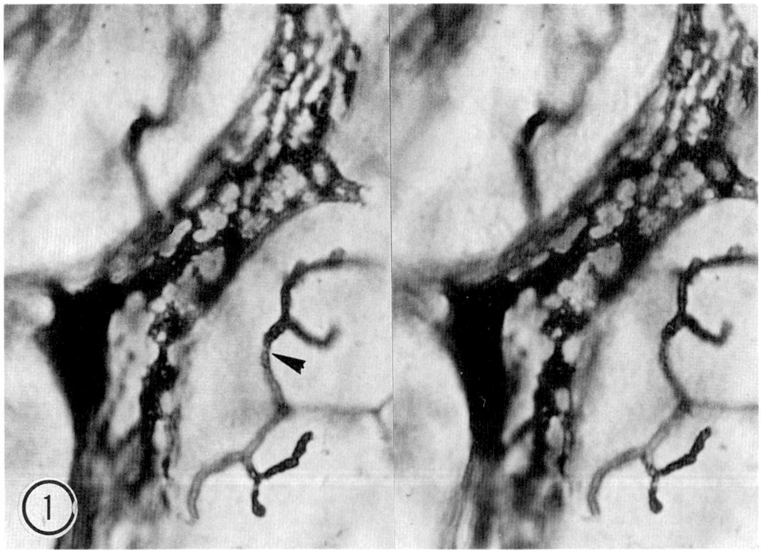

Figure 1. Stereo-pair micrographs showing an Golgi-stained endothelial cell from the rat liver, taken with a BH-2 Stereo-microscope. Arrow head, bile canaliculus.

thin portions, extend to cover the neighboring sinusoidal channels and dominate the branching sites of the sinusoids. The thick portions emit from the perikaryon, run obliquely or longitudinally towards the cell periphery, and taper off. The rest of the cytoplasmic processes is occupied by the presence of numerous pores or fenestrae. The unconventional slender processes extend from the outer surface of the endothelial cell. These processes proceed into the interhepatocellular clefts where some of them, referred to as a 'tail', terminate as small swelling. The other, called 'intersinusoidal processes or bridges', extend to the neighboring sinusoids. At an accelerating voltage of 200kV, TEM of the semithin Golgi sections reveals numerous electron-lucid fenestrations of the sieve plates in the endothelial cells (Fig. 3). The observed feature of individual endothelial cells varies from zone to zone. The thick portions of those cells in the periportal zone are more dominant than that of those found in the centrilobular zone. In contrast, the thin portions are more elaborate in the centrilobular zone as compared to the periportal zone. The area of distribution of a single endothelial cell amounted to 3471.3 ± 151.6 (SE) μm^2 periportally, and 5068.2 ± 309.0 μm^2 centrilobularly.

Perisinusoidal stellate cells

The three-dimensional entire forms of individual stellate cells differ notably from species to species and vary appreciably with the age of the animals. A Golgi-stained stellate cell of a six-month-old pig's liver is shown in Figure 2. Generally,

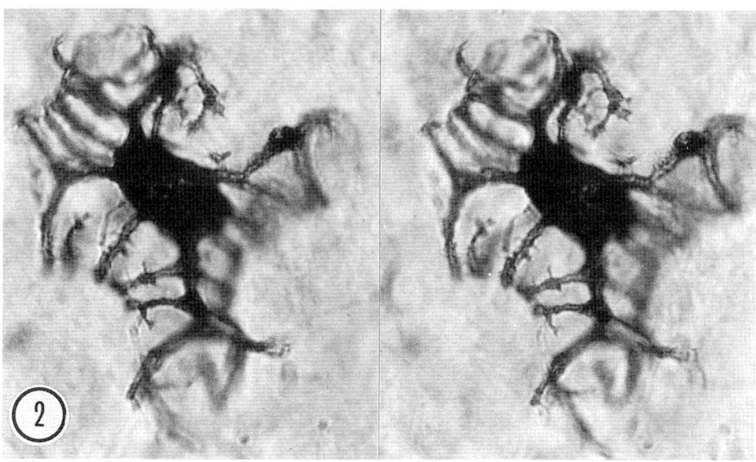

Figure 2. Stereo-pair micrographs of a perisinusoidal stellate cell from the porcine liver. Rapid Golgi method.

the cells in livers of lower vertebrates and of the young in higher vertebrates, are more simple in form. The cell bodies are somewhat spindle-shaped or angulated or even rounded. The lipid droplets, which contain retinol in the form of retinyl ester, may be recognized in the silver-precipitated cytoplasm. Following administration of excess vitamin A to the animals, the cells round up storing greater amount of lipid.

The cell bodies display two patterns of cytoplasmic processes, perisinusoidal and intersinusoidal. The former emerge laterally alongside the sinusoid with a series of side-branches regularly emitted to embrace the sinusoid (Figs. 2 and 4). The latter, much less in number, arise from the abluminal aspect of the cell body or of proximal part of the perisinusoidal processes, traverse the hepatic cell plate through the interhepatocellular space, and abut on the sinusoid nearby. The former are studded with thorn-like microprojections (Fig. 4), while the latter are of smooth surface.

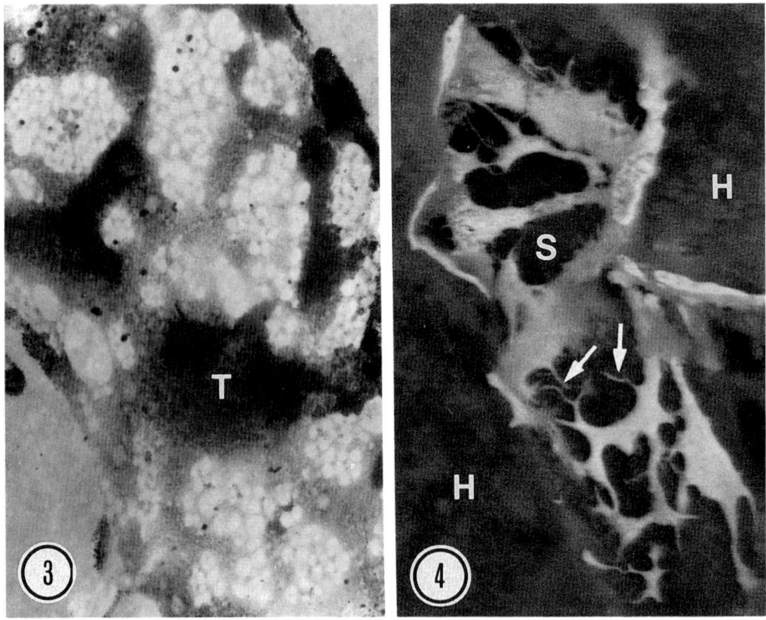

Figure 3. Electron micrograph of a semithin section from Golgi preparation of the rat liver,taken at accelerating voltage of 200kV, showing a part of cytoplasmic process with sieve plates. T, thick portion.
Figure 4. Backscattered electron imaging of a Golgi-stained stellate cell in the porcine liver. Arrows, thorn-like micro-projections; H, hepatocyte; S, sinusoid.

Stratification of the sinusoidal wall

Electron microscopic study of ultrathin sections reveals that the above-described cells together with other cellular and extracellular components of the hepatic sinusoids, are arranged in concentric layers. Considered from the lumen outwardly, the Kupffer cells and pit cells are lodged innermost. The endothelium lines an attenuated - yet discrete - layer with an underlying discontinuous basement membrane. The outermost is contributed by the perisinusoidal stellate cells whose cytoplasmic processes establish close apposition with the endothelial cells of the more central layer. Collagen fibrils and, at least in the porcine liver, oxytalan fibers occur facing the space of Disse to form a loose adventitial coat for the sinusoidal wall.

DISCUSSION

Zimmermann (1923) described as well the "Pericyten (pericytes)" as the endothelial cells in the liver based on the Golgi method. The present study re-evaluated the virture of Golgi method considerately and re-confirmed the result carefully by several sophisticated techniques of investigation. From the observed findings which ascertain the cells with their location and the capability of vitamin A-storage, it is concluded that Zimmermann's pericytes are identical to perisinusoidal cells which were earlier named as "Sternzellen (stellate cells)" by Kupffer (1876) (Wake, 1980), and were also referred to as "fat-storing cells" by Ito (1951).

The endothelial cells and the stellate cells are, by the presented Golgi staining, brought into view evidently in the entire forms, the cell bodies with their elaborate processes. These cells adhere together closely, separated by a heavily-interrupted discontinusous basement membrane. Neighboring sinusoids are united with intersinusoidal processes of both cells forming an interconnected network pervading throughout the liver.

The outstanding feature of the perisinusoidal processes of the stellate cells is the presence of numerous thorn-like microprojections. Though the significance of these microprojections is unknown, they are virtually neither microvilli nor filopodia in nature. Recent biochemical studies show that the stellate cells incorporate the retinol-RBP complex, that are secreted into the space of Disse from the hepatocytes, by the receptor-mediated endocytosis (Blomhoff et al. 1987). This metabolic role is supported in part by the complexity in cell shape and the largeness in cell size which result in increment of effective surface area responsible for the absorption.

The proportions of the thick and thin portions in the endothelial cells are variable, depending on the zone of the hepatic lobules. This finding coincides with the fact that the total area of the fenestrations of the sieve plates per unit square is larger in the centrilobular than in the periportal zone (Wisse et al. 1985). The predominance of the thin portion with greater numbers of fenestrae in the centrilobular zone can possibly facilitate transmural exchange of various substances between the space of Disse and the sinusoidal lumen, thus the decrement of PO_2 in the sinusoidal flow downstream being compensated by this situation.

ACKNOWLEDGEMENTS

The excellent technical assistance of Mr. Kiyoyuki Motomatsu and Mr. Akira Masuda is greatly acknowledged. The author thanks Dr. Kenji Kaneda and Dr. Chieko Dan for valuable discussions and Dr. Wichai Ekataksin for reviewing the English manuscript. This work was supported by a Grant-in-Aid (58370001) from the Ministry of Education, Science, and Culture, Japan.

REFERENCES

Blomhoff R, Berg T, Norum KR (1987). Absorption, transport and storage of retinol. Chemica Scripta 27: 169-177.

Kupffer C (1876). Ueber Sternzellen der Leber. Briefliche Mitteilung an Prof. Waldeyer. Arch Mikr Anat 12: 353-358.

Ito T (1951). Cytological studies on stellate cells of Kupffer and fat storing cells in the capillary wall of the human liver. Acta Anat Nippon 26: 2.

Wake K (1980). Perisinusoidal stellate cells (fat-storing cells, interstitial cells, lipocytes), their related structure in and around the liver sinusoids, and vitamin A-storing cells in extrahepatic organs. In Bourne GH, Danielli JF (eds) International Review of Cytology Vol 66, pp 303-353. (New York: Academic Press).

Wake K, Motomatsu K, Dan C, Kaneda K (1988). Three-dimensional structure of endothelial cells in hepatic sinusoids of the rat as revealed by the Golgi method. Cell Tissue Res 253: 563-571.

Wisse E, De Zanger RB, Charels K, Van Der Smissen P, McCuskey RS (1985) The liver sieve: considerations concerning the structure and function of endothelial fenestrae, the sinusoidal wall and the space of Disse. Hepatology 5: 683-692.

Zimmermann KW (1923). Der feinere Bau der Blutcapillaren. Z Anat 68: 29-109.

Cells and Tissues: A Three-Dimensional Approach by Modern Techniques in Microscopy, pages 263-267

THE MORPHOLOGY OF HEPATIC GLYCOGEN SYNTHESIS

Robert R. Cardell and Emma Lou Cardell

Department of Anatomy and Cell Biology, University of Cincinnati College of Medicine, Cincinnati, Ohio 45267

Permit us to join our distinguished colleagues in their recognition of the exceptional scientific contributions of Marcello Malpighi. As researchers who have studied the structure and function of the liver for most of our careers (Cardell, 1977; Garfield and Cardell, 1987), We are personally grateful to Dr. Malpighi for his valuable and scholarly study of this organ. In 1666, his description of the lobular organization of the liver was published (Malpighi, 1666) and in the 3 centuries since that writing, this concept has been the basis for numerous studies by a host of scientists attempting to relate the functions of the liver to its structure. In the 20th century the electron microscope has enabled us to finely "dissect" the liver so that we can now describe its ultra-structural morphology - the various cell types present, and the organelles within these cells. In the future the application of molecular biological approaches will clarify some of the functional and mechanistic mysteries which are not fully understood presently.

The liver, a vital organ with very diverse and essential functions, is a major site of glycogen deposition (Figure 1). In late gestation of the developing vertebrate, hepatocytes deposit glycogen which provides a source of glucose which is crucial to the survival of the animal immediately after birth and throughout life. Glycogen, an insoluble complex carbohydrate, is a reserve for glucose, the "fuel" of the body. So essential is glucose to the organism that the liver is capable of synthesizing glycogen, and thus glucose, from gluconeogenic precursors when circulating glucose levels become too low and dietary glucose is not available. Although a great deal is known about the biochemistry of glycogen metabolism, many questions remain about this process: why does glycogen deposition occur in limited areas, not throughout the cytosome, of hepatocytes? Why do hepatocytes which share the same sinusoid, and thus are bathed in the same milieu, sometimes differ in the quantity and stage of glycogen deposition? Where exactly are the enzymes that catalyze the biochemical reactions? Why are hepatocytes near portal veins morphologically different from those near central veins? What are the signals, the stimulators, and the regulators; and how do they relate to the morphology we observe in various natural and experimental situations? Research on the liver must continue in order to shed light on these and other questions.

Our studies (Cardell, 1977) of control-fed normal rats indicated that glycogen distribution within hepatocytes and across the lobule reflected the nutritional state of the animal: during glycogen deposition after feeding fasted rats, the glycogen masses in cells around the portal triad tended to be very compact, whereas those in cells near the central vein were more dispersed; SER was near the newly deposited glycogen in both cell types, however. At extremely high glycogen levels, even centrilobular cells could be observed with regions of compact glycogen.

Utilizing adrenalectomized rats which were fasted overnight to deplete their glycogen stores, we observed that shortly following the administration of a synthetic glucocorticoid, dexamethasone, and without refeeding, a few hepatocytes began to synthesize glycogen (Jerome and Cardell, 1983; Cardell et al., 1985) Earliest responding cells, however, were not necessarily side by side or located in a specific area of the lobule, although overall as glycogen deposition continued, more centrilobular cells had glycogen deposits earlier than periportal cells. Eventually all hepatocytes deposited glycogen but it is not clear why some cells responded earlier than others. Another consistent observation is that the early responding cells deposit glycogen in small localized regions of the cytoplasm (Figures 2 and 3). These regions have smooth endoplasmic reticulum (SER) and small glycogen particles which can be visualized in the electron microscope, but what other components are localized there -- enzymes? substrate? ions? We are presently working on ways to better characterize these SERGEs (Smooth Endoplasmic Reticulum, Glycogen or substrate, and Enzymes presumed to be present) our acronym for the specific cytoplasmic regions where glycogen is deposited (Cardell, et al., 1985).

Studies of diabetic animals which were not given insulin to control the diabetes, showed that the glycogen in most hepatocytes was dispersed, regardless of lobular localization. Thus during glycogen breakdown as in diabetes (Garfield and Cardell, 1979), in newborns shortly after birth (unpublished observations), and in fasting animals, a great deal of SER among the glycogen gives a dispersed pattern regardless of location within the lobule.

Autoradiographic studies enabled us to localize newly synthesized glycogen in which label from ^{3}H galactose had been

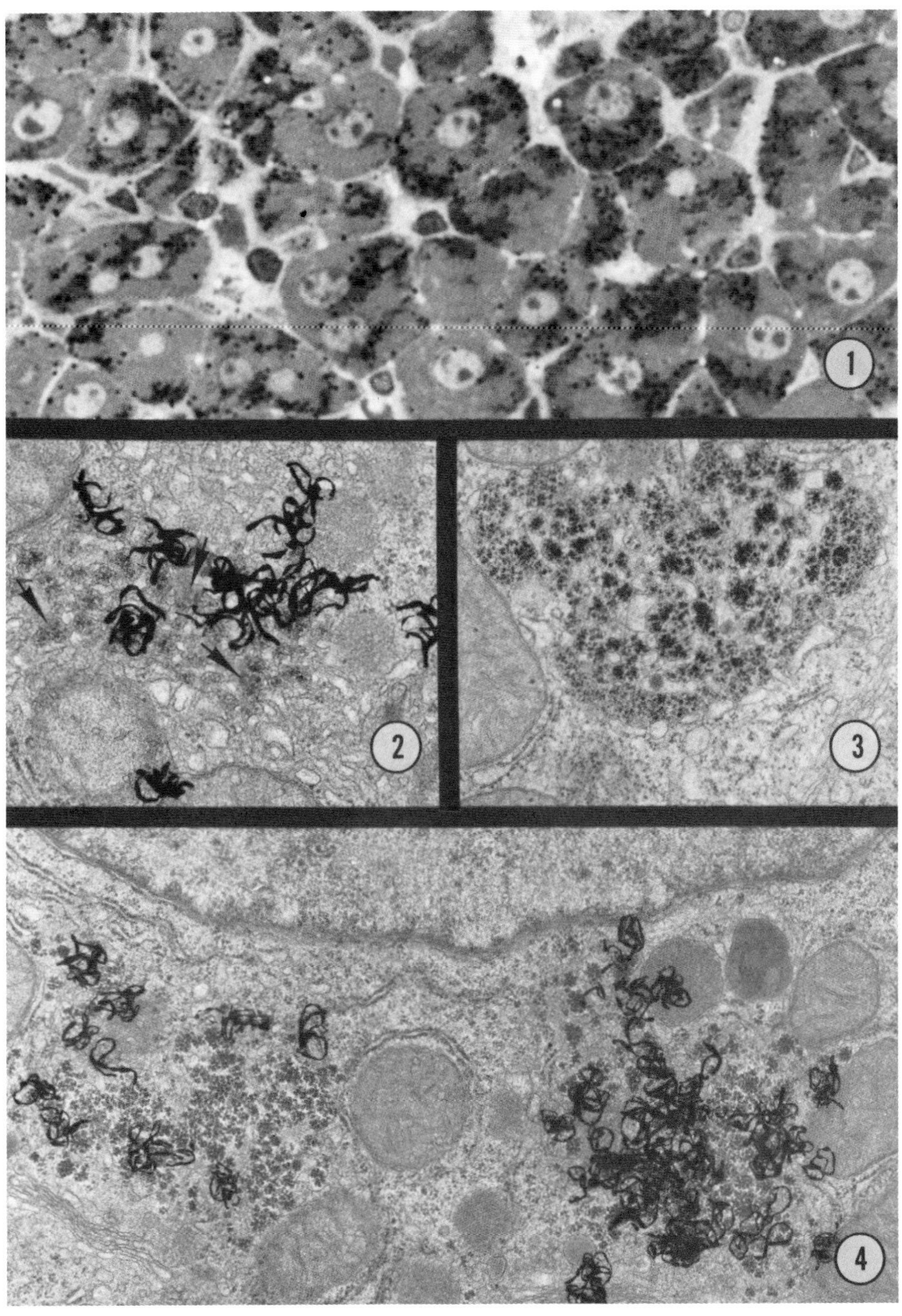
1
2
3
4

incorporated into the complex carbohydrate (Figures 1, 2, and 4). We observed interesting variations in the labeling patterns, especially at short intervals after administration of the isotope (Cardell, et al., 1985). Our specific interest was whether the newly synthesized and labeled glycogen was near SER. In liver from overnight fasted-adrenalectomized rats given dexamethasone and ^{3}H galactose, small focal areas of early glycogen deposits were heavily labeled (Figure 2). In these SERGEs, SER was evident and close to the glycogen. These morphological observations are corroborated by Margolis' (Margolis et al., 1979) studies of highly purified membrane fractions in which both synthase phosphatase (involved in glycogenesis) and phosphorylase phosphatase (involved in glycogenolysis), were found in rough and smooth microsomal fractions. At later intervals when glycogen patches were quite large (Figure 4), label was not necessarily over glycogen near SER, but occured also distant from the membranes over glycogen particles. This may reflect rapid translocation of newly synthesized glycogen, or nearby SER out of the plane of section.

We interpret our observations to affirm that ER, specifically SER, is involved in glycogen synthesis as well as glycogenolysis. The precise role for this organelle in glycogen synthesis is unclear. Perhaps the membranes act as a carrier for certain required enzymes in the glycogenesis process. The SER membranes may play a role in translocating the "glycogenin" molecule recently identified as important in the initial stages of glycogen synthesis (Smythe, et al., 1987). Alternatively, the organelle may function to regulate ionic calcium levels in subcellular compartments of the cell (Garfield and Cardell, 1987). Presently we are developing antibodies and probes for a more precise localization of enzymes related to glycogen metabolism. Because the liver is essential for the regulation of blood glucose levels, any irregularities in hepatic glycogen metabolism may affect the well-being of the individual. Our scientific interests in elucidating the minute details of the mechanisms of glycogen synthesis and breakdown may contribute to the understanding of such health problems as diabetes, glycogen-storage disease, and others. Hopefully, such studies may lead to new ways of treatment or even cures for metabolic diseases.

REFERENCES

Cardell RR (1977). Smooth endoplasmic reticulum in rat hepatocytes during glycogen deposition and depletion. Int Rev Cytol 48:221-279.

Cardell RR, Michaels JE, Hung JT, Cardell EL (1985). SERGE, the subcellular site of initial hepatic glycogen deposition in the rat: A radioautographic and cytochemical study. J Cell Biol 101:201-206.

Garfield SA, Cardell RR (1979). Hepatic glucose-6-phosphatase activities and correlated ultrastructural alterations in hepatocytes of diabetic rats. Diabetes 28:664-679.

Garfield SA, Cardell RR (1987). Endoplasmic reticulum: Rough and smooth. Int Rev Cytol 17:255-273.
Jerome WG, Cardell RR (1983). Observations on the role of smooth endoplasmic reticulum in glucocorticoid-induced hepatic glycogen deposition. Tissue & Cell 15:711-727.
Malpighi M (1666). De viscerum structura exercitatio anatomica. Bononiae.
Margolis RN, Cardell RR, Curnow RT (1979). Association of glycogen synthase phosphatase and phosphorylase phosphatase activities with membranes of hepatic smooth endoplasmic reticulum. J Cell Biol 83:348-356.
Smythe C, Pitcher J, Cohen P (1987). The initiation of glycogen biosynthesis. J Cell Biol 105:298a.

LEGENDS FOR MICROGRAPHS

Figure 1. This light microscopy autoradiograph of hepatocytes between a portal triad and a central vein show dark-stained glycogen regions which are restricted to specific areas of the cytoplasm. Silver grains, the small round black dots, are primarily over the glycogen masses. Notice that some cells have less glycogen than others, and some glycogen patches are more heavily labeled than others. 819X

Figure 2. When glycogen-depleted hepatocytes are stimulated to synthesize glycogen, very small cytosomic regions with glycogen can be found within the cell. This early SERGE is a localized region of SER, glycogen or its precursors, and probably enzymes involved in the process. Glycogen (arrows) in this micrograph is smaller than the typical α and β particles. Silver grains indicate the incorporation of ^{3}H galactose which had been provided this overnight fasted adrenalectomized rat treated with dexamethasone for 2 hrs. 37,200X

Figure 3. SERGE areas become larger as length of time after stimulus to deposit glycogen increases. Glycogen particles become more typical in appearance. 26,390X

Figure 4. Two regions within this hepatocyte show glycogen in rosettes (β particles). Notice that the glycogen patch on the right is more heavily labeled than that on the left. 25,480X

Cells and Tissues: A Three-Dimensional Approach by Modern Techniques in Microscopy, pages 269–275

MORPHOLOGICAL ASPECTS OF CHOLESTEROL STORAGE IN THE HUMAN GALLBLADDER.

Liliana Luciano

Laboratory of Cell Biology and Electron Microscopy, School of Medicine, Konstanty-Gutschow-Str. 8, 3000 Hannover 61, FRG

INTRODUCTION

A serious problem in morphological studies of the human gallbladder is obtaining satisfactory preservation of the structure through rapid fixation of the specimens. Cholecistotomized organs are considered the most suitable approach to this problem. On the other hand, such surgically removed gallbladders are all affected by pathology to different degrees.

Cholesterosis, first recognized by Virchow (1857), is a very common gallbladder disorder (Salmenkivi, 1964) which is characterized by an accumulation of cholesterol in the gallbladder wall. Usually this process is extended to the entire surface of the mucosa and is visible macroscopically (strawberry gallbladder, McCarty and Corkery, 1920). However, in several cases only some areas, or even one portion of the mucosa, are involved while the interposed regions have a normal appearance. The special interest in electron microscopical research of cholesterosis gallbladder is due to the possibility of investigating the features of these two regions with the purpose of collecting morphological data concerning the nearly normal gallbladder wall and of studying the poorly understood mechanism of cholesterol storage in foam cells which reside in the lamina propria (Luciano and Wolpers, 1973; rev. by Luciano and Reale, 1989).

In this investigation 8 gallbladders affected by cholesterosis at different degrees of severity obtained from patients (seven women and one man ageing from 27 to 72 years) undergoing cholecistectomy were examined.

The specimens were taken from the two above mentioned regions and immediately fixed either in OsO_4 solutions or prefixed in aldehyde solutions and then postfixed in OsO_4. The specimens were processed for scanning (SEM) and transmission (TEM) electron microscopy.

RESULTS

SEM of the regions of the mucosa with a macroscopically normal appearance (Fig. 1a) revealed numerous folds covered by a seemingly uniform type of epithelial cells. TEM demonstrates, however, that actually several types of cells compose this epithelium (rev. by Luciano and Reale, 1989).

The regions of the mucosa where the storage of the cholesterol occurred (Fig. 1b) revealed in SEM a completely different morphology when compared with the macroscopically normal regions.

Numerous folds of the epithelium, but not all, displayed gibbosities of very irregular size (Fig. 1b). Folds of intermediate size between the apparently normal and the large humps were also present (asterisks in Fig. 1b). The large humps nearly always appeared cracked at the summit in such a way that the epithelium retracted laterally and the underlying basement membrane or even the lamina propria was exposed. This phenomenon, probably occurring during dehydration and/or critical point drying, allowed the visualization and therefore the study of the inside of the mucosa. These aspects unmistakably showed that the site of the accomplished cholesterol storage implicated only the lamina propria and not the epithelium (Fig. 1b).

In en face view, the epithelial cells displayed mainly the same morphology as in the normal region, e.g., a well developed brush border. In the cracked regions the lateral infoldings of the epithelial cells became visible. Under the basement membrane, formed by a rich anastomosing network of unstructured soft material, the numerous foam cells revealed their characteristic roundish form and their surface covered by craters (Fig. 4).

Thin sections demonstrated that beside areals of the mucosa in which the enormous amount of foam cells with their cholesterol content completely filled the lamina propria until deformation - the humps in SEM -, other regions were present, in which the cholesterol storage is in progress - intermediate folds in SEM -. At these sites very numerous macrophages could be detected crossing the basement membrane (Fig. 3). These macrophages were characterized by large cisternae of smooth endoplasmic reticulum that could fill extensive areals of cytoplasm (Figs. 2,3a). In the lamina propria several roundish cells were seen with morphological characteristics of both the migrating macrophages (Fig. 3a) and the foam cells (Fig. 3b).

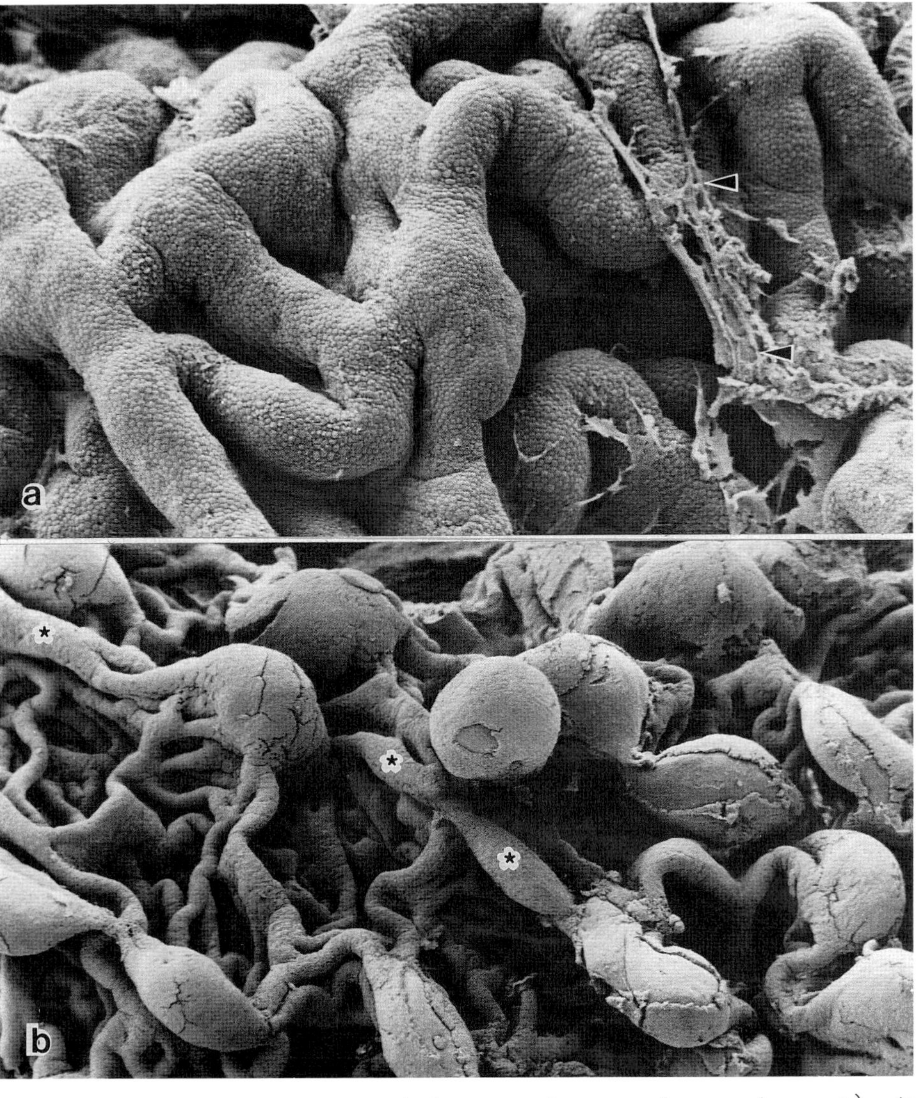

Fig. 1a,b. Gallbladder cholesterosis, surface view. a) An apparently normal mucosa with remnants of mucus (arrowheads). b) A "strawberry" region of the same specimen. In the large roundish humps of this region the process of cholesterol storage is accomplished; in the fusiform folds (some marked by asterisks) the process is in progress. a x 100, b x 28.

DISCUSSION

Cholesterosis of human gallbladder is due to an altered synthesis of hepatic cholesterol associated with the transfer

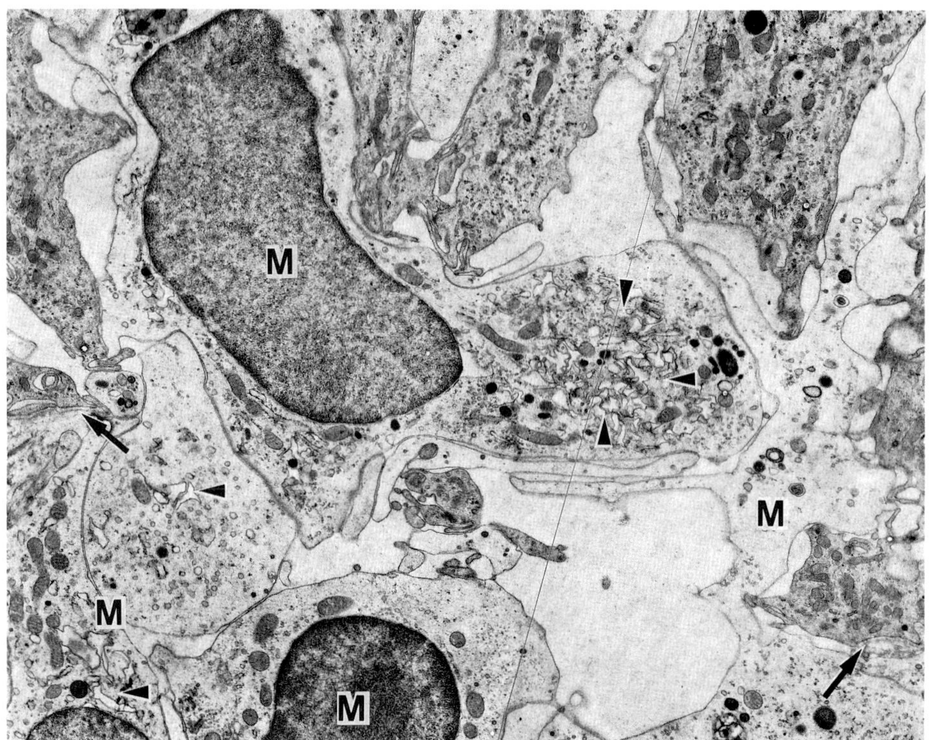

Fig. 2. Cross section of the mucosa in a region where cholesterol storage occurs. Several macrophages (M) crossing the epithelial basement membrane. The arrowheads point to the numerous cisternae of the smooth endoplasmic reticulum. The two arrows indicate the interruption of the basement membrane. x 6.000

of free cholesterol from the bile to the gallbladder mucosa (Tilvis et al., 1982a). As recently demonstrated (Jacyna et al., 1987) free cholesterol of the bile content can enter the epithelium through a passive as well as a partially active process. However, how the absorbed cholesterol reaches the lamina propria and how its storage occurs is not yet established. The results of the present study strongly suggest that the principal cells and the numerous macrophages crossing the epithelial basement membrane could both be involved in the mechanism of cholesterol storage: the principal cells in the absorption of free cholesterol from the bile, and the macrophages in the uptake of cholesterol from the intercellular space, as well as its storage and transport from the epithelium to the lamina propria. Here they become foam cells.

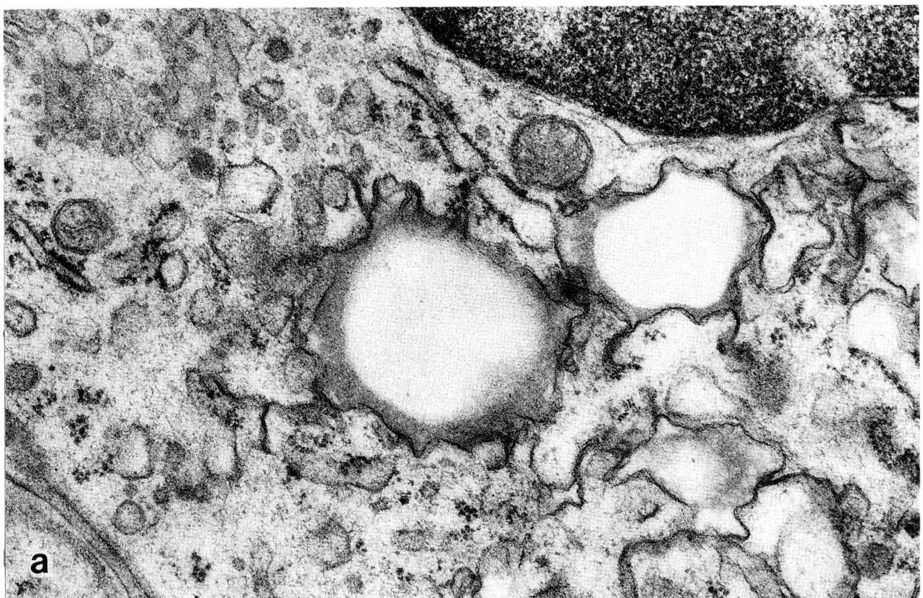

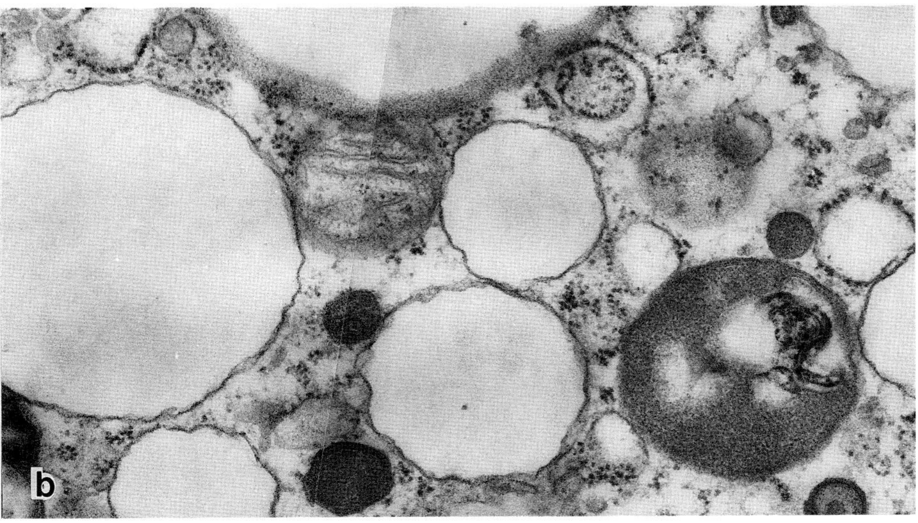

Fig. 3a,b. a) Part of a macrophage showing transitional forms from the smooth endoplasmic reticulum to granula. b) the cytoplasm of a foam cell containing several granules of different diameter. a,b x 37.000.

Studies in progress using digitonin as a marker for cholesterol reveal a large amount of digitonin-cholesterol-complexes not only inside the epithelial principal cells (Luciano and Reale, 1989) but also in the intercellular space (Luciano and Reale, in preparation). The macrophages that are

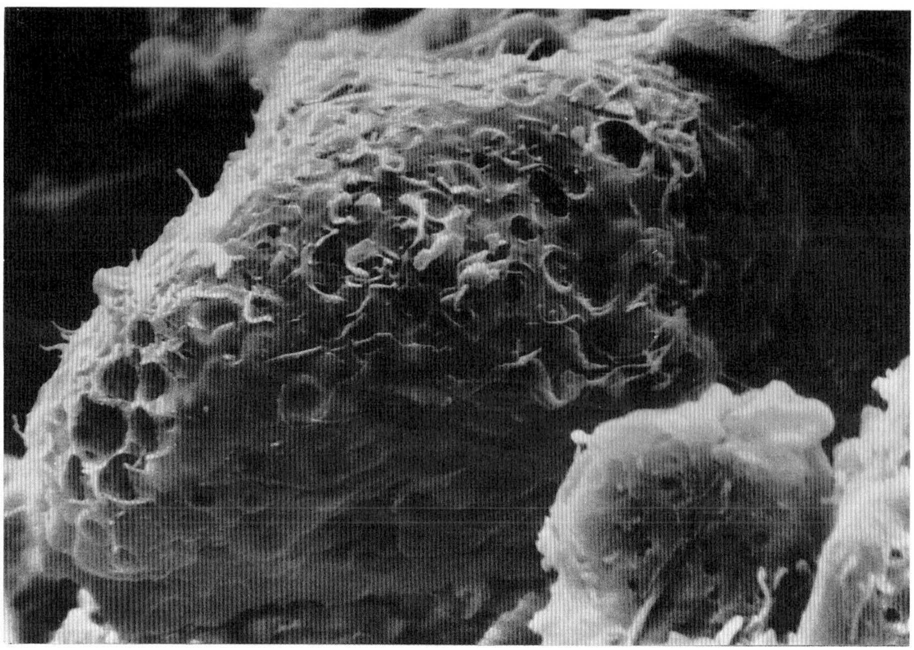

Fig. 4. Surface view of macrophages inside the lamina propria. Few microvilli, several large plicae (or lamellipodia) and numerous crater like openings which probably correspond to the granules which have lost theit content. x 4.150.

located inside this space possibly absorb the cholesterol. However, the significance of the large amount of smooth endoplasmic reticulum in the cytoplasm of the macrophages is not clear. The presence of transitional forms between the smooth cysternae of endoplasmic reticulum and the granula of the foam cells suggests that a transformation from one structure to the other occurs.

Absorption of cholesterol by the gallbladder epithelium takes place only in free form but not as cholesterol ester (Niederhiser et al., 1976). The gallbladder mucosa contains both unesterified and esterified cholesterol and this second form largely prevails (Tilvis et al., 1982a). Since previous studies show that free cholesterol could be demonstrated only in the epithelium of the gallbladder whereas in the foam cells it appeared bound i.e. in an esterified form (Luciano and Reale, 1989) it is conceivable that the esterification process occurs in the macrophages. The endoplasmic reticulum could represent the site of synthesis of fatty acids which are essential for the esterification of the cholesterol (Niederhiser et al., 1976).

This hypothesis is in agreement with previous _in vitro_ studies demonstrating that 1) peritoneal macrophages in suspension are able to synthetize phospholipids;2) the addition of cholesterol to the incubation medium causes a highly significant increase in the incorporation of phosphate by macrophages into phospholipid (Day et al., 1966). Finally, 3) triglycerides, that are not present in human bile, are effectively synthetized in vitro by the gallbladder mucosa with cholesterosis (Tilvis et al., 1982b).

REFERENCES

Day AJ, Fidge NH, Wilkinson GN (1966). Effect of cholesterol in suspension on the incorporation of phosphate into phospholipid by macrophages in vitro. J Lipid Research 7:132-140.

Jacyna MR, Ross PE, Bakar MA, Hopwood D, Bouchier IAD (1987). Characteristics of cholesterol absorption by human gallbladder: relevance to cholesterosis. J Clin Pathol 40:524-529.

Luciano L, Wolpers C (1973). Die Feinstruktur der Gallenblase und der Gallengänge. IV. Die Cholesterose der menschlichen Gallenblasenschleimhaut. Virchows Arch Abt B Zellpath 14:147-158.

Luciano L, Reale E (1989). The human gallbladder. In Riva A, Motta PM (eds.): "Ultrastructure of the Extraparietal Glands of the Alimentary Canal-Physiophatological Aspects", Vol VI, Boston: Nijhoff Publ (in press).

McCarty WC, Corkery JR (1920). "Early lesions in the gallbladder". Amer J Med Sci 159:646-753.

Niederhiser DH, Harmon CK, Roth HP (1976). Absorption of cholesterol by the gallbladder. J Lipid Res 17:117-124.

Salmenkivi K (1964). Cholesterosis of the gall-bladder: a clinical study based on 269 cholecystectomies. Acta Chir Scand 324 (Suppl):1-93.

Tilvis RS, Aro J, Strandberg TE, Lempinen M, Miettinen TA (1982a). Lipid composition of bile and gallbladder mucosa in patients with acalculous cholesterosis. Gastroenterol 82:607-615.

Tilvis RS, Aro J, Strandberg TE, Lempinen M, Miettinen TA (1982b). In vitro synthesis of triglycerides and cholesterol in human gallbladder mucosa. Scand J Gastroenterol 17:335-340.

Virchow R (1857). Über das Epithel der Gallenblase und über einen intermediären Stoffwechsel des Fettes. Virchows Arch path Anat 11:574-578.

Cells and Tissues: A Three-Dimensional
Approach by Modern Techniques in Microscopy,
pages 277–280

HUMAN FETAL LIVER CELLS AS SEEN BY SCANNING ELECTRON MICROSCOPY.

Sayoko Makabe*, Guido Macchiarelli and Pietro M.Motta.

*Department of Obstetrics and Gynecology, Toho University, Tokyo, Japan and Department of Anatomy, University of Rome "La Sapienza", via A. Borelli 50, 00161 Rome, Italy.

INTRODUCTION

During the last years scanning electron microscopy (SEM) allowed the best three-dimensional approach to the study of the liver's structure. In fact, many SEM studies on both healty and pathological liver samples have permitted to better understand the complex microstructure of this gland (Motta 1988). However, most of these studies regarded the adult liver, and, up to date, only a few SEM studies have been performed on human embryonal or fetal samples (Makabe and Motta 1980, Motta 1988, Macchiarelli et al. 1988). In addition, many problems concerning the nature and functions of liver cells during developmental stages, such as the differentiation of hepatocyte surface polarity, the formation of endothelial sieve plates,the structural relation of Kupffer cells to the sinusoidal wall, as well as the influence of hemopoiesis on liver cell structure, are still topics of debate (Bankston and Pino 1980, Macchiarelli et al. 1988).

In order to provide a better view of the fine three-dimensional architecture of developing hepatocytes, human liver biopsies obtained from embryos and fetuses ranging from 7-20 weeks post fertilization were observed by means of SEM. Special emphasis has been given to the hepatocytes and sinusoidal cells (endothelial cells and Kupffer cells), as well as to their microtopographical relationships to the hemopoietic cells.

MATERIALS AND METHODS

Human embryos and fetuses from 7 to 20 weeks' gestation were obtained from legal and/or spontaneous abortions. Whenever possible, the samples were fixed with glutaraldehyde by the puncture-perfusion method or by cannulation of the cut vessels;

otherwise, they were rapidly put in cold fixative.

For SEM observations, samples were post-fixed with 1% OsO_4, dehydrated with ethanol and critical point dried with liquid CO_2. Surface tissue exposure was performed in both wet and dried specimens by using a razor blade or by cracking liquid nitrogen-frozen samples. Observations were made through Hitachi HFS-2, or Jeol 35 microscopes operating at 10-30 kV. Specimens from the same livers were processed for transmission electron microscopy (TEM) in order to obtain comparative images. The ultrathin sections were stained with uranyl acetate and lead citrate, and observed in Zeiss EM 9A and Philips EM 400 microscopes.

RESULTS

During the developmental stages observed by SEM, neither the general architecture nor the cellular morphology of the liver cells showed notable changes. Therefore the results have been grouped by cell type instead of by chronologic order.

Liver parenchymal cells. (Figs. 1 and 2). Since early developmental stages the liver cells presented a typical arrangement in several cell-thick cords that seemed to anastomose large spaces resembling central veins and portal spaces. These cords delimited either large sinusoidal lacunae in which liver capillaries were present, or interstitial spaces, which were mainly occupied by rounded cells. By TEM, the latter corresponded to hemopoietic elements. The hepatocytes had an irregularly polyhedral shape, with several facets. Three different ultrastructural aspects of the hepatocyte facets could be recognized: a) surfaces showing a central groove and longitudinally placed double row of short and thick microvilli (corresponding to the typical bile hemicanaliculus) bordered by two flat and smooth areas; b) facets facing the sinusoidal wall and appearing evenly covered by long, thin microvilli; and c) facets facing hemopoietic elements appearing concave due to the close relation with these elements and were either smooth or, more frequently, covered by long and thin microvilli.

Sinusoidal cells. The sinusoidal wall appeared mainly composed of endothelial cells and of other cell types embedded in the sinusoidal lining.The endothelial cells possessed a bulging area, corresponding to the nucleus, and large, flat cellular extensions having isolated small fenestrae and characterized by numerous large gaps (Makabe and Motta 1980) These aspects were mainly characteristic of the early developmental stages observed (7-12 weeks). During later periods (15-20 weeks), small fenestrae were more numerous and showed a tendency to cluster,

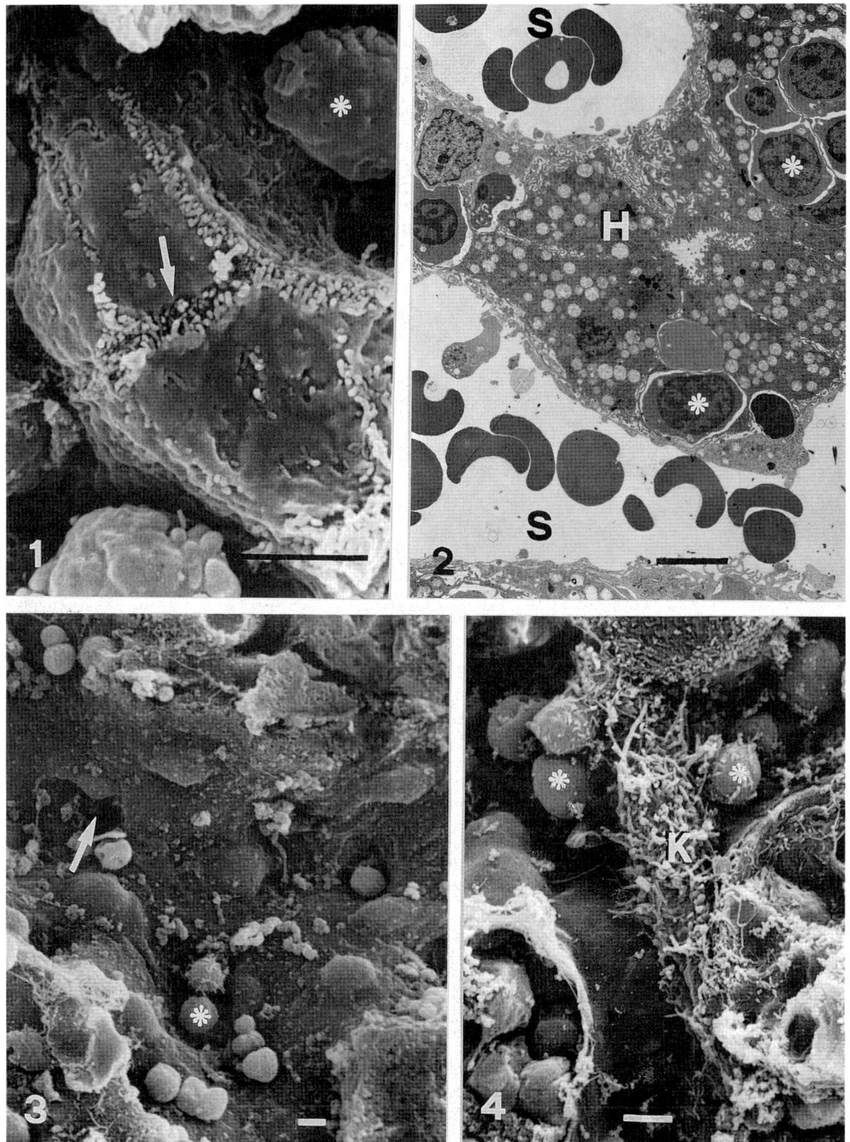

Fig. 1, SEM of human hepatocyte (15 weeks of gestation). Note the bile canaliculus (arrow) and the facet in contact with an hemopoietic cell (*). Fig. 2, TEM of human fetal liver (18 weeks of gestation). H = hepatocyte; S = sinusoids; * = hemopoietic cells. Fig. 3, SEM human sinusoid (9 weeks of gestation). Note the large gaps (arrow) and the transmural passage of hemopoietic cells (*). Fig. 4, SEM of human Kupffer cell (K) (20 weeks of gestation). Note blood cells (*) reached by Kupffer cell cytoplasmic prolongation. (Bar = 5 microns).

forming an arrangement more similar to that observed in normal adult liver sinusoids (Macchiarelli et al. 1988). Rounded cellular elements were often observed in the larger gaps (Fig. 3) and, by TEM, corresponded to hemopoietic cells (Fig. 2). Attached to the endothelium, and frequently in contact with blood elements, irregularly shaped gross cells having a rough surface, rich in microvilli and cell projections were seen (Fig. 4). By TEM, these elements could be easily identified as activated macrophages or Kupffer cells.

DISCUSSION

These morphological results allows to consider further the relation that the fetal liver hemopoiesis activity might have with the three-dimensional arrangement of human liver cells. As viewed by SEM, the hepatocytes presented biliary and vascular facets rather similar to those observed in the adult, suggesting that they may function as mature elements.However,the hemopoietic cells often seemed to altered hepatocyte shape and intercellular connections.The sinusoidal wall presented a high degree of polimorphism,likely due to the gradual development of endothelial sieve plates.However,the characteristic "sieving" endothelium of adult liver sinusoids was never recognized.In fact,large gaps caused by the transmural passage of hemopoietic elements were always present in the endothelium,even in the later developmental stages studied.Macrophages that morphologically resembled adult liver Kupffer cells, often appeared to be phagocytizing blood cells. In conclusion,the close microtopographical relation among the cell types studied,suggests that liver cell differentiation and function are strictly linked to the liver's principal role during the developmental stage we studied: hemopoiesis.

REFERENCES

Bankston PW, Pino RM (1980).The development of sinusoids of fetal rat liver: morphology of endothelial cells, Kupffer cells, and the transmural migration of blood cells into sinusoids. Am J Anat 159: 1-15.

Macchiarelli G, Makabe S, Motta PM (1988). Scanning electron microscopy of adult and fetal liver sinusoids. In P Bioulac-Sage and C Balabaud (Ed) "Sinusoids in Human Liver: Healt and Disease", Rijswijk, Kupffer Cell Foundation (Pbl),pages 63-85.

Makabe S, Motta PM (1980). Scanning electron microscopy of foetal liver sinusoids in humans. Folia Morphol Praha 28: 85-87.

Motta PM (1988). "Biopathology of the Liver. An Ultrastructural Approach" Boston, London, Kluwer Acad. Publishers, pages 37-58.

Cells and Tissues: A Three-Dimensional Approach by Modern Techniques in Microscopy, pages 281–286

DEVELOPMENT OF ADIPOSE TISSUE IN THE HUMAN FETUS

S. LABBE, J.P. BARBET, H. COPIN, M. MAILLET, B. SCHRAMM

Dept of Histology Embryology and Biology of Reproduction
Hôpital Fernand Widal, F. 75010 PARIS
Dept of Pathology, Hôpital Saint Vincent de Paul.
F. 75014 PARIS.

The development of adipose tissue in the human fetus was studied by Berg in 1911 and Dabelow in 1957 . Until recently, however, the moment of appearance of adipose tissue was not clearly established. Some authors like Wasserman noted a tight relationship between the development of blood and the appearance of adipocytes (Wasserman 1965). Wasserman concluded from this that adipocytes form from the external vessel wall and described the primitive organs of the white adipose tissue constituted of blood capillary vessels and connective cells. Others like Napolitano considered that the fibroblast is the adipocyte precursor (Napolitano 1968). This question is still open.

More recent studies describe the laying down and evolution of adipose tissue. Poissonet et al distinguish between adipogenesis (adipocyte differentiation) and lobulogenesis (laying down of densified connective tissue delimiting adipose lobules) (Poissonet et al 1983). For these authors, adipogenesis is carried out in five intricate but theoretically well defined stages:

-Young and loose connective tissue;
-Vascular invasion and mesenchymal condensation foci
-Formation of mesenchymal lobules constituted of a network of blood capillaries and stellate connective cells;
-Primitive fat lobules formed of a mixture of adipocytes and stellate cells;
-Definitive fat lobules limited by interlobular septa.

As to lobulogenesis, it would start with vascular invasion and would end up with the laying down of the adipose tissue during the second trimester of pregnancy (Poissonet et al 1983).

MATERIAL AND METHODS

Samples from shoulder, abdominal wall and gluteal region were obtained from 11 fetuses of either sex, aged from 15 to 40 weeks. After a fixation in 10% formalin, samples were divided and processed for light microscopy (LM) as well as for scanning electron microscopy (SEM). SEM was performed both on tissue blocks and on dewaxed tissue sections.

RESULTS

Our perfectly correlating histological and ultrastructural results show that adipose tissue develops according to the same morphological scheme in the three locations under study.

At 15 weeks the connective tissue located beneath the periderm is still loose. It displays elongated and often fusiform cellular elements and contains blood capillary vessels. Neither in LM nor in SEM have we observed any differentiation toward adipogenesis (Fig. 1).

At 17 weeks there are recognizable pileous formations. Mesenchymal lobules constitute rounded formations. These lobules are made of vascular elements, stellate cells well evidenced by SEM, and some small-sized adipocytic elements. They appear on the deep side of the connective tissue located beneath the periderm and are delimited by a dense fibrillar network (Figs. 2 and 3).

At 20 weeks, the lobules have increased in size. They seem well separated from the adjacent connective tissue by a collagenous condensation and contain a variable amount of cellular elements. LM shows that these are either typical adipocytes with a unique intracytoplasmic vacuole or adipocytes with one large-sized vacuole and several small-sized vacuoles, or undifferentiated elements. SEM confirms the presence of three cell types and demonstrates the often spherical morphology of undifferentiated cells. Adipocytes are closely piled up, connected by a connective felting. On the periphery the connective fibers delimiting the lobules wedge into the adipose cells (Figs. 4 and 5).

At 25 weeks, the adipose lobules are well defined and contain many piled up adipocytes. They form a deep continuous layer (Fig. 6). Outstanding is the appearance of new differentiation foci located above the deep fat layer at the peribulbar part of the pilosebaceous follicles. These foci are morphologically identical to those appearing deeper (Figs. 7 to 10).

At 27 weeks, the adipose tissue forms a wide deep pad. Perifollicular differentiation foci are numerous and new foci have appeared. At this stage a topographical difference is observed: the lobules of the abdominal wall contain more young elements than those in the deltoid and buttock region. In SEM it can be seen that the adipocytes lie on a fibrillar layer and are covered with a fibrillar meshwork (Fig. 11).

At 40 weeks two layers of adipose tissue can be clearly distinguished: a superficial and a deeper layer separated by connective fasciae. These two layers are arranged in lobules containing mature adipocytes as well as undifferentiated elements (Fig. 12).

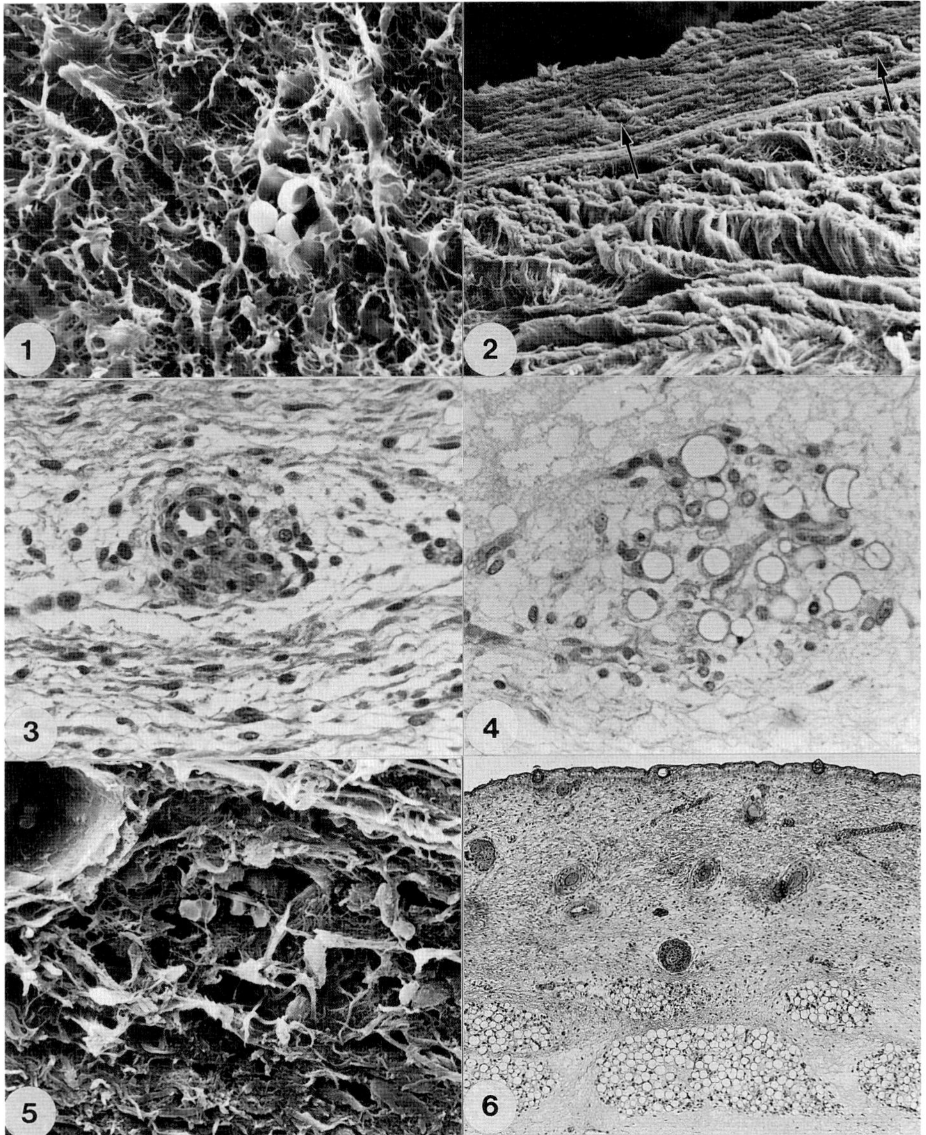

Fig. 1: 15 weeks old fetus. SEM. Loose connective tissue and blood capillary.
Fig. 2: 17 weeks old fetus. SEM. Condensed mesenchyme with recognizable pileous formations (arrows).
Fig. 3: 17 weeks old fetus. H&E. Group of connective cells close to a vessel wall.
Fig. 4: 20 weeks old fetus. H&E. Primitive fat lobule with young adipocytes and undifferentiated cells.
Fig. 5: 20 weeks old fetus. SEM. Mesenchymal condensation with stellate cells near a blood vessel.
Fig. 6: 22 weeks old fetus. H&E. Primitive fat lobules constitute a deep layer.

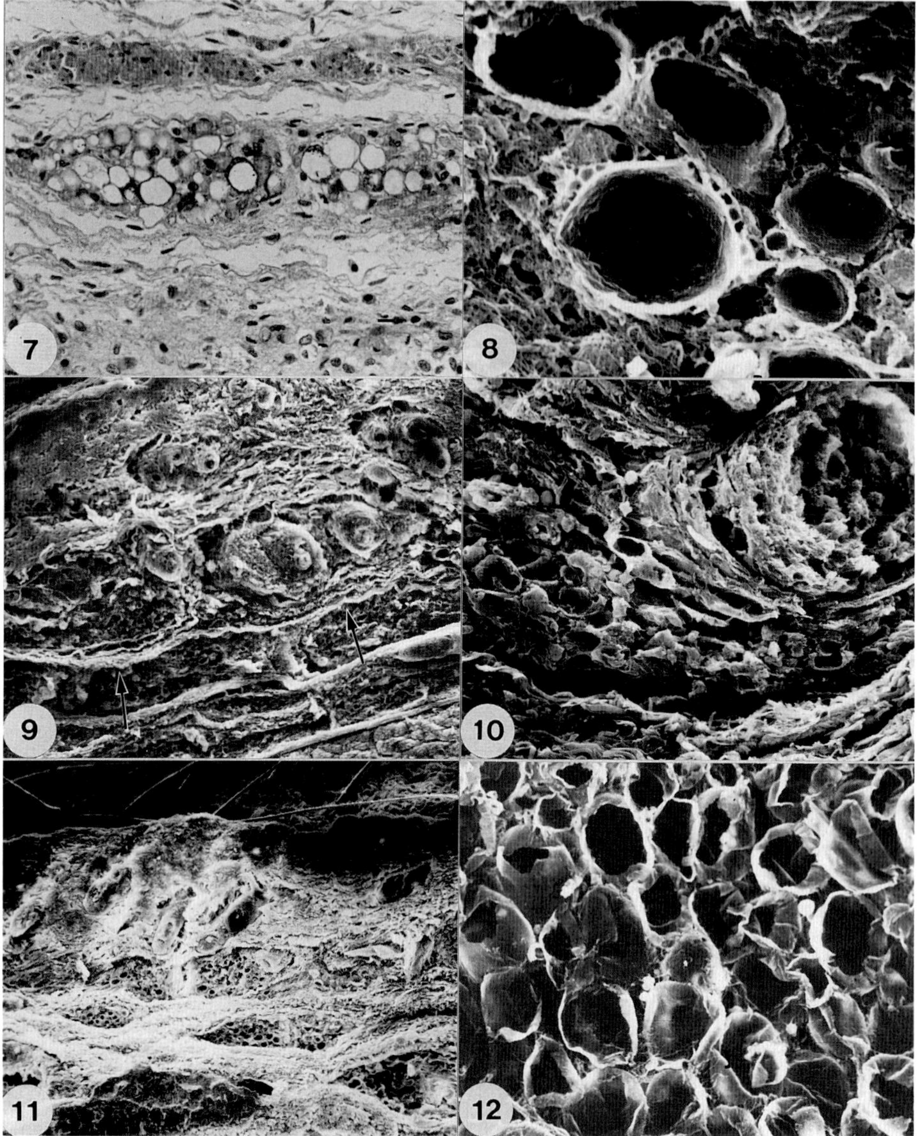

Fig. 7: 22 weeks old fetus. PAS. This primitive fat lobule is heterogenous and contains PAS+ undifferentiated cells.
Fig. 8: 22 weeks old fetus. SEM. Persistence of some multilocular adipocytes.
Fig. 9: 25 weeks old fetus. SEM. Fat lobules form a deep layer separated from the surface by a connective fascia (arrows).
Fig. 10: 25 weeks old fetus. SEM. A new focus of adipogenesis appears near the developing pileous follicles.
Fig. 11: 27 weeks old fetus. SEM. Hypodermic lobules are separated from the subcutaneous tissue by a well defined connective fascia.
Fig. 12: 40 weeks old fetus. SEM. Mature adipocytes.

DISCUSSION AND CONCLUSION

The second trimester of pregnancy is the essential period of adipose tissue laying down. Our results agree with those of Poissonnet et al who set the appearance and laying down of the adipose tissue between 16 and 26 weeks (Poissonet et al, 1983 and 1984). Location and sex do not affect this laying down, although the hormonal environment is very different in the boy or the girl during intrauterine life (Enzi et al 1980).

Our results show that the build up of subepidermal adipose tissue occurs in two steps:

-laying down of a deep adipose tissue which is always separated from the surface by a layer of connective tissue;

-secondary appearance of lobules located in between this zone and the epidermis.

Both zones evolve morphologically in the same way, the only differences being topographical and chronological. The most superficial zone corresponds to the hypodermis, the deepest to the proper subcutaneous tissue. It is indeed nosogically difficult to consider that the fusion of the hypodermis with this deep adipose tissue constitute the subcutaneous tissue since the hypodermis belongs by definition to the skin. Applying this terminology, our results thus allow to state that the subcutaneous tissue appears before the hypodermis.

As to adipogenesis stricto sensu two stages are classically recognized: a hyperplastic stage and a metabolic differentiation one (Poissonnet et al 1984). Our results show that the hyperplastic stage is very extended in time (beyond the 27th week) since the appearance of new lobules is observed for a rather long period as well as the persistence of undifferentiated elements. Dunlop et al consider that these elements are preadipocytes (Dunlop et al 1978). The problem of the adipocyte precursor cannot be solved by SEM. The obligatory presence of vascular element was an argument for Wasserman in favor of the origin of adipocytes in detached cells from the external vessel wall (Wasserman 1965). We indeed observed a close relationship between stellate cells and vessel wall, but this is no definitive argument.

Metabolic differentiation can be well observed by LM, but more so by SEM, displaying multi- and unilocular phases. The multilocular phase is very transient and shifted in time for the hypodermal lobules. This differentiation is not uniform , as each lobule is heterogeneous. The increase in size of the lobules is related to the increase in number and size of the adipocytes, but also to that of the non-adipocyte fraction. This growth takes place under the influence of multiple nutritional and hormonal factors (Enzi et al, 1980). We have seen that sex steroids do not determine any significant differences in the distribution of adipose tissue during fetal life. In contrast, insulin is an important growth factor of adipocytes in utero. This is why a better understanding of adipogenesis would allow to prevent macrosomies in cases of maternal diabetes and help elucidating how far the fetus contributes to child obesity.

REFERENCES

Berg W (1911). Uber die Anlage und Entwicklung des Fettgewebes beim Menschen. Z Morphol Anthropol 13:305-341.

Dabelow A (1957). Die Entwicklung der Fettorgane in subcutanen Gewebe menschlicher Feten. Verh Anat Ges (Jena) 54:83-96

Dunlop M, Court JM, Hobbs JB, Boulton JC (1978). Identification of small cells in fetal and infant adipose tissue. Pediatr Res 12:905-907.

Enzi G, Inelmen EM, Caretta F, Rubaltelli F, Grella P, Baritussio A (1980). Adipose tissue development "in utero". Relationships between some nutritional and hormonal factors and body mass enlargement in newborns. Diabetologia 18:135-140.

Napolitano L (1968). The differentiation of white adipose cells: an electron microscopic study. J Cell Biol 18:663-679.

Poissonnet CM, Burdi AR, Bookstein FL (1983). Growth and development of adipose tissue during early gestation. Early Hum Develop 8:1-11.

Poissonnet CM, Burdi AR, Garn SM (1984). The chronology of adipose tissue appearance and distribution in the human fetus. Early Hum Develop 10:1-11.

Wasserman F (1965). The development of adipose tissue. In Reynold AE, Cahill GF (eds):"Handbook of Physiology: adipose tissue", Washington: J Am Phys Soc, pp 87-100.

Cells and Tissues: A Three-Dimensional
Approach by Modern Techniques in Microscopy,
pages 287–294

ULTRASTRUCTURAL EVIDENCE OF RECEPTOR-MEDIATED ENDOCYTOSIS DURING EARLY MOUSE LIMB MORPHOGENESIS: A THREE DIMENSIONAL ANALYSIS USING DEEP ETCHING AND SCANNING ELECTRON MICROSCOPY

Robert O. Kelley

Department of Anatomy, University of New Mexico,
School of Medicine, North Campus, Albuquerque,
New Mexico. U. S. A. 87131

INTRODUCTION

Epithelial-mesenchymal interactions are essential to normal embryonic development and morphogenesis. One classic example, which is the subject of considerable experimental interest among developmental, cell and molecular biologists, is the vertebrate limb bud. These structures develop at their distal tips a thickened epithelium called the apical ectodermal ridge. The ridge functions as an inducer of limb development, since upon removal, further development ceases and only those limb parts established prior to ridge removal will form (Saunders, 1948; Zwilling, 1949). In addition, a necessary relationship exists between limb mesoderm and the overlying ectoderm. Mesodermal cells destined to form the limb induce the adjacent ectoderm to become active, i.e. to acquire properties of the apical ridge (Kieny, 1960). Once this is accomplished, mesodermal cells maintain the ridge in its active thickened form until all limb parts have been specified (Zwilling, 1964).

Although the phenomenon of morphogenetic tissue interaction is well known and thoroughly studied, little information is available about mechanisms which either effect or mediate embryonic induction. Wolpert (1971), in his hypothesis outlining the concept of positional information in development, included the possibility of morphogenetic signaling molecules, or "morphogens". Since it is known that interacting cells in limb mesoderm and ectoderm do not directly contact one another (they are separated by the complex molecular nature of a basal lamina; Kelley, 1973, Kelley and Fallon, 1976), it would seem that developmental signals would be required to traverse intercellular distance containing extracellular matrix, and furthermore, intersect the responding cell's plasma membrane before information (either instructive or permissive) is transmitted into the cell's interior.

In this regard, cell biologists have long recognized ligand-receptor interaction as a mechanism for intercellular signalling (see Goldstein, et al, 1985, for review) and receptor-mediated endocytosis is thought to be among the mechanisms used by cells to remove membrane bound signals once a cell has responded (Pastan and Willingham, 1985). The structures which participate in membrane receptor localization and internalization area clathrin-coated pits and vesicles (reviewed recently by Pearse and Crowther, 1987).

To investigate the possibility that morphogenetic signals involve cell surfaces during early ridge induction and thickening, and, further, that structural evidence of signal-receptor interaction might provide both temporal and spatial information on the nature of ridge induction, techniques of electron microscopy were applied to mouse embryos at stages prior to and during thickening of ridge ectoderm. This paper reports structural evidence of receptor-mediated endocytosis in pre-ridge limb ectoderm. The presence of coated pits and vesicles suggests receptor activity at cell surfaces, and, further, suggests a potentially important role for receptor-mediated events in limb morphogenesis.

MATERIALS AND METHODS

Swiss Webster mouse embryos at days 9, 9.5, 10, 10.5 and 11 (day 1 calculated as the day copulation plugs were observed) were fixed in 2.0% glutaraldehyde in 0.1M cacodylate buffer (pH 7.4, 4 C), postfixed in 2.0% osmium tetroxide in 0.1M cacodylate buffer (4 C), and either embedded in epon for thin sectioning, or critical point dried for scanning electron microscopy. Other limb buds were dissected, quick-frozen in either liquid freon or nitrogen and deep-etched in a Balzers BAF 301 apparatus (Heuser, 1981). Specimens and replicas were examined with Hitachi H-600 transmission and S-800 scanning electron microscopes.

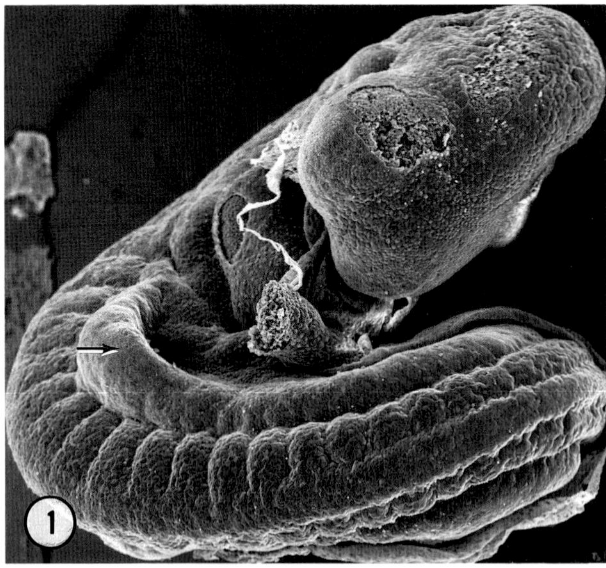

Fig. 1. Scanning electron micrograph of 9d mouse embryo. Note development of forelimb bud (arrow) and rudimentary presence of hindlimb. X 200.

OBSERVATIONS

The mouse embryo exhibits a prominent forelimb bud, but only a rudimentary hindlimb bud, at day 9 (fig. 1). At 9.5d (fig. 2), the hind limb bud enlarges, although both buds lack an apical ectodermal ridge at this stage. The ridge (fig. 3) is clearly developed in the forelimb bud at 10.5d and appears by day 11 in the hindlimb. The ridge regresses in both limbs by 12.5d. Figure 3 reveals that the murine ridge is a stratified epithelium in contrast to the simple cuboidal organization of adjacent, non-ridge ectoderm.

As ridge development begins in the forelimb (day 10-10.5), the relationship to underlying mesoderm is illustrated in figure 4. A thin, continuous basal lamina separates mesoderm from ectoderm at all stages. Numerous projections from mesenchymal cells explore the matrix at the mesodermal surface of the basal lamina (figs. 4 and 5), but do not penetrate the lamina to contact the basal membrane of adjacent ectoderm cells. Cell division and production of matrix are the prinicipal activities of subridge mesodermal cells (fig. 4).

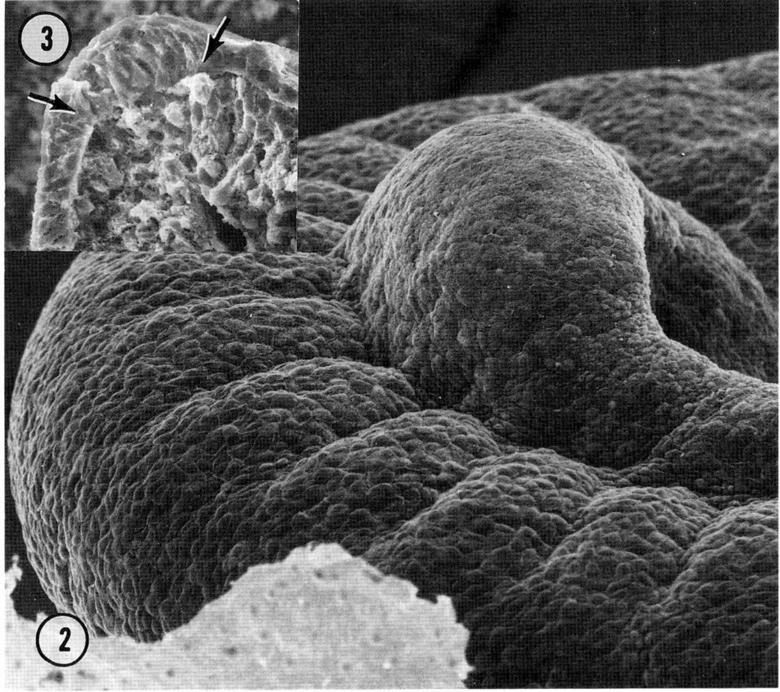

Fig. 2. Forelimb bud of 9.5d mouse embryo. The bud exhibits polarity in relation to other body parts and axes. Note the prominent development of the bud, but absence of apical thickening of ectoderm. X 500.
Fig. 3. By breaking the specimen open, the thickened apical ectoderm is clearly contrasted to the thinner dorsal and ventral ectoderm. In addition, subjacent mesoderm and investing extracellular matrix is revealed. X 450.

By 9.5d, numerous coated pits are observed at basal (fig. 6) and basolateral (fig. 7) surfaces of membranes of cells which abut the basal lamina. It is noted that few coated pits are observed in flank ectoderm at 8.5d and in cells of limb ectoderm which will not participate in the ridge by day 9. However, these pits are infrequently located at basal membrane surfaces and do not appear to be as numerous as in the distal limb tip prior to ectodermal thickening. Internalization of coated pit membrane is in evidence at 9.5d by the presence of numerous coated vesicles in the cytoplasm of cells destined to form ridge tissue (fig. 8). A replica of fractured membranes of pre-ridge ectodermal cells (fig. 9) illustrates the distribution of pits forming in adjacent cells. Similar membrane organization persists through day 12 in both ridge and non-ridge ectoderm, with the former exhibiting the more prominent distribution of coated pits and vesicles.

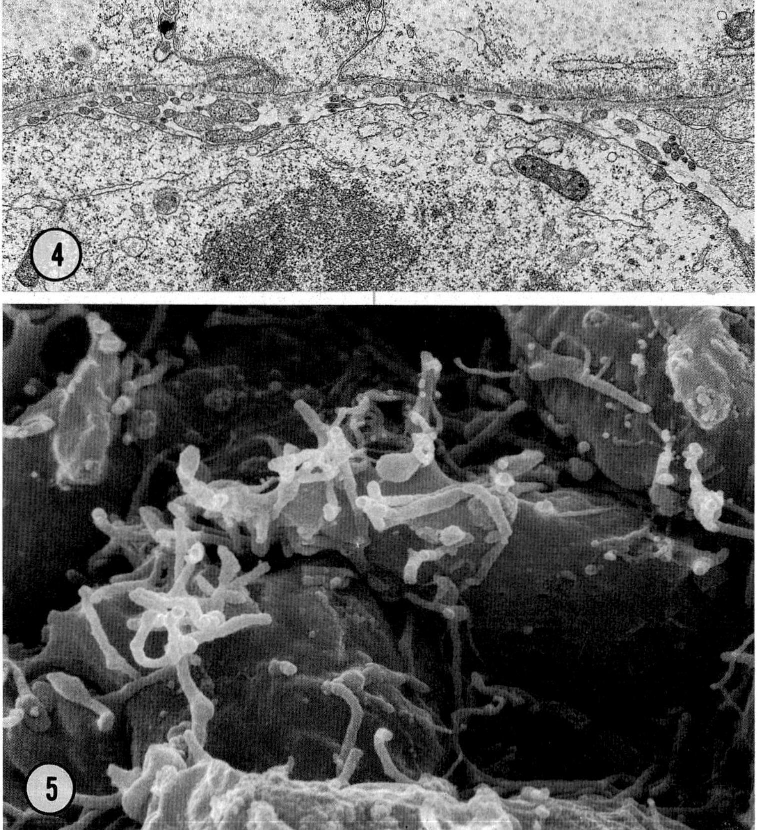

Fig. 4. Transmission electron micrograph of ectodermal-mesodermal interface. A dividing mesodermal cell extends projections which intersect a basal lamina subtending the ectoderm. X 10,000.
Fig. 5. Scanning electron micrograph of mesodermal cell surfaces immediately beneath the apical ridge. Processes extend both towards the epithelial surface and towards adjacent cells within the limb mesoderm. X 20,000.

Once internalized, coated vesicles appear to become intimately associated with the cytoskeleton in epithelial cells forming the ridge. Figure 10 illustrates two vesicles surrounded by filamentous elements of the cytoplasm, whereas figure 11 shows an extensive array of vesicles (not all coated vesicles) associated with the cytoskeleton. Current thought suggests that the cytoskeleton may participate in membrane traffic from the surface to selected sites within the cell (e.g. the Golgi apparatus, lysosomes, fusion with other vesicles, etc.) where spent receptors are processed and new membrane is returned to the cell surface.

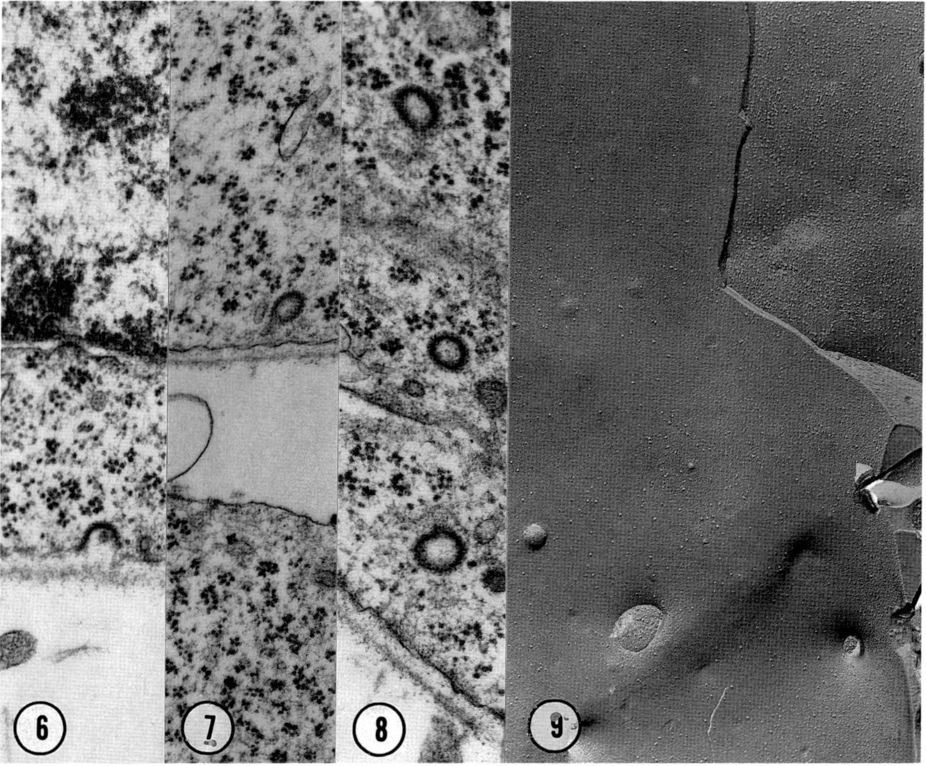

Figs. 6-9. Formation of coated pit in the basal membrane of apical epithelial cell at 9.5d. X 50,000 (fig. 6). Internalization of the coated segment of membrane yields a coated vesicle. X 50,000 (fig. 7). Multiple coated vesicles are observed at 9.5d prior to morphological appearance of apical thickening. X 50,000 (fig. 8). Freeze fracture exhibits membrane faces which show distribution of depressions and bumps which may be forming pits and vesicles. X 30,000.

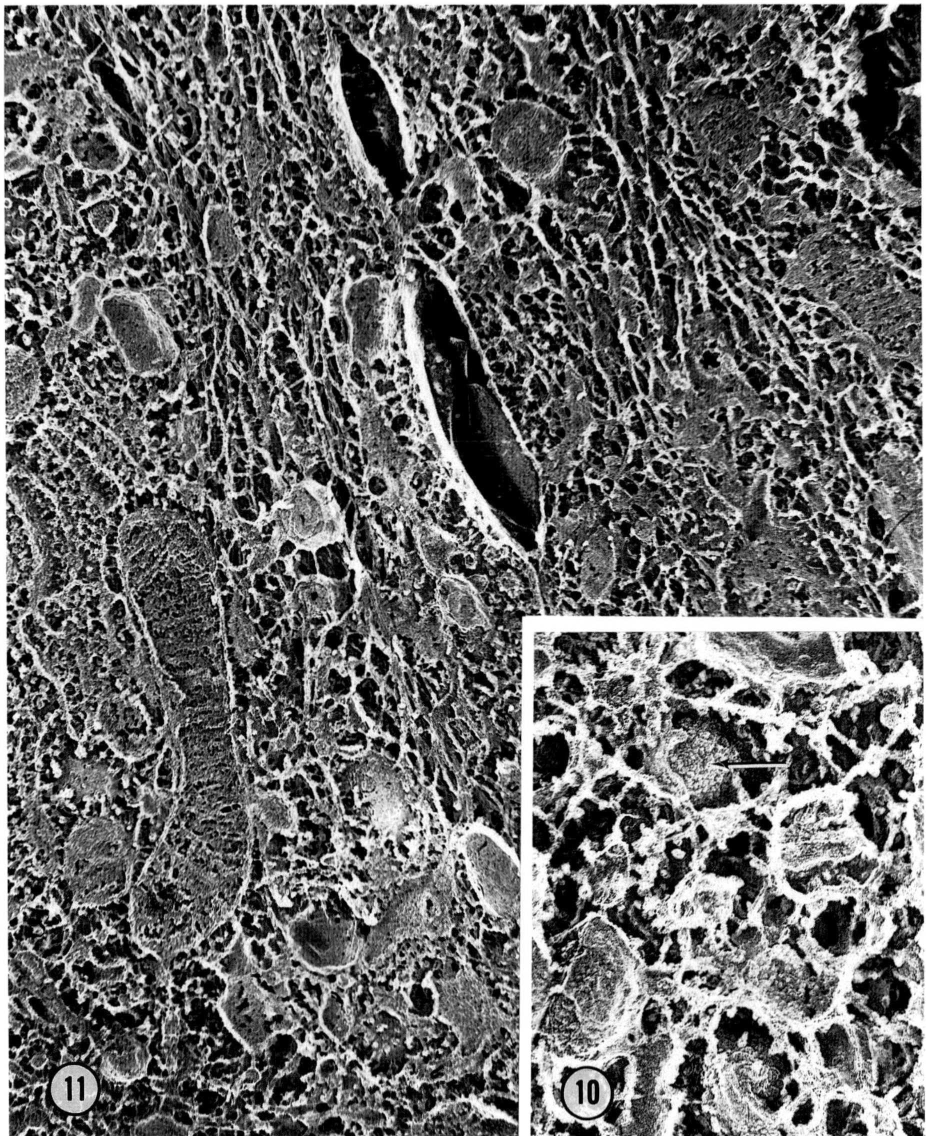

Fig. 10. Replica of quick-freeze, deep etch specimen revealing internalized vesicles (arrows) in close association with components of the cytoskeleton. X 80,000.
Fig. 11. Lower magnification image shows distribution of vesicles (arrows) along basolateral borders of epithelial cells. X 60,000.

DISCUSSION

It is reasonable to interpret experimental results obtained with the chick limb (e.g. Zwilling, 1964) to mean that the apical ectoderm of the limb bud is instructed and maintained in its active form only by limb mesoderm. The investigations of Milaire (1965) and Cairns (1965) suggest that the same is true in mammals. What remains unclear, however, is (1) the nature of gene products from mesoderm which may serve as "morphogens" to instruct ectoderm to thicken and, in turn, permit growth of underlying mesoderm, and (2) what structural mechanisms cells in intact embryonic tissues employ to effect the transduction of those signals across cell surfaces once they are intercepted by the responding cell.

The observations presented in this brief note suggest that cells engaged in the epithelial-mesenchymal interactions which result in the appearance of an apical ridge in limb buds employ membrane-mediated mechanisms for internalization of ligand-receptor complexes. Although it is not known what the nature of such a complex might be, it is clear from the structural features of interacting cell surfaces that the coated pit-coated vesicle mechanism is active in these tissues. Coated pits, which are present in a wide variety of cells and first described by electron microscopy in the early 1960's, are associated with the plasma membrane and are involved in endocytosis. Numerous studies have shown that many different ligands (such as epidermal growth factor, fibroblast growth factor, platelet-derived growth factor, and others) enter cells by way of the large coated pits of the plasma membrane (see Pearse and Crowther, 1987). The observation of receptor-mediated endocytosis in pre-ridge limb ectoderm provides structural evidence which permits the suggestion that (1) a growth-like factor may be transmitted from mesodermal cells; (2) this factor locates its receptor on competent pre-ridge ectoderm cells and (3) in turn, provides the initial signal for these cells to form the stratified layer of epithelium which has the active qualities of the apical ridge. This hypothesis is attractive in that regulation of receptor availability through gene action of cells in the limb ectoderm may serve as the basis for morphogenesis of ridge pattern (i.e. the apical position of the ridge and the presence of cells in limb ectoderm which do not respond to mesodermal influence by forming ridge tissue).

ACKNOWLEDGMENT

Grateful acknowledgment is extended to Judi Davis for technical assistance, and to Carol Corsaut for preparation of the manuscript. Use of instrumentation and supplies were supported, in part, by grants from the NIH (AG 00191 and RR 08139).

REFERENCES

Cairns J M (1965) Development of grafts from mouse embryos to the wing bud of the chick embryo. Devel Biol 12:36-52.

Goldstein J L, Brown M S, Anderson R G W, Russell D W, Schneider W J (1985) Receptor-mediated endocytosis:concepts emerging from the LDL receptor system. Ann Rev Cell Biol 1:1-39

Heuser J (1981) Quick-freeze, deep-etch preparation of samples for 3-D electron microscopy. Trends in Biochem Sci 6:64-68.

Kelley R O (1973) Fine structure of the apical ridge-mesenchyme complex during limb morphogenesis in man. J Embryol Exp Morphol 29:117-131.

Kelley R O, Fallon J F (1976) Ultrastructural analysis of the apical ectodermal ridge during vertebrate limb morphogenesis. I. the human forelimb with special reference to gap junctions. Devel Biol 51:241-256.

Kieny M (1960) Role inducteur du mesoderme dans la differenciation precoce du bourgeon de membre chez l'embryon de poulet. J Embryol Exp Morphol 8:457-467.

Milaire J (1965) Aspects of limb morphogenesis in mammals. in Organogenesis (Dehaan R L, Ursprung H, eds) Holt,Rinehart and Winston pp 283-300.

Pastan I, Willingham M (1985) The pathway of endocytosis. in Endocytosis (Pastan I, Willingham M, eds.) Plenum, New York, pp 1-68.

Pearse B M F, Crowther R A (1987) Structure and assembly of coated vesicles. Ann Rev Biophys Biophys Chem 16:49-68.

Saunders J W Jr (1948) The proximo-distal sequence of origin of the parts of the chick wing and the role of the ectoderm. J Exp Zool 108:363-403.

Wolpert L (1971) Positional information and pattern formation. Curr Top Dev Biol 6:183-224.

Zwilling E (1949) The role of epithelial components in the developmental origin of the wingless syndrome of chick embryos. J Exp Zool 111:175-187.

Zwilling E (1964) Development of fragmented and dissociated limb bud mesoderm. Dev Biol 9:20-37.

Cells and Tissues: A Three-Dimensional Approach by Modern Techniques in Microscopy, pages 295–300

THREE-DIMENSIONAL LOCALIZATION OF CONTRACTILE PROTEINS IN CULTURED CARDIAC MYOCYTES BY IMMUNOGOLD STAINING AND DEEP-ETCHING REPLICA ELECTRON MICROSCOPY

Yuji Isobe, Guan R. Hou, Dino A. Messina, and Larry F. Lemanski

Department of Anatomy and Cell Biology, State University of New York, Health Science Center at Syracuse, Syracuse, New York 13210.

INTRODUCTION

We describe here a combination of techniques that permit the three-dimensional localization of specific antigens inside the cytoplasm at high resolution. In brief, specific proteins that compose the cytoskeletal lattice are labeled with colloidal gold probes and then they are visualized in three-dimensions by deep-etching replica electron microscopy. The distributions of gold markers are seen superimposed on the spatial images of cytoskeletal elements as well as membranous structures. As examples, we show here the labeling of microtubules and intermediate filaments in developing hamster heart cells *in vitro* with the specific antibodies to their constituent proteins.

MATERIALS AND METHODS

Primary cultures were prepared from heart ventricles of Syrian hamster neonates as described previously (Lemanski and Tu, 1983) with slight modifications. They were grown on gelatin-coated glass coverslips for 1-3 days at 37°C. All of the following experimental procedures on these cultured cells were performed at room temperature. Cells were permeabilized by splitting them open with poly-L-lysine coated coverslips (Isobe and Shimada, 1986), which then were placed in HEPES buffer containing 70 mM KCl, 5 mM $MgCl_2$, 3 mM EGTA, 30 mM HEPES, 0.25 mM PMSF and 10 μM Taxol (supplied by the Drug Synthesis and Chemistry Branch, Division of Cancer Treatment, NCI, Bethesda, MD), pH 7.0.

Permeabilized cells were fixed for 5 min. with 2% formaldehyde in HEPES buffer and treated with 2% nonfat dry milk in the same buffer. Specimens were incubated with either a 1:40 dilution of rabbit anti-tubulin antiserum (Miles; Naperville, IL) or a 1:10 dilution of monoclonal anti-desmin or vimentin antibody (Amersham; Arlington Heights, IL) for 60-120 min. After rinsing in HEPES buffer, the samples were treated with secondary antibodies. A 1:4 dilution of 10 nm colloidal gold conjugated goat anti-rabbit IgG (Janssen Pharmaceutica; Beerse, Belgium) was applied to the samples incubated with antitubulin. The specimens stained with monoclonal antibodies were incubated with a 1:50 dilution of biotinylated sheep anti-mouse Ig (Amersham) for 60-120 min. The specimens treated with biotinylated secondary antibodies were further incubated with a 1:5 dilution of 5 or 10 nm colloidal gold coupled with streptavidin (Amersham) for 180 min. All gold labeled specimens were fixed in 2% glutaraldehyde in HEPES buffer.

After postfixation with 0.2% tannic acid and 0.5% osmium tetroxide (Isobe and Shimada, 1986), samples were plunged into liquid nitrogen cooled Freon 13 using 70% ethanol as a volatile cryoprotectant, then freeze-dried for 30-90 min. at -90°C and rotary replicated with platinum at a shadowing angle of 20° and carbon at 90° in a Balzers BA 360M apparatus. After the coverslips were dissolved with hydrofluoric acid, the replicas were cleaned in household bleach and picked up on 200-mesh formvar carbon-coated grids. Electron micrographs were taken using a JEOL 100CX-II Electron Microscope at 100 kV and photographically reversed.

RESULTS AND DISCUSSION

In previous studies we have described the specificity of the immunogold replica method including control studies as well as light microscopic evaluations (Isobe et al., 1988 a,b). In this report, we focus on the identification of major cytoskeletal elements within the complex spatial images which allow us to address the organization of myocyte cytoskeletons.

Low magnification stereo views of a myocyte cytoskeleton which has been exposed by the partial rupture of cell membranes and subsequently treated with anti-tubulin and gold probes, as shown in Fig. 1, allow ready discrimination of individual microtubules. These microtubules appear to be thickened with antibody complex and labeled with colloidal gold particles along their entire length, while very little gold markers are found on the other cytoskeletal elements. It has been suggested that during the early development of muscle, microtubules may play a role in the orientation and elongation of myogenic cells and the longitudinal arrangement of myofibrils (for review, see Ishikawa, 1983). It has been reported in adult heart muscle that microtubules are associated with myofibrils in a herical arrangement and form a network in the extramyofibrillar space (Goldstein and Entman, 1979). Although some microtubules appear to run parallel to nascent myofibrils, microtubule-myofibril associations are not

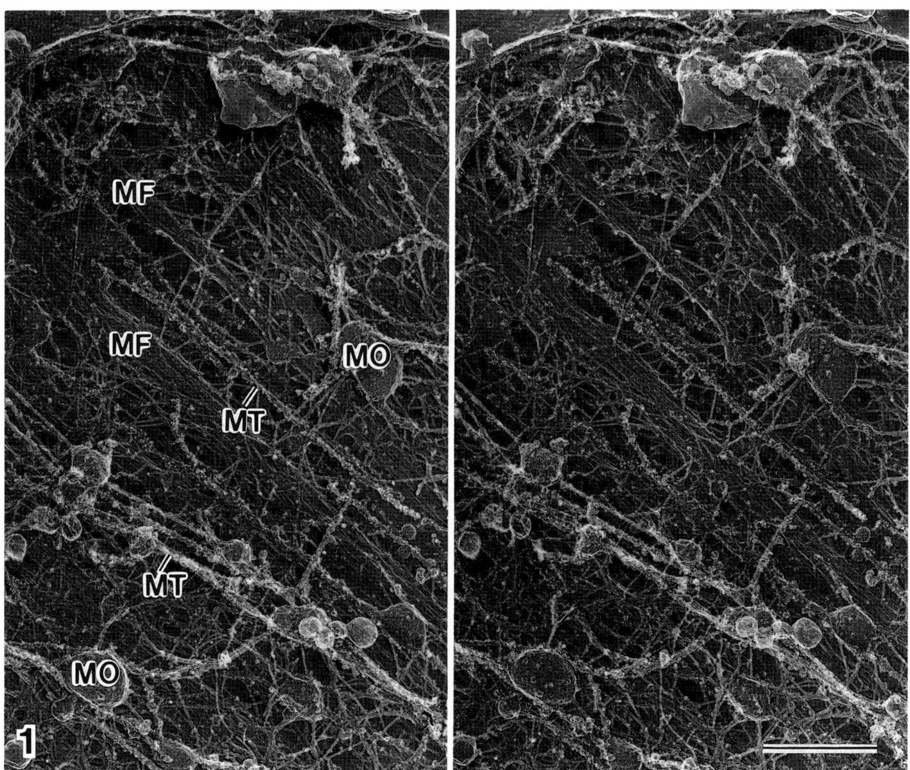

Figure 1. Stereo replica electron micrographs demonstrate a cytoskeletal network of a cardiomyocyte from a 3-day old culture. The cell was physically opened, stained with rabbit anti-tubulin antibody and tagged with 10 nm colloidal gold conjugated gold anti-rabbit IgG. Gold particles (seen as white spots) were localized on microtubules (MT). Myofibrils (MF). Membranous organelles (MO). Bar, 1 µm.

prevalent in our tissue culture system. Further developmental studies on the longer-term cultures or cultures from adult dissociated heart cells may lead to a better understanding of microtubule-myofibril relationship.

Intermediate filaments are seen in abundance in the cytoplasm of cardiomyocytes. It is clear now that desmin is a major component proteins of intermediate filaments in various muscle cells (Lazarides and Hubbard, 1976; Gard et al., 1979). When the present hamster cardiomyocytes are treated with monoclonal antibody to desmin, three-dimensional electron micrographs show that portions of intermediate filaments in these cells are heavily but intermittently, immunostained leaving others unstained (Fig. 2).

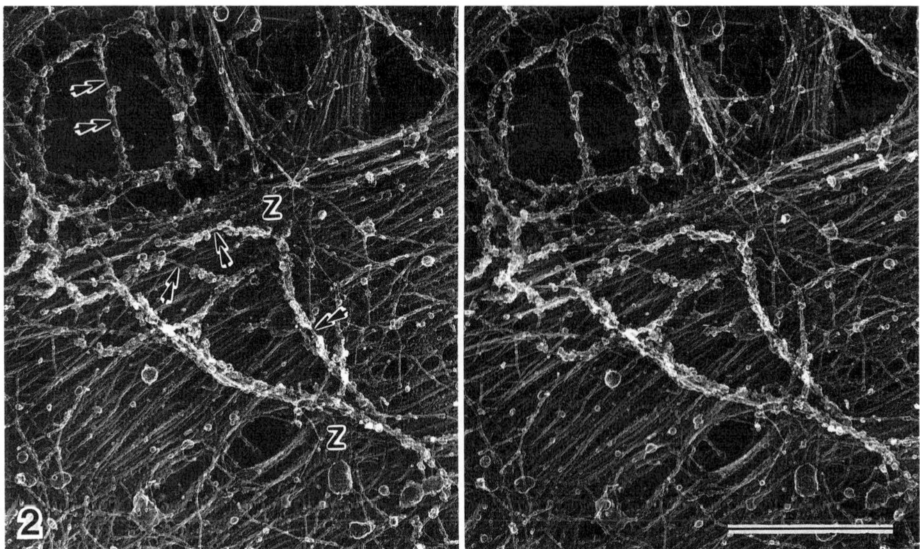

Figure 2. High magnification stereo micrographs exhibit a cytoskeletal network around nascent myofibrils in a 3-day old myocyte. The cell was physically opened, stained with monoclonal anti-desmin antibody and tagged with 5 nm colloidal gold by the biotin-streptavidin system. Antibody complexes and gold probes decorated the intermediate filaments. Note some bare segments (double arrowheads). Myofibril Z bands (Z). Bar, 1 μm.

Stained filament portions are thickened by the antibody complex and are labeled with gold particles. Bare segments, completely devoid of staining, are seen on these same filaments. In some instances, heavily stained filaments accumulate in the extra-myofibrillar space. Such a heterogeneous pattern of desmin staining on intermediate filaments in hamster heart cells with monoclonal antibody does not corroborate the previous pilot three-dimensional immunoelectron microscopic study (Ip et al., 1983), in which intermediate filaments of cultured chicken cardiac myocytes were uniformly labeled with polyclonal anti-desmin antibody. This may reflect a species difference or a difference in the specificity of monoclonal antibody we used. In the present study, the monoclonal antibody used was made against porcine stomach desmin. It is possible that this antibody recognizes only certain desmin isoforms present in the intermediate filaments of hamster cardiac myocytes. Another possibility is that in hamster myocytes, desmin and vimentin copolymerize to form intermediate filaments similar to that observed by Ip et al. (1983) in chicken fibroplast intermediate filaments, and by Tokuyasu et al. (1985) in developing chicken skeletal muscle cells. We hope to elucidate this important point with our

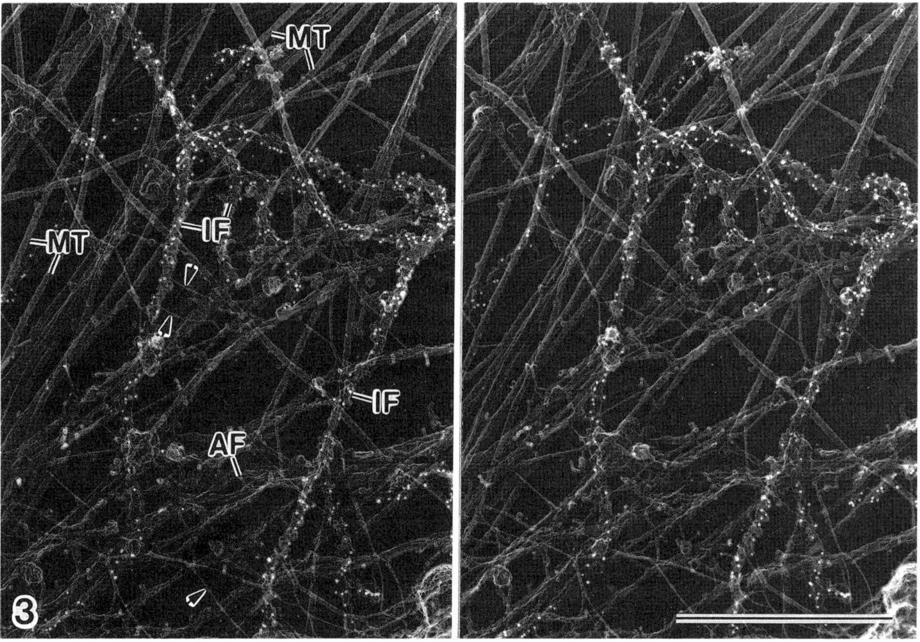

Figure 3. Replica of a fibroblastic cell after 4 days in culture. Stereomicrographs show a cytoskeletal network which was exposed by physical rupturing, incubated with monoclonal anti-vimentin antibody and tagged with 15 nm colloidal gold probes by the biotin-streptavidin system. Intermediate filaments (IF) were heavily labeled with antibody complexes and gold particles while microtubules (MT) and actin filaments (AF) remain unstained. 2-5 nm filaments (arrowheads). Bar 1 μm.

studies now in progress in which we use double immunolabeling with various combinations of monoclonal and polyclonal anti-desmin and anti-vimentin antibodies. We finally show the staining of a fibroblast by the monoclonal anti-vimentin antibody from our ongoing preliminary experiment (Fig. 3). By light microscopic observation, while this antibody stains all fibroblasts intensively making many individual intermediate filaments discernible, it stains myocytes heterogeneously. Some myocytes have accumulations of anti-vimentin staining in their cytoplasm similar to anti-desmin staining, but other myocytes show only a few short stained segments of filaments even in the same cultures. We have just begun the experiments using this antibody in combination with three-dimensional electron microscopic methods.

ACKNOWLEDGEMENT

We thank Dr. Fred D. Warner for allowing us to use his freeze-etching apparatus. We also thank Ms. Masako Nakatsugawa and Sharon Lemanski for their technical help, and Ms. Jo-Ann Pellett and Nancy Snyder for typing the manuscript. This work was supported by grants from NIH (HL32184 and HL37702) and the AHA to LFL.

REFERENCES

Gard DL, Bell PB, Lazarides E (1979). Coexistence of desmin and the fibroblastic intermediate filament subunit in muscle and non-muscle cells: Identification and comparative peptide analysis. Proc Natl Acad Sci USA 76:3894-3898.

Goldstein MA, Entman ML (1979). Microtubules in mammalian heart muscle. J Cell Biol 80:183-195.

Ip W, Danto SI, Fischman DA (1983). Detection of desmin-containing intermediate filaments in cultured muscle and non-muscle cells by immunoelectron microscopy. J Cell Biol 96:401-408.

Ishikawa H (1983). Fine structures of skeletal muscle. In Dowben RN, Shay JW (eds): "Cell and Muscle Motility," Volume 4, New York/London: Plenum Press, pp 1-84.

Isobe Y, Shimada Y (1986). Organization of filaments underneath the plasma membrane of developing chicken skeletal muscle cells _in vitro_ revealed by the freeze-dry and rotary replica method. Cell Tissue Res 244-47-56.

Isobe, Y, Warner FD, Lemanski LF (1988a). Three-dimensional immunogold localization of α-actinin within the cytoskeletal networks of cultured cardiac muscle and nonmuscle cells. Proc Natl Acad Sci USA, in press.

Isobe Y, Messina DA, Lemanski LF (1988b). Spatial immunolocalization of cytoskeletal proteins during cardiac myogenesis _in vitro_. In stockdale F, Kedes L (eds): "Cellular and Molecular Biology of Muscle Development. UCLA Symposia on Molecular and Cellular Biology, New Series Volume 93," New York: Alan R. Liss, in press.

Lazarides E, Hubbard BH (1976). Immunological characterization of subunit of the 100A filaments from muscle cells. Proc Natl Acad Sci USA 73:4344-4348.

Tokuyasu KT, Maher PA, Singer SJ (1985). Distribution of vimentin and desmin in developing chick myotubes _in vivo_. II. Immunoelectron microscopic study. J Cell Biol 100:1157-1166.

Digestive Tract

Cells and Tissues: A Three-Dimensional Approach by Modern Techniques in Microscopy, pages 303–308

COMPARATIVE STUDIES ON THE STEREO ARCHITECTURE OF THE CONNECTIVE TISSUE PAPILLAE IN SOME MAMMALIAN TONGUES

Kan Kobayashi and Shin-ichi Iwasaki

Dept. of Anatomy, School of Dentistry at Niigata, The Nippon Dental University, Niigata 951 Japan

INTRODUCTION

The connective tissue papillae (CP) have been regarded as adaptive structures which enlarge the epithelial-connective tissue interface. It is also conceivable that stereo structures of connective tissue papillae affect the differentiation of the epithelium as well as the formation of the exterior of lingual papillae. This study deals with the morphology of various types of lingual connective tissue papillae from comparative morphological point of view.

MATERIALS AND METHODS

Tongue tissues were fixed in formol or Karnovsky's fixative. Epithelial cell layers of the lingual papillae were exposed by long term hydrochloric acid treatment (8N HCl for 2-3 weeks at room temperature) and the surface of connective tissue papillae was observed by scanning electron microscopy.

RESULTS

A. Filiform papillae

Suncus murinus Filiform papillae of the tip of the tongue had connective tissue papillae seen as spherical protrusions on which a shallow groove runs in the antero-posterior direction. In the middle region, connective tissue of filiform papillae had one or two small round head on each spherical protrusion, while several branched structures were seen in the posterior region of the anterior tongue (Fig.1).

Mouse Filiform papillae distributed in the anterior part of the tongue had connective tissue papillae of conical shape and had a round depression that declined slightly antero-downward (Fig.2), and a long narrow depression running along the anterior

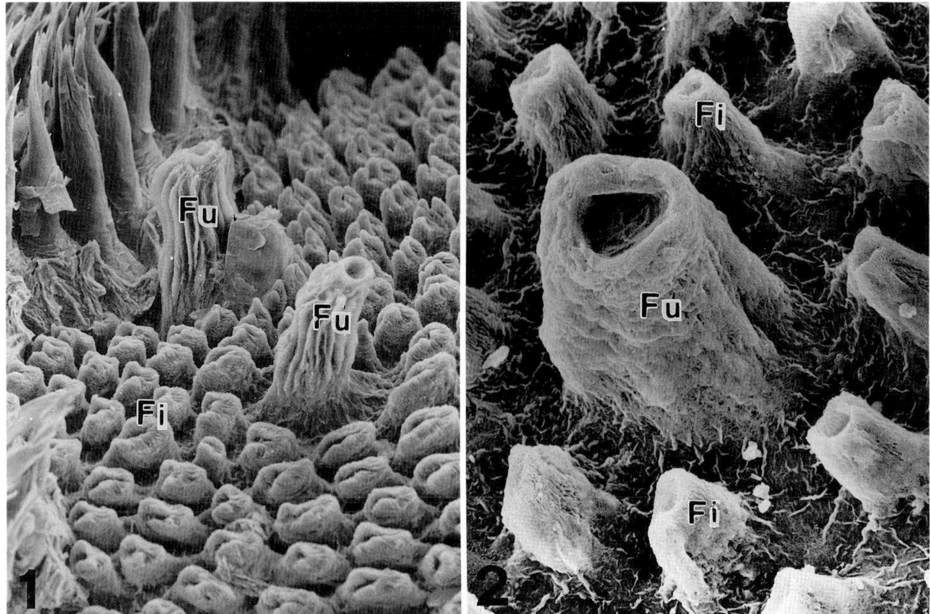

Scanning electron micrograph of CP of filiform (Fi) and fungiform (Fu) papillae of *Suncus murinus* (Fig.1) and mouse tongue (Fig.2). Note a round depression on the top of the fungiform papillae of both animals. Fig.1: x 140, Fig.2: x 580.

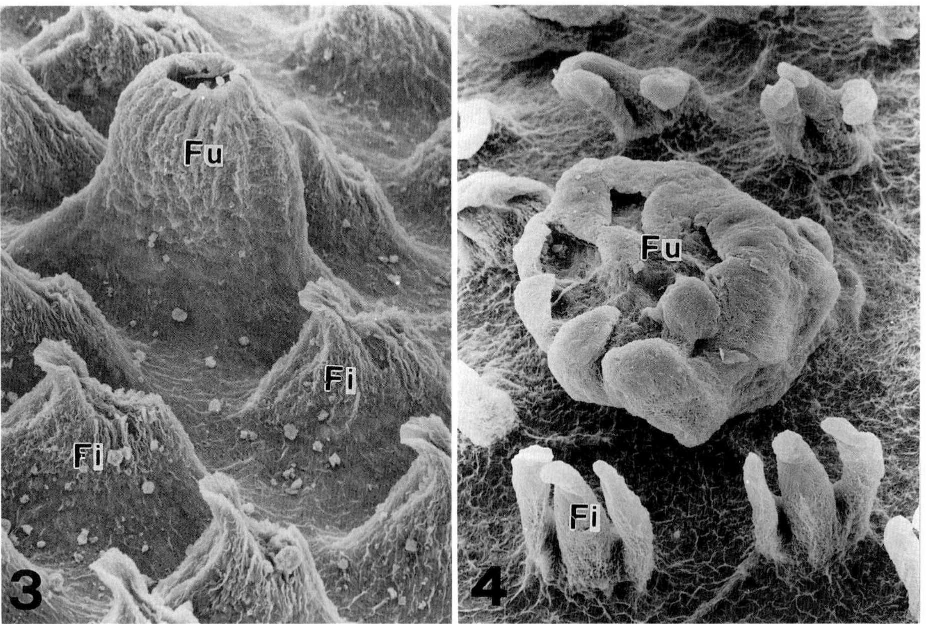

Scanning electron micrograph of CP of rat (Fig.3) and guinea pig (Fig.4) tongue. Note sea anemone-like appearance of fungiform (Fu) of rat and fist-like appearance of guinea pig. Fi:filiform papilla. Fig.3: x 350, Fig.4: x 270.

edge of each connective tissue papilla. Large conical papillae distributed at the anterior margin of the intermolar prominence had shovel-like connective tissue papillae with a depression on the posterior surface unlike that of the anterior filiforms. Branched papillae distributed in the posterior part of the prominence had a depression on the anterior surface.

Rat Filiform papillae distributed at the anterior part of the tongue have connective tissue of a roughly conical shape with the top slender depression running in an under-posterior direction (Fig.3). Concerning to the large conical papillae and branched papillae of rat, there was the same tendency to those of the mouse.

Guinea pig Filiform papillae that were distributed in the anterior part of the tongue were characterized by having several long connective tissue protrusions (Fig.4), while large filiform papillae (large conical papillae) which were distributed on the intermolar prominence had much more slender connective tissue protrusions (Fig.5).

Dog 1 week after birth: filiform papillae of ovoid shape contained a connective tissue papilla representing a smaller elliptical protrusion like a puppet. 4 weeks after birth: the head became somewhat longer and vertical slits occurred on the shoulder as well as the middle arm of the connective tissue of the filiform papillae. Adult: a spearhead-like external structure of the filiform papillae contained connective tissue papillae which consisted of one large main process and several small rod-shaped accessory processes (Fig.6).

Cat Large filiform papillae included connective tissue papillae consisting of one large posterior process (main process) and several small processes (accessory process) arranged circularly on the anterior basal margin of the primary connective tissue papilla (Fig.7).

Man The connective tissue filiform papilla contains ten or more rod-shaped protrusions from a round primary connective tissue papilla. Some of these protrusions showed further bifurcation.

B. Fungiform papillae

Fungiform connective tissue papillae of *Suncus murinus* appeared thick columnar in shape with a round depression on the top of each (Fig.1). Mouse fungiform connective tissue papillae distributed in the anterior part of the tongue showed a barnacle-like protrusions (Fig.2). Rat connective tissue papillae of each fungiform papilla had a sea anemone-like structure (Fig.3). At the top there was a small halo that may correspond to the taste bud. Guinea pig had connective tissue of the fungiform papillae in the form of a fist-like structure (Fig.4). Fungiform connective tissue papillae of dog had a round profile with several smaller round depressions on the flat surface. Connective tissue papillae of the fungiform papillae of cat were cup-like in shape with several round depressions on the top.

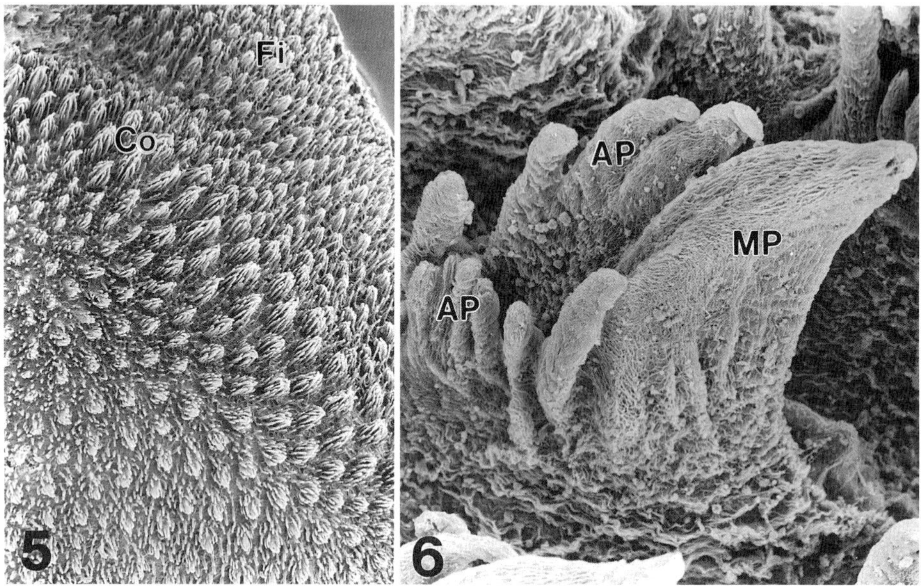

Fig.5. Guinea pig tongue showing CP of filiform (Fi) and large conical papillae (Co) consisted of numerous rod shaped projections. x 18. Fig.6. Adult dog tongue. CP of a filiform consists of one large main process (MP) and several pairs of small rod-shaped accessry processes (AP). x 350.

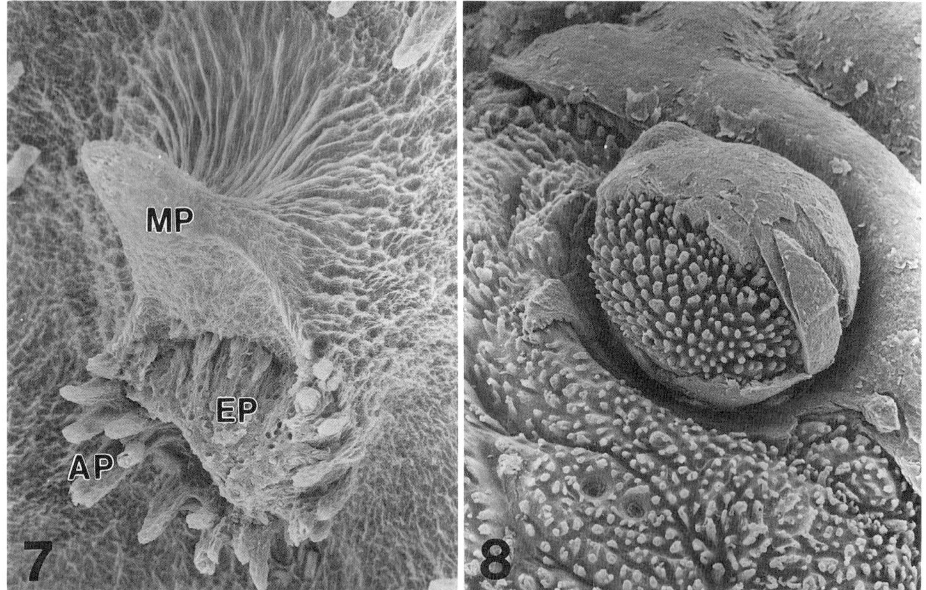

Fig.7. Cat tongue. CP of a large filiform consists of one large main process (MP) and several pairs of accessory processes (AP). EP:remaining epithelial cell mass. x 180.

Fig.8. Vallate papilla of man. Epithelial layer was removed at half part where rod-shaped secondary CP are prominent. x 28.

C. Vallate papillae

Vallate connective tissue papillae in *Suncus murinus*, mouse and rat had a beehive-like structure. In the dog vallate connective tissue papillae appeared as morning glory-like structures. The vallate papilla is surrounded by modified conical papillae arranged closely side by side. Connective tissue papilla of the vallate in man had numerous rod-shaped secondary connective tissue papillae on a large round primary papilla (Fig.8).

D. Foliate papillae

After removal of the epithelium foliate connective tissue papillae appeared as oval holes in accordance with the external several slits arranged in parallel on the posterior margin of the anterior tongue in mouse, rat and guinea pig.

DISCUSSION

The interdigitated structure between the epithelium and the underlying connective tissue increases the surface area over which the two tissue types are in contact with each other. This increases the strength of the junction between the two tissues. It also decreases the distance between the vascular supply and the epithelial cells. Blood vessels that remain inside the connective tissue nourish the epithelial cells. Blood vessels in the connective tissue papillae can carry more nutrients closer to more epithelial cells than can blood vessels that are located underneath a straight epithelium-connective tissue interface (Horstmann 1954).

Connective tissue papillae contain not only blood vessels but also nerve fibers that are related to sensory functions such as the sense of taste, touch, temperature, pain and pressure. Though it is not clear whether nerve terminals reach the epithelial cell layer or not, it may be reasonable to think that connective tissue papillae are developed in order to distribute nerve fibers as far as possible to the superficial layer of the epithelium.

The findings studied so far showed that each animal species has properly shaped connective tissue papillae of lingual papillae though there are some regional differences on the dorsal surface of the tongue (Kobayashi et al., 1987, 1988).

Transplantation experiments have suggested that the typical papillary architecture and the features of the epithelium covering the connective tissue papillae result from the action of unknown connective tissue inducers (Plagmann et al., 1974; Karring et al., 1975).

It has been established that epithelial mesenchymal interactions are fundamental in creating new and diverse differentiated tissues, and mesenchyme often plays a predominant role in directing the differentiation of epithelia (see Bissell et al., 1982). They proposed dynamic reciprocity between the extracellular matrix containing such substances as collagen, glycoproteins of glycosaminoglycans on the one hand and the cytoskeleton and the nuclear matrix on the other hand. The extracellular matrix is thought to exert physical and chemical effects on the geometry and the biochemistry of the cell via transmembrane receptors so as to alter the pattern of gene expression by changing the association of the cytoskeleton with mRNA and the interaction of the chromatin with the nuclear matrix.

Further detailed studies on the three dimensional architecture of connective tissue papillae of the tongue that consisted mainly of collagenous fibers may help to elucidate epithelium -connective tissue interaction in the differentiation and morphogenesis of lingual papillae.

SUMMARY

The epithelial cell layer was removed from the underlying connective tissue papillae by the method of long term hydrochloric acid treatment and the connective tissue surface was observed by scanning electron microscopy. Tongues of insectivora *(Suncus murinus)*, rodentia (mouse, rat, guinea pig), carnivores (dog, cat) and primates (man) were studied. Each animal species had proper architecture of connective tissue papillae as to four types of lingual papillae. There was a tendency for the stereo structure of connective tissue papillae of these lingual papillae to become more complicated from insectivora to primates.

REFERENCES

Bissell MJ, Hall HG, Parry G (1982). How does the extracellular matrix direct gene expression? J theor Biol 99:31-68.

Horstmann E (1954). Morphologie und Morphogenese des Papillarkörpers der Schleimhäute in der Mundhöhle des Menschen. Z Zellforsch 39:479-514.

Karring T, Land NP, Löe H (1975). The role of gingival connective tissue in determining epithelial differentiation. J periodont Res 10:1-11.

Kobayashi K, Miyata K, Iino T (1987). Three-dimensional structures of the connective tissue papillae of the tongue in newborn dogs. Arch histol jap 50:347-357.

Kobayashi K, Miyata K, Takahashi K, Iwasaki S (1988). Developmental and morphological changes in dog lingual papillae and their connective tissue papillae. In "Regeneration and development", Proc 6th Int M Singer Symp pp 609-617.

Plagmann H Chr, Lange DE, Bernimoulin JP, Howe H (1974). Experimentelle Studie über die Epithelneubildung bei heterotopischen Bindegewebstransplantaten. Deut zahnärz Z 29:497-503.

Cells and Tissues: A Three-Dimensional
Approach by Modern Techniques in Microscopy,
pages 309–316

EPITHELIAL CELL PROLIFERATION AND DIFFERENTIATION IN THE GASTRIC MUCOSA: COMPARISONS BETWEEN HISTOGENETIC AND CELL RENEWAL PROCESSES

Katsuko Kataoka, Akiko Kantani-Matsumoto and Yasuko Takeoka

Department of Anatomy, Hiroshima University School of Medicine, Hiroshima 734

CELL PROLIFERATION, DIFFERENTIATION AND MATURATION IN RENEWAL IN THE GASTRIC MUCOSA IN THE ADULT ANIMAL

The surface and foveola of the gastric mucosa are lined with surface mucous cells (Fig. 1). The gastric gland proper, which opens to the bottom of the foveola, is divided into 3 portions — isthmus, neck and base. The isthmus is also called the generative zone because of the presence of actively proliferating immature cells intermingled with parietal cells. The neck consists of mucous neck cells and parietal cells. Chief cells and some parietal cells form the base. The fine structure of each cell type has been well elucidated (Helander, 1981).

Cell kinetic studies have revealed that surface mucous cells are formed in the isthmus and migrate along the foveola to the mucosal surface, while other types of cells migrate downward from their generation sites (Hattori and Fujita, 1976). By panoramic electron microscope observation, it is proved that the cells are generally placed in the order of their fine structural maturation stages along the superficial-deep axis of the mucosa (Kataoka and Sakano, 1984). This can be observed by light microscopy with Feulgen hydrolysis-PAS-Con A-HRP-Bowie's staining after Hayama et al.(1987) (Fig. 2). Mucous granules of surface mucous cells, stained with periodic acid-Schiff (PAS) reaction, increase from the isthmus toward the mucosal surface. Mucous granules of mucous neck cells, stained with paradoxical Con A reaction, increase from the upper to lower part of the neck. Zymogen granules in chief cells,

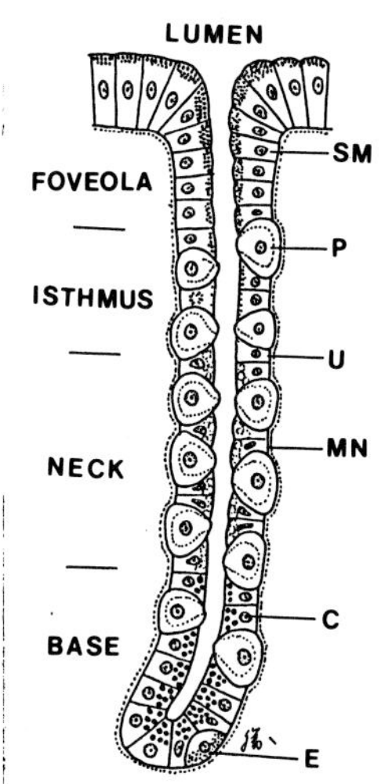

Figure 1. Schematic drawing of the gastric mucosa. SM:surface mucous cell, P:parietal cell, U:undifferentiated cell, MN:mucous neck cell, C:chief cell, E:endocrine cell.

stained with Bowie's solution, increase in number and size from the upper to lower part of the base.

By autoradiography 1 hour after the injection of ^{3}H-thymidine (^{3}H-TdR), most labeled cells are in the isthmus and identified as immature surface mucous cells containing a few mucous granules in the apical cytoplasm (Kataoka, 1970)(Fig. 3, 4). Surface mucous cells become ultrastructurally mature with migration along the foveola to the mucosal surface; mitochondria, rough endoplasmic reticulum and Golgi complex develop and secretory granules increase in number (Kataoka and Sakano, 1984)(Fig. 5). At the mucosal surface, surface mucous cells undergo in situ degeneration followed by extrusion (Kataoka et al., 1985).

Mucous neck cells uptake ^{3}H-TdR in both immature and mature stages (Kataoka, 1970)(Fig. 3). Ultrastructural maturation of mucous neck cells can be followed from the upper to lower part of the neck by panoramic electron microscopy (Kataoka and Sakano, 1984). Immature mucous neck cells contain a few mucous granules of moderate density, while mature mucous neck cells have abundant mucous granules with heterogeneous content (Fig. 6, 7). In the mucous neck cell, rough endoplasmic reticulum is not so well developed even in the mature stage, but Golgi complex is very large.

Immature chief cells occasionally appear at the transitional region between the isthmus and neck, and continue along the neck to the base (Kataoka et al., 1986). In addition, a small group of immature chief cells is found in the upper part of the base (Fig. 8). These immature cells uptake ^{3}H-TdR and undergo mitosis occasionally (Kataoka, 1970). At the junction between the neck and

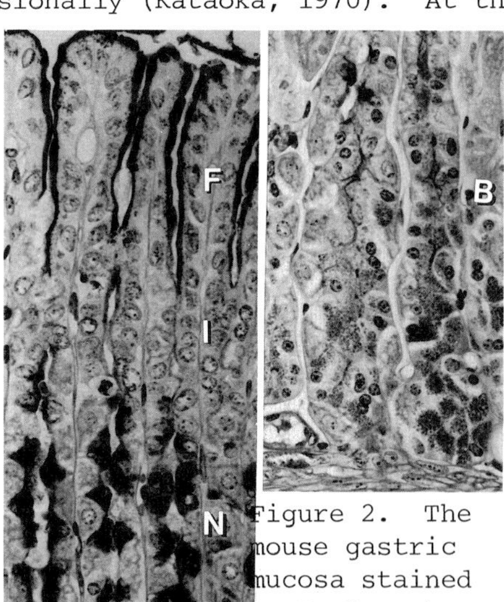

Figure 2. The mouse gastric mucosa stained by Feulgen hydrolysis-PAS-Con A-Bowie's method. The lower border of the left picture continues to the upper of the right. F:foveola, I:isthmus, N:neck, B:base.

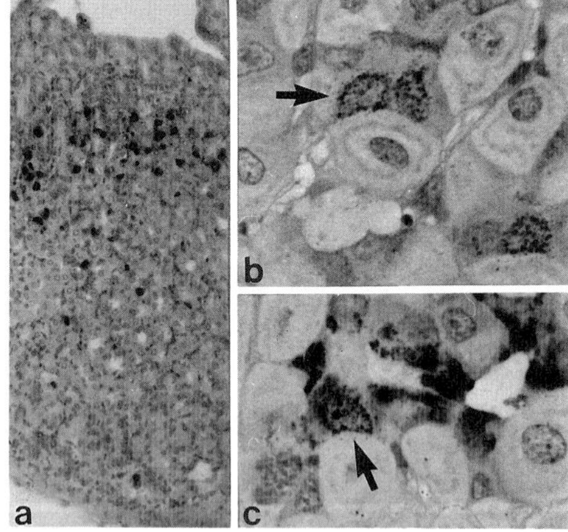

Figure 3. Autoradiograms 1 hour after an injection of ^{3}H-TdR. a:Labeled cells gather in the isthmus to form the generative zone. H.E.stain. b,c:Labeled immature surface mucous cells in the isthmus (b) and a labeled mucous neck cell in the neck (c). PAS stain.

base, the intermediate cell between the mucous neck and chief cells is seen. The cell contains heterogeneous secretory granules, which exhibit pepsinogen I (PG I)-immunoreactivity and bind lectins as mucous granules in the mucous neck cell (Suzuki et al., 1983; Suganuma et al., 1985)(Fig. 9). These findings suggest that chief cells are formed by direct differentiation from undifferentiated cells in the isthmus, occasional mitosis of immature chief cells and transition from mucous neck cells (Kataoka et al., 1986). Ultrastructural maturation of chief cells can be followed toward the bottom of the gland (Kataoka and Sakano,

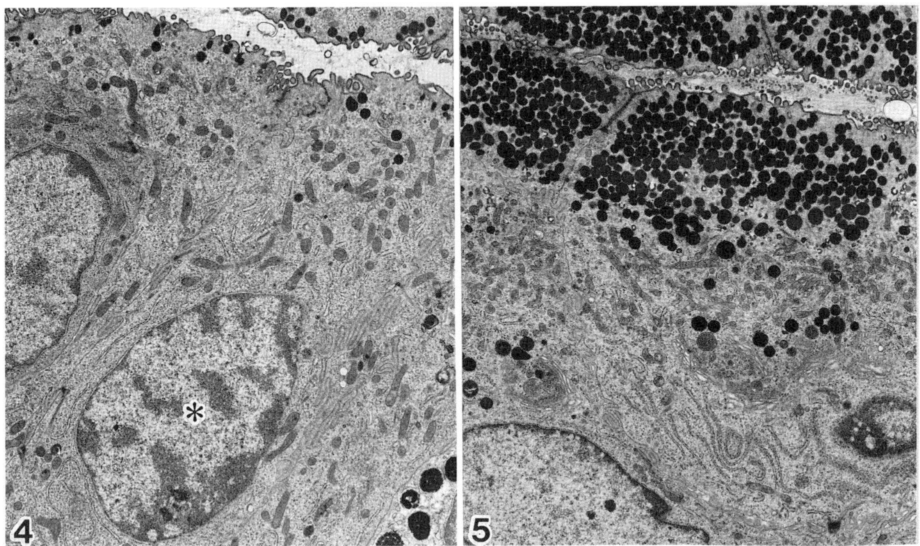

Figure 4. Immature surface mucous cells in the isthmus. The chromatin pattern indicates that the cell (*) is in the mitosis.
Figure 5. Mature surface mucous cells lining the foveola.

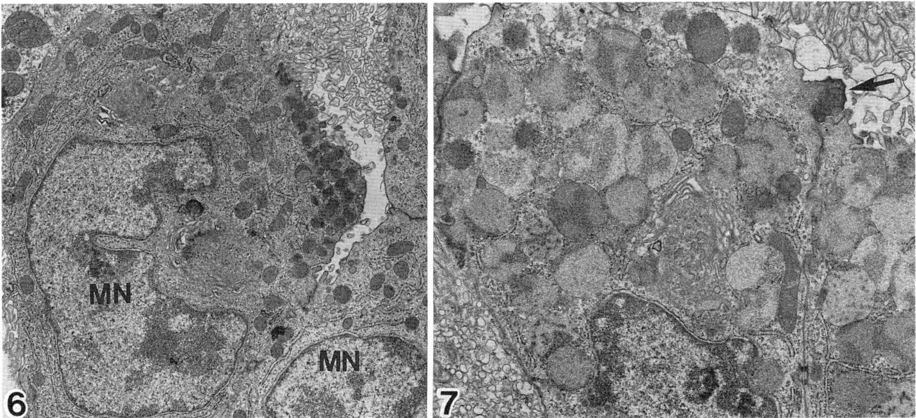

Fig. 6. Immature mucous neck cells (MN) at the uppermost part of the neck. The right cell contains only a few mucous granules.
Figure 7. Mature mucous neck cells interconnected by the midbody (arrow) in the terminal stage of mitosis.

1984). Mature chief cells contain numerous parallel arrays of rough endoplasmic reticulum and many zymogen granules (Fig. 10).

The very immature form of the parietal cell, identified by characteristic microvilli, is present at the lower part of the isthmus (Kataoka, et al., 1986)(Fig. 11). Then intracellular canaliculi are formed, and mitochondria and tubulovesicular elements develop to become mature parietal cells (Fig. 13). Immature parietal cells uptake ^{3}H-TdR very occasionally (Kataoka et al., 1984). Immature parietal cells sometimes contain a few granules resembling those in mucous neck cells (Kataoka et al., 1986)(Fig. 12). These findings suggest that parietal cells are formed by differentiation from undifferentiated cells, occasional mitosis of immature parietal cells and probable transition from immature mucous neck cells (Kataoka et al., 1986). Parietal cells migrate principally toward the bottom of the gland, but some parietal cells migrate toward the mucosal surface (Kataoka et al., 1986).

The probable cell lineage in the cell renewal process is summarized in Figure 14.

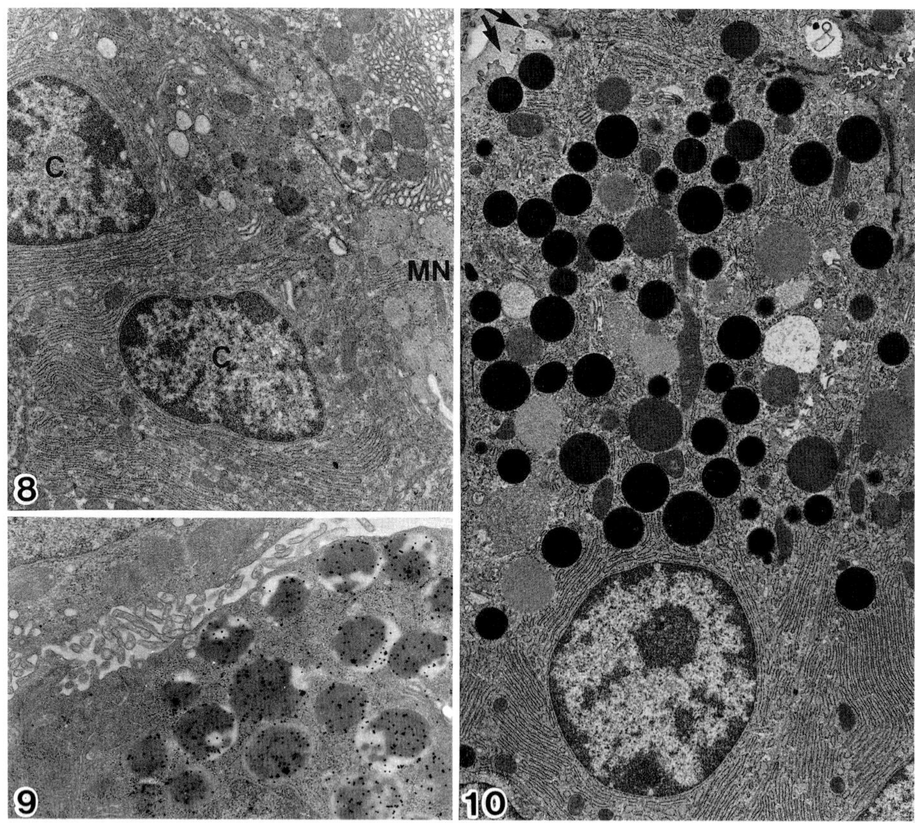

Figure 8. Secretory granules of immature chief cells (C) and a mucous neck cell (MN) exhibit PG I-immunoreactivity.
Figure 9. Secretory granules of an intermediate cell between mucous neck and chief cells have PG I-immunoreactivity.
Figure 10. Mature chief cells with exocytotic figures (arrows).

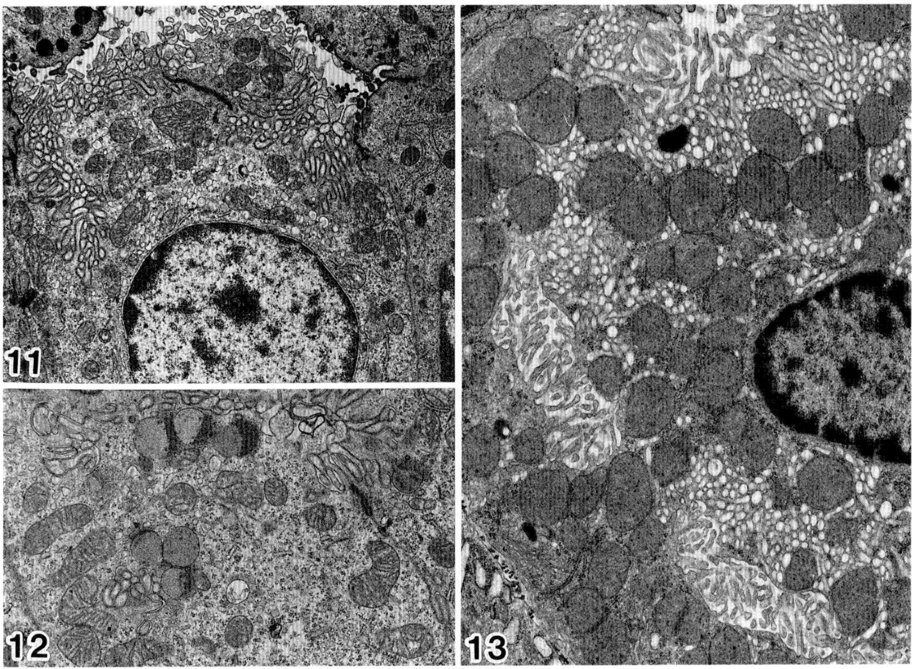

Figure 11. An immature parietal cell showing characteristic microvilli. Intracellular canaliculi begin to be formed.
Figure 12. An intermediate cell between mucous neck and parietal cells. Intracellular canaliculi and mucous granules are seen.
Figure 13. A mature parietal cell.

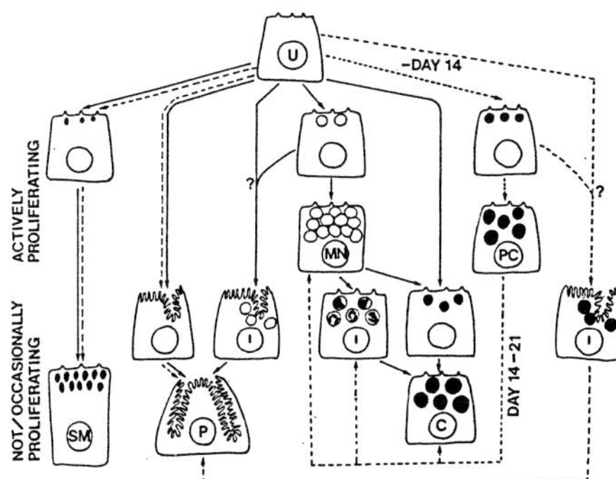

Figure 14. Schematic drawing showing the cell lineage of the gastric mucosa in cell renewal in the adult (solid lines) and in histogenesis in the developing mouse (broken lines). I:intermediate cell, PC:primitive chief cell. Refer Figure 1 for abbreviations for other cell types.

CELL PROLIFERATION, DIFFERENTIATION AND MATURATION IN HISTOGENESIS OF THE MOUSE GASTRIC MUCOSA

The gastric mucosa is lined with undifferentiated cells by day 15 of gestation (day E15)(Kataoka et al., 1984). Surface mucous cells, parietal cells and the only one type of pepsinogen-producing cells, tentatively called primitive chief cells, are identified on day E16 for the first time (Fig. 15, 16). Surface mucous cells and primitive chief cells are identified by dense secretory granules in the supranuclear cytoplasm. Parietal cells are identified by characteristic microvilli at the cell apex.

Other than these, the cytoplasmic structure looks very immature with numerous free ribosomes, scanty membranous organelles and accumulation of glycogen particles. Organelles participating in secretion develop to some extent until birth, and they increase rapidly within the postnatal first day (day 0) in each cell type.

On day 1, surface mucous cells contain mucous granules, mitochondria and rough endoplasmic reticulum as developed as those in

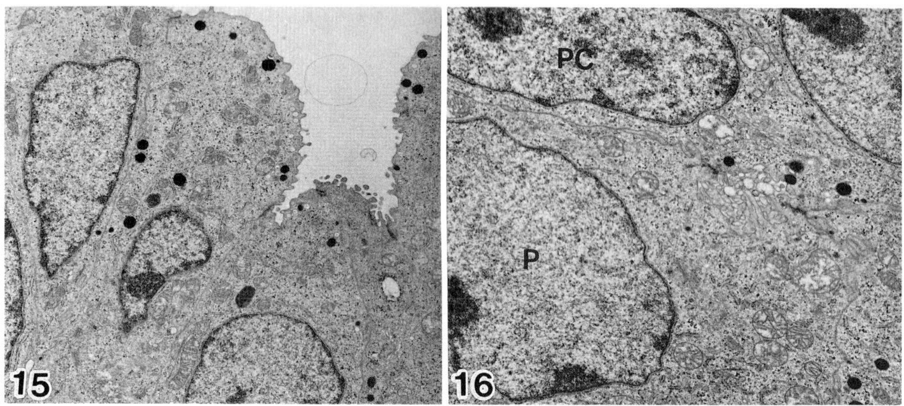

Figure 15. Immature surface mucous cells on day E16.
Figure 16. Immature primitive chief cells (PC) and an parietal cell (P) on day E16. The latter has characteristic microvilli.

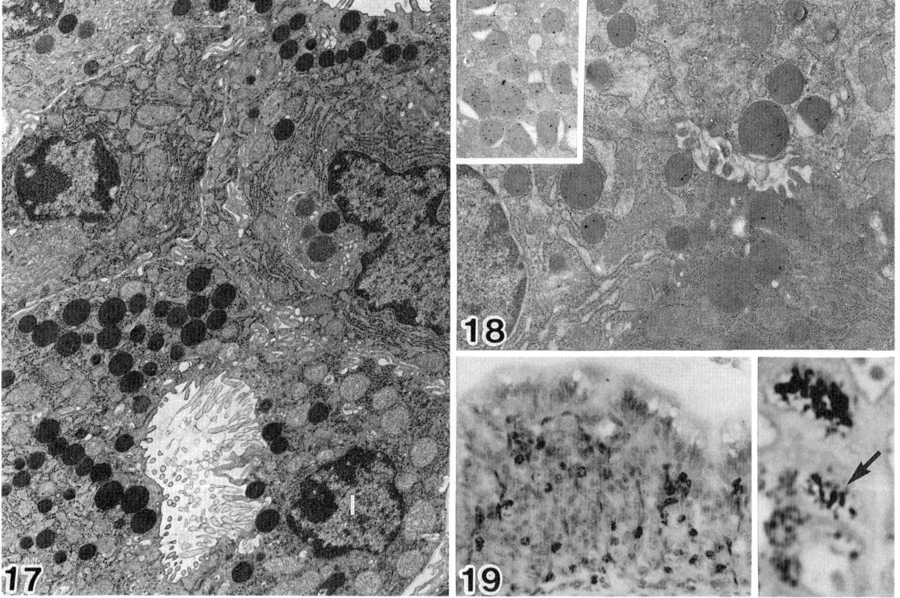

Figure 17. Primitive chief cells on day 14. An intermediate cell (I) projects parietal cell-like microvilli and contains dense secretory granules.
Figure 18. PG I-immunoreactivity in primitive chief cells.
Figure 19. ^{3}H-TdR-uptake cells are diffusely distributed throughout the gland. A labeled primitive chief cell is enlarged (arrow).

the adult. Results of carbohydrate histochemistry of mucous granules are consistent throughout the ontogenic development. They are stained by PAS and galactose oxidase-Schiff reactions, and bind lectins, such as SBA, RCA-I, WGA and UEA-I. Surface mucous cells uptake ^{3}H-TdR even at the mucosal surface in fetal mice. After day 1, ^{3}H-TdR-uptake is restricted in immature surface mucous cells lining the lower part of the foveola and the isthmus.

Mature parietal cells with enlarged intracellular canaliculi, numerous mitochondria and tubulovesicular elements appear on day 1. The number of parietal cells is relatively small by day 14. It increases rapidly from day 14 to 21. Parietal cells rarely uptake ^{3}H-TdR even in this period, but parietal cells containing granules resembling those in primitive chief cells are often seen (Kataoka et al., 1984). These facts suggest that parietal cells are formed by differentiation from undifferentiated cells and by probable transition from primitive chief cells but rarely by mitosis of parietal cells.

Mature primitive chief cells, which first appear on day 1, exhibit typical fine structure of protein-secreting cells with parallel arrays of rough endoplasmic reticulum and dense secretory granules (Kataoka et al., 1984)(Fig. 17). Occasionally the granules look biphasic and consist of dense portion and less dense cap. Primitive chief cells enlarge from the upper to the lower part of the gland. Rough endoplasmic reticulum and secretory granules increase as well. These fine structural features and positive stainability of the granules with Bowie's solution suggest that secretion of the primitive chief cell is rich in protein. However, secretory granules in the majority of primitive chief cells are immunoreactive with anti-PG I-serum very weakly (Fig. 18). Only a few cells, which are indistinguishable from other cells ultrastructurally, contain secretory granules immunoreactive as intensely as zymogen granules in the definite

TABLE. Pepsinogen I-Producing Cells in Mouse Gastric Mucosa

		Primitive Chief Cell	Mouse Neck Cell	Chief Cell
		∼Day 14	Day 21∼ Adult	
Structure	rER	++	+	+++
	Granules	dense≫biphasic*	heterogeneous	dense
^{3}H-TdR Uptake		+++	++	+
Histochemistry of Granules	Pepsinogen I	+≫+++	++	+++
	Bowie's stain	++	-	+++
	PAS reaction	-∼±	++	-
	(Fluorescence)	+∼++	+++	-
	SBA	+∼++	+++	-
	HPA, RCA-I, WGA	+∼++	+++	-∼±(rim)
	UEA-I	++∼±	-≫++	-
	Paradox Con A	±∼++	+++	-

*Dense≫biphasic means that the cells containing dense granules are far larger in number than the cells containing biphasic secretory granules. rER: rough endoplasmic reticulum.

chief cell in the adult. Secretory granules of the primitive chief cell seem to contain complex carbohydrates resembling mucus in the mucous neck cell in quality but less in quantity. The granules in the primitive chief cell are stained with PAS reaction so weakly that the stainability is confirmed only by fluorescence microscopy (TABLE). The granules are stained with various lectins, but the staining is generally weaker than staining of mucous neck cell-granules. Primitive chief cells uptake ^{3}H-TdR very actively throughout the gastric gland (Fig. 19).

On day 21, chief cells and mucous neck cells are clearly distinguished by ultrastructure, Bowie's staining and carbohydrate histochemistry (TABLE). PG I-immunoreactivity of either type of cells is far more intense than that in primitive chief cells.

The probable cell lineage during the histogenesis of the mouse gastric mucosa is summarized in Figure 14.

Acknowledgement. This study was supported in part by the grant provided by the Ichiro Kanehara Foundation.

REFERENCES

Hattori T, Fujita S (1976). Tritiated thymidine autoradiographic study in the gastric gland of the golden hamster. Cell Tiss Res 172:171-184.

Hayama M, Katsuyama T, Nakayama J, Akamatsu T, Honda T (1987). A combined stain for identifying epithelial cells of gastric mucosa. Stain Tech 62:35-40.

Helander HF (1981). The cells of the gastric mucosa. Int Rev Cytol 70:217-289.

Kataoka K (1970). Electron microscopic observations on cell proliferation and differentiation in the gastric mucosa of the mouse. Arch histol jap 32:251-273.

Kataoka K, Sakano Y (1984). Panoramic observation of the mouse gastric mucosa by superwide-field electron microscopy. Arch histol jap 47:209-221.

Kataoka K, Sakano Y, Miura J (1984). Histogenesis of the mouse gastric mucosa, with special reference to type and distribution of proliferative cells. Arch histol jap 47:459-474.

Kataoka K, Takeoka Y, Hirano S (1985). Electron microscopic observations of surface mucous cells in the mouse gastric mucosa during physiological degeneration and extrusion. Arch histol jap 48:327-339.

Kataoka K, Takeoka Y. Maesako J (1986). Electron microscopic observations on immature chief and parietal cells in the mouse gastric mucosa. Arch histol jap 49:321-331.

Suganuma T, Tsuyama T, Murata F (1985). Glycoconjugate cytochemistry of the rat fundic gland using lectin/colloidal-gold conjugates and Lowicryl K4M. Histochem 83:489-495.

Suzuki S, Furihata C, Tsuyama S, Murata F (1983). Immunocytochemical localization of pepsinogen in rat stomach. Histochem 79: 167-176.

Cells and Tissues: A Three-Dimensional Approach by Modern Techniques in Microscopy, pages 317–326

CORROSION CASTS IN LIVER AND STOMACH MICROCIRCULATION

Osamu Ohtani

Department of Anatomy, Okayama University School of Medicine, Okayama, Japan

INTRODUCTION

Scanning electron microscopy (SEM) of corrosion casts allows a three-dimensional view of blood and lymphatic vascular architecture, more clearly than previously employed techniques. Although injection of casting media into the parenchyma or into the blood vessels of the fetus or newborn animals reproduces the interstitial tissue spaces where macromolecules and tissue fluid flow, the organization of interstices cannot be fully examined by this method. This limitation can be at least partly overcome by recently introduced cell-maceration/SEM method (Ohtani, 1987; Ohtani et al., 1988) which can demonstrate organization of collagen fibrillar networks delimiting the interstitial compartment.

This paper describes the three-dimensional organization of blood and lymphatic vessels in the liver and stomach as revealed by SEM of corrosion casts, and also demonstrates collagen fibrillar networks with cell-maceration/SEM method.

MATERIALS AND METHODS

The animals used were healthy rats (Wistar, adults and newborns), rabbits, and young mongrel dog. The human liver tissues used for SEM of collagen fibrils were obtained in the autopsy of a Japanese individual died of lung cancer at 74 years of age.

The blood vascular corrosion casts were made according to the method by Murakami (1971). The lymphatic corrosion casts of the stomach were reproduced by either intraparenchymal puncture-injection (Kobayashi et al., 1976; Ohtani and Ohtsuka, 1985; Ohtani, 1987) or intra-arterial injection (Ohtani and Murakami,

1987) of methacrylate resin (Mercox, Oken Shoji, Tokyo) whose viscosity was adjusted to 1.5-2.0 centistokes with supplement of monomeric methyl methacrylate. The lymphatic corrosion casts of the liver were replicated by retrograde injection of resin into the common bile duct (Yamamoto and Phillips, 1986). The corrosion casts obtained were air-dried, mounted on metal stubs, coated with gold, and observed under the SEM (JSM-U3, JEOL) with an accelerating voltage of 5kV.

The collagen fibrillar networks of the liver and stomach were extracted by cell-maceration method (Ohtani, 1987). In brief, tissue pieces fixed with 2.5% glutaraldehyde were put in a 10% aqueous solution of NaOH for 3-7 days at 25°C, followed by washing in water. The pieces were conductive stained with 1% tannic acid and 1% OsO_4 (Murakami, 1973). The pieces were dehydrated through a graded series of ethanol, freeze-cracked with a razor blade in liquid nitrogen, and freeze-dried with t-butyl alcohol (Inoue and Osatake, 1988). The specimens were mounted on metal stubs with a double sticky tape, coated with gold, and observed under the SEM (JSM-U3, JEOL, or HSF-2, Hitachi) with an accelerating voltage of 15-20kV.

Some of the stomach tissue pieces that were perfusion-fixed with 2.5% glutaraldehyde were processed for SEM of tissues. The pieces were conductive stained with tannic acid and OsO_4 (Murakami, 1973), and dehydrated through a graded series of ethanol. They were freeze-cracked in liquid nitrogen, critical point-dried, mounted on metal stubs, coated with gold and observed under the SEM with an accelerating voltage of 15-20kV.

RESULTS AND DISCUSSION

Liver

The hepatic lobule (Kiernan, 1833) as seen by the corrosion casts is the mass of extensively interconnected sinusoids. There are some regional differences in the organization of the sinusoids. Near their origins from the interlobular (portal) venules the sinusoids are slightly narrower, more tortuous and more extensively interconnected than their terminations in the central venule (Fig. 1). However, in most mammals the peripheral boundaries of the lobule are not well defined except in pigs. The interlobular vein runs with the hepatic arterial branches in the portal tract. The interlobular vein gives off numerous short side branches or inlet venules which rather abruptly branch out into the hepatic sinusoids at the margin of the hepatic lobule (Fig. 1). The terminal arterioles fall into four categories depending upon their terminations; 1) the terminal arterioles draining directly into the hepatic sinusoids, 2) those connect with the interlobular venules, 3) those forming the peribiliary plexus surrounding the bile duct, and 4) those forming the

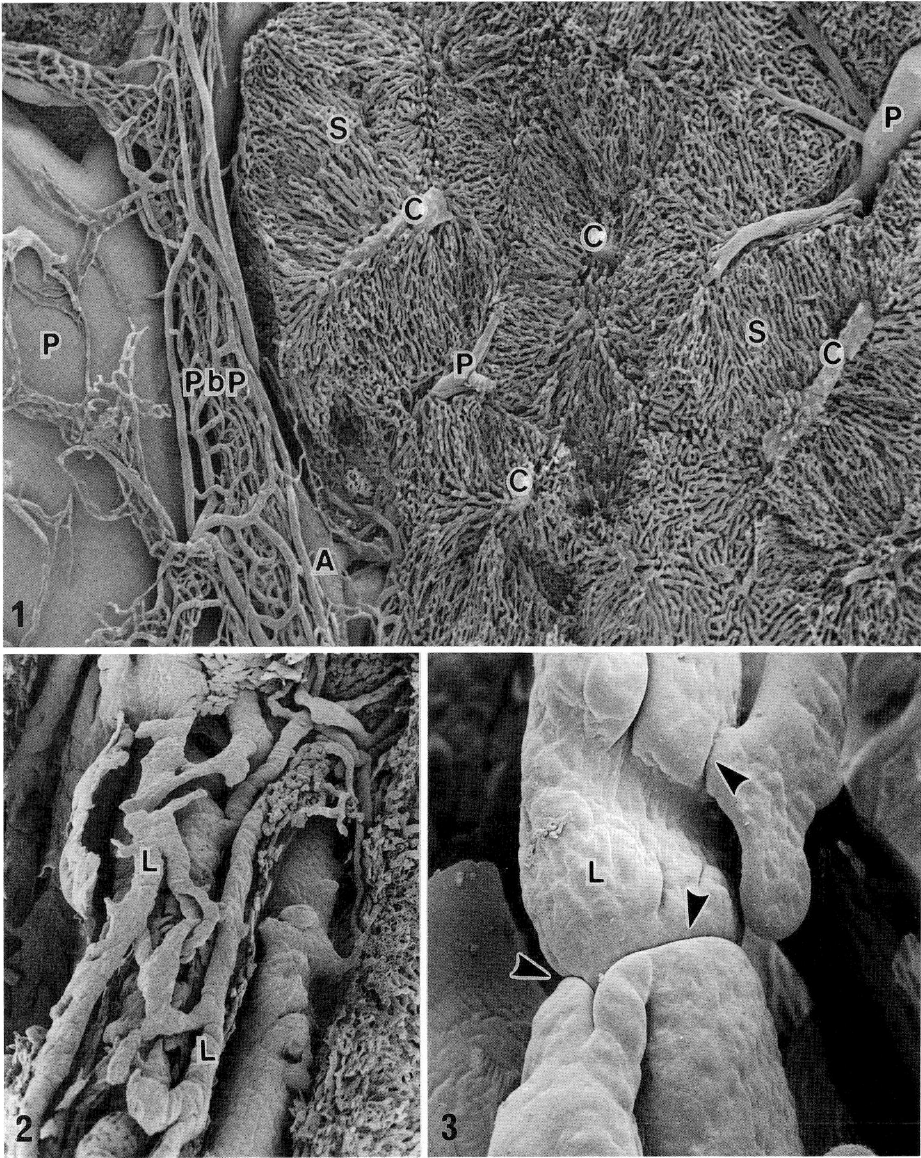

Fig. 1. SEM view of the blood vascular corrosion cast of the rat liver. The interlobular vein (P), hepatic arterial branches (A), and peribiliary plexus (PbP) run parallel in the portal tract. The extensively interconnected sinusoids (S) converge into the central vein (C). X50. Figs. 2,3. SEM views of the corrosion cast of the rabbit liver showing the network of lymphatic vessels (L) in the portal tract. There are many impressions formed by the endothelial nuclei, and marked notches (arrowheads) indicative of bicuspid valve locations. Fig. 2. X80, Fig. 3. X620.

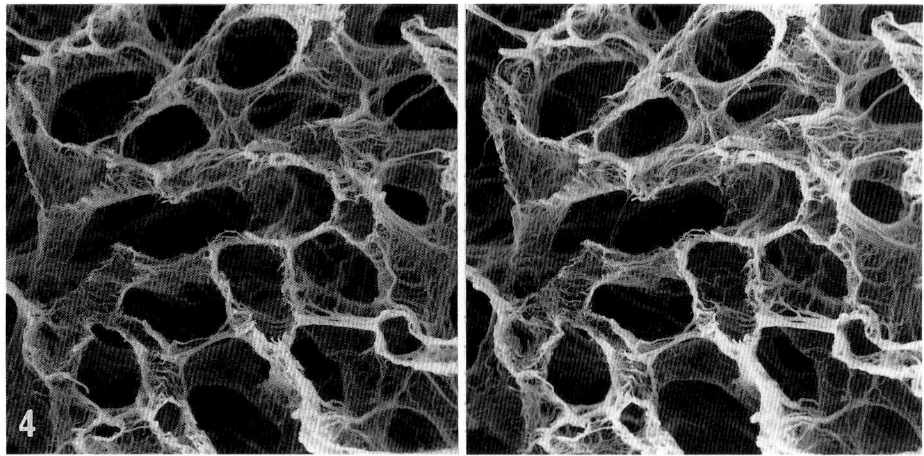

Fig. 4. Stereo SEM view of collagen fibrillar sheaths in the spaces of Disse (human liver). X 700.

periportal plexus to supply the tissues in the portal tract (Ohtani, 1979).

As noted by Mall (1906) and confirmed by later studies (Andrews et al., 1949), most of the terminal arterioles contribute toward the formation of the peribiliary capillary plexus. The efferent vessels of the plexus drain into the hepatic sinusoids either directly or via the interlobular venules (Ohtani and Murakami, 1978; Ohtani, 1979). Thus the vascular route from the peribiliary plexus to the hepatic sinusoids qualifies for the portal system which seems to play a role in transporting substances absorbed from the bile and any hormonal substances produced by the biliary system back to the hepatic lobule presumably to act there (Murakami et al., 1974).

The resin injected retrogradely into the common bile duct leaks at the periphery of the lobule and fills the lymphatic vessels in the portal tract (Yamamoto and Phillips, 1986) (Figs. 2,3). The lymphatic vessels repeatedly divide and anastomose, and form the network surrounding the portal triad (Fig. 2). Closer views of the cast show many impressions on its surface formed by the endothelial nuclei, and marked notches indicative of the bicuspid valve locations (Fig. 3). Biliary constituents are known to enter the lymphatic vessels following bile duct obstruction (Bloom, 1923). This mechanism is thought to be a route for the regurgitation of the bile into the blood. Herein demonstrated resin leakage from the bile into the lymphatic vessels indicates a possible route through which biliary substances enter the lymphatic vessels following obstruction of the bile duct.

SEM of the liver tissues macerated with alkali plus water has clearly shown the continuum of the collagen fibrillar network. There are collagen fibrillar sheaths in the spaces of Disse (Fig. 4). The collagen fibrillar bundles frequently stretch between adjacent sheaths for the sinusoids (Ohtani, in preparation). Such bundles seem to be located interhepatocellularly, because, as Motta and Porter (1974) reported, the spaces of Disse extend interhepatocellularly and form a continuous labyrinth system. Our preliminary TEM studies has also shown that the collagen fibrillar bundles exist interhepatocellularly. The collagen fibrillar framework reported here seems to play important roles not only in maintaining mechanical stability of the hepatic parenchyma, but also in providing the fluid and macromolecules with their passageways.

Stomach

The microvascular pattern of the mucosa shows some species-specific differences. In the rat, near the base of the glands the mucosal arterioles from the submucosal arterial plexus break up into capillaries which, in turn, ascend along the glands toward the luminal surface (Figs. 5,6). The mucosal capillary network is collected just beneath the surface epithelium into venules which descend perpendicularly to lead into the submucosal venous plexus (Gannon et al., 1982) (Fig. 5). In the rabbit, the submucosal arterioles give off two types of arterioles, short and long, to the mucosa (Ohtsuka and Ohtani, 1984). The short arterioles break up into capillaries near the base of the gastric gland, while the long ones ascend along the gland and supply the capillary network beneath the surface epithelium. In the glandular neck region, the superficial and deep capillary networks connect with each other and converge into common venules which descend along the glands to lead into the submucosal venous plexus. The microvascularization of the human gastric mucosa is essentially equivalent to that in the rabbit (Raschker et al., 1987). Arteriovenous anastomoses in the stomach have not been confirmed by SEM of corrosion casts. The microvasculature of the stomach is so organized that the blood in the glandular mucosal capillaries flows upward from the glandular bottom to the surface epithelium. This microcirculation pattern seems to have physiological significance in protecting the gastric mucosa from acid injury (Gannon et al., 1982, 1984).

Between the base of the gastric glands and the muscularis mucosa in the rat and dog stomach, there is a single-layered network of lymphatic capillaries (Figs. 7,8), which is in agreement with fluorescent _in vivo_ microscopy by Nagata and Guth (1984) and other injection studies (Donini, 1955). The beadwork-like casts of the interglandular tissue spaces connect with the lymphatic casts in the base of the mucosa (Fig. 7). Among the beadwork-like casts there are occasionally lymphatics-like

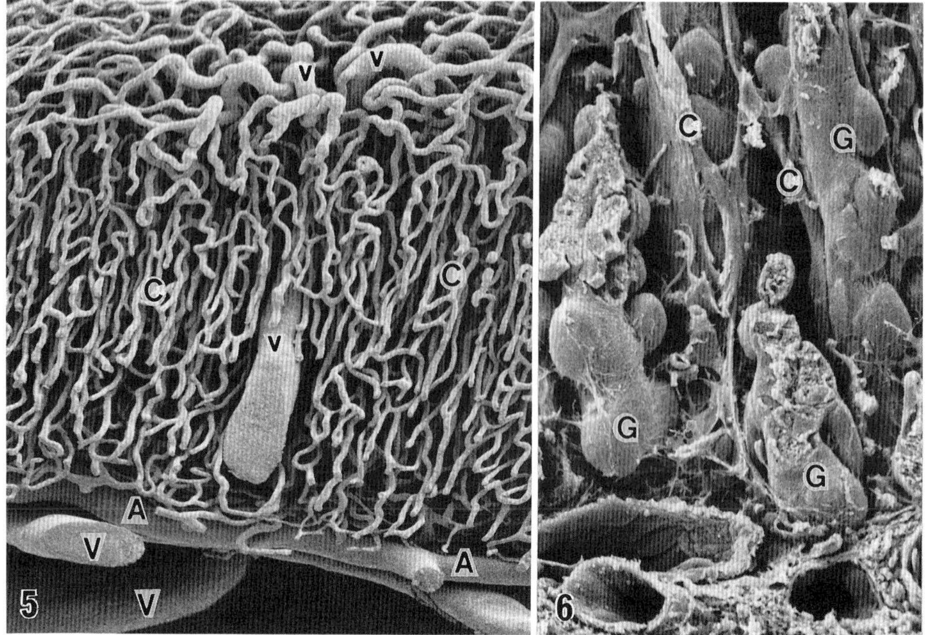

Fig. 5. SEM view of the corrosion cast of the rat stomach. The glandular capillaries (C) are collected just beneath the surface epithelium into venules (v) which descend perpendicularly to lead into the submucosal vein (V). A: submucosal artery. X 95. (Micrograph courtesy of Dr. A. Ohtsuka)
Fig. 6. SEM view of the rat gastric mucosa showing the capillaries (C) ascending along the glands (G). X720.

structures replicated (Fig. 7), which suggests the presence of lymphatic vessels in the interglandular region of the deep mucosa. The mucosal lymphatic capillaries gather into slightly thicker lymphatic vessels which descend through the muscularis mucosa and lead into the thicker submucosal lymphatic plexus (Fig. 9). This lymphatic plexus leads into the lymphatic plexus located between the inner and outer layers of the tunica muscularis (Ohtani and Murakami, 1987)(Fig. 10). The lymphatic casts in the submucosa and tunica muscularis show notches indicative of valve locations (Figs. 9,10).

SEM of fractured tissues shows the lymphatic capillaries located between the gastric glands and the base of the muscularis mucosa (Fig. 11). The lymphatic vessels lying between the outer and inner layer of the tunica muscularis are satisfactorily exposed by peeling off the serosa with the outer muscular layer (Fig. 12). The lymphatic vessels in the tunica muscularis possess bicuspid valves, each of which consists of two flat semilunar leaflets covered with the endothelial cells (Fig. 12).

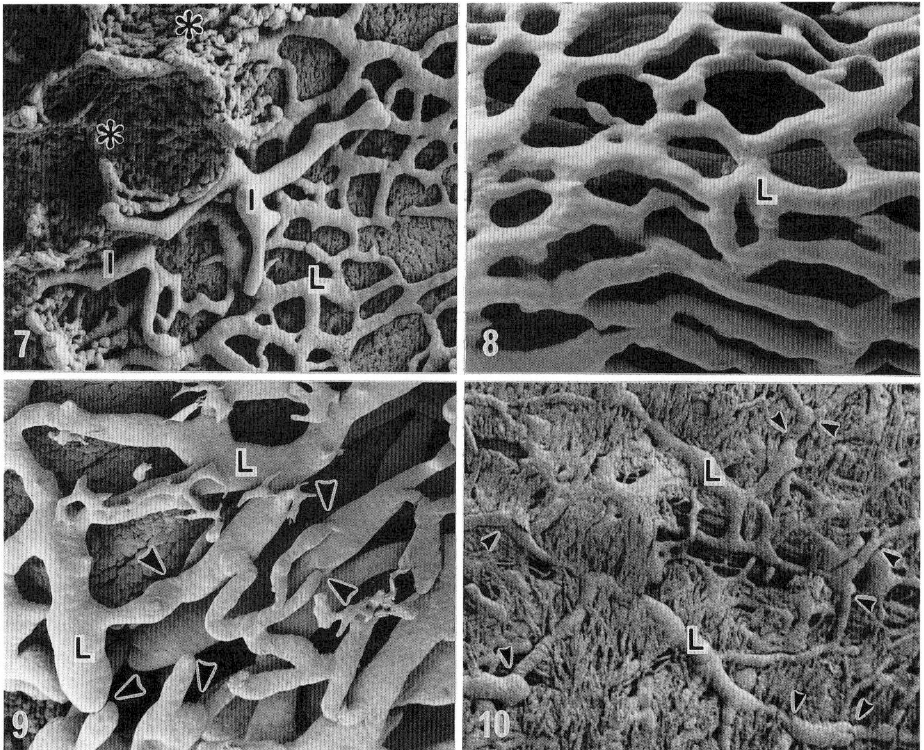

Figs. 7-10. SEM views of lymphatic corrosion casts of the stomach in the adult rat (Fig. 7), young dog (Figs. 8,9) and newborn rat (Fig. 10). Figs. 7,8: The lymphatic network in the base of the mucosa (L). The interstitial tissue spaces of the interglandular region are replicated as scrambled egg-like casts (*), among which tubular structures (l) probably of lymphatics are also reproduced. Fig. 7: X80, Fig. 8: X160. Fig. 9: The lymphatic network (L) in the submucosa. Arrowheads indicate notches indicative of valve locations. X50. Fig. 10: The lymphatic networks (L) in the tunica muscularis. These lymphatic casts also show many notches (arrowheads) indicative of valve locations. X60.

This, therefore, endorses that the notches observed on the lymphatic corrosion casts actually represent the locations of valves. All of the lymphatic vessels observed are lined with thin endothelial cells with nuclear bulges (Figs. 11,12). The endothelial cells project many columnar cytoplasmic leaves which interdigitate with those of adjacent cells (Fig. 12). These characteristics of the stomach lymphatic vessels are essentially the same as those of small intestinal ones (Ohtani, 1987).

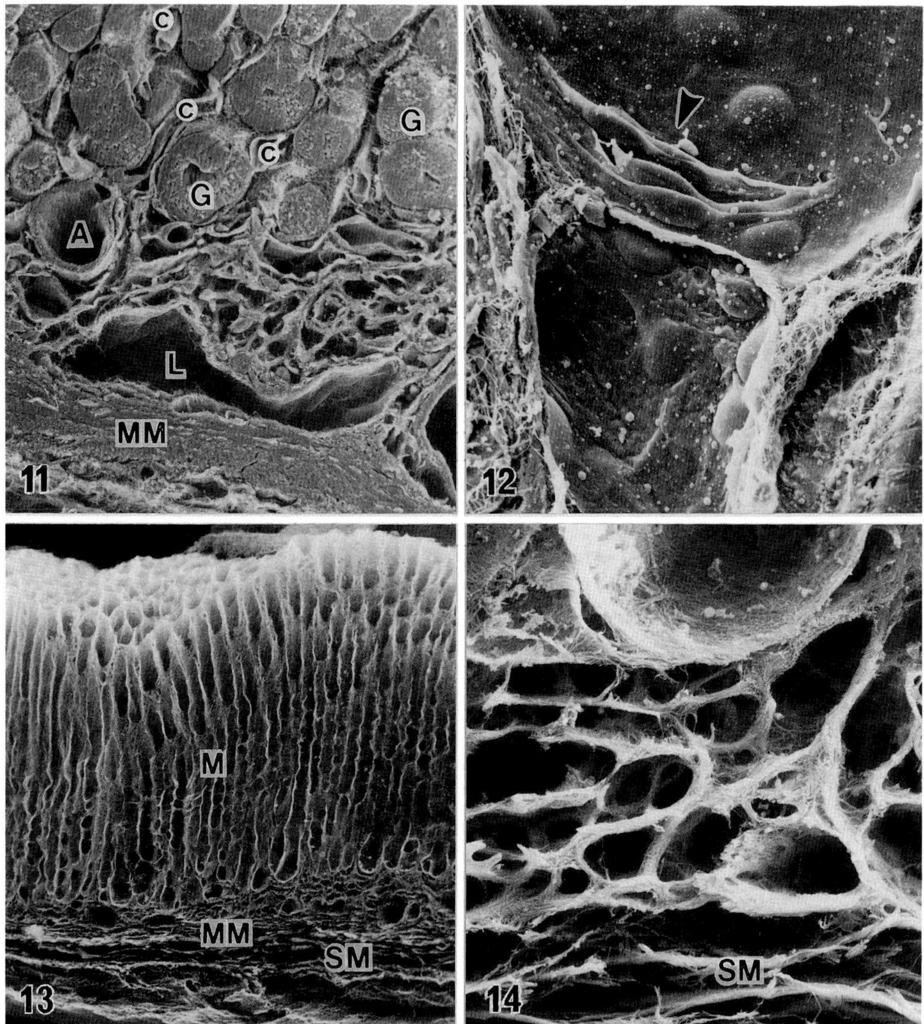

Figs. 11, 12. SEM views of the fractured rat stomach. Fig. 11: A lymphatic capillary is located immediately above the muscularis mucosa (MM). G: gastric gland, C: blood capillary, A: arteriole. X480. Fig. 12: The luminal view of the lymphatic vessel in the tunica muscularis. Nuclear regions of the endothelial cells show elliptical bulges, and flat cytoplasmic processes of the endothelial cells interdigitate with those of adjacent cells. There is a bicuspid valve (arrowhead) which corresponds to the notch seen on the lymphatic cast surface (Fig. 10). X1,200.
Figs. 13, 14. SEM views of collagen fibrillar network of the rat stomach. The collagen fibrillar network clearly delimits the interstitial compartment. M, MM, SM: mucosal, muscularis mucosal, and submucosal layer. Fig. 13: X100, Fig. 14: X1,600.

SEM of the stomach macerated with alkali plus water reveals the continuum of the collagen fibrillar network (Figs. 13,14) which provides the organ with mechanical stability. The interstitial compartment of the interglandular region is delimited by the dense collagen fibrillar network (Fig. 13). This compartment is continuous with that in the mucosal base where the collagen fibrillar bundles form a complicated network resembling that of the reticular tissue (Fig. 14). The interstitial compartment serves for the passageway of the interstitial tissue fluid and macromolecules.

CONCLUSION

The present study demonstrates the three-dimensional organization of blood and lymphatic vessels in the liver and stomach by SEM of corrosion casts. The peribiliary portal system in the liver has again been confirmed. Retrograde injection of resin into the bile duct replicates rich lymphatics in the portal tract. The microvasculature of the stomach is so organized that the blood in the mucosal capillaries flows from the glandular base to the surface epithelium. There are rich lymphatic capillaries in the base of the gastric mucosa. SEM of the NaOH macerated tissues demonstrates the collagen fibrillar networks which maintain the mechanical stability of the organs and also clearly delimit the interstitial compartment where tissue fluid flows.

REFERENCES

Andrews WHH, Maegraith BG, Wenyon CEM (1949). Studies on the liver circulation. II. The microanatomy of the hepatic circulation. Ann Trop Med Parasit 43:229-237.

Bloom W (1923). The role of the lymphatics in the absorption of bile pigments from the liver in early obstructive jaundice. Bull Johns Hopkins Hosp 34:316-320.

Donini I (1955). Sur la fine distribution des vaisseaux lymphatiques dans l'estomach humain (English abstr). Acta Anat 23:289-311.

Gannon B, Browning J, O'Brien P (1982). The microvascular architecture of the glandular mucosa of rat stomach. J Anat 135:667-683.

Gannon B, Browning J, O'Brien, Rogers P (1984). Mucosal microvascular architecture of the fundus and body of human stomach. Gastroenterology 86:866-875.

Inoue T, Osatake H (1988). A new drying method of biological specimens for scanning electron microscopy: The t-butyl alcohol freeze-drying method. Arch Histol Cytol 51:53-60.

Kiernan F (1833). The anatomy and physiology of the liver. Phil Trans Roy Soc London 123:711-770.

Kobayashi S, Osatake H, Kashima Y (1976). Corrosion casts of lymphatics. Arch Histol Jpn 39:177-181.

Mall FP (1906). A study of the structural unit of the liver. Am J Anat 5:227-308.

Motta PM, Porter RP (1974). Structure of rat liver sinusoids and associated tissue spaces as revealed by scanning electron microscopy. Cell Tiss Res 148:111-125.

Murakami T (1971) Application of the scanning electron microscope to the study of fine distribution of the blood vessels. Arch histol Jpn 32:445-454.

Murakami T (1973). A metal impregnation method of biological specimens for scanning electron microscopy. Arch histol Jpn 35:323-326.

Murakami T, Itoshima T, Shimada Y (1974). Peribiliary portal system in the monkey liver as evidenced by the injection replica scanning electron microscope method. Arch histol Jpn 37:245-260.

Nagata H, Guth PH (1984). In vivo observation of the lymphatic system in the rat stomach. Gastroenterology 86:1443-1450.

Ohtani O (1979). The peribiliary portal system in the rabbit liver. Arch histol Jpn 42:153-167.

Ohtani O (1987). Three-dimensional organization of lymphatics and its relationship to blood vessels in rat small intestine. Cell Tiss Res 248:365-374.

Ohtani O (1987). Three-dimensional organization of the connective tissue fibers of the human pancreas: A scanning electron microscopic study of NaOH-treated tissues. Arch histol Jpn 50:557-566.

Ohtani O, Murakami T (1978). Peribiliary portal system in the rat liver as studied by the injection replica scanning electron microscope method. Scanning Electron Microscopy/1978/II:241-244.

Ohtani O, Murakami T (1987). Lymphatics and myenteric plexus in the muscular coat in the rat stomach: A scanning electron microscopic study of corrosion casts made by intra-arterial injection. Arch histol Jpn 50:87-93.

Ohtani O, Ohtsuka A (1985). Three-dimensional organization of lymphatics and their relationship to blood vessels in rabbit small intestine. A scanning electron microscopic study of corrosion casts. Arch histol Jpn 48:155-268.

Ohtani O, Ushiki T, Taguchi T, Kikuta A (1988). Collagen fibrillar networks as skeletal frameworks: A demonstration by cell-maceration/scanning electron microscope method. Arch Histol Cytol 51:249-261.

Ohtsuka A, Ohtani O (1984). The microvascular architecture of the rabbit stomach corpus in vascular corrosion casts. Scanning Electron Microscopy/1984/IV:1951-1956.

Raschker M, Lierse W, van Ackeren H (1987). Microvascular architecture of the mucosa of the gastric corpus in man. Acta Anat 130:185-190.

Yamamoto H, Phillips MJ (1986). Three-dimensional observation of the intrahepatic lymphatics by scanning electron microscopy of corrosion casts. Anat Rec 214:67-70.

Cells and Tissues: A Three-Dimensional
Approach by Modern Techniques in Microscopy,
pages 327–332

GASTRIC MUCOSAE AS STUDIED BY SCANNING ELECTRON MICROSCOPY IN 3,242 HUMAN BIOPSIES

B.Foliguet, J.C.Guedenet, F.Vicari, L.Marchal and G.Jeanvoine , G.Grignon
Laboratoire de Microscopie Electronique. Faculté de Médecine de NANCY - BP 184, 54505 Vandoeuvre Cedex FRANCE

INTRODUCTION

The report in 1983 of the association of Campylobacter pylori (CP) and activ chronic gastritis gave a new field of development for scanning electron microscopy (SEM) in the study of the gastric mucosae (Marshall 1983, Warren 1983).

The aim of our work based on the systematic search for CP in 1,714 patients undergoing endoscopy it to show that scanning electron microscopy is a technique which is particulary well adapted to detect CP on the surface of gastric mucous membranes.

MATERIAL AND METHODS

The research was carried out on 3,242 samples taken from 1 714 adult patients who had undergone a gastroduodenal endoscopy (1,690 antral and 1 552 fundic biopsies).

Each biopsy was washed in physiological serum to remove the surface mucus, fixed (for 1 to 48 hours) in a 2,5 % glutaraldéhyde solution buffered in a cacodylate solution (pH : 7,2), washed in a buffer solution with saccharose added (7g/100ml), dehydrated in alcohol acetone baths, dried either through the technique of critical point of CO_2 (Anderson, 1951) or in air after passing through Hexamethyldisilazane (HMDS) (Nation, 1983). The samples were then orientated with a binocular microscope, attached to the support, gold coated through ionisation and observed under SEM (Autoscan Siemens Etec and Stereoscan Cambridge).

RESULTS

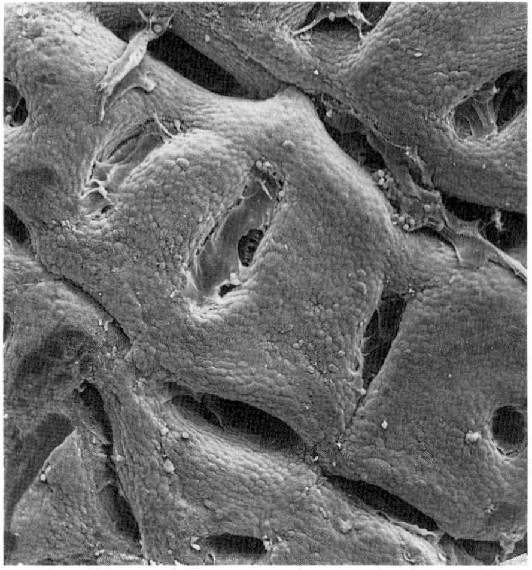

Figure 1

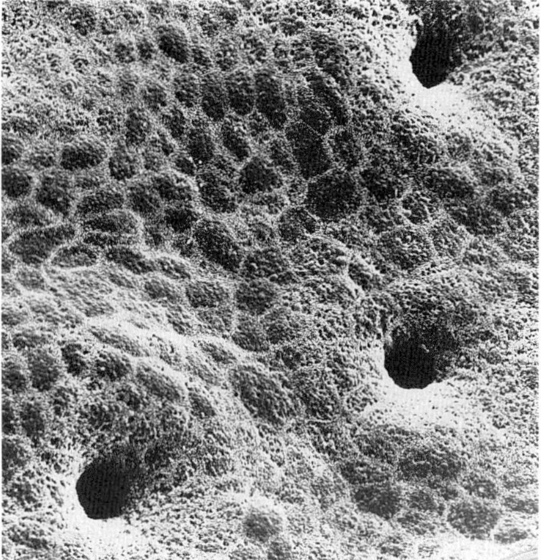

Figure 2

Normal gastric mucosae :

Figure 1 : Antral mucosa : In the area devoided of mucous secretions, the glandular pits are regularly spaced - X 200
Figure 2 : Fundic mucosa : Glandular pits and normal mucocytes - X 700

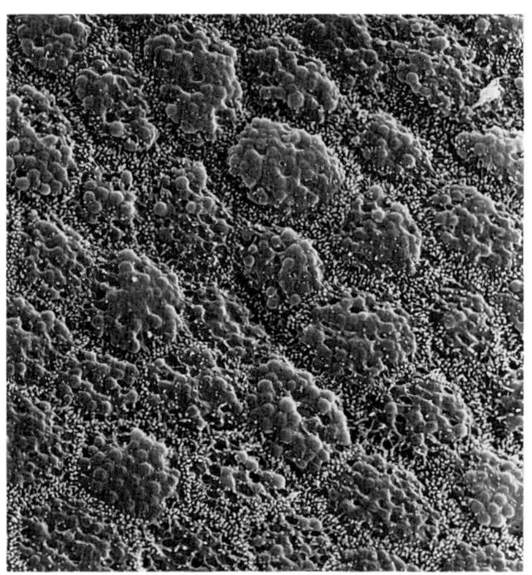

Figure 3

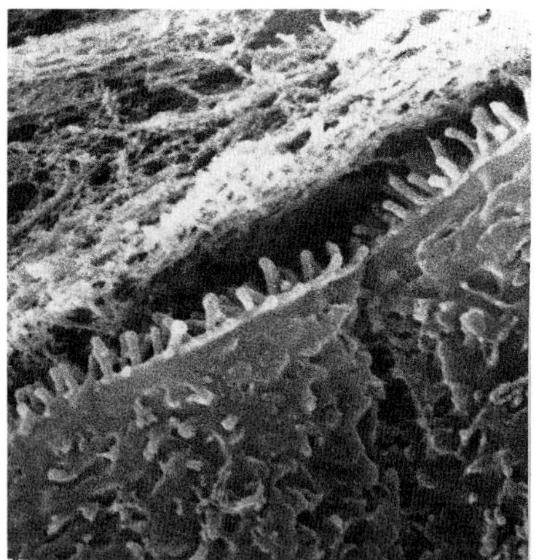

Figure 4

Epithelial secretions :

Figure 3 : Apical secretion of antral mucocytes - X 2,200
Figure 4 : Microvilli of fundic secretory cells and deep layer of mucus - X 10,000

Normal gastric mucosae :

At the surface of normal antral and fundic mucosae the glandular crypts are wide open and regularly spaced (Fig. 1 and 2). The secretion of mucocytes is usually of merocrine type (Fig. 3). In the area non devoïded of mucous secretions the gastric mucus exibits a reticular pattern in deep layer (Fig.4).

Gastroduodenal lesions :

Most of the biopsies show evidence of chronic gastritis. By place the mucocyte appeares hypertrophied with a smooth apical pole rounding in the lumen. The erosion of the apical pole cause a characteristic "honeycomb" appearance (Fig.5). Most phenomena of epithelial delamination leading to denudation of basal membranes are originating around the neck of glands (Fig.6 et 7). Some area of intestinal metaplasia were observed at the gastric mucosal surface (Fig.8).

Detection and quantification of CP :

CP was detected in 671 patients mainly at the fundic level and antral (Fig. 9 et 10)(in 515 and 582 biopsies). CP is characterized by it spiral appearance (the long form 2,5 to 6 µm) or by its curved shape (the short form 1,5 µm). The principal location is the neck of gastric glands, near intercellular spaces.

The high definition given by SEM allows to distinguish CP from other microorganisms observed on the surface of gastroduodenal mucosae (Fig. 11 : Spirillium, Candida albicans, Cocci, Bacteria...).

DISCUSSION

Our study using scanning electron microscopy confirms the focalisation of CP at the surface of antral and fundic epithelium, under the mucus (Goodwin et col.1985). CP has sometimes been detected on the surface of apparently normal mucosae but that does not imply that there is no interstitial lesion. The situation of CP near intercellular spaces may be due to the fact that gastric mucosafavorise growth through chemotactic factors (Marshall 1984) . Sometimes CP are so abundant that the covering of CP hides a large part of the epithelial surface. SEM confirms the frequent association of an infestation to CP with alterations of the surface of gastric mucocytes (Marshall 1984) and suggests a pathogenic action of this microorganism. The rate of positivity found in our series was almost identical to that of other published series studied through bacteriological techniques or optical microscopy (Marshall 1984). The preparation techniques are more rapid than bacteriological or histological techniques allowing interpretations within 2 hours following the arrival of the biopsies in the laboratory.

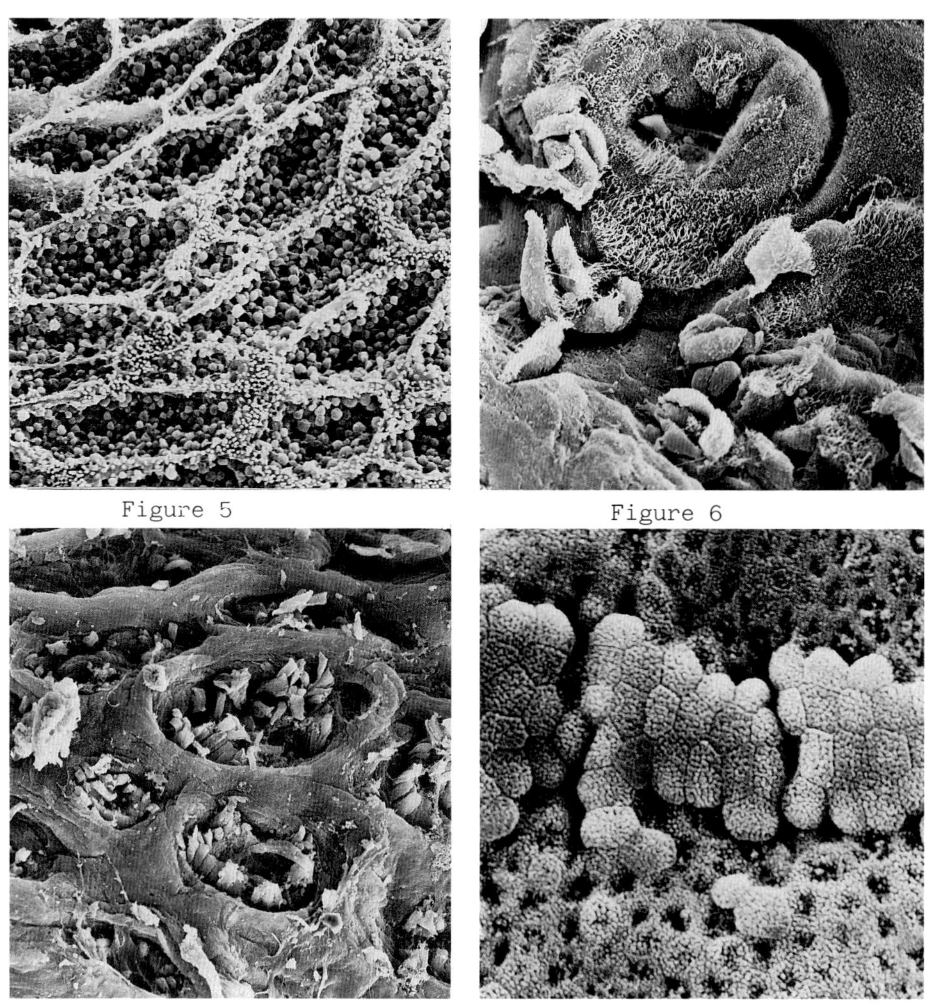

Figure 5

Figure 6

Figure 7

Figure 8

Elementary lesions of gastric mucosae :

Figure 5 : Antral mucosae - Apical erosion of mucocytes with image of vesicular excretion. The surface has a "honeycomb" appearance.

Figure 6 : Epithelial desquamation around the neck of antral glands - X 1,000

Figure 7 : Epithelial delamination leading to denudation of the basal membrane - X 350

Figure 8 : Intestinal metaplasia at the antral surface - X 2,000

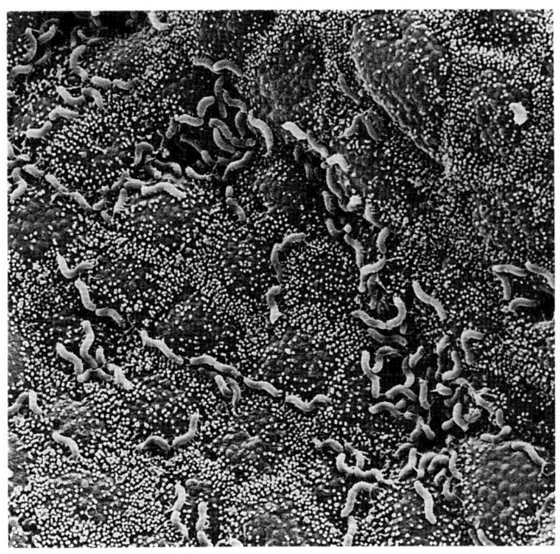

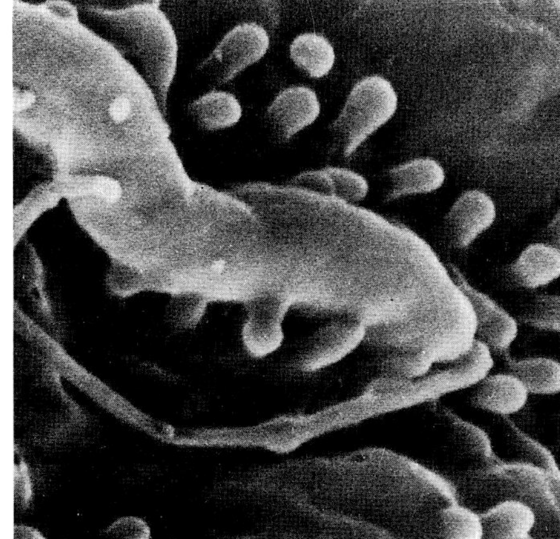

Figure 9 Figure 10

Campylobacter pylori :
Figure 9 : CP at the antral surface - X 3,000
Figure 10 : High magnification of bacterial body and flagellum of CP - X 38,000.

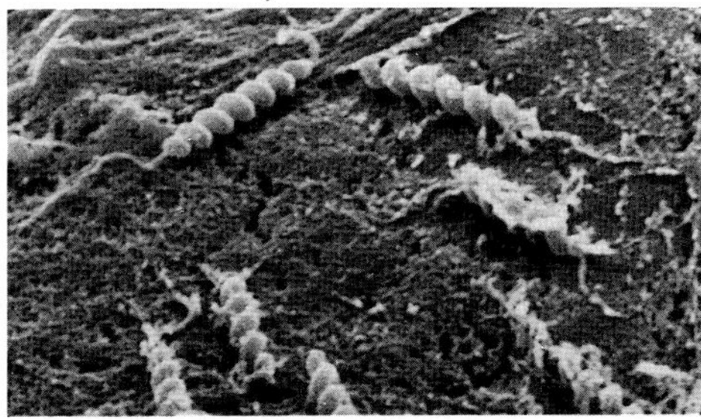

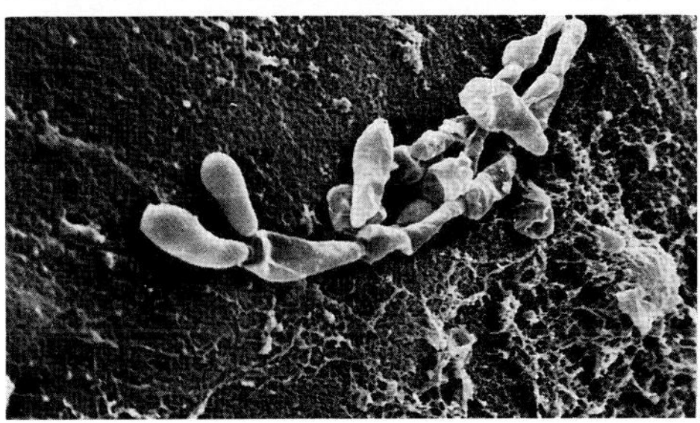

Figure 11 : Other microorganismes founds at the digestiv surfaces.

a) Spirillium - Antrum X 8,000
b) Candida albicans - Antrum X 3,000

CONCLUSION

SEM is a rapid and specific technique, well adapted to searching for and characterising Campylobacter pylori in large gastric biopsies (Steer 1984). It also allows us to appreciate the extent and seriousness of the associated lesions.

REFERENCES

Anderson T.F. (1951) Techniques for the preservation of the dimensional structure in preparing specimens for the electron microscope. Trans. N.Y. Acad. Sci., 13, 130-134.

Goodwin C.S., Mc Culloch R.K., Armstrong J.A. Wee S.H. (1985) Unusual cellular fatty acids and distinctive ultrastructure in a new spiral bacterium (campylobacter pyloridis) from the human gastric mucosa. J.Med.Microbiol., 19, 257-267.

Marshall B. (1983) Unidentified curved bacilli on gastric epithelium in active chronic gastritis Lancet, i, 1273-1275.

Marshall B.J., Royce H., Annaer D.I. (1984) Original isolation of campylobacter pyloridis from human gastric mucosa, Microbios. Letters, 25, 83-88.

Nation H.L. (1983) A new method using hexamethyldisilazane for preparation of soft insect tissues for scanning electron microscopy. Stain Technology. 58, 347-351.

Steer H.W. (1984) Surface morphology of the gastroduodenal mucosa in duodenal ulceration. Gut, 25, 1203-1210.

Warren J.R. (1983) Unidentified curved bacilli on gastric epithelium in active chronic gastritis. Lancet, i, 1273.

Cells and Tissues: A Three-Dimensional
Approach by Modern Techniques in Microscopy,
pages 333–341

STRUCTURAL AND ULTRASTRUCTURAL OBSERVATIONS ON RAT GASTRIC MUCOSA TREATED WITH CYTOPROTECTIVE AGENTS PRIOR TO ETHANOL EXPOSURE.

F.CARPINO°,G.PRINO*,E.GAUDIO°°,V.PETROZZA°,M.MANTOVANI*
D.BOSCO°,P.ALBERICO*,M.MELIS°
°DEPARTMENT OF HUMAN BIOPATHOLOGY
°°DEPARTMENT OF ANATOMY-UNIV. OF ROME "LA SAPIENZA",
ITALY, *CRINOS RESEARCH DEPT., COMO ITALY.

INTRODUCTION

Gastric cytoprotection can be defined as the ability of certain orally-administered pharmacological substances to prevent the development of acute hemorrhagic-necrotic lesions of the gastric mucosa commonly induced in vivo by a variety of necrotizing agents (Brendan J.R. 1985, Hollander D. 1985, Lacy E. 1982, Lacy E. 1985, Morris G. 1986, Morris G. 1987, Ohno T. 1985, Porta R. 1986, Robert A. 1985, Schmidit K.L. 1985, Tarnawski A. 1985, Tarnawski A. 1986, Trabucchi R. 1986). Such pharmacological substances have been termed "cytoprotective agents". Aim of this study was to verify the capacity of cyto-protective agents "per os" administered to prevent gastric mucosal damage from exposure to 80% ethanol.

The following agents were tested: sulcralfate, carbenoxolone, 16,16 -dimethyl prostaglandin E_2, (PGE_2) sulglycotide (GLPS) and Maalox TC®.

MATERIALS AND METHODS

GROUP	TREATMENT P. Os
A (negative control group)	H_2O 5 ml/kg for 5 days
B (positive control group)	FIRST: H_2O 5 ml/kg for 5 days 30'AFTER: ETOH 80% 1 ml
C	FIRST: Sucralfate 200mg/kg/5ml for 5 days 30' AFTER: ETOH 80% 1 ml
D	FIRST: Carbenoxolone 200mg/kg/5ml for 5 days 30' AFTER: ETOH 80% 1 ml
E	FIRST: 16,16 dm PGE_2 20mcg/kg/5ml for 5 days 30' AFTER: ETOH 80% 1 ml
F	FIRST: Sulglycotide 200mg/kg/5ml for 5 days 30' AFTER: ETOH 80% 1 ml
G	FIRST: Maalox TC 5 ml/kg for 5 days 30' AFTER: ETOH 80% 1 ml

One hundred sixtyfour male Sprague-Dawley rats whose body weight varied from 180 to 190 gr. were housed in separate wire-bottom cages and maintained on regular feed.

After fasting for 48 hrs the animals were divided into 7 groups. Each animal received by gavage, for 5 consecutive days, one of the following treatments:

Thirty minutes after the treatment administered on 5th day animals except those belonging to the negative-control, group (A), received 1 ml of 80% ethanol by intragastric instillation.

Six hrs after ethanol administration, each animal was anesthetized with a mixture of ether and oxygen and had its stomach lavaged with 0.9% saline solution. The stomachs were then fixed by intragastric perfusion of glutaraldehyde buffered to pH 7.35, and removed and processed for structural and ultrastructural study by scanning and transmission electron microscopy (SEM and TEM). For light microscopy (LM) conventional stains such as hematoxylin and eosin, Alcian blue, PAS, and Van Gieson's trichromic, were used. For ultrastructural study, current techniques for SEM and TEM were employed. All experimentation and observations were done according to the double-blind method.

RESULTS

In the negative-control group (A), all specimens observed showed normal histological morphology of the rumen, and normal morphology of the glandular region as well as the pyloric region. It is interesting to note that in one case microerosions of the apical portion of the superificial epithelium which might be correlated to pre-existent lesions or to phenomena related to the turnover of the gastric epithelial lining were observed by SEM. In another case, SEM demonstrated an erosion of the mucosa in the region of the body of the stomach, probably due to the stomach-tube trauma.

Histologic examination of the stomachs taken from the animals belonging to the second group, which had been treated only with distilled water and ethanol, showed classic, extensive acute necrotic ulcerative lesions of the glandular and pyloric mucosa, including disruption and exfoliation of the surface epithelium, gastric-gland damage, hyperemia and hemorrhage throughout the thickness of the mucosa. Such lesions never involved the submucosa.

SEM, performed on specimens from group B, confirmed these findings and was very useful in evaluating the extent of hemorrhagic-necrotic damage, the amount of mucus present on the surface epithelium and the alterations undergone by the epithelial cells.

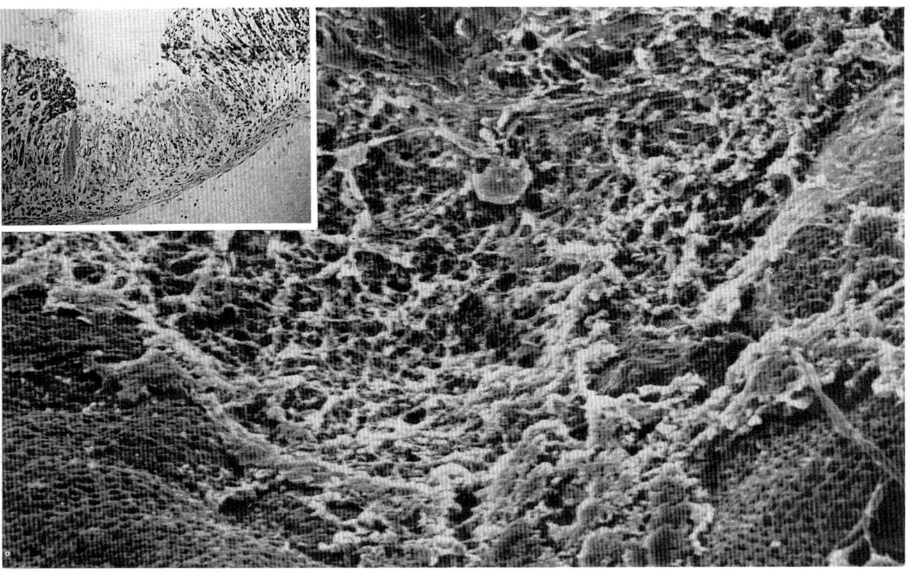

[Fig. 1: SEM (100 x) Positive control; an extensive necrotic-hemorrhagic lesion is evident. Insert: LM (125x), Van Gieson stain: The ulcerative lesion is extented throughout the mucosa, until to the muscolaris mucosae.]

High magnification of the necrotic areas revealed the presence of erythrocytes, cellular debris and fibrin filaments. In the rumen, hyperemia of the mucosa was always evidenced.

TEM ultrastructural observations of the necrotic lesions confirmed the presence of cells undergoing necrotic degeneration which were characterized by margination of the nuclear chromatin even to the point of pyknosis, rupture of the cytoplasmic membrane, vacuolization and rarefaction of the hyaloplasmic matrix, and mitochondrial swelling with rarefaction of the matrix and disappearance of the granules and mitochondrial cristae. In addition, the vessels showed necrosis of the endothelial cells. Free erythrocytes were scattered among the necrotic cells or within the intercellular spaces, which were notably enlarged.

Cells showed various degrees of ultrastructural alteration at the margin of the lesions. The superficial epithelial cells were reduced in height and showed partial or complet disappearance of the apical microvilli and depletion of secretory granules.

Numerous scarcely electron-dense vacuoles and heterophagosomes were seen in the cytoplasm. Diminution of the rough endoplasmic reticulum (RER) and other organalles as well as homogenization of the hyaloplasmic matrix were also noted.

The nucleus appeared indented, with the heterochromatin disposed mainly along the nuclear membrane. Some cells occasionally showed a characteristically clear cytoplasm in addition to dilation of the smooth endoplasmic reticulum (SER) cisternae, destruction of cytoplasmic organelles, rarefaction of the nuclear chromatin and condensation of the perinuclear matrix. Intercellular junctional complexes appeared well preserved. Immediately below the epithelial layer, congested vessels and conspicuos centers of platelet agglutination were present.

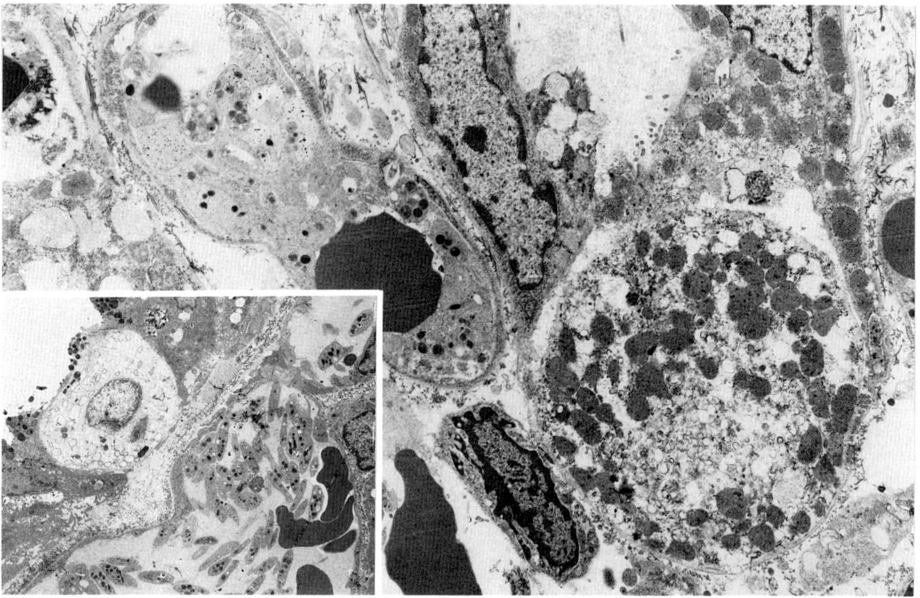

[Fig.2: TEM (2000x) Positive Control. Ultrastructure of necrotic lesion.
Insert: TEM (2000x) A cell with clear cytoplasm and a congested vessel with platelet agglutination are evident.]

In frank ulcerative lesions, the glandular epithelial cells showed nuclear pyknosis and cytoplasmic vacuolization, with disappearance of organelles and homogenization of the hyaloplasmic matrix; numerous free red blood cells were seen scattered among groups of these glandular cells. At the margins of the ulcers the residual glandular cells appeared more intact and showed RER disposed in concentric rings, as well as various degrees of cytoplasmic vacuolization and margination of the nuclear chromatin. In some of these cells, typical pepsinogen granules were also observed. In the more intact glandular regions, the secretory canaliculi appeared dilated and showed rarefied microvilli. Such lesions extended only as far as the muscolaris mucosae; the submucosae generally was apparently normal.

Observations relative to group C (sucralfate for 5 days, then exposure to ethanol), showed in 50% of cases, typical hemorrhagic-necrotic lesions. In 30% of cases, we noted erosions, while in the remaining 20%, only superificial damage was observed. The hemorrhagic-necrotic lesions were qualitatively similar to those seen in positive-control animals (group B). In other cases SEM showed erosions involving the surface of the tunica mucosa. In addition,exfoliative phenomena, with superficial erosion of the apical lining cells, were evident.

TEM ultrastructural observation of the regions containing typical lesions were equivalent to those described for group B in that cells undergoing necrosis, and congested and thrombotic capillaries, as well as lesions showing various degrees of degeneration involving either the lining cells or the glandular epithelium, reaching as far as, but non involving the muscolaris mucosae, were evident.

With regard to group D (carbenoxolone for 5 days,then ethanol exposure), only 10% of cases observed, whether at the structural or ultrastructural level of the specimens examined showed acute hemorrhagic-necrotic lesions similar to those commonly seen after ethanol insult. In 75% of cases, the lesions were limited to superficial epithelial damage. In the remaining 50% of cases, the lesions were comprised of microerosions and increased exfoliation, mainly at the level of the body of the stomach.

Ultrastructural TEM investigations of this group confirmed the absence of advanced stages of mucosal necrosis, as previously described. The superficial epithelium, largely well preserved, did show, in some areas, various degrees of cytoplasmic vacuolization and dilation of the intercellular secretory spaces. Although these cells showed a certain scarcity of microvilli on their surfaces, they were of normal height and were covered with numerous secretory granules. Intercellular junctions appeared intact. In zones showing greater damage to the superficial epithelium (reduction in cell height, nuclear condensation to the point of pyknosis, intense cytoplasmic vacuolization with heterophagosomes, extremely reduced numbers of microvilli and secretory granules), notable vascular congestion and platelet agglutination were seen.

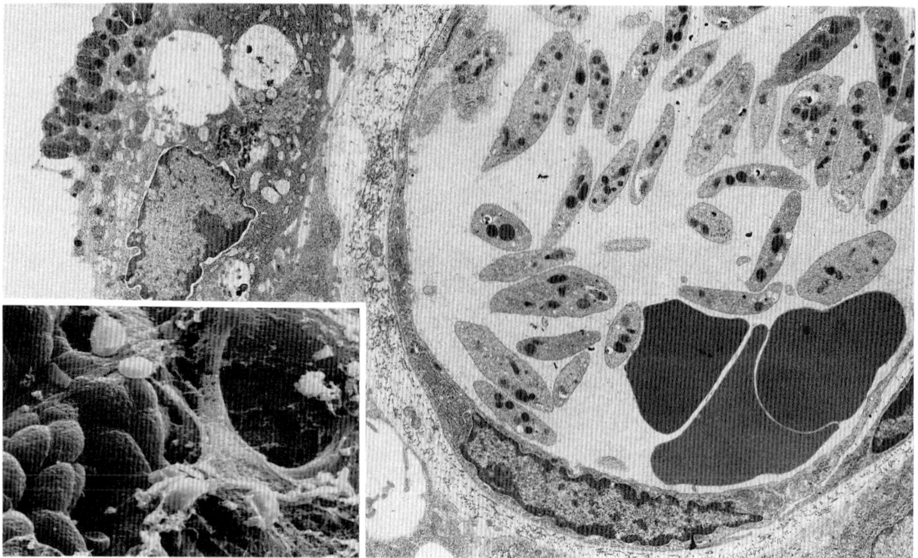

[Fig.3: TEM (2000x): Carbenoxolone pretreated group. Cytoplasmic vacuolization of superficial epithelium, with reduction of luminal microvilli and secretory granules. Vascular congestion and numerous platelets are also present.
Insert:SEM (1000x) The SEM micrograph shows typical superficial microerosion.]

In all specimens belonging to group E_2 examined (16,16 dimethyl PGE_2 for 5 days then ethanol exposure), we noted the absence of acute hemorrhagic-necrotic lesions. Some lesions that we noted were comprised of focal microerosions and ultrastructural alterations of the apical surface cells of the lining epithelium. Also, edema and subepithelial vasodilatation appeared after the acute administration of ethanol.

TEM of these specimens revelaed a good morphology of the superficial epithelium, although some degrees of dilation of the intercellular secretory spaces, areas showing sparse microvilli and various grades of cytoplasmic rarefaction, were noted. Some tracts of the glandular epithelium, which was also generally well preserved, showed shortened cells and swelling of the secretory capillaries. This group showed superficial damage in 70% of cases, while erosions were evident in the remaining 30%.

In the all group F (sulglycotide for 5 days then ethanol exposure) specimens observed, only in 2.5% of cases ulcerative mucosal lesions involving the lamina propria were noted. In 95% of cases, the lesions observed were microerosion of the mucosal surface with were accompainied by subepithelial edema and hyperemia.

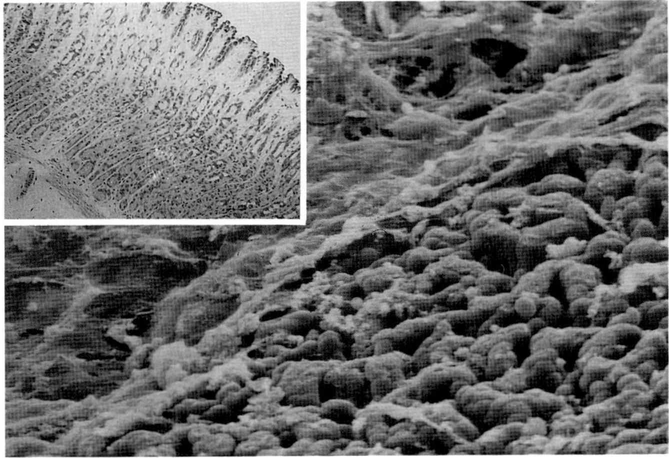

[Fig.4: SEM (500x):GLPS pretreated group. Small microerosion are present in generally well preserved mucosal surface. Insert LM (125x) Subepithelial edema and hyperemia are evident in the to- naca mucosa.]

Ultrastructural study of the specimens from this group showed also a good preservation of the superficial epithelium. The most significant and consistent findings were the rarefaction of the apical microvilli, notable enlargment of the intercellular secretory spaces, various degrees of degenerative vacuolization of the cytoplasm and dilatation of the SER cisternae. The intercellular junctions always appeared well preserved. In the lamina propria beside the well preserved glands, significant vascular congestion, without evidence of platelet activation, could be seen.

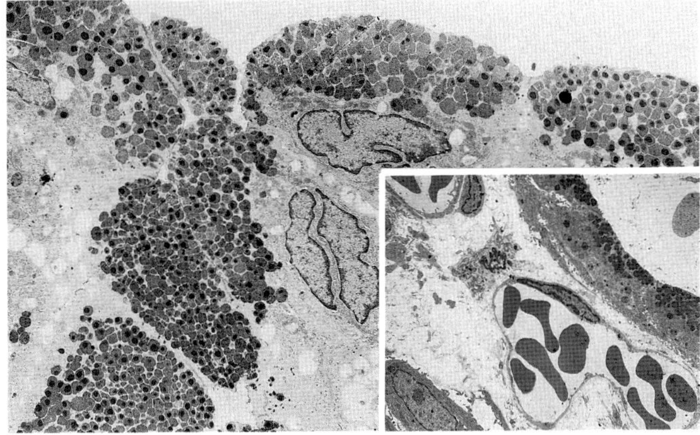

[Fig.5: TEM (2000x) GLPS pretreated group. Ultrastructural aspect of good preservation of superficial epithelium, with rarefaction of apical microvilli. Insert: TEM (2000x) In the enlarged capillaries, platelets activation is not present.]

Observations concerning group G (Maalox TC® for 5 days, then ethanol exposure) showed, in 40% of cases, the presence of acute hemorrhagic-necrotic lesions. In another 30% of cases, the lesions observed were limited to superficial erosions. In the remaining 30%, the only observable alteration was superficial damage.

Finally, TEM observation of these specimens showed necrosis of the superficial cells, vasodilation and aggregation of platelets. In addition at the margins of the ulcers the superficial mucosal cells show very scarce microvilli. Further, the number of granules is diminished, and in the cytoplasm of the mucous cells there are numerous and phagolysosomes. In the lamina propria the capillaries are dilated and are filled with platelets.

CONCLUSIONS

In concluding the present study we can sustain that rats treated with 16,16 PGE_2 (group E), carbenoxolone (group D) or sulglycotide (group F) show remarkably less ethanol-induced damage to the gastric mucosa than do positive-control animals or animals treated similarly with certain cytoprotective agents.

With regard to the remaining 2 agents tested in our experimental trial, sucralfate (group C), and Maalox TC®(group G), the evidence in a certain percentage of acute ulcers of the type commonly seen following ethanol exposure leads us to consider these less efficacious for gastric cytoprotection, at least when the doses cited are administered as described.

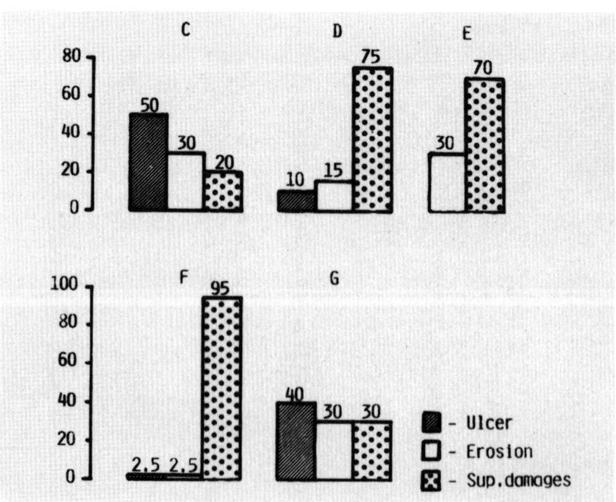

Tab. 2 : Percentage of various extension of damage in the different studied groups.

REFERENCES

Brendan J.R. Whittle and Graham Steel (1985). Evaluation of the protection of rat gastric mucosa by a prostaglandin analogue using cellular enzyme marker and histologic techniques. Gastroenterology 88: 315-27.

Hollander D., Tarnawski A., Karuse W.J. and Gergely H. (1985). Protective effect of sucralfate against alcohol-induced gastric mucosal injury in the rat. Macroscopic, histologic, ultrastructural and functional time sequence analysis. Gastroenterology 83: 619-25.

Lucy E.R., Susumu I. (1982). Microscopic analysis of ethanol damage to rat gastric mucosa after treatment with a prostaglandin. Gastroenterology 83: 619-25.

Lucy E.R. (1985). Prostaglandins and histological changes in the gastric mucosa. Digestive Disease and Sciences Vol. 30, n. 11: 835-945.

Morris G.P., Keenan C.M., Shriver D.A. (1987). Morphological and physiological effects of a cytoprotective prostaglandin analog (riprostil) on the rat gastric mucosa. Clinical and Investigation Medicine, vol. 10, n. 3: 121-131.

Ohno T., Ohtsuki H., Okabe S. (1985). Effects of 16,16-Dimethyl Prostaglandin E on ethanol-induced and aspirin-induced gastric damage in the rat. Scanning Electron Microscopy study. Gastroenteroly 88: 353-61.

Porta R., Niada R., Pescador R., Mantovani M., and Prino G. (1986) Gastroprotection and lyosomal membrane stabilization by sulglicotide. Arzeneimittel-Forschung/Drug Research 36 (II), 7: 1079-1982.

Robert A., Lancaster C., Davis J.P., Field S.O., Wickerema Sinha A.J. and Thornburgh B.A. (1985). Cytoprotection by prostaglandin occurs in spite of penetration of absolute ethanol into the gastric mucosa. Gastroenterology 88: 328-33.

Schmidt K.L., Henagan J.M., Smith G.S., Hilburn P.J. and Miller T.A. (1985). Prostaglandin cytoprotection against ethanol-induced gastric injury in the rat.Gastroenterology 88: 649-59.

Trabucchi E., Foschi D., Colombo R., Baratti C., Del Soldato P., Centemero A., Rizzitelli E. and Montorsi W. (1986) Cytoprotection by PGE , pirenzepine or vagotomy: a transmission and scanning electron microscopic study in rats. Pharmacological Research Communications Vol. 18, n. 4: 357-369.

Cells and Tissues: A Three-Dimensional
Approach by Modern Techniques in Microscopy,
pages 343–347

THREE-DIMENSIONAL VISUALIZATION OF THE GANGLIONATED ENTERIC NERVE PLEXUSES IN THE SMALL INTESTINE OF THE PIG.

Dietrich W. Scheuermann, Werner Stach and
Jean-Pierre Timmermans

Institute of Histology and Microscopic Anatomy, University of Antwerp, Belgium (D.W.S., J.P.T.) and Institute of Anatomy, Wilhelm-Pieck-Universität Rostock, GDR (W.S.)

INTRODUCTION

To clarify the histology of the complex neuronal network located intramurally in the gastrointestinal tract, various light- and transmission electron microscopical techniques have been used (for review, see Baumgarten, 1982; Furness and Costa, 1987). In most of these studies only two ganglionated networks are described, i.e. the myenteric plexus, recognized for the first time by Auerbach (1862, 1864), and the Meissner's plexus, attributed to the investigator who succeeded to demonstrate the presence of a ganglionated nerve plexus in the submucosal layer of the small intestine and stomach (Meissner, 1857).

Although it seemed indicated to use scanning electron microscopy (SEM) to elucidate the three-dimensional organization and topographical features of this part of the autonomic nervous system, it was not until recently by introducing digestion techniques for the removal of connective tissue components concealing the surface of the intramural enteric nervous system that the SEM-visualization of the enteric nerve structures became possible (Fujiwara and Uehara, 1980; Scheuermann et al., 1984, 1986, 1987a,b; Komuro, 1986; Stach et al., 1987). By using SEM, we aimed to reconsider the controversy which has been raised in the literature concerning the existence of two distinct submucosal nerve plexuses.

MATERIAL AND METHODS

One to six week old domestic pigs were killed by a blow on the head and duodenal, jejunal and ileal segments, about 20 to 30 cm in length were dissected out. The serosal and outer muscle layer were everted from the underlying tissue, so as to obtain two tubes, one consisting, from the outside inwards, of the submucosal and mucosal layer, and the everted tube composed of the circular smooth muscle layer, the longitudinal smooth muscle layer and the serosal layer. Both tubes were ligated, filled with phosphate buffer and immersed in a trypsin solution (2.5mg/ml; Sigma III) for 45 min at 37°C, sometimes followed by an additional enzymatic digestion in a mixture of glycosidases (0.1 mg/ml; Seikagaku Kogyo) for 1 hour at 37°C. After rinsing, both tubes are filled and immersion-fixed with 2.5% phosphate-buffered glutaraldehyde for 2 hours at 4°C. After fixation, the tubes were split open along the mesenteric border. The epithelial layer and lamina propria of the segment containing the submucosal plexuses are scraped off with a scalpel. The fibres of the circular or longitudinal smooth muscle layer are stripped off with the aid of watchmakers forceps. The segments are then treated with 8N HCl for 20-30 min at 60°C, rinsed in aq. dest. and further processed as for conventional SEM. The specimens were investigated using either an ETEC Autoscan or a Philips SEM 515 operating at 15-20 kV.

RESULTS AND DISCUSSION

After performing enzymatic digestion techniques SEM allows beyond doubts the visualization of the three distinct ganglionated plexuses in the small intestine of the pig. The myenteric plexus (Auerbach) (Fig. 1) can clearly be recognized in the everted part of the intestinal wall between the longitudinal and circular layer of the external smooth musculature. It appears in SEM as the known, wide-meshed network of ganglia with primary, secondary and tertiary connecting nerve strands. On examining the whole mounts from the remaining inner tube of the intestinal wall the surface texture of two submucosal plexuses can clearly be recognized each at a distinct level of the submucosal layer (Fig. 2). One plexus, i.e. the plexus submucosus externus (Schabadasch) (Fig. 2,3) located in the outer region of the submucosal layer above the vessel arcades, viewed from the abluminal side, appears as a very irregular meshwork with ganglia varying greatly in size and shape. A second submucosal meshwork, i.e. the plexus submucosus internus (Meissner) (Fig. 2,4), observed from the abluminal side, lies below the submucosal vessel arcades closely against the lamina muscularis mucosae. In this inner plexus the meshes and the ganglia are much smaller. Its neuronal perikarya are usually smaller in diameter compared to those of the other ganglionated plexuses. The three enteric ganglionated nerve plexuses are linked to each other by connecting strands which can be traced from a ganglion in one plexus to a ganglion in another plexus (Fig. 2). In contrast to other microscopical techniques, the high depth of focus characteristic for SEM enables a fair simultaneous visualization of both submucosal networks. (Fig. 2,3).

In conclusion, the procedure used for the present SEM-study allowed to extend previous light- and transmission elec-

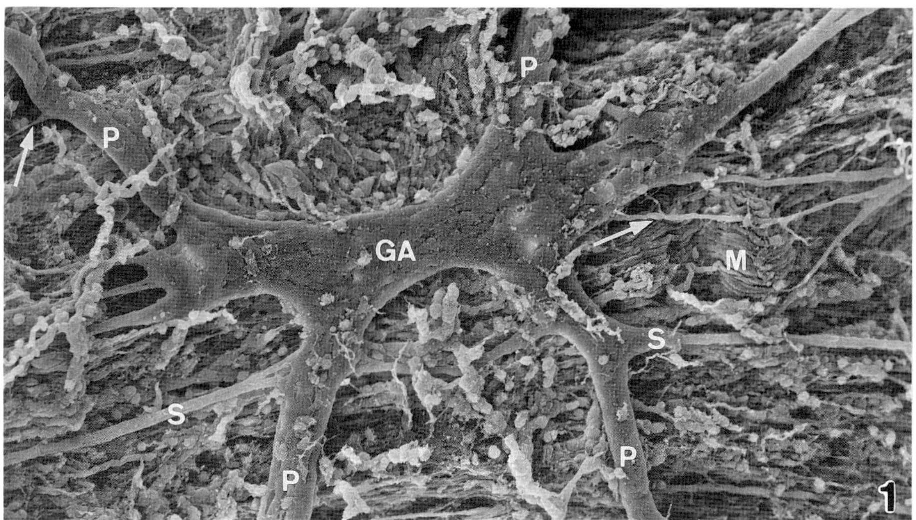

Fig. 1: Ganglion of the plexus myentericus (Auerbach). GA: ganglia; P: primary strand; S: secondary strand;→: tertiary strand; M: circular smooth muscle layer. x310.

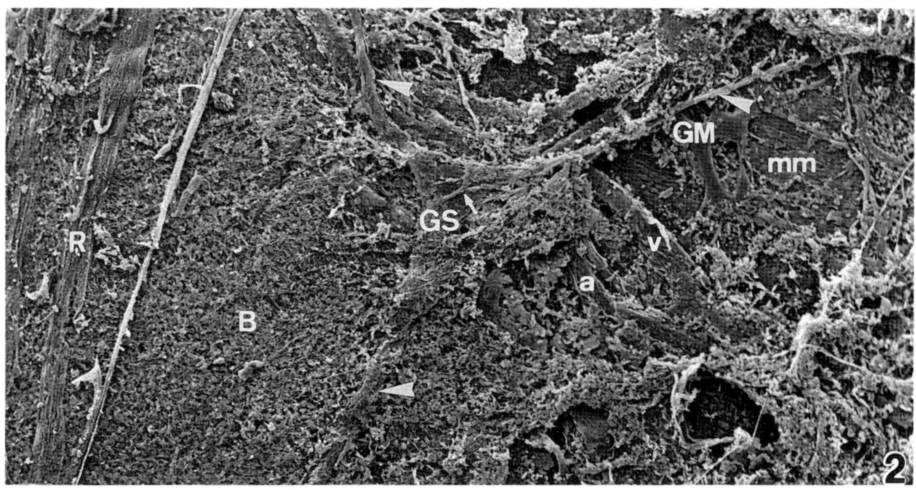

Fig. 2: Both submucosal nerve networks visualized simultaneously under SEM. GS: ganglion of the plexus submucosus externus (Schabadasch); GM: ganglion of the plexus submucosus internus (Meissner); a: small artery; v: small vene;→: capillary; B: connective tissue; R: remnants of the circular smooth musculature; mm: lamina muscularis mucosae; ►: connecting nerve strands to neighbouring ganglia of the plexus Schabadasch. x80.

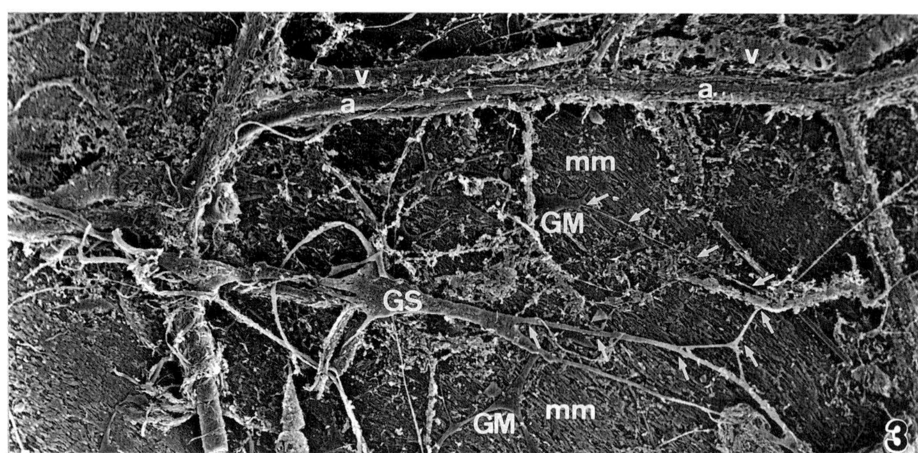

Fig. 3: Low power micrograph showing a connecting nerve strand (⟶) between ganglia of the two distinct submucosal nerve plexuses. GS: ganglion of the plexus Schabadasch; GM ganglion of the plexus Meissner; mm: lamina muscularis mucosae; a: artery; v: vene. x80.

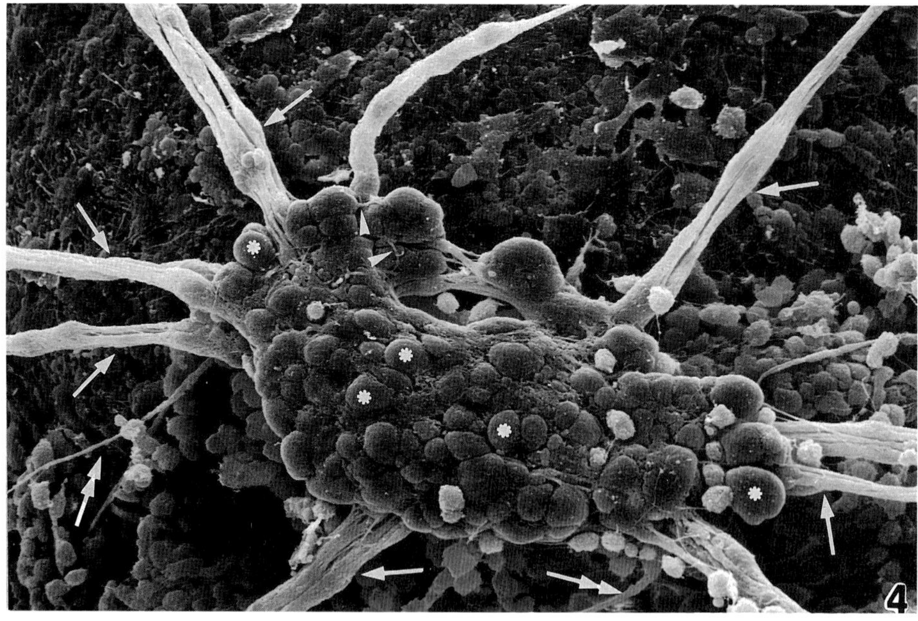

Fig. 4: Ganglion of the plexus Meissner. ✱: neuronal perikarya; ➤: neuronal process leaving a neuronal cell body; ⟶: connecting nerve strands to neighbouring ganglia of the Meissner's plexus; ⟶⟶: connecting strands directed to the mucosal layer. x1280.

tron microscopical findings and, in particular, illustrates in great detail the existence of two morphologically distinct and topographically separated submucosal plexuses.

REFERENCES

Auerbach L (1862). Über einen Plexus gangliosus myogastricus. 39er Jahr-Bericht u Abh d Schlesischen Gesells f vaterland Cult, 103-104.

Auerbach L (1864). Fernere vorlaufige Mitteilung über den Nervenapparat des Darmes. Arch Pathol Anat Physiol 30:457-460.

Baumgarten HG (1982). Morphological basis of gastrointestinal motility: structure and innervation of gastrointestinal tract. In Bertaccini G (Ed): "Mediators and Drugs in Gastrointestinal Motility. I.", Berlin: Springer Verlag, pp 7-53.

Fujiwara T, Uehara Y (1980). Scanning electron microscopy of myenteric plexus: a preliminary communication. J Electron Microsc 29:397-400.

Furness JB, Costa M (1987). "The Enteric Nervous System." Edinburgh, Churchill Livingstone, pp 1-25.

Komuro T (1986). Three-dimensional visualization of the intraganglionic structures of the rat myenteric plexus: scanning electron microscopy with the connective tissue digestion method. Neurosci Lett 72:49-53.

Meissner G (1857). Über die Nerven der Darmwand. Z ration Med, N F 8:364-366.

Scheuermann DW, Stach W, Timmermans J-P (1984). Scanning electron microscopy (SEM) of the plexus submucosus internus (Meissner) in the porcine small intestine. Scanning 6:96-102.

Scheuermann DW, Stach W, Timmermans J-P (1986). Three-dimensional organization and topographical features of the myenteric plexus (Auerbach) in the porcine small intestine: scanning electron microscopy after enzymatic digestion and HCl-hydrolysis. Acta Anat 127:290-295.

Scheuermann DW, Stach W, Timmermans J-P (1987a). Topography, architecture and structure of the plexus submucosus internus (Meissner) of the porcine small intestine in scanning electron microscopy. Acta Anat 129:96-104.

Scheuermann DW, Stach W, Timmermans J-P (1987b). Topography, architecture and structure of the plexus submucosus externus (Schabadasch) of the porcine small intestine in scanning electron microscopy. Acta Anat 129:105-115.

Stach W, Scheuermann DW, Timmermans J-P (1987). Licht- und Rasterelektronenmikroskopie der submukösen Nervengeflechte des Darmwandnervensystems. Verh Anat Ges 81:729-730.

Neuroendocrine, Cardiocirculatory and Immune Systems

Cells and Tissues: A Three-Dimensional
Approach by Modern Techniques in Microscopy,
pages 351-361

PINEALOCYTES - NEUROENDOCRINE UNITS RELATED TO THE VASCULAR AND CEREBROSPINAL FLUID COMPARTMENTS OF THE BRAIN

A. Oksche

Department of Anatomy and Cytobiology,
Justus Liebig University of Giessen, 123 Aulweg,
D-6300 Giessen, Federal Republic of Germany

INTRODUCTION : GENERAL CONSIDERATIONS

This treatise pays tribute to the great master Marcello Malpighi (1628-1694) whose ingenious scientific work contains, among a vast treasure of various microscopic observations, first data on the vascularization of the brain and several glandular structures. The pineal organ (Fig. 1),an evagination of the diencephalic roof, is an integral component of the brain (Bargmann, 1943). The neuroendocrine character of this organ is determined by the neural and secretory properties of its parenchymal cells, the pinealocytes (Oksche, 1988). The complex array of these cells is exposed, on the one hand, to a mighty vascular tree, on the other hand, to extensive compartments encompassing another circulatory medium, the cerebrospinal fluid. The analysis of this complex structural arrangement requires a three-dimensional approach, leading stepwise from a gross topographical survey to ultrastructural details.

Two discoveries have had a great impact on the development of pineal research: (1) The isolation and identification of the indoleamine melatonin (5-methoxy-N-acetyl tryptamine), a specific pigment-aggregating agent, from the mammalian (bovine) pineal organ (Lerner et al., 1959). (2) The neurophysiological demonstration of the photoreceptive character of the pineal sense organs ("third eyes") of lower vertebrates (Dodt et al.; cf. Dodt, 1973).

PINEALOCYTES : COMPARATIVE ASPECTS

In phylogeny, successive changes in the pineal organ have produced striking variations in its sensory and secretory apparatus. According to ultrastructural criteria the pinealocytes of sauropsids and mammals are derivatives of pineal photoreceptor cells of anamniotes (Collin, 1971; Oksche, 1971). All these subtypes (Fig. 2) can be regarded as members of a common class of receptor-like pinealocytes (Collin and Oksche, 1981). This concept is further substantiated by immunocytochemical evidence that pinealocytes of the receptor line share molecular similarities (immunoreactivity for opsin, α-transducin, retinal S-antigen = arrestin) with retinal photoreceptors (cf. Korf and Oksche, 1986; Foster et al., 1987). The problem of a multiplicity of pineal receptors, i.e., the existence of separate photoreceptor and photoneuroendocrine units (Meiniel, 1981), is still open to discussion. The former are to be regarded as primary sensory cells displaying synaptic ribbons and utilizing a conventional neurotransmitter (possibly glutamate or aspartate; cf. Meissl, 1986). The latter might possess the capacity to translate a light signal directly into a neuroendocrine response (Oksche, 1971). Pinealo-

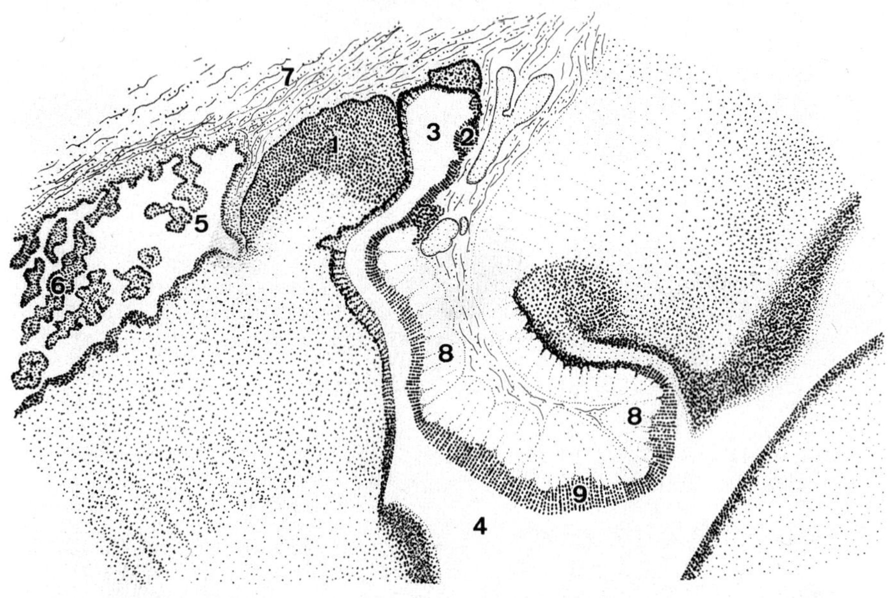

Figure 1. Midsagittal section through the pineal region of a human fetus, mens III (chromalum hematoxylin-phloxin stain, semidiagrammatic drawing). At this stage, the primordium of the pineal organ (1 anterior lobe, 2 posterior lobe) still has a diverticular appearance. Note the compartments containing cerebrospinal fluid: Pineal lumen (3) in open communication with the third ventricle (4); suprapineal recess (5) displaying a portion of the choroid plexus (6) of the third ventricle; leptomeningeal sheath (7) in which the subarachnoid space will develop. Posterior commissure (8) covered by the tall columnar ependyma of the subcommissural organ (9).

cytes endowed with characteristic outer segments may contain serotonin, a precursor of melatonin, as shown by use of microspectrofluorimetric and immunocytochemical (cf. Korf and Oksche, 1986) procedures. However, it is a matter of debate whether all pinealocytes bearing an outer segment are capable of melatonin synthesis. Unfortunately, to date, basic methodological handicaps have

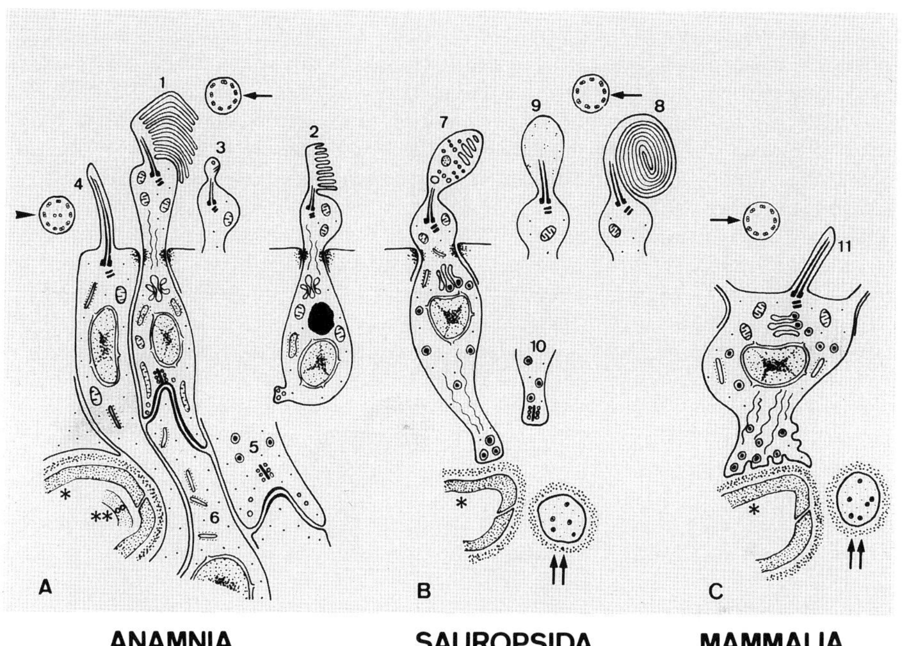

Figure 2. Diagrammatic representation of pinealocytes of the receptor line in Anamnia (A), Sauropsida (B) and Mammalia (C). 1 Typical pineal photoreceptor displaying regular outer segment, 2 serotonin-containing pinealocyte of the receptor line, 3 bulbous cilium of the 9 + 0 type (arrow), 4 supportive cell bearing cilium of the 9 + 2 type (arrowhead), 5 basal pedicle of pineal photoreceptor containing dense-core vesicles, 6 second-order neuron, 7 modified pineal photoreceptor (lacertilians) rich in dense-core vesicles, 8 irregular outer segment of modified pineal photoreceptor (birds), 9 bulbous cilium (9 + 0 type) of modified pineal photoreceptor (arrow), 10 basal process of modified pineal photoreceptor containing dense-core vesicles and occasional synaptic ribbons, 11 mammalian pinealocyte displaying cilium of the 9 + 0 type (arrow) and dense-core granules. Single asterisk leaky endothelium of pineal capillaries, two asterisks capillary endothelium endowed with tight junctions. Two arrows sympathetic nerve terminals. From Oksche et al. (1987), courtesy of John Libbey & Co Ltd.

limited the use of antibodies for unequivocal immunocytochemical demonstration of melatonin at the cellular level. On the other hand, it is beyond doubt that melatonin is produced by the pineal sense organs of lower vertebrates and, as has been strongly suggested, also by the retina (cf. O'Brian and Klein, 1986).

The molecules shared by retinal and extraretinal (pineal) photoreceptor cells have been highly preserved during evolution; thus immunoreactive S-antigen was shown to occur also in a representative number of typical, reduced and even uncertain photoreceptor organs of invertebrates (van Veen et al., 1986). Furthermore, there is immunocytochemical evidence of molecular photoreceptor markers in a definite class of brain tumors. In this respect, the neoplastic cells of retinoblastomas, pineocytomas and medulloblastomas share common features. S-antigen-like immunoreactivity was shown to exist in human pineocytoma (Korf et al., 1986). Furthermore, with the use of antisera against bovine S-antigen and opsin, immunoreactive tumor cells were demonstrated in approximately one third of the medulloblastomas investigated (Korf et al., 1987). These immunocytochemical results may speak in favor of a common diencephalic matrix of all S-antigen-immunoreactive cellular elements. However, since the occurrence of indoleamines in medulloblastomas is still uncertain, the present data do not allow to suggest a photoneuroendocrine character of medulloblastoma cells. Furthermore, it should be kept in mind that only certain pinealocytes and medulloblastoma cells possess the epitopes characteristic of photoreceptive elements.

AMPHIBIAN PINEALOCYTES: MODEL UNITS EXPOSED TO VASCULAR AND CEREBROSPINAL FLUID COMPARTMENTS

Due to anatomical and functional peculiarities the situation in anurans, especially in ranid frogs, deserves particular attention. It appears noteworthy that batrachians were favorite experimental animals of Marcello Malpighi. For the investigation of basic photoneuroendocrine mechanisms, the pineal complex of frogs offers a number of advantages. One of the advantages of this system is its anatomical position: it is easily accessible for all kinds of anatomical studies and experimental procedures. A century after the discovery of the pineal end-vesicle (frontal organ) of frogs, an eye-like body located in the skin of the parietal region (for references, see Bargmann, 1943), it was shown by electrophysiological recordings that this organ is capable of a direct response to light (cf. Dodt, 1973); for recent developments, see Meissl, 1986).

In close cooperation with Dodt and associates the fine structure and cytochemistry of the pineal complex of anurans have been explored in my laboratory (for details, see Oksche, 1971; Oksche and Hartwig, 1979; Korf and Oksche, 1986). Photoreceptor cells bearing a characteristic outer segment (see Fig. 2A) were revealed by electron microscopy, and indoleamines (other than serotonin) were shown to occur in such receptor elements by microspectrofluorimetric analysis; however, the immunoreaction for serotonin was - in marked contrast to teleosts - negative. Furthermore, pineal photoreceptor cells of frogs contain photopigments and several proteins involved in phototransduction by their interaction with the photopigment. The pineal complex of anurans is connected to the brainstem via conspicuous neural pathways. On the other hand, its lumen is in open communication with the third ventricle. Thus, Vigh-Teichmann and Vigh (1983) regard the pineal photoreceptor cells as a specialized form of cerebrospinal fluid-contacting neurons. Since anurans do not display a fully developed subarachnoid space, in the present context the relationship of the pinealocytes to the adjacent leptomeningeal sheath seems to be less important. The pineal complex of frogs possesses a complex intrinsic circuitry employing different neurotransmitters. It is highly vascularized, but so far - in contrast to the pineal organs of mammals, birds and some teleosts - the blood-brain-barrier in the pineal complex of frogs appears to be tight (cf. Oksche et al., 1987). Pineal circulation (Fig.3 A, C) can be observed over a period of several hours in live, anesthetized frogs (Mautner, 1965). A more detailed, three-dimensional picture of these vessels (Fig. 3 B) was revealed by scanning electron microscopy of Mercox casts (Syed Ali, unpublished observations). So far, there is no indication of an epithalamic or pineal portal vascular system.

In parallel, other laboratories have succeeded in showing by use of bioassay, enzyme biochemistry and radioimmunoassay that the pineal organ (and also the retina) of anurans are capable of melatonin synthesis (see Bagnara and Hadley, 1970; Pang et al., 1985). In larval amphibians melatonin is a hormone that regulates the blanching response in darkness; this response is independent of the function of the lateral eyes (retina).

All these results indicate that the pineal complex of anurans provides an excellent model of a functioning photoneuroendocrine system (cf. Oksche and Hartwig, 1979) that can translate a light signal into an endocrine response. In this context, light-dependent rhythmic phenomena involved in the periodic discharge and renewal of the pinealocyte outer segments deserve particular attention; they may be controlled by indoleamines acting in a paracrine or even autocrine fashion (Hartwig, 1986, and unpublished findings). The functional role of the photic information conveyed from pineal photoreceptor organs to the brain, either neural or humoral, is only partly understood.

This complexity of biological aspects is often overlooked in pineal research. Pineal sense organs of submammalian vertebrates provide the key for a more accurate and biologically significant analysis of the pineal body of mammals, which, in spite

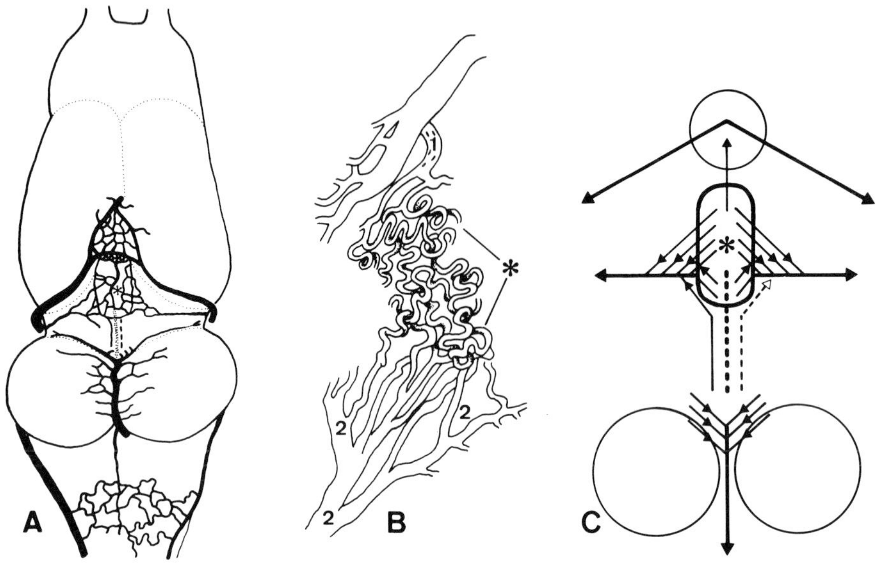

Figure 3. Vascular system of the pineal organ (asterisk) in frogs. (A) Specimen injected with India ink. Note the network formed by meandering capillaries. (B) Scanning electron-microscopic images of pineal vessels (Mercox cast) showing a supporting artery (1) and an abundant venous drainage (2). (C) The exact direction of the blood flow (arrows) in these vessels can be verified only by observations in live, anesthetized animals. For details, see text. After Mautner (1965) and Syed Ali (unpublished).

of all recent discoveries, still remains an enigmatic neuroendocrine structure of the brain.

NEURONAL FEATURES OF PINEALOCYTES

Recently, the exact nature of pinealocytes, the parenchymal cells of the pineal organ, has gained increasing attention. As already mentioned, pineal photoreceptors of anamniotes have the character of primary sensory cells, i.e., sensory neurons. This is emphasized by the presence of presynaptic vesicles, synaptic ribbons, pre- and postsynaptic membrane thickenings and conventional neurotransmitters. Such cells can be regarded as the first neuron in the pineal pathways to the brain.

According to our observations, all subtypes of pinealocytes of the photoreceptor line share certain molecular features characteristic of neurons. Thus, neuron-specific enolase was found not only in true and modified pineal photoreceptors but also in mammalian pinealocytes (cf. Oksche et al., 1987). Furthermore, mammalian pinealocytes display immunoreactions with antisera against neurofilament protein and synaptophysin (H.W.Korf, unpublished results; cf. Vollrath and Schröder, 1987). Even if several of these markers are not restricted to neurons but may occur also in various representatives of the diffuse neuroendocrine (Pearse) or paraneuron (Fujita) system, the above-mentioned findings give an indication of basic molecular similarities within this class of elements. Finally, both in pineal sense organs (anamniotes) and pineal glands (mammals) some pinealocytes display direct projections to the epithalamus, especially to its habenular region (cf. Korf and Ekström, 1987).

Recent immunocytochemical observations indicate that pinealocytes of the receptor line may contain, in addition to indoleamines, also proteins differing from the common structural proteins of the cell. Thus, a great variety of vertebrate pinealocytes of the receptor line (typical and modified pineal photoreceptors, mammalian pinealocytes) display a material immunoreactive with antisera against the glycoprotein secretion of the subcommissural organ, an ependymal structure adjacent to the pineal organ (Oksche et al., 1987; Rodríguez et al., in press). The production of peculiar proteins in avian and mammalian pinealocytes has already been discussed by Pévet and Collin (for references, see Oksche et al., 1987).

If the elaboration of specific proteins indeed represents a basic feature of a definite population of pinealocytes, then these pinealocytes may turn out to be closely related to secretory neurons or paraneurons. In fact, no rigid borderline can be drawn between neurosecretory nerve cells and paraneurons (cf. Oksche, 1988). It should be kept in mind that Bargmann (1943) already came to the conclusion that pinealocytes should be viewed in close context with neurons, especially those hypothalamic nerve cells that are capable of secretion (Scharrer). From our present point of view this was a highly prophetic prediction.

These considerations can be further substantiated by new observations concerning a close relationship between the pineal organ and the habenular ganglion. The existence of reciprocal neural pathways between the pineal organ and the habenular ganglion, in addition to other central projections (cf. Korf and Oksche, 1986), reflects the central nervous origin of the pineal organ. In the pineal photoreceptor organs this character is further underlined by the presence of an intrinsic neuronal apparatus. The successive phylogenetic and - in a number of cases - also ontogenetic replacement of this apparatus by autonomic projections is a further aspect of basic functional importance (for references, see Korf and Oksche, 1986). Even the interconnection of mammalian pinealocytes via gap junctions (Huang and Taugner, 1984) has its precursor in pineal sense organs of anamniotes where gap junctions exist between adjacent basal processes of the photoreceptor cells and probably enhance their electric communication (Omura, 1984). The bioelectric activity of mammalian pinealocytes, however, is not easily classified (cf. Vollrath and Schröder, 1987).

There is increasing evidence that the habenular ganglion of several mammalian species contains S-antigen-immunoreactive cells resembling pinealocytes; in a strain of pigmented mice, these cells have recently been identified at the electron-microscopic level (Korf et al., in press). Processes of these cells were found in close, occasionally even synaptic association with dendrites and perikarya of conventional habenular neurons, and also in juxtaposition to capillaries of the habenular region. Furthermore, in albino rats kept under constant illumination, the habenular region displayed patterned arrays of cells immunoreactive with antibodies against retinal S-antigen, and, although restricted in number, also against opsin. After sequential staining, some of the S-antigen-immunoreactive perikarya also showed a positive immunoreaction with an antibody against the secretory product of the subcommissural organ (Rodríguez et al., in press). Other habenular perikarya exhibited exclusively an immunoreaction with antisera against the subcommissural secretion; however, even then the distributional pattern of the S-antigen-immunoreactive elements in the same or subsequent sections was rather similar. Whether any of these perikarya also contain serotonin is open to discussion; serotonin-containing cells, however, have been shown to occur in the habenular region of the rat (Wiklund, 1974).

Thus, the habenular complex harbors cells sharing basic immunocytochemical and histochemical properties with pinealocytes. In this connection it should be mentioned that the pineal organ of ferrets contains an aggregation of nerve cells resembling habenular neurons (Herbert, 1971). The existence of an epithalamo-epiphyseal complex was already suggested by Roussy and Mossinger (1938), but this concept was not generally accepted at that time.

CONCLUSIONS

The pineal organ displays a very complex micro- and chemoarchitecture, not inferior to that of well-established neuroendocrine centers of the hypothalamus. The pineal organ is - like the hypothalamus - a derivative and integral component of the brain, i.e. diencephalon. This concept can be further extended if the habenular complex is taken into consideration. Pinealocytes are exposed to both hemal- and cerebrospinal fluid-dominated microenvironments (cf. Leonhardt, 1980); the functional role of these factors for sensory, synthetic and release mechanisms requires systematic exploration. On the other hand, the problem of melatonin receptors, especially their occurrence in defined target areas of the brain, is of major interest. In future investigations of the pineal complex the molecular type of analysis will be dominant. However, at the same time systemic, organism-dependent aspects deserve increasing attention.

REFERENCES

Bagnara JT, Hadley ME (1970). Endocrinology of the amphibian pineal. Am Zoologist 10:201-216.
Bargmann W (1943). Die Epiphysis cerebri. In von Möllendorff W (ed): "Handbuch der mikroskopischen Anatomie des Menschen", vol. IV/4, Berlin: Springer, pp 309-502.
Collin J-P (1971). Differentiation and regression of the cells of the sensory line in the epiphysis cerebri. In Wolstenholme GEW, Knight J (eds): "The Pineal Gland", Edinburgh London: Churchill Livingstone, pp 79-125.
Collin J-P, Oksche A (1981). Structural and functional relationships in the nonmammalian pineal gland. In Reiter RJ (ed): "The Pineal Gland", vol 1, Anatomy and Biochemistry, Boca Raton: CRC Press, pp 27-67.
Dodt E (1973). The parietal eye (pineal and parapineal organs of lower vertebrates). In Jung R (ed): "Handbook of Sensory Physiology", vol VII/3B, Berlin: Springer, pp 113-140.
Foster RG, Korf H-W, Schalken JJ (1987). Immunocytochemical markers revealing retinal and pineal but not hypothalamic photoreceptor systems in the Japanese quail. Cell Tissue Res 248: 161-167.
Hartwig H-G (1986). Turnover of pineal photoreceptive membranes in the frog, Rana esculenta complex. In O'Brian PJ, Klein DC (eds): "Pineal and Retinal Relationships", Orlando: Academic Press, pp 47-55.
Herbert J (1971). The role of the pineal gland in the control by light of the reproductive cycle of the ferret. In Wolstenholme GEW, Knight J (eds): "The Pineal Gland", Edinburgh London: Churchill Livingstone, pp 303-327.
Huang S-K, Taugner R (1984). Gap junctions between guinea-pig pinealocytes. Cell Tissue Res 235:137-141.
Korf H-W, Czerwionka M, Reiner J, Schachenmayr W, Schalken JJ, de Grip W, Gerry I (1987). Immunocytochemical evidence of molecular photoreceptor markers in cerebellar medulloblastomas. Cancer 60:1763-1766.
Korf H-W, Ekström P (1987). Photoreceptor differentiation and neuronal organization of the pineal organ. In Trentini GP, De Gaetani C, Pévet P (eds): "Fundamentals and Clinics in Pineal Research", New York: Raven Press, pp 35-47.
Korf H-W, Klein DC, Zigler JS, Gerry I, Schachenmayr W (1986). S-antigen immunoreactivity in a human pineocytoma. Acta Neuropathol 69:165-167.
Korf H-W, Oksche A (1986). The pineal organ. In Pang PKT, Schreibman M (eds): "Vertebrate Endocrinology. Fundamentals

and Biomedical Implications", vol 1, New York: Academic Press, pp 105-145.

Korf, H-W, Sato T, Oksche A (1988). Reflections on the nature and connectivities of pinealocytes. In Reiter RJ et al (eds) "Melatonin and the Pineal Gland (Satellite Symposium, 8th International Congress of Endocrinology)", London Paris: John Libbey (in press).

Leonhardt H (1980). Ependym und circumventriculäre Organe. In Oksche A, Vollrath L (eds) "Handbuch der mikroskopischen Anatomie des Menschen", vol IV/10, Berlin: Springer, pp 177-666.

Lerner AB, Case JD, Heinzelmann RV (1959). Structure of melatonin. J Am Chem Soc 81: 6084.

Mautner W (1965). Studien an der Epiphysis cerebri und am Subcommissuralorgan der Frösche. Mit Lebendbeobachtungen des Epiphysenkreislaufs, Totalfärbung des Subcommissuralorgans und Durchtrennung des Reissnerschen Fadens. Z Zellforsch 67:234-270.

Meiniel A (1981). New aspects of the phylogenetic evolution of sensory cell lines in the vertebrate pineal complex. In Oksche A, Pévet P (eds): "The Pineal Organ: Photobiology - Biochronometry - Endocrinology", Amsterdam: Elsevier, pp 27-48.

Meissl H (1986). Photoneurophysiology of pinealocytes. In O'Brian PJ, Klein DC (eds): "Pineal and Retinal Relationships", Orlando: Academic Press, pp 33-45.

O'Brian PJ, Klein DC eds (1986). "Pineal and Retinal Relationships". Orlando: Academic Press, pp 442.

Oksche A (1971). Sensory and glandular elements of the pineal organ. In Wolstenholme GEW, Knight J (eds): "The Pineal Gland", Edinburgh London: Churchill Livingstone, PP 127-146.

Oksche A (1988). Sensory and secretory potencies and differentiations of the central nervous system. Acta anat 132:216-224.

Oksche A, Hartwig H-G (1979). Pineal sense organs - components of photoneuroendocrine systems. In Kappers JA, Pévet P (eds): "The Pineal Gland of Vertebrates including man (Progr Brain Res, vol 52)". Amsterdam New York: Elsevier/North Holland Biomedical Press, pp 113-130.

Oksche A, Korf H-W, Rodríguez EM (1987). Pinealocytes as photoneuroendocrine units of neuronal origin: concepts and evidence. In Reiter RJ, Fraschini F "Advances in Pineal Research: 2", London Paris: John Libbey, pp 1-18.

Omura Y (1984). Pattern of synaptic connections in the pineal organ of the ayu, *Plecoglossus altivelis* (Teleostei). Cell Tissue Res 236:611-617.

Pang SF, Shiu SWY, Tse SF (1985). Effect of photic manipulation on the level of melatonin in the retinas of frogs (*Rana tigrina regulosa*). Gen Comp Endocrin 58:464-470.

Rodríguez EM, Korf H-W, Oksche A, Yulis CR, Hein S (1988). Pinealocytes immunoreactive with antisera against secretory glycoproteins of the subcommissural organ: A comparative study. Cell Tissue Res (in press).

Roussy G, Mosinger M (1938). Le complexe épithalamo-épiphysaire. Les corrélations avec le complexe hypothalamo-hypophysaire. Le système neuro-endocrinien du diencéphale. Revue neur 69:459-470.

Van Veen T, Elofsson R, Hartwig H-G, Gery I, Mochizuki M, Cena V, Klein DC (1986). Retinal S-antigen: Immunocytochemical and immunochemical studies on distribution in animal photoreceptors and pineal organs. Exp Biol 45:15-25.

Vigh-Teichmann I, Vigh B (1983). The system of cerebrospinal fluid-contacting neurons. Arch histol japon 46:427-468.

Vollrath L, Schröder H (1987). Neuronal properties of mammalian pinealocytes ? In Trentini GP, De Gaetani C, Pévet P "Fundamentals and Clinics in Pineal Research", New York: Raven Press, pp 13-23.

Wiklund L (1974). Development of serotonin-containing cells and the sympathetic innervation of the habenular region in the rat brain. Cell Tissue Res 155: 231-243.

Acknowledgements. The investigations of the author reported here were supported by research grants from the Deutsche Forschungsgemeinschaft and the Stiftung Volkswagenwerk. The author is grateful to Ms. Karola Michael for her skilled artwork, to Ms. Inge Lyncker for her aid in preparing the manuscript, and to Dr. R.L. Snipes for his comments.

Cells and Tissues: A Three-Dimensional
Approach by Modern Techniques in Microscopy,
pages 363–369

MAMMALIAN TANYCYTES : TRANSMISSION AND SCANNING ELECTRON MICROSCOPY

J. Flament-Durand

Department of Pathology, Hôpital Universitaire Erasme, Université Libre de Bruxelles, Brussels, Belgium.

Tanycytes are peculiar ependymal cells lining the ventral part of the 3rd ventricle.
The name "tanycyte" derives from the greek "tanus" which means elongated and was chosen to stress the shape of these bipolar cells (Horstmann 1954).
Their apical pole is in relation with the infundibular recess of the 3rd ventricle, and their distal pole is in contact with the portal vessels in the median eminence. They are in close relation with the hypophyseotrophic area.
Their functions are still a matter of debate.
I would like to present some morphological data of tanycytes obtained by transmission and scanning electron microscopy in mammals and to comment on these morphological grounds on their possible functions.

TRANSMISSION ELECTRON MICROSCOPY (TEM)

Because of their bipolar elongated shape, tanycytes are extremely difficult to observe in TEM from one pole to the other. However, in some favorable incidence, photographic recontructions allow such study (fig. 1).

The apical pole, in contact with the CSF shows microvilli, cytoplasmic expansions of various shapes, and only one or two cilia. These apical expansions increase the surface in contact with the CSF.

The distal process makes contact with the basement membrane of the portal vessels, at the level of the median eminence.

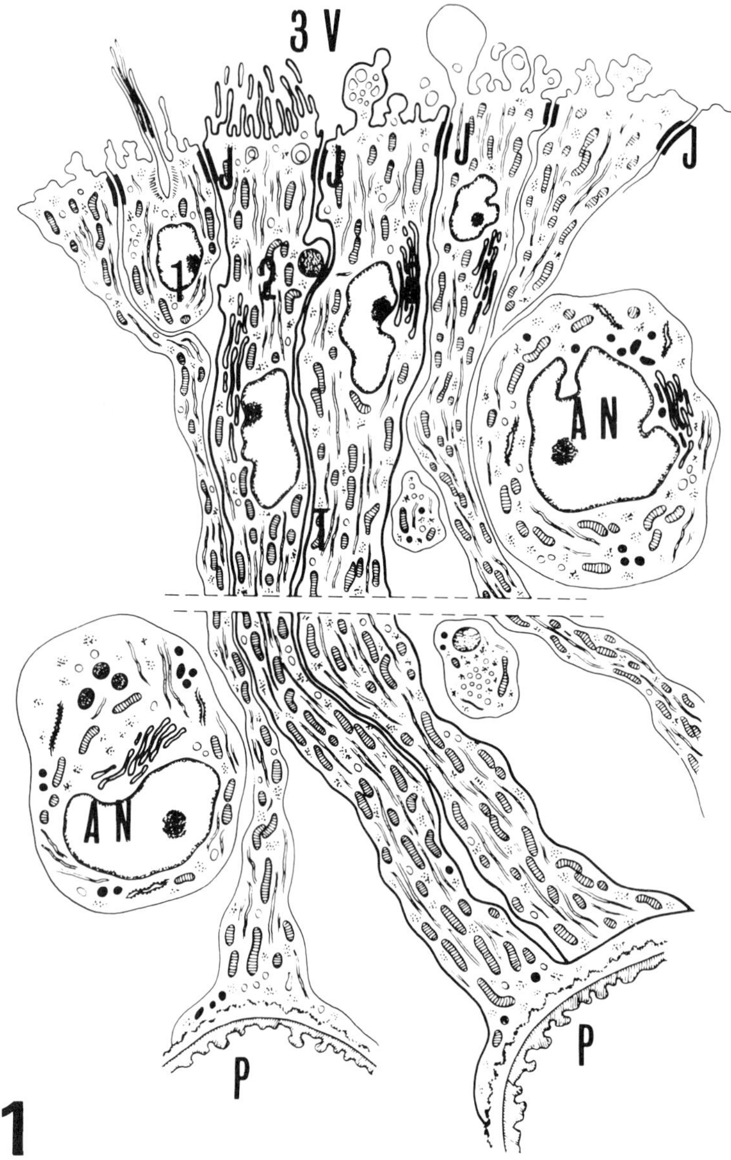

Fig. 1. Schematic drawing of the tanycytes facing the arcuate nucleus of the rat. T. Tanycyte; AN, arcuate neuron; P, portal capillary; 3 V, third ventricle; 1, oligociliated tanycyte; 2, microvilli decorating the apical pole of a tanycyte; 3 and 4, blebs of various shapes of the apical pole of tanycytes; J, tight junctions binding the apical neck of the tanycytes.

By courtesy of the International Review of Cytology.

Their Golgi apparatus is elongated. Their cytoplasm is poor in ribosomes, virtually devoid of granular endoplasmic reticulum and contain some lipid droplets. Their processes are full of microtubules, intermediary filaments and elongated mitochondria. The transition between the lining of the 3rd ventricle by classical ciliated ependymal cells is very sharp in the rat but more gradual in man. Whereas classical ciliated ependymal cells are linked by gap junctions, the apical zones of the tanycytes are linked by tight junctions.
The importance of these junctions and their possible functional significance will be discussed later.

SCANNING ELECTRON MICROSCOPY (SEM)

The surface of the ependymal lining of the 3rd ventricle has been studied by scanning EM.
The lower third of the 3rd ventricle is lined by polygonal cells, closely aligned, and their surface presents microvilli, blebs, and often a single centrally located cilium.
We have made correlations between SEM and TEM of the blebs. In our rat and human material, the blebs of the apical pole of the tanycytes are formed by cytoplasmic expansions containing free ribosomes and occasional lipid droplets, some are hollow with a cavity surrounded by a cytoplasmic expansion (fig. 2).
As already mentionned, the transition between the two types of cellular lining is very sharp in the rat but more gradual in man. The apical surface is very complex and exhibits either microvilli or bulbous protrusions also called blebs, which increase the surface of contact between tanycytes and the CSF.
The numerous microvilli already described in TEM increase the surface for possible absorption, whereas the significance of the blebs remains to be determined (Flament-Durand and Brion, 1985).
These morphological features easily observed by SEM have led to speculations about the possible functional significance of tanycytes, since they are mainly present in the ventricular wall overlying the hypophyseotrophic area (Halasz et al, 1962) and the median eminence which are of great neuroendocrine importance because they are the site of synthesis and of passage of the hypothalamic releasing hormones.
Let us consider the possible functions of these cells in relation with their morphology. This is a matter of great controversy and speculation. Because of their peculiar situation between the CSF contained in

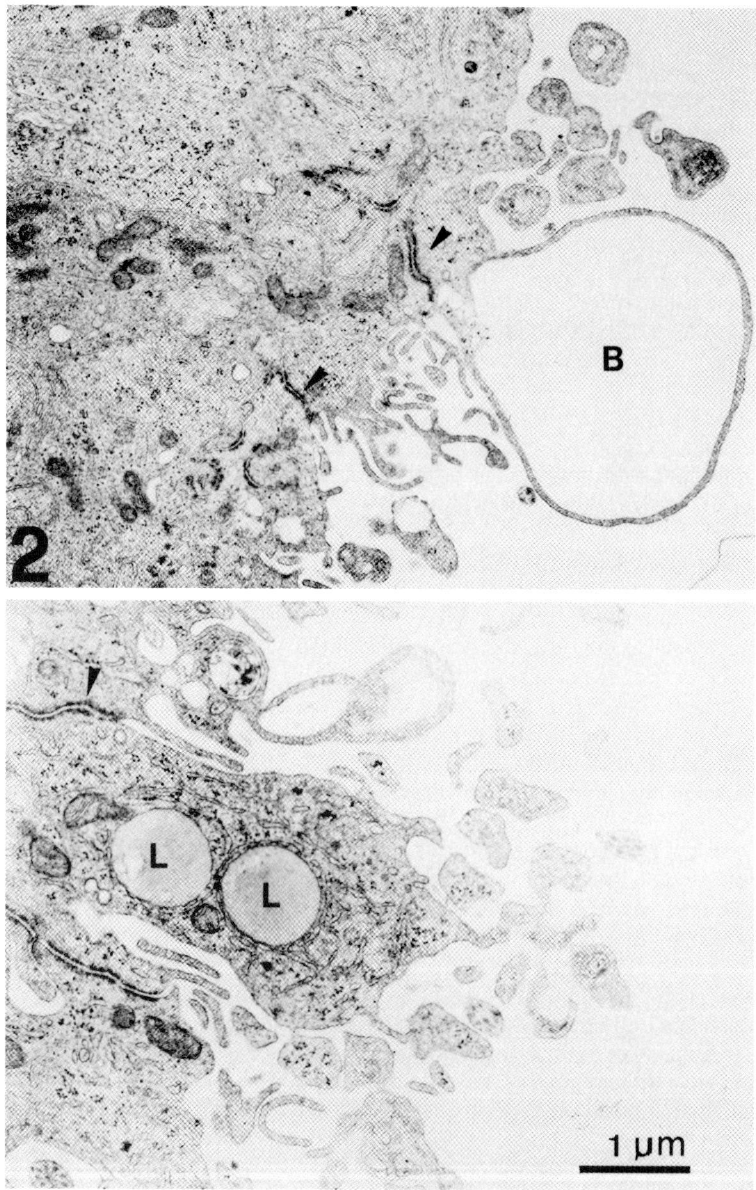

Fig. 2. Various aspects illustrating the complexity of the shapes of the apical surface of the tanycytes and their great diversity (rat). L: Lipid droplets. ▼: Point on the tight junctions. B: Bleb.

By courtesy of The International Review of Cytology.

the infundibular part of the 3rd ventricle and the portal vessels of the median eminence they have been assumed to play a role in neuroendocrine controls (Knigge and Scott, 1970; Knigge and Silverman, 1972; Knigge et al, 1976).
Morphological changes of the apical pole has been described by several authors (Knowles and Kumar, 1969; Gordon et al, 1972; Coates, 1974) during the estrous cycle in the rat and the menstrual cycle in the monkey. Results are sometimes contradictory but from these numerous studies, these cells appear sensitive to oestrogen levels. We have observed morphological modifications in rats after adrenalectomy, with accumulation of lipid droplets, increased number of blebs (Brion et al, 1982).
From our observations in rat and Man, the morphology of the tanycytes suggest a transport function. This is mainly supported by their richness in microtubules and mitochondria and by the scarcity of the rough endoplasmic reticulum, which pleads against a synthetic activity.
As far as transport is concerned, experiments studying the intraventricular injection of HRP are of great interest. The most important results were obtained by Brightman, Prescott and Reese in 1975.
They have shown clearly that the permeability of the ventricular lining presents regional differences. After intraventricular injection of HRP there is a rapid passage in the upper part of the 3rd ventricle where the lining is composed of classical ependymal cells joined by gap junctions. In the lower part of the 3rd ventricle lined by tanycytes joined by tight junctions, the intercellular passage is prevented and pinocytotic vacuoles labelled with the enzyme appear at the apical pole of the cells.
In contrast, intravascular injection of HRP is followed by a rapid passage of the enzyme in the median eminence, whereas the blood brain barrier prevents the passage of the enzyme in the upper part of the 3rd ventricle. Thus the enzyme is prevented from passing between the tanycytes which are linked by tight junctions in the area of the median eminence devoid of blood-brain barreer, as the endothelium of these capillaries is fenestrated.
This is very important to consider and has been also the object of many controvery and discussions.
We feel that the fact that tanycytes may absorb exogenous peroxidase injected in the CSF can be taken as an indication that transport may occur in tanycytes.
The nature of the material that might be transported through tanycytes in physiological conditions from

the CSF to the portal vessels and or vice versa remains to be discovered.
Hormone have been suggested.
More experimental work is needed for a better understanding of the possible functions of these interesting and intriguing cells.
To end, I would like to quote Pilgrim (1978) who wrote : "The shape and the topography of the tanycyte ependyma has stimulated the fantasy of numerous, especially morphologically orientated neuroendocrinologists". He concludes : "it remains to be proven that tanycytes have a function which can definitively be connected to neuroendocrine mechanisms".

REFERENCES

Brightman MW, Prescott L, Reese TS (1975). Intercellular junctions of special ependyma. In Knigge KM, Scott DE, Kobayashi H, Ishii S (eds) : "Brain-endocrine interaction II. The ventricular system in neuroendocrine mechanisms", Basel, S. Karger, pp 146-165.

Brion JP, Couck AM, Depierreux M, Flament-Durand J (1982). Transmission and scanning electron-microscopic observations on tanycytes in the mediobasal hypothalamus and the median eminence of adrenalectomized rats. Cell Tiss Res 221 : 643-655.

Coates PW (1974). Response of tanycytes in primate third ventricle to ovariectomy : scanning electron microscopic evidence. Anat Rec 178 : 330-331.

Flament-Durand J, Brion JP (1985). Tanycytes : morphology and functions. A review. In : "International Revue of Cytology", Acad. Press, vol. 96, pp 121-155.

Gordon JH, Bollinger J, Reichlin S (1972). Plasma thyrotropin responses to thyrotropin-releasing hormone after injection into the third ventricle, systemic circulation, median eminence and anterior pituitary. Endocrinology 91 : 696-701.

Halasz B, Pupp L, Uhlarik S (1962). Hypophysiotrophic area in the hypothalamus. J Endocr 25 : 147-154.

Horstmann E (1954). Die Faserglia des Selachiergehirns. Z. Zellforsch 39 : 588-617.

Knigge KM, Scott DRE (1970). Structure and function of the median eminence. Amer J Anat 129 : 223-244.

Knigge KM, Silverman AJ (1972). Transport capacity of the median eminence. In : Brain-endocrine interaction. Median eminence : structure and function. Knigge KM, Scott DE, Weindl A (eds), Basel : S. Karger, pp 350-363.

Knigge KM, Joseph SA, Sladek JR, Notter MF, Morris M, Sundberg DK, Holzwarth MA, Hoffman GE, O'Brien L (1976). Uptake and transport activity of the median eminence of the hypothalamus. Int Rev Cytol 45 : 383-408.

Knowles F, Kumar TCA (1969). Structural changes related to reproduction in the hypothalamus and in pars tuberalis of the rhesus monkey. Phil Trans Roy Soc London, 256 : 357-375.

Pilgrim Ch (1978). Transport function of hypothalamic tanycyte ependyma : how good is the evidence ? Neuroscience 3 : 277-283.

KEY WORDS

Tanycytes - TEM - SEM - Hypophyseotropic area - Infundibular recess - Portal system.

Cells and Tissues: A Three-Dimensional
Approach by Modern Techniques in Microscopy,
pages 371–376

A TECHNIQUE FOR STUDYING THE PERIKARYAL PROJECTIONS OF SPINAL GANGLION NEURONS BY SCANNING ELECTRON MICROSCOPY.

E. Pannese, M. Ledda, V. Conte, E. Marini, S. Matsuda

Institute of Histology, Embryology and Neurocytology,
University of Milan
Via Mangiagalli, 14 - I-20133 Milano Italy

INTRODUCTION

In electron micrographs of thin sections of sensory ganglia slender projections arising from the neuronal perikaryon have been observed by many authors. Shape and length of these projections, however, cannot be evaluated on single thin sections examined by transmission electron microscopy since they often extend beyond the plane of the section. As a consequence the entire outline of these perikaryal projections can only be evaluated in the transmission electron microscope by means of the serial section technique, a very laborious and time-consuming procedure, which limits the number of possible observations.

Recently, enzymatic digestion techniques have been successfully used to dissociate the spinal ganglia of adult mammals (Scott, 1977; Silberberg and Kim, 1979; Fukuda and Kameyama, 1980; Matsuda and Uehara, 1981; Smith and McInnes, 1986). We have modified these techniques so as to secure the removal of both the connective tissue and satellite cells of spinal ganglia thereby exposing the true neuronal surface. It has thus become possible to examine neuronal projections by means of the scanning electron microscope, which permits direct observation of the true neuronal surface. This procedure allows the examination of a large number of projections and evaluation of their quantitative parameters.

MATERIALS AND METHODS

Two adult rats and 2 adult rabbits were used as controls;

these animals were perfused with 3% buffered glutaraldehyde, their spinal ganglia were removed, postfixed with 2% buffered OsO_4 and routinely processed for examination under the transmission electron microscope. Ten adult rats and 6 adult rabbits were perfused with a Ringer solution, their spinal ganglia were removed and cleaned of the connective tissue capsule using fine needles; these ganglia were then incubated in a buffered solution containing 0.2 mg/ml $CaCl_2$ and 1 mg/ml collagenase and successively in a buffered solution containing 2.5 mg/ml trypsin under constant stirring. The ganglia were then fixed with 3% buffered glutaraldehyde and postfixed with 2% buffered OsO_4. Some of the ganglia were routinely processed for examination under a transmission electron microscope; the others were treated with 1% tannic acid followed by 2% OsO_4, then dehydrated in alcohol, immersed in isoamyl acetate, dried by the critical point method and observed under a scanning electron microscope.

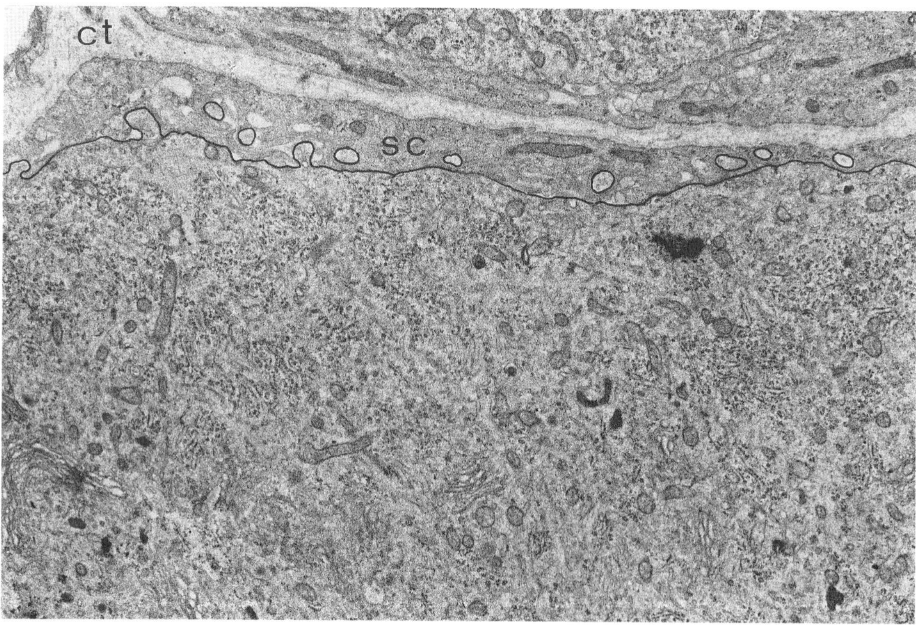

Figure 1. Rabbit spinal ganglion fixed in situ by perfusion. The satellite cell sheath (sc) is directly applied to the neuronal surface. The outlines of the neuronal perikaryon and of its projections arising at other levels and thus appearing as isolated entities were traced with ink to make them more easily visible. ct = connective tissue. Transmission electron microscopy, x11,000.

RESULTS

Transmission electron microscopy. In the ganglia fixed in situ by perfusion, the neuron-satellite cell boundary appears complicated by the presence of many projections from both neuron and satellite cells (Fig. 1). In single sections perpendicular to the surface of the nerve cell body neuronal projections may sometimes appear as evaginations continuous with the neuronal perikaryon; more frequently, however, they arise from the neuronal perikaryon at another level thus appearing as discrete entities (Fig. 1). In the latter case the projections may appear

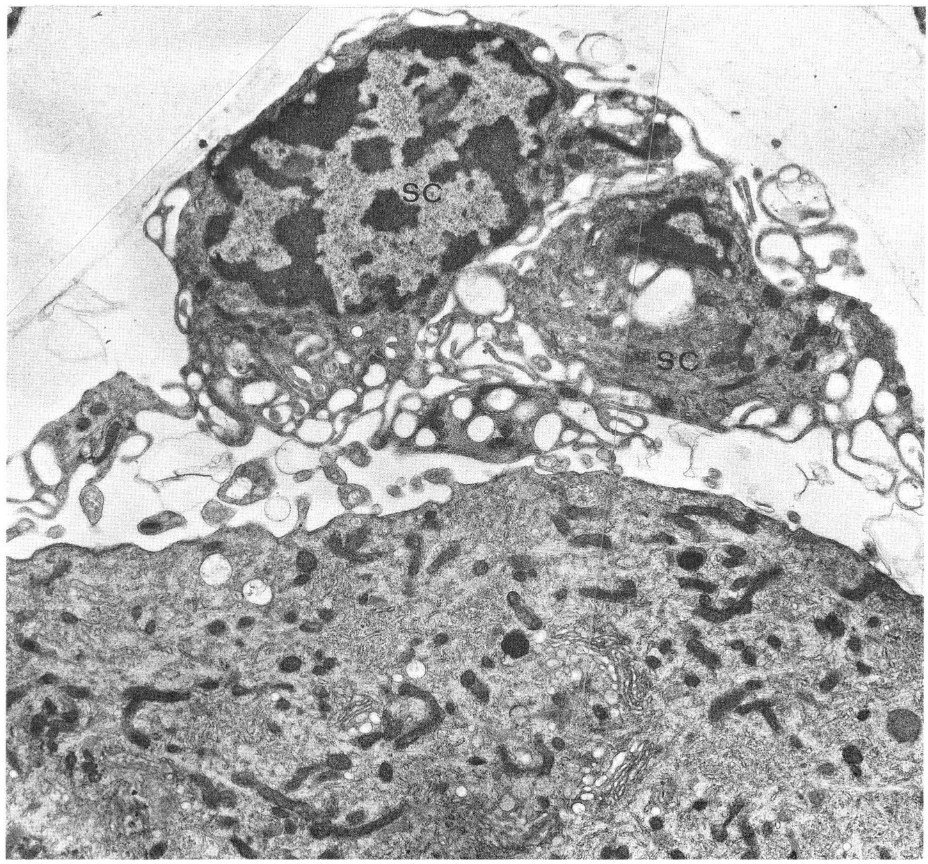

Figure 2. Rabbit spinal ganglion treated with collagenase and trypsin. A wide gap is evident between the neuronal surface and the satellite cells (sc); neuronal projections jut into this gap. Transmission electron microscopy, x11,000.

embedded in the satellite cell sheath or may be found along the gap between the neuron and its satellite cell sheath.

In ganglia treated with collagenase and trypsin some nerve cell bodies appear surrounded by their own satellite cells, but, unlike those observed in control ganglia, these cells are not directly applied to the neuronal surface, but are sharply separated from the latter. A wide gap is generally present between the neuronal surface and the satellite cells; projections arising from the neuronal perikaryon jut into this gap (Fig. 2). Other nerve cell bodies appear completely devoid of any enveloping sheath with projections continuous with the perikaryon. Moreover many isolated profiles of projections can be seen near the neuronal surface (Fig. 3). Usually, these projections have a similar appearance to that of their counterparts in the control ganglia; in a few neurons only their free ends are enlarged. Exceptionally one of these projections may appear frayed. The fine structure of nerve cell bodies isolated by enzymatic digestion is fundamentally similar to that of nerve cell bodies fixed in situ by perfusion (Figs. 1 and 3).

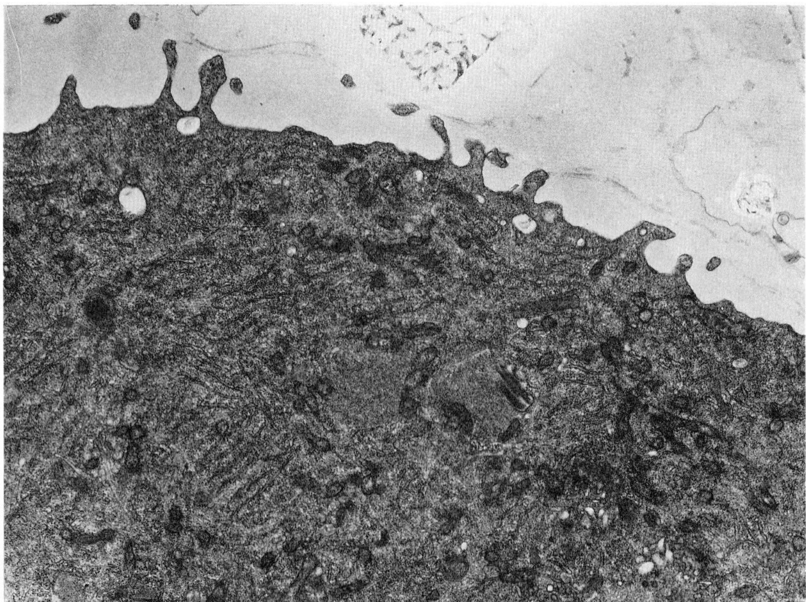

Figure 3. Rabbit spinal ganglion treated with collagenase and trypsin. A nerve cell body appears completely devoid of any enveloping sheath. Note profiles of neuronal projections continuous with the nerve cell body or isolated near the neuronal surface. Transmission electron microscopy, x11,000.

Scanning electron microscopy. After mechanical removal of the connective tissue capsule surrounding the whole ganglion and subsequent collagenase and trypsin digestion, many spherical or oval corpuscles become evident. A single process can sometimes be seen to arise from such corpuscles. Some corpuscles exhibit a smooth or finely wrinkled surface; each of these consists of a nerve cell body completely covered by satellite cells. Other corpuscles exhibit a highly irregular surface; these are nerve cell bodies devoid of satellite cells showing up their own true surface (Fig. 4). The latter exhibits a large number of projections, whose most common shapes are given in Figure 5. The projections range between 0.30 and 2.50 µm in length (average 1.32 ± 0.44 µm) in the rat and between 0.41 and 2.80 µm (average 1.30 ± 0.42 µm) in the rabbit. Average transverse diameter is 0.16 ± 0.03 µm in the rat and 0.19 ± 0.03 µm in the rabbit.

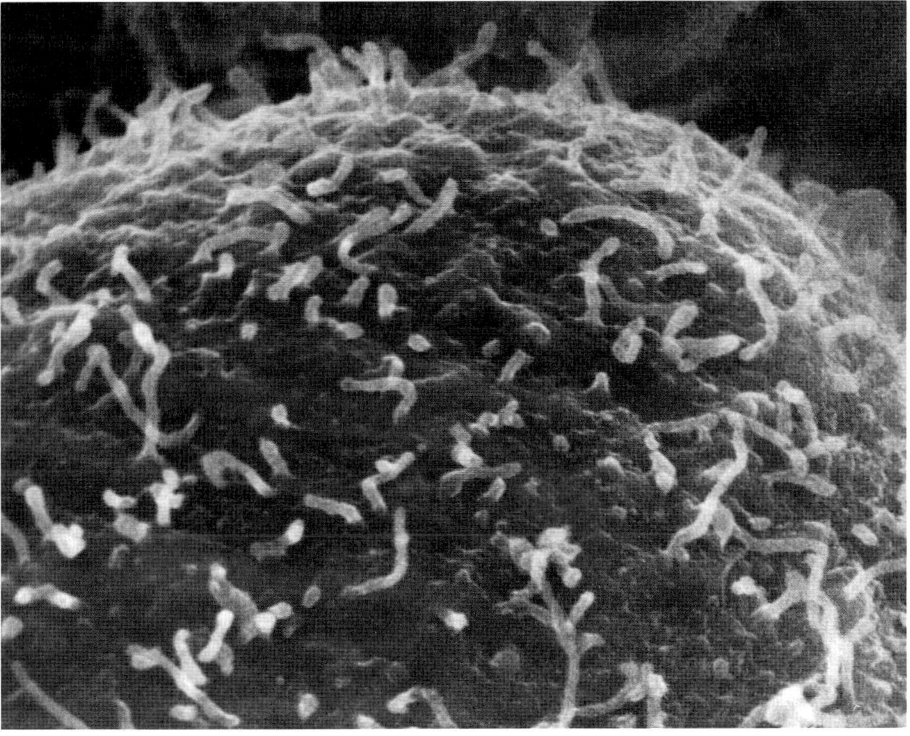

Figure 4. Rabbit spinal ganglion treated with collagenase and trypsin. The true exposed surface of a nerve cell body exhibits many projections. Scanning electron microscopy, x11,000.

DISCUSSION

To check that the enzymatic digestion technique does not alter the projections arising from the neuronal perikaryon, neuronal projections in the ganglia fixed by perfusion in situ and those to be seen in the dissociated ganglia were compared. This comparison reveals that shape, length and thickness of projections arising from neurons deprived of their envelopes correspond well to the shape, length and thickness of the projections found in ganglia fixed in situ by perfusion and to those observed in reconstructions prepared from serial sections (Pannese et al., 1983). It may therefore be concluded that this technique does not affect the nerve cell body projections in the spinal ganglia and is thus suitable for studying the quantitative parameters of these projections.

Figure 5. The most common shapes of neuronal projections are shown.

REFERENCES

Fukuda J, Kameyama M (1980). A tissue-culture of nerve cells from adult mammalian ganglia and some electrophysiological properties of the nerve cells in vitro. Brain Res 202: 249-255.

Matsuda S, Uehara Y (1981). Cytoarchitecture of the rat dorsal ganglia as revealed by scanning electron microscopy. J Electron Microsc 30: 136-140.

Pannese E, Gioia M, Carandente O, Ventura R (1983). A quantitative electron microscope study of the perikaryal projections of sensory ganglion neurons. I. Cat and Rabbit. J Comp Neurol 214: 239-250.

Scott BS (1977). Adult mouse dorsal root ganglia neurons in cell culture. J Neurobiol 8: 417-427.

Silberberg DH, Kim SV (1979). Studies of aging in cultured nervous system tissue. Int Rev Cytol Suppl 10: 117-130.

Smith RA, McInnes IB (1986). Phase contrast and electron microscopical observations of adult mouse dorsal root ganglion cells maintained in primary culture. J Anat 145: 1-12.

Cells and Tissues: A Three-Dimensional
Approach by Modern Techniques in Microscopy,
pages 377–382

3-D CHANGES IN NEUROBLASTOMAxGLIOMA HYBRID (NG 108-15) CELL DIFFERENTIATION AS STUDIED BY SEM AND TEM.

A.Iavarone,M.L.Eboli*, M.Osti**,
A.Redler**, M.Pocchiari* and M.A.Russo**.
**Dip.di Medicina Sperimentale, Universita' LaSapienza, and *Istituto di Patologia Generale, Universita'Cattolica Roma,Italy.*

INTRODUCTION.

Sodium Butyrate (**NaB**) is a natural four carbon fatty acid which induces differentiation *in vitro* of Freund erytroleukaemia (Kruh, 1982), melanoma (Nordenberg,1987) and neuroblastoma cells (Prasad, 1980),and *in vivo* of acute myelogenous leukaemia cells; until now it has not been used on NG 108-15 cells. Such a cell line constitutes one of the most easy and reproducible *in vitro* model for neuronal and tumor cell differentiation (Hamprecht, 1977); usually differentiation is achieved by other differentiating agents such as dibutyril-cAMP (**dBcAMP**) or retinoic acid and related compounds. By using SEM, TEM and assays of biochemical and biological markers of the differentiation, in this paper we have explored if the differentiating effect of the NaB on the NG 108-15 cells was advantageous in comparison to that of the dBcAMP. NaB appeared much more effective than dBcAMP in inducing a)-morphological modifications of neuronal type, b)-acetyl-cholinesterse (**AchE**) activity, with a peculiar pattern of induction; and c)-in drastically inhibiting cell growth.

MATERIALS AND METHODS.

A clonal hybrid cell line of neuroblastoma/glioma (**NG 108-15**) (kindly provided by dr.T.Costa from Max Planck Institute of Munich, WG) was cultured for 7 days in Dulbecco's modified medium, supplemented with 7% BCS and HAT, in the presence or absence of 1 mM of NadB or dBcAMPs. Cell counting was performed daily in Burker chamber; cell viability was evaluated by Trypan Bleu exclusion method. Samples were taken daily to determine AchE (EC 3.1.1.7) activity as a marker for neuronal differentiation and cell LDH activity as a marker for tumour transformation. Cells suspended in PBS Ca++/Mg++--free, were sonicated two times, 10" sec. each, and enzyme activity was assayed on total cell homogenate by the procedure described by Ellman(1961) for AchE and by Bergmeyer (1974) for LDH. Further samples were fixed at various times with 3.5% glutaraldehyde in phosphate buffer (pH 7.4) for at least two hours and then processed for optical microscopy, for SEM and TEM, following standard procedures.

RESULTS.

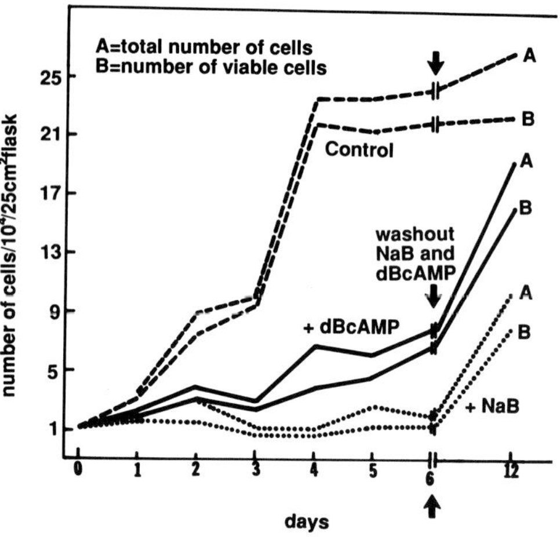

Cell proliferation was drastically inhibited by NaB, less by dBcAMP , when compared to control cells; the latter at seventh day were largely confluent (fig.1). At SEM, control cells appeared round in shape, with numerous microvilli and short protrusions on their surface, growing in irregular and dense multilayers (fig.2). At TEM, they showed prominent nucleus and nucleolus, irregular and mostly swollen mitochondria, a few short segments of endoplasmic reticulum, very few granules with dense core and various types of microfilaments irregularly organized (fig.6).

In the presence of NaB and dBcAMP cells displayed long processes, sometimes branched, often under/overlining similar processes of adjacent or distant cells (fig. 4-5). These and other features were evident especially with NaB: cell surface was mostly smooth, without microvilli; further, axonal-like long processes may show terminal patches which were similar to axo-axonic or axo-somatic synaptic terminals (fig.15).

At TEM, cytoplasm showed numerous important changes: organules, dense core vesicles and cytoskeletal components were largely increased (fig.7-13). In particular, mitochondria were more numerous (+60/80% per similar field), endoplasmic reticulum was abundant and mainly constituted of long segments of RER. Dense core granules were largely increased (+100% or more per similar field); sometimes also clear vesicles were present in NaB treated cells. Intermediate filaments were increased and vectorially organized as small bundles in the extended long processes; microtubules were also present, scattered all over the cell, but were more abundant and vectorially organized in the long processes (fig.7). A few clear core vesicles were in close association with microtubules. Adjacent cells with granules showed frequent coated caveolae, suggesting receptor-mediated endocytosis (fig. 11). AchE activity was increased with both agents, being much more evident with NaB (fig.14). All the above chan-

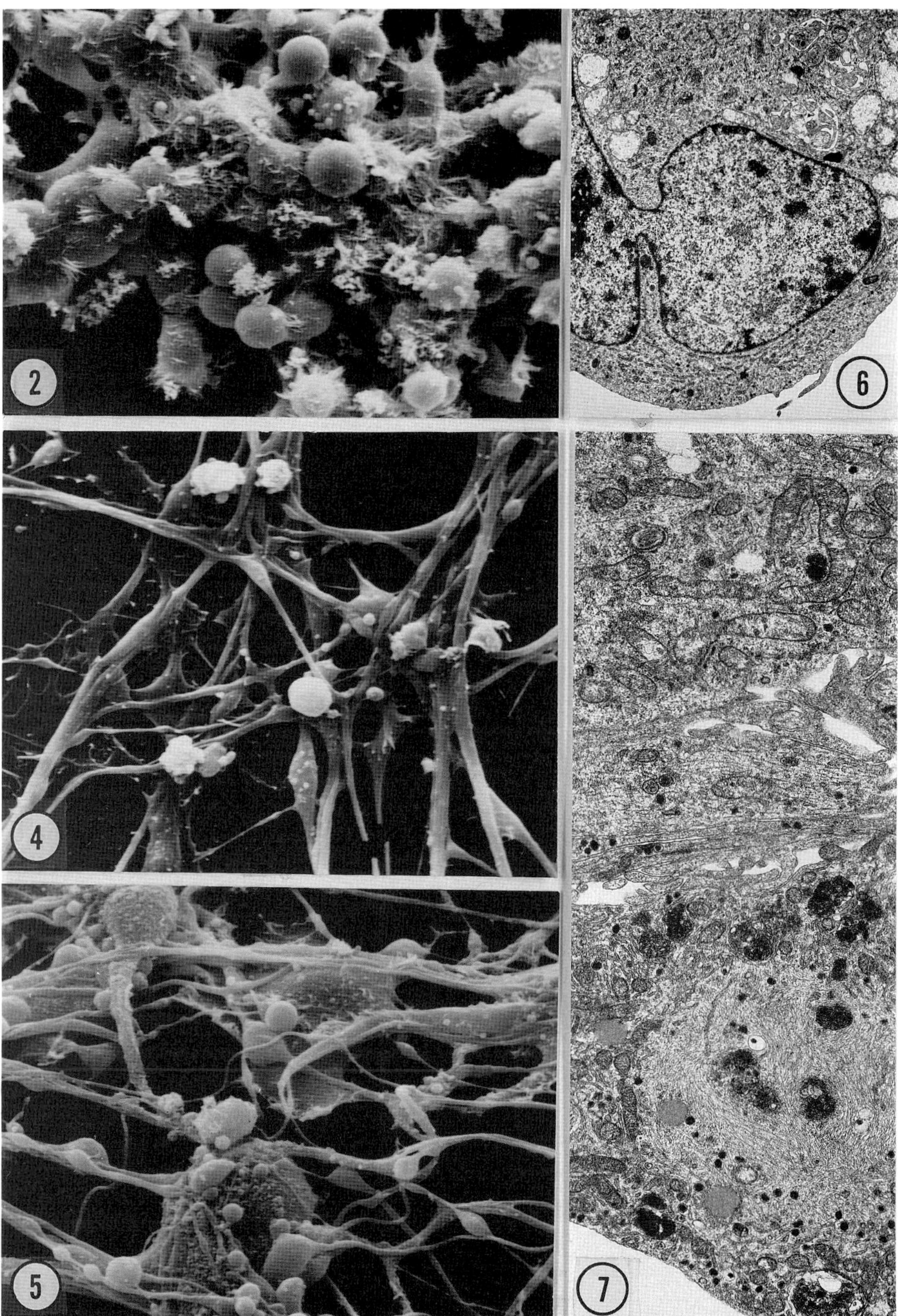

Figure 2: SEM, NG 108-15 control cells. ***Figure 4***: TEM, as figure 2. ***Figure 5***: SEM, treated with NaB for 7 days. ***Figure 6***: SEM, treated with dBcAMP for 7 days. ***Figure 7***: TEM, treated with NaB for 7 days.

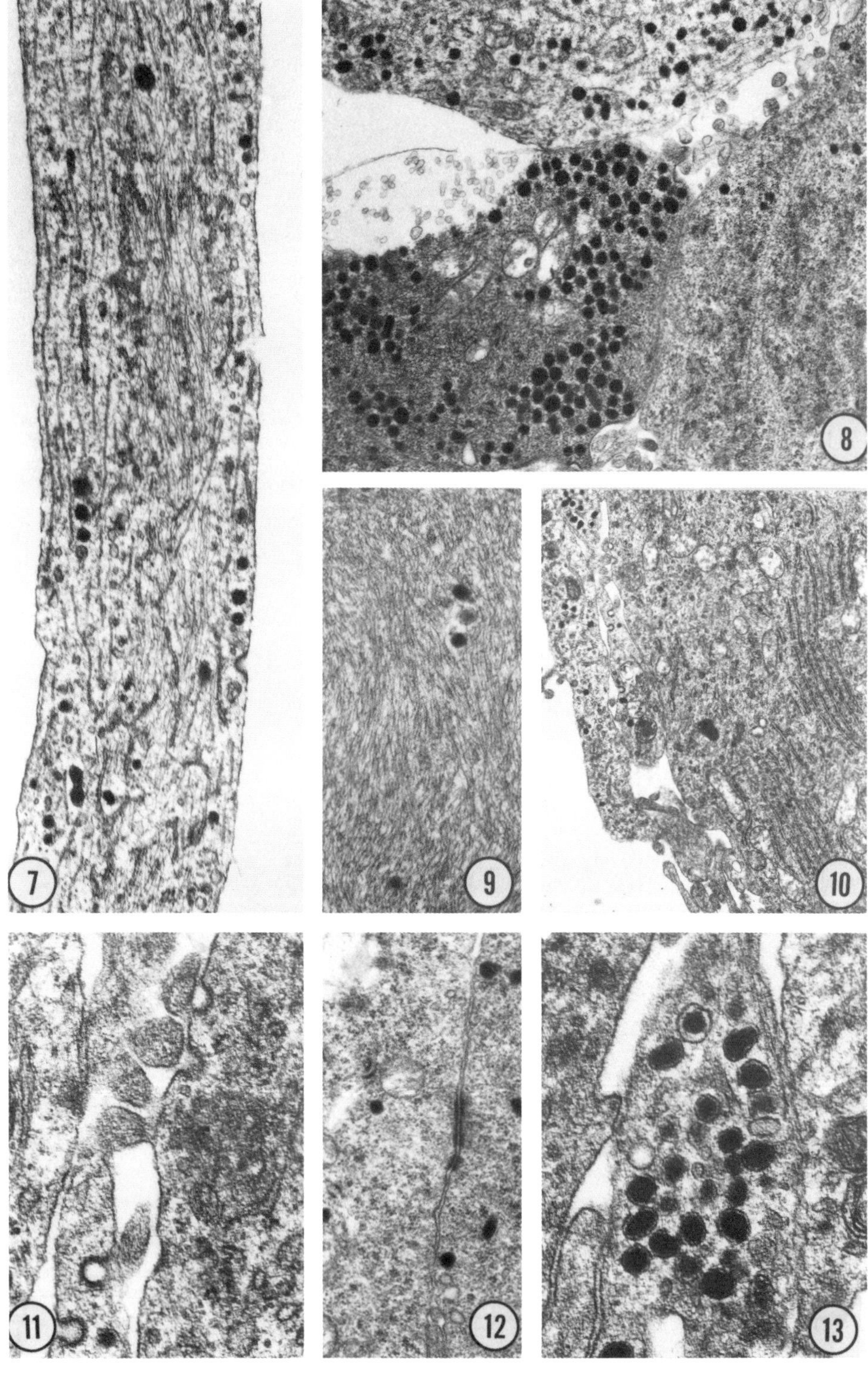
7
8
9
10
11
12
13

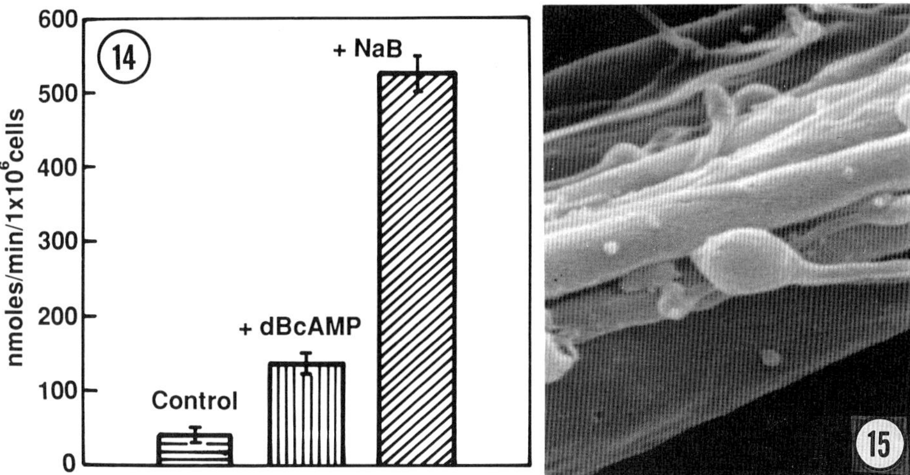

Figure 14: AchE activity of NG 108-15 cells at 6th day of culture. ***Figure 15***: Detail at SEM of an axo-axonic synaptic-like button.

ges were generally more pronounced in NaB treated cells, in comparison to cAMPdB treated cells. Preliminary results show that washout of both agents and a further incubation for 6 days without both agents dramatically restimulates cell growth (fig.1) and disappearance of ultrastructural and biochemical findings above described.

CONCLUSIONS.

Both agents have shown to be able to induce in NG 108-15 cells new biochemical and morphologic features, most of which are of neuronal type. The ultrastructural results obtained with dBcAMP were similar to that described by others (Daniels, 1974; Furuya, 1983). An advanced differentiation from tumoral phenotype is also suggested by cell growth inhibition, which is drastic with NaB, less pronounced with dBcAMP, and by cell LDH activity, which shows a 50% decrease in the presence of dBcAMP and a further 50% decrease with NaB.

The comparison between the two agents strongly suggests that, whatever is the feature considered, NaB is

Legends:
Details at TEM of NG 108-15 cells differentiated by NaB.
Figure 7: Detail of an axonal process.
Figure 8: Part of three cells with different granules.
Figure 9: Bundles of intermediate filaments.
Figure 10: Well organized granular endoplasmic reticulum.
Figure 11: Coated caveolae between adjacent cells.
Figure 12: Emidesmosomal junction between two cells.
Figure 13: Neuropeptide-like granules.

much more efficient in inducing both qualitative and quantitative changes of neuronal type. In particular, neurosecretory granules, RER, Golgi apparatus, coated caveolae, large bundles of neurofilaments and neurotubules are indicative of an advanced and coordinated intra/inter-cellular regulation.

In conclusion, from our results it seems that NaB, at least in this cell type, acts as a major potent and specific differentiative agent when compared to the classical dBcAMP. It remains to be established through which mechanism(s) NaB is able to accomplish such a better and coordinated differentiation of neuronal type.

REFERENCES.

Bergmeyer HU (1974). *Methods in enzymatic analysis,* New York: Academic Press, vol 2, pp 574-579.

Daniels MP, Hamprecht B (1974). *The ultrastructure of neuroblastoma glioma somatic cell hybrids.* J Cell Biol 63:691-699.

Ellman GL, Courtney KD, Andres V jr, Featherstone RM (1961). *A new and rapid colorimetric determination of acetylcholinesterase activity.* Biochem Pharmac 7:88-95.

Furuya S, Furuya K (1983). *Ultrastructural changes in differentiating neuroblastoma x glioma hybrid cells.* Tissue & Cells 15:903-919.

Hamprecht B (1977). *Structural, electrophysiological, biochemical, and pharmacological properties of neuroblastoma/glioma cell hybrids in cell culture.* Int Rev Cytol 49:99-170.

Kruh J (1982). *Effects of sodium butyrate, a new pharmacological agent,on cells in culture.* Mol Cell Bioch 42:65-82.

Nordenberg J, Wasserman L, Peled A, Malik Z, Stenzel KH, Novogrodsky A (1987). *Biochemical and ultrastructural alterations accompany the anti-proliferative effect of butyrate on melanoma cells.* Br J Cancer 55:493-497.

Prasad KN (1980). *Butyric acid: a small fatty acid with diverse biological functions.* Life Sciences 27:1351-1358.

*Partially supported by CNR funds from **Progetto Finalizzato Oncologia,** given to MAR.*

Cells and Tissues: A Three-Dimensional
Approach by Modern Techniques in Microscopy,
pages 383–387

PERIVASCULAR GLIA-ENDOTHELIUM RELATIONSHIPS IN THE DEVELOPING CEREBRAL VESSELS. A THREE-DIMENSIONAL COMPUTER AIDED STUDY

Luisa Roncali, Mirella Bertossi, Domenico Ribatti,
Beatrice Nico, Daniela Virgintino and Lucia Mancini
Institute of Human Anatomy & Histology and General
Embryology, University of Bari Medical School, Bari,
Polyclinics, 70124 Bari, Italy

INTRODUCTION

Three-dimensional computer assisted reconstructions (Computer PDP 11/23 Digital, Image Analyzer Tesak VDC 501) were made of the neural blood vessels in the developing chick embryo optic lobes in order to obtain as much information as possible on the spatial organization of endothelium, pericytes and perivascular glia.

On the basis of the results of recent studies, the astrocytic periendothelial expansions may be involved in the differentiation process of the tight interendothelial junctions responsible for the blood-brain barrier devices in the cerebral vessels (Arthur et al., 1987; Janzer and Raff, 1987; Tao-Cheng et al., 1987; Tao-Cheng and Brightman, 1988).

The supposition that perivascular glia controls the morphological features and the biochemical properties of the cerebral vessel endothelium is based on experimental evidence (*in vitro*) alone and the aim of this research is to corroborate it with further knowledge of the endothelium-glia relationships *in vivo*, during the normal embryonic development of the blood-brain barrier.

MATERIALS AND METHODS

Neural blood vessels from optic tecta of chick embryos of 14 and 20 days of incubation were serially cut with a LKB-V ultramicrotome after routine procedures of fixation (3% glutaraldehyde), post-fixation (1% osmium tetroxide) and embedding (Epon 812). The ultrathin sections, with an average thickness of 200 nm, were stained with uranyl acetate and lead citrate and observed under a Zeiss 9A electron microscope. For each of the two developmental stages one intratectal vessel was chosen and photographed in fields at 6000x magnifications, in each second or third section. Photographic reconstructions of the vessel sections were made, on which the outlines of endothelium, pericytes, endothelial nuclei and mitochondria, and perivascular glia were drawn by means of a pencil. The tracings of all vessel reconstructions were sequentially superimposed taking care that the boundaries of endothelium, pericytes and glia, and as many intraendothelial structures as possible, corresponded, in order to obtain final coincident x and y axes. For each section the traces of endothelium, pericytes and astrocytic glia were digitized and, simultaneously, displayed on the TV screen. The contour lines of each section, separately digitized, were then assembled in order to obtain the three-dimensional image of each of the vessel components, and, finally, photographed on slides. A computer aided evaluation of the contact region between endothelium-pericytic layer and, respectively, glial perivascular endfeet was also made.

RESULTS AND CONCLUSIONS

At 14th incubation day the astrocytic perivascular expansions are small and isolated, surrounding the endothelium-pericyte layer for 12.5% of its perimeter (Figs. 1 A, B).

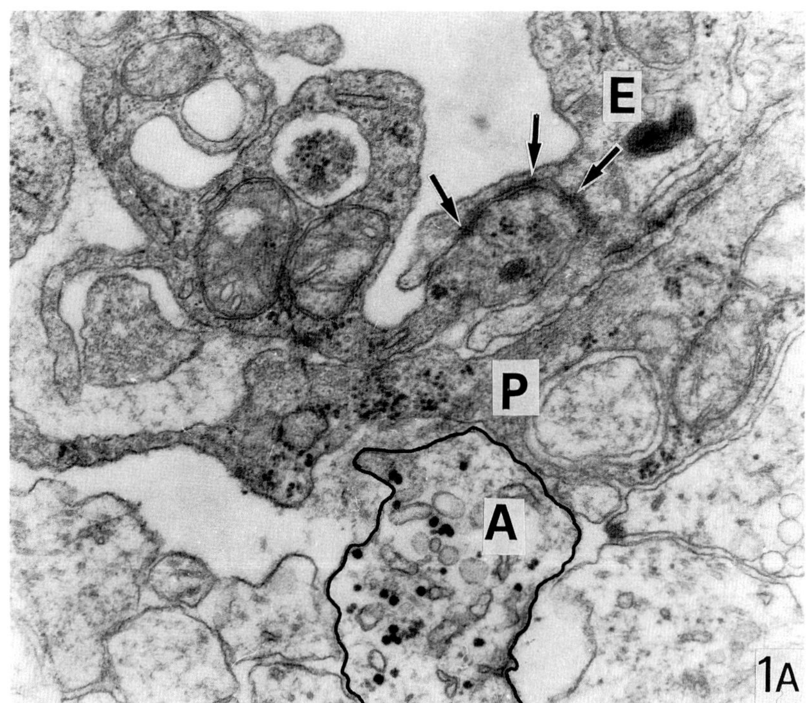

Figure 1 A. 14th incubation day. The endothelial cells (E) are coupled by junctions built of punctuate contacts between the facing plasmalemmas (arrows). Beneath the endothelium a slender pericyte process(P)and an isolated astrocytic endfoot(A)(51000x).

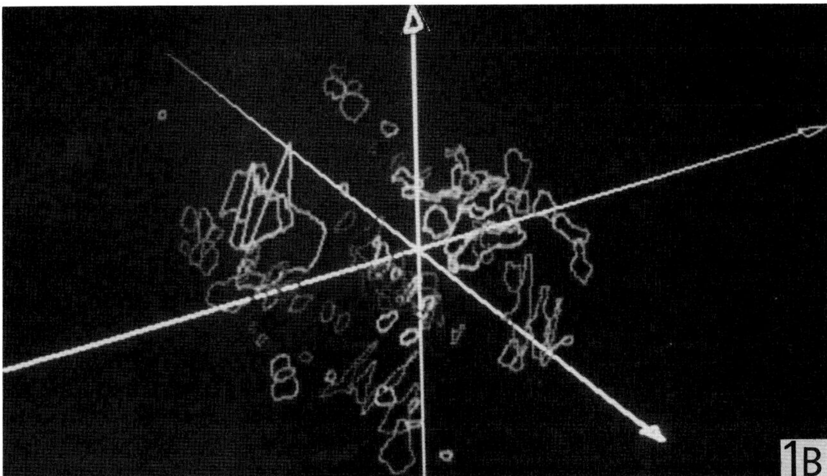

Figure 1 B. A three-dimensional view of the glial perivascular endfeet at 14th incubation day.

At 20th incubation day the glia forms an almost continuous sheath, 96% of the endothelium-pericyte layer being enveloped by its expansions (Figs. 2 A, B).

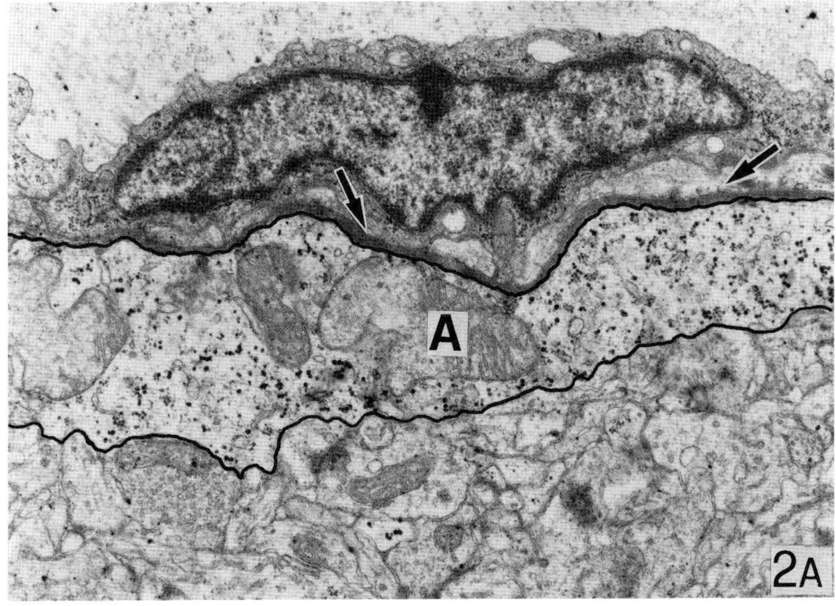

Figure 2 A. 20th incubation day. A continuous astrocytic sheath (A) underlies the endothelium-pericyte layer, encircled by a thick basal lamina (arrows) (18500x).

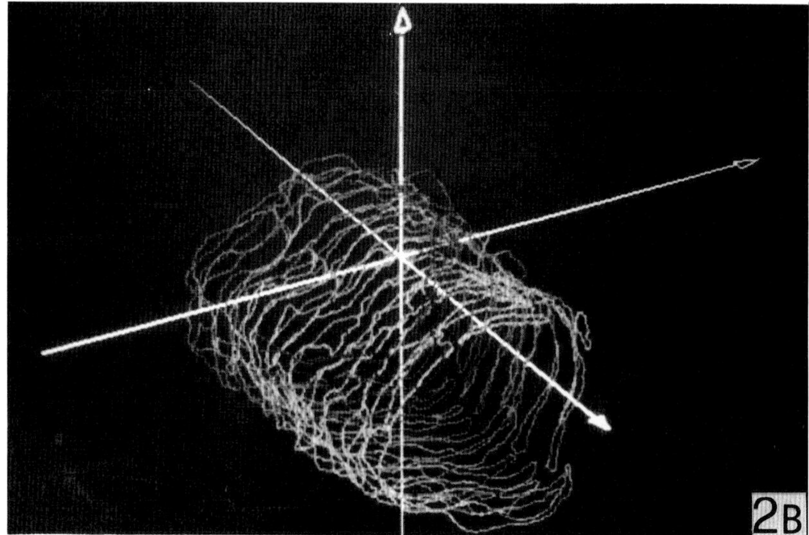

Figure 2 B. A three-dimensional view of the glial perivascular endfeet at 20th incubation day.

According to the results of our previous ultrastructural researches (Roncali et al., 1985, 1986) when the perivascular glia appears as circumscribed expansions, at 14th incubation day, a blood-brain barrier to the marker of vascular permeability horseradish peroxidase (hrp) begins to form in the tectum, and the contact regions in its vessels between endothelial cells are characterized by fusion points of their plasmalemmas. On the contrary, at 20th incubation day, when a continuous perivascular layer of glia encircles the vessels, the blood-brain barrier to hrp is complete and extensive tight junctions weld the endothelial cells.

On the whole, the previous ultrastructural studies and the present three-dimensional reconstructions seem to confirm that the perivascular glia, and therefore the microenvironment in which the neural blood vessels develop, may control the process of morphological and functional maturation of the cerebral endothelium.

REFERENCES

Arthur FE, Shivers RR, Bowman PD (1987). Astrocyte-mediated induction of tight junctions in brain capillary endothelium: an efficient in vitro model. Devel Brain Res 36:155-159.

Janzer RC, Raff MC (1987). Astrocytes induced blood-brain barrier properties in endothelial cells. Nature 325:253-257.

Roncali L, Ribatti D, Ambrosi G (1985). Ultrastructural basis of the vessel wall differentiation in the chick embryo optic tec-

tum. J Submicrosc Cytol 17:83-88.
Roncali L, Nico B, Ribatti D, Bertossi M, Mancini L (1986). Microscopical and ultrastructural investigations on the development of the blood-brain barrier in the chick embryo optic tectum. Acta Neuropathol 70:193-201.
Tao-Cheng JH, Nagy Z, Brightman MW (1987). Tight junctions of brain endothelium in vitro are enhanced by astroglia. J Neurosci 7:3293-3299.
Tao-Cheng JH, Brightman MW (1988). Development of membrane interactions between brain endothelial cells and astrocytes in vitro Int J Devl Neurosci 6:25-37.

Cells and Tissues: A Three-Dimensional Approach by Modern Techniques in Microscopy, pages 389–393

SEM OF SUBDURAL SPACE IN MAMMALS

Eduard Klika, Milan Richter

Department of Histology, Faculty of General Medicine, Charles University in Prague, Prague 2 Albertov 4, 128 00

INTRODUCTION

The subdural space of the skull and its existence was studied in rat /Rattus norvegicus var. alba/, rabbit /Oryctolagus cuniculus f. domestica/ and pig /Sus scrofa f. domestica/ under the SEM. The results are compared with findings obtained on membranous impression mounts in light microscope and ultrathin sections in TEM /Klika,1967/.

The dura mater of the brain and spinal cord lies upon the outer surface of the arachnoid. Between both layers the subdural space is supposed, the existence of which is in question. According to Clara /1953/ the subdural space appears postmortally following imbibition of cerebrospinal fluid. Pease and Schultze /1958/ observed in some locations contacts between the mesothelium of dura mater and outer surface of arachnoid. A fine granular material was found between both layers. They accept the existence of a potential subdural space. Klika /1967/ in a compararative study in light microscope and TEM demonstrated in a series of mammals, including man, the subdural space as discontinuous. The presence of islets of adhering arachnoid cells to dura mater is a proof of rather potential than a virtual, real subdural space.

The subdural space is a matter of dispute of the authors nowadays. Andres /1967/, Nabeshima et al /1975/, Allen and Didio /1977/ underline in mammals a close contact of dura mater and arachnoid. Mc Lone and Bondareff /1975/ in a developmental study in mouse did not demonstrate the subdural space or transitional layer between dura mater and arachnoid. Schachenmayer and Friede /1978/ studied the dura arachnoid junction in man. In well fixed specimen the arachnoid and dural border barrier form a uniform layer. It is to abandone the classical conception of existence of subdural space. The close contact of fibroblasts of inner surface of dura mater and arachnoid barrier in mammals was confirmed by Oda and Nakanishi /1984/, Angelov and al /1984/ and Zajícová /1987/.

METHODS

Using the method of membranous impression mounts and SEM analysis the subdural surface of the dura mater and the corresponding outer surface of arachnoid in rats /Rattus norvegicus var. alba/ rabbits /Oryctolagus cuniculus f. domestica/ and pigs /Sus scrofa f. domestica/ was performed by perfusion and immersion fixation with 3% glutaraldehyde solution in phosphate buffer. After fixation the upper segment of the skull was cut off by a coronal section and fixed with brain in situ for 24 hours in 2,5% solution of glutaraldehyde in phosphate buffer. The brain was luxated from the skull. In SEM the compartments of the inner surface of dura mater and arachnoid were studied. A part of material for light microscopical analysis was postfixed for 24 hours in 10% formaline. The specimen well washed in water were stained for 3 hours with Harris hematoxylin. After having washed the specimen in water the membranous, impression mounts were prepared /Klika, 1967/.

RESULTS

The application of the method of membranous impression mounts combined with the SEM techniques offers pictures completing one another and enabling to identify the structure of the inner surface of dura mater and arachnoid.

The findings obtained with method of membranous impression mounts demonstrate in some locations the inner surface of dura mater with an extremely thin discontinuous layer of fibroblasts. They incompletely coat a meshwork of collagen fibers /Fig. 1/. On the contrary it is possible in relatively numerous areas to show the islets of stratified arachnoid cells adhering to the inner surface of dura mater /Fig.2/. The round nuclei of arachnoid can be easely distinguished from elongated nuclei of fibroblasts. The layer of stratified arachnoid cells are torn off. Its margin is distinctly bordered from dural fibroblasts. The pictures demonstrated in rat are of the same character as seen in rabbit and pig.

Dural fibroblasts and stratified arachnoid lining distinctly distinguish from endothelium of sinus sagittalis as it was seen in pig.

Fig. 1 Dura mater of rat, inner surface lined with discontinuous layer of fibroblasts. Membranous impression mount stained with Harris´hematoxylin. 400 times.
Fig. 2 The same mount as in Fig.1; 1-dural fibroblasts, 2-adhering islet of arachnoid cells, 3-their round nuclei. 400 times

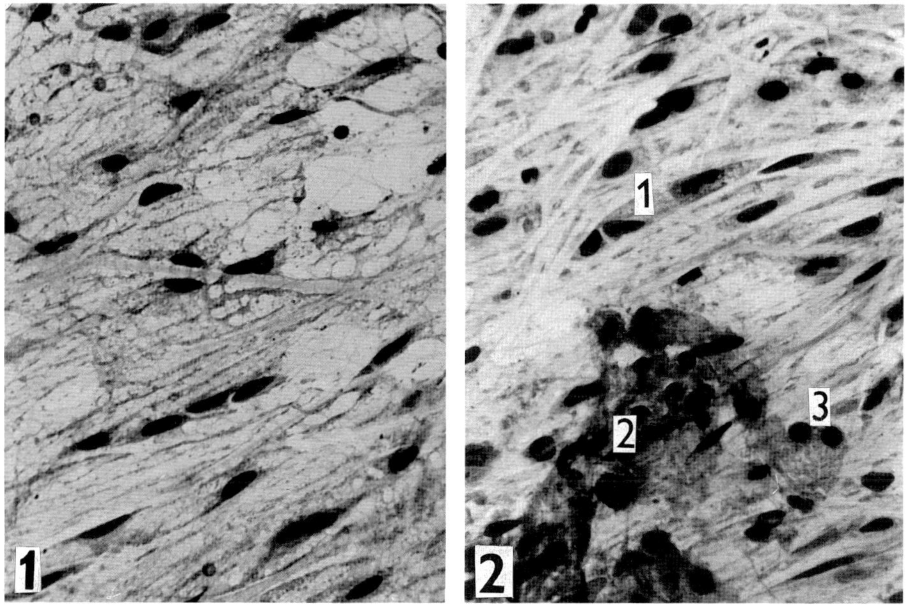

The three - dimensional aproach of SEM to topics under study offers more detailed informations and is also a proof of results obtained on membranous impression mounts. The places can be demonstrated in which the thin processes of fibroblasts coat the collagen fiber system the relief of which protrudes over the inner surface of dura mater /Fig.3/.

In some locations the layer of fibroblasts is discontinuous and collagen fibers meshwork is naked. On the contrary the SEM enables to demonstrate the overlapping of delicate processes which seem to be according to our opinion the most superficial layers of adhering arachnoid cells. In SEM pictures compact layers of cells torn off from arachnoid and adhering to the inner surface of dura mater are seen, too /compare Fig. 4 with Fig. 2/.

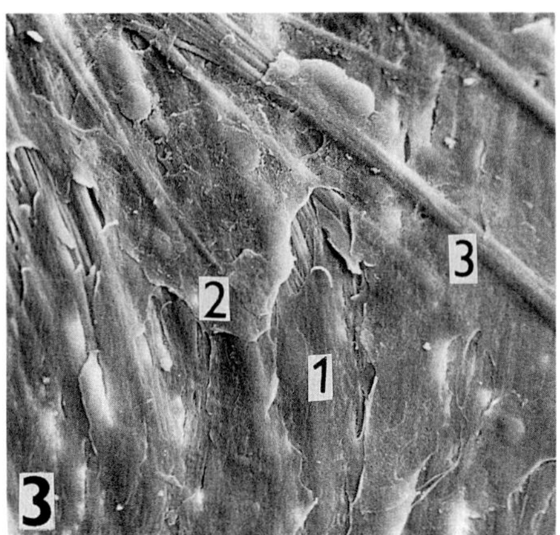

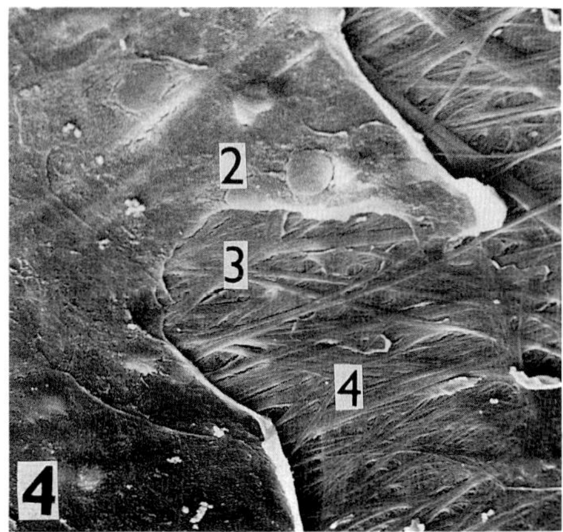

Fig. 3,4 Inner subdural surface of dura mater. SEM, 500 times. 1-dural fibroblast, 2-stripped off arachnoid cells adhering to dura mater, 3-coated, 4-naked collagenous fibers of dura mater.

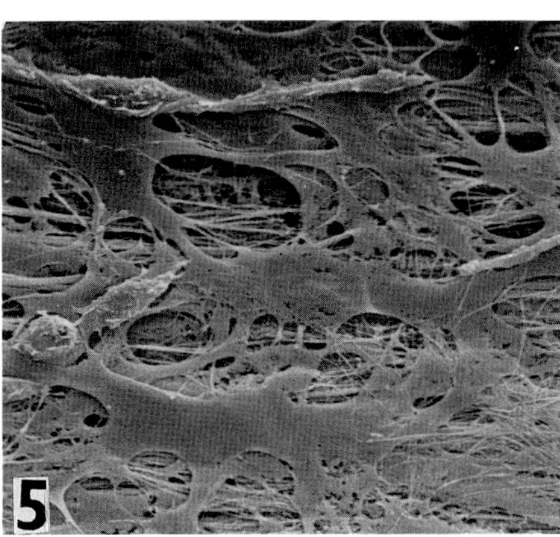

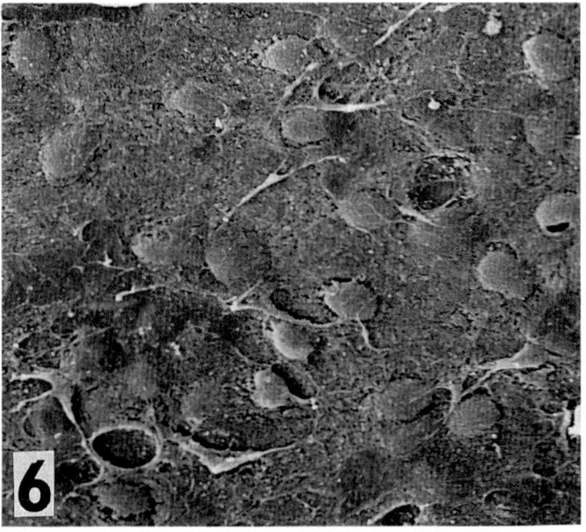

Fig. 5 Discontinuous fibroblast layer and naked collagen fibers meshwork of inner surface of dura mater in rabbit. SEM, 1000 times. Fig.6 Linning of the outer surface of arachnoid in rabbit. SEM, 750 times

The outer subdural surface of arachnoid, stretched on brain was also the subject of SEM analysis. The densely packed, squamous arachnoid cells are characterized by protruding numerous nuclei. Tiny cytoplasmic folds and fragments of cytoplasm are caused by separation of arachnoid during the preparation of the specimen /Fig. 6/.

The SEM analysis is a contribution to our hypothesis /Klika 1967/ of non existence of continuous subdural space in mammals. The intricated relief of the inner compartment of dura mater formed by collagen fibers meshwork is the site of the close contact with arachnoid which copies this relief. SEM confirms the general feature of dura arachnoid complex and its close connection.

REFERENCES

Allen DJ, Didio LJA / 1977/. Scaning and transmission electron microscopy of the encephalic meninges in dogs. J Submicr Cytol 9:1-22.

Andres KH /1967/. Über die Feinstruktur der Arachnoidalzotten bei Mammalia. Z Zellforsch 82:92-109.

Angelov DN, Vassilev VA, Vidinov NK, Chakalski KR /1984/. On the ultrastructure of dura mater in test animals. Comptes rendus de l´Académie bulgare des Sciences 37:395-398.

Clara M /1953/. Das Nervensystem des Menschen. Leipzig:Barth JA

Klika E /1967/. The ultrastructure of meninges in vertebrates. Acta Univ Carol Medica 13:53-71.

Mc Lone DG, Bondareff W /1975/. Developmental morphology of the subarachnoid space and contiguous structures in the mouse. Am J Anat 142:273-294.

Nabeshima S, Reese TS, Landish DMD, Brightman MW /1975/. Junctions in the meninges and Marginal glia. J Comp Neurol 164:127-170.

Oda Y, Nakanishi I /1984/. Ultrastructure of the Mouse Leptomeninx. J Comp Neurol 225:448-457.

Pease DC, Schultz RL / 1958/. Electron microscopy of rat cranial meninges. Am J Anat 102:301-321.

Schachenmayer W, Friede RL /1978/. The Origin of Subdural Neomembranes. I.Fine Structure of the Dura-Arachnoid Interface in Man. Am J Path 92: 53-62.

Zajícová A /1987/. Ontogenetic Development of the dura mater encephali et spinalis of the Laboratory Mouse Mus musculus v. alba. Folia Morphol 32:46-52.

Cells and Tissues: A Three-Dimensional
Approach by Modern Techniques in Microscopy,
pages 395–400

ULTRASTRUCTURAL MORPHOLOGY OF THE HYPOPHYSEAL CLEFT IN SOME MAMMALS .

Silvia Correr, Serena Petrillo, Marco Laureti and Fabrizio Barberini

Department of Anatomy, Faculty of Medicine, University "La Sapienza", Via A. Borelli 50, 00161 Rome, ITALY

INTRODUCTION

The ultrastructure of the hypophyseal cleft, the embryonic remnant of Rathke's pouch, was studied in a few species of mammals: rats and rabbits, in particular, and a small number of cats.

The hypophyseal (pituitary) cleft is situated between the pars distalis and the pars intermedia of the adenohypophysis. It appeared as a thin fissure in most of the animals studied. The cleft is lined on both sides by a continuous epithelium, made up of so-called "marginal cells". The lumen of the pituitary cleft is filled with a colloidal substance of variable density (Hanstrom, 1966; Vanha-Perttula and Arstila, 1970; Ciocca and Gonzales, 1978; Barberini and Correr, 1984). The ultrastructure of the cleft is quite different in the various mammals which we have examined in this study.

RESULTS AND DISCUSSION

Rat marginal cells seen by transmission electron microscopy (TEM) showed a voluminous nucleus with a scarse endoplasmic reticulum, many ribosomes, a small Golgi complex and mithocondria. The posterior marginal cells were very rich in cilia and, by TEM, appeared as a single layer of polygonal elements, held together by junctional complexes (Correr and Motta 1985). By scanning electron microscopy (SEM), the marginal cells appeared rather to be squamous, having a polygonal border and presenting numerous microvilli and a few cilia. The anterior side consisted of polygonal cells with borders rich in microvilli. There were very few ciliated cells. The posterior marginal cells (those lining the side of the pituitary cleft facing the pars intermedia) were generally similar although the majority of the cells possessed many cilia (Correr and Motta, 1981a). Only in the rat we did find an additional type of cell. These cells were located on both sides of the marginal epithelium. Because of their position and similarity to the "supraependymal cells" (Coates, 1973; Mestres and Breipohl, 1976) they have been called "supramarginal cells" (Correr and Motta, 1985). By SEM, these cells appeared as polymorphic elements, some of which possessed a voluminous cellular body with long and thin cytoplasmic processes; others were smaller cells with larger and longer cytoplasmic extrusions (Figs. 1 and 2). Based on these observations, the ultrastructural organization of the rat hypophyseal cleft seems to share similar features with the third ventricle (Scott et al;, 1974), especially with its infundibular recess, with which it is in close proximity (Correr and Motta, 1985). The third ventricle, in fact, appeared to be lined with cells possessing many cilia and microvilli and it appeared similar to the marginal cells lining the pituitary cleft. In addition, the third ventricle presented supraependymal cells on its surface, which were similar to the supramarginal cells of the cleft (Figs. 1 and 2). It has been suggested that the supramarginal cells (likely macrophagic elements) play a role in modulating the secretory activity of the parenchymal cells of the adenohypophysis (Correr and Motta, 1985).

In rabbits the hypophyseal cleft was not constant, often appearing similar to a cystic structure, located between the pars distalis and the pars intermedia (Young et al., 1965). By SEM, the anterior and posterior marginal cells showed characteristic polygonal boundaries. Both sides of the rabbit cleft were lined by ciliated and microvillous cells, but in the posterior wall the elements with cilia were more abundant (Figs. 3 and 4). The pituitary cleft of the rabbit was also similar to the third ventricle. In fact the ventricular tanicytes of the rabbit, located next to the infundibulum, were rather similar to the marginal cells. Whereas the anterior elements of the infundibulum were very rich in microvilli, the posterior cells were very rich in cilia (Figs. 5 and 6).

In the cat, the pituitary cleft showed an anterior epithelium

Fig. 1: A "supramarginal cell" lies on the marginal epithelium of the rat pituitary cleft. Ciliated cell (C), microvillous cell (M) (5000 X).

Fig. 2: A "supraependymal cell" on the epithelium lining the third ventricle (4000 X).

Fig. 3: Anterior side of the pituitary cleft of the rabbit. Observe the presence of microvillous (M) and ciliated (C) cells (4000 X).

Fig. 4: Posterior side of the pituitary cleft of the rabbit. There are more ciliated cells (C) than microvillous cells (M) (5000 X).

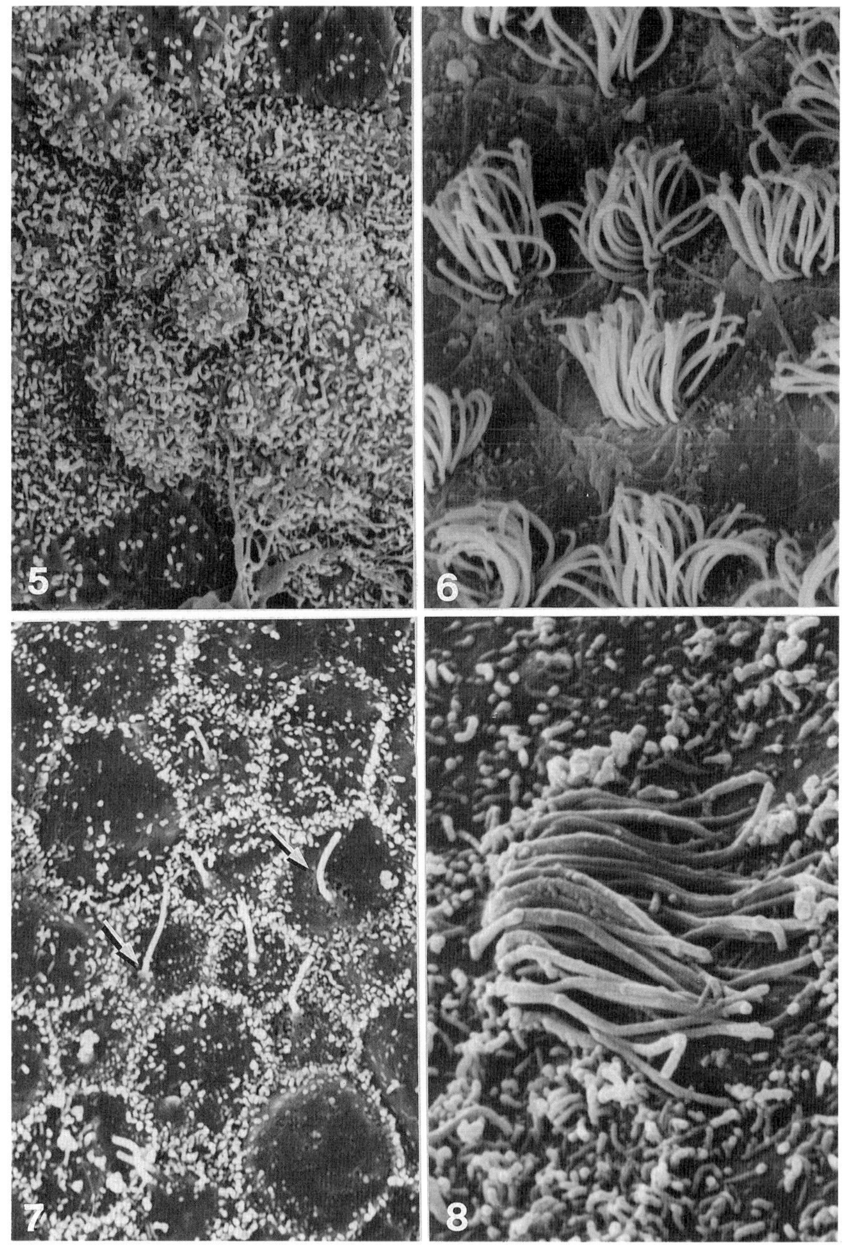
5
6
7
8

lined by polygonal cells with many microvilli and some with a single cilium. The posterior epithelium appeared covered by cells with microvilli and isolated ciliated cells (Figs. 7 and 8).

The hypophyseal cleft, considering its close topographical association with the hypothalamo-hypophyseal complex and the ventricular system of the brain, might possibly play some role in a neuroendocrine regulation of this tract (Correr and Motta, 1981b). Further, as demonstrated in part by this study, the pituitary cleft of the rat not only possesses an ependyma-like lining (marginal cells) but also shows free cells (supramarginal cells) similar to the supraependymal macrophagic cells found mostly in the third ventricle (Figs. 1 and 2). This morphological observation corroborates the suggestion that the third ventricle and the cleft may share some common origin. This might be related to an early and close neuroectodermic interaction during the formation of the hypothalamo-hypophyseal complex (Schechter, 1970; Takor Takor and Pearse, 1975; Correr and Motta, 1985). In considering the area in which the pituitary cleft and third ventricle are located, their cellular linings may assume a strategic role in the modification, recirculation and, possibly, the storage of hormonal hypophyseal products (Correr and Motta, 1981a; 1985).

Fig.5: Anterior side of the rabbit infundibulum. Microvillous cells. Note the different density of the microvilli above them (5000 X).

Fig.6: Posterior side of the rabbit infundibulum. Cells with cilia (5000 X).

Fig.7: Anterior side of the cat pituitary cleft. In the polygonal cells the majority of the microvilli are situated along the borders. Centrally, some isolated cilia are visible (arrows) (5000 X).

Fig.8: Posterior side of the cat pituitary cleft. A ciliated cell between microvillous cells (10000 X).

REFERENCES

Barberini F, Correr S (1984). Ultrastructure of Rathke's pituitary cleft. In: (ed PM Motta), Ultrastructure of Endocrine Cells and Tissues, Martinus Nijhoff Publishers, Boston/The Hague/Dordrecht/Lancaster pp, 57-63.

Ciocca DR, Gonzales CB (1978). The pituitary cleft in the rat: An electron microscopic study. Tissue and Cell, 10: 725-733.

Coates PW (1973). Supraependymal cells: Light and transmission electron microscopy extends scanning electron microscopic demonstration. Brain Res, 57: 502-507.

Correr S, Motta PM (1981a). The rat pituitary cleft: a correlated study by scanning and transmission electron microscopy. Cell Tissue Res. 215:515-529.

Correr S, Motta PM (1981b). The pituitary cleft and associated parenchymal tissue in the rat adenohypophysis as revealed by scanning electron microscopy. In:(ed DJ Allen, PM Motta and LJA DiDio), Three Dimensional Microanatomy of Cell and Tissue Surfaces, Elsevier North Holland Inc. pp 167-181.

Correr S, Motta PM (1985). A scanning electron microscopic study of "supramarginal cells" in the pituitary cleft of the rat. Cell Tissue Res. 241:275-281;

Hanstrom B (1966). Gross Anatomy of the Hypophysis in Mammals. In: (ed GW Harris, BT Donovan), The Pituitary Gland, Butterworths, London, 1: 1-57.

Mestres P, Breipohl W (1976). Morphology and distribution of supraependymal cells in the third ventricle of the albino rat. Cell Tissue Res, 168: 303-314.

Schechter J (1970). A light and electron microscopic study of Rathke's pouch in fetal rabbits. Gen Comp Endocrinol 14: 53-67.

Scott DE, Kozlowski GP, Sheridan MN (1974). Scanning electron microscopy in the ultrastructural analysis in the mammalian cerebral ventricular system. Int Rev Cytol 37: 349-388.

Takor Takor T, Pearse AGE (1975). Neuroectodermal origin of the avian hypothalamo-hypophyseal complex. The role of the ventral neural ridge. J Embryol Exp Morphol 34: 311-325.

Vanha-Perttula T, Arstila AU (1970). On the epithelium of the rat pituitary residual lumen. Z Zellforsch 108: 487-500.

Young BA, Foster CL, Cameron E (1965). Some observations on the ultrastructure of the adenohypophysis of the rabbit. J Endocrinol 31: 279-287.

Cells and Tissues: A Three-Dimensional
Approach by Modern Techniques in Microscopy,
pages 401–417

THE ENDOCRINE HEART

Wolf-Georg FORSSMANN

Department of Anatomy and Cell Biology,
University of Heidelberg
Im Neuenheimer Feld 307, D-6900 Heidelberg (FRG)

INTRODUCTION

Since William HARVEY's ingenious description of the cardiovascular system (Harveii, 1625) the heart has been considered as a pump and as the mechanic and dynamic center of blood circulation. Marcello MALPIGHI's discovery of the capillary circulation namley in the frog lung (Malpighi, 1684) completed a valid morphological concept of the cardiovascular system. Over centuries, many new discoveries about the morphology and functions of the heart were made. One of the most revolutionary discoveries, however, is the recent establishment that the heart is an endocrine organ.

The first steps in the discovery of cardiac hormones have their origin in morphological and functional studies of 1956 published simultaneously and independently. Soon after the introduction of electron microscopy, ultra-morphological studies on cardiac muscles gave evidence for differences between ventricular and atrial myocardiocytes. Bruno KISCH recognized peculiar inclusions in the cells of the atria which were described as round, osmiophilic and dark bodies (Kisch, 1956). Since these secretory granules resemble very much the catecholamine-storing granules of the adrenal medulla, several authors suggested that the specific function of these atrial myoendocrine cells is the production and storage of these biogenic amines. However, atrial catecholamines are now to be ascribed to the dense network of nerves which form a dense plexus around the atrial myocytes and the conductive system. In an exhaustive study, JAMIESON and PALADE (1964) postulated a relationship between the specific atrial granules and an unknown secretory function. Beside these morphological studies, investigations of HENRY, GAUER and REEVES (1956) showed that the atria played an important role in renal functions. The authors observed a strong diuresis when the atria were dilated by an inflatable balloon. They supposed the diuresis to be related to volume receptors of the atria. The same effect - increased diuresis after atrial distention - was maintained after cardiac denervation (Linden and Sreeharan, 1981). Thus it was evident that the observed cardio-renal axis was associated with both a neuronal and a so far unknown hormonal system.

ISOLATION AND CHARACTERIZATION OF CARDIAC HORMONES

In 1976, MARIE and collaborators made an interesting discovery concerning the nature of the atrial cells. They observed that the granulated atrial myoendocrine cells were influenced by changes in the water-electrolyte balance, and showed that a change in water-sodium uptake went along with a significantly altered granular index of the atrial myoendocrine cells, i.e. the number of granules per cell depends on its functional state. Only in 1981, the first experiments of DeBOLD and collaborators on acid atrial extracts revealed that the granules contained a substance with strong diuretic and sodiuretic activities. Almost independently, a vasorelaxant effect was subsequently found by DETH and coworkers (1982) and by NEEDLEMAN's, INAGAMI's and our groups (Currie et al., 1983; Grammer et al., 1983; Forssmann et al., 1983). Hence, the two main functions of the substance in the atrial myoendocrine cells had been ascertained by biological tests. Applying a biological test on the basis of the different degrees of vasorelaxation of renal (strongly sensible to atrial extracts) and mesenteric vessels (almost inert against atrial extracts), our research group, together with Viktor MUTT from Stockholm, isolated the porcine ***cardiodilatin*** (Forssmann et al., 1983), a new peptide which was postulated to be a cardiac hormone. At the same time, FLYNN and coworkers (from DEBOLD's group) published on the structural analysis of the sodiuretic ***cardionatrin*** of the rat. In 1984, KANGAWA and MATSUO isolated and characterized the ***alpha-atrial natriuretic polypeptide*** of the human heart.

A great number of studies on the isolation of new peptides of the cardiodilatin-ANP family was published in 1984. The comparison of the amino-acid sequences obtained for various mammalian species showed that all isolated polypeptides are homologous and belong to one family of peptide hormones (For review see Cantin and Genest, 1985; Forssmann, 1986). Almost simultaneously, several groups analyzed the gene structure, e.g. GREENBERG and collaborators the human one in 1984, who confirmed the above mentioned results. Now, this new polypeptide hormone is generally called **atrial natriuretic peptide (ANP or ANF)**. In our laboratory, we use the term **cardiodilatin (CDD).** The most recent findings in this field of morphology and biochemistry are reported in studies on the isolation and structural analysis of the circulating form of human cardiodilatin (K. Forssmann et al., 1986) and of a peptide of the same family from human urine (Schulz-Knappe et al., 1988) called **"urodilatin"**. Recently, MATSUO and coworkers (see Sudoh et al., 1988) isolated a new peptide of 26 amino acids from porcine brain, which they called **brain natriuretic polypeptide (BNP)**. This peptide is related to cardiodilatin but it is certainly not an expression product of the CDD gene.

MOLECULAR BIOLOGY OF THE ENDOCRINE HEART

Also the molecular biology of the processing of cardiac hormones is known in some detail, several major questions, however, are still open. The cleaving enzyme, which converts cardiodilatin-126 into the C-terminal circulating form of 28 amino acids, has not been definitely characterized and localized. Most probably it is localized in the membrane of the granules and activated during exocytosis when the membrane is exposed to a pH which parallels that of the interstitium. The structure and localization of the CDD gene has been studied: the CDD gene is found on chromosome 1 (p 3.6) and consists of three exons. The primary structure of the pre-prohormone can be derived from a number of cDNA studies. Posttranslational processing occurs during exocytosis, i.e. the circulating cardiodilatin is formed. Few data are available on the enzymatic degradation, namely that a specific renal endopeptidase cleaves cardiodilatin-99-126 between positions 105 and 106, where a cysteine-phenylalanine bond is located. This cleavage alters the ring structure of the molecule and puts an end to its biological activity (see Gagelmann et al., 1988).

Summarizing the results of the biochemical extraction we point out that the human polypeptide is processed as follows: according to the cDNA analysis a large molecular form of 151 amino acids is synthesized; this pre-prohormone contains a signal peptide of 25 amino acids which is removed during or shortly after translation. Then the prohormone of 126 amino acids is transferred to the Golgi apparatus and stored in the secretory granules where its presence is proven by extraction studies and the combined use of different segment-specific antibodies (Figs. 4 and 5). Finally, the circulating form of cardiodilatin (alpha-ANP or CDD-99-126) is formed during exocytosis. It is interesting that the 28-amino-acid-containing peptide is identical in the human, porcine, bovine or feline species whereas in rodents, the methionine residue in position 110 is replaced by an isoleucine.

MORPHOLOGY OF THE ENDOCRINE HEART

The basic morphological features (Fig. 7) of the atrial myoendocrine cells are detected by light-microscopical immunocytochemistry and corresponding electron microscopy. The immunoreactive cardiac hormones are localized close to the nuclei of the myoendocrine cells (Figs. 1-3). These areas at the two poles of the nuclei are called the *perinuclear regions*, where the cardiodilatin-immunoreactive material is represented by intensely stained patches. This area corresponds to the perinuclear zone in which electron microscopy reveals the secretory apparatus, including the GOLGI-complex and a high concentration of secretory granules (Fig. 3). The GOLGI-apparatus is usually surrounded by some progranules and mature secretory granules.

We have found that the thin-walled, trabecular, and extensible parts of the right and left atria are the major sites of the synthesis and storage of cardiodilatin. Macroscopically, these parts of the heart walls are well defined as those of the atrial appendages: the entire wall is exclusively built by myocytes of the myoendocrine type. These distinct topographical regions are illustrated in figure 7. Some scattered cells with myoendocrine features are also found in the rest of the atria, in some parts of the pulmonary vessels and along the cells of the ventricular conductive system. Nevertheless, the atrial appendages of the right and left heart can be considered as a genuine, compact endocrine organ with specific features, which are further demonstrated by ultrastructural means and by electron-microscopical immunocytochemistry. The granular storage of cardiodilatin in the human and in the rat myoendocrine cells

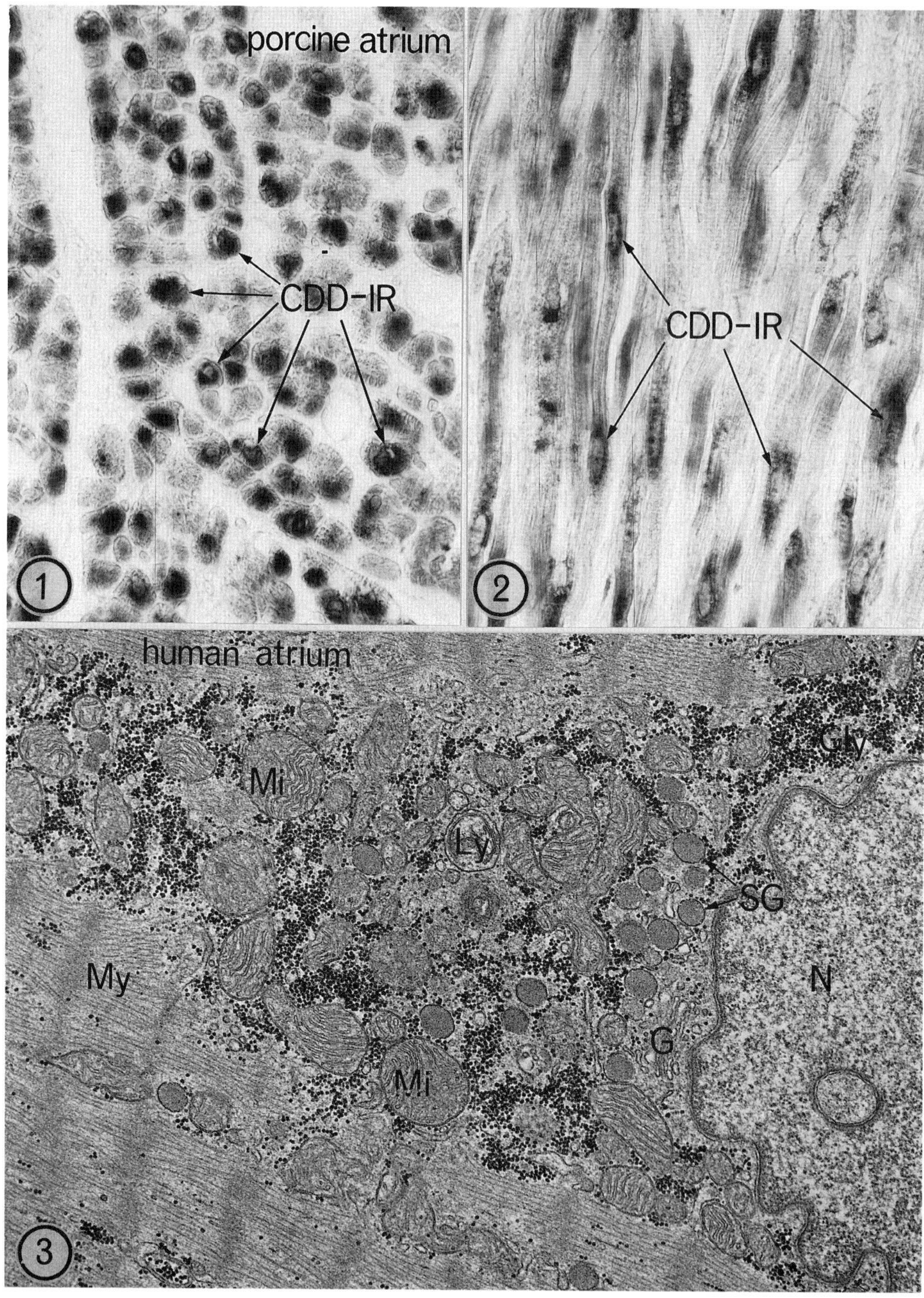
porcine atrium
CDD-IR
1
CDD-IR
2
human atrium
Mi
Gly
Ly
SG
My
N
G
Mi
3

is seen in figures 4 and 5. The markedly labelled mature secretory granules can be distinguished by the immunogold complexes imposed on tissue sections. It is noteworthy that little labelling is seen in the progranules and virtually no labelling in the GOLGI cisternae and the rough endoplasmic reticulum of the normal heart. This is probably due to the low hormone concentrations of unstimulated myoendocrine cells in these cellular compartments.

Fig. 1: Light microscopical demonstration of myoendocrine cells in the porcine atrium. Note the CDD-IR in the perinuclear regions stained by antibodies against CDD (here a C-terminal CDD-99-126 specific antibody is used, however, all epitope-specific antibodies against CDD-1-126 result in identical staining). Here in the transversal section, the spots of CDD-IR are located in the centre of the cells or as ring-like caps around the nuclei. x 530

Fig. 2: Immunohistochemical staining of CDD-IR under the same conditions as in Fig. 1, showing the longitudinally sectioned myoendocrine cells. The perinuclear regions are distinctly stained showing the CDD-IR Golgi complexes. x 530

Fig. 3. Electron micrograph of a myoendocrine cell from human heart showing the secretory apparatus: note adjacent to the Golgi complex (G) specific secretory granules (SG), mitochondria (Mi), lysosomes (Ly), myofibrils (My), and the cell nucleus (N). x 19,300

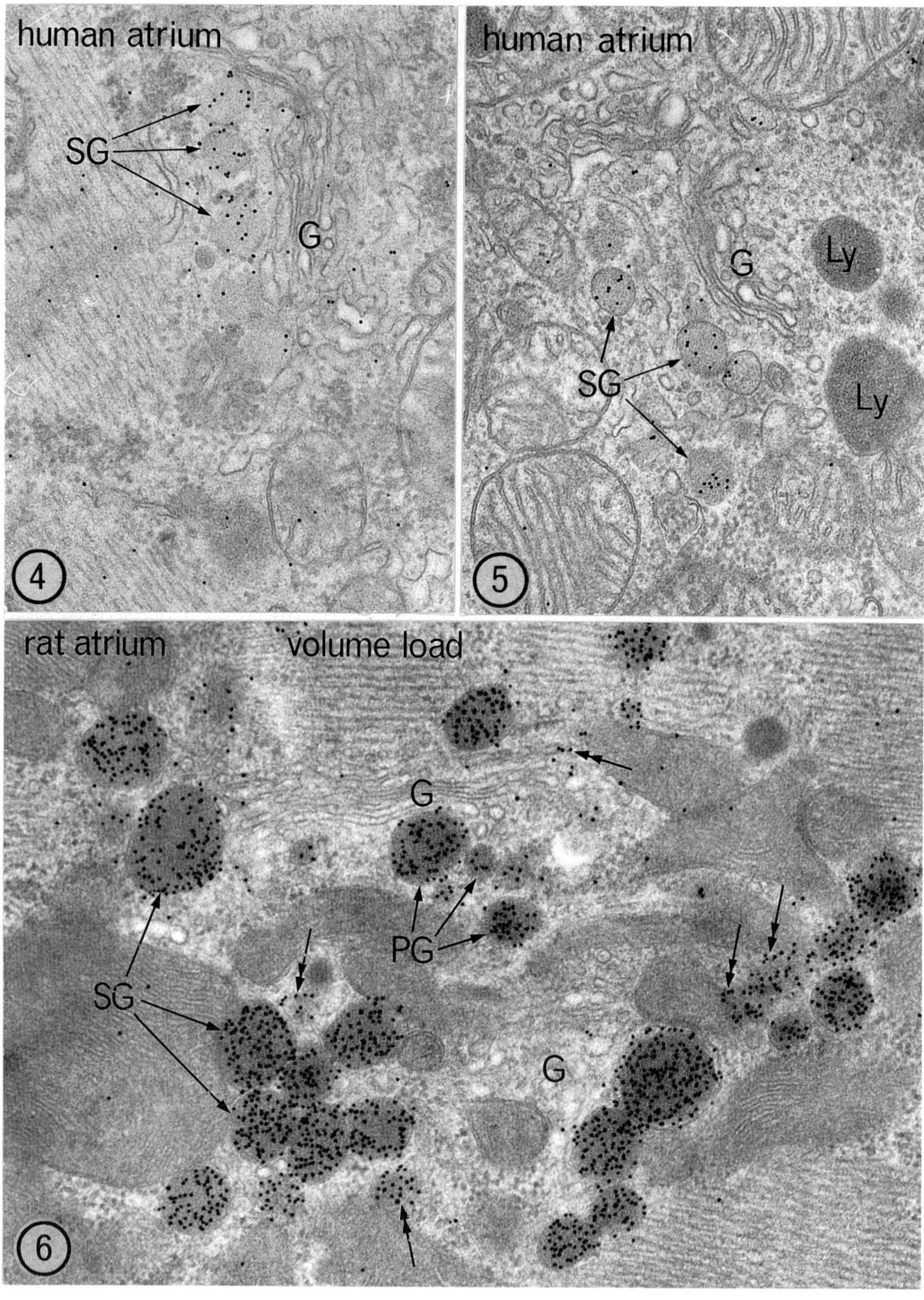
human atrium
SG
G
4
human atrium
G
Ly
SG
Ly
5
rat atrium
volume load
G
PG
SG
G
6

INNERVATION, MICROVASCULATURE, AND CONNECTIVE TISSUE OF THE ENDOCRINE HEART

Beside the myoendocrine cells the morphology of the endocrine heart also comprises the interstitial tissue including the myotendineous junctions, the microvasculature, and a specific innervation.

In our studies we also analyzed the innervation of the atrial myoendocrine cells. Beside a marked innervation by the adrenergic plexus we observed a great number of peptidergic nerves, which are found by immunohistochemistry and also by electron microscopy which confirms the great number of heteromorphic atrial nerves around the myoendocrine cells. The most important neuropeptides found in axons of atrial nerves are vasoactive intestinal polypeptide (VIP), neuropeptide Y (NPY), neurotensin (NT), calcitonin-gene-related polypeptide (CGRP) and substance P (SP).

Fig. 4: Human atrial myoendocrine cell: CDD-IR staining by the immunogold method using a segment specific antibody against pCDD-8-24. Note the specific staining of the secretory granules (SG) which indicates the presence of the prohormone within this storage site. Golgi complex (G). x 36,000

Fig. 5: Human atrial myoendocrine cell: CDD-IR staining by the immunogold method using a segment specific antibody against pCDD-99-126. Note the specific staining of the secretory granules (SG) which indicates the presence of the C-termial portion prohormone within this storage site. Golgi complex (G), lysosomes (Ly). x 36,000

Fig. 6: Rat atrial myoendocrine cell after stimulation by volume load: CDD-IR staining by the immunogold method shows an intense staining of the secretory granules (SG), but also a number of markedly stained progranules (PG) appear. Additionally some Golgi cisternae are laterally labeled (double head arrows) which is not the case in unstimulated control animals. Golgi complex (G). x 43,600

The microvasculature of the endocrine heart exhibits characteristic features. It is well known that the heart, as a rule, contains the continuous type of capillaries. Only WEIHE and KALMBACH (1978), however, detected a few fenestrated capillaries in the heart which are related to the conductive system. Recent studies on the ultrastructure of atrial capillaries show that they are frequently fenestrated in the endocrine atrium.

Another interesting feature of the endocrine heart is its specific relation to the connective tissue. A high number of elastic fibres are found which may be related to the high distensibility of this part of the atria. Also myotendineous junctions were characterized at the borders of the endocrine heart, where the atrial appendages insert into the annuli fibrosi.

SECRETORY CYCLE OF MYOENDOCRINE CELLS

In order to understand the secretory cycle of the endocrine heart the biochemistry of the processing of cardiac peptides of the cardiodilatin/ANP family must be emphasized: the final, biologically important posttranslational product, which is secreted from the atrial myoendocrine cells, is the circulating form of the CDD-molecule which is found in the blood plasma. Amino-acid sequencing of the isolated bioactive and immunoreactive peptide shows only one assayable product (K. Forssmann et al., 1986), which is identical to alpha-atrial natriuretic peptide or CDD-99-126. It represents the last C-terminal 28 amino acids of the stored prohormone cardiodilatin-126.

On the basis of this investigation, we have studied several experimental states of the cardiac-hormone production, i.e. we compared the stimulation and inhibition of the endocrine secretion after volume load or volume reduction. The latter is achieved by submitting animals to thirst or by acute hemorrhagia. Overload is achieved by an activation of the aldosterone mechanism, by exogenous sodium load, or by acute volume load. Acute changes in the body-fluid balance, induced by the injection of plasma-expanding saline into rats, or by the reduction of the blood volume by withdrawal of blood, significantly alter the plasma content of cardiodilatin. Using the combined radioimmunoassay/HPLC method, we can prove, in any of the experimental stages, that only one major molecular form exists in the plasma (Rippegather et al., 1987).

The activation of the secretory apparatus in studies of volume load occurs tremendously fast in the heart. Already three minutes after volume load the secretory apparatus is enormously stimulated in the rat, the GOLGI cisternae contain more progranules and they are easily stained by the immunogold method (Fig. 6). As a morphological sign of the activated secretion exocytotic events are observed more frequently. These are irrefutably proven by combined ultrastructural immunocytochemistry. Hence, the secretory cycle of the myoendocrine cells can be postulated. As in the classical concept of PALADE we distinguish the following steps: (1) the amino-acid uptake, (2) the precursor synthesis at the rough endoplasmic reticulum, (3) the transport and condensation of the secretory product in the GOLGI cisternae, (4) the storage and (5) the exocytosis of the secretory granules.

FUNCTIONAL ROLE OF THE ENDOCRINE HEART

Only few functional aspects of the endocrine heart can be given account to, which are of basic relevance for the cardiovascular regulation and comprise the main functions of this hormone. Among the important effects of cardiac hormones are the varying degrees of sensitivity of the smooth muscle of certain blood vessels. Vascular strips, which have been tested with identical concentrations of synthetic cardiodilatin, respond according to the degree of receptor activation, which is measured by an increase in cGMP and by dilatation of smooth muscle cells. Renal artery smooth muscle is highly sensitive to cardiodilatin-like substances whereas mesenteric and femoral arteries are almost inert. The dilatation of the renal blood vessels may result in an important redistribution of the blood stream in this organ depending on the functional state of the cardiovascular system. Also the distribution of systemic circulation is altered by cardiac hormones.

The functional role of cardiodilatin in renal circulation appears to be evident; the observed diuresis and sodiuresis may largely result from the altered glomerular filtration, which is due to the vasorelaxation of preglomerular arteries. Also an intra-renal redistribution of the blood stream as a cause of diuresis is possible. Furthermore, a tubular effect has to be considered (Schnermann and Briggs, 1988). The intra-renal effect of cardiodilatin on blood vessels can nicely be shown by means of microcinematographic methods. In these *in vivo* studies on the application of cardiodilatin the vasodilatation of a preglomerular cortical artery is seen. This effect has been investigated by STEINHAUSEN and collaborators (Marin-Grez et al., 1986) who found a dose-dependent vasodilatation in all preglomerular segments of the kidney vasculature. However, the vas ef-

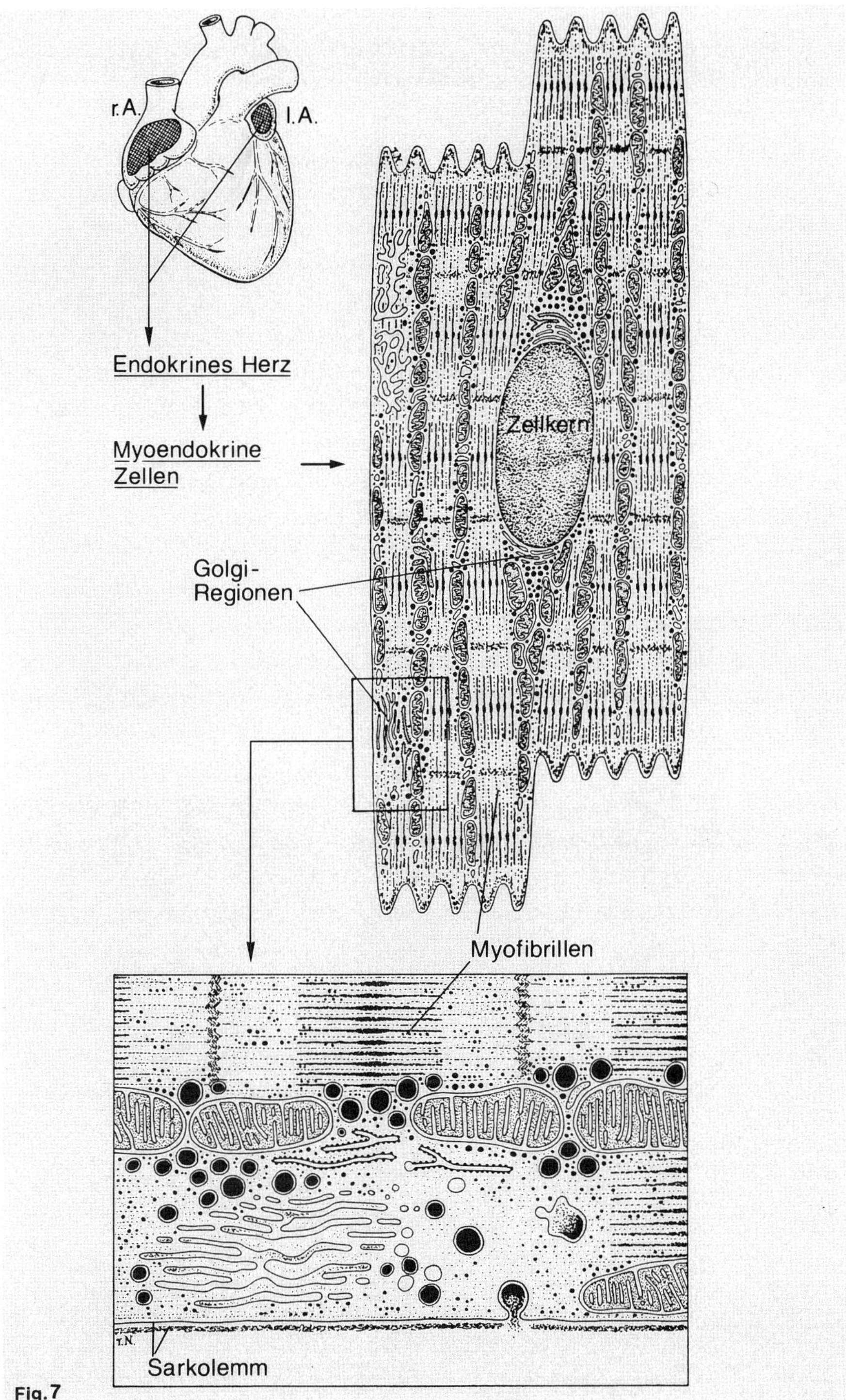

Fig.7

NOTE:Please turn to page 414 for Legend to Figure 7 shown above.

ferens is subject to vasoconstriction, thus increasing the intra-glomerular pressure and an increased glomerular filtration rate may occur.

A clinically important function of cardiac hormones is their effect on the systemic blood pressure. The antagonism with several vasoconstrictive substances, such as catecholamines, angiotensin and vasopressin, is evident (Christmann, 1986). Furthermore, a number of publications show that cardiac hormones inhibit the hypophyseal AVP (vasopressin) secretion, the adrenal aldosterone production, and furthermore the renin secretion from the juxtaglomerular apparatus. This may be a further clue to explain why CDD decreases the systemic blood pressure.

Thus the main function of CDD is of great clinical significance showing that the endocrine heart regulates the blood pressure and blood volume, by a decrease in the blood pressure via vasorelaxation and decreased resistance in the peripheral vasculature. The chronical effects on the kidney result in a decrease in the blood volume via an enhanced excretion of water, sodium and other electrolytes. Therefore the entire system may contribute to a decrease in cardiac overload and may protect the heart from acute overload.

Fig. 7: Schematic drawing of the endocrine heart on macroscopical, microscopical and ultrastructural scales. Note the compact endocrine organ is confined to the atrial appendages, where myoendocrine cells constitute the thin trabecular cardiac wall. Myoendocrine cells contain peri- and telenuclear Golgi regions including the secretory machinery. Exocytosis of the processed hormone is observed at the sarcolemma. Further description see text. (From Forssmann, 1988)

REFERENCES

Cantin M, Genest J (1985). The heart and the atrial natriuretic factor. Endocr Rev 6: 107-127

Christmann M (1986). Die Bedeutung des endokrinen Herzens und der cardialen Hormone für die Regulation des systemischen Blutdruckes der Ratte und der renalen Ausscheidungsfuntion am wachen Hund. Inauguraldissertation, Heidelberg

Currie MG, Geller DM, Cole BR, Boylan JG, YuSheng W, Holmberg SW, Needleman P (1983). Bioactive cardiac substances: potent vasorelaxant activity in mammalian atria. Science 221: 71-73

DeBold AJ, Borenstein HB, Veress AT, Sonnenberg H (1981). A rapid and potent natriuretic response to intravenous injection of atrial extracts in rats. Life Sci 28: 89-94

Deth RC, Wong K, Fukozawa S, Rocco R, Smart JL, Lynch CJ, Awad R (1982). Inhibition of rat aorta contractile response by natriuretic-inducing extract of rat atrium. Fed Proc 41: 983

Flynn TG, DeBold ML, DeBold AJ (1983). The amino acid sequence of an atrial peptide with potent diuretic and natriuretic properties. Biochem Biophys Res Comm 117: 859-865

Forssmann K, Hock D, Herbst F, Schulz-Knappe P, Talartschik J, Scheler F, Forssmann WG (1986). Isolation and structural analysis of the circulating human cardiodilatin (alpha ANP). Klin Wochenschr 64: 1276-1280

Forssmann WG (1986). Cardiac hormones. I. Review on the morphology, biochemistry and molecular biology of the endocrine heart. Eur J Clin Invest 16: 439-451

Forssmann WG (1988). Morphological review: immunohistochemistry and ultrastructure of the endocrine heart. In: W.G. Forssmann, D.W. Scheuermann, J. Alt (eds.), Functional Morphology of the Endocrine Heart, pp. 13-42. Steinkopff, Darmstadt

Forssmann WG, Hock D, Lottspeich F, Henschen A, Kreye V, Christmann M, Reinecke M, Metz J, Carlquist M, Mutt V (1983). The right auricle of the heart is an endocrine organ: Cardiodilatin as a peptide hormone candidate. Anat Embryol 168: 307-313

Gagelmann M, Hock D, Forssmann WG (1988). Urodilatin (CDD/ANP-95-126) is not biologically inactivated by a peptidase from dog kidney cortex membranes in contrast to atrial natriuretic peptide/ cardiodilatin (alpha-hANP/CDD-99-126). FEBS Lett 223: 249-254

Grammer RT, Fukumi H, Inagami T, Misono KS (1983). Rat atrial natriuretic factor: purification and vasorelaxant activity. Biochem Biophys Res Comm 116: 696-703

Greenberg BD, Bencen GH, Seilhamer JJ, Lewicki JA, Fiddes JC (1984). Nucleotide sequence of the gene encoding human atrial natriuretic factor precursor. Nature 312: 656-658

Harveii G (1625) De motu cordis et sanguinis en animalibus. Anatomica exercitatio. Lugduni Batavorum. Ex officina Ioannis Maire

Henry JP, Gauer OH, Reeves JL (1956). Evidence of the atrial location of receptors influencing urine flow. Circulation Res 4: 85-90

Jamieson JD, Palade GE (1964). Specific granules in atrial muscle cells. J. Cell Biol 123: 151-172

Kangawa J, Matsuo H (1984). Purification and complete amino acid sequence of alpha-human atrial natriuretic polypeptide (alpha-hANP). Biochem Biophys Res Comm 118: 131-139

Kisch B (1956). Electron microscopy of the atrium of the heart. I. Guinea pig. Exp Med Surg 14: 99-112

Linden RJ, Sreeharan N (1981). Humoral nature of the urine response to stimulation of atrial receptors. Q J Exp Physiol 66: 431-438

Malpighi M. Opera omnia (Londra, 1686-7; Leida, 1687). cited from: L. Belloni (ed.) Opere scelte di Marcello Malpighi. pp. 75-99 Tipografia Torinese S.p.A., Torino

Marie JP, Guillemot H, Hatt PY (1976). Le degré de granulation des cardiocytes auriculaires. Étude planimétrique au cours de différents apports d'eau et de sodium chez le rat. Pathol Biol 24: 549-554

Marin-Grez M, Fleming JT, Steinhausen M (1986). Atrial natriuretic peptide causes pre-glomerular vasodilatation and post-glomerular vaso-constriction in rat kidney. Nature 324: 473-476

Rippegather G, Maldonado CA, Schulz-Knappe P, Hock D, Lang RE, Forssmann WG (1987). Morphologie der myoendokrinen Zellen des Rattenherzens bei akuten Veränderungen des Blutvolumens. Verh Anat Ges 81: 109-113

Schnermann J, Briggs JP (1988). The physiological importance of atrial natriuretic peptide. In: W.G. Forssmann, D.W. Scheuermann, J. Alt (eds.) Functional Morphology of the Endocrine Heart, pp. 113-122. Steinkopff, Darmstadt

Schulz-Knappe P, Forssmann K, Herbst F, Hock D, Pipkorn R, Forssmann WG (1988). Isolation and structural analysis of "urodilatin", a new peptide of the cardiodilatin-(ANP)-family, extracted from human urine. Klin Wochenschr 66: 752-759

Sudoh T, Kangawa K, Minamino N, Matsuo H (1988). A new natriuretic peptide in porcine brain. Nature 332: 78-81

Weihe E, Kalmbach P (1978). Ultrastructure of capillaries in the conductive system of the heart in various mammals. J Mol Cell Cardiol 192: 77-78

Cells and Tissues: A Three-Dimensional
Approach by Modern Techniques in Microscopy,
pages 419–425

SURFACE MORPHOLOGY OF MYOCARDIUM, PURKINJE FIBERS, AND TRANSITIONAL CELLS

Tatsuo Shimada and Tsuyoshi Noguchi

Department of Anatomy, Medical College of Oita, Oita 879-56, Japan

INTRODUCTION

The ventricular wall of the mammalian heart consists of three main layers: the endocardium, the myocardium and the epicardium. Purkinje fibers act as a rapid pathway for the transition of impulses throughout the ventricles and predominantly located in the subendocardium. THe ungulate heart has a particularly dense network of Purkinje fibers which can be observed in the subendocardium with a dissecting microscope. In order to examine more closely the Purkinje network and the working myocardium, the following methods for scanning electron microscope (SEM) were employed.

After the endocardial endothelium was removed by mechanical abrasion (Izumi et al.,1981; Canale et al.,1983;1983) or by chemical digestion (Shimada et al.,1983; 1984; 1985; 1986), the subendocardial connective tissue was digested with hydrochloric acid (HCl). In the present study, the stromal surface of cells was revealed under SEM following treatment with sodium hypochlorite (NaClO) and then HCl.

Tissues studied in this way were ordinary myocytes, Purkinje fibers, cells of the Purkinje-myocyte junctional region (P-M region) and their associated capillary networks and nerve fibers. Architecture of the P-M region was of particular interest because although described by light microscopy (Tawara,1906) it has not been examined at the electron microscope level.

MATERIALS AND METHODS

Hearts were excised from 10 adult sheep and 5 adult goats under Nembutal anesthesia and immersed in a solution containing 2.5% glutaraldehyde and 2% paraformaldehyde in a 0.1M cacodylate buffer (Karnovsky's fixative) for 1h for SEM. Another group of 3 sheep were used to study the microcirculation and then were fixed by retrograde aortic perfusion with Karnovsky's solution and subsequent immersion in the same fixation for 1hr. Large specimens (approximately 5 x 5 cm) of the left and right ventricular walls were then cut from both groups of hearts, immersed in fresh Karnovsky's fixative for at least 2 h and washed in cacodylate buffer for a further 1 to 3 hr.

In order to uncover the stromal surface of cells, a two steps digestive process was followed. The tissue specimens were first immersed in 3-5% NaClO, or detergent containing NaClO for 40-90 sec at room temperature. This treatment removed the endocardial endothelium. After three rinses in cacodylate buffer, each of 5 min duration, the specimens were immersed in 8N HCl for 20-30 min at 60°C followed by 1 to 2 h at 37°C. Treatment with HCl removed subendocardial connective tissues and basal lamina materials. The tissue specimens were given a final wash in saline solution. They were then postfixed in buffered 1% osmium tetroxide for 2h, dehydrated through a graded series of ethanol and dried by the critical point method. After the specimens were sputter-coated with gold, they were viewed in a JEM-25 or HFS-II scanning electron microscope and photographed at original magnifications of 10-10,000 times.

For gross-anatomical observation, modified periodic acid schiff (PAS) reaction (Otsuka and Hara, 1965) was applied to tissue from 2 sheep hearts.

RESULTS AND DISCUSSION

A dense subendocardial network of Purkinje fibers was observed directly under a dissecting microscope in the sheep ventricles prior to treatment (Fig. 1a). The Purkinje network was clearly visible after PAS reaction (Fig. 1b). The cytoarchitecture of the Purkinje network and the P-M regions was examined in detail under SEM.

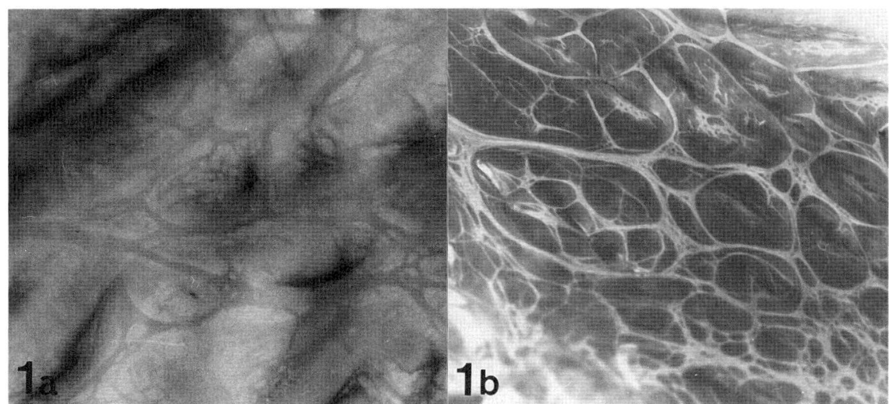

Fig. 1 Gross anatomical images of the Purkinje network in the right ventricles of sheep heart. a:untreatment, b:PAS reaction

Fig. 2 shows a low power SEM image of the Purkinje network in the subendocardium of the right ventricle. In this material the endocardial endothelium was digested with the NaClO treatment. When the fixed heart tissue was briefly immersed in a NaClO solution, only the endothelium was digested-away. The endothelium was affected little by the treatment with HCl, NaOH or trypsin, but following removal of the endothelium, subendocardial connective tissue elements are easily digested by HCl (Evan et al., 1976). HCl also digested basal lamina materials, exposing the stromal surface of cells to study with SEM.

The ventricular myocytes were generally cylindrical in form, arranged in parallel but bifurcated and connected with adjacent cells (Fig. 3). Purkinje strands consisted of a bundle of oval cells which were larger and shorter than ventricular myocytes (Fig. 3). The strands varied in size and appeared as a delicate network resembling a fishing-net (Fig. 2).

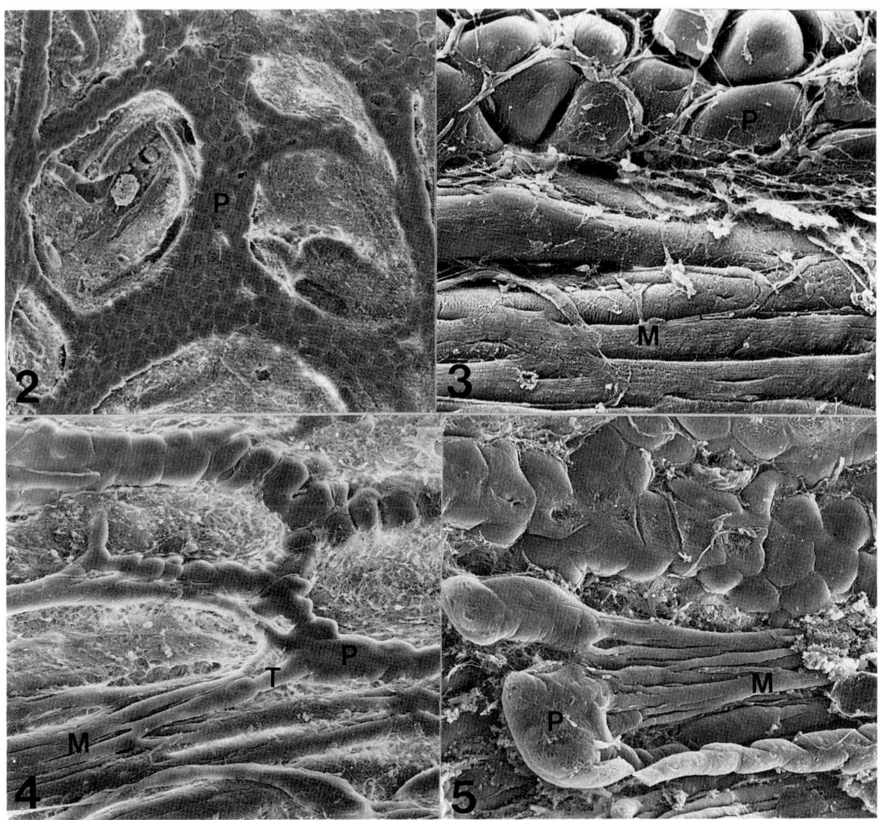

Fig. 2 SEM image of a network of Purkinje strands (P) of the subendocardium of the goat right ventricle. X 65

Fig. 3 Stromal surface of working myocytes (M) and Purkinje fibers (P) in the goat right ventricle. X 430

Fig. 4 P-M region of goat heart. Purkinje fibers (P) are indirectly connected with working myocytes (M) via transitional cells (T). X 130

Fig. 5 Purkinje fibers (P) are directly continuous with working myocytes (M) X 200

The P-M regions showed complex three-dimensional architectures. Purkinje fibers were indirectly connected with working myocytes via intermediary transitional cells (Fig. 4), or directly connected via a complex of anastomosing muscle fibers (Fig. 5).

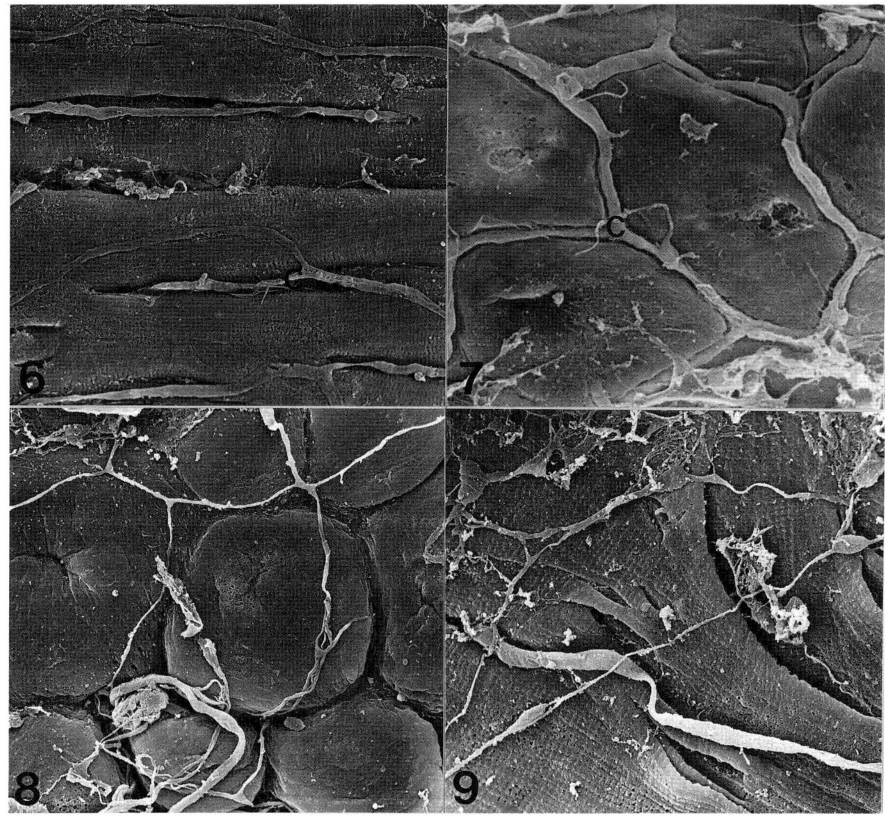

Fig. 6 Blood capillary networks (C) running along the long axis of ventricular myocytes of sheep heart. X 560

Fig. 7 Blood capillary networks (C) surrounding Purkinje fibers of sheep heart. X 760

Fig. 8 Networks of nerve fibers running close to terminal Purkinje fibers of sheep heart. X 1,700

Fig. 9 Networks of nerve fibers over the ventricular myocytes of sheep heart. X 1,400

SEM of vascular casts has been very useful in reconstructing the three-dimensional architecture of blood capillaries (Murakami, 1971), but does not provide information about anatomical relationships between blood capillaries and adjacent cells. In this study, chemical digestion techniques were used to simultaneously expose muscle cells and blood capillaries for observation with SEM. The capillaries of the ventricular myocardium were arranged mainly parallel to the long axis of the muscle cells and formed dense elongated networks (Fig. 7). In contrast, those of the terminal Purkinje fibers formed loose circular networks surrounding the fibers (Fig. 7).

For the first time, nerve fibers have been demonstrated surrounding Purkinje fibers of the subendocardial networks (Fig. 8). These are similar in appearance to those previously observed in false tendons (Canale et al.,1983;1986). Fig.9 illustrates networks of nerve fibers over myocytes.

CONCLUSION

Myocytes, Purkinje fibers and transitional cells of adult ungulate heart were investigated by SEM. The endocardial endothelium was digested with NaClO treatment, and the subendocardial connective tissue elements and basal lamina materials removed with HCl treatment. The myocytes were generally cylindrical in form, arranged in parallel but bifurcated and connected with adjacent cells. In contrast, terminal Purkinje strands were composed of bundles of larger oval cells in a delicate network resembling a fishing-net. These cells were indirectly connected with working myocytes via transitional cells, or were directly continuous with myocytes. Blood capillaries and nerve fibers surrounding myocytes and terminal Purkinje fibers were also observed in close proximity by SEM.

Acknowledgments. We thank Prof. Mitsuo Nakamura (Medical College of Oita), Dr. Gordon R Campbell (University of Melbourne) and Dr. Enlico Canale (University of melbourne) for pertinent advice.

REFERENCES

Canale ED, Fujiwara T, Campbell GR (1983). The demonstration of close nerve-Purkinje fiber contacts in false tendons of sheep heart. Cell Tissue Res 230:105-111

Canale ED, Campbell GR, Uehara Y, Fujiwara T, Smolich JJ (1983). Sheep cardiac Purkinje fibers; configurational changes during the cardiac cycles. Cell Tissue Res 232:97-110

Canale ED, Campbell GR, Smolich JJ, Campbell JH (1986). Cardiac muscle. In Oksche, Vollrath (eds):'Handbook of Microscopic Anatomy' Berlin, New York: Springer-Verlag, pp 60-142

Evan Ap, Dail WG, Dammrose D, Palmer C (1976). Scanning electron microscopy of cell surfaces following removal of extracellular material. Anat Res 185:433-446

Izumi T, Miura K, Hattori A (1981). Myofiber branching in hypertrophic human heart. Biomed Res 2 Supple:265-271

Murakami T (1971). Application of the scanning electron microscope to the study of the fine distribution of blood vessels. Arch Histol jap 32:445-454

Otsuka N, Hara T (1965). Gross demonstration of the mammalian atrioventricular bundle by a periodic acid-Schiff procedure. Stain Technol 40:305-308

Shimada T, Nakamura M, Kitahara Y, Sachi M (1983). Surface morphology of chemically-digested Purkinje fibers of the goat heart. J Electron Microsc 32:187-196

Shimada T, Nakamura M, Notohara A (1984). The Purkinje fiber-myocardial cell region in the goat heart as studied by combined scanning electron microscopy and chemical digestion. Experientia 40:849-850

Shimada T, Kitamura H, Fujimori O, Itose M (1985). Light and electron microscopic studies on capillaries of the goat cardiac muscle with special reference to the topographical relationship between the vessels and Purkinje fibers. Acta Anat 124:127-132

Shimada T, Noguchi T, Asami I, Campbell GR (1986). Functional morphology of the conduction system and the Myocardium in the sheep heart as revealed by scanning and transmission electron microscopic analyses. Arch histol jap 49:283-295

Tawara S (1906). Das Reizleitungssystem des Saugetierherzens. Jena: Fisher.

Cells and Tissues: A Three-Dimensional Approach by Modern Techniques in Microscopy, pages 427–433

MORPHOMETRY OF CORROSION CASTS

A. Lametschwandtner, T. Weiger and G. Bernroider
University of Salzburg, Institute of Zoology (Head: Prof. Dr. H. ADAM), Department of Experimental Zoology, Hellbrunnerstrasse 34, A-5020 Salzburg, Austria (Europe)

Corrosion casts - also termed microcorrosion casts (Hodde and Nowell, 1980) or injection replicae (Murakami, 1975) - represent fillings of hollow space systems with hardening material which withstands the corrosive action of lyes and acids one has to use for maceration and decalcification of tissues and final cleaning of the cast surfaces.

Although quite a number of studies is done with Scanning Electron Microscopy (SEM) of corrosion casts (for bibliography see Lametschwandtner et al., 1984) since the very first report on this technique (Murakami, 1971) we still have to learn much about it.

Morphometry - very simply defined - is a method which helps to solve practical problems in microscopy using stereological principles (Weibel, 1979).

Dealing with morphometry of corrosion casts - in particular with that of vascular corrosion casts, i.e. replicated blood vessels - several basic questions concerning the corrosion casting procedure itself have to be answered before applying morphometric approaches to casted structures.

So one first has to ask What practical problems can be solved with morphometry of vascular corrosion casts at all? This is Which functional variables of the blood vascular system actually can be studied by this method? Which details of these variables then can be investigated with sufficient reliability leading to questions summarizable under What precautions, caveats and corrections have to be considered before doing measurements on corrosion casts. If we then finally know advantages, but moreover disadvantages and pitfalls of the casting technique we then finally may ask What morphometric approach is best suited for a particular cast anatomy.

Remembering that circulation - beside a series of other important functions - has to serve the demands of respiratory gas exchange - 6 variables (see table 1) out of a list are important in closed and open circulatory systems.

Table 1: Functional important variables of the blood vascular system in open and closed circulatory systems.

1. Total volume
2. Blood flow
3. Total number of vessels
4. Ratio of different vessels
5. Spatial arrangement
6. Sinuses, lacunae, interstitial spaces

Because corrosion casting is a static method we cannot receive direct information on variable 2, but by the arrangement of the vascular bed we indirectly get valuable information how blood flow in a particular area is likely to be.

To calculate vessel capacities - or vessel surfaces - we have to use vessel diameter and vessel lengths. Both variables can be affected greatly by injection pressure and shrinkage to name just two factors out of several ones.

So intravasal pressures while flushing the system with appropriate solutions to free it of blood and pressures occuring while injecting casting medium - in generally polymerizing resins (Gannon, 1979) - have to be within the physiological range. Otherwise vascular distentions, leakage or rupture will occure creating evasates. Evasates by their characteristical morphology and the lack of any endothelial imprint patterns clearly can be distinguished from existing vascular structures.

Shrinkage of the injected medium results from polymerization. So the grade of prepolymerization of the injection medium at the time of injection defines the extent of the final shrinkage. The use of very low viscous resins as needed to cast very delicate structures - like bile duct systems - will result in maximal shrinkage. Testing two injection media, methylmethacrylate (MMA) and Mercox-CL-2B (M) we found 20 % volume shrinkage in MMA, but 6 % in M due to the initial differences in viscosity , i.e. the grade of prepolymerization (Weiger et al. 1986). Today detailed studies on the linear shrinkage in horizontal and vertical directions are not yet done. But, knowing the extent of shrinkage we have a correction factor availiable to work with in doing any calculations on casted structures.

To define vascular densities beside number and capacity of vessels the ratio of different types of vessels, i.e. arteries, arterioles, capillaries, venules or veins is of great interest. Information on this variable can be received from fully filled casts only. Fully filled casts - with the exception of avascular regions and areas with angiogenesis - are those casts which reveal no blind ending vascular structures. Vascular sprouts and broken vessels are clearly to be differentiated from unsufficiently filled vessels having characteristic rounded tips.

In corrosion casts arteries and veins can be clearly differentiated by characteristic endothelical imprint patterns (Miodonski et al. 1978). These patterns, however, can be influenced by several factors, whereby a) fixation of vessels, b) replication quality of injection medium, and c) injection pressure are the three most important ones. In a previous study (Weiger et al., 1986) we used a microchip as a model system to test the replication quality of M and MMA. The replication was equally good with both media used and was good to replicate structural details down to a quarter of a micrometer. By using low voltage-high resolution scanning electron microscopy to corrosion casts it should be possible to study this topic in more detail. More interestingly we also could replicate early events of reactive vasodilation occuring in a host vessel approaching the tumor periphery in the case of the Lewis lung carcinoma (GRUNT et al.; 1986a, 1986b).

The preservation of the 3-D arrangement is another problem since the spatial arrangement of vessels defines intervascular distances and branching angles of vessels just to name two variables out of several ones depending upon structural preservation.

Distortion of the 3-D-arrangement is likely to occure by a) application of hot solutions during various preparation steps, b) air-drying of casts, c) heat created by evaporating or sputtering of casts or d) electron bombardment of the casts during SEM inspection. While a, c and d challenge thermal stablity, b) - if no surface tension lowering agents are used - creates high surface tensions to the casts. We therefor recommend to freeze properly orientated casts simply in distilled water and then to cut them either in each desired direction or thickness and/or to freeze-dry them (Lametschwandtner et al. 1984).

If we are finally aware of the variables which can be influenced by the casting procedure, then - and to our opinion then only - it is reasonable to quantify casted structures.

Quantification of corrosion casts can be done by

1. Planimetry
2. Stereophotogrammetry
3. Point counting methods
4. Image analysis systems

Planimetry is the simplest method. Fig. 1 shows the casted subepidermal capillary network of the hagfish skin (Lametschwandtner et al. 1989). Vessel lengths are calculated by tracing them along their midline on a micrograph overlaid to a digitizing pad, which gives the data to a computer which according to a specific software handles them.

Stereophotogrammetry (Boyde, 1973, 1974) is another method. This method however needs a specific stereoscope equipped with a paralax measuring system as well as a specific way of taking micrographs in the SEM (M nnig et al., 1988).

Point counting methods using the principles of stereology (Weibel, 1979) can be applied to flat vascular networks or to cut surfaces (Fig. 2). Then appropriate test grids can be overlaid either to the micrograph or the TV-screen of the SEM directly and calculations can be made by using appropriate stereological formulas.

Image analysis systems are modern tools for quantification of all sorts of structures. In quantitative microvascular casting they are presently used by several groups (Nelson, 1987; Schraufnagel 1987). We sofar have studied the respiratory surfaces of teleost gills (Pohla et al., 1987). These gills consist of gill filaments supplied by a filamental artery giving rise to the secondary lamellae vascular bed (Fig. 3). Having removed these secondary lamellae

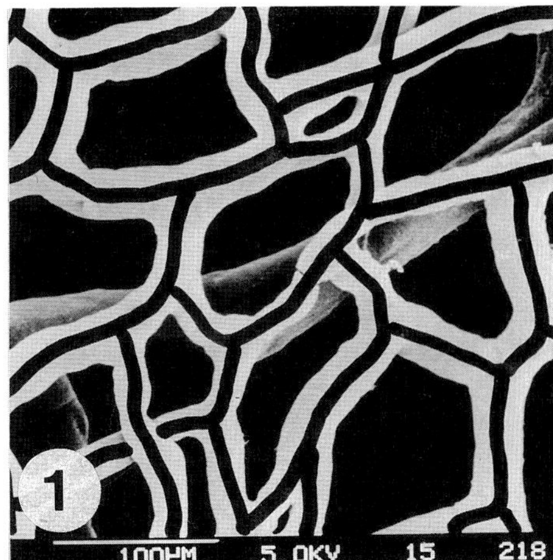

Figure 1. Subepidermal capillary network of the skin of the Atlantik Hagfish, Myxine glutinosa L. Vascular corrosion cast. Black lines along the midline of capillaries result from length measurements by planimetry.

Figure 2. Subepidermal capillary network of the skin of the Atlantik Hagfish, Myxine glutinosa L. Vascular corrosion cast overlaid with a stereological testgrid.

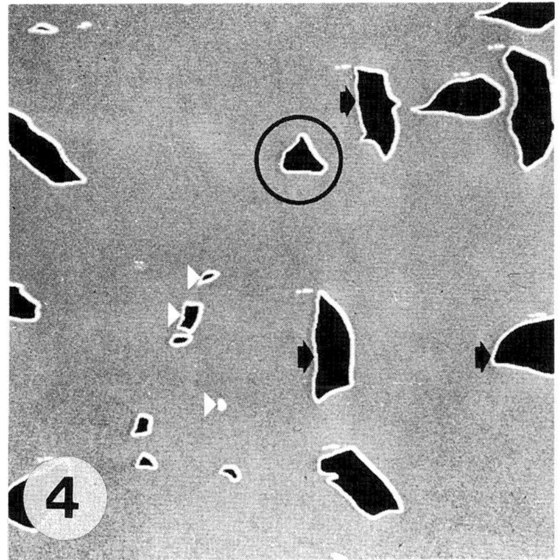

Figure 3. Vascular bed of an isolated secondary lamella of a teleost fish gill. Vascular corrosion cast.

Figure 4. Individual capillary beds of secondary lamellae (black arrows) and broken beds and detritus (white arrows) after segmentation from the background with a "Quantimet 720" image analyzer.

and having suspended them in a drop of water on a microscope slide size, surface area and volume of them were calculated, whereby broken lamellae were discriminated and excluded from measurement (Fig. 4).Another approach is grey level image analysis (Schraufnagel,1987). Here the SEM-image is digitized and structures right in focus are designed white those out of focus black. Thus by the ratio of total area measured to black area an index of vascular density is received one can work with (Schraufnagel, 1987).

Concluding this rather short and fragmentary review we once more stress the need to know more about the methodical procedure of casting, of physical and chemical properties casting media have and on the effects these media exert upon the vessel during the casting procedure, a topic not yet investigated in any detail up to now. If we have this knowledge - then only we can get reasonable data by the application of morphometry to corrosion casts.

REFERENCES

Boyde A (1973). Quantitative photogrammetric analysis and qualitative stereoscopic analysis of SEM images. J. Microc.98:452-471.

Boyde A (1974). Photogrammetry of stereopair SEM images using separate measurements from the two images. Scanning Electron Microsc.1974:101-108.

Gannon BJ (1979). Pre-polymerization of methacrylate corrosion casting media for microvascular replication using ultra violet light. J. Anat 120:665.

Grunt TW, Lametschwandtner A, Karrer K, Staindl O (1986a). The angioarchitecture of the Lewis lung carcinoma in laboratory mice. Scanning Electron Microsc.1986:557-573.

Grunt TW, Lametschwandtner A, Karrer K (1986b). The characteristic structural features of the blood vessels of the Lewis lung carcinoma. Scanning Electron Microsc.1986:575-589.

Hodde KC, Nowell JA (1980). SEM of micro-corrosion casts. Scanning Electron Microsc.1980:88-106.

Lametschwandtner A, Lametschwandtner U, Weiger T (1984). Scanning electron microscopy of vascular corrosion casts - technique and applications. Scanning Elelctron Microsc.1984:663-695.

Lametschwandtner A, Weiger T, Lametschwandtner U, Patzner RA, Adam H (1989). The vascularization of the skin of the Atlantic Hagfish, *Myxine glutinosa* L. Scanning 3:(in press).

Miodonski A, Hodde KC, Kus J (1978). SEM of the

cochlea vasculature. Arch Otolaryngol 104:313-317.

Mönnig B, El-Gammal S, Witsch P, Stanka P (1988). Computer-aided three-dimensional reconstruction of terminal blood spaces in the proximal tibia metaphysis of the growing rat. VIII Intern Symp Morphol Sciences. Roma. p 274.

Murakami T (1971). Application of the scanning electron microscope to the study of the fine distribution of blood vessels. Arch histol jap 32:445-454.

Murakami T (1975). Injection replica scanning electron microscope method for studying the fine distribution of the blood vessels. Jap J. 7:11-18.

Nelson AC (1987). Study of rat alveoli using corrosion casting and freeze fracture methods coupled with digital image analysis. Scanning Microsc.1:817-822.

Pohla H, Bernroider G, Lametschwandtner A, Goldschmid A (1987). Computerized measurement of gill respiratory area with corrosion casts of gill vasculature. Proc 5th Cong Europ Ichthyol, Stockholm 1985.

Schraufnagel DE (1987). Microvascular corrosion casting of the lung. A state-of-the-art review. Scanning Microsc.1:1733-1747.

Weibel ER (1979)." Stereological methods". Vol.1. Practical methods for biological morphometry. London-New York-Toronto: Academic Press, pp 415.

Weiger T, Lametschwandtner A, Stockmayer P (1986). Technical parameters of plastics (Mercox CL-2B and various methylmethacrylates)used in scanning electron microscopy of vascular corrosion casts. Scanning Electron Microsc.1986:243-252.

ACKNOWLEDGEMENTS

This work was supported in parts by the the Stiftungs- und Förderungsgesellschaft der Paris Lodron Universität Salzburg and the "Jubiläumsfonds der Oesterreichischen Nationalbank (Project No 1868)". The authors thank Dr. Arno Laminger and Missis Karin Bernatzky for photographic work.

Cells and Tissues: A Three-Dimensional Approach by Modern Techniques in Microscopy, pages 435–441

MICROVASCULAR ASPECTS OF MYOCARDIAL LYMPHATIC VESSELS IN THE DOG BY SCANNING ELECTRON MICROSCOPY

J.A. Esperança-Pina

Department of Anatomy, Faculty of Medical Sciences, New University of Lisbon, 1198 Lisbon, Portugal

INTRODUCTION

The lymphatic vessels of the heart in the dog consist of three intercommunicating plexuses.

The existence of a subendocardial and subepicardial plexus has been confirmed by every author, but the myocardial plexus remains controversial.

Whereas some authors consider that the lymph circulates in the fissurae of the connective tissue, forming a sort of lymphatic sponge, others deny the existence of lymphatic vessels in the myocardium. We agree, however, with a third group of authors who believe that the myocardium has true lymphatic vessels.

MATERIAL AND METHOD

Ten hearts of dog were studied and injected after the coronary arteries were catheterized and the arterial tree washed with a saline solution at 37°C until the liquid coming out through the coronary sinus was transparent. The specimens were then fixed with 2% glutaraldehyde and, then, injected with Mercox very slowly during 2 to 4 minutes. Polymerization was performed in a water basin to avoid distortions in the vascular mould. Corrosion was done in a 20% sodium hydroxide solution.

The vascular cast was observed with a dissectin microscope in order to choose the part of the specimen for Scanning Electron Microscopy (SEM) study.

The vascular cast was mounted in a metallic stub using colloidal silver paint, then, it was sputter coated with a thin layer of gold, and afterwards examined and photographed with a ISI SX-30 SEM.

OBSERVATIONS

ORIGIN OF LYMPHATIC VESSELS IN THE INTERSTITIAL SPACES

In all our cases the Mercox perfused through the coronary arteries appeared in the interstitial spaces.

The passage to the interstitial spaces was done through the blood capillary wall and these capillaries had calibres between 5 and 8 µm. In the interstitial spaces we could observe two or three lymphatic capillaries converging to one another with calibres between 1 and 3 µm (Fig.1).

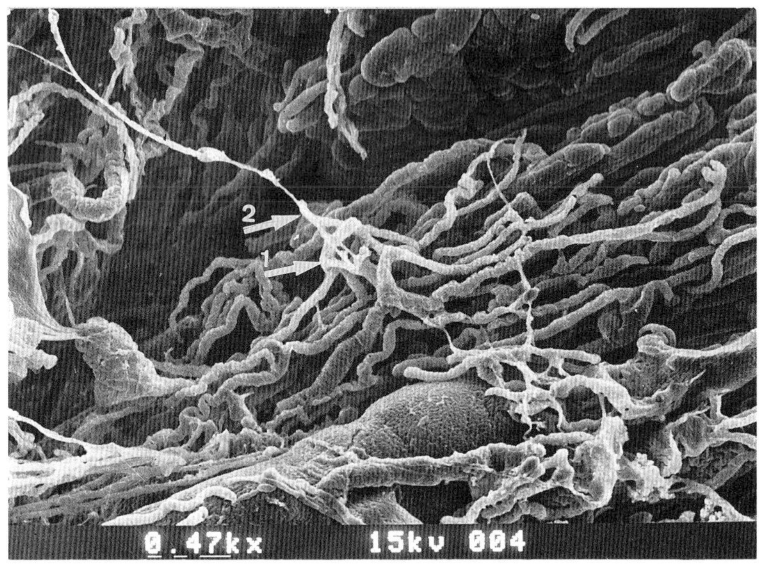

Fig.1. The origin of lymphatic capillaries in the interstitial space (arrow 1) and their convergence to one another (arrow 2) can be seen.

In some cases, the lymphatic capillaries do not exist in the interstitial spaces and the lymphatic vessels get immediately filled up. In other cases, lymphatic nodes in the lymphatic vessels were near the interstitial spaces and immediately after their origin.

The origin of the lymphatic capillaries in the interstitial spaces showed a great variability.

CONFIGURATION OF THE LYMPHATIC VESSELS

The myocardial lymphatic vessels exhibited a cylindroid shape.

CALIBRE OF THE LYMPHATIC VESSELS

Each lymphatic capillary had a calibre between 1 and 3 µm.

RELATION OF THE LYMPHATIC VESSELS

The lymphatic capillaries of the myocardium were perpendicularly disposed to the blood capillaries, where we could observe an arterial, intermediate and venous part.

The lymphatic capillaries were in relation with arterioles, venules and blood capillaries and perpendicularly disposed to them; further they were smaller and fewer (Fig.2)

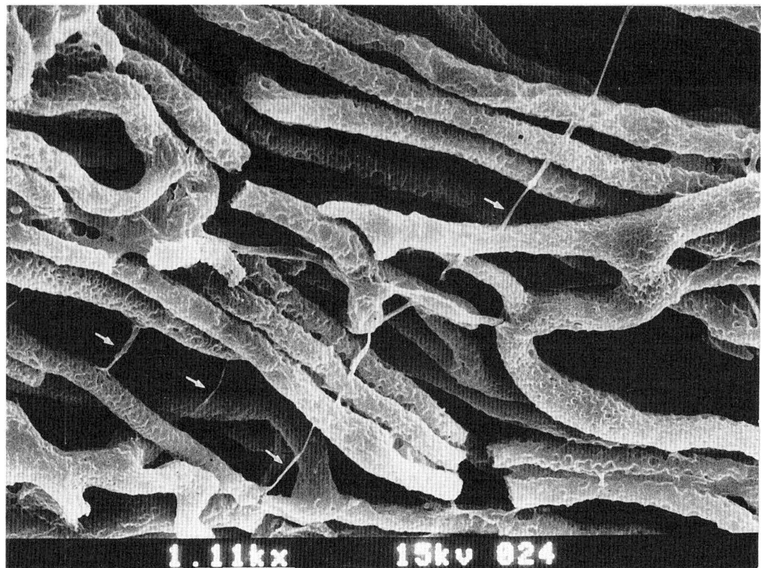

Fig.2. The lymphatic and blood capillaries have a close relation and a perpendicular position (arrows).

LYMPHATIC VESSELS AND CAPILLARIES OF VASCULAR WALLS

Direct relations between lymphatic vessels and the venules of second and first order were frequent.

The lymphatic vessels had an irregular cylindrical shape and their endothelial cells showed an extremely irregular shape.

In some cases, the lymphatic capillaries were in relation with the postcapillary venules and they were perpendicularly disposed to them. These relations as well as the presence of lymphatic nodes, could be clearly observed in high magnifications.

ANASTOMOSIS BETWEEN THE LYMPHATIC VESSELS

The existence of lymphatic capillaries in the myocardium of the dog was confirmed, we could observe that these vessels were perpendicularly disposed to the blood capillaries and parallel to each other. Several transversal anastomoses between adjacent lymphatyc capillaries could be noted (Fig. 3).

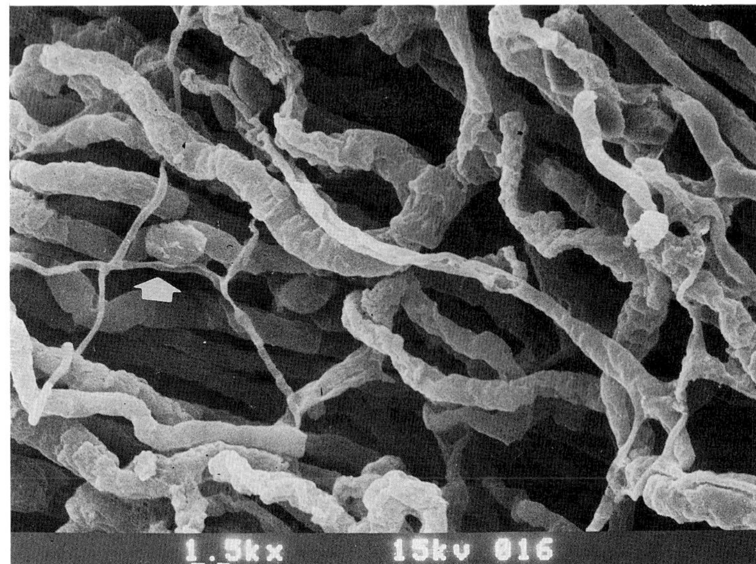

Fig. 3. A transversal anastomosis is seen between two adjacent lymphatic capillaries (arrow).

LIMPHATIC NODES

The lymphatic vessels of the myocardium always presented enlarged zones. These zones corresponded probably to lymphatic nodes separated by regular intervals (fig.4).

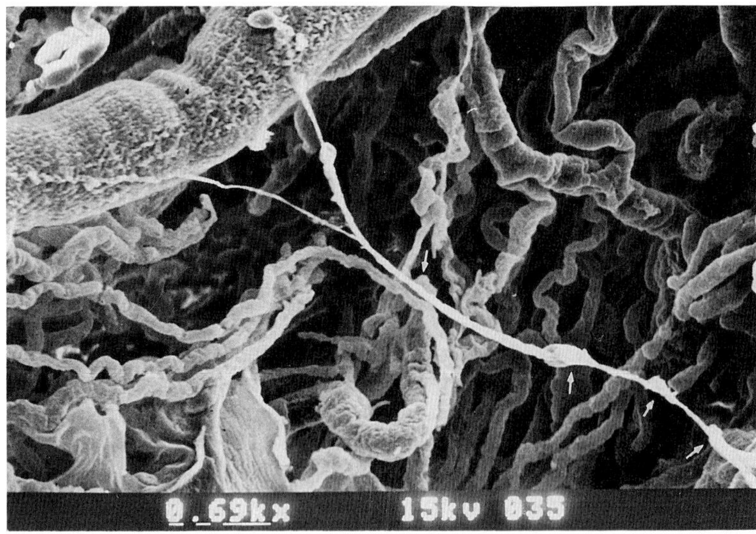

Fig.4. Some lymphatic nodes are separated by regular intervals (arrows).

The venules of 2nd and 1st order were very often in relation with the lymphatic nodes. The lymphatic nodes were well developed almost spherical, and we could not observe more than four or five; they formed all together, the lymphatic nodal groups. They were adjacent to the vascular walls and their axis measure 7 to 12 µm. The lymphatic nodes, in high magnifications could be classified in afferent and efferent vessels (Fig.5).

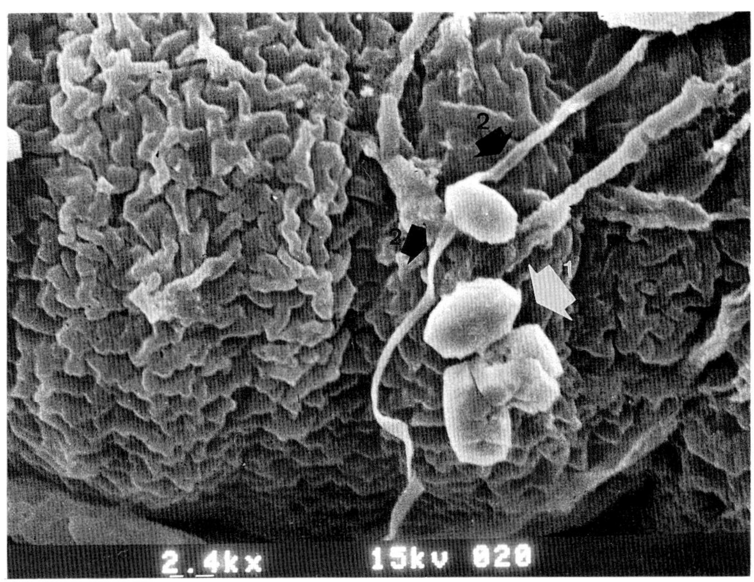

Fig.5. Spherical lymphatic nodes, adjacent to the vascular wall (arrow 1) are noted. One of the lymphatic nodes has an afferent and efferent vessel (arrow 2).

DISCUSSION

It has been proved by several authors that the difficulty to demonstrate the lymphatic vessels of the heart is due to the limitation of the technique used. The injection, after catheterization, is difficult owing to the small calibre and frailty of the vessels, and the existence of valves. The demonstration of the vessels by direct injection in several epicardial zones is also rather limitative.

We used the most physiological technique, as every lymphatic vessel is filled with Mercox injected through the coronary arteries.

In all our cases, we could observe that there was substance in the interstitial spaces coming from the lymphatic plexus.

We could thus confirm the existence of the lymphatic vessels in the myocardium of the dog observing in SEM the vascular casts obtained after corrosion.

The lymphatic vessels had their origin in the interstitial spaces, they showed an irregular cylindrical shape and a calibre between 1 and 3 μm.

The lymphatic vessels were in relation with arterioles, venules and blood capillaries, and they were perpendicularly disposed to them.

The lymphatic vessels were in close relation with the venules and there were many transversal anastomoses between them.

The existence of enlarged segments, corresponding probably to lymphatic nodes, is one of the characteristics of the lymphatic vessels.

The results here obtained were not found in the literature we reviewed.

The existence of lymphatic vessels in the myocardium of the dog is evident, contradicting the authors who deny it, with other techniques.

REFERENCES

Aagard OC. (1925). Les vaisseaux lymphatiques du coer chez l'homme et chez les mammiferes. Copenhagen: Levin-Munksgaard.

Bullon A , Huth F. (1972). Fine structure of lymphatics in the myocardium. Lymphology 5:42-48.

Eberth CJ, Belajeff A (1866). Ueber die lymphgefasse des herzens. Virchows Arch. 37: 124-130.

Eliskova M & Eliska O. (1973). Lymph drainage of the dog heart. Folia Morphologica 22:320-323.

Esperança-Pina JA, Sampaio-Tavares A. (1980). Comparative morphological study of epicardial ventricular lymphatics in the dog before and after ligature of the superficial veins. Acta Anat. 107: 72-79.

Esperança-Pina JA, Sampaio-Tavares A. (1974). Linfaticos sub-epicardicos ventriculares do cao, Folia Anatomica Universitatis Conimbrigensis 43: 1-31.

Johnson RA. (1969). The lymphatic system of the heart. Lymphology 2:95-108.

Johnson RA , Blake,M.T. (1966). Lymphatics of the heart. Circulation 33:137-142.

Kline IK. (1969). Lymphatic pathways, Arch. Path. 88: 638-644.

Miller AJ. (1963). The lymphatics of the heart. Arch. Inter. Med. 112: 501-511.

Patek PR. (1939). The morphology of the lymphatics of the mammalian heart. Amer.J.Anat. 64:203-249.

Rodrigues A, Carvalho R, Pereira S. (1935). Irrigacao (sanguinea e linfatica) do coraçao do cao. Arq.Anat.Antropol.

17: 317-327.

Rouviere R. (1981). Anatomie des lymphatiques de l'homme. Paris, New York, Barcelona, Milan, Mexico, Rio de Janairo: Masson, pp. 224-230.

Rusznyak I, Foldi M, Szabo G. (1967). Lymphatics and lymph circulation. Oxford, London, Edinburgh, New York, Toronto, Sidney, Paris, Braunschweig: Pergamon Press, pp. 79-80.

Salvioli G. (1978). Sulla struttura e sui linfatici del cuore. Arch. per le scienze mediche 2:379-386.

Schmidtova K, Gregor A, Munka V. (1974). Note on lymph drainage of the heart in the dog. Folia Morphologica 22: 405-407.

Uhley HN, Leeds SE. (1976). Lymphatics of the canine papillary muscles. Lymphology 9: 72-74.

Ullai SR, Kluge TH, Kerth WJ, Gerbode F. (1972). Anatomical studies on lymph drainage of the heart in dogs. Ann. Surg. 175: 305-310.

Zhemtchuzhnikova LY. (1953). The lymphatic system of the human heart. Trudy Leningr. san.-gig.med.in-ta. 17-63.

Cells and Tissues: A Three-Dimensional Approach by Modern Techniques in Microscopy, pages 443–450

CORROSION CASTS IN THE MICROCIRCULATION OF SKELETAL MUSCLE

E. Gaudio, L. Pannarale and G. Marinozzi

Department of Anatomy, Universities of Rome "La Sapienza" and of L'Aquila - Italy.

INTRODUCTION

The first studies on skeletal muscle microcirculation were conducted by Ranvier (1874) and Spaltenholz (1888). spaltenholtz's contrast injection (diaphanization) technique provided much interesting information, but it could cause some confusion due to the difficulty in distinguishing vessels on different focal planes.

More recently, intravital microscopy made possible joining morphological and dynamic observations. This provided most interesting information concerning microvascular bed subdivisions and blaod flow in the different microvascular sections (Zweifach and Metz, 1955; Stingl, '76; Myrhage and Hudlickà, '76). But, by this technique, only the superficial part of some muscles of a laboratory animal can be studied, in paraphysiological conditions.

Morphometry applied to histological sections has tried to provide numerical data on the capillary bed. The quantitative approach is the only one that allows the study of the influence of different factors on capillarization. However, sometimes, these studies have been biased by a false structural knowledge of the microvascular bed. For this reason many authors, believing that only longitudinal, mostly straight capillaries are present in the skeletal muscles, have kept relying on transverse sections only (Zumstein et al. 1983).

As we will discuss here, the contribution of corrosion casts has been that of providing an unbiased concept of the microvascular pattern at any level in the skeletal muscle.

Different problems make the study of skeletal muscle microcirculation interesting and all of them must be taken into consideration while one is investigating the features of the microvascular bed. First, not all the muscles show the same metabolism: oxygen consumption and metabolism, even within the same muscle group are variable. Second, work load is variable in different activities. Third, microcirculation is influenced or regulated differently by the nervous system and by local factors. Fourth, microcirculation is influenced by the variation of the muscle belly morphology and by the heightening of interstitial pressure during contraction. For these reasons, we performed our studies on the tibialis anterior and soleus muscles of the rat. These two are very well typified histochemically and physiologically muscles. The tibialis anterior is a fast twitch muscle. It is mainly anaerobic and sustains less prolonged work load than the soleus. On the other hand, the soleus is a slow twitch mainly oxidative muscle (Baldwin and Tipton, '72). We obtained casts from muscles that had been injected both while mantained in situ fully shortened position or fully extended.

Casts made of Mercox CL 2 R resin were prepared by a standard technique, following Miodonski et al.'s (1976) suggestions. We performed observations at different depths and in different zones of the muscle belly microvascular beds.

RESULTS

In this paper we will deal with the terminal network of the vascular bed. For this reason, we will not consider the supplying artery and vein and their branches since in the human being they are not a part of the microcirculation.

The arterioles (fig.1) up to 40 micrometer wide are the largest vessels taken into consideration and are what we called primary arterioles. They run mostly parallel to capillaries. Secondary arterioles (20 to 15 um in diameter) stem from those 40 micrometers wide. The secondary vessels give off side branches, 10 micrometers wide or narrower, along their course mainly transverse to the capillary trend. These side branches directly give origin to capillaries or divide into two short trunks, 55 micrometers long, that, in turn, give rise to the capillary bed. At the origin of the 10 micrometers wide arterial section, the casts show a reduction in diameter or even a groove, like the imprint of a sphincter. All of these vessels show spindle shaped nuclear imprints.

A special feature of the soleus muscle is the possible presence of a different type of configuration at the origin of the capillary bed. The most downstream part of the secondary arteriole gives origin, sometimes,to 6-8 micrometers wide vessels that do not show on their surfaces the typical spindle shaped imprints of the arterious endothelial nuclei. On the other hand, they do show (fig. 2) the round imprints generally attributed to

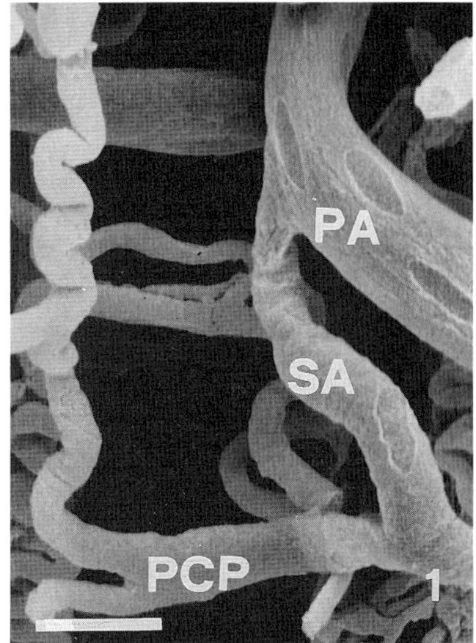

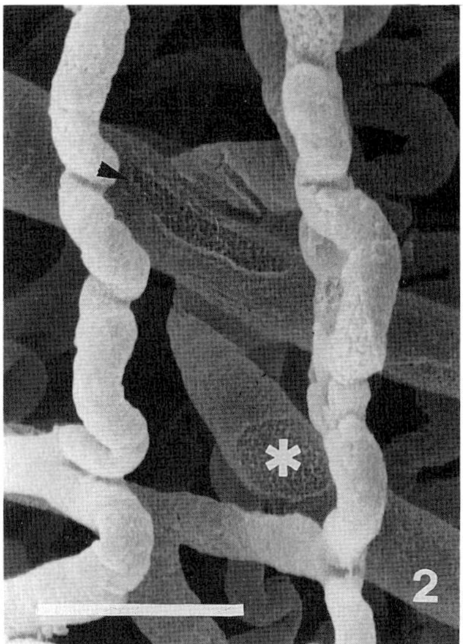

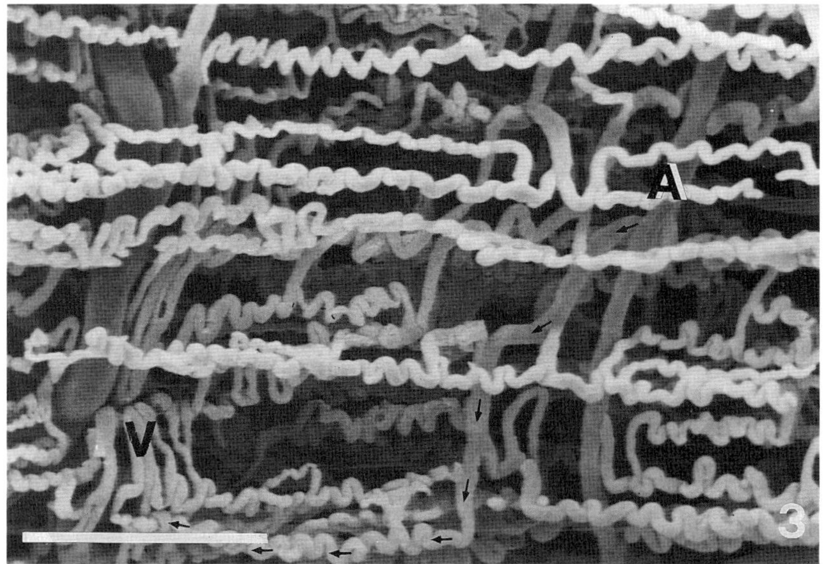

Fig. 1.: soleus muscle. Arteriolar subdivisions. PA= primary arteriole; SA= secondary arteriole; PCP= precapillary arteriole. Bar= 20 micrometers.

Fig. 2.: soleus muscle. Origin of a metarteriole. Notice round venous type endothelial imprint (rosette) on the cast of the metarteriole and spindle shaped endothelial imprint on the arteriole (arrowhead). At the arterious and of the metarteriole a reduction in the cast diameter is present. Bar= 20 micrometers.

Fig. 3.: soleus muscle. A metarteriole (arrowheads) showing an oblique course can be followed from its arteriolar origin (A) to its venular end (V) Bar= 100 micrometers.

the venous endothelium. These vessels are connected at one end with an arteriole and at the other end with a venule. At the origin of the casts of these vessels it is possible to see the same reduction in diameter as at the origin of the 10 micrometers wide arterioles. These hybrid vessels can take two different kinds of courses:

1) they can follow (fig. 3) a longer oblique course running from the arteriole to the venule; in this case they give origin to a number of capillaries that arise only from one side of the vessel;

2) they can run parallel (fig. 4) to the capillaries, anastomosing with them on both sides and getting progressively thinner toward their venular end, where they can be confused with capillaries.

In both cases, at the origin of capillaries and capillary anastomoses one can observe imprints that, as we will see later, are related to the presence of pericytes. The trends and capillary connections of these vessels are very similar to those vessels named "thoroughfare" channels and "metarterioles" described by Zweifach and Metz (1955), but are different in that they do not originate from arterial arcade (Stingl '76).

In the first configuration, they represent not only a preferential channel but, due to the large number of capillaries depending on them and singly controlled by the presence of pericytes at their origins, they represent the only precapillary selective control of blood flow ever described. Precapillary arterioles, in fact, are known to regulate blood flow in a group of vessels (segmental control). The presence of metarterioles, together with the presence of pericytes along the capillary bed, can explain differences in blood flow and perfusion in two adjacent capillaries as found by Eriksson and Myrhage ('72) and Kayar and Banchero ('85).

In the second configuration they certainly represent mainly a preferential channel that is able to supply selectively additional vessels.

In both muscles the capillary bed develops from subsequent tuning fork divisions.

Both in the contracted and in the extended positions the overall capillary patterns of the two muscles appear different. In the tibialis anterior we recognize longitudinal capillaries with a straight or slightly tortuous trend (fig. 5). The tortuosity is generally accentuated in the contracted muscles. Transverse anastomoses of different lengths connect capillaries at different distances. They are more frequent in the venous halves of vessels. In the soleus we (fig. 6) recognize a more complex pattern, because of the presence of a special arrangement of capillaries around spaces without vessels. This is

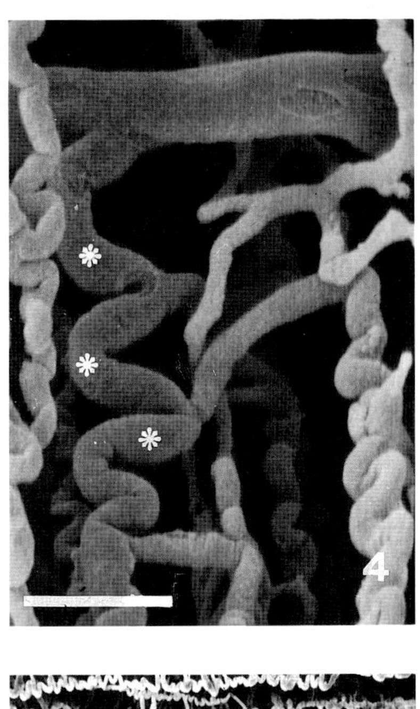

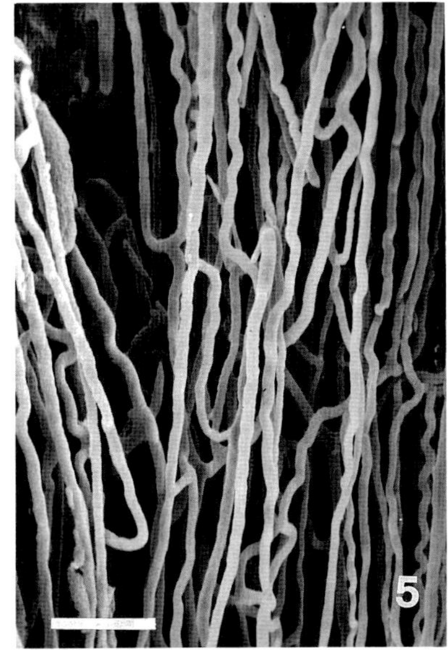

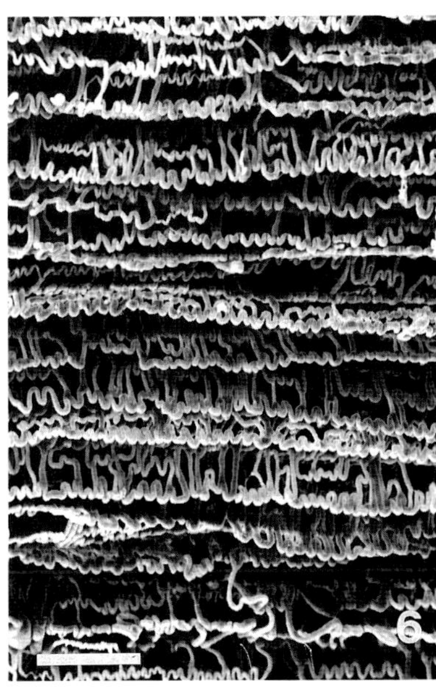

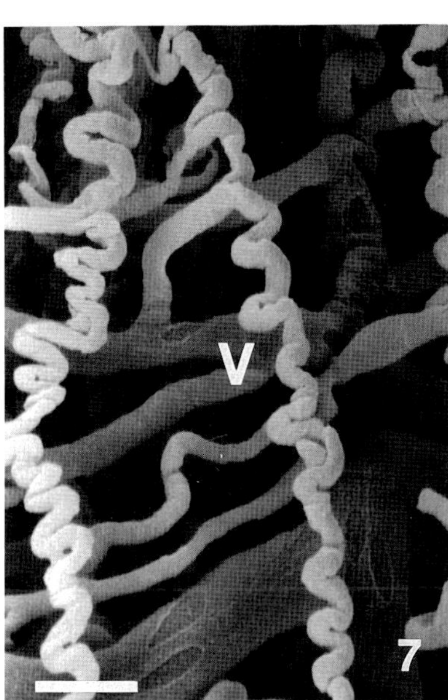

Fig. 4.: soleus muscle. The arterious origin of a metarteriole (rosette) running parallel to capillaries. Notice capillary connections. Bar= 30 micrometers.

Fig. 5.: tibialis anterior. Notice the straight or slightly tourtuous trend of capillaries. Bar= 50 micrometers.

Fig. 6.: soleus muscle. Notice the complex arrangement of capillaries. Bar= 100 micrometers.

Fig. 7.: soleus muscle. The venular tree (V) paralleling the arteriolar pattern. Bar= 20 micrometers.

what we called the capillary cage of the muscle fiber. The cage is composed of three or more tortuous longitudinal capillaries (that, also for their possible function, have been called "main capillaries") and their side branches. The side branches are transverse anastomoses and capillary loops connected at both ends with the same longitudinal capillary. Sometimes, the loops give rise to transverse anastomoses. Only in the very superficial soleus the capillary cage pattern is much less evident. But, even in the distended soleus, the tortuosity of capillaries and the frequency of side branches, in a word, the density of the capillary cast, is higher than in the tibialis anterior.

How can the different arrangements of the capillary bed be explained?

Metabolism is certainly the most important determinant of the capillary density, or, better, capillary volume. For this reason the mainly oxidative soleus is more densely capillarized than the mainly glycolytic tibialis anterior.

Nevertheless, mechanical factors due to the shape and the anatomical situation can cause "squeezing and pinching" of the vessels that certainly must be compensated for,... possibly structurally. If this hypothesis is taken into consideration, one can easily explain the superficial pattern of the soleus microvasculature. Perhaps, the functional complexity of the capillary cage is less needed at that level, because the possibility of sliding under the gastrocnemius minimizes compressive forces. On the other hand, superficial capillaries are not shared on the outer side by oxygen consuming fibers, but simply feed the muscle fascia.

The particular pattern that we have seen in the soleus indicates that the blood supply to the muscle fibers can not, from any point of view, be studied quantitatively, by merely considering the longitudinal capillaries. In the soleus so many side branches of the capillary cage cooperate with the main capillaries in delivering oxygen and metabolites almost all around the fiber. A viable model should take into consideration not the distance from the supplying vessel but the distance from the surface of the structure, in this case the muscle fiber, supposed to be immersed in an homogeneously oxygenate media. From this point of view the Hill's model seems to fit. The same morphology of the soleus microvascular bed, makes it evident that capillarity can not be studied by transverse histological sections.

Postcapillary venules, 12 micrometers in diameter, take rise from two or more capillaries joining together. A single capillary can reach a postcapillary venule or also a larger venous trunk (fig. 7). The venous tree can parallel the arterial pattern or can show a wide range of different arrangements.

As we have mentioned before, both in the tibialis anterior and in the soleus muscle, imprints related with the presence of pericytes, can be observed at the level of longitudinal capillaries branching points and side branches, as well as at the origin of the venular tree. The same kind of feature is present at the point where a capillary is connected to or stems from a metarteriole. Correlation of these imprints on the cast surface with the presence of pericytes have been discussed elsewhere. Here, we will mention that their localizations correspond with those found by others through the use of other techniques.

CONCLUSIONS

As shown by Tilton et al. (1979), pericytes cannot close a vascular lumen, but can reduce its diameter.

In vivo studies have shown that a certain number of capillaries remains intermittently nonperfused by red cells, even during central or periferal nerve stimulation. The influence of precapillary arteriole vasomotion on these situations is only marginal: as observed by Eriksson and Myrhage (1972), they perform only a segmental control of the capillary bed (or an undiscriminating control of blood flow through all the vessels downstream at the same time).

We can hypothesize regulation of red cell flow by pericyte contraction if these cells can reduce the vascular diameter under the threshold of 2.45 micrometers, that is the minimum diameter of a tube through which a normal rat eritrocyte can pass. Most of the imprints at the level of branching points, and at the level of venular and metarteriolar ends, make the cast show that diameter. That is like saying pericyte contraction could stop red cell flow. Contraction at the level of tuning fork divisions, and at the metarteriolar origin of capillaries, influences perfusion and speed of flow in the single capillary; contraction at the venular end could slow down and even stop red cell column; a transient diameter reduction at the level of transverse anastomoses could cause the flow of the plasma only.

Summing up the microvascular pattern of the soleus muscle (presence of the capillary cage and of metarterioles) and the supposed action of pericytes, we develop the idea of an extremely dynamic adaptability of the microvascular bed in this muscle. Metarterioles, where present, can regulate blood flow through every single capillary; thoroughfare channels can spread their high flow through capillaries at their sides; finally, the main capillaries can act as thoroughfare channels if access to their side branches is reduced below a certain level. The extreme flexibility of the vascular features of the soleus muscle would enable it to respond to a very wide range of extra metabolic demands.

REFERENCES

Baldwin KM and Tipton CM (1972): Work and metabolic patterns of fast and slow twitch skeletal muscle contracting in situ. Pflugers Arch. 334: 345-356.

Eriksson E and Myrhage R (1972): Microvascular dimensions and blood flow in skeletal mucle. Acta Physiol. Scand. 86: 211-222.

Gaudio E., Pannarale L., Marinozzi G. (1985): An SEM corrosion cast study on pericyte localization and role in microcirculation of skeletal muscle. Angiology 36 (7): 458-464.

Kayar SR and Banchero N (1985): Sequential perfusion of skeletal muscle capillaries. Microvasc. Res. 30: 298-305.

Miodoński AJ, Hodde CK, Backker C (1976): Rasterelektronenmikroskopie von Plastik-Korrosion-Prapareten: Morphologische Unterschiede zwischen Arterien und Venen. Beitr Elektronenmikroskop Direktabb Oberfl (Munchen) 9: 435-442.

Myrhage R and Hudlickà O (1976): The microvascular bed and capillary surface area in rat extensor hallucis proprius muscle (EHP). Microvasc. Res. 11: 315-323.

Pannarale L., E. Gaudio, G. Marinozzi (1986): Microcorrosion casts in the microcirculation of skeletal muscle. SEM III : 1103-1108.

Ranvier L. (1874): Note sur les vaisseaux sanguins et la circolation dans les muscle rouges. CR Soc. Biol. 26: 28-31.

Spaltehotz W (1888): Die Vertheilung der Blutgefasse im Muskel Abhandl math phys Cl sachs. Gesellsch Wissench 14: 507-535.

Stingl J (1976): Fine structure of precapillary arterioles of skeletal muscle in the rat. Acta Anat. 96: 196-205.

Tilton RG, Kilo C, Williamson JR, Murch DW (1979): Differences in pericytes contractile function in rat cardiac and skeletal muscle microvasculatures. Microvasc. Res. 18: 336-352.

Zweifach BW and Metz DB (1955): Selective distribution of blood through the terminal vascular bed of the mesenteric structure and skeletal muscle. Angiology 6: 282-290.

Zumstein A, Mathieu O, Howald H, Hoppeler H. (1983): Morphometric analysis of the capillary supply in skeletal muscles of trained and untrained subjects - Its limitations in muscle biopsies. Pfleugers Arch. 397: 277-283.

Cells and Tissues: A Three-Dimensional
Approach by Modern Techniques in Microscopy,
pages 451–456

MORPHOLOGICAL ASPECTS OF NASAL BLOOD VESSELS

Gerhard Grevers, Ulrich Heinzmann, Joana Grevers-Colorian
Dept. of Otorhinolaryngology, the University of Munich,
Klinikum Großhadern, Marchioninistr. 15, 8 München 70
Federal Republic of Germany

During the past 20 years, electron microscopic evaluations have been of fundamental importance for a variety of new findings in the morphology of both normal and pathologic nasal mucosa. Several investigators have studied the appearance of this tissue under various conditions using either Transmission (TEM) - or Scanning (SEM) - Electronmicroscopy. However most authors working on ultrastructural aspects of the nasal mucosa restricted themselves to a description of the respiratory or olfactory epithelium, hardly mentioning the underlaying lamina propria.
During the last years we have studied the vasculature of the nasal mucosa using different techniques in order to evaluate the morphological peculiarities of this tissue, which is responsible for a variety of functional behaviours of the nose.
The specimens - septal mucosa of adult rabbits - were prepared for TEM and SEM; furthermore vascular corrosion casts were received (for reference see Grevers and Herrmann 1987, Grevers and Heinzmann 1988a,b).
Evaluation of the specimen prepared for TEM revealed significant differences in the appearance of the endothelial lining of capillaries and - occasionally - large venous sinuses, i.e. capacity vessels. Capillaries with fenestrated endothelium were found mainly in the respiratory part (Figs. 1); in the olfactory mucosa only a few fenestrated capillaries could be encountered displaying very few diaphragmed fenestrations (Figs. 2a,b). Additionally, fenestrated endothelia could be verified in the subepithelial walls of some of the large veins of the respiratory mucosa (Figs. 3a,b,c) - an untypical feature in muscularized veins of this size. These fenestrations might be transient specialisations for certain physiological needs such as increased vascular permeability.

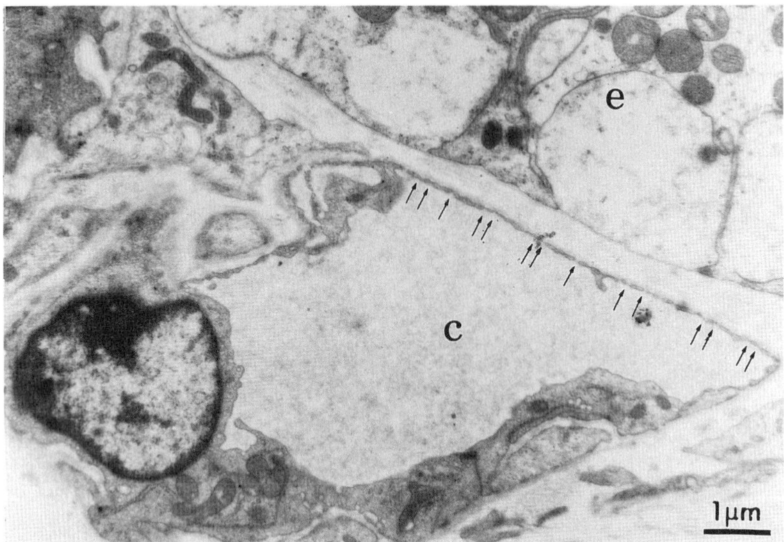

Fig. 1. Capillary (c) of the respiratory mucosa facing the adjacent epithelium (e) with its attenuated, fenestrated parts (arrows). Magnification x 8960

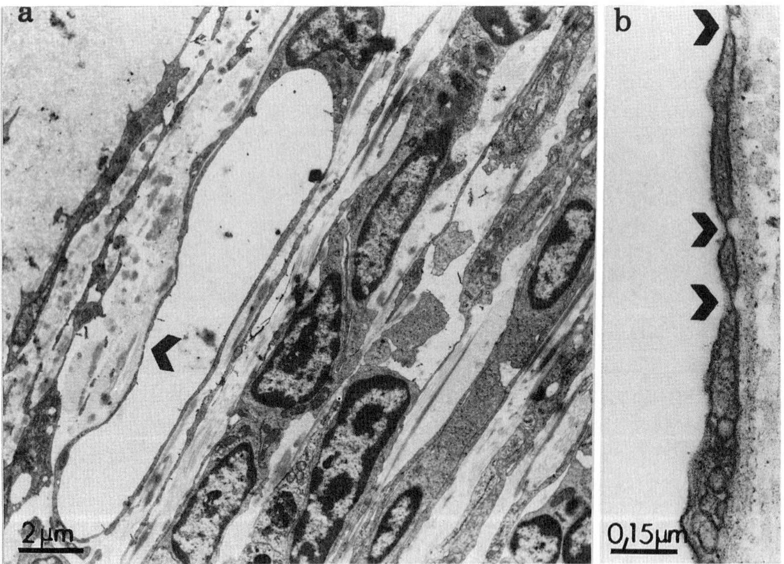

Fig. 2a. Fenestrated capillary of the olfactory mucosa located close to the osseous nasal septum. Note the very few fenestrations (arrowhead) in the endothelial lining. Magnification x 5000

Fig. 2b. Higher magnification reveals a better impression of the diaphragmed fenestrations (arrowheads). Magnification x 63000

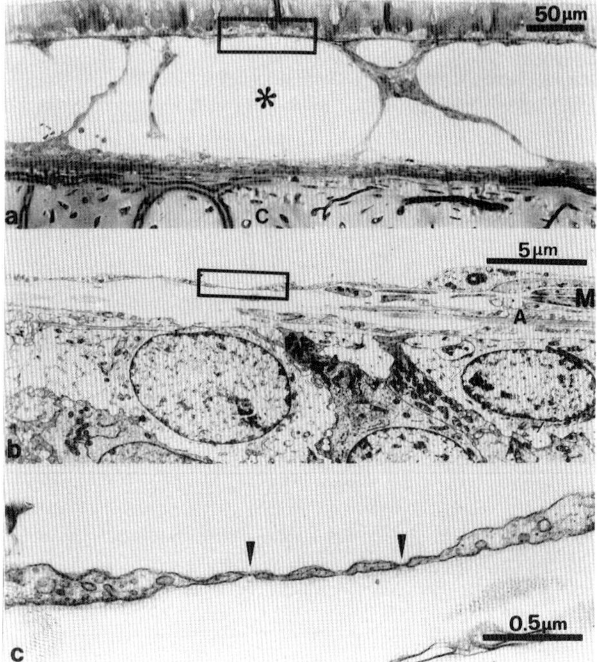

Fig.3a Semi-thin section of the respiratory mucosa showing several large veins which span almost the entire thickness of the lamina propria (star). cartilage (c). Magnification x 320

Fig. 3b Low power electron micrograph of an area corresponding to that boxed in the preceding micrograph. It illustrates a venous wall segment which faces the respiratory epithelium (below) and consists of an extremely thin, partly fenestrated endothelium. The latter, however, is underlaid by a three-layered media (m) at the far right end of the micrograph. Note also the closely adjacent axons (a.). Magnification x 5500

Fig. 3c High power view of the boxed area in figure b. Typical diaphragmed fenestrae (arrows) can be seen. Magnification x 55000

Vascular corrosion casts enabled us to explore further aspects of endonasal angioarchitecture. In the perichondral area, a dense capillary network could be demonstrated (Fig. 4), not allowing to examine the underlaying vasculature. Contrary to these findings the subepithelial capillaries were less in number and density (Fig. 5); therefore one could distinguish easily the vasculature in the deeper layers. The arterial or venous character of larger vessels could be ascertained by the special appearance of the endothelial cell nuclei (Fig. 6) as well as the course of the vessels.

Finally SEM-studies were performed. Using this technique, intraluminal aspects of the endonasal vasculature could be visualized (Fig. 7), as well as junctions between vessels of different sizes (Fig. 8).

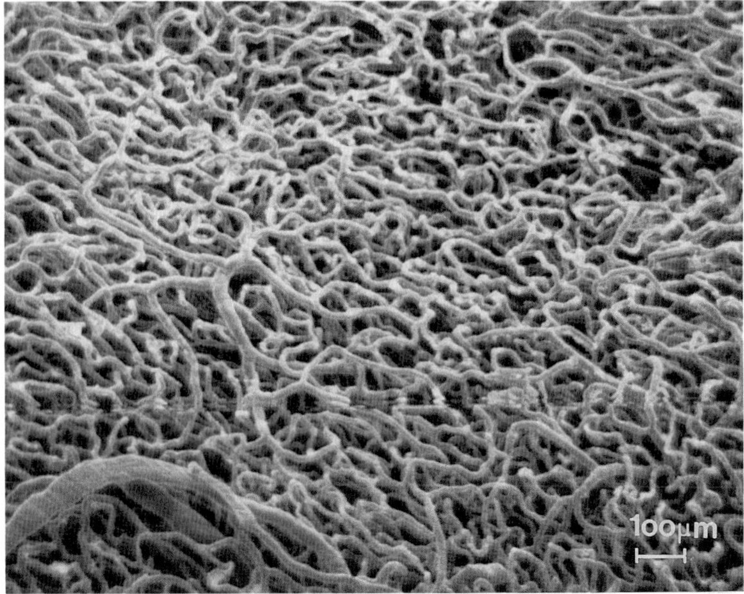

Fig. 4 Dense capillary network in the perichondral area of the nasal septum. Magnification x 70

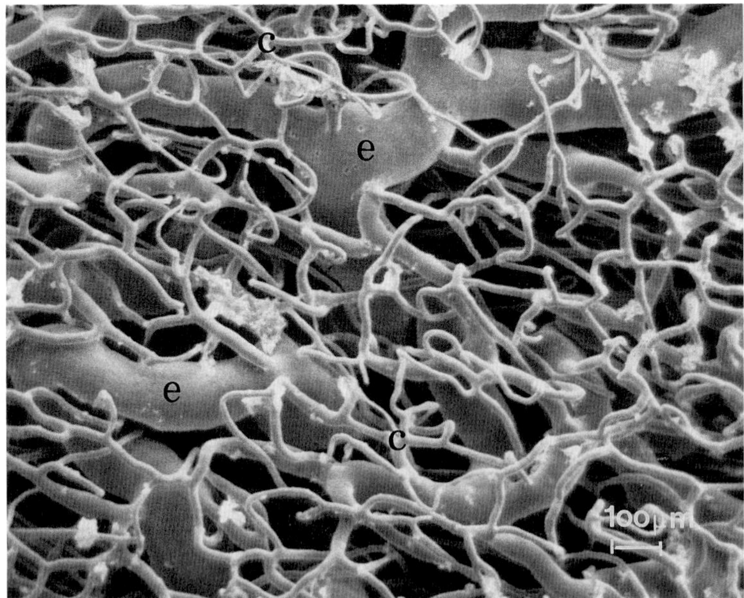

Fig. 5 Subepithelially the capillaries (c) are less in number and density. Venous vessels (e). Magnification x 70.

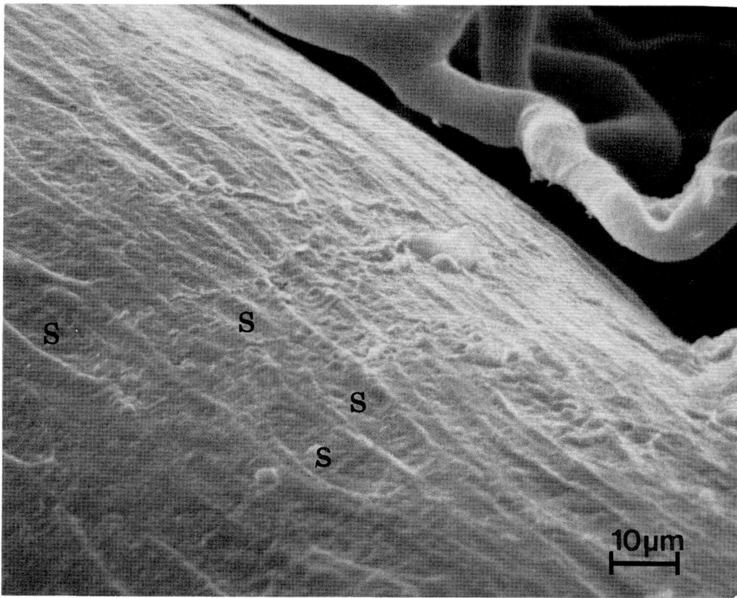

Fig. 6 Arterial blood vessel of the nasal septum. Note the spindle-shaped endothelial cell nuclei (s). Magnification x 1000

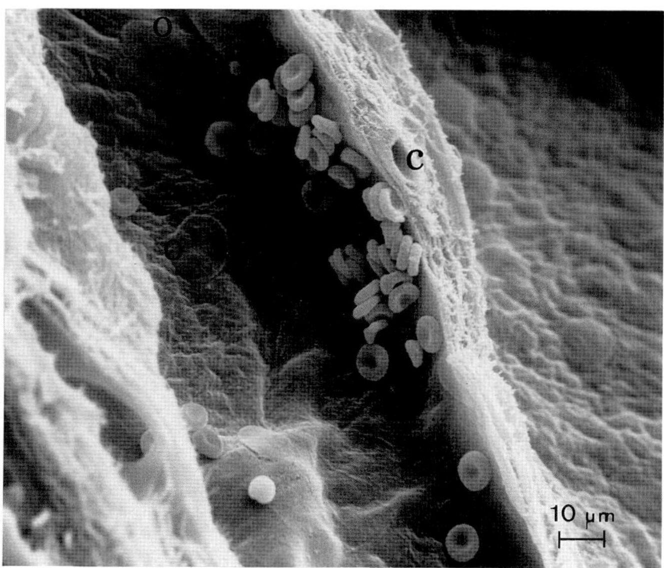

Fig. 7 SEM-micrograph of the vasculature of the nasal septum; note the capillary (c) between two venous blood spaces. The veins can be identified by the appearance of their endothelial cell nuclei (o). This technique offers a unique chance to examine the intraluminal aspects of blood vessels. Magnification x 615

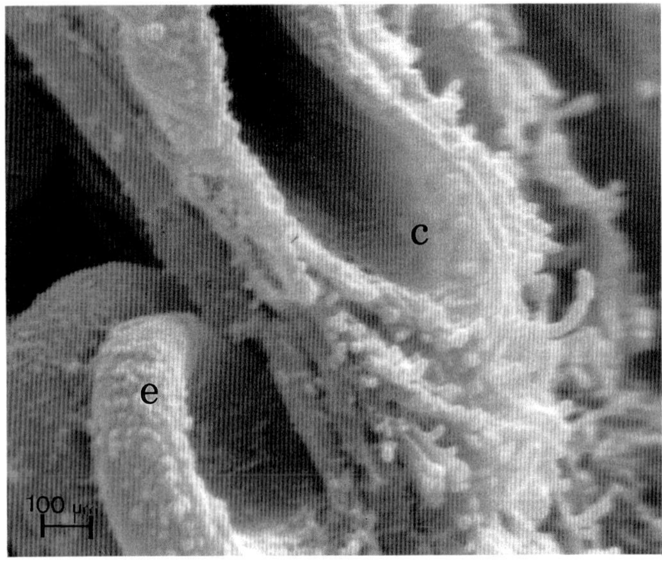

Fig. 8 Higher magnification of Fig. 7. The diameter of the capillary vessel (c) equals the size of an erythrocyt (e). Magnification x 3660

Still - neither TEM nor SEM will be available for routine clinical use in the nearest future for a variety of reasons. Nevertheless these morphological techniques might be of further importance for basic research in all fields of medicine.

Suming up our findings it can be pointed out that the use of different morphological techniques is necessary to understand the complicated morphology of the endonasal vasculature. Some of our results might be of functional significance - like the subepithelial fenestrated capillaries which have already been described in vessels of the small intestine by Horstmann (1966). The appearance of fenestrated capillaries in the subepithelial layers of the tissue mentioned above support the hypothesis of a functional differentiation of the endothelium for improved vascular permeability and fluid exchange.

References

Grevers G., Herrmann U. (1987). Fenestrated endothelia in vessels of the nasal mucosa. Arch Otorhinolaryngol 244: 55-60

Grevers G., Heinzmann U. (1988 a). Scanning electron microscopic studies of the nasal blood vessels. Arch Otorhinolaryngol 244: 363-366

Grevers G., Heinzmann U. (1988 b) Zur Aussagefähigkeit sog. "corrosion casts" bei der Beurteilung endonasaler Gefäßverläufe. Arch Otorhinolaryngol (Suppl.) in press

Horstmann E. (1966). Über das Endothel der Zottenkapillaren im Dünndarm des Meerschweinchens und des Menschen. Z. Zellforsch 72: 364-369

Cells and Tissues: A Three-Dimensional
Approach by Modern Techniques in Microscopy,
pages 457–467

STATISTICAL, MORPHOMETRIC AND STEREOLOGICAL BASICS IN THE MORPHOLOGICAL SCIENCES AND THEIR POSSIBLE APPLICATIONS IN CORROSION CAST STUDIES.

NIGEL T. JAMES.
Department of Biomedical Science,
University of Sheffield,
Sheffield, U.K.,
S10 2TN.

Morphometry

Morphometry is quantitative anatomy. Stereology is that branch of morphometry which, for anatomists, usually consists of obtaining information on the three dimensional nature of test objects using only two dimensional test sections. An essential principle embdedded within stereological practice is the dimensional reduction in the preparation on the test sections or projections upon which measurements are carried out and the estimation using appropraote estimating formaulae of values in a higher dimension. In addition to the commonly used sectioning techniques, other less widely known techniques involving dimensional reduction involve projection stereology (Underwood. 1970). Numerous standard textbooks are available on morphometry and stereology (DeHoff and Rhines, 1968; Underwood, 1970; Weibel, 1979; 1980; Steer, 1981; Serra, 1982; Elias and Hyde, 1983)

Occasionally, the theory of the dimensional reduction from two dimensional sections to one dimensional lines is needed but in anatomy its use is rare. Biology (including anatomy) used to be one of the least numerate of the natural sciences but within the last 20 years the advantages of quantitaion has become clear and has transformed the techniques of experimental investigation.

Quantitation in general, and particularly that provided by stereology especially, is capable of providing both relative and sometimes absolute values both to know more of the structure which interest investigators and more readily make comparisons which are objective rather than subjective. The basis of modern practice in morphometry and stereology is that of geometrical probability. A clear understanding of the principles of deriving stereological relations and formulae from the basics of geometrical probability and the subsequuent analysis using correct probabilitic techniques is vital. The mode of generating results of an entirely probabilitic nature and the correct analysis using sophosticated statistical techniques can hardly be more relevant and interwined than in any other branch of biology. Understanding of the basic principles in order to avoid spurious conclusions is essential. The use of such principles and their specific implementation by making assumptions and using any series of formulae is intellectually dishonest if used as a series of recipies without understanding.

Many types of measure of size or quantity can be estinated using stereological techniques. For example, the volume fraction (V_V) and their numerical density (N_V) of organelles such as nuclei and mitochondria; unfolding of the size distribution of populations of regular particles from two dimensional section profiles; membrane surface densities (S_V), the lengths of capillaries per unit volume of tissue (L_V) and quantitative estimates of their degree of orientation (Ω & K). Standard textbooks on morphometry list all the commonly used stereological symbols.

An essential feature of stereology is that of dimensional reduction. This indicates that three dimensional tissues are sectioned to use two dimensional sections for analysis and that two dimensional plane images can be projected onto one dimensional line length for measurements. With the approprtiate stereological formulae, the data from two dimensional sections is used to compute information on the three dimensional structues of the original test tissue.

Practical measurements are either carried out on the two dimensional section images by used randomly placed overlying transparent test screen bearing relevant patterns of points or test lines (probes) or by the use of instrumental measuring devices if appropriate. Often, however, test screens are superior due to their ready availability, their simplicity and cheapness and also due to their greater flexibility of use in that new screens for new tasks can be constructed in minutes by drawing and xeroxing the test probes onto plastic sheets. The large amounts of data collected by expensive instruments is often entirely unnecessary on statistical grounds and tends to give spurious confidence to the numerical values obtained.

The information derived from the test sheets consists either of measuring the proportion of test points in a test lattice which fall upon the profiles of interest (P_P) or counting the number of intersections per unit length of line that test lines make with profile boundaries (I_L). The numerical values are then used to evaluate measures of size using expressions of the type

$$P_P = A_A = V_V$$

$$S_V = 2I_L$$

Derivations of these relations are to be found in several standard textbooks of stereology. Special stereological formulae are used for anisotropic tissue either to allow for the special sectioning conditions which may apply or to derive additional information abount tissues which would not be available with standard formulae.

An elegant and concise way of descibing the underlying theory of stereology isth use the standard notation of set theory. Consider a three dimensional specimen Y intersected by a section T. Part of the specimen is contained in section T as the intersection of Y with T and is represented symbolically by the relation

$$Y \cap T$$

where $\cap$ is the symbol for '...the intersection of..' If the specimen Y contains a test structure, X, of particular interest, then a profile of X may appear in T. In this case the intersection of X and T is denoted by the relation

$$X \cap T$$

Inferior number notation (subscipt) is used to denatote the dimensions of test structures and probes (points lines and sections) so that Y_3 is a three dimensional specimen containing X_3 and T_2 is the two dimensional test section. The area (A) of specimen intersection between Y_3 and T_2 is given by

$$A\,(Y_3 \cap T_2)$$

whilst the area of intersection of X_3 and T_2 can be written as

$$A\,(X_3 \cap T_2)$$

Clearly the relation of the two intersection areas is an areal fraction, A_A, which is expressed by the relation

$$A_A = \frac{A\,(X_3 \cap T_2)}{A\,(Y_3 \cap T_2)}$$

However, there is a finite probability that the probe T does not intersect with X_3 with the consequence that an empty set, $\emptyset$, is generated. This is symbolised by the relation

$$T \cap X = \emptyset$$

The probability that T hits the specimen Y or its contained test structure X is expressed by the symbol $\uparrow$ and is the probability that a set, $\emptyset$ is **not** generated. Hence the probability of T intersecting the test structure X is given by

$$Pr\{T \uparrow X\}$$

More complicated statements such as

$$Pr\{T \uparrow X \mid T \uparrow Y\}$$

can also be constructed using the standard conditional statistical conditional symbol '|' for conditional probabilities. This statement can therefore be read as "...the probability of the probe T hitting the test structure X given that T hits the specimen Y. Additional symbols include

T_3	=	Tissue section (thick)
T_2	=	Plane surface
T_1	=	Lines on plane surface
T_0	=	Lattice of dimensionless points

Clearly, various types of intersection between T and X can be expressed according to empirical conditions, for example,

$$X_2 \cap T_2 = R_1$$

expresses the intersection of a a two dimensional plane X_2 with a two dimensional section which results (R) in a one dimensional line, R_1. The expression

$$X_1 \cap T_2 = R_0$$

denotes a one dimensional line structure, X_1, (such as a capillary network) intersecting a two dimensional section, T_2 resulting in dimensionless points, R_0. An areal fraction, A_A, of a test structure, X_3, seen in a section, T_2, of a whole specimen, Y_3, is expressed by

$$A_A = \frac{A(X_3 \cap T_2)}{A(Y_3 \cap T_2)}$$

Using similar notation for the probabilities of intersection using $\uparrow$, the probability that a test structure X_3 is hit by dimensionlesss points, T_0given that the specimen Y_3 intersects with the section, T_2, is given by

$$Pr\{T_0 \uparrow X_3 \mid T_0 \uparrow Y_3\} = \frac{\mid A \mid_X}{\mid A \mid_Y} = V_V$$

which is the volume fraction of X in reference volume Y. Clearly, $\mid A \mid_X$ and $\mid A \mid_Y$ are the absolute areal values obtained from $A(X_3 \cap T_2)$ and $A(Y_3 \cap T_2)$, respectively, measured in sections.

Projection and Thick Section Stereology

The theory of obtaining stereological data from measurements obtained from projected images of thick sections has long been available (Underwood, 1970; Miles, 1976). Given that speciment preparation conditions are adequate and that the basic assumptions of projection theory are not violated (as in any statistical analysis) then corrosion casts may be analysed using these techniques. Two types of technique are possible. In the first, single sections of known thickness, t, are prepsred and areal fraction and boundary density counts obtained from projected images. In the second approach, due to Miles (1976), two sections of differing but **unknown** thickness are used and additional tangent count measurements are required.

Areal fraction and boundary density measurements, denoted by A'_A and B'_A respectively are made on the projected images or micrographs for both methods whilst surface tangent counts per unit area T'_A made at test profile boundaries by sweeping a test line the images in a constant orientation are also nedded for the technique of Miles.

For single section thickness, t, some simple relations for estimating line lengths (L), surfaces (S), surface densities (S_V), volume fractions (V_V) and numerical densities (N_V) in three dimensional space can readily be obtained from the various measurements L', A', and N'_A on projected images. The notation (') denotes measurements on projected images. For line lengths the following derivations are obtained:

$$L = 4/\pi.L'; L_V = \frac{4/\pi.L'_A}{t}$$

For surfaces we obtain the well known Cauchy result for convex surfaces without re-entrant (invaginating) surfaces $S = 4A'$ whilst for re-entrant convex surfaces the inequality $S \geq 4A'$ is applicable. For surface densities we obtain the corresponding relations for non re-entrant and re-entrant surfaces

$$S_V = \frac{4A'_A}{t}; S_V \geq \frac{2A'_A}{t}$$

In order for the above equality to hold the additional conditions of no truncation and no test object overlapping must be met. For numerical densities, $N_v = N'_A/t$ but for volume fractions, the measure of size most commonly and readily used in thin section studies, the estimating equation is more complex and requires independent measurements of the mean random intercept length (L_3) unless the test object shape can be specified accurately, e.g. spheres. The estimation equation is given by

$$V_V = \frac{A'_A L_3}{t}$$

The mean random intercept length is related to test volume and surface by the well known equation of Tomkieff

$$L_3 = 4\frac{V}{S}$$

from which the volume fraction can be derived by substitution.

The alternative approach of Miles (1976) of obtaining estimates of V_V and S_V from two differing but **unknown** thicknesses requires a completely different set of estimating equations and also an additional surface tangent density count (T'_A) on the projected image although measurements of projected areal fractions A'_A and projected profile boundary densities B'_A are made as for the previous technique. The estimating equations are given by

$$V_V = 1 - e^{-u}; S_V = 4\nu a_{21} e^{-u}$$

where values for u and ν are obtained from the relations

$$u = \frac{\{a_{21}(c_1 a_2 - c_2 a_1) - b_{21}(a_1 b_2 - a_2 b_1)\}}{b_{21}^2 - a_{21} c_{21}}$$

and

$$\nu = \frac{c_1 b_2 - c_2 b_1}{b_{21}^2 - a_{21} c_{21}}$$

Values for a_i, b_i and c_i (i = 1,2) are obtained from the two unknown section thicknesses t_1 and t_2 and are calculated from

$$a_i = -ln(1 - A'_A)$$

$$b_i = \frac{B'_A}{\pi(1 - A'_A)}$$

$$c_i = \frac{T' - A}{2\pi(1 - A'A)} + \frac{\pi b_i}{4}$$

where $a_{21} = a_2 - a_1$, $b_{21} = b_2 - b_1$ and $c_{21} = c_2 - c_1$

Statistical Experimental Design

In any subject where specimen variability exists allowances for this variability must be made using standard statistical techniques. Unlike the more exact physical sciences and experimental value of, say 17 units, will be recognised as being clearly greater than 14. If, however, in the biological sciences the same values are obtained as arithmetic means, then in the presence of a large variance, say 12 units, then an empirical mean of 17 cannot be regarded as being significantly greater than a control mean of 14. The prime role of statistical theory is to control and allow for such biological variability. That branch of statistics responsible for the construction of appropriate models of experiments is termed experimental design.

Analysis of variance (ANOVA) for univariate statistics and multivariate analysis of variance (MANOVA) occupy central positions in statistical experimental design. Each carefully designed experiment which is concerned with altering controlling variables to determeine the alterations is response variables must contain this element of experimental design. Analysis of variance is a powerful technique and specific models can be represented by an equation of the experimental design such as a two way random effects model which is given by

$$Y_{ijk} = \mu + \alpha_i + \beta_j + (\alpha\beta)_{ij} + \varepsilon_{ijk}$$

where Y_{ijk} is the dependent variable, μ is the parametric mean of the population, α_i is the fixed treatment effect for the ith group of treatmant A (e.g. fixative concentration), β_j the treatment of the jth group of treatment B (e.g. fibre type) and $(\alpha\beta)_{ij}$ is the interaction between treatment effects, A and B. The error term is given by ε_{ijk} for the kth item in each group. Tukey's test for additivity was used to indicate whether the additive model is an adequate fit for the empirical data. Values for volume fractions or non-normal data will be transformed using the expression arcsin $\sqrt{Vv}$ $(0 < Vv < 1)$ prior to normality and homogeneity of variance testing.

So often in reasearch involving morphometric or stereological techniques statistical proceedures are either neglected or inappropriate ones area used. Typical examples of statistical misuse include the the statistical analysis of volume fraction estimates, inappropriate multiple comparisons and the direct violation of important statistical assumptions for the statistical tests used.

Under most conditions of morphometric data collection, areal or volume fractions should not normally be analysed directly since the values may have different weight and their variances and means may not be independent. Transformations, such as the simple arcsin transformation found on most simple scientific pocket calculators, are need prior to analysis to ensure more adequate analysis.

In larger empirical studies where many measures of size or quantity are being analysed (note that parameters are constants of a statistical distribution) spuriously significant significance levels may be obtain by incorrect statistical tables for critical values for multiple comparisons. Suppose a large number of measures of size are being compared statistically at significance levels of either 5 or 10 per cent levels. Then for a hundred such measures there is a 50 per cent chance that at least 5 or 10 statistical results will be significant in hypothesis testing by chance alone. For a large input of effort, the finding of such a large group of statistically significant values which have arisen by chance is difficult for most enthusiastic investigators to ignore - particularly if, on reflection, some "plausible" explaination linking these chance findings can be found. The significance of a set of observations (S) to the significance level (α) for k individual univariate statistical hypothesis tests is given by the Bonferroni inequality

$$\mathbf{S \geq 1 - (1 - \alpha)^k}$$

Formally, the Bonferroni inequality for constructing **simultaneous** confidence intervals is given by the expression

$$\mathbf{Pr}\left(\bigcap_{i=1}^{k} \mathbf{E_i}\right) \geq 1 - \sum_{i=1}^{k} \mathbf{Pr}\,(\mathbf{E_i})$$

whilst the Bonferroni inequality for **simultaneous** hypothesis testing is given in the form

$$\mathbf{Pr}\left(\bigcup_{i=1}^{k} \mathbf{E_i'}\right) \leq \sum_{i=1}^{k} \mathbf{Pr}\left(\mathbf{E_i'}\right)$$

where events are denoted by $E_1, E_2, \ldots, E_t$ and the corresponding complements are denoted by $E_1', E_2', \ldots, E_t'$.

Where appropriate critical values for the t test are not available they may be approximated from the generally available tables of the normal curve using the relation

$$t^* = z + (z + z^3)/4n$$

where n is the number of degrees of freedom and z is the critical normal curve value for P/k. Alternatively, as an approximate conservative approach the critical Student t value required for a single comparison can be divided by the value for k, the number of comparison to obtain the appropriate multiple comparison value.

The Bonferroni procedure is conservative as it assumes each test is independent and overestimation of the probability of a false positive test may result. If we assume the parameters to be compared are normally distributed and the tests nonindependent then less conservative critical values are available. Some comparative values for the different probabilities according to independence or nonindependence of the multiple tests at the 5 % level are given in the table below:

Number of comparisons	Independent tests	Nonindependent tests
1	0.05	0.05
3	0.14	0.11
6	0.26	0.21
10	0.40	0.30
15	0.54	0.39
21	0.64	0.47

For example, the **experimentwise** significance level for 15 comparisons of measures of size is in reality 54 per cent where individual tests at the 5 % level are repeatedly applied. The probabilty that the conclusions reached by the investigator that they are true is therefore considerably less than half (and more than likely to be untrue). The use of appropriate statistical tables (Bonferroni or Sidak's) are required for all tests involving multiple comparisons.(Rohlf and Sokal, 1981; Kres, 1984). Different probability values are required for results which are non-independent rather than independent. Alternatively, the use of multivariate statistics is to be recommended.

Statistical assumptions needed to be valid for a statistical analysis should never be assumed to be valid but should be specifically **tested** to prevent their violation with the consequence of spuriously significant results. For example, normality of data can be tested either graphically or by using the Shapiro and Wilk test. The commonly used parametric Student's t test requires both normality of data and similar variances for control and experimental data (homoscedastic populations). Test for homoscedasticity (homogeneity of variance) are carried out using Snedcor's F test. Whilst the Student t test is fairly robust to lack of normality, at least for equal sample sizes, the F test requirement for normal data is extremely strong.

Other common violations include the use of statistical tests at levels of data other than that assumed for the test, incorrect use of non-parametric statistics and spurious correlation. For example, empirical data can be classified in the progressively ascending levels of nominal, ordinal, hedonic, interval and ratio data. In hypothesis testing using the student t test empirical data is required to be of at least interval level.

Substitution of non-parametric statistics for the often more powerful parametric tests is often fraught with difficulties. Many nonparametric tests are tests of **differences** between two populations rather than specific differences between means or some other parameter. Heterogeneity of variances always needs to be tested using a nonparametric test but with different means such tests loose power untill when there is no overlap of the two test populations the test power falls to zero. Rescaling is then required. The use of nonparametric statistics is based on a series of assumptions such as symmetrical popuations (as for the most powerful Walsh test) or that apart from the parameter under test the populations shoule be **otherwise identical**. Hence the term **pleistoparametric** is to be preferred to the term nonparametric (James, 1986).

The nature of the test structure under investigation is also of critical importance. The presence of a significant degree of either anisotropy in which structural elements possess a preferred average orientation or of spatial autocorrelation in which nearest neighbour test elements occur at specifically correlated distances with each other may significantly alter or even invalidate estimates of measures of size and their associated statistical significance levels. In the case of anisotropy failure to use the appropriate equations may yield over or underinflated estimates and and best results in loss of information. The failure to recognise the presence of spatial autocorrelation may result in the estimation of false variances which preclude both statistical estimation and hypothesis testing (Rogers, 1974).

The techniques of sampling are of major importance as in all analyses involving statistical estimations and hypothesis testing. Not only must samples be representative of the sampled structure but each sample must be identically and independently drawn such that the basic assumptions of subsequent testing must not be violated. Calculations to determine the appropriate sizes of samples needed to obtain **a priori** defined significance levels are also of considerable importance. Various sampling formulae have been given in the literature. For example, Hally (1964) suggested that for a given relative standard error, the number of test points required (n) can be given by the relation

$$R.S.E. = \sqrt{\frac{(1 - V_V)}{n}}$$

DeHoff (Underwood, 1970) has suggested that the number of test points required is given by

$$\frac{200\sigma(P_P)}{(\%accuracy).P_P)}$$

where σ is the standard deviation. Gladman and Goodman (Underwood, 1970) suggest that the number of test points is given by

$$\frac{P_P(1 - P_P)}{\sigma^2(P_P)}$$

Weibel (1979) has indicated that the minimal size can be given by the relation

$$P_T = \frac{t_\alpha^2}{\Delta^2}\left(\frac{1 - V_V}{V_V}\right)$$

where Δ is the magnitude of the **a priori** difference required to br detected and where t_α is the appropriate critical value obtained from tables of the Student t distribution. Preliminary estimates of either V_V or P_P are required either from pilot studies or from previous experience. However, for multiple comparisons it is safer to use the appropriate critical values for either independent or non-independent comparisons from Bonferroni multiple comparison tables. The reader may notice the general nature of these formulae is similar to that of the binomial distribution.

Particular care should also be given to the number of animals used in empirical studies, the numbers of blocks and tissue sampling fields required. In general, it is more important to maintain as large a number of anaimals as is either convenient, posssible or appropriate and that, surpisingly, the number of points used beyond a basic minimum has relatively little effect on the estimating procedure compared with the numbers of the animals used. This can be illustrated by the general variance, (V(x)), relation for a measure of size is given by

$$\mathbf{V(x)} = \frac{\mathbf{S_1^2}}{\mathbf{n_1}} + \frac{\mathbf{S_2^2}}{\mathbf{n_1.n_2}} + ... + \frac{\mathbf{S_4^2}}{\mathbf{n_1.n_2.n_3.n_4}}$$

where S_i^2 is the variance respectively of values obtained from animals, blocks, micoscopical fields and point counting procedures and n_i the specific item numbers of each category. The largest component for variance is normally given by interanimal variance, and the influence of the number of points is manifested only in the the most right hand term. A general rule for experimental design, therefore, is to

...do more less well...

indicating that high values for $n_1$and to a lesser extent for n_2 are required whilst those for subsequent stages, particularly values for n_4 should be as low as is consistent with the experimental aims.

As in all empirical investigations considerable care must be exercised in the selection of suitable statistical design procedures and the implementation of the measurement and analysis procedures without violation of basic underlying assumptions.

References

DeHoff, R.T. and Rhines, F.N. (1968) Quantitative Microscopy. McGraw-Hill, New York.

Elias, H. and Dallas, M.H. (1983) A Guide to Practical Stereology. S. Karger, Basel.

James, N.T. (1986) Hyperbaric oxygen in multiple Sclerosis British Medical Journal, **292**, *829*, London.

Kres. H. (1983) Statistical Tables for Multivariate Analysis. Springer, Berlin.

Miles. R.E. (1976) On estimating aggregate and overall characteristics from thick sections by transmission microscopy. In: Proceedings of the Fourth International Congress for Stereology, National Bureau of Standards Special Publication 431. U.S. Government Printing Office, Washington.

Rohlf, F.J. and Sokal, R.R. (1981) Statistical Tables. 2nd. Edition, W.H. Freeman and Co., San Francisco.

Serra, J. (1982) Image Analysis and Mathematical Morphology. Academic Press, London.

Steer, M (1981) Understanding Cell Structure. Cambridge University Press, Cambridge.

Underwood. E.E. (1970) Quantitative Stereology. Addison-Wesley, Mass., U.S.A.

Weibel, E. (1979) Practical Stereology, Vol. I Academic Press, London.

Weibel E. (1980)Practical Stereology, Vol. II. Academic Press, London.

Cells and Tissues: A Three-Dimensional
Approach by Modern Techniques in Microscopy,
pages 469–474

SCANNING ELECTRON MICROSCOPY OF MICROVESSELS AND PERIVASCULAR CELLS IN DIFFERENT ORGANS AFTER KOH DIGESTION

Andrea Maggioni, Alberto Caggiati, Guido Macchiarelli

Department of Anatomy, Faculty of Medicine, University of Rome "La Sapienza", Via A. Borelli 50, 00161 ROME

INTRODUCTION

In the last decades numerous studies on the ultrastructural features of smooth muscle cells, pericytes and endothelium were carried out by means of transmission electron microscopy (TEM). These investigations, however, could not easily elucidate the form, dimension and three-dimensional arrangement of the periendothelial cells. Scanning electron microscopy (SEM) provided the evidence of the microtopographic relationship between periendothelial cells and microvessels and other structures occupying the interstitial space. In order to perform a SEM study of the microvascular district and perivascular cells it is essential to remove the overlying connective tissue elements. In the past, several techniques for the digestion of cellular and intercellular elements obscuring the objects of interest have been tried. All these techniques were based on the combined use of strong acids or bases, enzymes or detergents; e.g., HCl, NaOH, KOH, NaClO, collagenase and trypsin (Evan et al. 1976; Miller, 1982; Takahashi-Iwanaga and Fujita, 1986).

With this study we have elaborated an experimental protocol which calls for KOH alone, and not in combination with any enzymes.

MATERIAL AND METHODS

Male adult Wistar rats, weighing about 150-200g, were anesthetized and perfused with physiologic solution followed by 2.5 % buffered glutaraldehyde in 0.1 M phosphate buffer at pH 7.4. The study has been carried out on tissue samples from skeletal muscle, liver and pancreas. These organs were chosen because of their different content in connective tissue. Tissue fragments were excised and cut into small cubes, about 1 mm each side, and immersed in the same fixative for 24 hrs. at room temperature. The samples were rinsed for 2-4 hrs. under running tap water. At the same time a solution of 30% 5M KOH was prepared and pre-heated (60°C) in an oven for at least 30 minutes. Each segment was quickly placed into the pre-heated KOH and then subjected to digestion, at 60°C, for 3-60 minutes (the period of time varied in relation to the kind of tissue and the amount of collagen fibers). The time of digestion had to be determined by trial and error. During maceration the tissue fragments were briefly vortexed every 2-3 minutes until they began to break up, and were then placed under running tap water in order to arrest their digestion. The KOH solution was prepared just before use . After the digestion the tissue fragments were rinsed for two hours under running tap water at mid flow in order to perform a mechanical dissociation after the chemical digestion. The specimens were postfixed in OsO_4, dehydrated in a graded series of alcohol, dried by the critical point drying method, sputter coated with gold and examined in a Cambridge 150 stereoscan electron microscope operating at 10-20 kV. After critical point drying the tissue fragments were manually cracked with a needle and mounted using a double-face tape placed on a stub.

RESULTS

The shape and arrangement of the periendothelial cells vary markedly in different vascular segments, although there always exists a gradual intersegmental transition from a smooth muscle cell to an intermediate cell to the pericyte. Smooth muscle cells consist of a central bulge and umbranched tapered ends. The SMC are oriented circularly around the vessels being offset with respect to one another so that the thick central portion of one cell is generally juxtaposed to the tapered end of the adjacent one (Fig.1) . In the transitional zone between a terminal arteriole and a precapillary arteriole SMC adopt branched ends that are usually oriented around the vessel. These cells are a hybrid between regularly aligned SMC and irregularly oriented pericytes and are identified as"intermediate cells" (Fig.2) because they are intermediate in position and in structure between the pericyte and the mature SMC. In most tissues, pericyte structure is characterized by three general features: 1) a cell body containing the nucleus; 2) primary processes extending from the cell body on the long axis of the vessel; and 3) numerous secondary processes of variable shape and size extending from the primaries to partially encircle the capillary

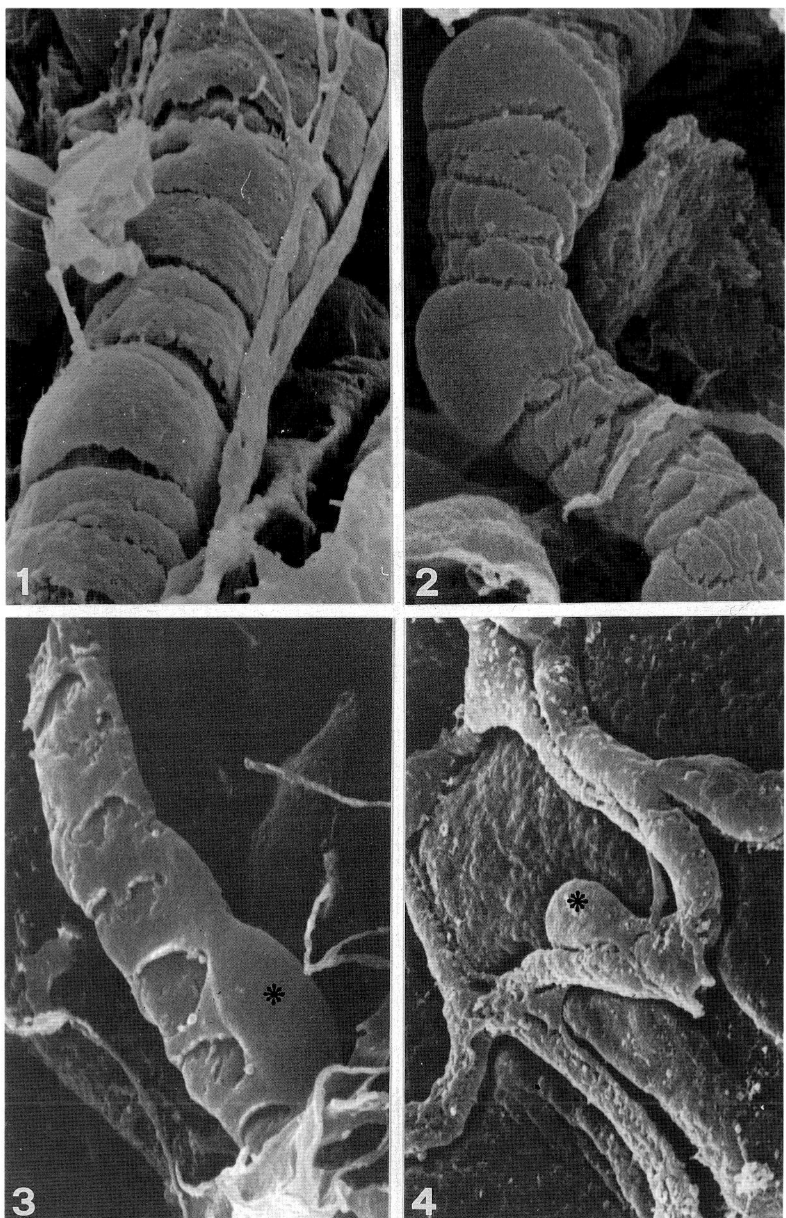

Fig.1: Skeletal muscle. Arteriole with its exposed SMC.

Fig.2: Skeletal Muscle. Precapillary arteriole with intermediate cells along the vessel.

Figs.3-4: Skeletal Muscle. Different aspects of pericytes at capillary level (*).

(Figs. 3,4). Pericyte processes demonstrated considerable variation in shape, and frequently formed bridges between neighboring capillaries (Figs. 5,6).This phenomenon was frequently observed in skeletal vessels and confirm earlier observations (Zimmermann, 1923; Tilton et al., 1979).

SEM observations of pancreas after KOH digestion demonstrated a great number of vessels, e.g., arterioles with typical smooth muscle cells and capillaries with pericytes studded along them. Nerve fibers entwining the blood vessels were well preserved (Fig.7). Numerous fibroblast-like cells with a stellate form were observed on the pancreatic acini (Fig.8). In the liver treated with KOH digesting medium it was possible to obtain a good intercellular separation and a satisfactory digestion of the reticular fibers around the sinusoids which allowed visualization of perisinusoidal cells (Figs.9,10). The perisinusoidal cell presented a round central body from which arose a pattern of branching characterized by primary, secondary and tertiary processes, which surrounded the whole sinusoids. Because of their position the perisinusoidal cells, can, in a sense, be considered as transformed pericytes (Motta, 1988).

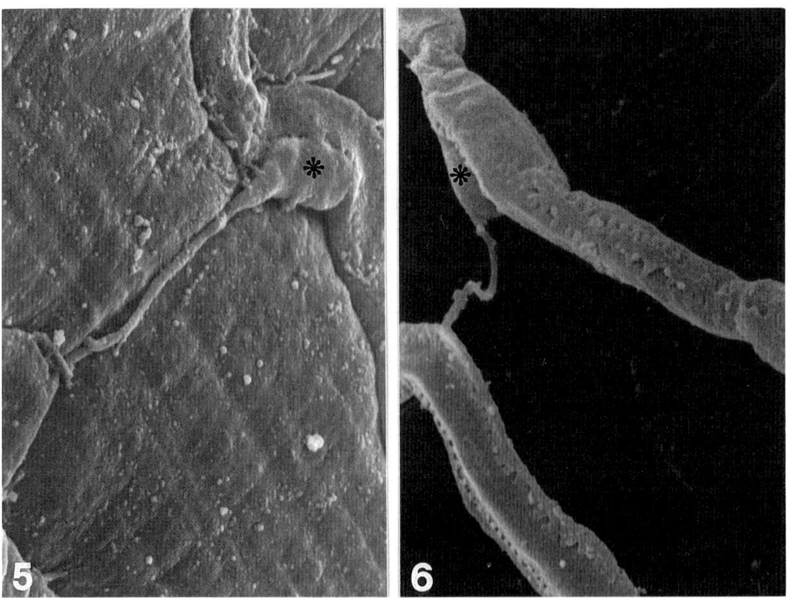

Figs.5-6: Skeletal Muscle. Pericytes (*) extending lateral processes that "bridge" to the neighboring capillary.

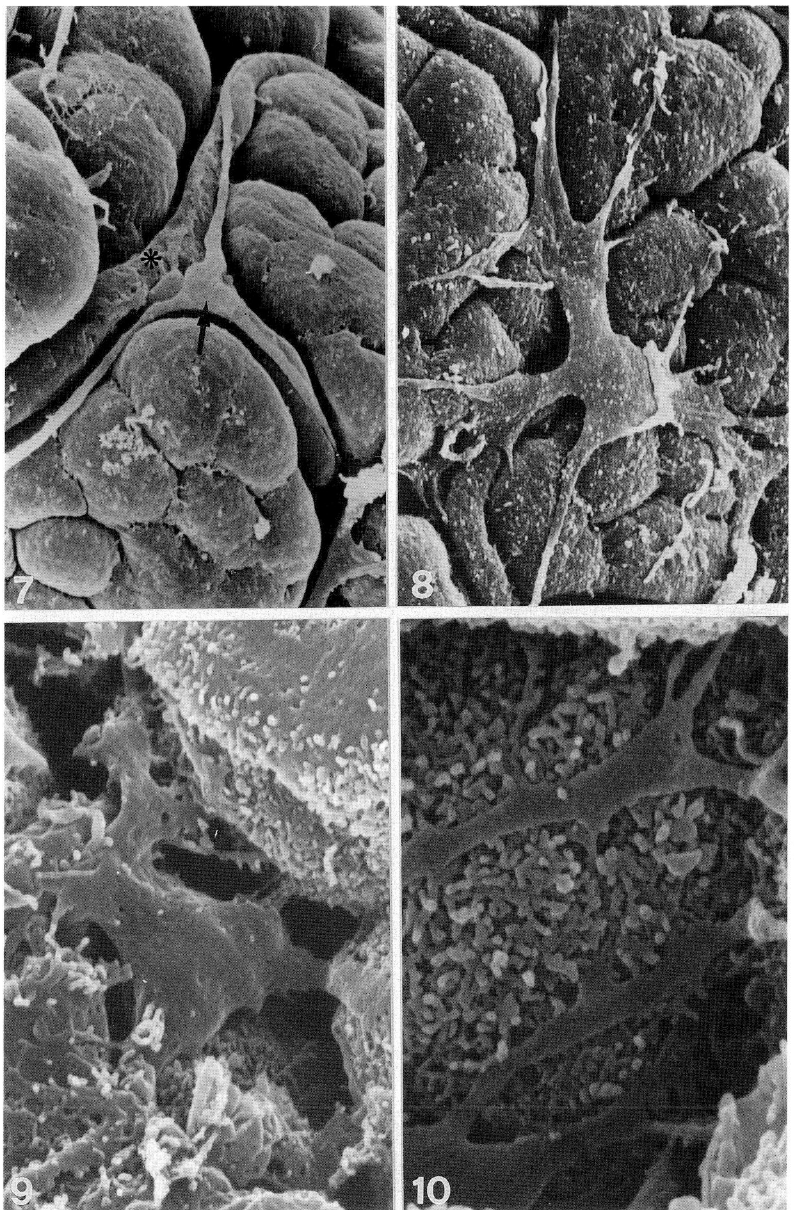

Fig.7: Pancreas. Pericyte (*) studded along a capillary entwining the pancratic acini, with nerve fibers associated (arrow).

Fig.8: Pancreas. Fibroblast-like cell among the pancreatic acini.

Fig.9: Liver. Perisinusoid cell body (Ito's cell).

Fig.10: Liver. Primary and secondary processes of Ito's cell.

CONCLUSION

Digestion with KOH allowed a satisfactory exposure of the microvascular bed and a good visualization of the periendothelial cells in all tissues examined. This two-step dissociation protocol was used to remove as much overlying connective tissue and other interstitial cellular elements from the vessels as possible, while minimizing the damage to the remaining tissues. Further a great number of vessels could be recognized. Compared to the digestion with HCl, the KOH solution seems to be more effective in removing the interstitial elements and bundles of fibers. In fact, it requires shorter exposure times, and seems to induce only little damage to the cell membrane and less loss of capillary and venular perivascular cells.

REFERENCES

Evan AP et al (1976). Scanning electron microscopy of cell surfaces following removal of extracellular material. Anat Rec 185:433-446.

Miller BG et al (1982). A new morphological procedure for viewing microvessels: a scanning electron microscopic study of the vasculature of small intestine. Anat Rec 203:493-503.

Motta PM (1988). "Biopathology of the Liver. An ultrastructural approach" Dordrecht, Boston, London: Kluwer Academic Publishers.

Takahashi-Iwanaga H , Fujita T (1986). Application of an NaOH maceration method to a scanning electron microscopic observation of Ito cells in the rat liver. Arch Histol Jap 49(3):349-357.

Tilton RG et al (1979). Pericyte-endothelial relationship in cardiac and skeletal muscle capillaries. Microvasc Res 18: 325-335.

Zimmermann K (1923). Die feinere Bau der Blutcapillaren. Z Anat Entwickl 68:29-109.

Cells and Tissues: A Three-Dimensional Approach by Modern Techniques in Microscopy, pages 475–480

SCANNING AND TRANSMISSION ELECTRON MICROSCOPIC STUDIES ON THE VASCULAR SYSTEM OF XENOTRANSPLANTED HUMAN TUMORS ON NUDE MICE

Moritz A. Konerding* and Fritz Steinberg**

*Institute of Anatomy, ** Institute of Med. Radiation Biology, University Essen, Hufelandstraße 55, D-4300 Essen 1, F.R.G.

INTRODUCTION AND PROBLEMATICS

It has been shown that grading and staging, both of which to a large extent form the basis for therapeutic procedures, cannot do justice to the interindividual variability of tumors (Vaupel et al., 1986). As a result, there has been an increasing demand for an individual biological assessment (Streffer, 1987).

A major factor influencing the efficacy of tumor therapy, particularly in the case of radio- and chemotherapy as well as hyperthermia, is the oxygen tension in tissues (Vaupel et al., 1987). For this reason, apart from morphologists, above all pathologists, physiologists and oncologists have been interested in the importance of tumor vascularisation since the last century (Gassmann, 1899).

Ultrastructural studies of the vascular system are necessary in order to obtain more detailed information on the sites affected by tumor therapy. Grunt et al. (1986 a, b) have already demonstrated conclusively the development and the characteristic structural features of the angioarchitecture of the Lewis lung carcinoma using microcorrosion casting. Walmsley et al. (1987) have attempted to define differences in tumor vascularisation in old and young hosts. However, such systematic studies have only been carried out on a few experimental tumors (Hammersen et al., 1983; Warren, 1979b) but not in primary tumors or in the more suitable tumor model involving nude mice. In context to the above mentioned, the angioarchitecture of xenotransplanted tumors was investigated in this study by means of scanning and transmission electron microscopy.

MATERIALS AND METHODS

For our studies 126 male and female congenital thymusaplastic nude mice (NMRI-strain), 8-10 weeks old, weighing 22-35 g, were used for transplantation. The tumor tissue from 40 human melanoma and 37 sarcoma-bearing mice was processed for scanning electron microscopy. The vascular system was exsanguinated with up to 40 ml 0.9% NaCl given by means of an olive-tipped cannula inserted into the left ventricle. The perfusion pressures were between 70-90 mm Hg and the solution temperatures ranged from 35-38°C. The animals were then fixed with a total of 100 ml 2.5% cacodylate-buffered glutaraldehyde (pH 7.40, 860 mosmol). After dissection the tumors were washed with cacodylate-buffer, broken with a scalpel, dried by the critical point method and sputtered with gold in an argon atmosphere. From the same specimens prepared for SEM, samples were also processed for TEM. After being washed several times in Sörensen phosphate buffer (pH 7.44, 460 mosmol) they were postfixed for 3 hours according to Dalton and rinsed in phosphate buffer prior to dehydration in ascending alcohol. Ultrathin sections with a thickness of 500-700 A were examined using an EM 109 T (Zeiss, F.R.G.) operated at 80 kV. Vascular corrosion casts were made from 29 melanomas and 20 sarcomas mainly (cf. Lametschwandtner et al., 1984). After perfusion with saline and fixation as described above, up to 35 ml of undiluted Mercox CL-2B (Japan Vilene Comp. Ltd., Ja-pan) mixed with 1.5% catalyzing substance was injected. After complete maceration the casts were sputtered with gold in an argon atmosphere and examined with a Stereoscan 180 scanning electron microscope (Cambridge, U.K.) at an acceleration voltage of 10 kV.

RESULTS

The subcutaneous tissue formed a capsular-like structure, through which the main supplying and draining vessels pass to and from the tumor. These arteries and veins, showing normal structural features outside the tumor, lead into intratumoral

---

Fig. 1: Freeze-broken sarcoma. Note the thin (↑) and, in places, incomplete vessel endothelium (↑↑), whereby the marked cell (*) cannot be clearly identified as an endothelial cell. tc = tumour cells. Bar = 5 µm.

Fig. 2: Freeze-broken specimen of a poorly differentiated xenotransplanted leiomyosarcom. Peripheral vessel with more prominent, oval to round endothelial cell nuclei shapes (↑) corresponding to venous type of endothelium. No medial layer, little perivascular tissue. Bar = 10 µm.

Fig. 3: Elongated, tortuous course of a peripheral vessel with numerous venous endothelial cell nuclei imprints (↑). Vascular corrosion cast of a melanoma. Bar = 50 µm.

Fig. 4: Corrosion cast specimen demonstrating the heterogeneity in vascular distribution (leiomyosarcoma). The periphery is located in the foreground and the central sections in the background. Note the differing vascular densities and the lack of an hierarchial order. * = flattened peripheral major vessels. Bar = 300 µm.

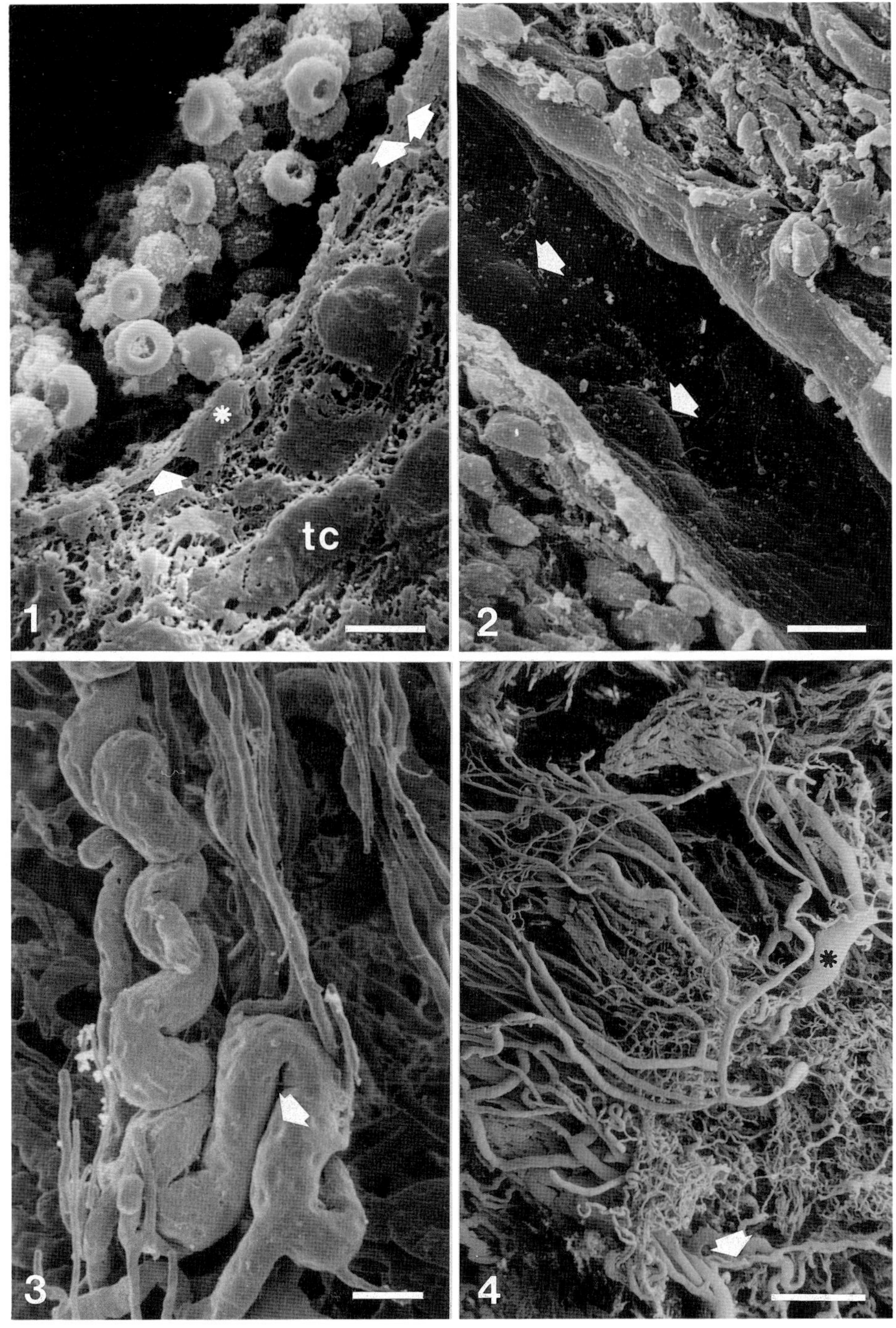
1
*
tc
2
3
4
*

vessels characterized by a widely different architecture. Regular vessels consisting of intimal, medial and adventitial layers are neither to be seen in the periphery nor in the center. Even the largest peripheral vessels with diameters of 50-150 µm consist mainly of an endothelial layer and some perivascular connective tissue only (figs. 1 and 2).

Vascular corrosion casts display the course of these vessels: adjacent to the xenograft they follow normal courses and branching modes, thus partially forming a "vascular envelope" as described by Grunt et al. (1986 a). Frequently these vessels follow a tortuous course when reaching the tumor tissue (figs. 3 and 4), where features such as glomeruloidal arrangement or widening and dilatations - already excellently described by Grunt and coworkers (1986 a, b) - are visible.

In the tumor itself the vascular system shows a much more chaotic arrangement. Very often regions with clusters of vessels could be seen just beneath areas almost free of vessels (fig. 4). In almost no case a relationship to the topographical localisation could be established. The size and extent of these avascular regions varied considerably from tumor to tumor as well as in the individual xenotransplants. In many sarcomas and melanomas, these vessels form a sinusoidal system with numerous blind ends without clearly discernible endothelial cell impressions (fig. 5). The diameters vary between 5 and 50 µm. Thus, they clearly differ from vessels in normal tissues. Sinusoids originating or confluencing from large calibre vessels could regularly be shown, even for large vascular densities (fig. 5), and must be assigned to venous vessels because of their endothelial cell nuclei impressions. Frequently, large calibre endothelialised vessels were found in the direct vicinity or in the center of non-vital zones, which could be seen as loose, partially unstructured areas under scanning electron microscopy (fig. 6).

In agreement with Hammersen et al. (1983), the transmission electron microscopic investigation shows that even large ca-

---

Fig. 5: Vascular corrosion cast of the sinusoidal system in a leiomyosarcoma. Note the flattened (↑) and partially tortuous main vessels (*), respectively draining and supplying the sinusoidal system. Bar = 300 µm.
Fig. 6: Large calibre vessels (*) in the vicinity of a partially necrotic, loose region (↑) in a leiomyosarcoma. Bar = 50 µm.
Fig. 7: Vessel in the central section of a xenotransplanted leiomyosarcoma with variously high endothelium. Note the extreme thinning in places (↑↑). Pericytic ramifications (*) are surrounded by the basal membrane, which is not clearly demarcated in places (↑). Magnification: x 3310.
Fig. 8: Excentric, conelike position of the perikaryon (p) in a vessel from an undifferentiated sarcoma. Note the varying electron density of the endothelial cells. Magnification: x 5200.

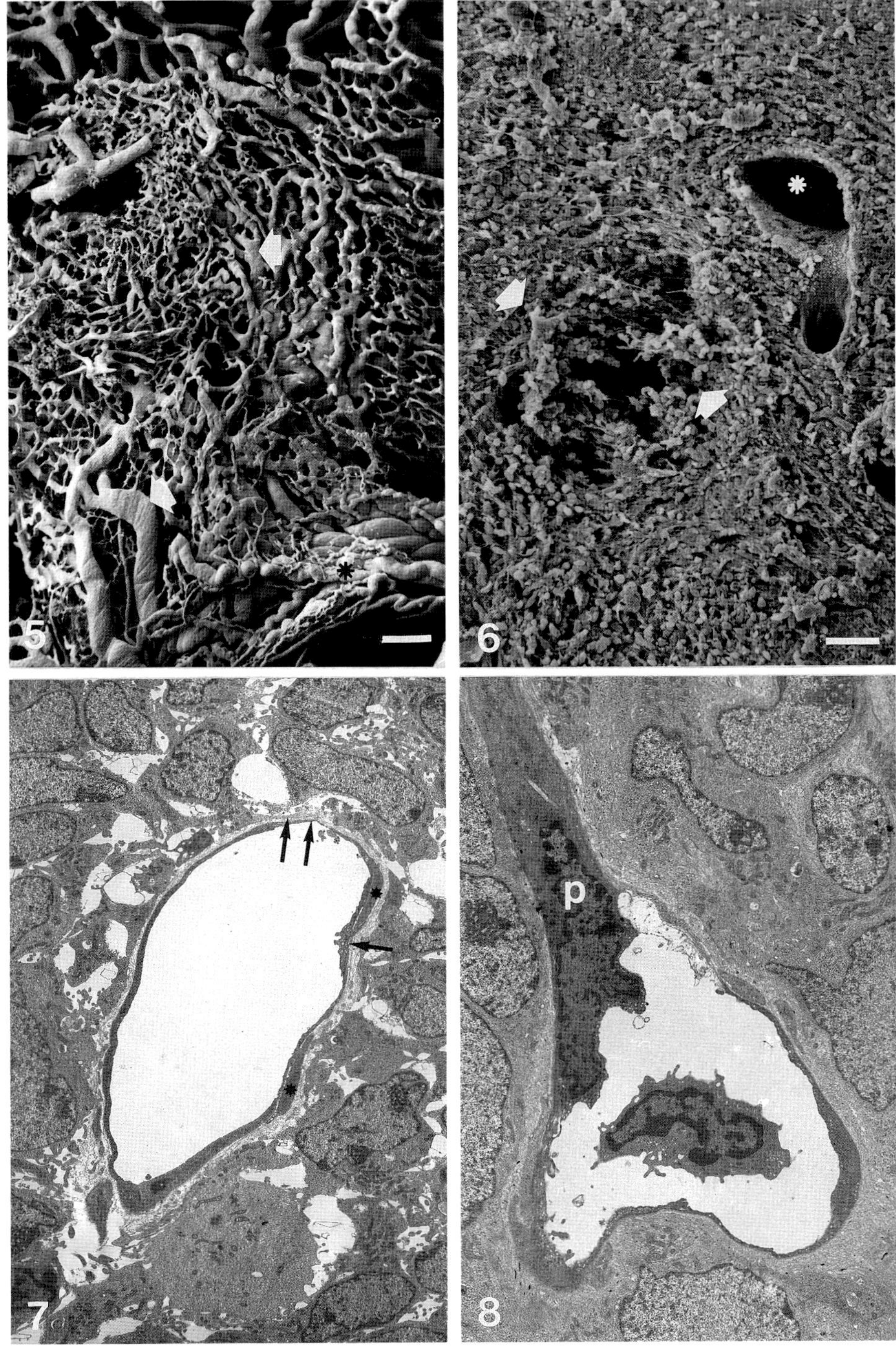
5
6
7
p
8

libre vessels basically have a capillary wall structure (fig. 7). The endothelia show a very alternating height and electron density, whereby frequently different cell types are apparently involved in the formation of the vessel wall. Frequently, a basal membrane cannot be observed at all or only incompletely (fig. 8). Instead, a very close association of tumor cells to the wall of the vessel is apparent.

REFERENCES

Gassmann, A.: Zur Histologie der Roentgenulcera. Fortschr. Roentgenstr. 2, 199-207 (1899).

Grunt, T.W., Lametschwandtner, A., Karrer, K., Staindl, O.: The angioarchitecture of the Lewis lung carcinoma in laboratory mice. Scanning Electron Microsc. II, 557-573 (1986a).

Grunt, T.W., Lametschwandtner, A., Karrer, K.: The characteristic structural features of the blood vessels of the Lewis lung carcinoma. Scanning Electron Microscopy II, 575-589 (1986b).

Hammersen, F., Osterkamp-Baust, U., Endrich, B.: Ein Beitrag zum Feinbau terminaler Strombahnen und ihrer Entstehung in bösartigen Tumoren. In: Mikrozirkulation in Forschung und Klinik. Vol. 1: Structure and Function of Endothelial Cells. Meßmer K., Hammersen F. (eds) Karger, Basel-München-Paris-London-New York-Tokyo-Sydney, 15-51 (1983).

Lametschwandtner, A., Lametschwandtner, U., Weiger, T.: Scanning electron microscopy of vascular corrosion casts - techniques and applications. Scanning Electron Microsc. II, 663-695 (1984).

Streffer, C.: Predictive assays of human tumour response. Introductory review . Proc. of the 8th Int. Congress for Radiation Research. Taylor & Francis, London, New York, Philadelphia, Vol. 2, 825-830 (1987).

Vaupel, P., Gabbert, H.: Evidence for and against a tumor type-specific vascularity. Strahlenther. Onkol. 162, 633-638 (1986).

Vaupel, P., Fortmeyer, H.P., Runkel, S.: Blood flow, oxygen consumption, and tissue oxygenation of human breast cancer xenografts in nude rats. Cancer Res. 47 (13), 3496-3503 (1987).

Walmsley, J.G., Granter, S.R., Hacker, M.P., Moore, A.L., Ershler, W.B.: Tumor vasculature in young and old hosts: scanning electron microscopy of microcorrosion casts with microangiography, light microscopy and transmission electron microscopy. Scanning Microsc. I, 823-830 (1987).

Warren, B.A.: The vascular morphology of tumors. In: Tumor Blood Circulation: Angiogenesis, Vascular Morphology and Blood Flow of Experimental and Human Tumors, Peterson H.I. (ed), CRC-Press, Boca Raton, Florida, 1-47 (1979).

Acknowledgements: The authors wish to thank B. Gobs for skilful technical assistance and D. Kittel for photographic work.

Cells and Tissues: A Three-Dimensional
Approach by Modern Techniques in Microscopy,
pages 481–485

CADMIUM TOXICITY ON THE AORTIC MEDIA OF PREGNANT RATS: TRANSMISSION AND SCANNING ELECTRON MICROSCOPIC STUDY

Mitsuaki Yoshizuka, Takeshi Maruyama, Naoki Mori,
Hiroshi Ueda and Sunao Fujimoto
Department of Anatomy, University of Occupational and
Environmental Health, School of Medicine
Kitakyushu 807, Japan

INTRODUCTION

It has been well known that parental and oral exposure to cadmium salts may cause hypertension (Baranski, et al., 1983, Perry, et al., 1971, Schroeder, 1964, Thind, et al., 1973), and the accumulation of the labeled cadmium in the vascular walls has been determined by the previous autoradiographic studies (Berlin and Ullberg, 1963, Shibata, 1977). However, no detailed ultrastructural observations as to toxic effects of this metal on various kinds of arterial wall have been made, except our previous report demonstrating the endothelial degeneration and increased endothelial permeability of the rat thoracic aorta after acute administration of cadmium sulphate (Yoshizuka, et al., 1987). Furthermore, we have several data that the cadmium toxicity becomes much pronounced in various organs during pregnancy.

On these grounds, the present study was planned for elucidation of morphological consequences of the aortic media of pregnant rats induced by cadmium. Ultrastructural localizations of cadmium in the medial smooth muscle cells were also determined with the X-ray microanalysis.

MATERIALS AND METHODS

Pregnant Wistar rats received intraperitoneal injection of 1.8mg/kg/day of cadmium sulphate from fifteen to nineteen days of gestation and were sacrificed at the third and fifth day of administration. The thoracic aortic rings were fixed in 4% paraformaldehyde saturated with 8-hydroxyquinoline (oxine) in 0.1M phosphate buffer for 24 hours. For transmission electron microscopy, specimens were postfixed in 2% osmium tetroxide in the same buffer and embedded in epoxy resin. Ultrathin sections were examined with a JEOL 100CX electron microscope and analyzed with a HITACHI H-500H electron microscope equipped with a KEVEX 7000 *energy dispersive X-ray microanalyzer. For scanning electron microscopy of elastic elements of the media, specimens were incubated in 88% formic acid at 45°C for 96 hours. Isolated elastic tissues were rapidly frozen in liquid nitrogen, freeze-dried in a JFD 7000 vacuum evaporator and examined with a HITACHI S-700 scanning electron microscope.*

RESULTS AND DISCUSSIONS

On the third day after cadmium administration, various degenerative changes were observed in the aortic endothelia: These include degeneration of mitochondria and appearance of inter- and intracellular vacuoles (Fig. 1). These morphological changes are thought to be related to the increase in permeability of the endothelia as previously observed by the marker experiment using horseradish peroxidase (Yoshizuka, et al., 1987). The innermost layer of elastic lamina also showed several morphological changes such as thinning, bending and fragmentation, possibly caused by the subendothelial edema after a massive influx of plasma through the endothelia (Fig. 1).

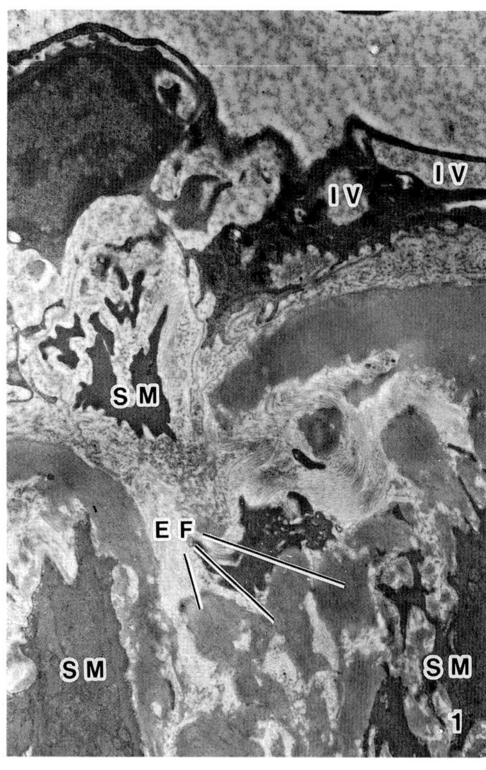

Fig. 1. The degenerating endothelial cells include many intracytoplasmic vacuoles (IV). The innermost elastin becomes a marked bending and fragmentation (EF). SM: smooth muscle cells in the media. At the third day after cadmium administration. X 6,700

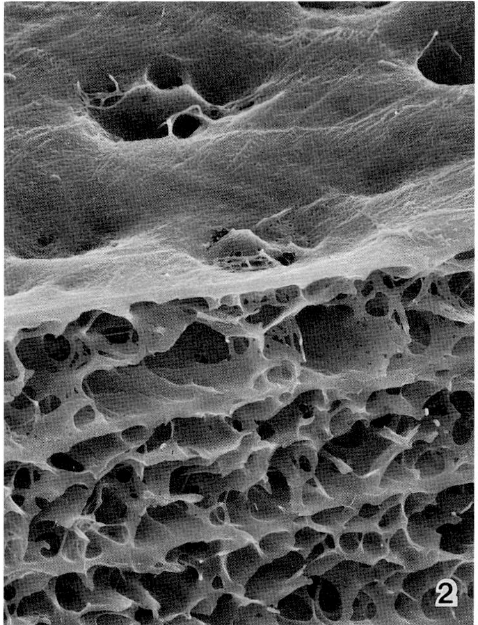

Fig. 2. A scanning electron micrograph showing elastin framework of the aortic wall after treatment with formic acid. The appearance of large fenestrations of each elastin layer is noted. At the third day after cadmium administration. X 2,300

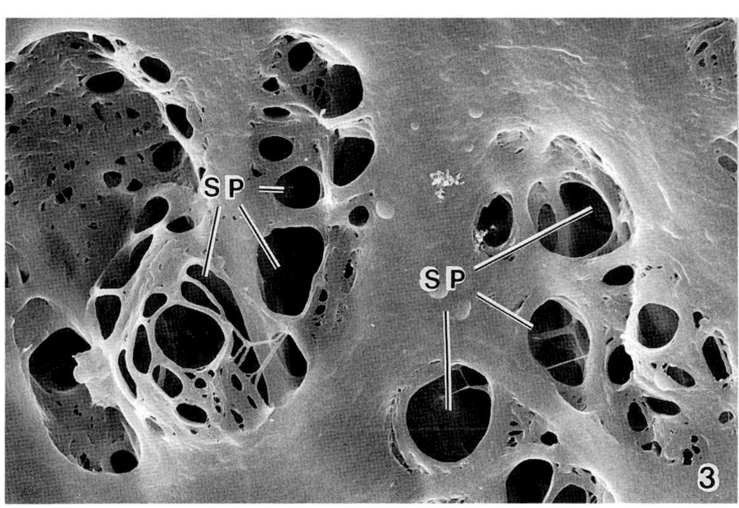

Fig. 3. A scanning electron micrograph showing the surface of the innermost layer of elastic lamina. Enlarged fenestrations consist of irregularly-arranged smaller pores (SP) and the intervening elastin meshwork. X 4,000

The scanning electron microscopy showed the presence of abundant enlarged fenestrations in the innermost layer of elastic lamina (Fig. 2). These fenestrations included irregularly-arranged smaller pores among elastin meshwork (Fig. 3).

Smooth muscle cells of the media also underwent significant degenerative changes. Some altered cells extended knobby edematous cytoplasmic projections which were frequently detached from the main cell body. Mitochondria under cytolysosomal degradations were pronounced in most altered muscle cells as shown in Fig. 4, and cadmium-oxine complexes were detected out from these cytolysosomes with X-ray microanalyses (Fig. 5).

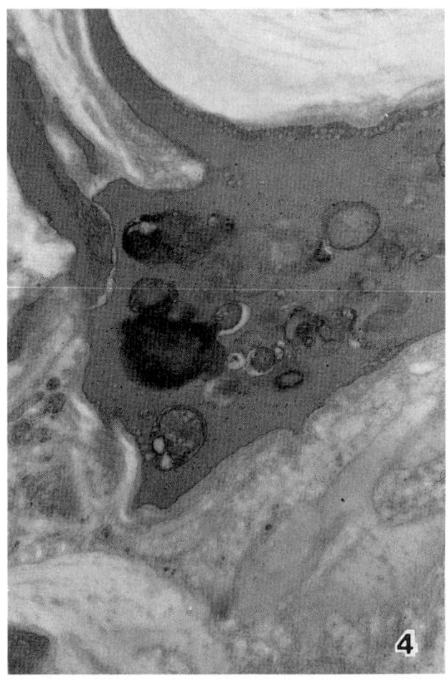

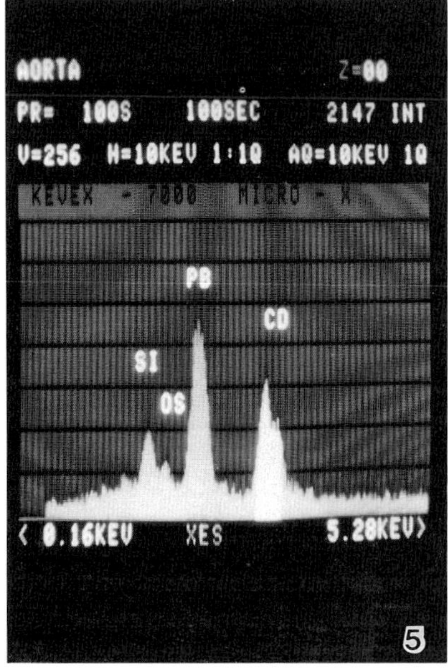

Fig. 4. Mitochondria of a smooth muscle cells in the aortic media show cytolysosomal degradations of varying degree. X 20,000

Fig. 5. By the X-ray microanalysis, cadmium-oxine complex was detected out from the cytolysosome of degenerating mitochondria as shown in Fig. 4.

REFERENCES

Baranski B, Opacka J, Wronska-Nofer T, Trzcinka-Ochocka M, Sitarek K (1983). Effect of cadmium on arterial blood pressure and lipid metabolism in rats. Toxicol Let 18:245-250.

Berlin M, Ullberg S (1963). The fate of Cd^{109} in the mouse. An autoradiographic study after a single intravenous injection of $Cd^{109}Cl_2$. Arch Environ Health 7:686-693.

Perry HM Jr., Erlanger M (1971). Hypertension and tissue metal levels after intraperitoneal cadmium, mercury, and zinc. Am J Physiol 220: 808-811.

Schroeder HA (1965). Cadmium as a factor in hypertension. J Chron Dis 18:647-656.

Shibata H (1977). Whole body autoradiographic studies on the fate of zinc, cadmium and mercury in mice. Bull Fac Agric Yamaguchi Univ 28:83-97.

Thind GS, Biery DN, Bovee KC (1973). Production of arterial hypertension by cadmium in the dog. J Lab Clin Med 81:549-556.

Yoshizuka M, Tanaka I, Miyazaki M (1987). Cadmium toxicity on the thoracic aortae of pregnant rats: structure and permeability. In Tsuchiya M et al. (eds): "Microcirculation - an update vol. 1" Amsterdam: Elsevier, pp 129-130.

Cells and Tissues: A Three-Dimensional Approach by Modern Techniques in Microscopy, pages 487–492

MORPHOLOGICAL CHARACTERS OF THE ABSORBING PERIPHERAL LYMPHATIC VESSEL BY TEM, SEM AND THREE-DIMENSIONAL MODELS

Giacomo Azzali, Guido Orlandini and Giovanna Bucci

Institute of Anatomy, University of Parma, via Gramsci 14 - 43100 PARMA - ITALY.

INTRODUCTION

Recent investigations have ascertained that the peripheral lymphatic absorbing vessel (initial or capillary lymphatic) is closely related to the interstitium and plays a pivotal role in lymph formation. Electron transmission (TEM) and scanning microscopies (SEM), togheter with three-dimensional models (TDM), have contributed, to different extent, to improve the morpho-functional knowledge about the lymphatic vessel, but many uncertainties and differences of opinion still exist. The aim of our participation is to examine some of the morphological characters of the endothelial lymphatic wall, which have represented a particular object of debate in the last decade. We have studied deep and superficial lymphatics in different organs (stomach, small intestine, urinary bladder, kidney) from different animals (micromammalian, man, sea and earth turtle) in experimental and physiological conditions.

There is an almost complete agreement about the fact that the absorbing peripheral lymphatic lacks a basal lamina and that it is formed by a monolayer of flat cells joined each other by different types of intercellular junctions: end to end, overlapping and interdigitating. Furthermore each cell is composed by a central globular part containing nucleus, common cytoplasmic organelles and lysosomes (nuclear area), and by a laminar peripheral part lacking fenestration whose cytoplasm is continuous. The cytoplasmic matrix of this portion contains free ribosomes, scarce mitochondria, some dense bodies, thin actin-like filaments, free uncoated vesicles and vesicles linked with luminal and abluminal plasma membranes.

The SEM shows that the luminal endothelial surface is slightly wavy and it is composed of flat cells, polyedric or rectangular in shape, lacking porès or fenestrations. The endothelial cells are joined each other by only two types of junctions: end to end and overlapping. The eventually occurring interdigitating junctions can not be correctly identified by SEM. In the proximity of the end to end contacts the surface shows small grooves with an irregular course. When an overlapping contact occurs the peripheral edge of the cell continues over the junctional complex forming an anfractuosity. The detachment between limited segments of the peripheral edge of two adjacent cells forms an intercellular pocket. This morphological feature is peculiar of the overlapping intercellular contacts. In this way two adjacents cells delimit an orifice introducing a space which deepens between the endothelial surfaces. Thus the rim of the pocket (luminal orifice) is delimited (Fig. 1) by the detached abluminal edge of a cell

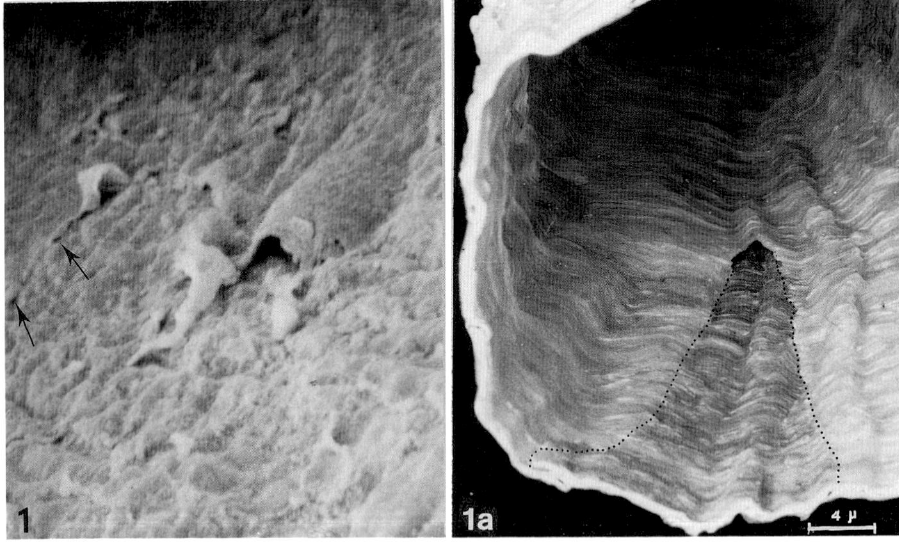

Figs. 1 and 1a: Turtle - lacteal vessel.
Wavy luminal endothelial surface with short detachment (pocket) of the peripheral edges of two contiguous cells. The arrows point to an anfractuosity. The above morphological picture as it appeare in a three-dimensional model (1a) from serial ultrathin slices. X 2,500.

arching over the luminal flat surface of an adjacent cell. Both "anfractuosities" and "pockets" (Castenholz, 1987) can not be adequately explored by SEM in order to value a possible continuity with the interstitium. The three-dimensional models (built up by serial thin slices) allowed to establish that only a limited number of intercellular pockets communicate with the interstitium (Fig. 1a). The anfractuosities are always dead-end structures; they are formed by limited protrusion of the peripheral cytoplasm after the intercellular contact which is always fixed by tight and gap junctions. The abluminal endothelial surface can not be properly observed with SEM because of a thick net of connectival fibers. In some areas these fibers are linked to the abluminal wall of the endothelial cell (anchoring filaments - Leak and Burke, 1968). The areas of the abluminal surface free of fibers show slight and wavy grooves connected with end to end contacts alternating with more or less deep anfractuosities connected with overlapping contacts. Furthermore cytoplasmic processes (generally 12 - 28 μm long and 1.5 - 4 μm wide), overcoming the border with the adjacent cell, overlap the abluminal surface of the latter, sticking to it only with the lateral edges. Each process shows different shape and lenght and their number varies in relation with particular physiological conditions (fasting, lymphatic or venous stasis). Abluminal cytoplasmic projections have been described with TEM and TDM in small intestine, urinary bladder and kidney lymphatics from different mammalians (Azzali, 1982 a, b; Azzali and Werner, 1976; Azzali et al., 1983; Niiro et al., 1986). A similar morphological behaviour was assessed with SEM in corrosion casts obtained by neoprene injection.

A comparative analysis of the data collected with TEM (serial slices), SEM and TDM allows a correct interpretation of the morphology of this structure that we define as "<u>intraendothelial channel</u>". This channel takes origin from the cytoplasmic process which, sticking to the abluminal surface of the adjacent cell only with the lateral edges, defines a space between the two cells. This structure is provided with two orifices, luminal and abluminal. The orifice formation is due to the missed adhesion of the initial and terminal tract of the process. The "intraendothelial channels" must be considered dynamic morphological structures (Azzali, 1988 a, b) representing a fast and unselective pathway in lymph formation. The luminal and abluminal orifices must not be confused with the "open junctions" or with the valve-working open intercellular contacts (Palay and Karlin,

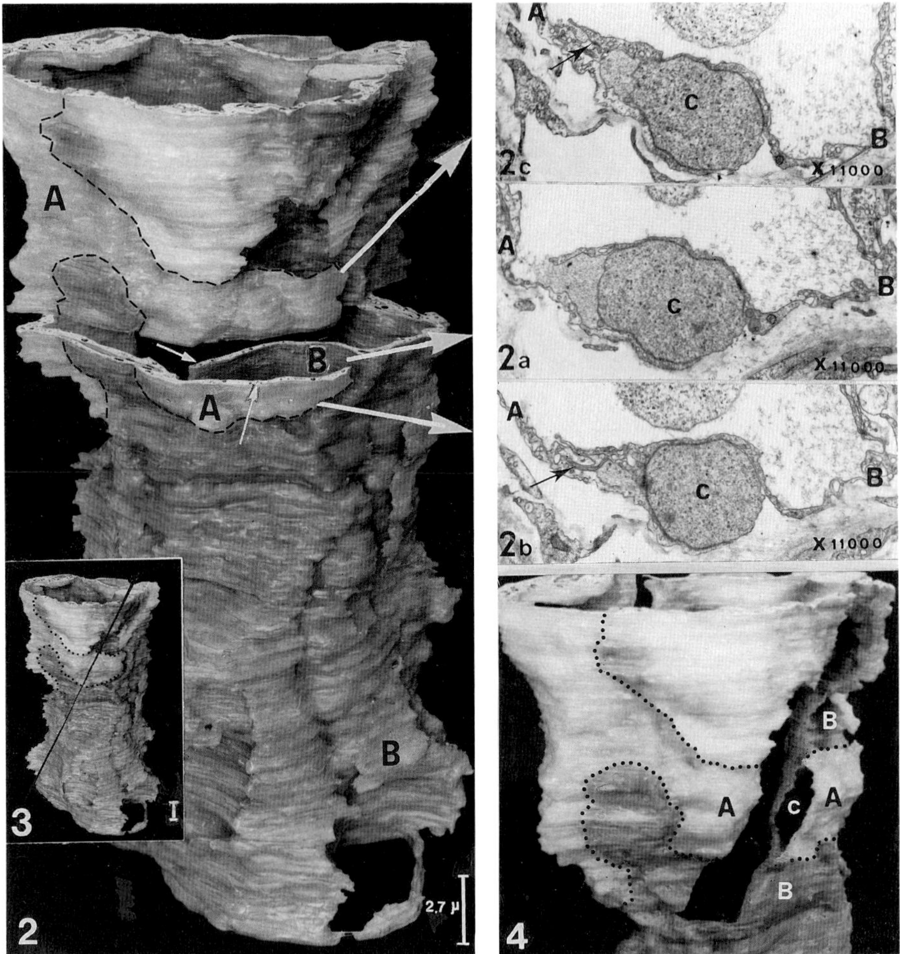

Fig. 2 = Three-dimensional model of a segment of human lacteal vessel. The cleft shows a so-called "open junction" determined by the "missed-adhesion" of the peripheral cytoplasmic expansion of two contiguous cells, A and B. Fig. 2a = Ultrastructural picture of cleft in Fig. 2 showing a globule (c) encircled by the peripheral cytoplasmic expansion. Fig. 2b and 2c = Upper and lower limits of the so-called "open junction". Fig. 3 = Same three-dimensional model. The black oblique line shows the section plan of Fig. 4. Fig. 4 = The new section plan shows the intraendothelial channel delimited by cells A and B. There are no open junctions. c = cavity of the channel.

1959; Leak, 1971; Kalima, 1973; Kalima et al., 1977; Casley-Smith, 1965, 1988) which are considered, by most Authors, the main pathway of the transendothelial transport.

We also found the "missed adhesion" (open junction) of peripheral cytoplasmic edges between two adjacent cells (with no trace of tight or gap junctions) in lymphatics from small intestine submucosal and subserosal net, trachea, urinary bladder and (although very rarely) lacteal vessels. A TEM analysis of 300-400 serial ultrathin slices showed (Figs. 2a, b and c) that the "missed adhesion" concerns 12, at most 24, slices. The endothelial wall immediatly above and below of this "opening" resumes a continuous course with intercellular overlapping contacts. Furthermore if we consider a lymphatic three-dimensional model reproducing the above described picture (open junction), but we choose a section plan unaltering the morphology of the vessel itself (Fig. 3), we obtain a morphological picture strikingly similar to the above mentioned "intraendothelial channels" (Fig. 4). Thus we believe that the frequency and the report of the so-called "open junctions" closely relate to the section plan of the lymphatic vessel. SEM pictures also support this theory. We think it is possible to assert that the "open junctions" must not be considered a morphological reality but the result of a longitudinal section of the "intraendothelial channels".

This report points out the pivotal role played by the comparative analysis of TEM, SEM and TDM morphological pictures in order to interpret and understand the controversial aspects of the mechanisms involved in the transendothelial lymphatic transport.

This work was supported by a grant from the Ministero Pubblica Istruzione 40% - Rome, Italy.

REFERENCES

Azzali G (1982 a). The ultrastructural basis of lipid transport in the absorbing lymphatic vessel. J. Submicrosc. Cytol. 14: 45-54.

Azzali G (1982 b). Transendothelial transport of lipids in the absorbing lymphatic vessel. Experientia 38: 275-276.

Azzali G (1988 a). Ultrastructural and seasonal aspects of the kidney lymphatic system of hibernating animals. Experientia 44: 441-444.

Azzali G (1988 b). The "intraendothelial channels" of the peripheral absorbing lymphatic vessel. In Partsch H (ed): "Progress in Lymphology. XI", Amsterdam: Elsevier Science Publishers B.V., pp 187-192.

Azzali G, Werner G (1976). Ultrastruttura del capillare linfatico e sua ricostruzione tridimensionale. Atti Soc.Ital.Anatomia: Ed. Grafica Toscana, Firenze, pp 42-43.

Azzali G, Romita G, Gatti R (1983). Ultrastruttura dei vasi linfatici della vescica orinaria. Acta Biomed. Ateneo Parmense 54: 105-115.

Casley-Smith JR (1965). Endothelial permeability II. The passage of particles through the lymphatic endothelium of normal and injured ears. Br.J.Exp.Pathol. 46: 35-49.

Casley-Smith JR (1988). The two modes of initial lymphatic filling: colloidal osmotic pressure, hydrostatic pressure, or both? In Partsch H (ed): "Progress in Lymphology. XI", Amsterdam: Elsevier Science Publishers B.V., pp 173-177.

Castenholz A (1987). Structural and functional properties of initial lymphatics in the rat tongue: scanning electron microscopic findings. Lymphology 20: 112-125.

Kalima TV (1973). Ultrastructure of the intestinal lymphatics in regards to absorption. Scand. J. Gastroent. 8: 193-196.

Kalima TV, Collan Y, Kalima SH (1977). Variations of lymphatic endothelial cell junctions in experimental conditions. In Mayall RC, Witte HM (eds): "Progress in Lymphology", Plenum Press, pp 7-12.

Leak LV (1971). Studies on the permeability of lymphatic capillaries. J. Cell Biol. 50: 300-323.

Leak LV, Burke JF (1968). Ultrastructural studies on the lymphatic anchoring filaments. J. Cell Biol. 36: 129-149.

Niiro GK, Jarosz HM, O'Morchoe PJ, O'Morchoe CCC (1986). The renal cortical lymphatic system in the rat, hamster and rabbit. Am. J. Anat. 177: 21-34.

Palay SL, Karlin KJ (1959). An electron microscopic study of the intestinal villus. II. J. Biophys. Biochem. Cytol. 5: 373-383.

Cells and Tissues: A Three-Dimensional Approach by Modern Techniques in Microscopy, pages 493–500

SEM OF IMMUNOHEMATOPOIETIC TISSUES

Tsuneo Fujita
Department of Anatomy
Niigata University School of Medicine
Asahimachi, Niigata 951

The lymphoid and hematopoietic tissues comprise complex labyrynthine structures which could be first elucidated by the aid of scanning electron microscopy (SEM). This paper will review what SEM has revealed in this field of histology, concentrating on the findings obtained by our own research group. The tissues usually are perfused with aldehyde fixatives and conductive-stained with OsO_4. Dehydrated in ethanol, they are freeze cracked in isoamylacetate for adequate fracture surfaces, critical point-dried, and evaporation-coated with gold palladium. Observations in this study have been mainly made with a field emission SEM (HFS-2, Hitachi).

THE LIVER

The sinusoidal and perisinusoidal tissue in the liver has been included in Aschoff's reticuloendothelial system. In fetal life hematopoiesis occurs here. The SEM study of this tissue started by Motta (1975) in Italy and by Muto of our research group (1975). These independent studies were soon united into an international collaboration, resulting in publication of a monograph, "The Liver" (1978). The first result thus reached was that the sinusoidal endothelium in mammals is perforated intra- and intercellularly with pores which are mostly small in size (less than 1 µm) and partly large reaching a few micra of diameter (Motta, 1975; Muto, 1975; Muto et al., 1977; Motta et al., 1978). The Kuppfer cells is characterized by filopodial and granular microprojections densely covering its surface and by the ability of recognizing and internalizing foreign bodies of large sizes (several micra in diameter) (Muto and Fujita, 1977); SEM has revealed that there are no intermediate forms between this cell and the endothelial cell, contradicting the classical view of the reticuloendothelial system of Aschoff (Motta et al., 1978; Fujita, 1978).

THE BONE MARROW

Our SEM studies, using mainly rat femur, demonstrated that the sinus endothelium comprises a thin cytoplasmic plate which shows perforations when maturing blood cells come close to it to pass through (Muto, 1976). SEM also visualized the process of platelet formation by megakaryocytes; the cells extend, through the sinus endothelium, beaded ribbons of cytoplasm, which later become separated into platelets (Ihzumi et al., 1977).

The sinus endothelium is incompletely covered by adventitial cells which extend processes connecting with those of reticular cells. Supported by the reticular cells are hemopoietic cells. Among interesting structures demonstrated by SEM, erythroblastic islands are noted. A group of immature erythrocytes, characterized by irregular indentations on the cell surface, gather around a macrophage nursing them. The three-dimensionally complex intercellular relation in the erythroblastic islands may be more precisely visualized by SEM than by transmission electron microsocpy of thin sections (Fujita and Kashimura, 1981).

The three-dimensional structures of reticular cells extending wing-like processes are suitable objects of SEM. Intermediate forms of the reticular cell and a rounded adipose cell may be recognized, presenting an interesting subject of study on the relationship and transformation between immunohematopoietic and adipose tissues.

THE SPLEEN

Successful visualization of the red pulp has been attained by complete arterial perfusion of the spleen with warmed saline followed by a fixative. The controversial structure of the sinus wall composed by transversely connected rod cells and supported by ring fibers and reticular cell processes has been unequivocally demonstrated by SEM, among others in human spleen (Fujita, 1974) (Figs. 2 and 3). Thread-like microprojections of unknown functions have been discovered (Fujita, 1974) and pathological changes in the morphology of rod cells have been demonstrated (Fujita et al., 1985).

The nature of macrophages independent of and untransformable to reticular and endothelial cells has been proved by our SEM studies (Miyoshi and Fujita, 1971; Fujita, 1974; Fujita, 1978). SEM unequivocally reveals where a macrophage occurs and how it extends delicate microprogections and filopodia to unexpected sites on the reticular and endothelial cells. Contact of a macrophage with a lymphocyte was often encountered and interpreted as the image of antigen presentation.

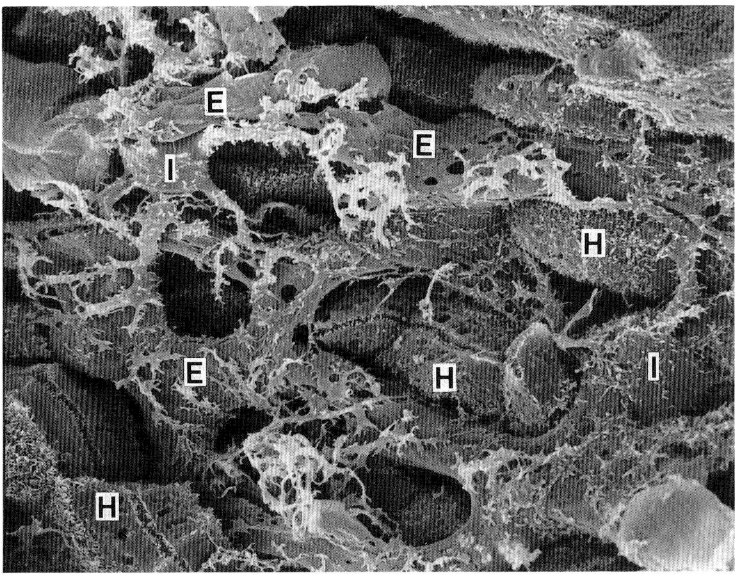

Figure 1. Ito cells (I) extending long branched processes. E: endothelium, H: hepatocyte. Rat liver after maceration with NaOH. X1,500. (Takahashi-Iwanaga and Fujita, 1986).

The perisinusoidal space of Disse regularly contains fat-storing cells of Ito. Combined use of secondary and back-scattered electron beams enabled us to visualize fat droplets (osmificated) in the cytoplasm together with the surface structure of Ito cells (Ushiki and Fujita, 1986). We also could demonstrate elaborate extensions of fine processes of Ito cells circularly embracing the sinuoid (Takahashi-Iwanaga and Fujita, 1986) (Fig.1).

The sinusoid under the SEM frequently reveals a lymphocyte-like round cell with irregular-shaped microvilli of a moderate number. This cell, which is attached by the endothelium even after vigorous perfusion, corresponds to the pit cell designated by Wisse et al. (1976) which recently has turned out to be natural killer (NK) cell (Kaneda et al., 1983). It is desired that identification under the SEM of this cell of immunological importance will be established.

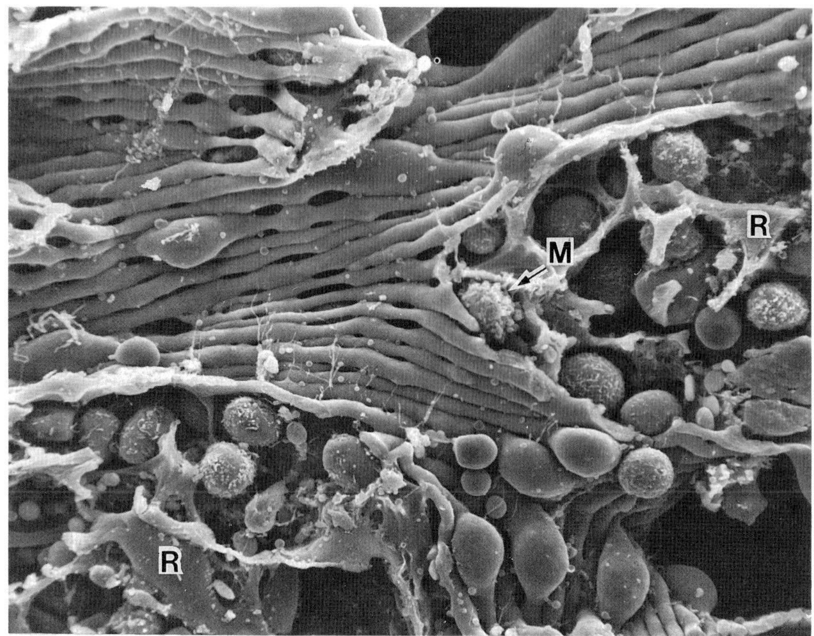

Figure 2. Human spleen revealing the lattice work of rod cells forming the sinus and the sponge-like cord tissue supported by reticular cells (R). M: macrophage. X1,000.

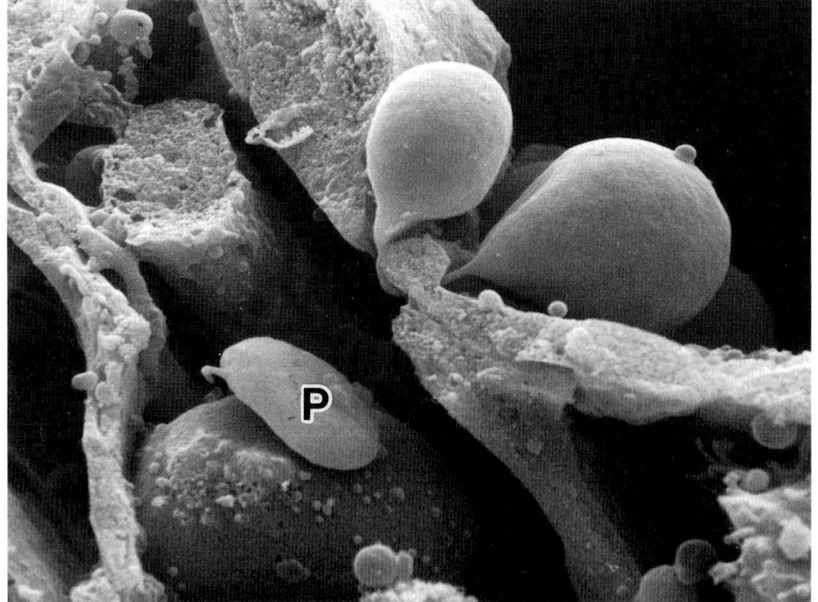

Figure 3. A red cell hanging on the rod cell lattice. P: blood platelet. X7,000.

The long disputed problem on the mode of arterial termination in the spleen, i.e., whether penicillar arteries open into the meshwork of splenic cords (open theory) or they are directly connected to the sinuses has been investigated by SEM. In addition to the observation of freeze-cracked tissues, resin casts of blood vessels were examined under the SEM. We first demonstrated evidence indicating the open circulation in human red pulp, and detailed description was made on the fine structure of arterial termination (Fujita, 1974; Irino et al. 1977; Fujita and Kashimura, 1983; Fujita et al., 1985). Recently, we recognized that at certain sites in human red pulp there are nests of labyrynthine blood vessels which are connected to penicillar arteries on one hand, and to thin sinuses on the other hand (Kashimura, 1985). At least in human spleen, the closed circulation occurs only in a small portion of the organ, while the open circulation is at the base and serves the function of the red pulp as the filter of blood.

The white pulp and marginal zone were also studied by SEM. The marginal zone receives many arterial terminals and also shows migrating lymphocytes (Fujita and Kashimura, 1983; Fujita et al., 1985). SEM is expected to serve elucidation of lymphocyte migration within the spleen and to other organs (Pabst, 1988).

LYMPH NODES AND TONSILS

The fundamental constructions of the lymph nodes and tonsils have been clearly visualized by SEM (Fujita et al., 1972; He, 1985). Lymphatic labyrynths originating from the vicinity of germinal centers and of high-endothelial venules (HEV) and pouring into sinuses were discovered in rat lymph nodes by SEM (He, 1985) (Fig. 4).

The HEV with unique stellate endothelial cells interdigitating with each other has been investigated by SEM in lymph nodes and tonsils. Lymphocytes with few or moderate numbers of microvilli have been observed to pass either intercellularly or intracellularly through the endothelium, and this image has been interpreted as the migration T lymphocytes from the blood to the thymus-dependent area surrounding the HEV (Umetani, 1977; Fujita and Kashimura, 1981; He, 1985) (Fig.5).

THE THYMUS

The three-dimensional architecture of the thymus was recently clarified by Ushiki (1986). The fine surface structures and mutual relations of the epithelial cells, lymphocytes, macrophages and interdigitating cells (IDCs) were demonstrated by SEM. IDCs often embraced some lymphocytes. Passage of lymphocytes into the general circulation was suggested to take place via a particular perivascular space and the endothelium of postcapillary venules (Ushiki, 1986).

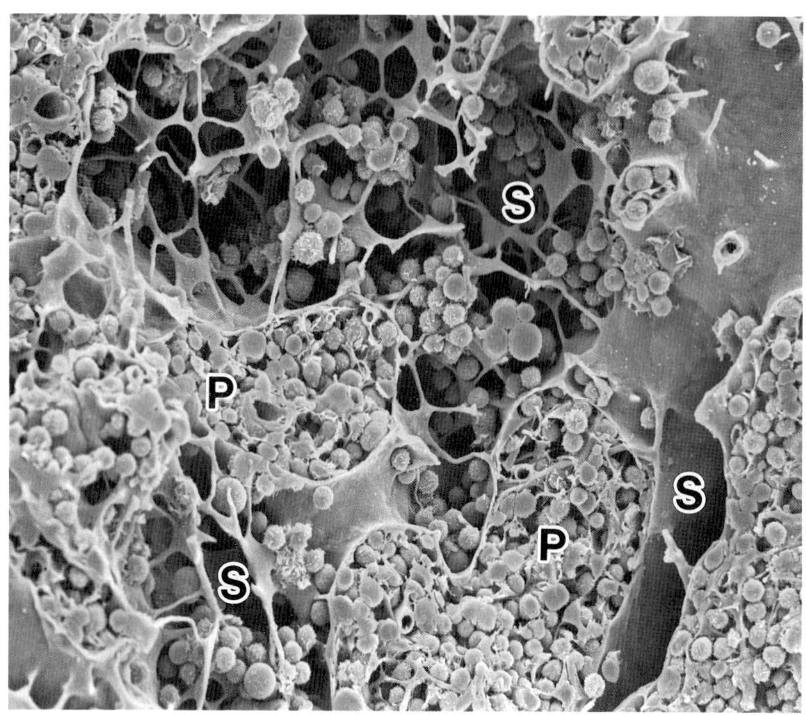

Figure 4. Rat mesenteric lymph node, showing medullary sinuses (S) and pulp (P). Note the flat endothelium separating both tissues. X660.

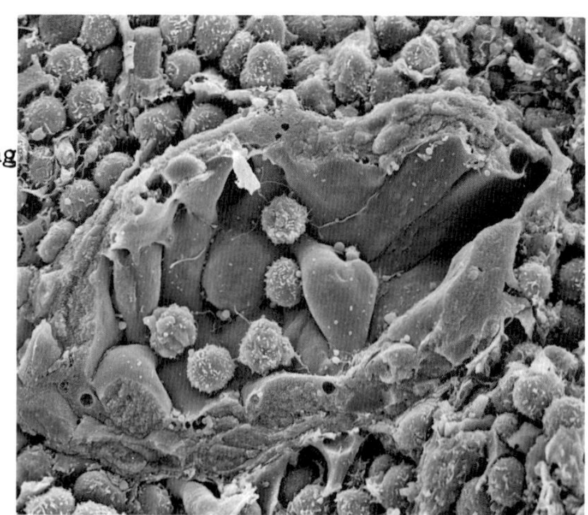

Figure 5. High-endothelial venule with five lymphocytes presumably immigrating through the endothelium. Rat lymph node. X1,300. Figs.4 and 5 were reproduced from He, 1985.

REFERENCES

Fujita T (1974). A scanning electron microscope study of the human spleen. Arch Histol Jap 37:187-216.

Fujita T (1978). Microarchitecture of reticular tissues. Reevaluation of the RES by scanning electron microscopy. Recent Advances in RES Research 18:1-19.

Fujita T, Kashimura M (1981). The "Reticulo-endothelial system" reviewed by scanning electron microscopy. Biomed Res 2 suppl: 159-171.

Fujita T, Kashimura M (1983). Scanning electron microscope studies of human spleen. Surv Immunol Res 2:375-384.

Fujita T, Kashimura M, Adachi K (1985). Scanning electron micro scopy and terminal circulation. Experientia 41:167-179.

He Y (1985). Scanning electron microscopy studies of the rat mesenteric lymph node with special reference to high-endothelial venules and hitherto unknown lymphatic labyrinth. Arch Histol Jap 48:1-15.

Ihzumi T, Hattori A, Sanada M, Muto M (1977). Megakaryocyte and platelet formation: a scanning electron microscope study in mouse spleen. Arch Histol Jap 40:305-320.

Irino S, Murakami T, Fujita T (1977). Open circulation in the human spleen. Dissection scanning electron microscopy of conductive-stained tissue and observation of resin vascular casts. Arch Histol Jap 40:297-304.

Kaneda K, Dan C, Wake K (1983). Pit cells as natural killer cells. Biomed Res 4:567-576.

Kashimura M (1985). Labyrinthine structure of arterial terminals in the human spleen, with special reference to "closed circulation". A scanning electron microscopy study. Arch Histol Jap 48:279-291.

Miyoshi M, Fujita T (1971). Stereo-fine structure of the splenic red pulp. A combined scanning and transmission electron microscope study on dog and rat spleen. Arch Histol Jap 33:225-246.

Motta P, Muto M, Fujita T (1978). "The Liver. An Atlas of Scanning Electron Microscopy." Tokyo: Igaku-Shoin.

Muto M (1976). A scanning electron microscopic study on endothelial cells and Kupffer cells in rat liver sinusoids. Arch Histol Jap 37:369-386.

Muto M, Nishi M, Fujita T (1977). Scanning electron microscopy of human liver sinusoids. Arch Histol Jap 40:137-151.

Pabst R (1988). The role of the spleen in lymphocyte migration. In Husband AJ (ed): "Migration and Homing of Lymphoid Cells." Boca Raton: CRC Press, pp 63-84.

Takahashi-Iwanaga H, Fujita T (1986). Application of an NaOH maceration method to a scanning electron microscopic observation of Ito cells in the rat liver. Arch Histol Jap 49:349-357.

Umetani Y (1977). Postcapillary venule in rabbit tonsil and entry of lymphocytes into its endothelium: a scanning electron microscope study. Arch Histol Jap 40:77-94.

Ushiki T (1986). A scanning electron-microscopic study of the rat thymus with special reference to cell types and migraion of lymphocytes into the general circulation. Cell Tiss Res 244:285-298.

Ushiki T, Fujita T (1986). Backscattered electron imaging. Its application to biological specimens stained with heavy metals. Arch Histol Jap 49:139-154.

Wisse E (1977). Ultrastructure and function of Kupffer cells and other sinusoidal cells in the liver. In Wisse E, Knook DL (eds): "Kupffer Cells and Other Liver Sinusoidal Cells." Amsterdam: Elsevier/North-Holland Biomedical Press, pp 33-60.

Eye and Ear

Cells and Tissues: A Three-Dimensional
Approach by Modern Techniques in Microscopy,
pages 503–510

SCANNING ELECTRON MICROSCOPY IN OPHTHALMOLOGY

P. VERSURA

Institute of Clinical Electron Microscopy,
University of Bologna, Italy
Via Massarenti, 9 - 40138 Bologna

INTRODUCTION

Scanning Electron Microscopy (SEM) plays an important role in studying ocular anatomy and physiology as it renders the complex interrelationships among the different structures more comprehensive. The ocular tissues have constituted a classical domain of SEM, since the eye, as a virtually hollow apparatus, exposes several free surfaces. The complexity of the structures in the eyeglobe has been clarified by SEM, thus yealding to new interesting morpho-functional correlations.

The present paper briefly reviews the major results obtained by SEM in this field, both from our personal experience and from the literature. It is subdivided into small sections dealing with the cornea, the conjunctiva, the trabecular meshwork, the uveal tract, the zonular apparatus, the lens, the retina. Because of the space the list of the references is forcedly limited but anyway essential. The micrographs derive from post-mortem human eyes, enucleated for corneal transplantation within 5 hours from the death.

Cornea. The clinical investigation of the corneal epithelium is usually accomplished by the use of a biomicroscope, which provides a three-dimensional view at 8-10 times magnification. Therefore, SEM appears the more appropriate consequent approach for the structural study of this tissue, as it enhances at higher magnification the 3-D images already familiar to the clinicians.

Human corneal epithelium consists of a mosaic of roughly polygonal cells which display a variable degree in size and

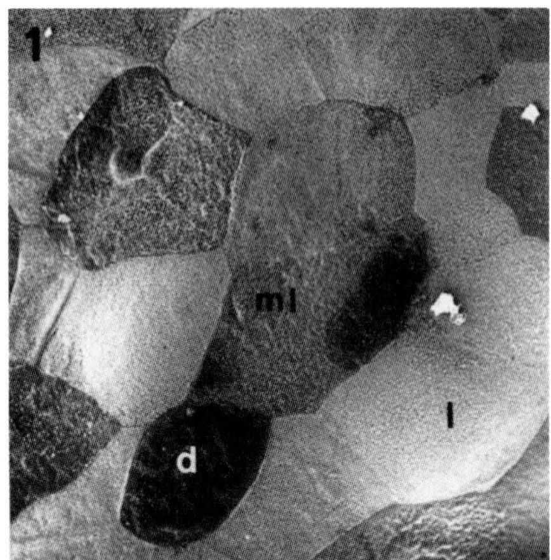

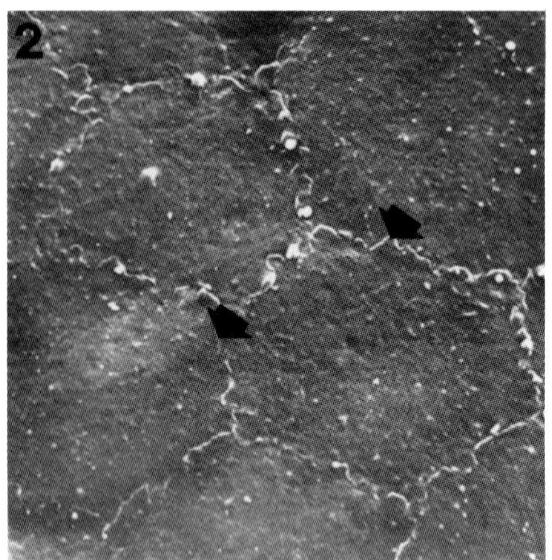

shape. These cells are usually classified, according to their brightness, into light, medium light and dark cells (fig.1). The brightness appears to be related to the number and type of microprojections present on the cell plasmamembrane, being the light and medium light cells the vital cells, while the dark cells are the older ones immediately before death. According to the "x,y,z hypothesis" (Thoft and Friend, 1983) the integrity of the corneal epithelium is maintained by the coordinated work of three independent movement of cells: basal cells from peripheral cornea move centripetally (y movement), then proliferate and mature while moving outward (x movement), and finally exfoliate (z movement). Therefore, in any case of epithelial surface abnormality SEM offers the opportunity to evaluate the wrong movement of the cells involved, by quantitatively analysing the relative presence of the three cell types.

The monolayered corneal endothelium faces the anterior chamber. As it is believed that any damage to this fragile tissue is healed by the spreading of the adjacent cells more than by their proliferation, the quantitative analysis of the number, size and morphology of endothelial cells is clinically carried out by the specular microscopy. SEM, again, represents the enhancement at ultrastructural level of the images to which the clinician is used. In man, four zones are observed, from the central Zone I, in which many intercellular overlaps are located (fig.2), to Zone IV at the border with the trabecular meshwork, in which irregular endothelial cells connected by some less overlapping are present (Svedbergh and Bill, 1972). The endothelial cell plasmamembrane is actively involved in maintaining the corneal deturgescence, therefore corneal edema can result due to mechanical or chemical injuries to this

Fig. 1. Light (l), medium light (ml) and dark (d) cells of the corneal epithelium. SEM 570 x

Fig. 2. Endothelial cells in the central zone show prominent overlapping (arrows). SEM 2,590 x

---

tissue. The literature on SEM studies in this respect is wide and is summarized in (Versura and Maltarello, 1988).

Conjunctiva. The various zones of the human conjunctiva show a different morphology in SEM (Versura et al., 1985). The upper tarsal zone consists of polygonal cells with fine and regular microvilli on the surface which represent (as the microvilli on the corneal cells), a mechanical support for the preocular tear film. The mucus-producing goblet cells are located in different functional stages among the epithelial cells. Another source of mucus, the Second Mucus System (SMS), consists of subsurface secretory granules located in the non-goblet epithelial cells. At SEM these cells show microvilli grouped in tufted structures cemented by the extruding mucus. This feature appears particularly represented in the contact lens-related conjunctivitis. As the SMS vesicles produce a mucus different from that of the goblet cells, an indirect evaluation on the mucus composition is provided by the SEM observation of the microvillar pattern rearrangement.

Trabecular meshwork. The observation of the whole-mounted anterior chamber by SEM provides a gonioscopic study at ultrastructural level of the angle, which is of particular relevance, both from clinical and research point of view. The trabecular meshwork (TM) acts as a sieve through which the aqueous humour runs in its way from the anterior chamber to the Schlemm's canal (SC). A progressive differentiation both from a morphological and a functional standpoint occurs in the endothelial cells from the peripheral cornea to the trabecular area (fig.3). From the endothelial cells of the zone IV still firmly interconnected, going towards the angle, the intercellular edges become enlarged and some cytoplasmic bridges are present between the cells. These structures border the initial openings located between the two plasmamembranes. The trabecular endothelial cells have elongated shape and interconnect each other by cytoplasmic connections which delimitate the openings through which the aqueous flows away (Grierson and Lee, 1975). The spindle-shaped cells of the Schlemm's canal walls are oriented following the aqueous flow. The cells of the inner wall show openings of up to a size of

3–5 µm , which represent the major route for the bulk outflow of the aqueous humor from the trabeculum to the lumen of the canal (Tripathi, 1971). The frequency of these pores is rather low and therefore a great amount of serial ultrathin sections would be required to make a quantitative analysis. SEM overcomes the technical problems associated with Transmission Electron Microscopy (TEM) observation, but potential artefacts during collection and preparation of the specimens should be equally taken into account.

Uveal tract. The iris offers two free surfaces, one towards the anterior and the other towards the posterior chamber, which can be easily investigated by SEM. At the anterior border, elongated and flattened cells are located, intermixed among bundles of collagen fibers (Dickson, 1975). As this layer is discontinuous, also for the presence of holes and pores which deepen into the underlying connective tissue, the aqueous humor penetrate freely the iris. The stroma consists of a network of collagen fibers running parallel to the major axis and forming canals, the arrangement of which depends on the contraction of the iris muscles. In addition, layers of tightly woven collagen fibers surround the vessels, probably to prevent a kinking during mydriasis or to counterbalance the vessel pressure against the intraocular pressure (Van der Zypen, 1978). The posterior border is runned by both circularly and radially oriented furrows and at high magnification excrescences due to the underlying melanin granules of the pigmented epithelial cells are well visible.

The ciliary body (CB) offers free surfaces as well, but, due to its embryological development, what we see at SEM is the basement membrane of the non-pigmented cells and not their apices. The relationships between CB surface and zonular attachment will be discussed later on.

To speak of choroid means to speak of blood supply and therefore the observation of the vascular corrosion casts in SEM has yealded to important functional correlations (Hayreh, 1975), mainly as to the choriocapillaris arrangement in independent vascular districts.

Zonular apparatus. The zonule represents the main effector of accomodation and its anatomy directly reflects how this mechanism works. In man, two groups of fibers have been described: one group comes from the pars plana and goes to the anterior surface of the lens, the other group originates from the ciliary valleys and inserts at the anterior surface of the lens. Fibers from both the two groups insert at the equator

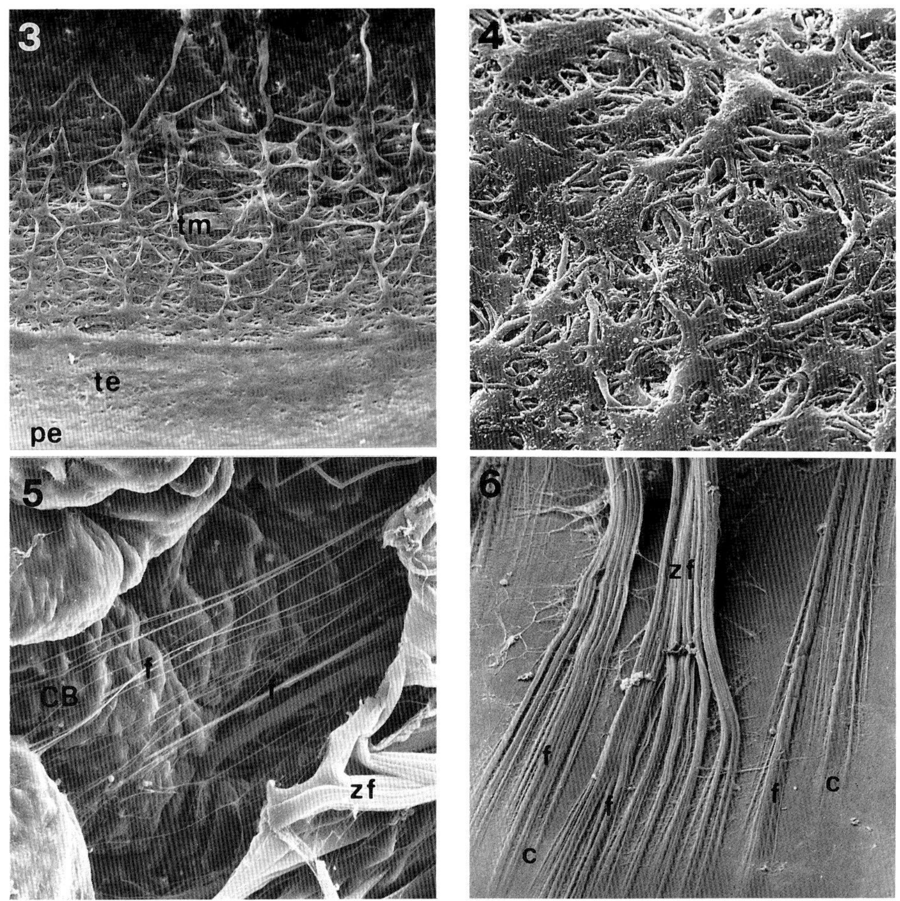

Fig. 3. Complete front-view of the angle. pe= peripheral endothelium; te= transitional endothelium; tm= trabecular meshwork. SEM 100x

Fig. 4. A discontinuous layer of cells with prominent interdigitating processes lines the anterior surface of the iris. SEM 836 x

Fig. 5. The zonular fibers (zf) adhere intimately to the ciliary body surface (CB) by thin fibrils (f). SEM 400 x

Fig. 6. The zonular fibers (zf) attach at the lens surface by dividing and thinning progressively into smaller fibrils (f), which apparently fuse with the capsule (c). SEM 325 x.

(Bornfeld et al., 1974; Farnsworth et al., 1976). In addition, short zonular fibers connect the posterior part of the CB processes with pars plana, anchoring the processes against the zonular pull during accomodation. A new theory of accomodation has been proposed on the basis of SEM observations (Rohen and Rentsch, 1969). This model assumes that the pars plana does not move forward during the accomodation.

The fibers thin progressively into fibrils which penetrate into the lens capsule approaching the epithelium.

Lens. The lens consists of a single type of cells which undergo a maturation process while moving from the germinative, to the cortical, to the nuclear zone. This organ actually is an ordered system of fibers connected by specific devices and only a minimal extracellular space. The presence and differential distribution of 3 distinct types of anchoring devices has been unequivocally shown by the three-dimensional appreciation of the fiber arrangement throughout the lens (Dickson and Crock, 1972). Ball-and-socket junctions are bulges and corresponding invaginations that fit into each other and are located on the lateral fiber surfaces. The interlocking protrusions are extroflessions of the apical and lateral edges that embrace analogous structures of the neighboring fibers. The microplicae are infolding and ridges of the lateral fiber membrane. As the density of ball-and-socket decreases from the cortical to the nuclear zone and there is a transitional region in which ball-and-socket and microplicae are simultaneously present, it is believed that these junctions are transient structures which can change during development. Such a complex of interconnecting junctions prevent any type of fiber movement and it is therefore believed that the lens shape changes during accomodation are accomplished by changes in the lenght of the fibers due to unfolding of bends and folds (Willekens and Vrensen, 1981).

Retina. As yet, the SEM studies relating to this organ have confirmed, clarified, highlighted and summarized previous knowledge (Borwein, 1985) and, apart from few works, the amount of really new information is not very large. The application of the methods of cracking to expose the internal structures and the subsequent combination with immunocytochemical techniques will be extremely useful to join neurophysiology and ultrastructure. This in order to reconstruct in 3-D functionally related neurons at a level not always allowed by the usually employed flat preparations of intact retinas observed in Light Microscopy.

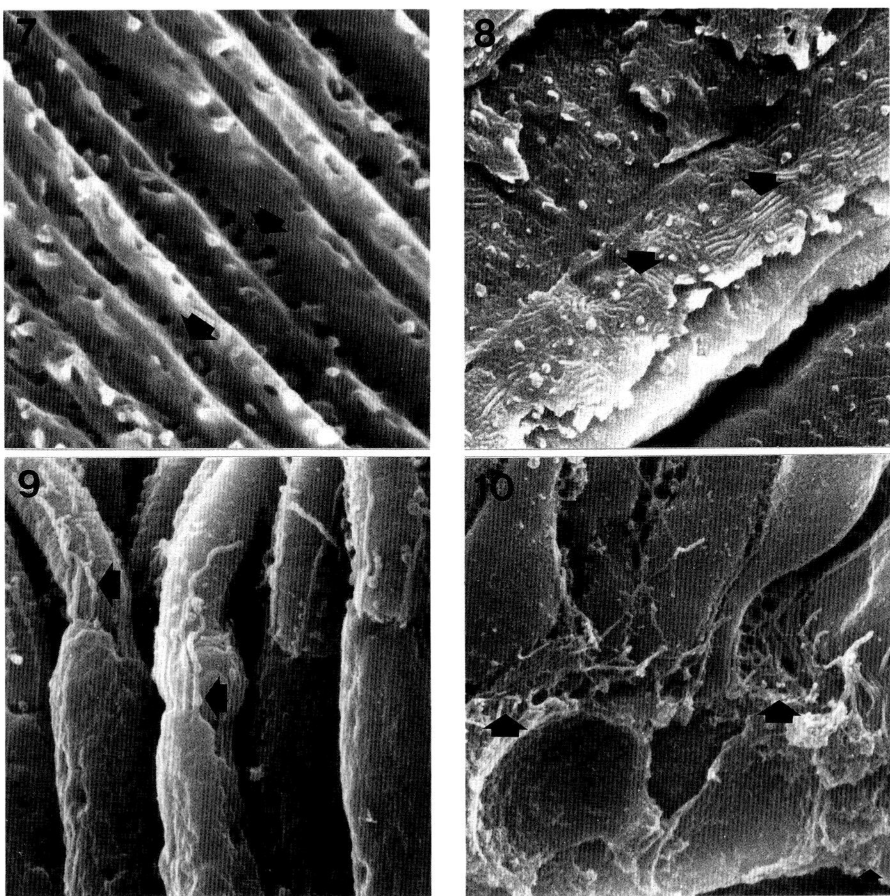

Fig. 7. Interlocking protrusions (arrows) in the deep cortex of the human lens. SEM 6,380 x

Fig. 8. Microplicae (or tongue-and-groove interdigitations, arrows) in the nucleus of the human lens. SEM 4,520 x

Fig. 9. Calyceal processes (arrows) encircle the base of a photoreceptor outer segment. SEM 8,200 x

Fig. 10. Mueller cell microvilli (arrows) surround the base of the photoreceptor inner segment, at the level of the external limiting membrane. SEM 4,520 x

REFERENCES

Bornfeld N, Spitznas M, Breipohl W, Bijvank G (1974). Scanning electron microscopy of the zonula of Zinn. I. Human eyes.Graefe's Arch Clin Exp Ophthalmol,192:117-129.

Borwein B (1985). Scanning electron microscopy in retinal research. Scanning Electron Microsc, 1985; I:279-301.

Dickson DH (1975). Fine structure and mechanics of the anterior border of the primate iris: a scanning and transmission electron microscopic study. Can J Ophth, 10:227–238.

Dickson DH, Crock GW (1972). Interlocking patterns of primate lens fibers. Invest Ophthalmol, 11:809–815.

Farnsworth P, Mauriello J, Burke P, Kulyk T, Cinotti A (1976). Surface ultrastructure of the human lens capsule and zonular attachments.Invest Ophthalmol Vis Sci,15:36–40

Grierson I, Lee WR (1975). The fine structure of the trabecular meshwork at graded levels of intraocular pressure: a qualitative analysis by light microscopy and scanning electron microscopy. Exp Eye Res, 19:505–521.

Hayreh SS (1975). Segmental nature of the choroidal vasculature. Br J Ophth, 59:631–648.

Rohen JW, Rentsch FJ (1969). The construction of the human zonular apparatus and its functional significance. Morphological basis for a new accomodational theory. Graefe's Arch Clin Exp Ophthalmol, 178:1–32.

Svedbergh B, Bill A (1972). Scanning electron microscopic studies of the corneal endothelium in man and monkeys. Acta Ophthalmol, 50:321–336.

Thoft RA, Friend J(1983).The "x,y,z" hypothesis of corneal epithelial maintenance.Invest Ophthalmol Vis Sci, 24:1442–1443.

Tripathi RC (1971). Mechanism of the aqueous outflow across the trabecular wall of Schlemm's canal. Exp Eye Res, 11:116–121.

Van der Zypen E (1978). The arrangement of the connective tissue in the stroma iridis of man and monkey. Exp Eye Res, 27:340–357.

Versura P, Bonvicini F, Caramazza R, Laschi R (1985). Scanning electron microscopy study of human cornea and conjunctiva in normal and various pathological conditions. Scanning Electron Microsc. 1985; IV:1695–1708.

Versura P, Maltarello MC (1988). The role of Scanning electron microscopy in ophthalmic science. Scanning Microscopy, in press.

Willekens B, Vrensen G (1981). The three-dimensional organization of lens fibers in the rabbit. A scanning electron microscopic reinvestigation. Graefe's Arch Clin Exp Ophthalmol., 216:275–289.

Cells and Tissues: A Three-Dimensional Approach by Modern Techniques in Microscopy, pages 511–514

SCLEROCOCORNEAL TRABECULA AND GLAUCOMA. A SCANNING ELECTRON MICROSCOPIC STUDY

Domingo Ruano-Gil and Jesus Costa-Vila

Department of Human Anatomy, Faculty of Medicine,
University of Barcelona
Avda, Diagonal s/n. (Anexo Fac. Farmacia),
Barcelona, A-Vila 08028 Spain

INTRODUCTION

Since Sondermann (1933) indicated the existence of morphological channels communicating the chamber fundus with Schlemm's canal, many authors have studied this subject (Ashton, 1951; Spencer, 1968; Bill, 1970; etc.). In previous studies we were able to show the presence of these channels between the anterior chamber and Schlemm's canal (Costa-Vila and Ruano-Gil, 1988). According to our studies, the spaces between the trabecular meshwork form Sondermann's interstices, and so it is clear that ciliary muscle contractions produce the dilatation and confluence of these interstices, aiding drainage of the aqueous humour from the anterior chamber to Schlemm's canal.

In cooperation with the Glaucoma Unit of the Department of Ophthalmology of our Faculty (Prof. Dr. D. Pita Salorio and Dr. C. Verges Roger), we studied the sclerocorneal trabecula in patients with open-angle glaucoma, where there was a blockage of the flow of the aqueous humor and consequently increases the intra-ocular pressure. After a detailed examination of the trabeculae obtained from patients with glaucoma, by means of therapeutic trabeculectomy, we noted the existence of magmatic plates covering the intrabecular interstices like a blanket, thus preventing communication between the chamber funds and Schlemm's canal.

After these observations, we thought that the most useful anatomical and physiological treatment would be to reconstruct these channels and restore their permeability. This required a cutting tool that would permit us to make small orifices in the sclerocorneal trabecula, which was accomplished with the Nd. Yag Laser.

In this study we describe the surgical technique used to perform the "trabeculotomy" with the Yag Laser and point out the results obtained by means of Scanning Electron Microsopy (SEM).

MATERIAL AND METHOD

We used ten normal eyeballs from the Eye Bank of the Hospital Clinico in Barcelona and ten trabeculae from patients with open-angle glaucoma, without any other relevant ocular or extraocular disease. These patients underwent "trabeculotomy" with the Yag Laser prior to filter surgery or "trabeculectomy". The "trabeculotomy" was performed with the Lasag Yag Microswitch Laser, using the multimode Q-switched system, with a spot diameter of 50 micra, wavelenqth of 1064 nm, exposure time of 15 nanoseconds and the Lasag CGA Goniolens.

After the cornea has been shaped into a "button" 9 mm in diameter, the normal eyes underwent Laser action, with the beam focused directly on the internal face of the sclerocorneal trabecula, in the middle area. In all cases, the convergence of the beam was 16°, with tree study groups formed according to the firing strength applied. Thus, one group was given a strength of 10 mJ. per impact, the second 12 mJ. and the third 14 mJ. They were all given a total of 40 coalescent impacts on an area of 30°.

After the trabeculotomy with the Yag Laser, both the eyeballs and the trabeculae of the operated patients were fixed according to Karnovky's method. They were dehydrated by critical point and coated with gold. To analyse them we used the Jeol SEM (between 10 and 20 Kv).

RESULTS

This study enabled us to analyse the morphology and arrangement of the sclerocorneal trabecular system that underwent the Yag Laser impacts.

In Fig. 1, a view with the SEM of the chamber fundus, we could see innermost fibers of the sclerocorneal trabecula of the patient with open-angle glaucoma who underwent filter surgery (trabeculectomy) without having previously been submitted to Laser action. The Laser impacts with a firing. strength of 10 mJ, and further how the innermost layers of the sclerocorneal trabecula are sectioned without exceeding 3 or 4 layers (Fig. 2).

In those cases where the trabeculotomy with the Yag Laser was performed using firing strengths of 14 mJ., the study by

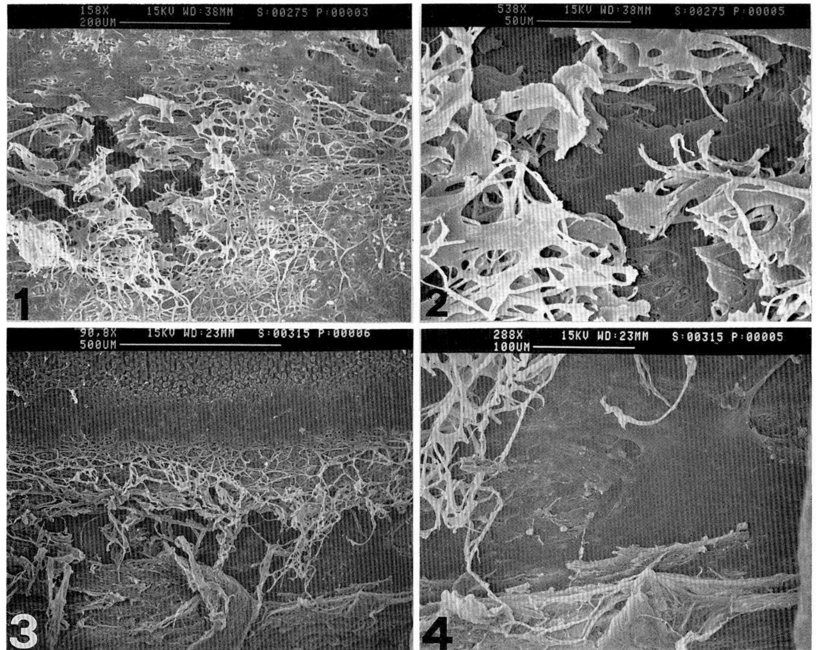

Figs. 1 and 2: Results with the Yag Laser, using low and-medium intensity impacts. The points of impact only penetrated into 2 or 4 layers of the inner surface of the sclerocorneal trabecula, and did not re-establish communication between the anterior chamber and Schlemm's canal.

Figs. 3 and 4: Yag Laser using high impacts. The sclerocorneal trabecula is totally sectioned, and communication is re-established between Schlemm's canal and the anterior chamber. This was accompained by a normalisation of intraocular pressure.

SEM of the chamber angle fundus of a piece from the filter surgery undergone by a patient with open angle glaucoma, following the trabeculotomy (Fig. 3), shows the complete section of the sclerocorneal trabecula situated between the scleral spur and septum, thus converting Schlemm's canal into open channel. In this case we could see the anterior endothelium of Schlemm's canal complete and without significant neighbouring impairment (Fig. 4).

DISCUSSION

The development of a new technique involves an evaluation

of the effects achieved. In the case of the "trabeculotomy" with the Yag Laser, the intention is to obtain the linear opening of an area of greater or lesser extend in the sclerocorneal trabecula, in order to establish a direct connection between the anterior chamber and Schlemm's canal. We have shown that with this technique we achieve these objectives, aiding free passage between the chamber funds and Schlemm's canal through the "trabeculotomy". Thus it is reasonable to belive that a certain degree of optimism may be permitted with regard to the technique of trabeculotomy with Yag Laser, since both the experimental and clinical results demonstrated its effectiveness immediately and in the short and medium term, with few complications.

In fact, detailed analysis of the preparations allowed us to see that, after undergoing the effects of the impacts, the structure of the slerocorneal trabecula breaks down to a greater or lesser depth depending on the amount of energy applied. Thus, in the cases where strengths of 10 and 12 mJ were applied, it was impossible to penetrate beyond 3 or 4 layers, without at any time reaching the interior of Schlemm's canal. On the other hand, where the firing strengths were sufficiently powerful (14-16 mJ.), we succeeded to section all the layers of the sclerocorneal trabecula until the interior of Schlemm's canal was reached, thus connecting the chamber funds with the canal, and so permitting the aqueous humour to circulate freely. Moreover, from a study of the side effects that the Laser impact may have caused both in the endothelium of the anterior wall of Schlemm's canal and in adjacent tissues, significant impairment was not observed.

REFERENCES

Ashton N (1951). Anatomical study of Schlemm's canal and aqueous veins by means of Neoprene casts. Br J Ophthalmol 35: 291-303.

Bill A (1970). Scanning electron microscopy studies of the canal of Schlemm. Exp Eye Res 10: 214-218.

Costa-Vila J, Ruano-Gil D (1988). Scanning electron microscope study of the communication between the anterior chamber and the sinus venous sclerae in primates. Acta Anat 131: 342-345.

Sondermann R (1933). The formation morphology and function of Schlemm's canal. Acta Opthalmol 14: 1-8.

Spencer WH, Alvarado J, Haeyes TL (1968). Scanning electron microscopy of human ocular tissues: trabecular meshwok. Invest Ophtalmol Vis Sci 7: 651-654.

Cells and Tissues: A Three-Dimensional
Approach by Modern Techniques in Microscopy,
pages 515–519

SCANNING ELECTRON MICROSCOPIC STUDY OF THE MICROCIRCULATION OF THE RABBIT'S ANTERIOR UVEA

J. Goyri O'Neill, J. Esperança-Pina

Department of Anatomy, Faculty of Medical Sciences
New University of Lisbon, 1198 Lisbon, Portugal

INTRODUCTION

The study of microvascularization of the anterior uvea acquired increased importance owing to different pathological conditions, the most significant being glaucoma, in order to clarify the trophic changes in the iris, and also the formation of new vessels.

The iris is formed by an anterior layer, mesodermic in origin, and a posterior one with ectodermic origin, forming two layers of pigmentd epithelial cells continuous with the two layers of retinal epithelium that line the ciliary body. Between both, the muscular fibers of the iris are derived from the anterior cells of the pigmented epithelial layer, the constrictor of the iris a compact layer of circular smooth muscle fibers near the pupilar margin. The dilator of the iris is formed by myo-epithelial cells and among them nervous fibers, with Schwann cells surrounding the muscle.

The ciliary processes consist chiefly of capillaries supported by connective tissue, covered with two epithelial layers, being the deeper one pigmented. They play an important role in the acommodation and nutrition of the anterior chamber, and capillaries of the ciliary processes produce most of the aqueous humor that passes from the posterior chamber between the lens and the iris to enter the anterior chamber, moving towards the angle of the iris, and entering the trabecular spaces being absorbed into Schlemm's canal.

In serious pathological conditions of severe ocular hypertension the iris suffers trophic morphological changes that are in most of the cases irreversible, with atrophic zones with angiographic translation.

RESULTS

We performed our study in 35 New Zealand rabbits owing to the fact that the non pigmented iris allows a clear morphological study of iris vessels, being their distribution similar to the human. (Fig.1)

The vascularization of the iris and ciliary processes derives from the posterior long ciliary arteries that run through the horizontal axis of the globe dividing into two vessels of lesser caliber superior and inferior, which surround the margin of the iris anastomosing with each other and forming the major arterial circle of the iris, the caliber of which varies between 100 and 150 µm (Fig.2)

The major arterial circle receives the anterior ciliary arteries. From the major arterial circle arise different branches, namely the arterioles of the ciliary processes, the anterior recurrent arteriole to vascularize the choroid, and the anterior arteriole of the iris.

Each of the arterioles of the ciliary processes runs to a ciliary process. They are eighty in number forming the vascular axis of the ciliary processes and an elongated vascular net, their caliber varies between 35 anf 70 µm, being the greater caliber near their emergency.

The radiating arterioles (fig.3) arise from the concavity of the major arterial circle and run towards the pupil anastomosing and forming near the constrictor of the iris the incomplete minor arterial circle of the iris (fig.4). Their caliber gradually diminishes as they give different arteriolar branches which vascularize the dilator of the iris.

In our cases, we observed that the arterioles of the Major Circle in 18 cases varied between 150 and 130 µm. In 30 cases their caliber varies between 120 and 110 µm. In 22 of the cases they presented a caliber between 110 and 100 µm. The vessels of larger caliber were observed more frequently at 3 o'clock and 9 o'clock.

As to the radiating arterioles at 12 o'clock the values of their caliber was between 32 and 50 µm. At 3 o'clock the values of their caliber was between 25 and 40 µm. At 6 o'clock the values of caliber varied between 35 and 48 µm and at 9 o'clock between 25 and 45 µm.

The arterioles of the ciliary processes at 0 o'clock presented a medium caliber of 45 µm. at 3 o'clock presented a medium calber of 42 µm, at 6 o'clock presente a medium caliber of 40 µm and at 9 o'clock presente a medium caliber of 45 µm.

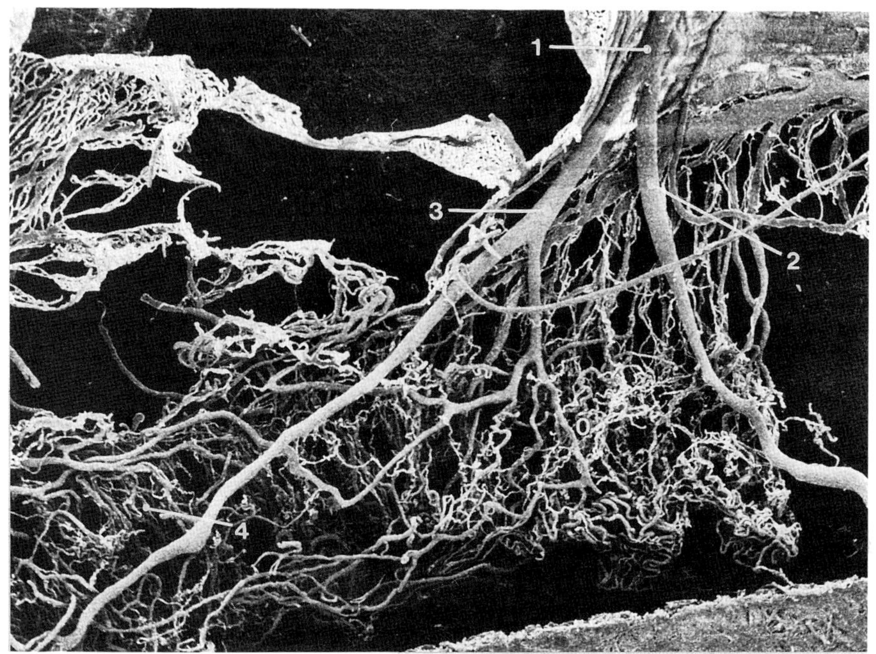

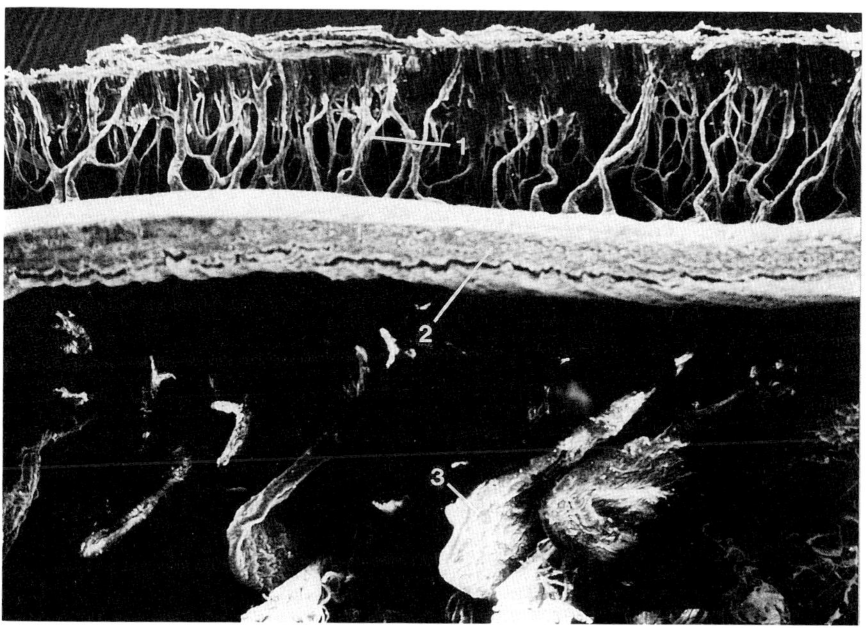

Fig.1 - Scanning electron microscopy - 34x. 1 -Iris processes; 2 -Iris:; 3 -Ciliary processes

Fig.2 -Scanning electron microscopy on corrosion casts-24x. 1-Ciliary posterior artery; 2-Inferior branch of the Major Arterial Circle; 3-Superior branch of the Major Arterial Circle.

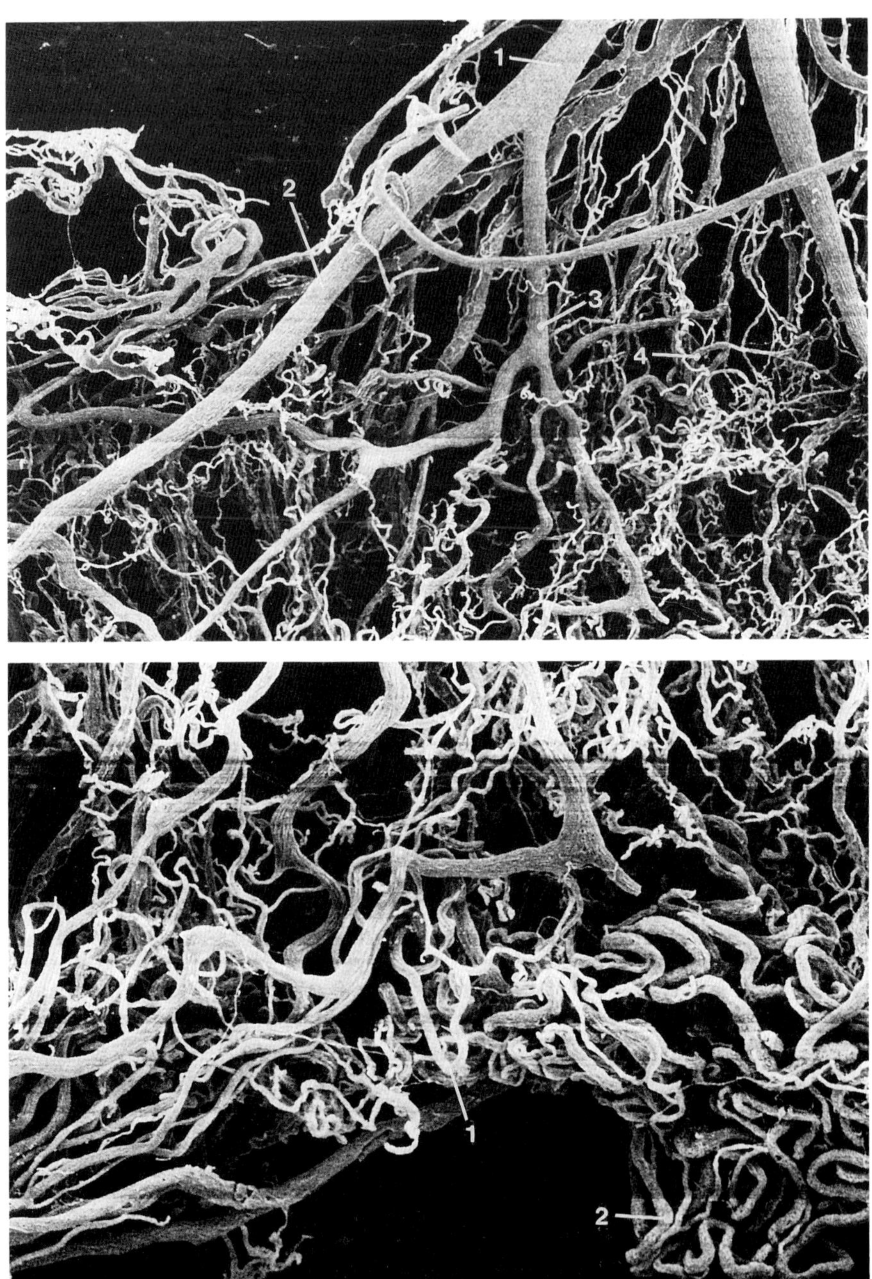

Fig.3 - Scanning electron microscopy on corrosion casts - 45x. 1 -Superior branch of the Major Arterial Circle; 2 -Major Arterial Circle; 3 -Radiating arteriole; 4 -Stroma vessels.

Fig.4 - Scanning electron microscopy on corrosion casts - 89x. 1 -Arterioles of the constrictor of the iris.

REFERENCES

Allen L, Burian HM, Braley AE. (1959). A new concept of the anterior chamber angle. Arch. Ophthalmol. 62: 966-971.

Ashton N, Brini A, Smith R. (1956). Anatomical studies of the trabecular meshwork of the normal human eye. Br.J.Ophthalmol. 40: 247-251.

Bellhorn RW.(1980) Control of blood vessel development. Trans. Ophthalmol. soc. UK. 3: 328-31.

Bill A. (1975). The drainage of aqueous humor. Invest. Ophthalmol. 14: 1-10.

Flocks M. (1965). The anatomy of the trabecular meshwork as seen as seen in tangential section. Arch. Ophthalmol. 56: 708-712.

Francois J, Duke-Elder S. (1955). A symposium, Springfield, Ill, Charles C. Thomas, Publishers.

Freddo TF, Raviola G. (1982). The homogenous stucture of blood vessels in the vascular tree of Macaca mulata iris. Invest.Ophthalmol. Vis. Sci. 3: 279-91.

Fujita H, Kondo K, Sears M. (1984). A new function of the non-pigmented epithelium of ciliary processes in the formation of aqueous humor. Klin. Monatsbl. Augenheilkd. 1:28-34.

Grierson L, Lee WR. (1974). Junctions between the cells of the trabecular meshwork. Albrecht v. Graefes Arch. Klin. Exp. Ophthalmol. 192: 89-93.

Hogan MJ, Alvarado JA, Weddell JE. (1971). Histology of the human eye. Philadelphia. W.B. Saunders Co.

Morrison JC, Van Buskirk EM. (1983). Anterior collateral circulation in the primate eye. Ophthalmol. 6: 707-15.

Moses RA. (1977). The effect of intraocular pressure on resistance to outflow: a review, Surv. Ophthalmol. 22: 88-99.

Risco JM, Nopanitaya W. (1980). Ocular microcirculation. Scanning electron microscopic study. Invest. Ophthalmol. Vis. Sci. 1: 5-12.

Rohen JW, Lutjen-Drecoll E, Futa R. (1979). Fine structure of trabecular meshwork in normal and glaucomatous eyes as seen in tangential sections. Invest. Ophthalmol. Vis. Sci. 18: 240-1990.

Salzamann M. (1980) The anatomy and histology of the human eyeball. Chicago Press.

Scremin OU, Sonnenschein RR, Rubinstein EH. (1982). Cerebrovascular anatomy and blood flow measurements in the rabbit. J. Cereb. Blood Flow Metab. 1: 55-66.

Ueda A, Nishida T, Otori, Fujita H. (1987). Electron-microscopic studies on the presence of gap junctions between corneal fibroblasts in rabbits. Cell. Tissue. Res. 2: 473-5.

Cells and Tissues: A Three-Dimensional
Approach by Modern Techniques in Microscopy,
pages 521–525

SEM-STUDIES OF THE NICTITATING MEMBRANE OF THE PIGEON

Wolfgang Kühnel and Uda Schramm

Institut für Anatomie, Medizinische Universität zu Lübeck, Ratzeburger Allee 160, 2400 Lübeck 1, F.R.G.

INTRODUCTION

The eyelids or palpebrae are essentially movable folds of skin which protect the cornea and sclera both from injury and excessive light. Each lid is covered by a thin skin, which is modified to form a transparent membrane, the conjunctiva on the posterior surface. The inner surfaces of the lids are lined with this conjunctiva which is in contact with the eyeball except near the medial angle, where a vertical fold of conjunctiva intervenes, also called the plica semilunaris conjunctivae, the third eyelid, or the nictitating membrane, containing a plate of cartilage in many animals.

Birds and reptiles posses a very well developed nictitating membrane. The fast and continual movements of the nictitating membranes of chickens was observed in photograms by De Angelis and coworkers (1985). The movements occur through the contraction of special muscles which rapidly stretch the membrane and the passive retractions which take place as a consequence of its elastic structure.

Kolmer (1923) as well as Kajikawa (1923) investigated the nictitating membrane of the pigeon and some reptiles. On the inner surface of the nictitating membrane they found a special form of epithelium which they called "gefiedertes Epithel" which could be translated in english as "looking like feather dusters" (Fig. 1). The long and pointed extensions of the superficial cells look like feathers or resemble the needles of a christmas tree. TEM observations showed a pseudostratified epithelium consisting of dark basal cells and light and sparce dark superficial cells. The light microscopically striking and impressive apical processes are formed by cytoplasmatic extensions which are provided with microvilli and long branched cytofila (Bielek, 1976; Schramm and Kühnel, 1986).

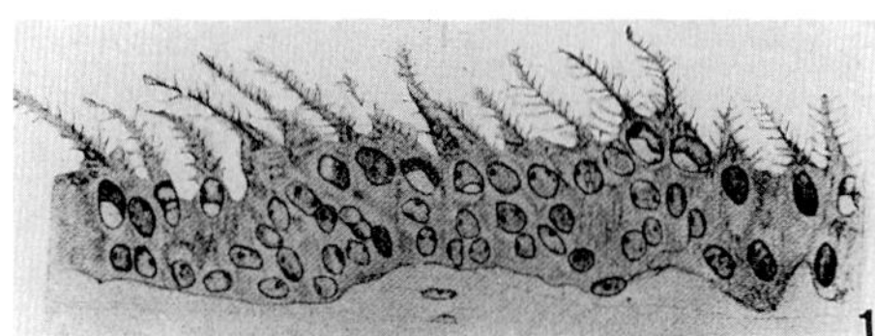

Figure 1. The so called "gefiedertes Epithel" of the nictitating membrane of the pigeon. Original drawing of Kolmer (1923).

However, the arrangement of apical featherlike projections of the nictitating membrane is far from being fully understood and, in particular, their three-dimensional organization still has to be analyzed. The present study reports on the three-dimensional surface relief of the third eyelid of the pigeon, as revealed by scanning electron microscopy. A preliminary account has been presented elsewhere (Schramm and Kühnel, 1986).

RESULTS AND DISCUSSION

The third eyelid of the pigeon is a transparent membrane demonstrated in figure 2 in its entirety. The membrane is pulled over the bulbus. As one can see, it is a thin, nearly transparent curtain. The outer surface or skin is covered by stratified squamous epithelium. The stroma consists of compact collagen fibers and abundant elastica including fibrocytes, blood vessels and myelinated nerve bundles. The inner surface is covered with the feathered epithelium (Fig. 3).

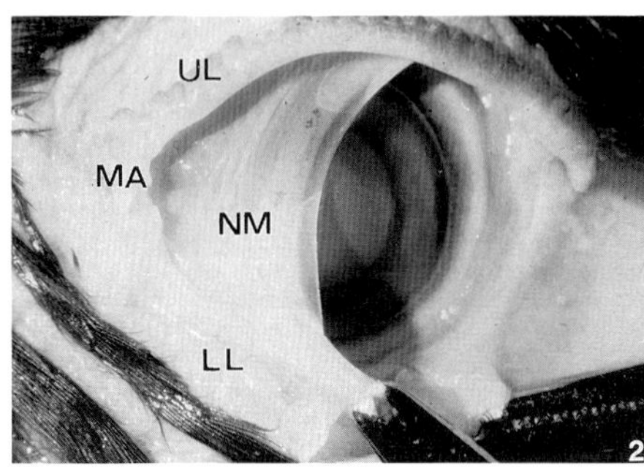

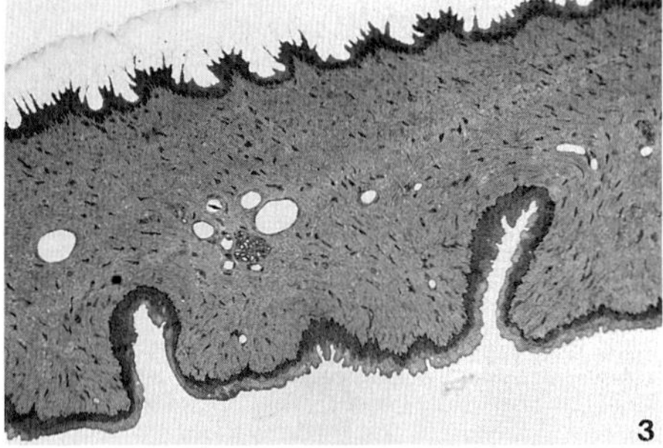

Figure 2. Lids and eye of the pigeon. UL = Upper lid, LL = Lower lid, NM = Nictitating membrane (third eyelid) pulled over the bulbus with a pincette. MA = Medial angle of the palpebral fissure.

Figure 3. Part of the nictitating membrane, vertical section. The feathered epithelium at the top, the outer stratified squamous epithelium below. Big blood vessels and nerves are located approximately in the middle of the membrane. Semi-thin section stained with azure II-methylenblue. x 110

The lid margin marks the transition between skin and inner surface. The keratinized stratified squamous ectoderm of the former becomes the nonkeratinized stratified squamous epithelium of the latter and is called the limbus region (Fig. 4). In the transition zone between the limbus and the actual feathered epithelial surface the epithelial cells have long, slim tentacle-like processes (microvilli) that are bent and superficially covered with small spherical structures. The ends of these long microvilli are often spherically distended. These processes resemble a chain of pearls (Fig. 5).

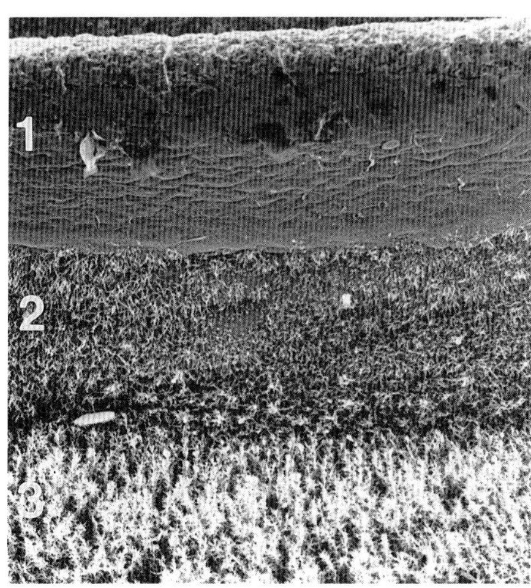

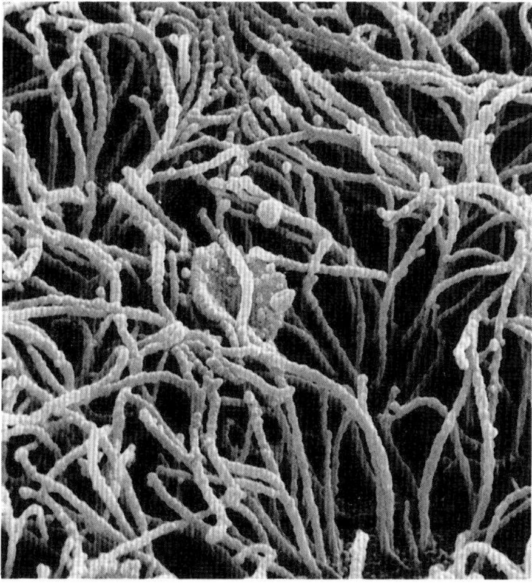

Figure 4. A scanning electron micrograph showing the inner surface of the nictitating membrane. 1 = Limbus, 2 = Transition zone, 3 = Feathered epithelium. x 400
Figure 5. A part of the epithelium of the transition zone with long, slim tentacle-like processes. x 3 700

The following surface area of the nictitating membrane is not only equipped with microvilli-like thin precesses. Here there are very differently structured apical protrusions. There are occasional thick outfoldings of the plasmalemma which fork several times resembling the branches of a brush or a small tree (Figs. 6, 8). At other points protrusions structurally reminiscent of long feather dusters appear (Fig. 7), or club and dumbbell shaped protrusions from which thin processes again extend (Fig. 9). In all cases spherical or button shaped bulges can be seen on the surfaces of these extensions.

Compared with the SEM-pictures figure 10 shows an overview of the corneal side of the nictitating membrane in a TEM-photograph. The epithelium is single layered at points and pseudostratified columnar at others. The lower basal cell surface is remarkable for its irregular protrusions into the underlying connective tissue looking like the dermal-epidermal junction.

Figure 6. Outfoldings of the plasma membrane which fork several times. x 4 600
Figure 7. Protrusions looking like feather dusters. x 2 300

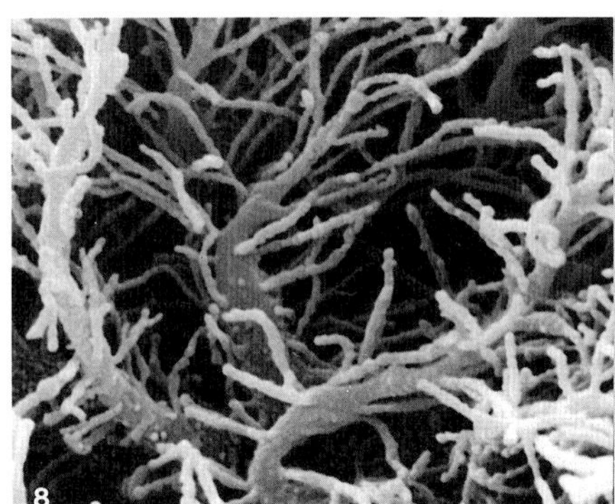

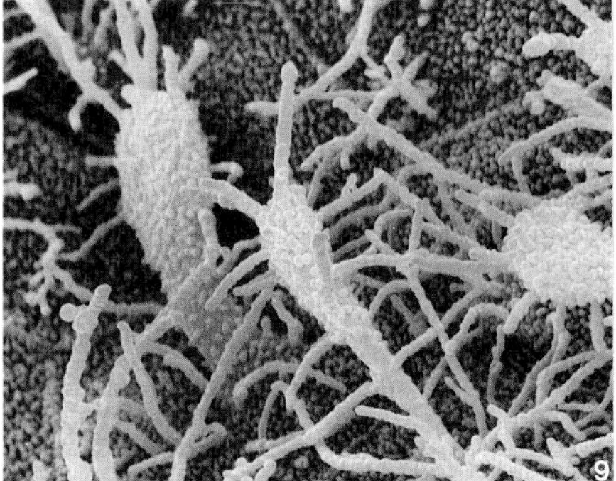

Figure 8. Processes resembling the branches of a tree. x 5 000
Figure 9. Club and dumb-bell shaped protrusions. x 5 900

The low cuboidal epithelial cells have superficially short and stumplike microvilli while the high columnar epithelial cells are thin and pointed. They are overlaid with microvilli and thin cytofila as seen in the SEM-photographes. The varying development of cell organelles and vacuoles indicates a secretory process releasing complex mucopolysaccharides and lipids, an assumption supported by histochemical staining tests. The spherical structures seen in SEM-photographs are seen in the TEM depiction to be small vesicles and vacuoles lying directly under the plasmalemma or bulging the plasmalemma. These vacuoles contain the secretory products which are extruded on the surface.

Figure 10. Vertical section through the nictitating membrane, area of the feathered epithelium. x 3 000

Our results show that the nictitating membrane of the pigeon is provided with a highly differentiated epithelium which serves in the care and cleansing of the cornea. At the same time the epithelium is secretory. The secretory products are spread over the cornea by the feathered epithelium. In this way, the cornea is not only cleansed but kept moist by the secretions. The foremost portions of the bulbus are prevented from drying out, a danger which is especially present due to the wind currents in flying.

REFERENCES

Bielek E (1976). Zur Feinstruktur des "gefiederten" Epithels der Nickhaut der Taube. Verh Anat Ges 70: 977-985.

De Angelis MA, Smith RL, Ribeiro EC (1985). Anatomia funcional da membrana nictitante de Gallus gallus domesticus. Rev Bras Cienc Morfol 2: 9-15.

Kajikawa J (1923). Beiträge zur Anatomie und Physiologie des Vogelauges. Arch Ophthal 112: 260-346.

Kolmer W (1923). Über eine bisher noch unbekannte Form von Epithelzellen (gefiedertes Epithel) in der Nickhautinnenfläche der Vögel. Anat Anz 57: 122-125.

Schramm U, Kühnel W (1986). Untersuchungen über den Bau der Nickhaut der Taube. Anat Anz 161: 167.

Cells and Tissues: A Three-Dimensional
Approach by Modern Techniques in Microscopy,
pages 527–535

APPLICATION OF THE CORROSION CASTS METHOD TO THE ANURAN AMPHIBIAN EYE AND MAMMALIAN COCHLEA

Adam J. Miodoński, Jan Kuś, Thomas Bar /+/

SEM Laboratory of the ENT Department of the N. Copernicus Academy of Medicine, Kopernika 23a, 31-501 Kraków, Poland
/+/ Max-Planck-Institut f. Systemphysiologie, Rheinlanddamm 201,4600 Dortmund, West Germany

INTRODUCTION

Each organ has its own characteristic microcirculatory bed that is the angioarchitecture and three-dimensional organization of the microcirculation varies from organ to organ. The angioarchitecture of microvascular beds in many organs and tissues have been already extensively studied in sectioned tissues both with and without contrasting of blood vessels (Rhodin, 1974, 1981; Michaelson, 1954; Axelsson, 1968). Although three centuries have already passed since Malpighi discovered capillaries, there still remains much to be investigated about microcirculation.

The invention of the scanning electron microscopy of vascular corrosion casts by Murakami in 1971 has significantly advanced our knowledge and understanding of angioarchitecture of different organs and tissues (Murakami, 1971, 1983; Hodde and Nowell, 1980; Lametshwandtner et al., 1984; Miodoński et al., 1981). Besides all known advantages but also disadvantages, vascular corrosion casts studied by scanning electron microscope (SEM) show as a rule the presence of surface features. As it has been ascertained by Miodoński et al., (1976, 1981) these features are distinctly different for arteries and veins. So it is possible to distinguish with high accuracy the type of vessel down to the capillary level.

The possibility offered by SEM of vascular corrosion casts which allowing more perfect resolution of details of the angioarchitecture within different organs and tissues, especially at the capillary level, were used by us to analyze the microcirculation of the frog eye and the rat cochlea.

It is generally accepted in frogs that the choriocapillaris is supplied only by two arteries i.e. arteria ciliaris nasalis and temporalis, outcoming from the ophthalmic artery (Virchow,

1881; Gaupp, 1904; Prince, 1956; Francois and Neetens, 1962). Therefore it was surprising to find in Rana temporaria and Rana esculenta presence of a second arterial source supplying the choriocapillaris. From the other side we were able, for the first time, to obtain in toto and then to analyze in detail the intra-ocular system of the membrana vasculosa retinae (Virchow, 1881; Gaupp, 1904; Michaelson, 1954; Miodoński and Bar, 1987a, 1987b). Similarly, previous extensive studies on the vascularization of the cochlea (Smith, 1951, 1954; Axelsson, 1968, 1974; Wustenfeld and Kuhnert, 1964) have faild in showing both the overall three-dimensional picture as well as spatial organization of the micro-circulation. This basic shortcoming could be again omitted by the use of SEM of vascular corrosion casts to the mammalian cochlea.

RESULTS

The choroid of anuran amphibians is composed mainly of choriocapillaris. In both species studied, an independent arterial supply to the choriocapillaris supplements that from the ciliary arteries (Fig. 1). This additional vascular route arises from the optic artery a separate branch of the arteria infundibularis superficialis (Fig. 2).

The optic artery, accompained by its vein within the vascular sheath of the optic nerve joins the rich, arterial in character, capillary network of the choriocapillaris and supplies the posterior pole of the ocular globe (Figs. 1, 3). The superficial capillary network displays a dense collar around exit of the optic nerve from the eye and is built of circular meshwork of capillaries arranged in several layers deep. More peripherally howener it becomes single layered (Fig. 3). This arterial capillary network extends nearly to the equator of the eyeball, and as a whole, establishes numerous connections with the underlying choriocapillaris at the posterior pole of the ocular globe (Fig. 4). The both arteriae ciliares start to give off secondary branches to the choriocapillaris at a certain distance from their proximal segments, therefore the territory of choriocapillaris around the exit of the optic nerve from the eye is supplied by the above mentioned superficial capillary network, which is fed by the optic artery. This situation resembles somewhat that encountered in mammals, in which branches of the short posterior ciliary arteries form an anastomosing circle (Zinn-Haler) around the optic disc, and the offshoots of this circle join the arterial network of the optic nerve.

The second analyzed system i.e. the superficial vascular hyaloid bed (membrana vasculosa retinae) of the frog extends on the vitreal surface of the avascular retina. The single-layered sheath follows the concave surface of the retina, so that in practice, one can compare it with a hemisphere measuring 6000 µm in diameter (Fig. 5). This vascular system, being a permanent

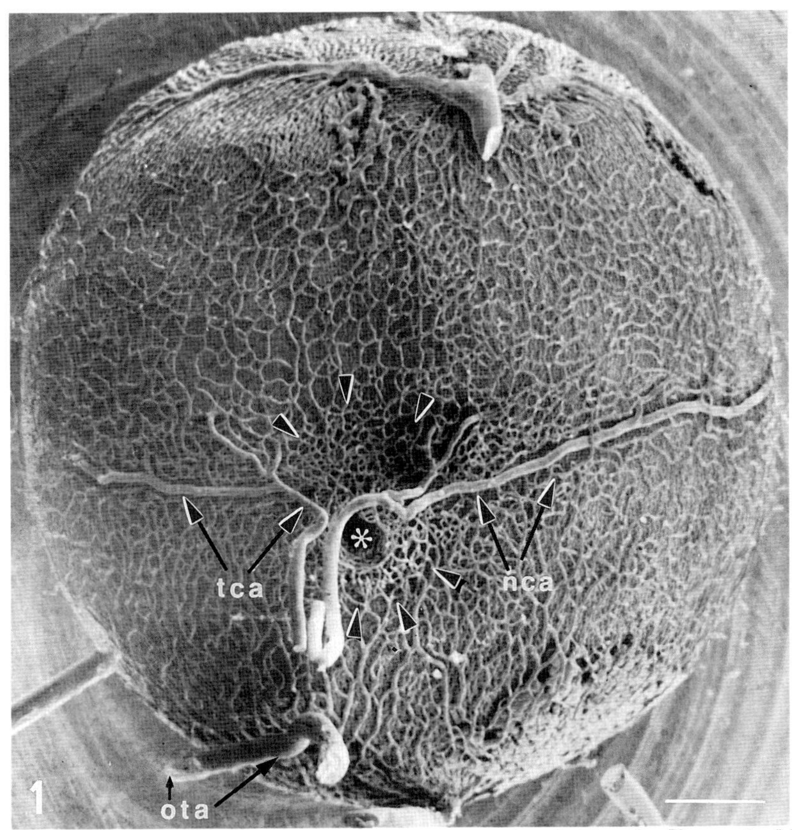

Fig. 1. SEM micrograph of a vascular cast of the left eyeball of the frog, viewed from the posterior pole. Note capillary network (arrowheads) extending over choriocapillaris, arteriae ciliares nasalis and temporalis, vascular collar around the optic nerve exit. (rosette) Bar = 1000 µm.

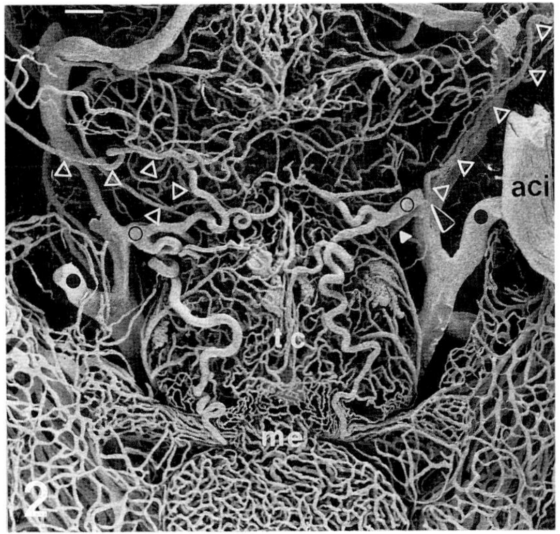

Fig. 2. SEM micrograph of 6a vascular cast of the brain of the frog, ventral aspect. Note the common trunk of the arteriae infundibulares (arrowhead) and the arteria optica (white triangles). Bar = 100 µm.

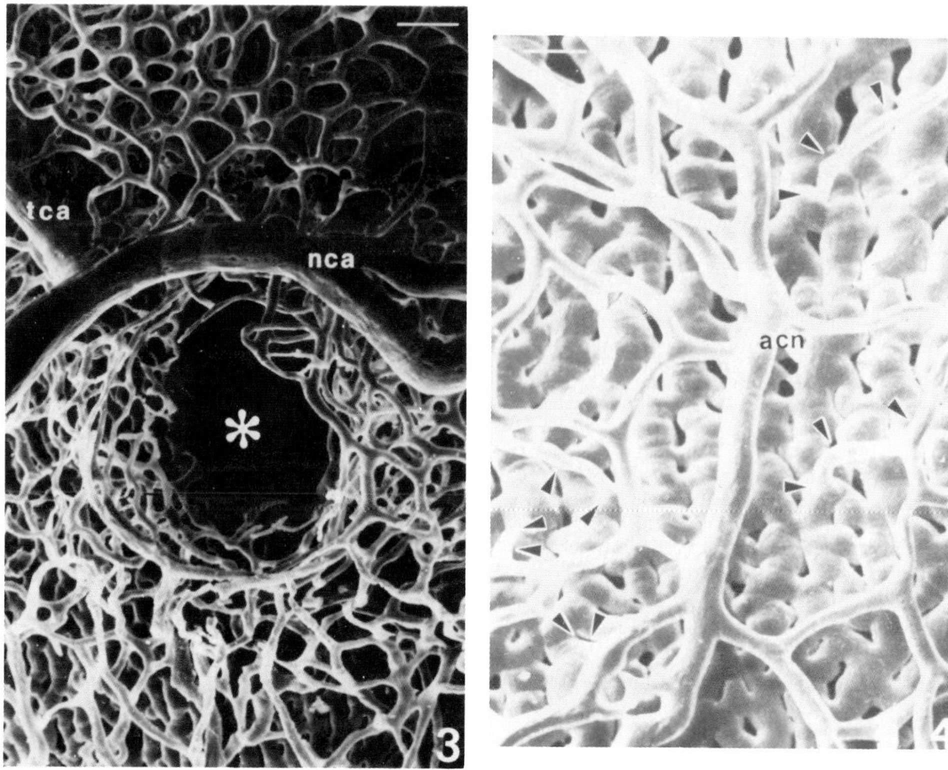

Fig. 3. Exit of the optic nerve (rosette) encircled by vascular collar supplied by the optic artery. Bar = 100 µm

Fig. 4. Anastomosing connections (arrowheads) of the superficial arterial capillary network, fed by the arteria optica. Bar = 25 µm.

structure, differs markedly in this respect from the vitreous vessels which are present only temporarily during embryonal development (Waele De, 1905; Francois and Neetens, 1962; Michaelson, 1954). The superficial hyaloid system is subdivided by the ventral venous trunk into three central areas: the dorsal, the temporo-ventral and the naso-ventral area. Toward the ora serrata this system is bordered by an fully closed arterial ring as well as by nasal and temporal venous branches constituting more or less hemicircle (Fig. 5). A vascular zone composed of several tonguelike sectors, which are most pronounced within the dorsal area of the system, establishes an interconnection between the peripheral vascular ring and the central areas of the fundus (Fig. 5). The whole system is supplied with blood by the terminal segments of the ophthalmic artery, the so-called hyaloid arteries. They form together a closed arterial peripheral ring, the only source of the meridional arterial twigs. These twigs, measuring 65-70 µm in diameter, branch from the parent vessels at nearly 90° angles (Fig.5, 6). On their course they split

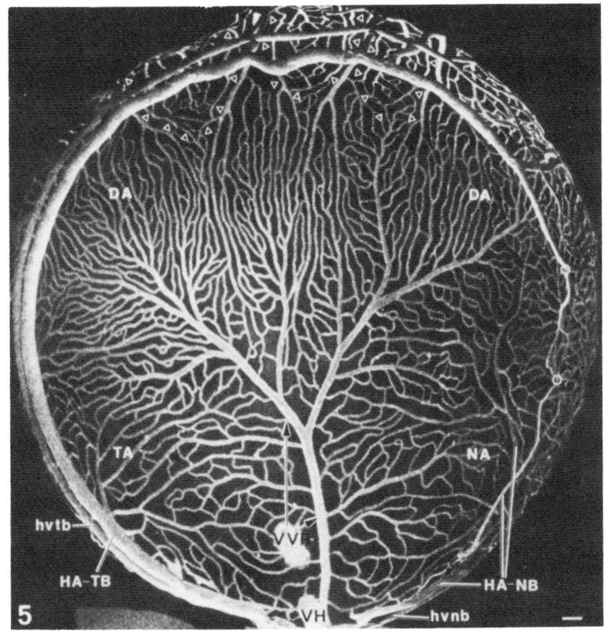

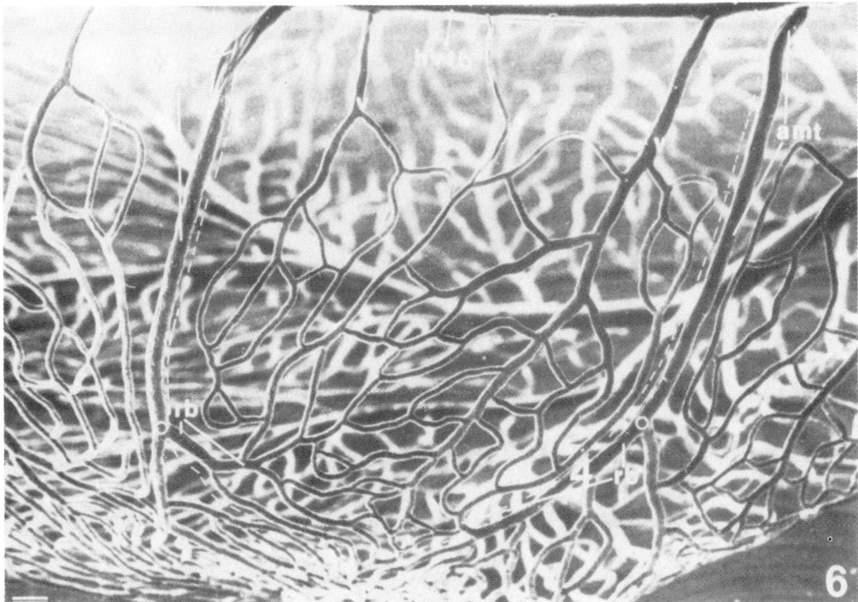

Fig. 5. SEM micrograph of a vascular cast of the superficial hyaloid system, aspect from the vitreous body. Note the peripheral arterio-venous ring and the Y-shaped venous ventral trunk draining the central areas. Bar = 100 µm

Fig. 6. Angioarchitecture of the tongue-like sector in the pheripheral zone. Note arterial recurrent branches (r) originating from arterial meridional twigs, veins of the sector discharging into temporal branch of the hyaloid vein. Bar = 100 µm

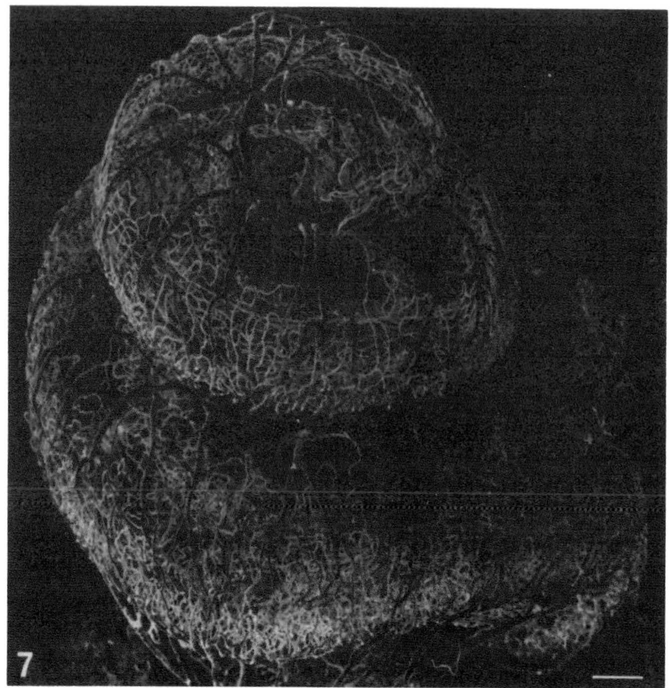

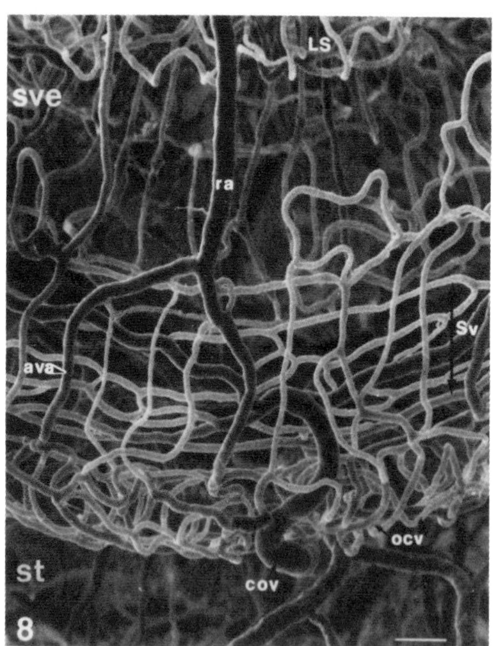

Fig. 7. Cochlear vasculature in toto, viewed slightly from above. Bar = 200 µm.

Fig. 8. Fragment of second turn showing vascularization of the stria vascularis. Note arterio-venous anastomoses, capillaries of stria vascularis and radiating arterioles Bar = 50 µm.

dichotomously. Secondary branches emerging from these points of bifurcation are principally directed toward the posterior pole of the eye globe and are arranged alternantly with venous twigs discharging into the temporo dorsal and naso-dorsal branches of the venous ventral ramus. The drainage of the superficial hyaloid system is provided via two routes: a) the Y-shaped ventral trunk, which is emptying into a short venous stem, the so-called hyaloid vein, collects blood from the central areas, b) the two peripheral venous branches i.e. nasal and temporal, drain the tongue-like sectors which form a circular band measuring 1560-1600µm in width. The single tongue like sectors are supplied either by two or by one recurrent small artery, outcoming from the main stem of arterial meridional twig (Fig. 6), or from one of its secondary offshoots. The territory of the tongue-like sectors are drained by a single central venous vessel or by two, varying in size, venous twigs (Fig. 6). These veins are discharging only into one of the two peripheral venous branches. Besides tongue-like sectors are connected by few capillary anastomoses with the vascular areas drained by the venous ventral ramus. Within the territory of the dorsal area practically all venular twigs, which alternate with the meridional arterial branches, are directed parallel to the dorso-ventral meridian of the ocular globe (Fig. 5). Numerous longitudinally oriented capillary loops display a simmilar pattern. The capillaries are more densely packed within the dorsal area in comparison with the ventral regions. This densely capillarized area, which estimated surface attains ca. 2,300 mm2 and the total length of vessels reaches 62,400 µm, correspond in location to the area centralis retinae. It is suggested therefore, that the peculiar array and the high density of capillaries are related functionally to the more specialized visual functions of the area centralis.

The application of the same technique i.e. SEM of vascular corrosion casts in studies on the rat cochlea has made it possible to obtain by us, for the first time (Miodoński et al., 1978) a three-dimensional picture of the vascular bed of this organ as a whole (Fig. 7). Closer study, especially of the scala media, has not only provided confirmation of other authors observations (Smith, 1951, 1954; Agazzi, 1948; Axelsson, 1968, 1974; Charachon, 1961; Wustenfield and Kuhnert, 1964; Hawkins, 1976) but allowed also to firm the presence of arterio-venous anastomoses (Fig. 8). These, which are vessels of different calibre, are usually secondary branches given off by the radiating arterioles. Anastomoses formed by a vessel of greater diameter costituting a continuation of radiating arteriole are also encountered. These anastomoses, running externally in respect to the vessels of the stria vascularis, join the loops omega-shaped capillaries discharging into the collecting venules of the scala tympani. Besides all types of blood vessels which have beeb described by Axelsson can be easily recognized and further analyzed. The quasi three-dimensional pictures give much better insight and informations about spatial relations between the particular groups of vessels.

aci= internal carotid artery; acn= arterial capillary network; amt= arterial meridional twigs; ava= arterio-venous anastomisis; cov= collecting venules; DA= dorsal area; HA-NB= hyaloid artery nasal branch; HA-TB= hyaloid artery temporal branch; hvnb= hyaloid vein nasal branch; hvtb= hyaloid vein temporal branch; ls= spiral limbus; NA= naso-ventral area; nca= nasal ciliary artery; ocv= omega shaped capillary venules; ota= ophthalmic artery; ra= radiating arteriole; rb= recurrent branches of meridional arterial twigs; St= scala tympani; sv= stria vascularis; Sve= scala vestibuli; TA= temporo-ventral area; tca= temporal ciliary artery; VH= vena hyaloidea; VVR= venous ventral ramus.

REFERENCES

Agazzi C (1948). Angioarchitettonica del ligamento spirale della cavia. Biol Lat 1:59-74.

Axelsson A (1968). The vascular anatomy of the cochlea in guinea pig and in man. Acta Oto Laryngol, Suppl, 243, pp 5-134.

Axelsson A (1974). The blood supply of the inner ear in mammals. In Keidel WD Neff WD (eds) "Handbook of sensory Physiology: Auditory System", Berlin, Heidelberg, New York, Springer Verlag, pp 213-260.

Charachon R (1961). Anatomie de l artere auditive interne chez l'homme. Lyons, Imprimerie Bosc Freres, pp 1-20.

Francois J, Neetens A (1962). Comparative anatomy of the vascular supply of the eye in vertebrates. In Davson H/ed/ "The Eye: Vegetative Physiology and Biochemistery". New York, London, Academic Press, pp 369-416.

Gaupp E (1904). Lehre von den Eingewiden, dem Integument und den Sinnersorganen. Eckers und Widersheims Anatomie des Frosches. Braunschweig, Vieweg, pp 762-901

Hawkins Jr. JE(1976). Microcirculation in the labyrinth. Arch Oto Rhino Laryng 212:241-251.

Hodde CK, Miodoński AJ, Backker C, Veltman WAM (1977). SEM of microcorrosion casts with special attention on arterio-venous differences and application to the rat cochlea. Scanning Electron Microscopy, Vol II, Chicago, IIT Research Institute, pp 477-484.

Hodde CK, Nowell JA (1980). SEM of microcorrosion casts. Scanning Electron Microscopy, Vol II, AFM O-Hare (Chicago), SEM Inc., pp 89-106.

Lametschwandtner A, Lametschwandtner U, Weiger T (1984). Scanning electron microscopy of vascular corrosion casts-technique and application. Scanning Electron Microscopy, Vol II, AMF O-Hare (Chicago), SEM Inc., pp 663-695.

Michaelson IC (1954). "Retinal Circulation in Man and Animals". Springfield, C.C. Thomas Publisher.

Miodoński AJ, Hodde CK, Backker C (1976). Rasterelektronenmikroskopie von Plastik-Korrosion-Prapareten: Morphologische Unterschiede zwischen Arterien und Venen. Beitr Elektronenmikroskop Direktabb Oberfl (München) 9:435-442.

Miodoński AJ, Kuś J, Tyrankiewicz R (1981). SEM blood vessel casts analysis. In Allen DJ, Motta PM, DiDio LJA (eds) "Three Dimensional Microanatomy of Cells and Tissue Surfaces", New York, Amsterdam, Oxford, Elsevier/North Holland, pp 71-87.

Miodoński AJ, Bar Th (1987a). Arterial supply of the choriocapillaris of anuran amphibians (Rana temporaria, Rana esculenta): SEM study of microcorrosion casts. Cell and Tissue Res 249:101-109.

Miodoński AJ, Bar Th (1987b). The superficial vascular hyaloid system in the eye of the frogs, Rana temporaria and Rana esculenta: SEM study of vascular corrosion casts. Cell and Tissue Res 250:465-473.

Murakami T (1971). Application of the SEM to the study of distribution of the blood vessels. Arch Hist Jap 32:445-454.

Murakami T (1983). Injection replication and scanning electron microscopy of blood vessels. In Hodges GM, Carr KE (eds) "Biomedical Research Applications of Scanning Electron Microscopy", Vol 3, London, Academic Press, pp 1-30.

Prince JH (1956). "Comparative Anatomy of the Eye", Springfield, C.C. Thomas Publisher.

Rhodin JAG (1974). "Cardiovascular system. In Rhodin JAG (ed) "Histology. A Text and Atlas", London and New York, Oxford Univ Press.

Rhodin JAG (1984). Anatomy of microcirculation. In Abramson DI, Dobrin PB (eds) "Blood Vessels and Lymphatics in Organ System", Orlando, London, Academic Press, pp 97-106.

Smith CA (1951). Capillary areas of the cochlea in guinea pig. Laryngoscope 61:1073-1095.

Smith CA (1954). Capillary areas of membranous labyrhint. Ann Otol Rhinol Laryngol 63:435-447.

Virchow H (1881). Ulber die Gefasse im Auge und der Umgebung des Auges beim Frosche. Zeirt Wiss Zool 35: 247-281.

Waele De H (1905). Notes sur l embryologie de l oeil des Urodeles. Intern Monatsschr f Anat u Physiol, Bd 22.

Wustenfeld E, Kuhnert D (1964). Experimenteller Beitrag zur Frage der Gefassversorgungen der Meerschweinchencochlea. Z Mikrosk Anat Forsch 71:172-184.

Cells and Tissues: A Three-Dimensional Approach by Modern Techniques in Microscopy, pages 537–541

MORPHOLOGICAL CORRELATIONS OF HEARING IN THE PHYLOGENETIC SCALE: MAUREMYS CASPICA.

J. Morales, V. García-Martínez, D. Sánchez-Quintana and A. Ambel.

Departamentos de Otorrinolaringología (J.M., A.A.) y Ciencias Morfológicas (V. G.-M., D.S.-Q.), Universidad de Extremadura. Badajoz. 06071 Spain.

INTRODUCTION

Although a number of extensive studies have been published dealing with inner ear morphology throughout the phylogenetic scale (Baird, 1.970; Wever and Gans, 1.973; Vinnikov, 1.982) as well as with functional aspects of hearing in different animal species (Gans and Wever, 1.972, Wever and Gans, 1.976), we consider a valid approach with wich to gain a better understanding of the functional significance of each of the elements comprising the Organ of Corti in man is through a correlational examination of morphology and function in the phylogenetic scale.

MATERIAL AND METHODS

A total of seven animals (Mauremys caspica) of both sexes between 4 and 6 years of age and 400-600 grams in body weight were studied. Surface specimens technique, microdissection and celloidin embedded sections were examined with dissecting microscope, light microscope and transmission and scanning electron microscope.

RESULTS

The middle ear is represented by an elongated cavity whose anterior third is partially stenotic at the level of its union with the posterior two thirds. The anterior part of the cavity forms the tympanic cavity proper. The tympanic membrane is formed from a thick circular cartilaginous layer encircled by a ligamentous ridge joining it to the boney edges of the outer wall of the tympanic cavity. Both internally and externally this cartilaginous layer is covered with mucous and poorly keratinized epithelium. The columella is an element of sound transmission in the middle ear. Its cartilaginous external portion is a continuation of the tympanic membrane cartilage (Fig. 1), whereas its inner part is boney. The crura of the columella is excentrically anchored to the base, which presents a ligament of insertion along the edges of the oval window.

The inner ear is separed from the middle ear by the periotic cistern, which is in communication with the round window. The later is covered by the fibrous secondary tympanic membrane. The cochlear duct, located within the vestibule, is situated dorsal to the saccule (Fig. 1) and cranial to the macula of the lagena, located in the caudalmost part of the cochlear duct. A wide communication exits between the cochlear duct and the saccule (Fig. 1). The basilar papilla located within the cochlear duct, is elongated and can be divided into three parts in craniocaudal direction: the head, neck and body (Fig. 2). The basilar papilla extends throughout the length of the duct on the basilar membrane, which separates the papilla from the scala tympani (Fig. 3), except in the head portion, where the space corresponding to the scala tympani is occupied by connective tissue showing areas of myxoid degeneration (Fig. 4).

Two types of the cells make up the basilar papilla: supporting cells and neurosensory cells (Fig. 5). The two populations are in intimate contact via juncture complexes joining cells of the same line as well as supporting and neurosensory cells. In supporting cells the nucleus is located basally, and their lower pole rests upon the basilar membrane. Between the basilar membrane and the basal pole of the neurosensory cells is a virtual space through which the nerve fibers of the acoustic portion of the VIII cranial pair rum. A noteworthy cytoarchitectural feature of the supporting cells is the presence of numerous microtubules arranged predominantly near the apical pole. The apical pole of neurosensory cells is covered with a cuticle from which a bundle of cilia arises (Fig. 5). The bundles from each neurosensory cell are arranged in strict geometric order from smaller to larger, so that each group of cilia produce a pyramid-like formation (Fig. 2). A kinocilium (Fig. 5) occupies the apex of each bundle, the rest of the pyramid being made up of shorter stereocilia. Ciliar bundles are oriented in the same direction in the body and neck of the basilar papilla, with their kinocilia pointing toward the dorsal edge of the papilla. In the head portion, kinocilia are located centripetally in relation to the central point. The neurosensory cells are innervated by a compact bundle of myelinated nerve fibers running ventral and medial to the cochlear duct. Fibers supplying the duct are arranged radially, reaching the anterior edge of the basilar papilla at the level of the basilar membrane. At this point they lose their myelin sheath, and a short distance later end in synaptic button near the basal pole of the neurosensory cells. These nerve endings can take two different forms: i) large structures containing numerous vesicles and ii) small dilations with few vesicles (Fig. 6). Membrane specializations are evident in both types in the nerve ending itself as well as in the cell membrane.

The tectorial membrane is joined to the cochlear duct at the level of the anterior wall of the later (Fig. 3), and can be divided into three parts on the basis of structural, morphological and topographical features: i) the peripheral part, serving as the union to the cochlear duct wall and made up of finely granular, highly compact material, ii) the intermediate portion, consisting of fibrillar material, and iii) a central portion lying upon the basilar papilla and made up of the same material as the peripheral part (Fig. 3,7).

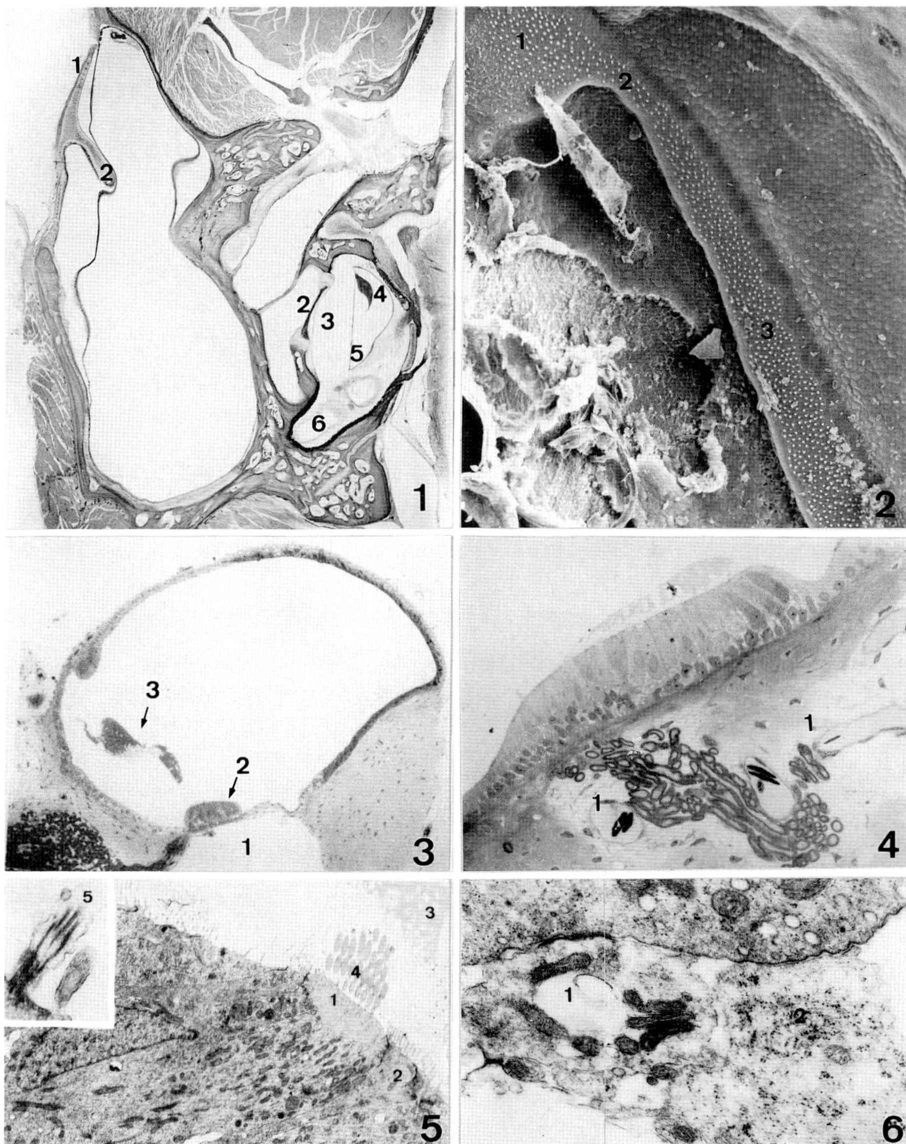

Fig. 1.- Celloidin transversal section. 1. Tympanic membrane. 2. Columella. 3. Vestibule. 4. Saccule. 5. Cochlear duct. 6. Posterior semicircular duct.

Fig. 2.- Scanning electron micrograph of the basilar papilla. 1.Head. 2.Neck. 3.Body.

Fig. 3.- Semithin section of cochlear duct. 1. Scala tympani. 2. Basilar papilla. 3. Tectorial membrane.

Fig. 4.- Semithin section of the head. 1. Scala tympany.

Fig. 5.- Transmission electron micrograph of the basilar papilla. 1. Hair cell. 2. Supporting cell. 3. Tectorial membrane. 4. Stereocilia. 5. Kinocilia.

Fig. 6.- Transmission electron micrograph of the nerve endings. 1.Afferent. 2.Efferent.

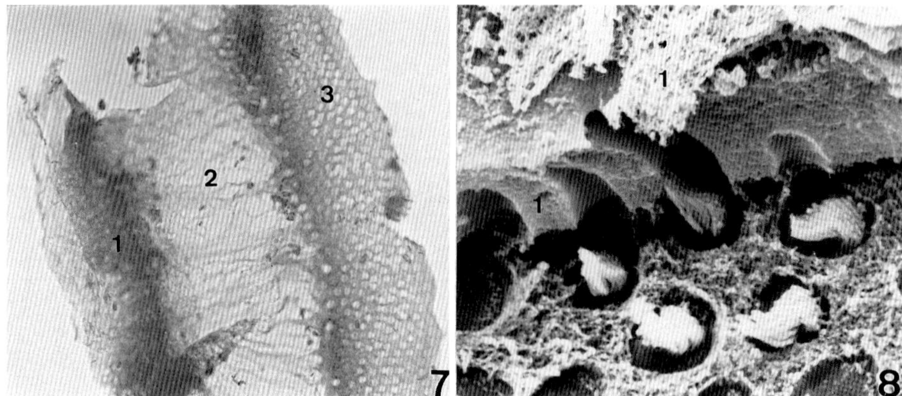

Fig. 7.- Surface specimen. Tectorial membrane. 1. Peripheral part. 2. Intermediate portion. 3. Central portion.
Fig. 8.- Scanning electron micrograph of the basilar papilla. 1.Central portion of the tectorial membrane.

The peripheral and central portions are connected by numerous, widely anastomosing minute canals of variable diameter which occasionally contain non-membrane bound vesicular structures made up a material similar in appearance to that seen in the peripheral and central portions of the tectorial membrane. The surface of the tectorial membrane in contact with the basilar papilla forms a series of more or less regular battlements which give rise to the ciliar bundles. In the intervals between the battlements is a finely fibrillar, highly lax tissue (Fig. 8).

DISCUSSION

The morphological pattern described thus far is suggestive of the functional roles noted below, althought further studies will be necessary to confirm these functions.

The filamentous structure which joins the columella to the anterior surface of the tympanic cavity may participate in the bone conduction. The vibrations detected in the region of the animal's head would be transmitted via the aforementioned ligament to the columella, hence air and bone conduction have a common route to the inner ear.

The wide communication between the saccule of the cochlear duct, together with the arrangement described above of the tectorial membrane and the absence of the tympanic ramp at the level of the head of the basilar papilla may be related to the following functional progression of the sound wave in the inner ear. Neurosensory cells in the basilar papilla are stimulated both by the sound wave and kinetic energy.

In the former case, sound energy is transmitted from the columella to the vestibule and to the cochlear duct. Following a kinetic stimulus, the turbulence in the endolymph caused by the movement of the statoconial membrane in the saccule would reach the cochlear duct via the wide communication connecting the two cavities. Either of these two mechanisms would lead to movement of the tectorial membrane upon the basilar membrane, thus triggering neurosensory cell stimulation without requiring vibration of the basilar membrane via the tympanic ramp. This would give functional significance to the head of the basilar papilla.

In Mauremys caspica the scala tympani may be involved in hearing by permitting vibration of the endolymph in the labyrinth, which leads to deformation of the secondary tympanum. The basic difference which distinguishes the double mechanism of neurosensory cell stimulation in the basilar papilla is the disproportionate amplitude of tectorial membrane movement in response to a given acoustic or kinetic stimulus. To offset this quantitative difference and fundamentally to protect young auditory cells from a kinetic energy overload, the peripheral part of the tectorial membrane may act as a shock absorber, reducing the movement of the central portion of the tectorial membrane in response to kinetic stimuli, even at the expense of a loss of sensivity.

The series of events described above for sound transmission seems a reasonable mechanism in the species studied, given its position on the phylogenetic scale. When auditory receptors first appear, gravitational receptors are fully constituted and functional. As the former represents a specialization of the latter, both may well respond to the same kind of stimuli.

ACKNOWLEDGEMENTS

This study was supported by a Research Grant from the Excelentísima Diputación of Badajoz.

REFERENCES

Baird IL (1970). The Anatomy of the Reptilian Ear. In Gans C, Parson TS (eds): "Biology of the Reptilia" Vol 2b, London: Academic Press, pp 192-275.

Gans C, Wever EG (1972). The Ear and Hearing in Amphisbaenia (Reptilia). J exp Zool 179: 17-34.

Vinnikov YA (1982). "Evolution of Receptor Cells. Cytological, Membraneous and Molecular Levels". New York: Springer Verlag, pp 78-89.

Wever EG, Gans C (1973). The ear in Anphisbaenia (Reptilia); further anatomical observations. J Zool 171: 189-206.

Wever EG, Gans C (1976). The caecilian ear: Further observations. Proc Natl Acad Sci USA 73: 3744-3746.

Cells and Tissues: A Three-Dimensional
Approach by Modern Techniques in Microscopy,
pages 543–547

THE ENDOLYMPHATIC SAC. A SCANNING ELECTRON MICROSCOPIC STUDY IN DIFFERENT ANIMAL SPECIES.

Maurizio Barbara, Masaya Takumida and Roberto Filipo

Departments of Otolaryngology, University "La Sapienza", Rome, Italy (M.B., R.F.) and Karolinska Hospital, Stockholm, Sweden (M.T.)

INTRODUCTION

The endolymphatic sac (ES) represents the blind appendage of the membranous labyrinth, with which is communicating via the endolymphatic duct (ED). Several investigations have been carried out in order to study its structure and function, formerly considered merely rudimental (Siebenmann, 1919). In particular, fluid and ion transport (Bagger-Sjöbäck and Rask-Andersen, 1986), pressure and volume regulation for the entire inner ear fluid compartment (Friberg et al., 1987), as well as secretory activity (Friberg et al., 1986; Rask-Andersen et al., 1987; Takumida , 1988; Barbara, in press) have been taken into consideration. Most of these theories have been derived from the ultrastructural findings related to the ES epithelial cell population, which is composed of *light* and *dark cells*. The *light cells* are believed to be involved in the resorption of fluid and ions, whereas to *the dark cells* a phagocytotic activity has been attributed (Lundquist, 1965).

Several investigations have also been carried out by using scanning electron microscopy (SEM) both in humans (Schindler, 1979; Galey and House, 1980) and in animal models (Lundquist, 1965; Mitani et al., 1973; Harada and Gaafar, 1976; Takumida , 1988; Barbara et al., 1988). This procedure has, moreover, become more valuable since the introduction of newly developed techniques, which allow a three-dimensional view and hence also the observation of the intracellular structures, such as the one proposed by Tanaka and Mitsushima in 1983.

In the present study we have compared SEM data of the intermediate portion of the ES in different animal species, giving particular emphasis to the advantages offered also by three-dimensional SEM procedures.

MATERIAL AND METHODS

Healthy, adult guinea-pigs, mice and mongolian gerbils have been used in this study. While in the mongolian gerbil a standard SEM procedure was carried out, the guinea-pig and murine ES underwent the SEM A-O-D-O method described by Tanaka and Mitsushima (1983). The specimens were observed and photographed by both JSM-880 and Philips 505 scanning electron microscopes with a back scatter electron detector.

RESULTS

Gerbilline ES

The intermediate portion of the ES appears to be quite different from both the proximal and distal portions. Its epithelial surface appears uneven, with rugosities made of small crypts and ridges. A distinct variation in cell typology can be observed. Two main groups of cells are recognizable, both having a pronounced convexity. Microvillar protrusions cover both type of cells, being taller in one cell and more rounded and short in the other. A third cell type can also be distinguished, having its surface covered by rounded and/or ovoid material (Fig. 1).

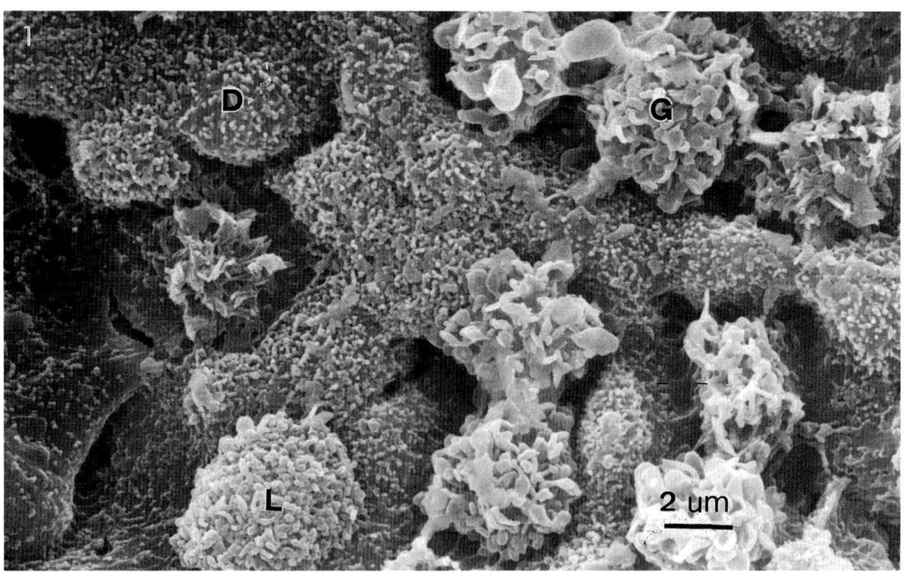

Fig. 1: Intermediate portion of the gerbilline endolymphatic sac. The surface appears uneven with crypts and microridges and cell population is mainly composed of microvilli-rich (L) and microvilli-devoid (D) cells. A third cell type (G) is

characterized by a cumulation of rounded and/or ovoid material over its luminal surface.

Murine ES

The rugose portion presented an uneven appearance. Moreover, along with a majority of flat cells, with short microvilli, part of the cell population is composed of cells presenting with a convex surface and a great number of short, thick microvillar protrusions (Fig. 2). The lateral intecellular spaces (LIS) are also easily recognizable, being generally widened in this portion of the ES.

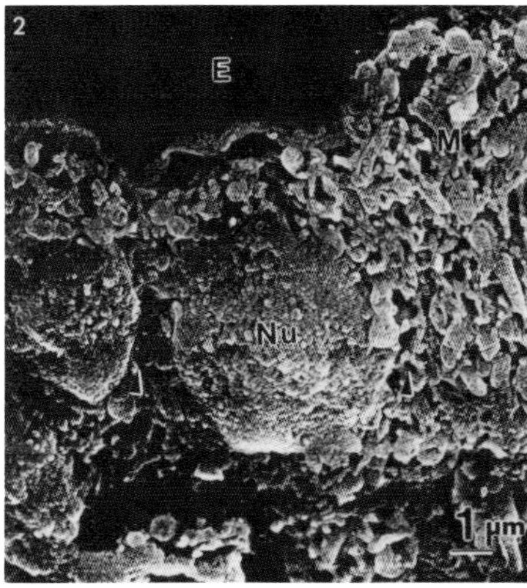

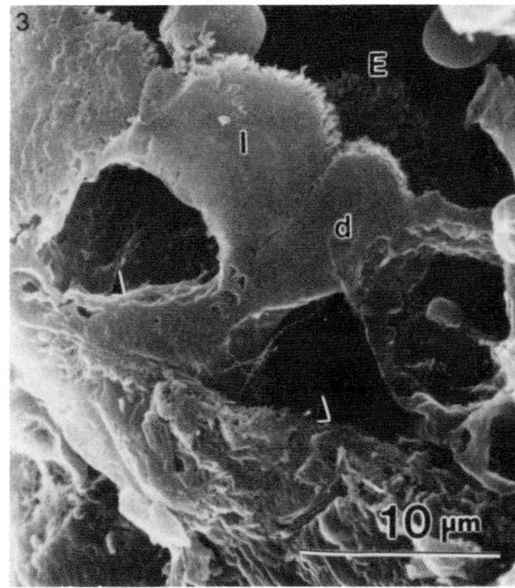

Fig. 2: ES epithelium of the guinea-pig (AODO method). The epithelial cells are cracked and macerated by 0.1% OsO_4. The intracellular structures such as nucleus (Nu) and mitochondria (M) are clearly revealed. The lateral intercellular spaces (arrowheads) are narrow in general. E= endolymphatic space.

Fig 3: Murine ES epithelium. The epithelial cells are cracked and show the light cell with microvilli (l) and the dark cell without microvilli (d). The lateral intercellular spaces (arrowheads) are widened in general. E= endolymphatic space.

Guinea-pig ES

Both the intra- and the extraosseous portion were easily recognizable with a cell population of two main types, i.e. light and dark cells. The light cell appeared very rich in mitochondria and apical microvilli (Fig. 3). An extensive network of rough endoplasmic reticulum (rER) was seen in close proximity of the apical plasmalemma, with the appearance of tubular structures or thin lamellae baso-apically directed. Connections between the inner surface of apical plasmalemma and rER were also noticed at some places. Also Golgi complexes were frequently noted in the supranuclear region. The dark cell, on the contrary, was seen to have a wedge shape, with fewer mitochondria and an indented nucleus. Both smooth and rugose endoplasmic reticulum can be recognized, rather evenly dispersed throughout the cytoplasm. The LIS are in this region quite narrow.

DISCUSSION

According to the criteria first proposed by Guild (1927), in all the three animal species studied it was possible to distinguish a proximal, and intermediate and a distal portion. The intermediate portion of the ES has morphologically revealed to be the most complex portion of the entire ES, thus confirming TEM findings as well as human studies. In the *mongolian gerbil* , apart from *light* and *dark cells*, also a third cell-type could be distinguished, which presented its luminal surface lined by solid material. This cell-type has been related to a presumed secretory activity, probably involved in pressure and volume regulation of the entire endolymphatic compartment (Barbara, in press). This cell has been named granular ccell. Another finding which seems to be specific for the *mongolian gerbil* is that the ES is completely covered by the bony operculum, and no true extraosseous ES can be observed, contrary to humans and other animal species.

The three-dimensional SEM method used in this study has permitted to obtain detailed informations at subcellular level and to distinguish the characteristics of the single cell type. *Light* and *dark cells* have easily been recognized according to the description earlier presented by Lundquist (1965). Moreover, the close topographic relationship existing in the *light cells* between the plasmalemma and the rough endoplasmic reticulum confirmed the possibility of an alternative pathways for water and solutes transport across the ES epithelium. The utilization of high-resolution SEM techniques can also be helpful for more advanced ultrastructural studies, as shown by the cytochemical analysis of the murine ES components carried out by Takumida (1988), who used scanning, other than light and transmission microscopy.

It can then be concluded that the newly-developed SEM techniques can offer a valuable contribute for the evaluation of dynamically-acting epithelia, such as ES is presumed to be. Although TEM still remains the main source of informations at subcellular level, the afore mentioned SEM techniques turned out to be particularly useful especially when a precise topographical identification of the structure under study is also needed.

REFERENCES

Bagger-Sjöbäck D, Rask-Andersen H (1986). The permeability barrier of the endolymphatic sac. A hypothesis of fluid and electrolyte exchange based on freeze fracturing. Am J Otol 7:134- 140.

Barbara M, Rask-Andersen H, Bagger-Sjöbäck D (1988). The surface morphology of the endolymphatic sac of the Mongolian gerbil (Meriones unguiculatus). A scanning electron microscopic study. J Laryngol Otol 102:308-313.

Barbara M (in press) Carbohydrate content of the endolymphatic sac. A histochemical and lectin- labelling study in the mongolian gerbil. J Laryngol Otol.

Friberg U, Wackym PA, Bagger-Sjöbäck D, Rask-Andersen H (1986) Effect of labyrinthectomy on the endolymphatic sac: a histologic, ultrastructural and computer-aided morphometric investigation in the mouse. Acta Otolaryngol (Stockh) 101:172-182.

Galey FR, House WF (1980) Scanning electron microscopy of the human endolymphatic sac. A preliminary report. Am J Otol 4:218-220.

Guild SR (1927) Observations upon the structure and normal contents of the ductus and saccus endolymphaticus in the guinea-pig (Cavia cobaya). Am J Anat 39:1-66.

Harada Y, Gaafar H (1976) Scanning electron microscopy of the endolymphatic sac epithelium. ORL 38:257-266.

Lundquist P-G (1965) The endolymphatic duct and sac in the guinea-pig. An electron microscope and experimental investigation. Acta Otolaryngol (Stockh) suppl. 201:1-108.

Mitani Y, Kosaka N, Konishi S, Ogura Y (1973) Observation of luminal surface of the endolymphatic sac in the guinea-pig. Acta Otolaryngol (Stockh) 76:149-156.

Rask-Andersen H, Erwall C, Steel KP, Friberg U (1987) The endolymphatic sac in a mouse mutant with cochleo-saccular degeneration. Electrophysiological and ultrastructural correlations. Hear Res 26:177-190.

Schindler RA (1979) The ultrastructure of the endolymphatic sac in Meniere's disease. Adv Oto- Rhino-Laryngol 25:127-133.

Siebenmann F (1919) Anatomische Untersuchungen uber den Saccus und Ductus endolymphaticus beim Menschen. Beitr zur Anat 13-59.

Takumida M (1988) Dynamic properties of the endolymphatic sac. A morphological and histochemical analysis of secretory and absorptive mechanisms. Thesis, Stockholm.

Tanaka K, Mitsushima A (1983) A preparation method for observing intracellular structures by scanning electron microscopy. J Microsc 133:213-222.

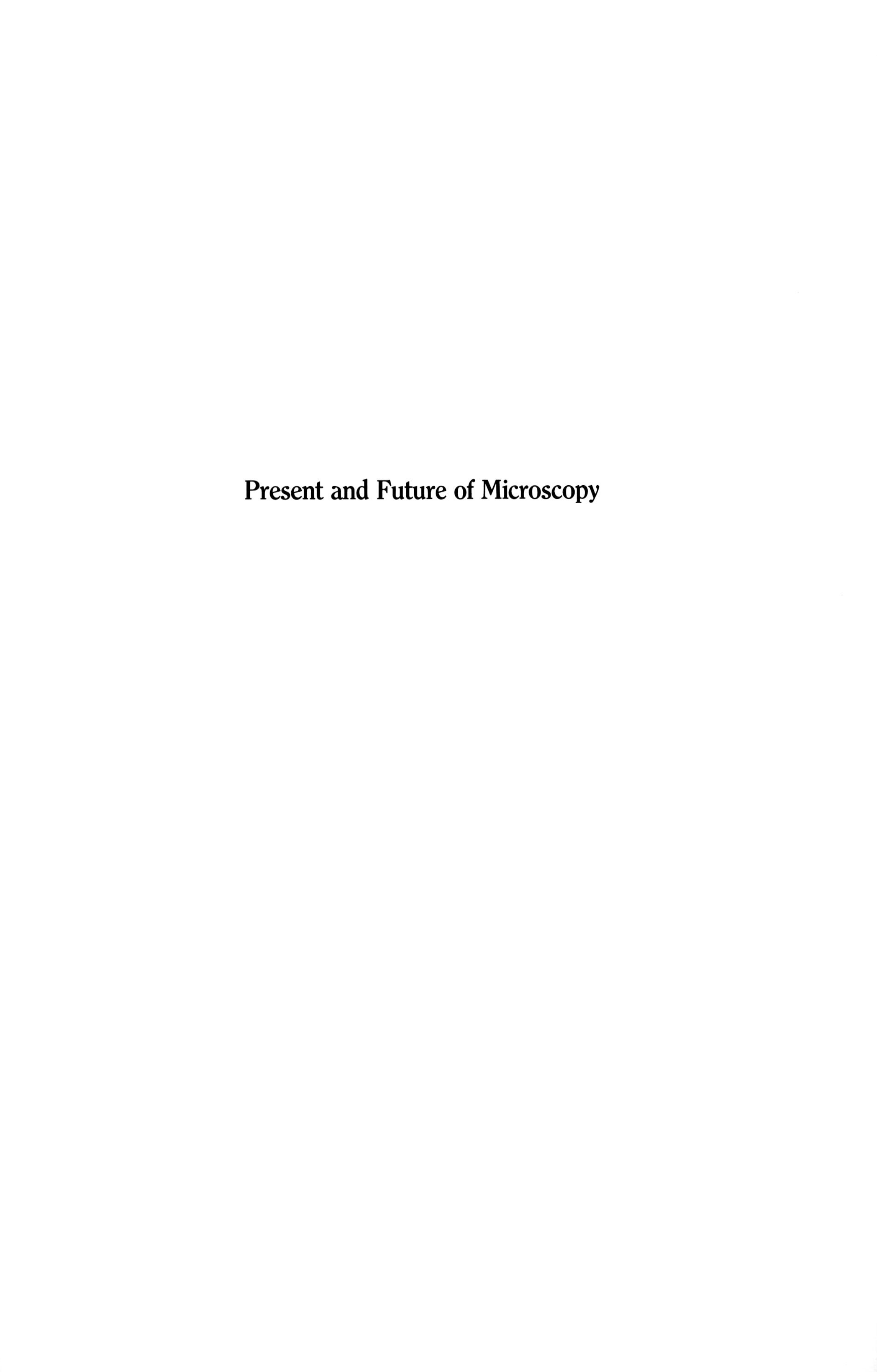

Present and Future of Microscopy

Cells and Tissues: A Three-Dimensional
Approach by Modern Techniques in Microscopy,
pages 551–558

TANDEM SCANNING MICROSCOPE - A NEW TOOL FOR THREE-DIMENSIONAL MICROANATOMY

Mojmír Petráň and Milan Hadravský

Institute of Biophysics, Charles University
Faculty of Medicine in Plzeň, 48 Karlovarská
Str., 301 66 Plzeň, Czechoslovakia

INTRODUCTION

As the microanatomical studies are conducted using a microscope as the principal tool, they were made difficult from the very beginning by the ominous contradiction between the three- (or even four-) dimensionality of the living tissue and the two-dimensionality of the microscopic image. The founding fathers - Marcello Malpighi (1628 - 1694), whose anniversary we celebrate today, our famous countrymen Georgius Prochaska (1749 - 1820) and Johannes Purkinje (1787 - 1869) and a score of other classics of microanatomy solved their problems by combined use of microdissection, blood vessel injection and observation by simple (and later also by compound) microscopes.

After the invention of microtome and paraffine embedding and staining of slices and introduction of photomicrography, three-dimensional models of microscopical organelles began to be forged of beeswax. Their fabrication was very laborious but the results were often inadequate because usually only one surface could be represented, not the minute inner structure contained in the depicted body.

Attempts to make stereoscopic photomicrographs of three-dimensional objects or even of thick slices were abandoned soon because only one thin layer could give a sharp image. To be able to get sharp images of a deeper space there were four prerequisites to be fulfilled: (1) To photograph on one photographic frame a stack of overlapping sharp images of thin layers lying in the object one upon the other; (2) To stack these layers in exact register;

(3) To be able to make photographs of such stacks with desired small lateral shifts between the photographs of individual levels (layers). And an additional condition was (4) that empty space be black to preserve contrast regardless of the number of superposed exposures. Theoretically it was possible to get such pairs of exactly but differently stacked photographs by microphotographing successively serial paraffine slices. But it was impossible to fulfill the fourth requirement in transmitted light microscopy and the fulfilment of the other three was also difficult.

METHODS AND RESULTS

Our new method of Tandem Scanning Reflected Light Microscopy (Petráň et al., 1968; Petráň et al. 1985) was originally contrived for an entirelly different purpose: to see only the focused-on structure, i. e. to prevent the access of unwanted light (i. e. of light coming from non focused-on parts of the object) into the image. Or as we put it aphoristically: to see structures (e. g. cells and their components) in unprepared, unstained bulk pieces of meat. It was Allan Boyde from University College London who first tried our Tandem Scanning Microscope in stereoscopy of fine structures in microanatomy and succeeded (Boyde 1985).

First aim is to limit the illumination as much as possible only to these parts of objects which lie in the focused-on object plane of the microscope objective, i. e. which will yield a sharp image. The second aim is to suppress as much as possible all light which does not emerge from the focused-on object plane but has been reflected elsewhere. Generally, both these aims can be fulfiled the more perfectly the greater is the aperture of the microscope objective which is used for illumination and for image formation.

Suppose a point like light source e. g. a strongly illuminated small hole (Fig. 1) situated in the centre of the image plane of the microscope objective. Light enters the object beneath the objective and converges in a broad cone to a small spot, which we will call a light point. So the illumination light density along the optical axis above and below the focal plane falls off rapidly. Now suppose a reflecting plane object is moving along the axis in the object space towards the objective. Its illumination first rises slowly and then quickly reaches a maximum when the object passes the focal plane of the ob-

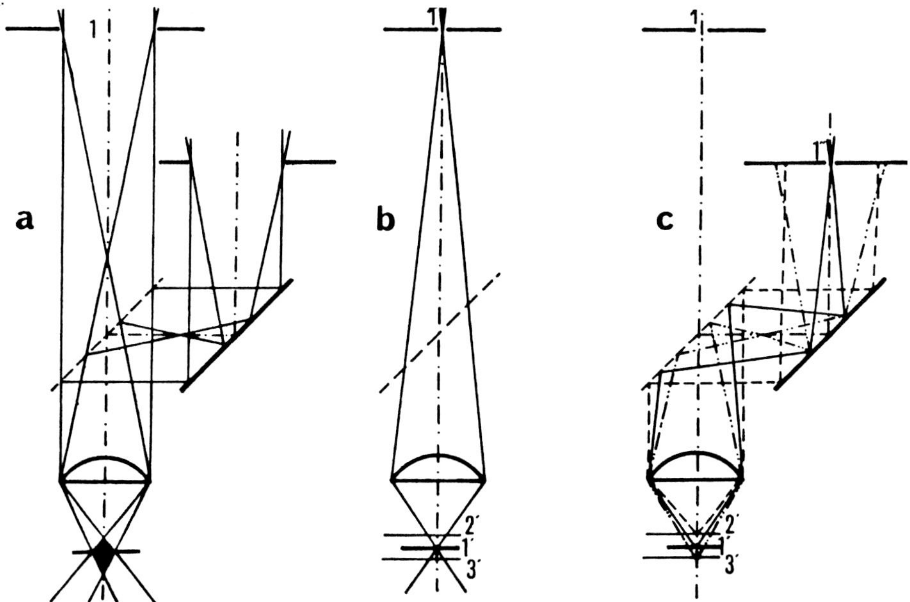

Figure 1. (a) When illuminating field of normal size in living matter we actually illuminate a volume of considerable depth nearly equally strongly. (b) Illuminating by means of a pinhole (1) a very small field (a "point" 1′), the illumination of object layers lying outside focus (2′, 3′) drops with the square of defocus. (c) Only the light which was reflected in focus (in the plane 1′) gets through the pinhole in the image plane unattenuated (1″). Only a small fraction of light which was reflected not in focus (2′, 3′) gets through the pinhole to be attenuated once more and so it impairs the image only negligibly.

jective and then the illumination falls as quickly as it has risen. But let us consider the spatial distribution of light reflected by the object: only when the light was reflected by an object lying in the focal plane, the reflected rays (after passing the objective) converge to the same small hole through which they entered. By this arrangement it should be possible to treat the reflected light selectively according to the place where the light was reflected: only the light which was reflected in focus, i. e. that able to form a sharp image of an focused-on object point in the small hole through which it entered the microscope is hence allowed to participate in image formation.

Figure 2. An enlarged detail of the Nipkow disc pattern. Area of the holes is one percent of the total active area of the disc.

To deal not with one object point but with extended object and image fields, it is necessary to scan simultaneously the illumination of the object and the formation of the image of the object. To shorten the time necessary to form the complete image and to repeat the process as frequently as necessary for a virtually steady image acceptable by a human eye, the scanning is performed not by one but by a set of several holes regularly distributed in a rotating scanning disc, a modification of the scanning disc by Nipkow (1884). To limit as much as possible the amount of any scattered light, i. e. light not carrying information but entering the image, it is necessary to have the openings in the Nipkow disc as small as possible (say about 20 micrometres in diametre) and occupying together less than one percent of the eyepiece field.

To give images of highest contrast it is necessary not to allow any illuminating light to enter the image space via a "short circuit", mainly via reflections on lens surfaces and tube walls. Hence follows that there must be not one scan but two scans coupled together: Illuminating light passes holes of the first scanning device and then a beam splitter to enter the objective. The light reflected in the focused-on plane of the object after passing the objective in reverse direction is reflected on the beam splitter and passes holes in the second scanning device lying in the front focal plane of the eyepiece.

Now the real optical setup is contrived so that both scans are performed by means of a single device, a modification of a Nipkow disc (Fig. 2) carrying two sets of holes arranged in Archimedean spirals, any hole in one set having its twin in the other set. The central symmetry of both sets of holes is transformed

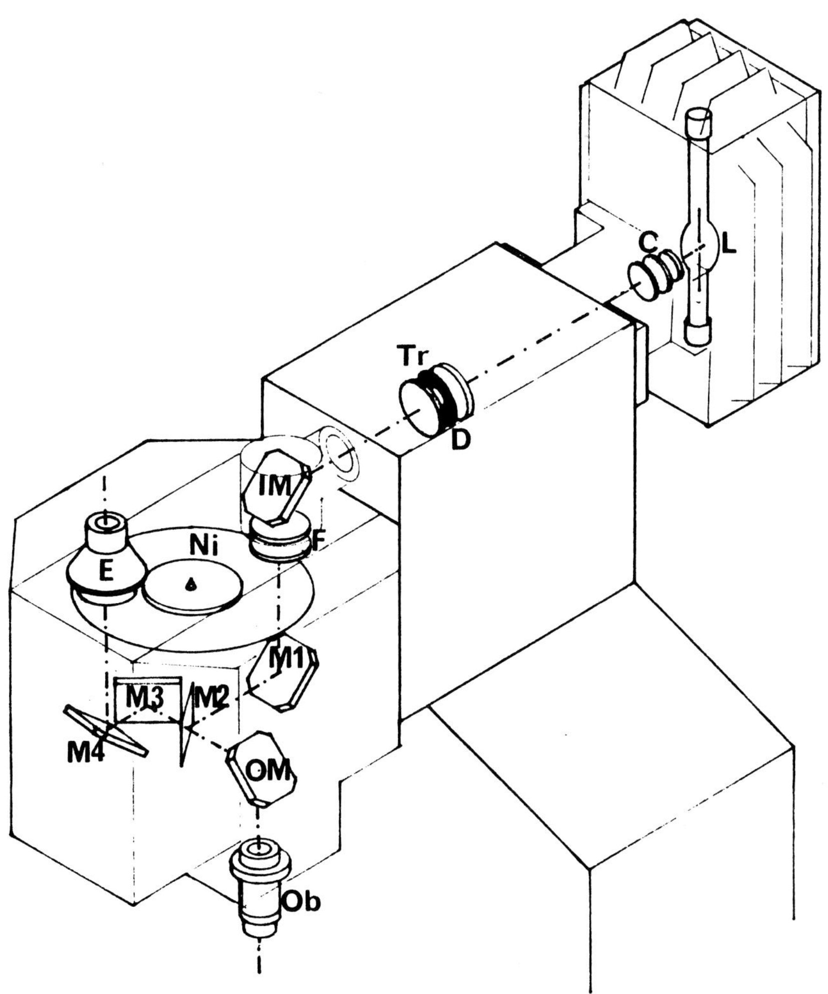

Figure 3. Disposition of the microscope. From right to left: Arc lamp (L), transfer lens (Tr) with aperture diaphragm (D), input mirror (IM), field lens (F), Nipkow disc (Ni), inverting mirror system (M 1, M 2, M 3, M 4) with semitransparent mirror (M 2), output mirror (OM), objective (Ob), eyepiece (E).

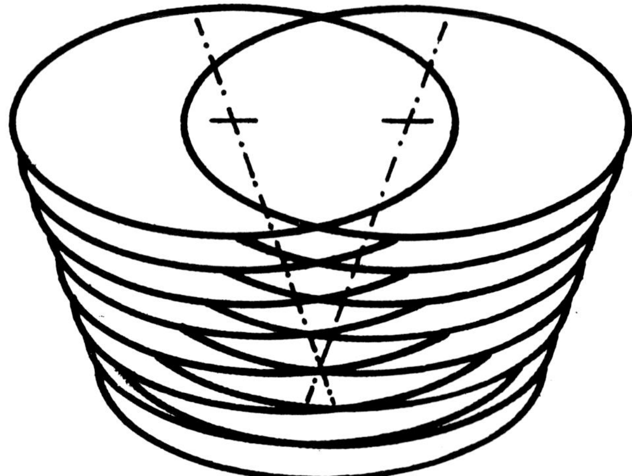

Figure 4. Stacking of exposures into one stereoscopic pair comprising a space model of the photomicrographed object.

into virtual identity by means of a set of four mirrors (Fig. 3), the above mentioned beam splitter being one of them. Needless to say, demands on exact localization of the holes, on stability of disc rotational axis, on positions of mirrors etc. are considerable. Nevertheless, this microscope enables us to pursue observations in microanatomy in vivo and in microphysiology virtually in real time monocularilly without any other additional device.

But we can also develop this microscope into a device yielding three-dimensional (stereoscopic) representations of living and also of fixed preparations (Boyde 1985). For this purpose it is necessary to make two frames, each consisting of one stack of many (e.g. twenty) exposures (Fig. 4). In one of these stacks any following exposure is taken after a small and equal shift in focus and at the same time a small lateral shift in one direction. In the second stack (comprising the second frame) after each exposure the focus is shifted exactly as it was shifted in the first stack (first frame), but the horizontal shifts are made in the opposite direction than in the first frame. By this means we get the paralax necessary for binocular observation to give stereoscopic perception. In this way impressive three-dimensional images of ramified neurons (e. g. Purkinje cells in the cerrebelum) can be obtained.

It is also possible to see three-dimensional images directly by using the method with two colour spectacles and a big memory for image storage and a closed circuit television displaying red image for one eye armed with red glass and a green image for the other eye with green glass. We can fancy also other approaches to this problem. Some of them are being tried on their feasibility.

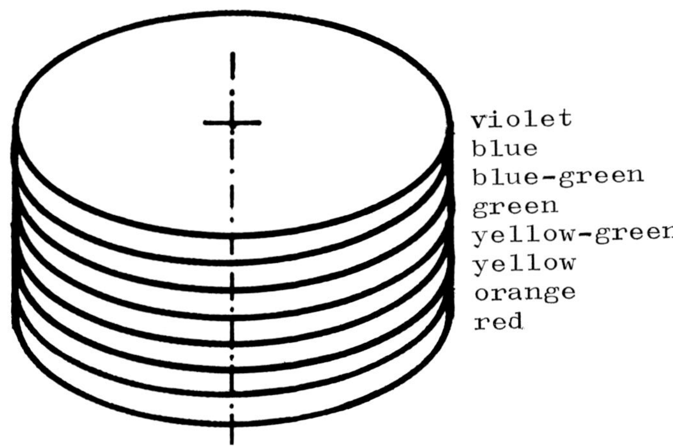

Figure 5. Stacking of exposures into one frame showing colored map of a thick object.

Even more simple methods can be used to demonstrate space distribution of constituents in a thick specimen: We photograph (Fig. 5) layer after layer as above but all exposures on one frame only (i. e. also without any sidewise shift) but we "denote" the third coordinate by colour using colour filters in illumination (Boyde 1987): First exposure (of the top layer) will be violett, the second one blue, then blue-green, green, yellow-green, yellow, orange, red. (In this order, because of Rayleigh scattering of light: The colour used for bottom layers should have longest wave length to be scattered least!)

It should also be noted that even the best objectives have noticeable chromatic aberrations of focal length (their aberrations are doubled because they are used in autocollimation). This residual chromatism can be used for exact measurement of small differences in elevation, e. g. in integrated circuits: The images look like maps which denote altitudes by means of different colours because foci for different colours do not lie exactly in the same distance from the front lens. By the differences in hue between different details the elevations of steps (e. g. in integrated circuits) can be estimated with the accuracy of about 0,1 micrometre or even better.

SUMMARY:

Tandem Scanning Microscope enables us to observe in real time one thin layer under the surface of a highly complex semitransparent object (e. g. living tissue), make photographs of thick layers in such objects and observe (by means of a stereoscope) three dimensional images of such objects. Important results can be obtained with moderate expenditure on basic hardware and the equipment can be expanded steadily according to the state of our budget.

REFERENCES

Boyde A. (1985). Stereoscopic Images in Confocal (Tandem Scanning) Microscopy. Science 230: 1270-1272

Boyde A. (1987). Colour-coded stereo images from the tandem scanning reflected light microscope. Journal of Microscopy 146: 137-142.

Petráň M., Hadravský M., Beneš J., Kučera R., Boyde A. (1985). The Tandem Scanning Reflected Light Microscope Part I - the principle, and its design. Proc Roy Microsc Soc 20: 125-129.

Petráň M., Hadravský M., Egger M. D., Galambos R. (1968). Tandem-Scanning Reflected-Light Microscope. Journal of the Optical Society of America 58: 661-664.

Cells and Tissues: A Three-Dimensional
Approach by Modern Techniques in Microscopy,
pages 559–562

APPLICATION OF OPTICAL DIFFRACTOMETRY IN THE ANALYSIS OF CYTOLOGICAL AND HISTOLOGICAL IMAGES

Kazimierz OSTROWSKI

Department of Histology, Institute of
Biostructure, Medical Academy,
Warsaw, Poland.

Frauenhofer optical diffraction has been applied in non biological disciplines of science for more than 35 years, although the first suggestion to use the method was made by Bragg as early as in 1939. The availability of coherent laser light as well as the remarkable development of electronics in the late sixties resulted in progress of optical diffraction pattern analysis and its use in biology. A number of sophisticated optical diffractometers coupled with analogue-digital detectors were designed and constructed.

Objective methods for quantitative image analysis are required in optical as well as in electron microscopy of biological objects as cells and tissues in order to compare two or more cell populations, evaluate the effects of pharmacological or other experimental treatment on cell or tissue fine structure. Stereology represents one approach to this problem. Optical diffractometry is another approach. In combination with optical filtering and three-dimensional reconstruction by Fourier synthesis, it has become a valuable tool in studies of macromolecular structures (Klug, 1971). Fourier analysis has been used in analysis of cytological and histological structures (eg. Pernick et al. 1978; Dziedzic-Goclawska et al. 1982; Ostrowski et al. 1983, 1988, Tedde et al. 1986). The strength and beauty of the discussed technique, as applied in the cytological and histological research, is connected with the numerical description of differences in structure which until now one could only describe.

The diffraction pattern obtained by the optical Fourier transform represents an integrated view of both, the size of structures and interstructural distances. Morover, the information on the direction of arrangement of the structures contained in the analysed image is obtained, even tough such an arrangement cannot be sometimes visually observed. Furthermore the diffractogram integrates information from the whole illuminated fragment of an object and the signal/noise ratio is thus much better than in the original image, where structural regularity might be masked by local impurities.

In principle each convergent lens achieves the Fourier transform. However, to apply diffraction pattern analysis in quantitative terms, an optical system of high sophistication as well as a high quality analogue-digital transformation system have to be used. In practice diffractometers contain high quality optical, rather expensive system of lenses. Diffractograms of the analysed images appear in the back focal plane of the transforming lens in the form of the light spots distributed in different distances and directions from the central spot of the nondiffracted light beam. Assumption is made that every image of biological object as cell or cell fragments, can be considered from the analytical point of view as being the sum of superimposed diffaction gratings, differing in their spatial frequencies (amount of lines per unit length) or directions (angles between the direction of lines and the arbitrary chosen NS axis). The higher spatial frequence the longer is the distance of the spot from the central point of diffractogram. The distribution of spots in relation to the diffractogramÕs center is called radial distribution. It contains the information about the size of the structures and interstructural distances. The distribution of spots of light in relation to the NS axis is called angular distribution. It contains the information concerning the spatial distribution of the structures present in the analysed image. The size of spots (their diameter) is measured by their angular size, which is determined by the higher or lower ordering of structures causing their appearence.

The light intensity and distribution of diffraction pattern can be measured or recorded on a film, photographic plate, TV video-camera or photoelectric detector.

The data obtained from the diffraction pattern are numerical ones and can be used for more or less complicated statistical analysis. Discrimination analysis is usually used for evaluation of differences between the two populations of images differing from each other by detectable structures.

The diffraction pattern obtained by the Fourier optical transform represents the image of an object imprinted in the microphotograph, as far as the number of bits of information are carried by both, the image and the diffaction pattern concerned. The image of the object may be reconstituted from the diffraction pattern by application of an additional optical Fourier transform system. Moreover, the diffraction pattern may be spatially modulated in order to cut off the unwanted structures from the reconstituted image. This is achieved by application of spatial filters or masks and is called spatial filtering. The modulation of the diffraction pattern eg. the masking of some of its fragments can change seriously the reconstituted image. Some structures characterized by given spatial frequency and given directional arrangement might disappear by masking of the appropriate fragments of the diffraction pattern. The famous example of spatial filtering is the vizualization of the dextra-laevorotary spiral of viruses (Klug, 1971).

More refined, altough technically difficult analysis of diffractograms can be done by the method called optical correlation. In this technique spatial matched filters are used. These filters are prepared by the holographic means, in which the hologram of the diffraction pattern of a reference structure is obtained. When the hologram of the reference structure is localized in the Fourier plane and subsequently the diffraction pattern of an examined image is formed in the same plane, the light passing through the additional optical Fourier transform system is converged in the back focal plane of the optical system into three separate areas which represent : the reconstructed image of the object, the convolution of the examined and the reference images and the correlation of the examined and the reference images, respectively. The latter is used to discriminate a given structure among many others in an examined object. The bright spots in the correlation area represent the autocorrelation between the reference structure and identical structures in the examined object. Morover, the localization of the autocorrelation spots reveals the spatial situation of the structures in the examined object.

Since the microphotographs of the biological objects taken under light, scanning or EM microscopes are the usual objects of the diffractometric analysis, some features of the photographic material should be taken into account and shortly discussed. They can and usually affect the accuracy of Fourier transform. The photographic films used for analysis should record with high fidelity informations on the distribution of light intensity and they should not deform the phase of the passing light. The deformations of the image imprinted on the film may be caused by : a). nonlinearity between the optical density of the photographic emulsion and the time of exposure, b). nonlinearity between the contrast on the microphotograph and the contrast of the object for a given spatial frequency, c). heterogeneity of the photographic material. The grain size of the photographic emulsion is another important feature which can, when unappropriatly chosen, increase the over-all noise of the system.

The tickness of photosensitive material and of film support vary. The phase of light passing through such emulsion and support is modulated and overall noise of the system is thereby increased. The input film gate has to ensure, appart from fixation in the gate of the film containing the analysed image, a uniform refractive index across the gate to prevent noise increase. This is achieved by placing the films in a glass chamber filled with liquid of the same refractive index as that of the film support.

Apart from the importance of optical diffractometry in the field of cytological and histological research as a tool for numerical teransformation of informations contained in analysed images this technique has big potency in future application in histopathological and cytological diagnosis. Automatization of histopatological diagnosis is already a necessity in the situation when the amount of experienced pathologists is relatively small in comparison to the growing demands of clinicians. Nobody will claim nowaday that any automatic devise could substitute the eye and brain of experienced histopathologists, but one can already use the optical diffractometry for screening the smears or tissue sections (Guerin et al. 1984).

REFERENCES

Guerin JF, Rozycka M, Raffat A, Dziedzic-Goclawska A, Ostrowski K, Czyba JC. (1984). Appreciation de la fibrose peritubulaire dans les biopsies testiculaire, par la technique de diffractometrie optique. In: La Biopsie Testiculaire ed. Arvis G. et Dadoune JP., SIMEP, Bruxelles - Villeurbanne France.

Klug A. (1971). Optical diffraction and filtering and three dimensional reconstructions from electron micrographs. Philos. Transactions R.Soc. Lond B. 261: 173-179.

Ostrowski K, Thyberg J, Dziedzic-Goclawska A, Rozycka M, Lenczowski S. (1983). Application of optical diffractometry in studies of cell fine structure. Comparison on arterial smooth muscle cells in contractile and synthetic state. Histochemistry 78: 435-449.

Ostrowski K, Wojtowicz A, Dziedzic-Goclawska A, Rozycka M (1988). Effect of 1-hyroxyethylidene-1,1-bisphosphonate (HEBP) and dichloromethylidene-bisphosphonate (Cl_2MBP) on the structure of organic matrix of heterotopically induced bone tissue. Histochemistry 88: 207-212.

Pernick B. Kopp RE. Lisa J. Mendelsohn J. Stone H. (1978). Screening of cervical cytological samples using coherent optucal processing. Appl. Opt. 17: 21-42.

Tedde G, Ostrowski K, Rozycka M, Dziedzic-Goclawska A, Daszkiewicz M, Montella A. (1986). Optical diffractometry of the developing or desaggregating Reinke crystals in the Leidig Cells of the human testis. Arch. Ital. Anat. Embriol. 91(n.3) 231-252.

Cells and Tissues: A Three-Dimensional Approach by Modern Techniques in Microscopy, pages 563–570

OSMIUM IMPREGNATION AND MICROMANIPULATION. THEIR ASSOCIATION IN STUDIES USING SECONDARY OR BACKSCATTERED ELECTRONS.

Gebhard Reiss and Enrico Reale

Laboratory of Cell Biology and Electron Microscopy, School of Medicine, Konstanty-Gutschow-Str. 8, D 3000 Hannover 61, FRG

INTRODUCTION

The conventional scanning electron microscopy (SEM) is used to detect the signal of secondary electrons (SE) investigating the surface of biological specimens. This surface normally is sputtered with gold or gold-palladium. The examination of structures lying under this primary surface may be unbared by fracturing the tissue. In this case charging artifacts may occur, which make a second or third sputtering procedure necessary. But every further sputtering reduces the resolving power of the primary surface by covering details.

In 1973 Kelley et al. introduced the ligand-mediated osmification, the OTOTO-method. The heavy metal coat on the surface made the tissue conductive like sputtered material. We will show, that in fracturing this OTOTO-treated tissue, the freshly unbared surfaces show no charging artifacts and permit a resolution of details at least like in sputtered probes. The further fracturing will be unlimited, even within the scanning electron microscope.

Ushiki and Fujita in 1986 demonstrated, that biological structures lying some micrometers below the surface of a specimen but stained with heavy metals like silver or osmium can be seen by SEM using the backscattered electrons, the BSE-mode. This procedure has been applied in the present study to OTOTO-treated specimens of the guinea pig inner ear.

MATERIAL AND METHODS

In our study we used the modification of the OTOTO-method

by Malick et al. from 1975. Tissue of the rat kidney and guinea pig inner ear was first prepared for scanning electron microscopy (fixation with aldehydes - vascular perfusion in rat kidney, perilymphatic perfusion in inner ear tissue -; OTOTO-method; graded alcchol series; critical point drying; mounting on stubs; observation in Philips SEM 505 without sputtering at accelerating voltages of 8 - 30 kV), and then - after observation in SEM - embedded in part for comparison in TEM.

To reveal in SEM structures, lying under the primary surface of the tissue, we developed a micromanipulator to be used inside the chamber of the SEM without interruption of vacuum and electron beam. It consists of two electrically driven motors allowing x/y-movements of a needle and two mechanical shifters for rotation and z direction. The stainless steel-needles have a diameter of about 1/10 mm and gradually taper at one extremity to a diameter of less than 10 micrometers.

To investigate the inner ear structures by SEM recording the backscattered electrons, a Philips "Multi - Function - Detector - System" with four BSE-detectors was used.

RESULTS

The precision of fracturing a tissue under light microscopic sight is limited by the dexterity of the operator and the resolving power of a stereoscopic light microscope. The desire of microsurgery within the chamber of a scanning electron microscope is obvious. The normally existing limits of this intention like probe charging and covering of details by several sputtering procedures were described above. Using the OTOTO-method we were able to remove these problems. Preliminary successful results on specimens treated with the OTOTO-method and dissected under the light microscope induced us to consider the construction of the micromanipulator.

In the following we want to demonstrate firstly the suitability of the association between OTOTO-procedure and micromanipulation; secondly, the application of the micromanipulator to the solution of one biological problem, for instance, to identify cellular structures; thirdly, the extension of the OTOTO-method and its association with micromanipulation to studies with backscattered electrons.

Firstly, it was necessary to demonstrate, that fracturing of OTOTO-treated tissue within the electron microscope is possible, and that it is not hindered by unexpected or undesired effects. Fig. 1a shows the tip of a needle inside of the urinary space of a renal corpuscle, close to a glomerular lobule. In Fig. 1b the lobule is detached from the glomerulus,

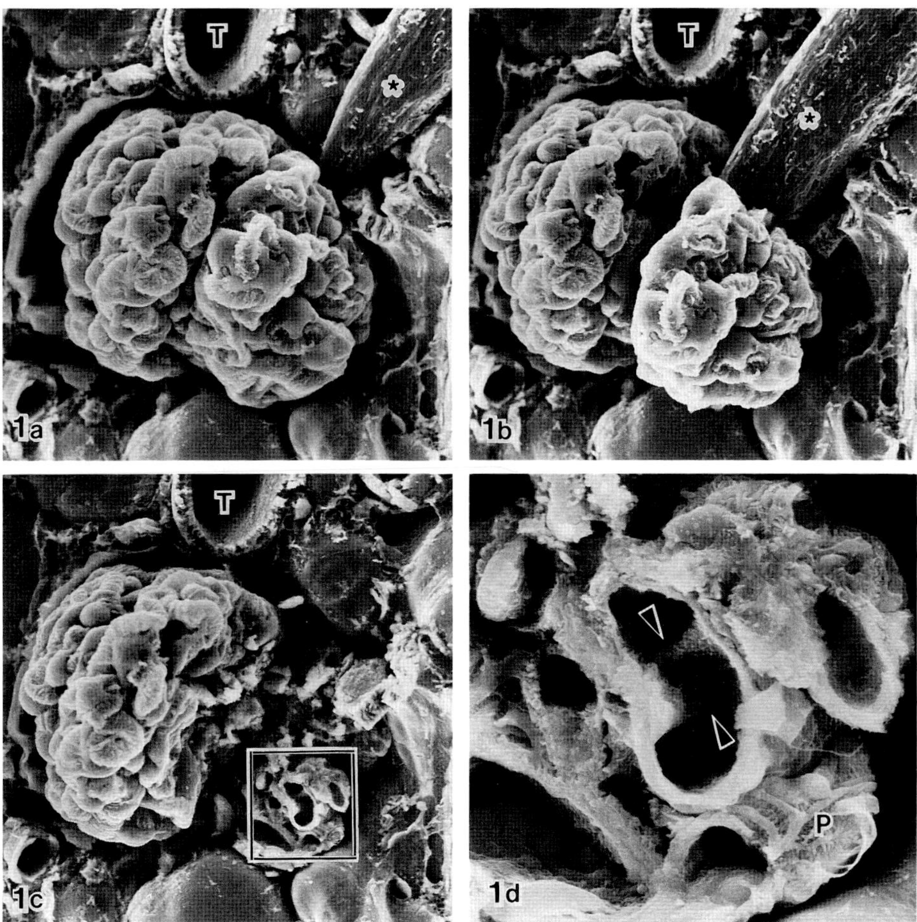

Fig. 1a-d. Rat kidney. Removal of a glomerular lobule from a renal corpuscle by microdissection. T, proximal tubule; asterix, tip of the micromanipulator-needle. In d, the fractured anchoring structures of the lobule - the capillary loops (framed in c) - become visible at higher magnification. P, podocyte; arrowheads, inner endothelial surface of a capillary (enlarged in Fig. 1e). a-c, 370 x; d, 1700 x.

in Fig. 1c it is removed by micromanipulation, and its fractured anchoring structures (framed in Fig. 1c) become visible. These structures - the capillary loops - are seen in Fig. 1d by higher magnification, the outer epithelial surface of one loop with a podocyte (arrow in Fig. 1d), and the endothelial inner surface (arrowheads). We can enter into this capillary loop (Fig. 1e) and observe the surface of the endothelial cells with numerous

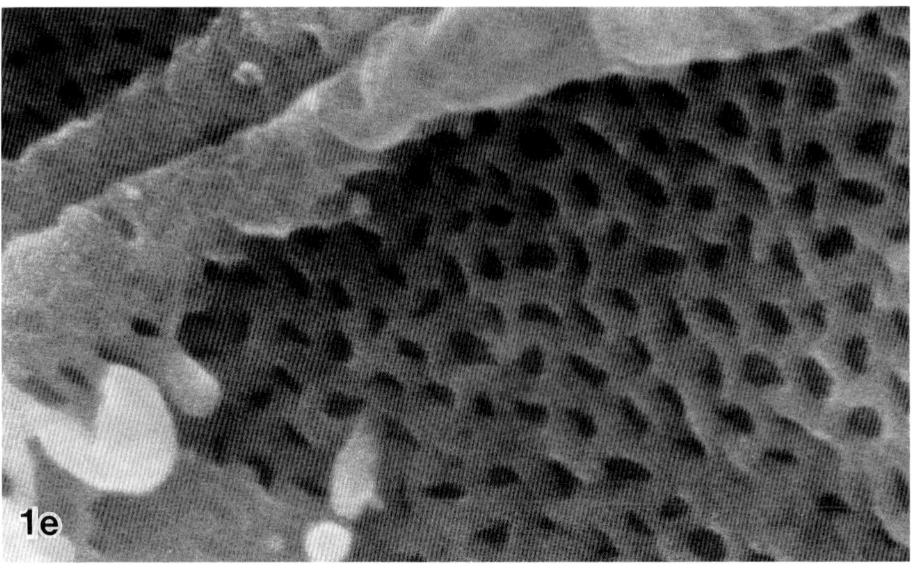

Fig. 1e. The inner surface of a capillary loop with endothelial pores. 52000 x.

irregularly shaped pores. Also higher magnifications can be seen without charging effects. This means, that the specimen has been sufficiently permeated by both, the osmiumtetroxide and the thiocarbohydrazide solutions.

By using the micromanipulator, in the following we tried to identify one tubule (arrows in Fig. 2a) of a medullary ray of the rat renal cortex. The tubule just showed its basement membrane. The needle extremity was put near to the tubule, and - see Fig. 2b - pushed laterally in order to break the tubular wall. The brush border (arrows) and the lateral interdigitations of the cells indicated, that the fragment was part of the proximal straight tubule.

As frequently described by thin section electron microscopy, a podocyte may lie with its body and the primary processes on parts of a capillary loop covered by secondary processes and by pedicles. To verify this arrangement the extremity of the needle was brought close to a podocyte body (Fig. 3a) and this was pushed away (Fig. 3b). Below the podocyte, the capillary wall was actually covered by processes of the removed podocyte (arrows indicate their fractured surfaces) and of a neighboring podocyte (P2).

In the organ of Corti, after microdissection of the

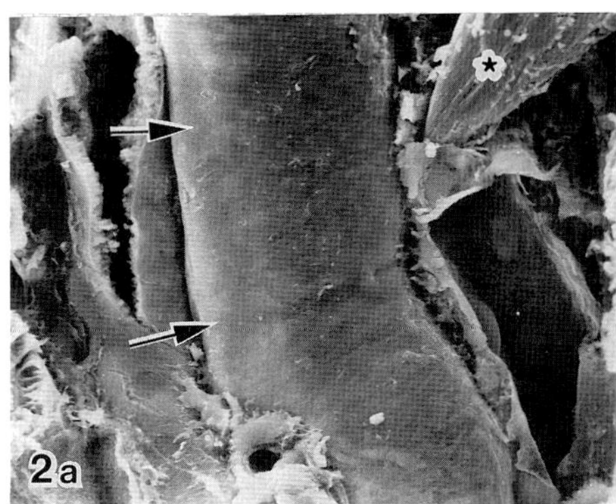

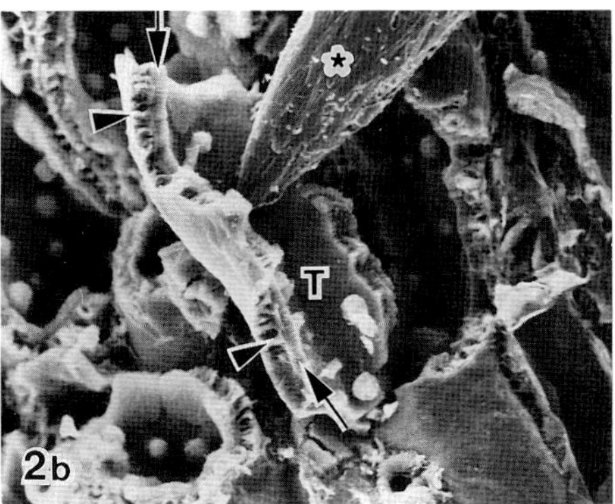

Fig. 2a,b. Rat kidney. Microdissection and identification of a proximal tubule (T) from cortical medullary ray. Asterix, tip of the micromanipulator-needle; arrows, brush border; arrowheads, lateral interdigitations. a 532 x, b 360 x.

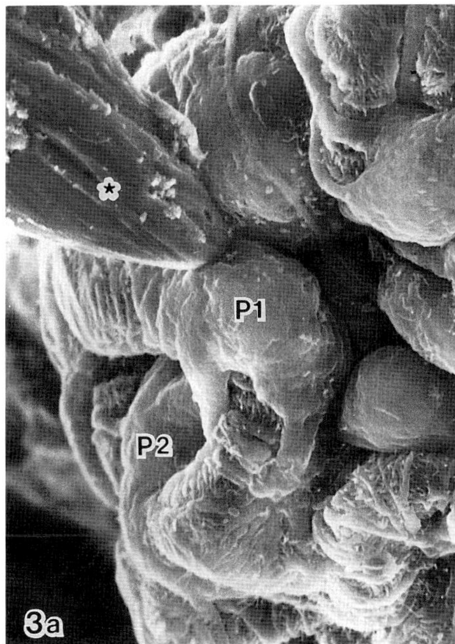

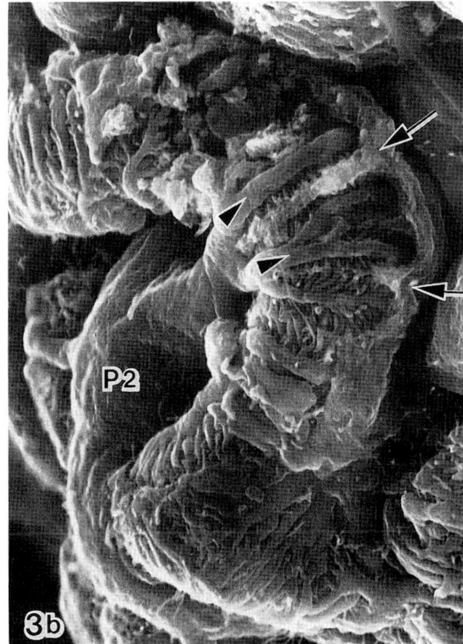

Fig. 3a, b. Rat kidney. Removal of a podocyte body (P1) by microdissection. Asterix, tip of the micromanipulator-needle. In b the fractured processes (arrows) of the removed P1 podocyte and those (arrowheads) of the neighboring podocyte (P2) cover the capillary wall. a 1730 x, b 2780 x.

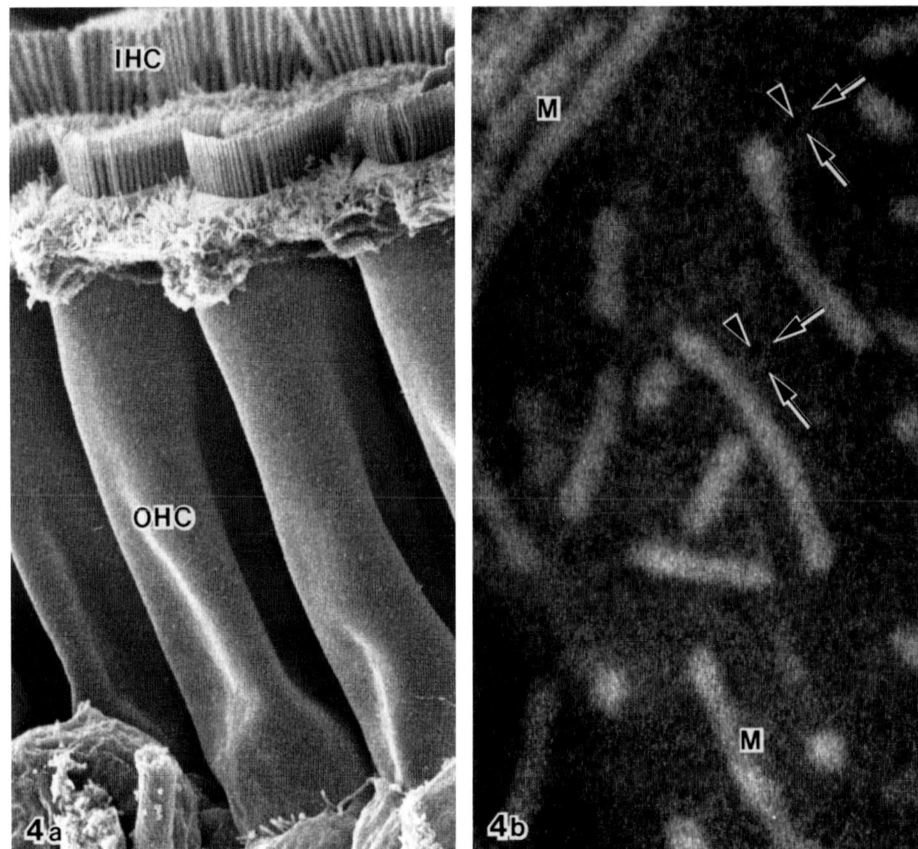

Fig. 4a, b. Guinea pig, organ of Corti. a, secondary emission. The lateral surface of the first row of outer hair cells (OHC). IHC, hairs of inner hair cell. b, backscattered electrons. Single outer hair cell at higher magnification. Below the plasma membrane withish mitochondria (M) and the network of the endoplasmatic reticulum (arrows) become visible. a, 3200 x; b 20000 x. (from: Reiss, 1988)

reticular membrane, the lateral surface of the first row of outer hair cells became visible (Fig. 4a). No charging of the probe was observed by using secondary emission. This again demonstrates the suitability of the OTOTO-procedure to impregnate the specimen deeply and uniformely and to render a sputtering unnecessary.

If we now changed to the BSE-mode and observed a single cell (Fig. 4b), we could recognize whitish elongated structures below the hair cell plasma membrane. Thin sections of the same

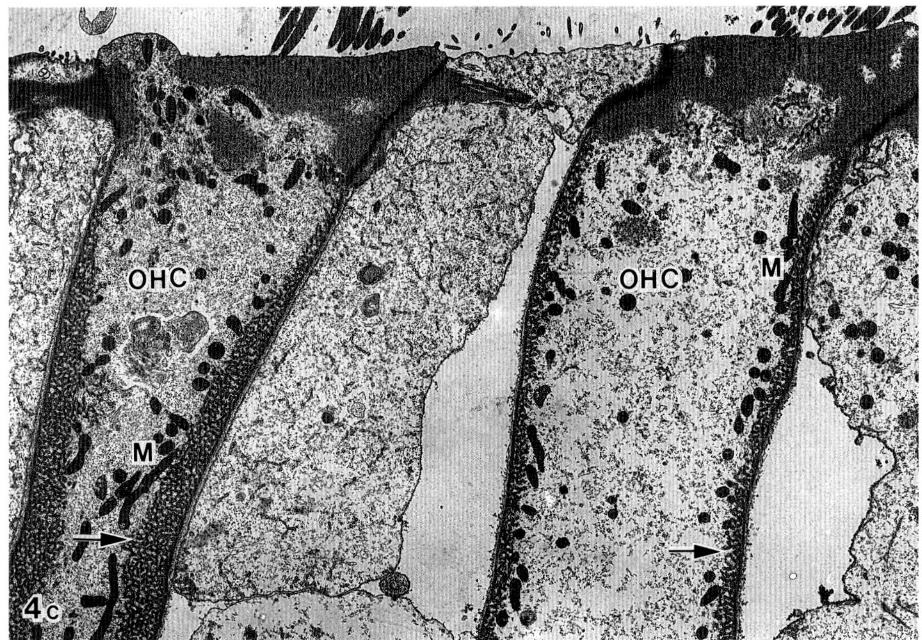

Fig. 4c. TEM of guinea pig organ of Corti embedded after observation in SEM. OTOTO-method. The lateral arranged mitochondria (M) and the network of the endoplasmatic reticulum (arrows) are heavily impregnated by osmium in two outer hair cells (OHC). 7600 x. (from: Reiss, 1988)

---

specimen embedded in Epon after observation by SEM (Fig. 4c), showed, that the whitish elongated structures were lateral arranged mitochondria (M in Fig. 4c). The SEM examination (Fig. 4b) also revealed the presence of a delicate reticular structure over and between the mitochondria. The thick meshes (arrows) were white, more electrons were backscattered; the small interposed spaces (arrowheads) were dark. This network corresponded to the flattened cisternae of smooth endoplasmic reticulum occurring beneath the hair cell plasma membrane (see arrows in Fig. 4c). Thin sections further demonstrated, that both - mitochondria and cisternae of the endoplasmic reticulum - were heavily impregnated by osmium. This explains the possibility to detect them by backscattered electrons.

CONCLUSIONS

In this short review, we demonstrated, whenever still necessary, the noticeable advantages of the OTOTO-procedure.

With two simple examples - identification of a renal tubule and removal of a podocyte body - we showed the utility and the possibilities of micromanipulation inside the scanning electron microscope. Of course, there are commerciably available micromanipulators; they, however until now, seldom have found biological applications. Therefore, we considered, that it is useful to remind them, especially in association with the OTOTO-procedure. This association, in addition, also allows investigations detecting backscattered electrons.

ACKNOWLEDGEMENTS

This work was done with the help of the DFG (Deutsche Forschungsgemeinschaft: Grant RE. 257/8-1). We are indebted to Mr. R. von der Fecht (Dept. of Mechanic, Central Workshop of the Hannover Medical School) for the construction of the micromanipulator.

REFERENCES

Kelley RO, Decker RA, Bluemink JG (1973). Ligand-mediated osmium binding: Its application in coating biological specimens for scanning elctron microscopy. J. Ultrastruct. Res. 45: 254-258.

Malick LE, Wilson RB (1975). Evaluation of a modified technique for SEM examination of vertebrate specimens without evaporated matal layer. SEM 1975: 259-264.

Reiss G (1988). Rasterelektronenmikroskopische Darstellung intracellulärer Strukturen an Haarzellen der Meerschweinchencochlea mittels Rückstreuelektronen. HNO 36: 102-105.

Ushiki T, Fujita T (1986). Backscattered electron imaging. Its application to biological specimens stained with heavy metals. Arch. Histol. Jap. 49: 139-154.

Cells and Tissues: A Three-Dimensional
Approach by Modern Techniques in Microscopy,
pages 571–580

MICRODISSECTION BY ULTRASONICATION FOR SCANNING ELECTRON MICROSCOPY

Frank N. Low

Department of Anatomy, Louisiana State University Medical Center, New Orleans, Louisiana 70112, USA.

INTRODUCTION

Experiments aimed at determining the usefuless of selective microdissection of biological tissues by ultrasound were initiated in the author's laboratory in the University of North Dakota at Grand Forks in 1978. Continuing work demonstrated the vascular network of the respiratory portion of the rat lung to advantage (Highison and Low, 1982). Further experiments indicated that simple epithelium in general was readily dissociated by ultrasound but that collagen and collagen containing tissues resisted the "vibratory insult". Later investigations ranged from dental tissues (Johnson and Highison, 1985) to Thebesian ostia in the mammalian heart (Rosinia and Low, 1986). These were recently reviewed by Highison et al. (1988). The present contribution summarizes the current status of certain phases of this technique with emphasis on recent attempts to achieve selective microdissection of the central nervous system.

METHODOLOGY

Various technical procedures have been utilized in search of useful and reliable preparations. These are briefly summarized below.

Frequency of Vibration

Commercial instruments capable of delivering ultrasonic mechanical vibrations are available in frequencies varying from 20 to 80 kilohertz (kHz; thousands of cycles per second). The lower frequencies (20 kHz) tend to break up the tissue samples without selectivity. The higher frequencies, notably 80 kHz, are gentler and more selective. The manufacture of ultrasonicators at frequencies greater than 80 kHz is prohibited by the Federal Communications Commission of U.S.A. Ultrasonic instruments commonly known as "tank cleaners' are ideal for microdissection when the tissue samples are immersed in small beakers containing pure acetone. The beakers are suspended in the water jacket of the tank by soft wire with the surface of acetone and water at the same level. Three to five minutes exposure usually suffices. Very little further microdissection occurs after 10 minutes.

Fixers

Fixation procedures usually utilize two percent OsO_4, buffered in the usual manner with immersion prolonged to 24 hours or longer. This produces brittleness in the tissue which then responds to the action of ultrasound. While interesting results are obtained with OsO_4 a satisfactory substitute is available in one percent aqueous boric acid (Vial and Porter, 1975). This compound (H_3BO_3) does not fix the tissue in the ordinary sense but softens it to a jelly-like mass that responds readily to ultrasonic vibration. Basement membranes retain their histological organization. Economy and ease of manipulation (overnight immersion) recommend this procedure.

The "hardeners" common in the fixing solutions of light microscopy can also be useful. One percent aqueous $HgCl_2$ or 2% $K_2Cr_2O_7$ yield useful samples after overnight immersion (McClugage and Low, 1984). Simple epithelia tend to peel off their basement membranes in a single sheet rather than undergo dissociation of individual cells.

Aldehyde fixation by cardiac perfusion is so commonly used that its special effects on subsequent microdissection by ultrasonication must be noted. In general, the stronger the aldehyde perfusate the more the tissue samples subsequently resist ultrasonics. The concentrations customarily used for transmission electron microscopy, such as Karnovsky's fixer diluted to circa 500 mOsm/l or the perfusate recommended by Palay and Chan-Palay (1974) for the central nervous system (1% formaldehyde and 1% glutaraldehyde) do not prevent subsequent microdissection. But perfusion of the living animal with ten percent formalin or any stronger aldehyde solution produces such strong cross linking that subsequent ultrasonication is futile. At the other extreme, when working with central nervous tissue and substitution of H_3BO_3 for OsO_4 at least mild prefixation with aldehydes is necessary. Otherwise, the samples become completely dissociated by the ultrasound. In general the concentrations of glutaraldehyde and paraformaldehyde in a cardiac perfusate are best kept at concentrations of less than two percent and the tonicity at circa 500 mOsm/l.

Pretreatment with Enzymes

Enzymatic digestion of fresh tissue samples prior to fixation affects the selectivity of subsequent microdissection. The effects of bacterial collagenase (Clostridium histolyticum; Sigma Type I) on dental tissues produces a variety of results depending on the duration of incubation (Highison et al., 1988). The use of collagenase, hyaluronidase, trypsin and pronase E (Grenier et al., 1985) in the analysis of basement membranes reveals a substructure limiting the edges of pores. Pretreatment with enzymes, although only briefly explored, appears to present unlimited possibilities for microdissection by ultrasonication.

Action of Detergents

Detergents added to certain of the solutions used are reported by Highison and Tibbitts to produce improved results (1986). Their best preparations result from tissues initially fixed in OsO_4. These are washed in cacodylate buffer and immersed in OsO_4 containing one-tenth percent detergent. Both ionic (Triton X-100; SDS) and non-ionic (NP40; Tween 20) detergents are useful but no real distinction can be made among the results obtained. The author uses saponin in similar and somewhat stronger concentrations with improved selectivity in the results.

RESULTS

The results obtained by various technical procedures used on different tissues prior to ultrasonication has been reviewed by Highison et al. (1988). Epithelial removal by variations of the technique has been reported by King and Hossler (1988) and after greatly prolonged immersion in OsO_4 by Jollie (1986). The present account is limited chiefly to certain variations (alimentary) and constancy (kidney) in basement membrane structure as revealed by ultrasonication. Some current experiments with the central nervous system are touched upon.

Alimentary Canal

The epithelial surface of the epithelial basement membrane is clearly revealed by ultrasonication. It is geometrically continuous throughout the lumen of the gut (Low & McClugage, 1984; McClugage and Low, 1984). Where it is covered by stratified squamous epithelium, as in the cardiac portion of the rat stomach, it is difficult to demonstrate. This is because keratinized tissue is somewhat resistant to ultrasonic vibration (Fig. 1). However, where a simple epithelium lines the gut, the basement membrane becomes clearly visible after

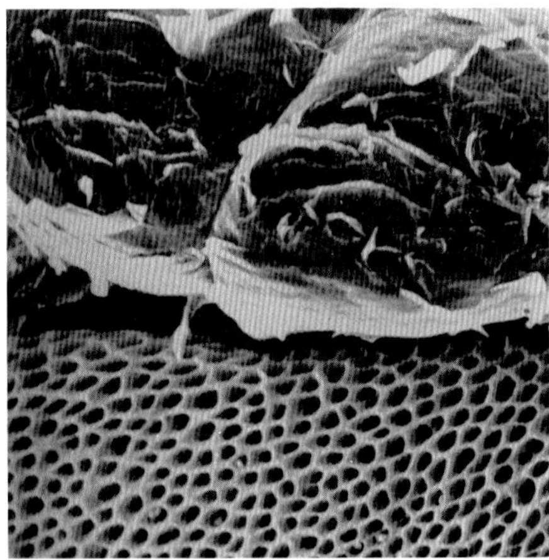

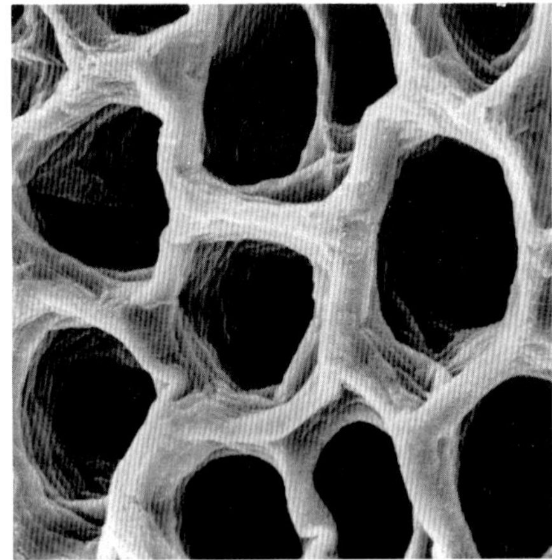

Figure 1. Stomach, Rat. Stratified squamous epithelium (above) is largely undisturbed but the epithelium has been removed from the gastric glands (below). 80X

Figure 2. Rat stomach. After epithelial removal the ostia of gastric glands are outlined by basement membrane. 850X

---

removal of the epithelial cells (Figs. 1-4). In the small intestine, in particular, the basement membrane may contain pores (McClugage and Low, 1984; McClugage et al., 1986). These vary in size and regional distribution.

The ostia of cul-de-sacs of basement membrane containing gastric and pyloric epithelial cells during life are revealed by removal of these cells (Figs. 1,2). In the small intestine the remaining connective tissue core of each villus is covered by its basement membrane (Figs. 3,4). In rats these are clavate but are cylindrical in humans. The lateral surfaces of these villous cores in rats possess a system of pores that accomodate lipid absorption (Dearing et al., 1984) and sometimes contain basal projections of columnar epithelial cells (McClugage and Low, 1984). Between the bases of the villous cores the ostia of crypts of Lieberkuhn are conspicuous. Where lymph nodules are located directly beneath the epithelium, whether singly or in Peyer's patches, villi are absent and the epithelium is specialized for antigen detection. Here the basement membrane is extremely porous (Low and McClugage, 1984; McClugage and Low, 1984; McClugage et al., 1986). In the colon the absence of villi emphasizes ostia of the glands of Lieberkuhn. In general appearance, the colonic basement membrane closely resembles that of the stomach. Its response to tumorigenesis has been recorded by Gallinghouse et al. (1986).

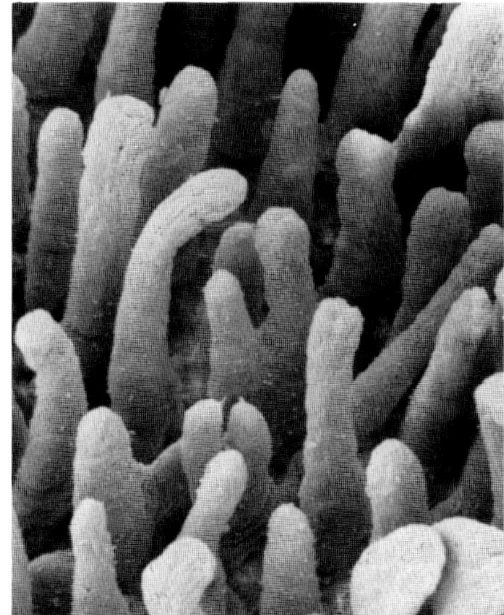

Figure 3. Small Intestine, Rat. The connective tissue cores of villi denuded of epithelium are clavate but vary in size. 110X

Figure 4. Small Intestine, Human. Villous cores in the human form tend to be cylindrical. 95X

Throughout the alimentary canal the epithelial basement membrane is structurally modified in a manner that accomodates food absorption (villous pores). It also protects the organism against invasion by antigens (extreme porosity over lymphatic nodules).

Kidney

The basement membrane of the kidney differs from that of the alimentary canal in that no pores are demonstrable. This applies to the entire extent of the nephric tubules as well as to the renal vasculature. The technique removes podocytes from glomeruli revealing the thick and poreless basement membrane well known in light microscopy and transmission electron microscopy.

Central Nervous System

The brain and spinal cord present a challenge. This is because collagen, the chief inhibitor of vibratory microdissection, is absent except in the walls of blood vessels.

The cerebellum was subjected to microdissection by methods described in this paper (Arnett and Low, 1985). Two responses

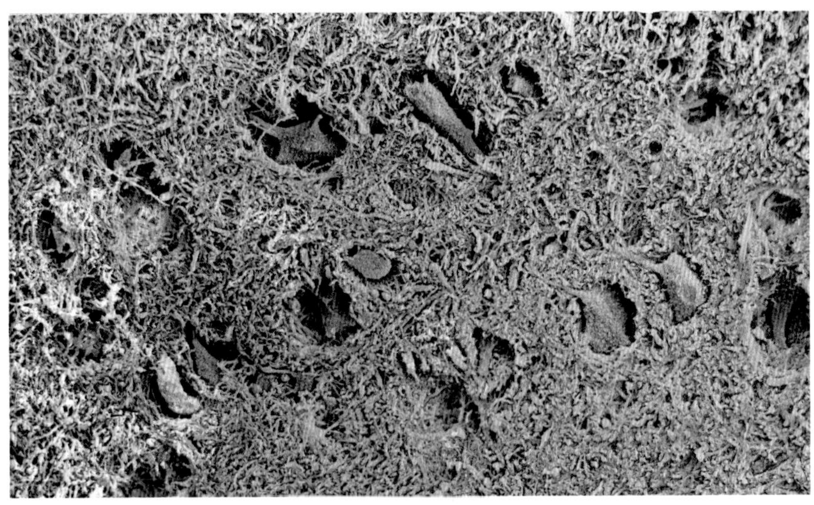

Figure 5. Anterior horn motor cells, Rat. Varying degrees of cavitation occur around cell bodies. 220X

of tissue from the central nervous system became evident during these experiments: (1) cavitation appeared in the neuropil surrounding cell bodies when ultrasound was applied and (2) the degree of microdissection depended on the concentration of aldehydes used in the initial cardiac perfusion of the experimental animal. Cavitation made it possible to identify all cell types of the cerebellum by means of their position and general morphology. In similar preparations of the spinal cord and brainstem it was observed that cavitation also occurred around anterior horn motor cells (Fig. 5). There was, however, a remarkable distinction. Synaptic vesicles remained on the surfaces of these cells (Figs. 6-8). There was clear distinction between the heavily studded surfaces of anterior horn motor cells, whether on the cell bodies or dendrites, and the comparatively smooth surfaces of the axon of the same cell (Figs. 6,7). The cell bodies of the anterior horn, with their many synapses, contrasted with the relatively smooth surfaces of cell bodies in the cochlear nucleus (Fig. 9).

The one percent concentrations of aldehydes of Palay and Chan-Palay (1975) microdissected well but anterior horn motor cells tended to drop out of the preparations. Reduction of the aldehyde content of the perfusate to one third or one quarter of one percent prevented this. Processes afferent to the

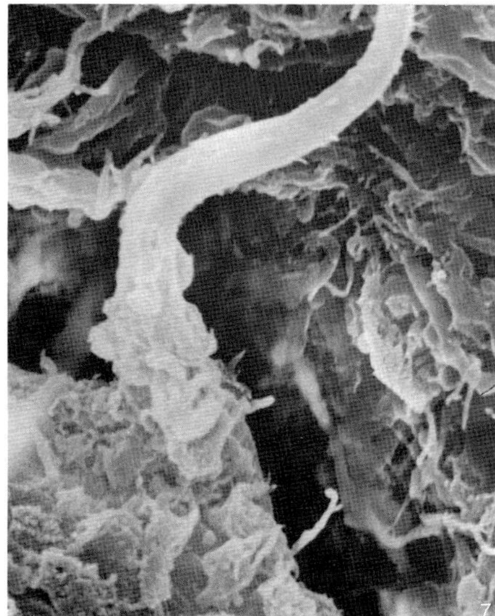

Figure 6. Anterior horn motor cell, Rat. The axon of this neurone is clearly visible. Detail in Figure 7, 740X

Figure 7. Anterior horn motor cell, Rat. Detail of cell in Figure 6. The axon (above) is smooth surfaced compared with the cell body which is studded with synaptic vesicles. 5310X

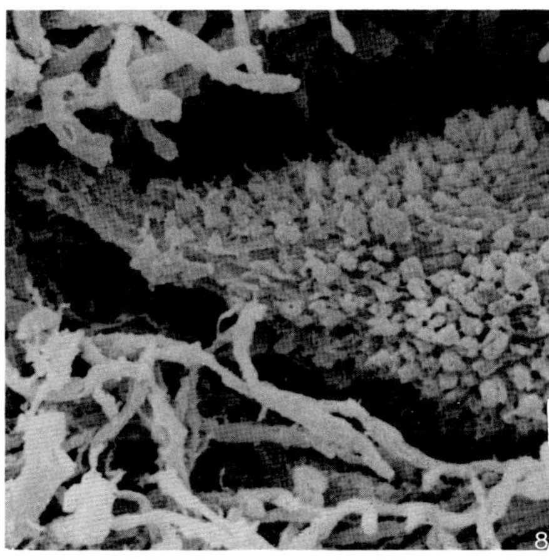

Figure 8. Anterior horn motor cell, Rat. The cavitation reveals synaptic vesicles and a few afferent axons. Compare with Figure 9. 2200X

Figure 9. Cell body, cochlear nucleus, Rat. This cell body is smoother than an anterior horn motor cell (Figure 8). 2000X

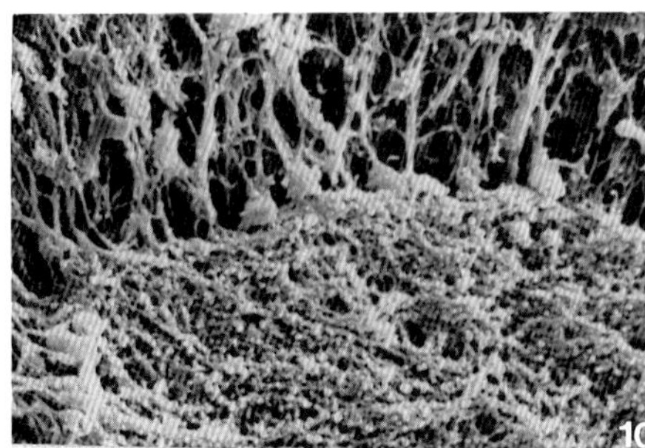
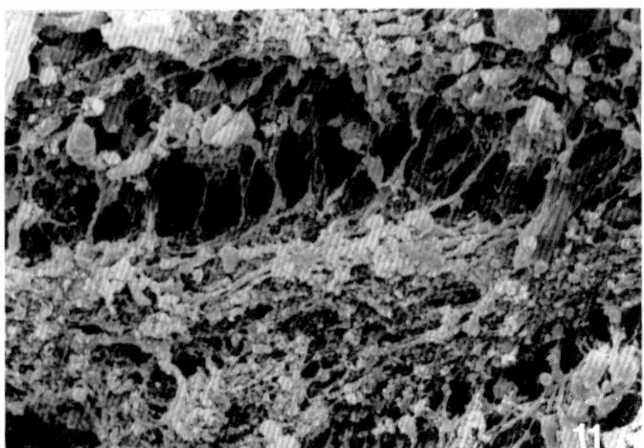

Figures 10 and 11. Anterior horn motor cells, Rat. Weaker prefixation with aldehydes produces lesser cavitation and preserves afferent processes. The artifically cut cell body occupies the lower half of each field. 2300X and 3700X

anterior horn motor cells were better preserved (Figs. 10,11).

DISCUSSION

Experiences thus far obtained with selective microdissection by ultrasonication indicate considerable potential for further investigation. The technique makes available, chiefly for examination by the scanning electron microscope, naturally occurring surfaces hitherto hidden from direct examination of their surface contours. This is particularly significant in the case of the basement membranes of epithelia. Epithelial cells are easily removed (although technical problems may arise in areas of cornification) and the underlying membranes readily examined. It is interesting to note that modifications of the membranes thus exposed reflect known physiological functions. In the alimentary canal the pores in the basement membrane of villous cores reflect an accomodation of digestive absorption at two separate levels; (1) the chylomicra of fat through pores and (2) transfer to the blood stream of molecules of carbohydrates and proteins. The extensively porous basement membranes overlying lymphatic nodules accomodate the passage of lymphocytes in combatting entry of antigens into the organism. Conversely the basement membranes of nephric tubules possess no similar pores; reflecting the unwanted nature of the contained waste products.

Enzymatic predigestion of living tissue samples presents a wide field of investigative possibilities, the range of which has been only slightly touched. Basement membranes act as a selective filter through which must pass the active ingredients and many of the products of essential metabolic processes. The possible effects of carefully selected enzymes on normally hidden surfaces could reveal much information concerning their molecular organization. At the risk of overspeculating, mention may be made of the "tip techniques" (Martin et al., 1988), now revealing surface organization at resolutions never before achieved (Abelson, 1988). Instruments involved include the scanning tunneling microscope (Binnig and Rohrer, 1985), the atomic force microscope (Marti et al., 1985) and related instruments (Martin et al., 1988). These items of equipment are now commercially available but are still in the process of refinement. They are able to examine biologically organized surfaces, whether electrically conductive or not, at very high resolution. Examination of the surfaces of untreated or enzymatically digested surfaces, whether plasmalemmal or those of basement membranes, could become a significantly productive venture.

The present status of microdissection of the central nervous system is admittedly imperfect. It is not yet possible to control the outcome of individual technical procedures. However, the possibilities for future development are interesting. The extraordinary complexity of physical relationships among neurones and neuroglia are presently clear to us. But their apprehension and interpretation are limited by the relatively complex analyses of thin sections (Conradi, 1976). A three dimensional view of the same areas would not only supplement present knowledge but would very likely reveal new information.

ACKNOWLEDGEMENTS

Special thanks are due to Drs. G. J. Highison, R. B. Johnson, S. G. McClugage, Jr. and to Dr. C.E. Arnett, III for pioneer work in the development of microdissection by ultrasonication. Angela Pepin typed the manuscript.

REFERENCES

Abelson PH (18 March 1988). Phenomena at interfaces. Science 239 (#4846):Editorial 1357.

Arnett CE III, Low FN (1985). Ultrasonic microdissection of rat cerebellum for scanning electron microscopy. Scanning Electron Microsc I: 247-255.

Binnig G, Rohrer H (1985). The scanning tunneling microscope. Scientific Amer 253(2):50-56.

Conradi, S (1976). Functional anatomy of the anterior horn motor neurone. In Landon DN (ed): "The Peripheral Nerve," London: Chapman and Hall, pp 279-329.

Dearing BD, McClugage SGJr, Low FN (1984). A SEM study of the epithelial basal lamina of small intestine during fat absorption. Anat Rec 208:43A.

Gallinghouse, GJ, McClugage SGJr, Low FN (1986). A SEM study of the intestinal epithelial basement membrane during tumorigenesis. Anat Rec 214:40A.

Grenier CP, McClugage SGJr, Low FN (1985). A SEM study of the fibrous microskeleton of pores within epithelial basement membranes. Anat Rec 211:73A.

Highison GJ, Johnson RB, McClugage SGJr, Low FN (1988). Ultrasonic microdissection techniques for scanning electron microscopy. CRC Crit Rev Anat Sci 1:193-227.

Highison GJ, Low FN (1982). Microdissection by ultrasonication after prolonged OsO_4 fixation: a technique for scanning electron microscopy. J Submicrosc Cytol 14:161-170.

Highison GJ, Tibbitts FD (1986). Ultrasonic microdissection of immature intermediate human placental villi as studied by scanning electron microscopy. Scanning Electron Microsc II:679-685

Johnson RB, Highison GJ (1985). Ultrasonic microdissection of the mouse mandible: exposure of the vasculature of alveolar bone and myelinated axons of the pulp. Anat Rec 211:96-101.

Jollie WP (1986). Changes in absorptive surfaces of rat visceral yolk sac with increasing gestational age. Scanning Electron Microsc II:661-669.

King, JAC, Hossler FE (1988). The gill arch of the striped bass, Morone saxatilis. III. Morphology of the basal lamina as revealed by various ultrasonic microdissection procedures. J Submicrosc Cytol Pathol 20:371-377.

Low FN, McClugage SGJr (1984). Microdissection by ultrasonication: scanning electron microscopy of the epithelial basal lamina of the alimentary canal in the rat. Amer J Anat 169:137-147.

McClugage SGJr, Low FN (1984). Microdissection by ultrasonication: porosity of the intestinal epithelial basal lamina. Amer J Anat 171:207-216.

McClugage, SGJr, Low FN, Zimny ML (1986). Porosity of the basement membrane overlying Peyer's patches in rats and monkeys. Gastroenterology 91:1128-1133.

Marti O, Ribi HO, Drake B, Albrecht TR, Quate CF, Hansma PK (1985). Atomic force microscopy of an organic monolayer. Science 239(#4846):50-52.

Martin Y, William CC, Wickramasinghe HK (1988). Tip-techniques for microcharacterization of materials. Scanning Microsc 2(1):3-8.

Palay SL, Chan-Palay V (1974). Cerebellar Cortex: Cytology and Organization. Springer-Verlag, pp 322-326.

Rosinia FA, Low FN (1986). Scanning electron microscopy of Thebesian ostia (microdissection by ultrasonication: enzymatic digestion). Scanning Electron Microsc IV:1363-1369.

Vial J, Porter KR (1975). Scanning electron microscopy of dissociated cells. J Cell Biol 67:345-360.

Cells and Tissues: A Three-Dimensional
Approach by Modern Techniques in Microscopy,
pages 581–587

MICROMORPHOLOGY BY CRYO-HVEM

Mircea Fotino

Department of Molecular, Cellular and Developmental Biology
University of Colorado, Boulder, CO 80309, U.S.A.

The study of ultrastructures in cells and tissues unavoidably depends on the methodology used for specimen preparation and on the visualization technique. Although in common use, it is widely recognized that the required preparation protocols usually introduce modifications and artifacts that alter the original configuration in a way that is difficult to identify or evaluate. This is especially the case when pursuing the highest levels of resolution currently attainable by scanning and transmission electron microscopy.

It is widely recognized that specimen alterations may unfortunately occur all along the various steps taken during the chemical protocols traditionally required for satisfying the limitations imposed by the instrument vacuum, by beam-induced damage and by low image contrast. Over the last several decades they have been the subject of careful study in search of unaltered structures. It is thus known that fixation by aldehydes that may take up to several seconds (Hackenbrock et al., 1971) is in fact a slower process than ideally required for instantly arresting cell movement and for preventing intracellular components from undergoing known distortions or rearrangements (Fineran, 1970; Fitzharris et al., 1972; Bretscher & Whytock, 1977; Green, 1981, 1982) as well as chemical alterations (Wood, 1973; Hayat, 1981). The use of alcohol, acetone or propylene oxide for dehydration leads to some uncontrollable loss of lipids (Korn & Weisman, 1966; Zingsheim & Plattner, 1976), to modifications of proteins (Hopwood, 1969; Ellar et al., 1971) and to overall shrinkage (Hayat, 1981). From a mechanical perspective, resin embedment is usually accompanied by some change in volume (Hayat, 1981), and sectioning disrupts specimen integrity in the vicinity of the advancing knife edge (Peachey, 1958). Finally, the last step in ultrastructural specimen visualization, namely imaging by exposure to the electron beam, produces complex radiation-induced damage that affects structures and matrix alike (Cosslett, 1970; Glaeser, 1971; Reimer, 1975) by such deleterious mechanisms as mass loss (Bahr et al., 1965; Reimer, 1965; Stenn & Bahr, 1970; Wall, 1972), heating, ionization, radiolysis and others. Some of these alterations may actually happen under the eye of the observer.

The approach described below combines preparative and operational procedures made available in electron microscopy during the last few years and provides an advantageous alternative to standard techniques (Fotino & Giddings, 1985; Fotino, 1986). In this approach each preparative and imaging step is simply reduced to its least distorting form currently available. Consequently, what may be considered at the present time the least disruptive method for microscopic investigation of biological specimens can be attained by combining the following steps:

1) **Physical specimen preparation by fast freezing.** It is preferable to the traditional chemical protocols because it achieves simultaneously three major goals:

a. preventing cellular rearrangement by fast completion in milliseconds rather than seconds as usually required by chemical fixation (Moor, 1964; Costello & Corless, 1978; Schwabe & Terracio, 1980),

b. avoiding use of reagents and thus chemical alteration (Chandler & Heuser, 1979),

c. eliminating structural damage deriving from slowly growing ice crystals during slow freezing by the direct formation of amorphous ice (Staub & Storey, 1962; Meryman, 1974; Brüggeller & Mayer, 1980; Dubochet & McDowall, 1981).

Good and undistorted preservation of ultrastructures by simple fast freezing may routinely be achieved to depths of up to 10-20μm.

2) **Direct microscopic observation at low temperature.** Major temperature-related benefits derive from such features as:

a. native structures viewed directly in unmodified, hydrated (frozen) form (Fernández-Morán, 1952; Heide & Grund, 1974; Taylor & Glaeser, 1976; Hutchinson et al., 1978; Taylor, 1978; Zierold, 1982; Lepault et al, 1983a; McDowall et al., 1983).

b. dehydration by sublimation carried out inside the microscope itself, if so desired, for avoiding possible alterations by exposure to the atmosphere during transport to and insertion into the instrument column, and

c. beam-induced damage reduced to an appreciable extent by comparison with the corresponding damage at room temperature (Kobayashi & Sakaoku, 1965; Lepault et al, 1983b; Siegel, 1972). The hydrated environment, however, renders the ice-embedded structures more radiation sensitive.

3) **Imaging by high-voltage electron microscopy (HVEM).** The main features of this type of microscopic imaging derive from the direct interaction of the high-energy electron beam with the specimen. They consist primarily of improved resolution and penetration as well as reduced beam-induced damage by comparison with the corresponding characteristics of conventional microscopes (Hama & Porter, 1969; Ris, 1969; Favard & Carasso, 1973). At the highest accelerating voltages of about 1 MV commonly accessible today,

a. the resolution in specimens of any thickness is the best attainable in these specimens by any instrument (Fotino, 1981),

b. thicker specimens than normally viewed by CTEM can be imaged satisfactorily with suitable depth of field, so that more information can be extracted from a single thick and thus mechanically undistorted sample, making this feature ideal for 3D reconstructions from serial sections, and

c. the energy transferred by the electron beam to the specimen is at a minimum and so is therefore the beam-induced damage (Bethe, 1933; Landau, 1944).

The advantageous characteristics afforded by these steps are simultaneously embodied in the low-temperature, high-voltage electron microscopy imaging mode (cryoHVEM) that has become available at state-of-the-art performance levels through the development since the late 1970s of an appropriate cryostage for the Boulder HVEM installation (Model JEM-1000).

Because of the cylindrically symmetric geometry of its top-entry configuration and of the large - and thus stable - cryostat that can be accommodated in the unusually large space available within the instrument's pole pieces, this cryostage has good mechanical and thermal stability, and therefore state-of-the-art performance. The original 0.3nm resolving power of the installation is preserved isotropically at arbitrary temperatures between 78K and room temperature. Frozen specimens immersed uninterruptedly in liquid nitrogen are inserted into the cryostage through a vacuum lock mechanism that eliminates specimen exposure to the atmosphere (Fotino, 1986).

Of the different types of uses for which the cryoHVEM mode may provide a unique and interesting experimental potential, only two that present a clear morphological interest, albeit in very different ways, are briefly illustrated here.

One of them, perhaps the most stimulating and radically differing from the customary procedures for ultrastructural visualization, is the direct viewing of a biological sample in its frozen natural environment (Fernández-Morán, 1952; Heide & Grund, 1974; Taylor & Glaeser, 1974; Hutchinson et al., 1974; Dubochet et al., 1982). Such condition is displayed in Fig. 1. It shows a suspension of entire flagellar axonemes isolated from sea-urchin sperm and

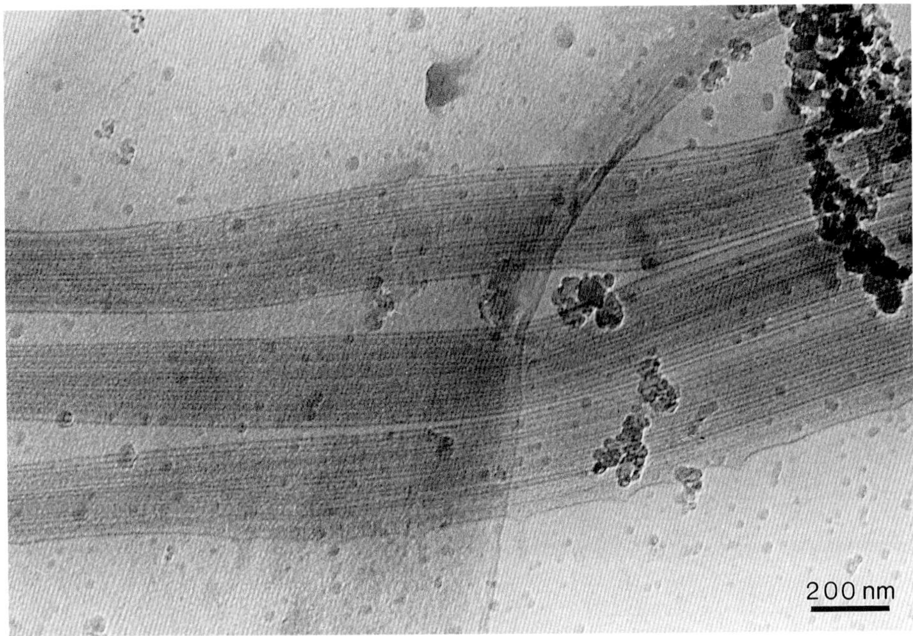

Fig. 1 - Phase-contrast image by cryoHVEM at 120K of a suspension of flagellar axonemes isolated from sea-urchin sperm and embedded in a thin layer of amorphous ice stretched across a holey carbon film. The flagella were neither fixed nor stained, but only fast frozen in liquid ethane. Good preservation was assured through minimum exposure (∿100-300 e/nm^2). Accelerating voltage 750 kV. (Specimen courtesy of Dr. J. Murray, University of Pennsylvania).

embedded in a thin layer of amorphous ice supported by a holey carbon film. These flagella are neither fixed nor stained. Nevertheless their internal structure is visible by phase contrast to a level of resolution that is determined by the amount of defocus. In addition to the unaltered, intrinsic information so obtained, this image is doubly interesting in itself for it shows that phase contrast can both occur in relatively thick layers of ice (in this case presumably about 0.2μm) and become visible at high voltage.

The other example of unique cryoHVEM resourcefulness consists of the view of massive whole mounts, such as whole cultured cells, imaged at some arbitrary temperature after complete freeze-drying has been accomplished **in situ** in a reasonable length of time and at an appropriate temperature, say 160K to 180K. The structures of whole cultured NRK cells seen in Fig. 2 not only have not been modified by any chemical treatment, but have also not been exposed to the rehydration and oxidation that usually occur by exposure to the atmosphere. It is also somewhat surprising that by this procedure all freeze-dried structures appear to be quite stable for imaging even though they were not stabilized for drift or charging effects by the usual carbon coating of traditional preparations.

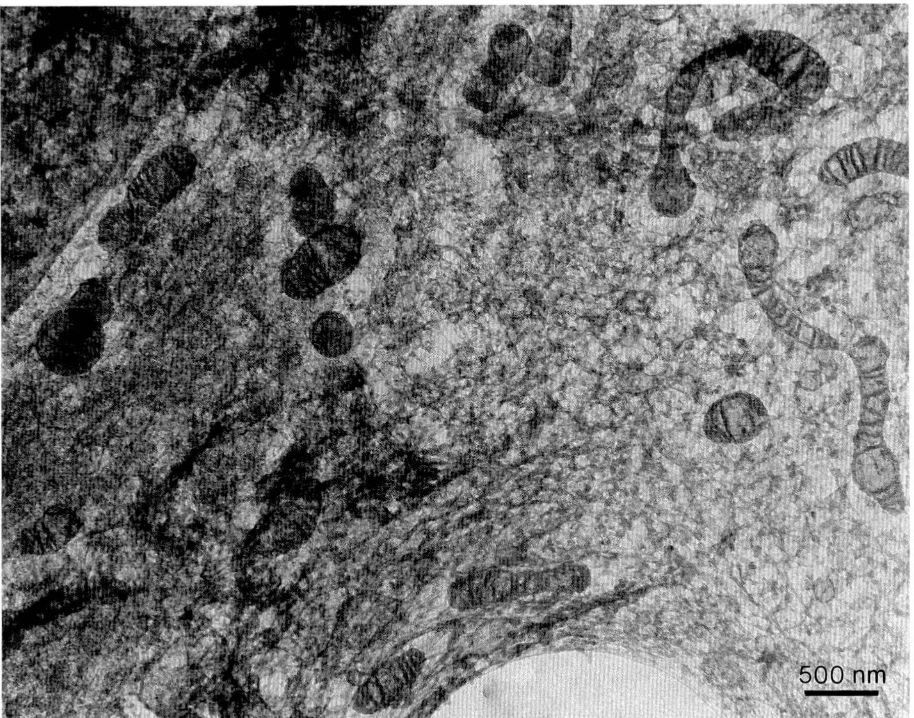

Fig. 2 - Portion of a fast-frozen NRK cell, unfixed and unstained, freeze-dried **in situ** *at 183K. Both the cell edge and the nuclear envelope (upper left) appear intact. Cristae are visible in most mitochondria. Accelerating voltage 1 MV.*

Even the thickest portions toward the center of the cell, where the freezing rate is slowest, and in particular the nuclear envelope and mitochondria appear well preserved by this procedure. It should be noted that the cytoplasmic matrix also appears to have an identifiable and non-random structure that is preserved by this chemically inert imaging procedure and that may therefore be interpreted as a real cellular feature (Wolosewick & Porter, 1979; Porter & Anderson, 1982) rather than as an artifact of preparation (Ris, 1980, 1985).

The limitation imposed on microscope use by a lengthy sublimation period can be avoided with specimens for which there is no evidence of structural alteration by exposure to the atmosphere after freeze-drying. The benefits of cryoHVEM for these can still be enjoyed after separate freeze-drying outside the microscope.

The advantages of cryoHVEM discussed above can be further enhanced by improved contrast attainable through increased illumination coherence and by stereo imaging that should become available in the near future.

In conclusion, visualizing cellular ultrastructure by fast freezing and subsequent imaging by cryoHVEM offers a combination of beneficial features that are currently accessible. Imaging by cryoHVEM is of unique interest both with hydrated-frozen as well as with freeze-dried specimens. The former are thought of being closest to their native condition, but delicate to explore. The latter do not seem to display significant differences with their critical-point-dried or freeze-dried counterparts, but their preparation is simple and the absence of chemical alteration may provide a subtle differential handle for assessing cellular features.

This work was supported in part by NIH grant No. RR-00592 and by NSF grant No. PCM-79-05631.

REFERENCES

Bahr GF, Johnson FB, Zeitler E (1965) in Bahr GF, Zeitler E (eds): "Quantitative Electron Microscopy", Baltimore, Williams and Wilkins, pp 377-395.

Bethe HA (1933) in Geiger H, Scheel K (eds): "Handbuch der Physik", 2nd ed., vol. 24/1, Springer, Berlin, pp 491-523.

Bretscher MS, Whytock S (1977). J Ultrastruct Res 61:215-217.

Brüggeller P, Mayer E (1980). Nature 288:569-571.

Chandler DE, Heuser J (1979). J Cell Biol 83:91-108.

Cosslett VE (1970). Ber Bunsenges Phys Chem 74:1171-1175.

Costello MJ, Corless JM (1978). J Microscopy 112:17-27.

Dubochet J, McDowall AW (1981) J Microscopy 124:RP3-RP4.

Dubochet J, Lepault J, Freeman R, Berriman JA, Homo JC (1982). J Microscopy 128:219-237.

Ellar DJ, Munoz E, Salton MRS (1971). Biochim Biophys Acta 225: 140-150.

Favard P, Carasso N (1973). J Microscopy 97:59-81.

Fernández-Morán (1952). Arkiv for Fysik IV/3:471-483.

Fineran BA (1970). J Microscopy 92:85-97.

Fitzharris TP, Bloodgood RA, McIntosh JR (1972). Tissue Cell 4:219-225.

Fotino M (1981) in Griffith JD (ed):"Electron Microscopy in Biology", vol. 1, New York: Wiley-Interscience, pp 89-138.

Fotino M & Giddings TH (1985). J Ultrastruct Res 91:112-126.

Fotino M (1986). J Microscopy 143:283-298.

Glaeser RM (1971). J Ultrastruct Res 36:466-482.

Green CR (1981). J Ultrastruct Res 75:11-22.

Green CR (1982). J Microscopy 125:201-206.

Hackenbrock CR, Rehn TG, Weinbach EC, Lemasters JJ (1971). J Cell Biol 51:123-137.

Hama K, Porter KR (1969). J Microscopie 8:149-158.

Hayat MA (1981). "Fixation for Electron Microscopy", New York, Academic Press.

Heide HG, Grund S (1974). J Ultrastruct Res 48:259-268.

Hopwood D (1969). Histochemie 17:151-161.

Hutchinson TE, Bacaner M, Broadhurst J, Lilley J (1974). Rev Sci Instr 45:252-255.

Hutchinson TE, Johnson DE, Mackenzie AP (1978). Ultramicroscopy 3:315-324.
Kobayashi K, Sakaoku K (1965). Lab Invest 14:1097-1114.
Korn ED, Weisman RA (1966). Biochim Biophys Acta 116:309-316.
Landau L (1944). J Phys USSR 8:201-205.
Lepault J, Booy FP, Dubochet J (1983a). J Microscopy 129:89-102.
Lepault J, Dubochet J, Dietrich I, Knapek E, Zeitler E (1983b). J Mol Biol 163:511.
McDowall AW, Chang J-J, Freeman R, Lepault J, Walter CA, Dubochet J (1983). J Microscopy 131:1-9.
Meryman HT (1974). Ann Rev Biophys Bioeng 3:341-363.
Moor H (1964). Z Zellforsch 62:546-589.
Peachey LD (1958). J Biophys Biochem Cytol 4:232-242.
Porter KR, Anderson KL (1982). Eur J Cell Biol 29:83-96.
Reimer L (1965). Lab Invest Suppl 14:344-358.
Reimer L (1975) in Siegel B, Beaman R (eds): "Physical Aspects of Electron Microscopy and Microbeam Analysis", New York: Wiley, pp 231-245.
Ris H (1969). J Microscopie 8:761-766.
Ris H (1980). Proc EMSA 38:812-813.
Ris H (1985). J Cell Biol 100:1474-1487.
Schwabe KG, Terracio L (1980). Cryobiology 17:571-584.
Siegel G (1972). Z Naturforsch 27A:325-332.
Staub NC, Storey WF (1962). J Appl Physiol 17:381-390.
Stenn K, Bahr GF (1970). J Ultrastruct Res 31:526-550.
Taylor KA (1978). J Microscopy 112:115-125.
Taylor KA, Glaeser RM (1974). Science 186:1036-1037.
Taylor KA, Glaeser RM (1976). J Ultrastruct Res 55:448-456.
Wall J (1972). Proc EMSA 30:186-187.
Wolosewick JJ, Porter KR (1979). J Cell Biol 82:114-139.
Wood JG (1973). Biochim Biophys Acta 329:118-127.
Zierold K (1982). Ultramicroscopy 10:45-53.
Zingsheim HP, Plattner H (1976) in Korn ED (ed): "Methods in Membrane Biology", New York: Plenum, vol 7, pp 1-46.

Cells and Tissues: A Three-Dimensional Approach by Modern Techniques in Microscopy, pages 589–595

REVIEW OF SCANNING TUNNELING MICROSCOPY - NEW BIOLOGICAL FRONTIER?

Branislav Vidić

Department of Anatomy and Cell Biology,
Georgetown University School of Medicine,
Washington, D.C. 20007

BACKGROUND

Scanning tunneling microscopy (STM) is relatively new technology. It was initially developed for imaging the structural and/or electrical properties of insulating layers of such diameter as to allow electron tunneling (Binnig and Rohrer, 1982). Further improvement was made by introducing vacuum as the separating medium between the sample and the scanning probe: 1. damage to sample was minimized to negligible; 2. surface of sample was accessible for treatment and/or investigation; and 3. free movement of the tip over the investigated surface was established. Bound electrons in the sample are the only source of radiation required for the operation of the STM. Hence, there is no additional need for free particles, lenses, light and/or special electron sources. Because condensation of decayed electrons, electron cloud, falls exponentially with their distance from the surface, the tunneling current (I) is very sensitive to the variation of this distance. In fact, a variation equal to an atomic diameter modifies the tunneling current by a factor of about 1,000 (Binnig and Rohrer, 1985a). Voltage (V) applied across the tunneling gap by means of scanning probe, as one electrode, and the sample, as the other electrode, is the third variable in the operation of STM, in addition to the tip to sample distance (s) and the tunneling current. The polarity of electrodes in general is irrelevant in this operation.

The principle on which the STM operates is relatively simple. The two electrodes, scanning tip and investigated sample, are approached to ca. 10 A distance so that two electron clouds, decayed electrons, overlap. Application of a voltage across the tunneling gap establishes the flow of electrons from both clouds, the tunneling current. Assuming that the two functions (current and voltage) are constant, the gap width (s) would remain also in stabilized mode as the tip is scanned along either X and/or Y axes over an electronically homogeneous surface. The tip will accurately reproduce, as a consequence, the surface topography of a sample by a number of scanning lines. If, however, a heterogeneous surface is investigated, the interpretation of results becomes somewhat complicated. The tunneling current, even under constant voltage, depends on the tip to sample distance and on surface electronic density, degree of decay or electron cloud, of each atomic element. The latter parameter, work function (required energy to remove electrons from a matter), undoubtedly influences stability of the gap width in order to preserve constant tunneling current and, hence, may result in a false topographic replica. Additional functions of the STM were subsequently developed, however, to circumvent such uncertainties in the graphic interpretation of data. Assuming that the tunneling current is kept in stabilized mode, for instance, s-V characteristics are derived from either work function test or scanning tunneling spectroscopy. Surface electron density of sample may also be ascertained at an atomic scale, in addition to surface topographic features at a subatomic resolution, by appropriate combination of tests. Essential capabilities of the STM, therefore, are in providing: 1. topographic reproduction of surfaces where basic variables (applied voltage, tunneling current and tip to sample distance) are kept constant; 2. work function test where the tip to sample distance is modulated; and 3. scanning tunneling spectroscopy where the applied voltage is modulated to keep the tunneling current in stable mode. Combination of these functions, first applied to hard materials, revolutionized the concept of atomic configuration, interatomic topography and atom to atom bounds.

BIOLOGICAL APPLICATION

Since the inception of the STM and its initial application it became evident that this novel technology provides unprecedented capabilities for topographic and compositional investigation of a variety of surfaces. Some functions of the STM were enumerated in previous reports by Binnig and Rohrer (1985b; 1986):

1. Reproduction of 3-dimensional, spatial orientation of surfaces at subatomic resolution in all directions;
2. Besides the field ion microscopy, it is the only technology not requiring lenses, due to utilization of bound particles from its electrodes;
3. Because of a low voltage used (mV and 10^{-4} V/cm range) the procedure is non-invasive and causes no compositional damage to physical structure and/or internal relationship of specimen;
4. Structural and chemical characteristics of periodic and non-periodic elements could be evaluated. Conductors and semiconductors may be studied in natural state, without fixation and/or dehydration; non-conducting samples are, however, first coated in natural state and subsequently investigated;
5. Evaluation of surface work function, electron density, on an atom by atom basis, in addition to topographic characterization of surfaces at an atomic level. Combination of these two criteria, topography and work function, is especially useful for differentiation of molecular subunits; and
6. Operation is also possible, although with some loss in resolution, at ambient pressure and in liquid media because a distance of only ca. 10A from the tip to the specimen (low atomic and molecular density across the tunneling field) is used.

From the very onset of the STM's application in the hard material sciences an attempt was made to assay its usefulness in the biological work as well. However, already initial trials identified several deficiencies in the nature of biological materials for the analyses with this technology: poor conductivity and hence insufficient amount of radiation for tunneling current; prone to deformation during regular STM's operation; large Brownian motion and other thermal fluctuations.The methodology of the STM, to circumvent such inherent disadvantages of biological materials, was modified in several respects. As a consequence, the atomic force microscopy, a branch of the STM, was developed by Binnig at al. (1986). Replicas of natural or freez-fractured surfaces were also found useful substitutes for subsequent investigation of topographic characteristics with the STM (Zasadzinski et al., 1988).

Although the instrument had been originally intended for the operation under vacuum, it was soon recognized that the STM could be successfully performed also at atmospheric pressure (Hansma et al., 1985; Miranda et al., 1986), in oil and liquid nitrogen (Drake et al., 1986) and in water (Sonnenfeld and Hansma, 1986). Furthermore, Gerber et al. (1986) developed an application of the STM in conjunction with the scanning electron microscopy (SEM) in order to facilitate and/or improve: 1. guiding of the STM's probing tip by the SEM's control; 2. controlling the surface deformities of the STM's electrodes; and 3. correlating STM and SEM data. Along with these technical improvements in the general operation of the STM its biological application was concomitantly expanded in recent years. Air-dried bacteriophage particles were first comparatively investigated by transmission electron microscopy (TEM) and the STM (Baro et al., 1985; Baro et al., 1986). Graphical images of the head to tail axis of DNA-containing phage particle, depicted by the STM, accurately corresponded to views obtained by the TEM. Subsequent studies by Travaglini et al. (1987) of air-dried bare samples, and of freez-dried and rotary shadowed samples of recA-DNA complexes demonstrated under atmospheric condition positive images of DNA molecules, filament fragments and the hypothetical placement of recA monomers along a recA-DNA filament. Some indication was also provided by this study that dI/ds test, work function, produced noticeable corrugation over a DNA molecule, or a recA-DNA filament fragment. Amrein et al. (1988) established, later on, on samples coated with a conducting film that recA-DNA complex is a right-hand helical filament of 10 nanometers in diameter. Six recA monomers were observed per helical turn by the STM, whereas the number of monomers was only approximated in the conventional TEM by averaging analysis. A new variation in the STM's technology was more recently developed by Lindsay and Barris (1988) to investigate topographic and electronic (dI/ds) properties of DNA molecules in a liquid medium. Aggregates of DNA molecules, in this instance, are placed on gold surface under a buffer solution in the electrochemical cell by the plating and stripping manipulation. Under this condition, which somewhat resembles an in vitro state, the matter is characterized by the application of either constant-current topography test, or the work function (dI/ds) criteria.

Lipid bilayers, stearic acid, purpur membrane and other biomembranes were also used as a biological model for the STM. The lipid film, consisting of two monomolecular layers of cadmium arachidate, was prepared by Langmuir-Blodget technique and transformed into the graphical mode by the STM. Due to sufficiently high electrical conductivity of fatty acid molecules, the intermolecular spacing of 4.9A (tail to tail) was measured, a value which is in agreement with earlier electron diffraction data (Rabe et al., 1986). Similarly conclusive results were obtained while comparing observations by TEM and STM on the ripple phase of dimyristoylphosphatidylcholine bilayer (Zasadzinski et al., 1988). Platinum-carbon replica of freez-fractured profiles of biomembranes was prepared for the STM in order to avoid thermal fluctuations and structural alterations of intermolecular relationship. The STM was proven once again superior to the other techniques (TEM, SEM, x-ray diffraction) in providing more detailed amplitude, asymmetry and configuration of surfaces of the biological matter.

The most recent assays strongly indicate a possibility of manipulating and/or altering the structure and chemical composition of some biological media directly by a modified STM's procedure (Foster et al., 1988; Pethica, 1988). Di(2-ethylhexyl) phthalate was used in these experiments as the adsorbate layer between the tunneling needle and the single crystal (highly oriented pyrolytic) graphite as the conductive surface. Application of a 100 nanoseconds, 3.7V pulse to the tip of needle results in pinning ("writing") of an adsorbate molecule, or a part of molecule, to the graphite. A similar pulse to the needle scanned over an already pinned molecule, inversely, results in a partial "erasure" of the latter. These processes are assumed to occur when enough energy is generated by tunneling electrons to pin a molecule against the graphite background, in case of "writing", or to cleave already pinned molecule by electric pulse, in case of "erasure".

FUTURE PERSPECTIVE

General properties of the STM inspire a great expectation for the future application of this methodology in biological sciences. This technology is indeed unique in combining a number of unprecedented qualities: subatomic resolution, operation in different media, no requirement for chemical alteration of samples, non-invasive nature, 4-dimensional capability (X, Y and Z axes and electronic state of surface), chemical manipulation of matter and molecular dissection. Clear, much more experimental data are needed to fully assess the facts about either technical capabilities and/or limitations of the STM in the biological work. Although many obstacles along this effort (conductivity, deformation, thermal drift, positioning of sample on background electrode, etc.) are yet to be resolved by researchers from physical and biological disciplines, the results obtained with the STM so far strongly indicate that the biological science may soon be in possession of a powerful tool for topographic and electronic analyses of substrates at a molecular and, even, submolecular levels.

REFERENCES

Amrein M, Stasiak A, Gross H, Stoll E, Travaglini G (1988) Scanning tunneling microscopy of recA-DNA complexes coated with a conducting film. Science 240:514-516.

Baro AM, Miranda R, Alaman J, Garcia N, Binnig G, Rohrer H, Gerber Ch, Carrascosa JL (1985) Determination of surface topography of biological specimens at high resolution by scanning tunneling microscopy. Nature 315:253-254.

Baro AM, Miranda R, Carrascosa JL (1986) Application to biology and technology of the scanning tunneling microscope operated in air at ambient pressure. IBM J Res Develop 30:380-386.

Binnig G, Quate CF, Gerber Ch (1986) Atomic force microscope. Physic Rev Lett 56:930-933.

Binnig G, Rohrer H (1982) Scanning tunneling microscopy. Helvet Physic Acta 55:726-735.

Binnig G, Rohrer H (1985a) The scanning tunneling microscope. Sci Am 253(2):40-46.

Binnig G, Rohrer H (1985b) Scanning tunneling microscopy. Surf Sci 152/153:17-26.

Binnig G, Rohrer H (1986) Scanning tunneling microscopy. IBM J Res Develop 30:355-369.

Drake B, Sonnenfeld J, Schneir J, Hansma PK (1986) Tunneling microscope for operation in air or fluids. Rev Sci Instrum 57(3):441-445.

Foster JS, Frommer JE, Arnett PC (1988) Molecular manipulation using a tunneling microscope. Nature 331:324-326.

Gerber Ch, Binnig G, Fuchs H, Marti O, Rohrer H (1986) Scanning tunneling microscope combined with a scanning electron microscope. Rev Sci Instrum 57(2):221-224.

Hansma PK, Sonnenfeld R, Schneir J, Drake B, Hodzicki J (1985) A scanning tunneling microscope can be operated in air. Bull Am Physic Soc 30:309.

Lindsay SM, Barris B (1988) Imaging deoxyribose nucleic acid molecules on a metal surface under water by scanning tunneling microscopy. J Vac Sci Technol 6(2):544-547.

Miranda R, Garcia N, Baro AM, Garcia R, Pena JL, Rohrer H (1985) Technological applications of scanning tunneling microscopy at atmospheric pressure. App Physic Lett 47(4):367-369.

Pethica JB (1988) Atomic-scale engineering. Nature 331:301.

Rabe J, Gerber Ch, Swalen JD (1986) An image of a lipid bilayer at molecular resolution by scanning tunneling microscopy. Bull Am Physic Soc 31:289.

Sonnenfeld R, Hansma PK (1986) Atomic-resolution microscopy in water. Science 232:211-213.

Travaglini G, Rohrer H, Amrein M, Gross H (1987) Scanning tunneling microscopy on biological matter. Surf Sci 181:380-390.

Zasadzinski JAN, Schneir J, Gurley J, Elings V, Hansma PK (1988) Scanning tunneling microscopy of freez-fracture replicas of biomembranes. Science 239:1013-1014.

Cells and Tissues: A Three-Dimensional Approach by Modern Techniques in Microscopy, pages 597–604

SCANNING ELECTRON MICROSCOPY IN BIOMEDICINE

Karlheinz A. Rosenbauer

Department of Anatomy, University of Düsseldorf, Faculty of Medicine,,
D-4000 Düsseldorf 1, F.R.G.

INTRODUCTION

Over the past three decades biomedicine has been one of the most expanding sciences in medicine. Not only modern techniques in surgery and the construction of new medical devices, but also the development of numerous biomaterials initiated the tremendous progress in this field.

While in 1973 in the United States about 45000 persons were living with artificial heart valves, it is suggested that nowadays this number has increased worldwide up to more than 900000.

The term biomaterials includes all kinds of materials used in biomedicine including materials for dentistry. Biomaterials frequently used are: ceramics, stainless steel, titanium, precious metals, various alloys, pyrolite carbon, silicone rubber, polypropylene, polyurethan, teflon, dacron, silver- and heparin-coated polymers or glass. These materials must be non-toxic, should resist platelet deposition and growth of thrombi, should be compatible with blood and tissues and finally should be virtually indestructible through fatigue and wear.

One has to keep in mind, that for instance the moving element of a prosthetic heart valve, the poppet or occluder, must withstand more than 40 millions cycles per year in vivo.

Besides this, medical devices contacting blood should have smooth surfaces and should be designed not to induce blood stream turbulence.

Scanning electron microscopy is the most useful umethod to qualify smoothness of biomaterials, to detect contamination of surfaces and to discover fatigue and wear in microscopic dimensions.

As in cell and tissue research, the samples which are to be investigated by SEM must have the following characteristics: They must be free of particles, not belonging to the sample and vacuum stable. They must remain stable after exposure to the electron beam and emit a sufficient number of secondary electrons. Finally they should develop as few surface charges as possible.

But, according to the heterogeniety of the samples, preparative techniques in SEM of biomaterials are often quite different from those used for cells and tissues (Rosenbauer et al. 1978, 1981, 1985).

Prior to SEM investigation on biomaterials one has always to test whether the samples are resistant against the solutions used in the different preparation steps, such as ethanol, acetone and DMP for dehydration, HMDS for stabilization prior to air drying or amylacetate as an intermedium in critical point drying.

If biomaterials are not resistant against ethanol, acetone or amylacetate, then the samples can be probably dehydrated with DMP (Maser and Trimble, 1977).Dehydration with acidified DMP is a chemical reaction. Adding water or water containing tissues to DMP, DMP dissociates into methanol and aceton. Since Biomaterials contain usually only small amounts of water, methanol and acetone-concentration is very low and not harmful to the samples. In addition dehydration with DMP requires only a few minutes.

If medical devices should be investigated, it is imperative to compare the used materials with unused ones. It is absolutely essential to coat the native materials with the same defined layer, for instance of gold, also, to avoid false conclusions due to misinterpretation of decoration effects after sputter coating (Rosenbauer and Kegel, 1978).

RESULTS

One of the problems to resolve was why gas-transfer through oxygenator membranes is reduced as a function of time during operation. Studying various oxygenator membranes, we observed that in different types of membranes, the design leads to blood flow turbulences which can be demonstrated particularly well in the spiral membrane of the Kolobow artificial lung.

The turbulences depend on the arrangement of the supporting meshwork stabilizing the oxygenator membrane (Fig. 1).One can see, that settling of platelets and fibrinization starts inside the excavations and grow out to other parts of the membrane. Finally all parts of the membrane are covered with deposits of denatured plasma proteins or fibrin.

Heparin-coated membranes showed slightly better results but up till now there are difficulties in manufacturing membranes properly coated with heparin. SEM reveals that there are numerous manufacturing defects, causing blood clotting also.

In preliminary studies we investigated the surface morphology, the tensile force at breaking point, the quality of manufacturing in different kinds of cava catheters as well as the influence of the design on blood traumatisation (Rosenbauer and Herzer, 1981).

Settling of formed elements of the blood was always related to roughness of the catheter wall caused by the design or by manufacturing defects. Thrombi, grown on the outer surface of the catheters were very rare. In those catheters where radio-opaque material at the head of the catheters is not covered by non radio-opaque material (Fig. 2) thrombus formation is related to the radio-opaque area (Fig. 3).

In these pilot studies we suspected, that most of the thrombi were stripped off during withdrawal of the catheters.

Therefore, we took the heads of the cathers directly out of the opened heart during cardiosurgery(Rosenbauer et al., 1985). By this time the patients were heparinized for 15 to 30 minutes. Out of 60 catheters we saw big thrombi already without any microscope in 66 %. In additional 5 % of the cases we could demonstrate thrombotic occlusions of the catheters. The surface of the majority of the catheters was covered by denaturated plasma proteins and we saw outgrowth of thrombi also starting from the radio-opaque materials.

As in our studies on oxygenator membranes, we also observed adherance of bacteria on the catheter-surfaces. (Rosenbauer et al. 1981). In response to our findings that roughness of the catheter heads is often related to the manufacturing process, one of the manufacturers

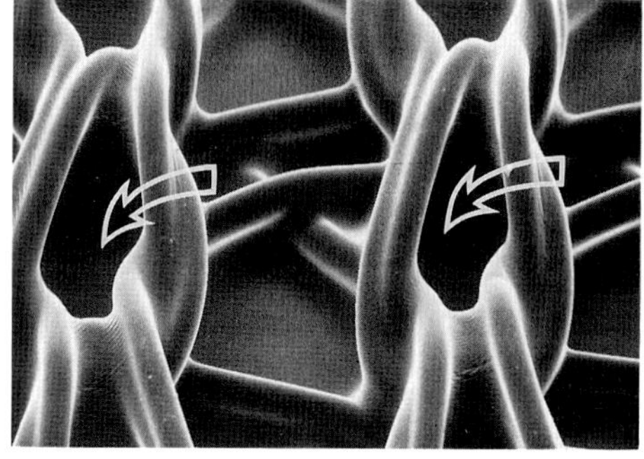

Fig. 1 Kolobow spiral lung. Supporting meshwork. Places where blood stream turbulences occur (arrows). x 70

Fig. 2 Cavafix catheter. Radio-opaque material(rom.) X 1000

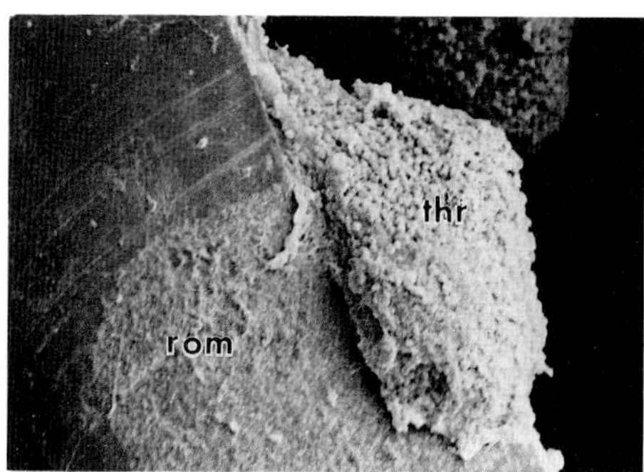

Fig. 3 Cavafix catheter. Radio-opaque material(rom.) Thrombus (thr.) x 140

Fig. 4 Surface of a native CAPD catheter. x 1500

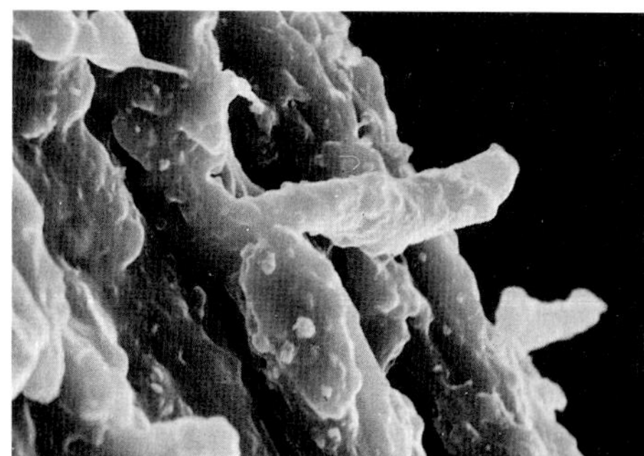

Fig. 5 CAPD catheter. Surface of the silastic disc. x 750

Fig. 6 Starr-Edwards mitral valve. Cracks in the fibrils of the sewing ring. x 560

reformed the manufacturing conditions. After that the catheter-heads became much smoother.

Recently we started to investigate catheters used in continuous ambulatory peritoneal dialysis (CAPD) because some of the surgical complications are canalicular infections and the so called CAPD-peritonitis. Already during the investigation of native catheters of different charges, we observed extreme contamination of the surface which may lead to aseptic inflammations of the peritoneum (Fig. 4 and 5).

Prosthetic heart valve failure is another problem which may be resolved by scanning electron microscopy. Since Rodman in 1974 started SEM research on prosthetic heart valves, obtained at post mortem, we have had opportunities to study specimens obtained in the course of reoperation for valve malfunction. They were in place between one month and ten years and eight months (Rosenbauer et al. 1978, 1980).

In some of the other cases we could demonstrate fatigue of the materials (Fig. 6) as multiple irregular cracks in the fibrils of the sewing ring or defective sewing rings. In other heart valve prostheses we saw manufacturing defects on the cage struts or poppets and microthrombi (Fig. 7) or fibrinization blocking the function of the occluder.

Unused needles for venipuncture often show rough tips or splinters of metal on the surface (Fig. 8). These manufacturing defects as well as rough surfaces of the introducers for catheters may cause endothelial lesions.

For about three years we are working on minimizing particle contamination of injection fluids to reduce particle stress in patients (Rosenbauer, 1986, Rosenbauer et al. 1986, 1988).

As the majority of the particles are splinters of glass we first studied the devices for opening the ampoules. Comparing the surfaces of unused (Fig. 9) and used diamond files we could demonstrate, that after using the files only three times the surface of the files is completely covered by splinters of glass (Fig. 10) which leads to an additional contamination. Settling of splinters on the broken edges of ampoules may cause supplementary contamination also, when during aspiration of the contents these glass particles are stripped off. In addition the content of the ampoules may become contaminated by crystals from the in-

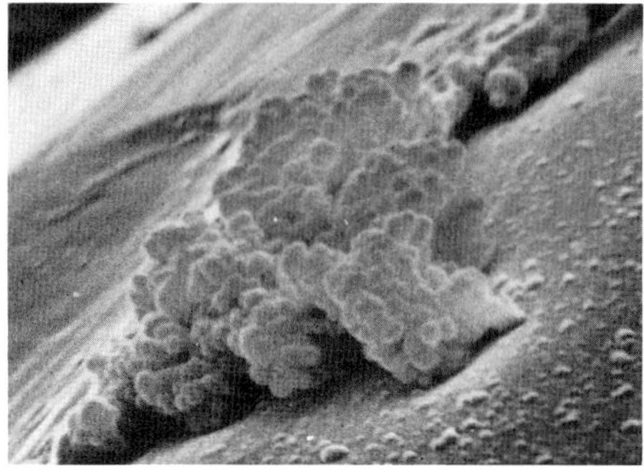

Fig. 7 Lillehei-Kaster mitral valve. Small thrombus, blocking the occluder x 3000

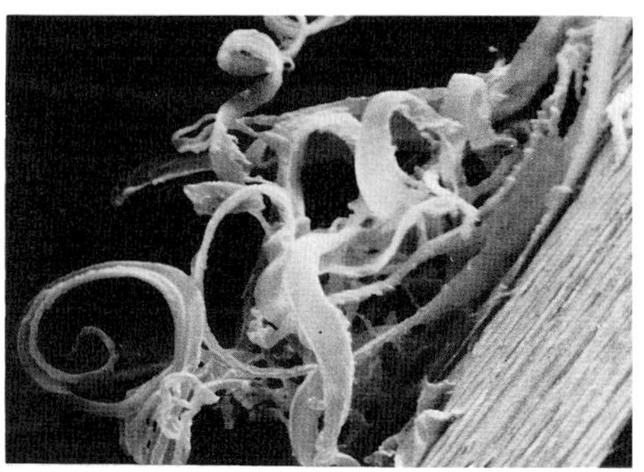

Fig. 8 Tip of a needle for venipuncture with splinters of metal. x 360

Fig. 9 Unused diamond file. x 90

Fig.10 Diamond file used three times, totally contaminated with glass. x 90

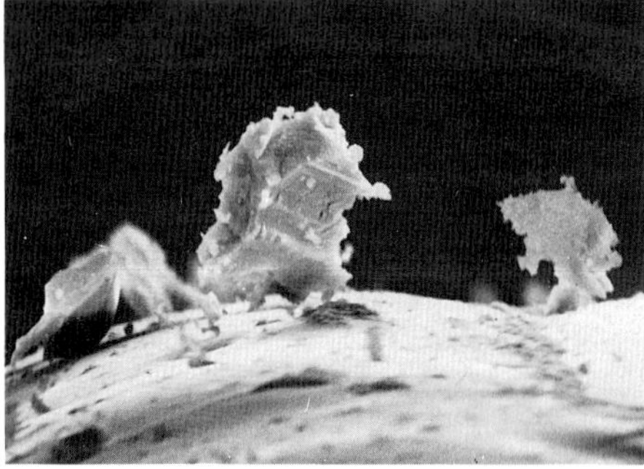

Fig. 11 Glass particles on the edge of an opened ampoule. x 600

Fig.12 Splinters of glass from the content of a 2 ml ampoule. x 560

jection solution, growing on the surface of the broken edges of the ampoules.

We have collected particles aspirated from the contents of 2 ml ampoules on Nuclepore filters with a pore size between 0.2 and 10 µm and found numerous glass particles (Fig. 12), often filiform structures and occasionally keratinocytes. Additionally a variety of particles not identifiable by SEM was observed (Rosenbauer et al. 1988).

In the contents of carpules we detected contaminations also. Filiform structures originate from the package and metal and particles of paint from the metal caps.

REFERENCES

Maser MD, Trimble JJ (1977). Rapid chemical dehydration of biologic samples for scanning electron microscopy. J. Histochem. Cytochem. 25: 247-251.

Rodman NF, Doty DB, Caughey RC (1974). Surface ultrastructural studies of mitral valve prosthetic devices. Scanning Electron Microscopy: 729-736.

Rosenbauer KA (1986). Glassplitter gespritzt: Emboliegefahr? Ärztl. Praxis 38: 377-378.

Rosenbauer KA, Bircks W, Becker U (1985). Rasterelektronenmikroskopische Untersuchungen an Spitzen zentraler Venenkatheter. Klinikarzt 11: 769-774.

Rosenbauer KA, Herzer JA (1980). Further scanning electron microscopic studies of different types of prosthetic heart valves which were in place between two and more than ten years. Scanning Electron Microscopy 1980/III: 219-225.

Rosenbauer KA, Herzer JA (1981). Surface morphology and tensile force at breaking point of different kinds of intravenous catheters before and after usage. Scanning Electron Microscopy 1981/III: 125-130.

Rosenbauer KA, Herzer JA, Kegel BW (1978). Scanning electron microscopic studies of four different types of prosthetic mitral valves which were in place between one and fifty months. Scanning Electron Microscopy 1978/II: 459-463.

Rosenbauer KA, Jansen B (1986). Untersuchungen über die Herkunft von Fremdpartikeln in Injektionslösungen. 1. Licht- und rasterelektronenmikroskopische Befunde an den Oberflächen von Ampullenfeilen. Labor-Medizin 9: 207-210.

Rosenbauer KA, Jansen B (1986). Untersuchungen über Herkunft von Fremdpartikeln in Injektionslösungen. 2. Rasterelektronenmikroskopische Befunde an den

Oberflächen von Halsstücken ungeöffneter und·geöffneter Ampullen. Labor-Medizin 9: 559-562.
Rosenbauer KA, Jansen B (1988). Untersuchungen über die Herkunft von Fremdpartikeln in Injektionslösungen. 3. Licht- und rasterelektronenmikroskopische Befunde an den Oberflächen ungeöffneter und geöffneter O.P.C.-Ampullen sowie deren Inhalt. Labor-Medizin 11: 170-174.
Rosenbauer KA, Kegel BH (1978). "Rasterelektronenmikroskopische Technik. Präparationsverfahren in Medizin und Biologie". Stuttgart: Thieme, pp 139-140.

Cells and Tissues: A Three-Dimensional
Approach by Modern Techniques in Microscopy,
pages 605–621

SCANNING ELECTRON MICROSCOPY IN CLINICS

R. LASCHI, G. PASQUINELLI, P. VERSURA, F. BONVICINI

Institute of Clinical Electron Microscopy,
University of Bologna, Italy
Via Massarenti, 9 - 40138 Bologna

INTRODUCTION

The application of scanning electron microscopy (SEM) in clinic has been underestimated at the beginning. This was mainly due to the general doubts that always are related to the introduction of new technologies in biomedical research. A transitional period in which SEM has been used associated to the other morphological techniques to provide correlative information can now be considered overcome. As far as our personal experience is concerned, at the moment SEM has begun to be an invaluable tool in some selected clinical problems, since it offers intrinsic advantages which will be discussed in detail.

In this paper we would like to point out some fields in which we have achieved a personal experience in the past years. In particular, the cardio-vascular, gastrointestinal and haematological fields will be dealt with, while the ophthalmological one is presented in another paper published in this book. The conclusive part will be devoted to the improvement of tissue processing methods, which appear important in order to obtain more complete information from human specimens.

CARDIOLOGY

SEM investigation of the heart presents some problems related to the fact that myocardium fibers are not naturally accessible to SEM observation. However, the introduction of specific techniques, as cryofracturing and osmium maceration associated to high resolution SEM, is foreseeing new and exciting possibilities in the study of the myocardium biopsy (Siew, 1985; Laschi, 1988; Dalen et al., 1987a, b).

Our experience concerns the physiopathology of myocardium ischemia in the "in vitro" perfused rat heart. In this model fiber structural changes induced either by a short rate of ischemia or by standard procedures of reperfusion can be well studied. SEM observation of conventionally processed specimens showed changes in fiber sizing and in the capillary network as well as interstitial oedema (Fig.1). Waving fibers, foci of necrosis and sarcolemmal foldings related to fiber contraction can be clearly documented (Fig.2). The use of fracturing techniques allows to observe extensive loss of myofilaments, subsarcolemmal accumulation of differently sized mitochondria. Moreover, the osmium maceration associated to high resolution SEM mode provides relevant information on the muscle cell membrane components. In particular the distortion of T-tubules and the leakage of intercalated discs are depicted. The prevention of these injuries by the parenteral administration of drugs linked to the myocardial cell metabolism (L-carnitine) can be also investigated (Figs.3,4), thus suggesting a promising clinical application.

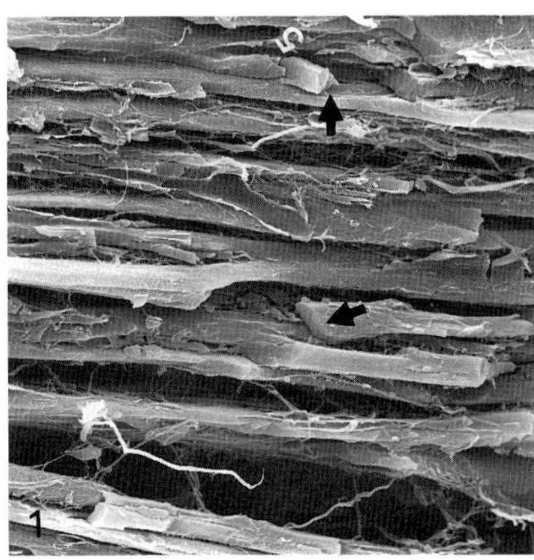

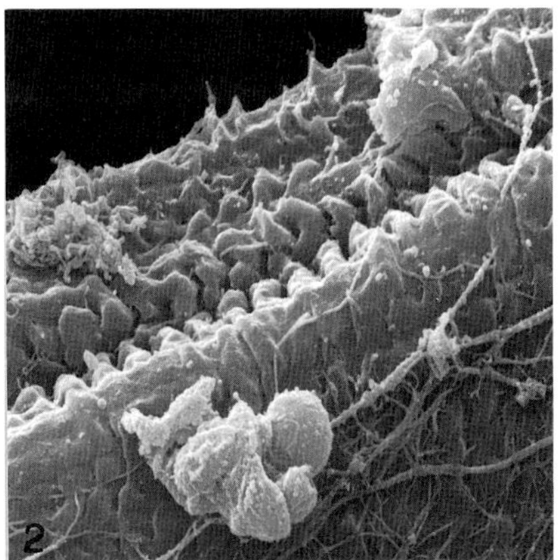

Fig.1 - Rat heart after ischemia. SEM allowed to observe ruptures in fibers (arrows) as well as interstitial edema.
SEM 340 X

Fig.2 - Rat heart after ischemia. At higher magnification, sarcolemmal foldings are evident. SEM 3,500 X

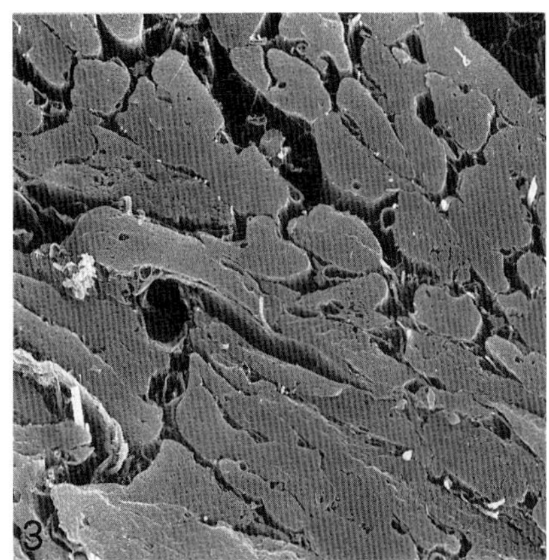

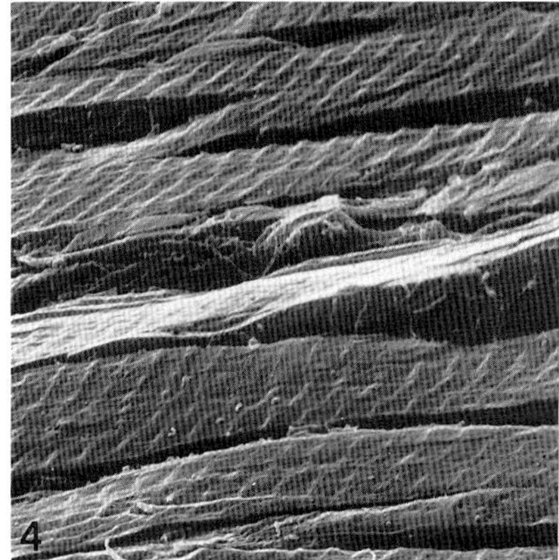

Fig.3 - Effects of reperfusion on ischemic rat heart after L-carnitine treatment. The miofibers retain their normal appearance. SEM 450 X

Fig.4 - Effects of reperfusion on ischemic rat heart after L-carnitine treatment. Cryofractured miofibers show no miofilament loss. SEM 2,100 X

ATHEROSCLEROSIS

For many years we have studied the atherosclerotic disease on human material. The human model presents of course many difficulties and limitations. Anyway, it is well known that deep differences exist as to the onset and evolution of animal and human atherosclerotic lesions, and this may impair the assumption of the proper conclusions (Laschi, 1982).

Relatively few satisfactory ultrastructural studies have been carried out on atherosclerosis in man (for a review see Laschi et al., 1987a). Most of these have been performed on tissue obtained at autopsy. However, the poor preservation of the specimens has made precise investigation of the lesion very difficult. In our opinion, the most recent and relevant contribution appeared in the literature was by Ross and coworkers (1984) who made a systematic analysis on the fibrous plaques from occluded femoral arteries during by-pass surgery. The lesions appeared to be fibroproliferative and contained numerous senescent pancake-shaped smooth muscle cells embedded in lacuna-like spaces at the level of the plaque fibrous cap. Moreover, immunocytochemical analysis performed on advanced plaques showed the presence of complex layers of smooth muscle cells and macrophages with considerable variation from region to region (Gown et al., 1986). It is quite obvious that specimens coming from the surgery room generally refer to the advanced stage of the disease. Therefore the early events are studied by inducing the lesion in experimental models.

There are different methods to generate experimental atherosclerosis. Dietary-induced hypercholesterolemia, ischemic and toxic injury and mechanical removal of endothelium have been proposed. In particular, the dietary-induced atherosclerosis model appear to be the most stimulating since we can assume it resembles what it happens in humans. Faggiotto et al.(1984a, b) described the early changes in the primate aorta and iliac arteries wall leading to fatty streak formation. These lesions appear as focal elevations on the vascular lumen due to the subendothelial accumulation of multiple layers of foam cells. Later on, the fatty streak conversion to fibrous plaque was observed at the same anatomical site. Worthly of mention is that loss of endothelial lining continuity as well as platelet adherence to exposed subendothelial matrix were detected only at this level. The authors stress the importance of these features for the subsequent evolution of the lesion. However, the evidence for conversion of fatty streaks into plaques remains a controversial issue. In fact, it has been claimed that fatty streaks in swine abdominal aorta develop from small intimal cushions of smooth (Thomas and Kim, 1983) muscle cells which are most probably present at birth. As to the endothelium injury, SEM has greatly contributed in establishing that endothelial cells are retained as a confluent monolayer over the developing early lesion in different animal models. In the light of few but relevant contribution (see Hansson et al., 1987) the presence of a functionally damaged endothelium, not yet detectable at the morphological level, should be taken into account. As for platelets, their role in the development and complications of atherosclerosis has been discussed by Packham and Mustard, 1986, and SEM contributed satisfactorily to these studies (Spurlock and Chandler, 1987). New exciting views on the pathogenesis of atherosclerosis arose from the observation of a focal adhesion of circulating leukocytes over the confluent endothelium. The role of monocytes in the onset of the disease was originally suggested by Leary (1941) and, most recently, stressed by numerous other authors (see Laschi et al., 1987a). Focal sticking of monocytes to the endothelium, followed by migration into the intima and by progressive loading with lipidic material, appears to be the sequence leading to fatty streak formation. Finally, modification of the elastic lamina in early

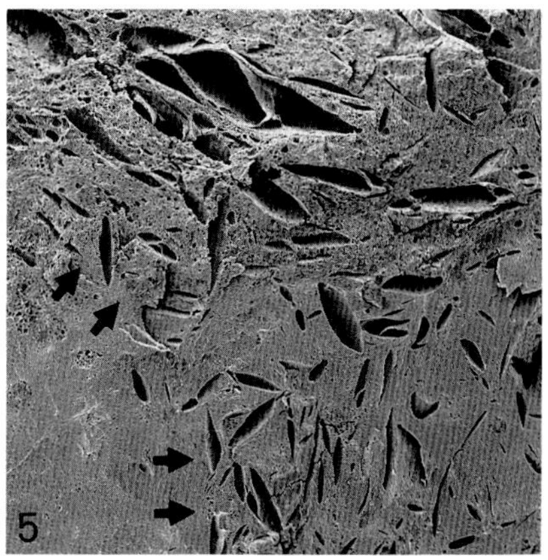

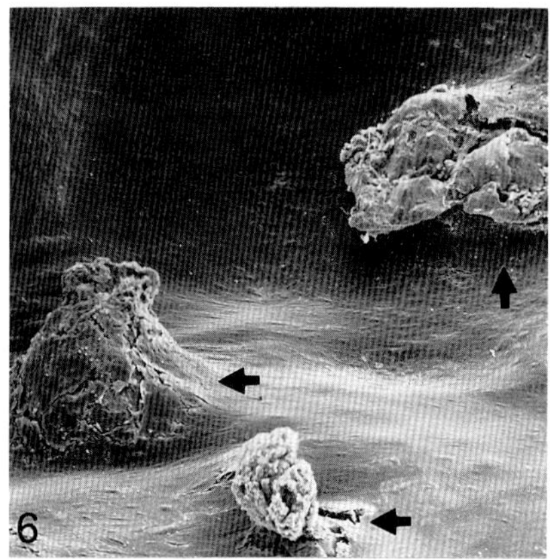

Fig.5 - Atherosclerotic carotid artery. Cryofractured arterial wall exhibits numerous newly formed vessels (arrows) at the periphery of the plaque central core. SEM 110 X

Fig.6 - Atherosclerotic carotid artery. The surface of the plaque presents foci of splitting (arrows) from which ateromatous debris bulge into the lumen. SEM 40 X

atherosclerosis has been finely described by Roach and Song (1988) who interpreted it as a defense mechanism intended to prevent succesive lesions. The ordered arrangement of vascular smooth muscle cells has been made possible by the development of new methodologies such as blunt dissection, digestion and microdissection (Miller et al., 1987).

In these years we have focalized our attention on human ateromatous plaques removed from the internal carotid artery at surgery (Laschi, 1985 a, b). In order to better preserve the fine morphology, the specimens were fixed directly in the operating room. The lesions were mainly fibroproliferative with significant stenosis (70%) associated to different degree of complications such as thrombosis, calcification and intramural hemorrhage. We found that hemorrhage was frequently associated with focal neurologic defects. In these cases SEM observation of cryofractured specimens showed newly formed vessels located at the periphery of the plaque core (Fig.5). By transmission electron microscopy (TEM) it was possible to demonstrate the occasional presence of inflammatory cells. In particular, lymphocytes, mainly of the T-subset, were seen in the capillary lumen. SEM observation showed mural thrombi which had not been

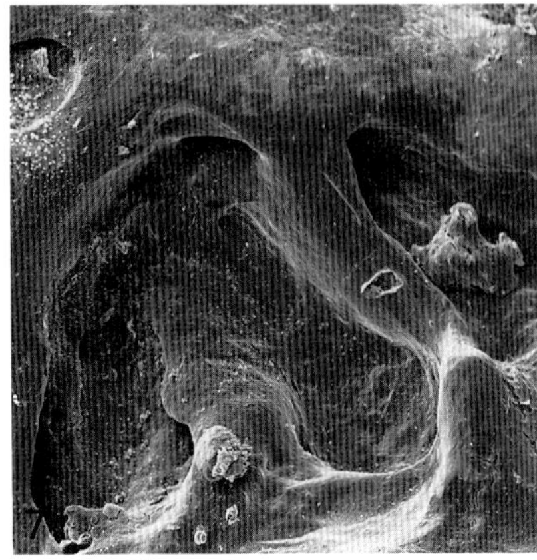

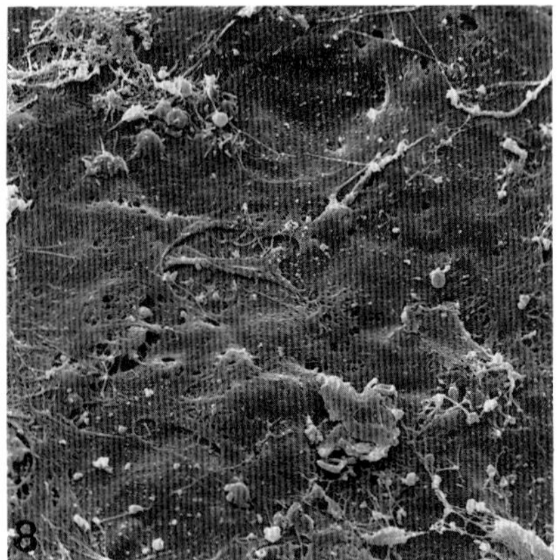

Fig.7 - Atherosclerotic carotid artery. SEM detects an healed ulceration. SEM 20 X

Fig.8 - Atherosclerotic carotid artery. SEM view of denuding endothelial injury. The surface of the plaque is covered by fibrin strands, amorphous material and matrix fibrils. SEM 1,000 X

disclosed at angiography. Moreover, splitting of the surface plaque (Fig.6) as well as recent and healed ulcerations (Fig.7) were observed. These findings well correlate with the presence of transient ischemic attacks in patients without evidence of any gross complications. Patches of calcification along the wall thickness were detected in SEM/backscattered electron (BSE) mode as well as by X-ray microanalysis. The use of digitonin and filipin followed by a controlled osmication localized lipid accumulation in the subsurface side of the plaque and in the basal core. At the core level, TEM showed numerous cholesterol crystals. SEM observation of the luminal side of the plaque revealed wide areas of denuding endothelial injury. The lining was composed of atheromatous debris, fibrin strands, amorphous material as well as matrix fibrils (Fig.8). By TEM, the plaque revealed the classical arrangement in an organoid pattern. The subsurface peripheral zone contained few cells and in particular lacuna-like smooth muscle cells (fibrous cap). Below this area, an highly cellular zone delimitating the basal core is located. The former was composed of multiple layers of smooth muscle cells exhibiting different phenotypes : contractile, synthetic and lypophagic smooth muscle cells. The cells were embedded in

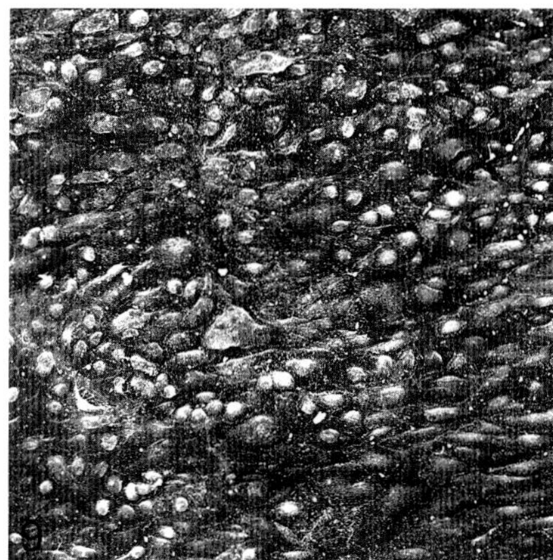

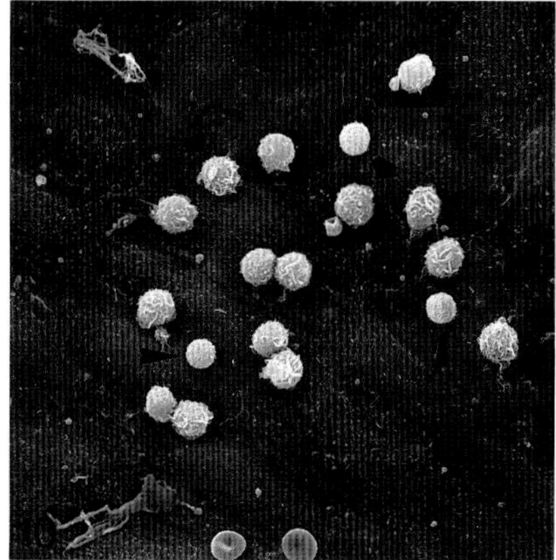

Fig.9 - Atherosclerotic carotid artery. Non-denuding endothelial injury. SEM reveals endothelial cells with loss of polarity and differences in shape and size. SEM 110 X

Fig.10 - Atherosclerotic carotid artery. SEM depicts monocytes (arrows) and lymphocytes (arrowheads) while adhering to the endothelium. SEM 650 X

an aboundant extracellular matrix (fibroproliferative zone) which consists of fragmented elastic fibers, flower-like collagen fibers and thick proteoglycan filaments. The basal pool of the plaque was composed of necrotic cells and lipids and, in some cases, it appeared to be completely calcified (central core). These data, as a whole, strongly support that no possibilities of recover can exist when the plaque reaches such a florid degenerative pattern.

The identification of the early lesions which are still susceptible of regression remains the goal to be achieved. For this reason, we are now focussing our attention on the areas adjacent to the main human carotid plaque, which are supposed to be affected by the disease in an earliest stage (Laschi et al., 1987b, 1987c). In fact the lesions vary quite a lot in appearance, ranging from fatty streak to uncomplicated plaque. SEM observation showed that the endothelial cells cover continuously the arterial wall, although evidence of non-denuding endothelial injury can be always recognized. In particular, endothelial cells show an uneven orientation, focal loss of polarity as well as differences in shape and size (Fig.9). A zonal sticking of mononuclear cells, monocytes and lymphocytes, to the endothelium was a common finding (Fig.10).

By TEM, inflammatory cells were detected just below the endothelial barrier. Monocytes were involved in the uptake of lipid material thus turning into a foam cell phenotype. Lymphocytes characterized by immunohystochemical techniques appeared to be of T-subset. They were arranged as single cell or as clusters sometimes in an indian file appearance. Filipin incubated specimens revealed the presence of a band of vesicles, sometimes with lamellar structures at the periphery. Finally, some modulated smooth muscle cells appeared to take on new phagocytic properties, as suggested by the presence of acid lipase positive vacuoles within their cytoplasm.

HEMATOLOGY

The role of SEM in the study of blood cell disorders has been widely assessed. SEM observations have improved the classification of red blood cell pathology based on alteration in size, shape and surface morphology. In particular, the thalassaemic syndromes as well as the hereditary abnormalities (acanthocytosis, elliptocytosis or stomatocytosis) have been well appreciated by the 3-D investigation. Polliack (1981) claims : "There is no doubt that in respect to red blood cell pathology, SEM contributes a vivid three-dimensional image of the cell deformities...and has led to a better understanding of erithrocytes disorders".

The study of the surface characteristic of leukocytes has, at the beginning, led to the inaccurate differentiation of the different cell types based only upon the arrangement of the membrane specializations. At present, it is evident that the presence of microvilli, microridges or blebs can reflect more the functional state than the origin of the cell. Therefore, conventional SEM morphology as alone is not sufficient to distinguish the various types of leukaemia, other than the hairy cell subtype. However, with the advent of cytochemical and immunocytochemical techniques associated to SEM we are now able to better characterize white series cells. For this purpose, different cytochemical rections, like esterase, peroxidase and toluidine blue, have been used to identify the azurophilic granules, while tantalum was employed as a marker for phagocytosis. The final product of these reactions, containing high Z metals, can be detected in backscattered electron mode (BEI) and in X-ray microanalysis and correlated to the surface morphology of the cell. Soligo et al.(1985) attempted the same approach. In particular, they applied the classical cytochemical reactions (alkaline and acid phosphatase, OsO_4-DAB) to investigate human leukaemic cells. Still, the complete characterization of a leukaemic leukocyte is not fully achieved only on the basis of these cytochemical reactions. The final improvement of these investigation has been carried out by applying immunocytochemical methods and in particular the gold-immunolabelling technique at SEM level. Many monoclonal antibodies used to characterize blood cells (see the complete list in Soligo et al., 1987) can recognize their epitope even after a light prefixation of the cell by 0.1 % glutharaldehyde. This allows to obtain a fine correlation between the pattern of positivity and the cell surface morfology, enhanced when the immunolabelling is visualized by the signal mixing mode in SEM. The more recent step in this technology is the introduction of the silver enhancement for the visualization of very small gold beads (5 nm) not otherwise detectable by the conventional instruments. Double labelling performed by applying antisera coupled with differently sized gold particles can be also carried out. A semi-quantitative evaluation of the immunolabelling has been recently proposed by De Harven and Soligo(1987) to provide cellular labelling index determinations.

The platelets are an interesting cellular model, which has been widely studied in these years by SEM (see for references White, 1987). The cell surface characteristics of resting and activated platelets can be well asessed. In particular, the morphologic alterations that accompany aggregation as well as the various stages of platelet spreading after contact activation are well documented in correlative SEM/TEM studies. These studies have made possible the correlation between shape changes and functional stages. Colloidal gold conjugated-fibrinogen and -fibronectin probes allow to resolve specific receptor sites (Albrecht et al., 1985). Monoclonal antibodies against glycoprotein IIb/IIIa as well as against von Willebrand factor have been successfully utilized. Thus, surface platelet receptor mapping is, at present, possible. Stereo high voltage electron microscopy on whole mount platelets contribute to simultaneously correlate the pattern of distribution of the specific receptor sites with the internal cytoskeleton. In addition, the application of specific preparative procedures visualizes, also in conventional SEM, the platelet cytoskeleton (Zobel, 1988).

Our experience mainly concerns platelet from patients affected by type IIa heterozigous hypercholesterolemia (Laschi et al., 1987b). The platelet activity was indirectly evaluated

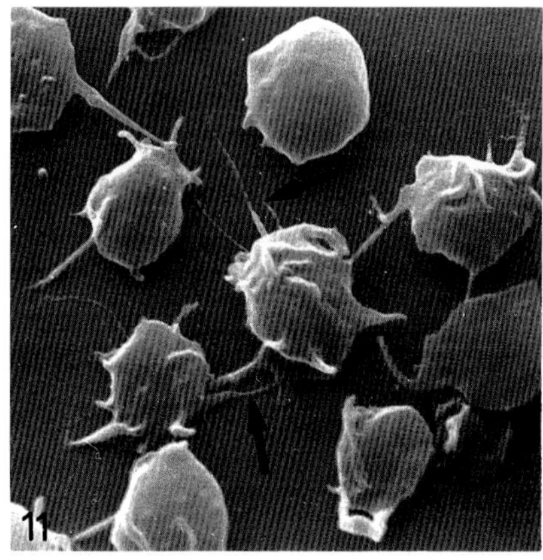

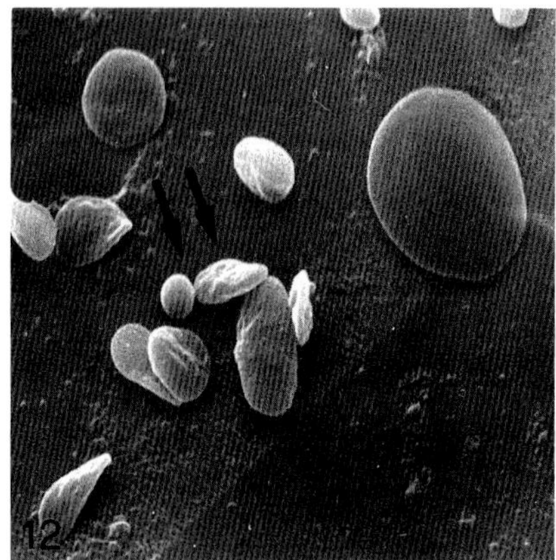

Fig.11 - Platelets from type IIa heterozigous hyper-cholesterolemic patients. Platelets show minimal shape changes. Pseudopodes (arrows) bulge from the cell periphery.
SEM 6,000 X

Fig.12 - Platelets from type IIa heterozigous hyper-cholesterolemic patients. SEM shows small platelets (arrows) in conjunction with normal sized platelets. SEM 3,000 X

by the SEM observation of platelet size and shape. SEM of peripheral blood platelets showed only minimal shape changes, namely the emission of long, thin pseudopodes bulging from the cell periphery (Fig.11). As to the size changes, an increased frequency of small platelets was detected (Fig.12), thus suggesting that hypercholesterolemia may influence the platelet production from megacaryocytes at the bone marrow level.

GASTROENTEROLOGY

Gastritis and peptic ulcer . We applied SEM to the study of some particular aspects of gastritis and peptic ulcer, such as the bacterial colonization, the extension and the healing of the mucosal lesions.

As concerns the first point, SEM has contributed to the knowledge of the morphological features of Campylobacter pylori (CP), a germ which is now considered a major etiological agent of type-B gastritis (Marshall et al., 1984). Campylobacter pylori has a spiral or kidney shaped morphology and bears polar flagella with bulbous tips (Goodwin et al., 1985). It is most frequently found at the level of the intercellular junctions (Fig.13). The distribution of CP on the gastric mucosa is extremely patchy, and the entity of the bacterial colonization

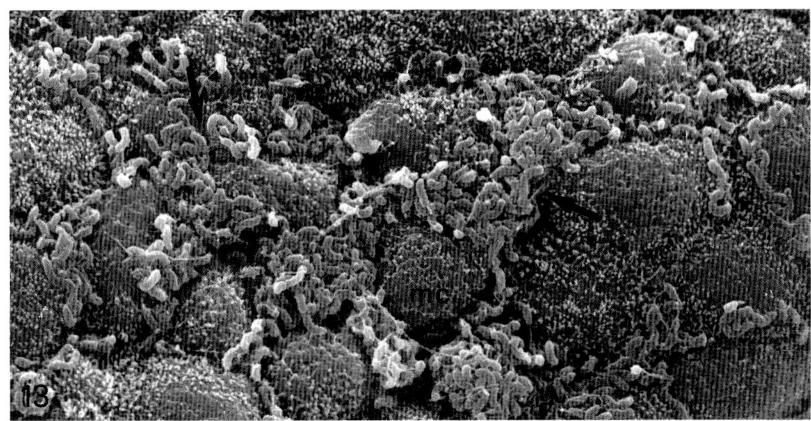

Fig.13 - Gastric mucosa. Campylobacter pylori are localized at the level of intercellular junctions (arrows). mc= surface mucous cells. SEM 2,400 X

varies within the same biopsy. SEM can easily detect even single and sparse bacteria. This accounts for the higher sensitivity of SEM with respect to histology and urease test, as we recently reported (Bonvicini et al., 1988).

As to the second point, the gastroduodenal mucosa far from inflammatory and/or ulcerative lesions may appear normal upon endoscopy and even histology, while SEM can detect minimal alterations of the surface structure such as blebs of the microvilli of the enterocytes in cases of duodenal ulcer. Blebs were also found by us to be the early mucosal lesions and, at the same time, the last to disappear after therapy, despite clinical and endoscopic remission. We also reported the persistence of microvillar blebs as predictive of recurrence.

The real extension of the mucosal lesions as well as the grade of healing are so better defined by SEM which can be considered a valuable tool in patient follow-up (Bonvicini et al., 1985).

The same applies to the gastric and duodenal lesions induced by non-steroidal antiinflammatory drugs and to the efficacy of H2 receptor antagonists in preventing them (Zoli et al., 1986a,b).

Malabsorption syndromes . As far as the small intestine mucosa is concerned, surface structure and function are closely correlated. The villi constitute an extension of the mucosal surface which favours the absorptive functions. Coeliac disease, a gluten-sensitive enteropathy which represents the major cause of primary malabsorption, is characterized, from a histopathological point of view, by a flat mucosal surface due to the loss of the villi associated with a compensatory crypt hypertrophy. The response to gluten-free diet is the most important parameter both for the diagnosis and the management of the patients. Histology has always been considered a reliable technique for assessing the mucosal repair. However, it does not permit to early detect the initial mucosal changes after gluten free diet nor to exactly know the grade of the repair.

In a follow up of 170 coeliac patients we identified by SEM the different stages of the mucosal reconstruction which can be described as follows: - a simple "rising" of the mucosa above the crypt plane, - the formation of convoluted ridges giving the mucosa a cerebriform pattern, - the formation of convoluted villous structures (intermediate mucosa), - the appearance of regularly shaped villi. The same biopsy fragment was utilized both for light microscopy and scanning electron microscopy (Fig.16). The latter after having dewaxed and reprocessed the histological block (Bonvicini et al., 1985).

Inflammatory bowel diseases . Crohn's disease of the colon and ulcerative colitis have in most cases well differentiated clinical, endoscopic and histopathological features.

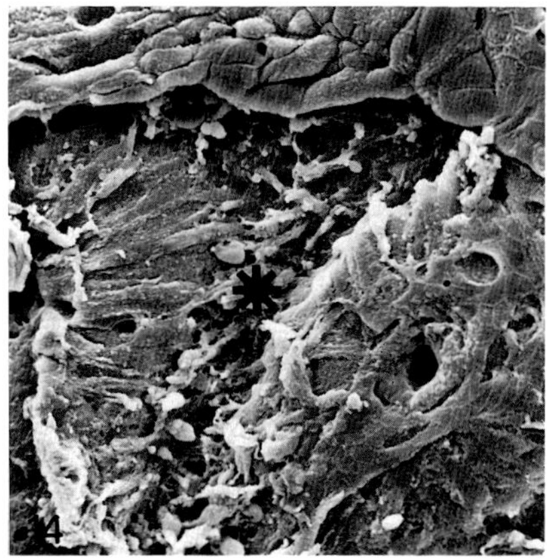

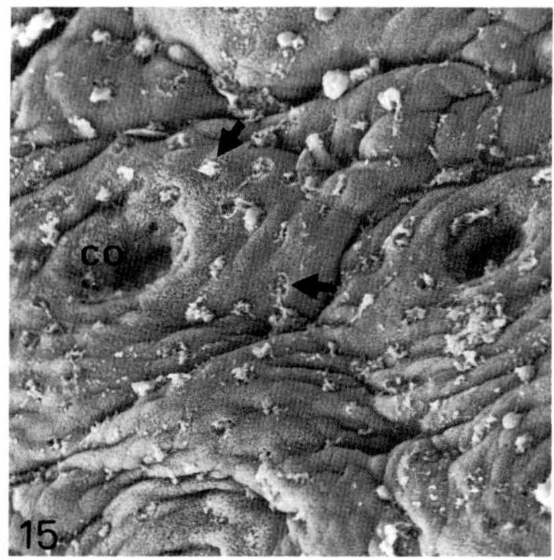

Fig.14 - Colonic mucosa in ulcerative colitis. Loss of the crypt unit pattern. * = loss of epithelial cells. SEM 800 X

Fig.15 - Colonic mucosa in Crohn's disease. Preserved surface mucosal architecture with regular crypt units. Increased number of goblet cells (arrows). co= crypt opening. SEM 800 X

Nevertheless problems in the differential diagnosis are not infrequent due to the fact that the mucosal biopsies available for the histopathological evaluation are small and include only the superficial mucosal layer above the muscolaris mucosae. We were interested in searching by SEM possible mucosal changes which could help in the differentiation of the two diseases (Bonvicini et al., 1985).

The colonic mucosal surface in normal conditions is flat and subdivided into polygonal units outlined by shallow furrows with the opening of a crypt orifice at their center. In active ulcerative colitis the mucosal surface architecture is completely subverted with loss of the crypt unit pattern (Fig.14), disepithelialized areas and decrease in the number of goblet cells. The colonic mucosa of Crohn's disease, even on ulcer edges, has a quite normal surface structure. The number of goblet cells opening on the mucosal surface is increased (Fig.15).

On the basis of these findings we suggest that SEM enhance the diagnostic sensitivity of the endoscopic biopsy.

CORRELATIVE MICROSCOPY

The application of correlative studies on the same human specimen is often a real necessity when the dimensions of the fragment are very small, like in the case of human conjunctival biopsies. In other cases correlative microscopy represents a practical advantage to optimize and extend the information coming from the same sample. Moreover, it allows to perform retrospective studies on histological blocks stocked in archives.

The reprocessing of specimens after SEM observation for LM and TEM by an improved schedule of embedding has been worked out in our lab (Versura and Maltarello, 1987). This common procedure permits to gain a better submicroscopic morphology of the inner structures and therefore correlative information can be obtained.

The possibility of recovering paraffin blocks for SEM observation allows retrospective studies on large series of biopsies stored in the histological archives. Moreover, it permits a fine correlation of the morphology of the transected edge with surface topography of the same specimen. This has been proved to be particularly useful in the study of the pattern of the jejunal mucosa at the beginning of the gluten-free diet in coeliac patients. In some cases, the recovered material may retain some antigenicity (Bonvicini et al., 1986) (Figs.16, 17).

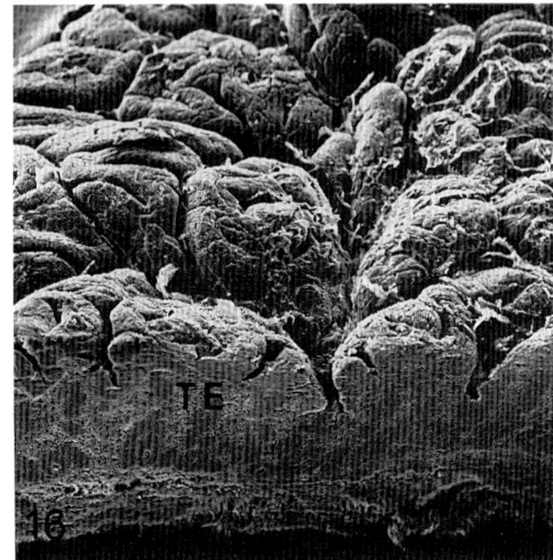

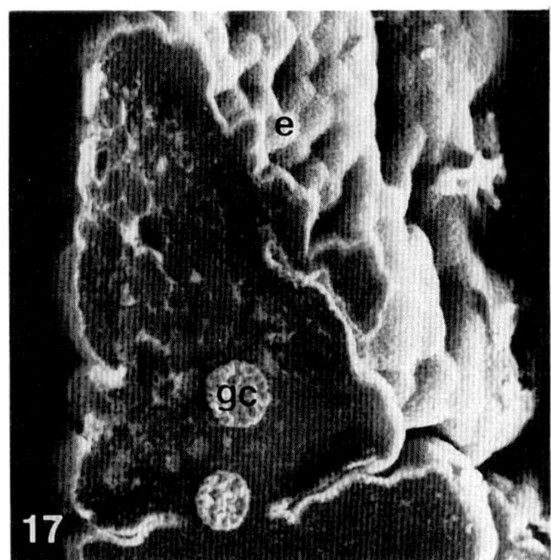

Fig.16 - Small intestine mucosa in coeliac disease after 1 month of gluten free diet. Cerebriform mucosa. Dewaxed histological block. TE= transected edge. SEM 50 X

Fig.17 - Small intestine mucosa. Receptors for Wheat germ agglutinin-colloidal gold are detected on enterocyte (e) luminal surface and in goblet cells (gc). Dewaxed paraffin block. SEM, backscattered mode 1,000 X

Improved techniques of tissue processing have made possible the observation, in sequence, of the same field of semithin sections first by LM and later by SEM. At the same time, cytochemical and immunocytochemical investigation can be performed. In some cases, this mode of SEM observation allows the identification of some morphological features, undisclosed at LM level, useful for diagnostic purposes (Pasquinelli et al., 1985).

REFERENCES

Albrecht RM, Oliver JA, Loftus JC (1985). Observation of colloidal gold labelled platelet surface receptors and the underlaying cytoskeleton using high voltage electron microscopy and scanning electron microscopy. In: Science of Biological Specimen Preparation for Microscopy and Micro analysis. M Muller, RP Becker, A Boyde, JJ Wolosewick eds; SEM,Inc;,AMF O'Hare,Chicago,IL 60666; 185-193.

Bonvicini F, Zoli G, Maltarello MC, Bianchi D, Pasquinelli G, Versura P, Gasbarrini G, Laschi R (1985). Clinical applications of scanning electron microscopy in gastrointestinal diseases. Scanning Electron Microsc. 1985;III:1279-1294.

Bonvicini F, Maltarello MC, Versura P, Bianchi D, Gasbarrini G, Laschi R (1986). Correlative scanning electron microscopy in the study of human gastric mucosa. Scanning Electron Microsc.,1986;II: 687-702.

Bonvicini F, Versura P, Pretolani S, Gasbarrini G, Laschi R (1988). Scanning electron microscopy in the study of Campylobacter pylori associated gastritis. Scanning Microsc., 1988, in press.

Dalen H, Odegarden S, Saetersdal T (1987a). The application of various electron microscopic techniques for ultrastructural characterization of the human papillary heart muscle cell in biopsy material. Virchows Arch A,410:265-279.

Dalen H, Saetersdal T, Odegarden S (1987b). Some ultrastructural features of the myocardial cells in the hypertrophied human papillary muscle. Virchows Arch A,410:281-294.

De Harven E, Soligo D (1987). Immunogold labeling of human leukocytes for scanning electron microscopy and light microscopy: quantitative aspects of the methodology. Scanning Microscopy,1987;II:713-718.

Faggiotto A, Ross R, Harker L (1984a). Studies of hypercholesterolemia in the non human primate. I. Changes that lead to fatty streak formation. Arteriosclerosis,4:323-340.

Faggiotto A, Ross R (1984b). Studies of hypercholesterolemia in the non human primate. II. Fatty streak conversion to fibrous plaque. Arteriosclerosis,4:341-356.

Goodwin CS, McCulloch RK, Armstrong JA, Wee SH (1985). Unusual cellular fatty acids and distinctive ultrastructure in a new spiral bacterium (Campylobacter pylori) from the human gastric mucosa. J Med Microbiol, 19: 257-267.

Gown AM, Tsukada T, Ross R (1986). Human Atherosclerosis. II. Immunocytochemical analysis of the cellular composition of human atherosclerotic lesions. Am J Pathol,125:191-207.

Hansson GK, Bondjers G (1987). Endothelial dysfunction and injury in atherosclerosis. Acta Med Scand,suppl,715:11-17.

Laschi R (1982). Atherosclerosis in man. Atlas of electron microscopy. Ed. Skema/Compositori,Bologna,1-71.

Laschi R (1985a). Studio ultrastrutturale dell'ateroma della carotide nell'uomo. In: Ateroma della carotide e ischemia cerebrale reversibile. M D'Addato, A Stella, L Pedrini eds. Editrice Compositori, Bologna, 31-39.

Laschi R (1985b). Contribution of scanning electron microscopy and associated techniques to the study of atherosclerotic disease. Scanning Electron Microsc. 1985;III:1215-1222.

Laschi R, Pasquinelli G, Versura P (1987a). Scanning electron microscopy application in clinical research. Scanning Microscopy 1987;IV:1771-1795.

Laschi R, Cenacchi G, Pasquinelli G, Preda P, Scala C (1987b). Correlative electron microscopy in the study of atherosclerosis. In:"Atherosclerosis and Cardiovascular Diseases". S Lenzi and GC Descovich eds.,MTP Press Limited, London,331-336.

Laschi R, Pasquinelli G, Preda P, Pileri S, Rivano MT, Stella A, D'Addato M (1987c). Morpho-functional correlation in the study of human atheromatous arterial wall. In: Proc Intern Symp on "Cholesterol Control and Cardiovascular Diseases: Prevention and Therapy";15-18.

Laschi R (1988). L-Carnitine and ischaemia. A morphological atlas of the heart and muscle. Ed. Biblioteca Scientifica/ Fondazione Sigma-Tau,Pomezia,1-117.

Leary T (1941). The genesis of atherosclerosis. Arch Pathol, 32:507-555.

Marshall BJ, Warren JR (1984). Unidentified curved bacilli in the stomach of patients with gastritis and peptic ulceration. Lancet i:1311-1314.

Miller BG, Evan AP, Glenn Bohlen H (1987). Exposure of vascular smooth muscle cells for analysis with the scanning electron microscope. Scanning Microscopy,III:1295-1313.

Packham MA, Mustard JF (1986). The role of platelets in the development and complications of atherosclerosis. Sem in Hematol,23:8-26.

Pasquinelli G, Scala C, Borsetti GP, Martegani F, Laschi R (1985). A new approach for studying semithin sections of human pathological material: intermicroscopic correlation between light microscopy and scanning electron microscopy. Scanning Electron Microsc. 1985;III:1133-1142.

Polliack A (1981). The contribution of scanning electron microscopy in haematology: its role in defining leukocyte and erythrocyte disorders. J Microsc 123:177-187.

Roach MR, Song SH (1988). Arterial elastin as seen with scanning electron microscopy: a review. Scanning Microscopy, in press.

Ross R, Wight TN, Strandness E, Thiele B (1984). Human Atherosclerosis. I. Cell constitution and characteristics of advanced lesions of the superficial femoral artery. Am J Pathol,114:79-93.

Siew S (1985). Scanning electron microscopy of the human myocardium. Scanning Electron Microsc. 1985;III:1295-1304.

Soligo D, de Harven E, Pozzoli E, Nava MT, Polli N, Lambertenghi-Deliliers G (1985). Scanning electron microscope cytochemistry of blood cells. Scanning Electron Microsc. 1985;II:817-825.

Soligo D, Lambertenghi-Deliliers G, de Harven E (1987). Immuno-scanning electron microscopy of normal and leukemic leukocytes labeled with colloidal gold. Scanning Microscopy 1987;II:719-725.

Spurlock BO, Chandler AB (1987). Aderent platelets and surface microthrombi of the human aorta and left coronary artery: a SEM feasibility study. Scanning Microscopy;III:1359-1365.

Thomas WA, Kim DN (1983). Biology of disease. Atherosclerosis as a hyperplastic and/or neoplastic process. Lab Invest, 48:245-255.

Versura P, Maltarello MC (1987). An improved processing method for electron microscopy investigation of conjunctival biopsies. Curr Eye Res,6 (7): 943-946.

White JG (1987). An overview of platelet structural physiology. Scanning Microscopy;IV:1677-1700.

Zobel CR (1988). The platelet cytoskeleton: evidence for its structure from interactions with $ZnCl_2$. J Submicrosc Cytol Pathol, 20 (2):269-275.

Zoli G, Pasquinelli G, Bonvicini F, Gasbarrini G, Laschi R (1986a). S.E.M. study I : gastric and duodenal lesions induced by non-steroidal anti-inflammatory drugs (Aspirin, Piroxicam) in man. Int J Tiss Reac, VIII: 47-53.

Zoli G, Pasquinelli G, Bonvicini F, Gasbarrini G, Laschi R (1986b). S.E.M. study II: protective effect of Ranitidine against gastric and duodenal lesions induced by non-steroidal anti-inflammatory drugs. Int J Tiss Reac, VIII: 71-77.

Cells and Tissues: A Three-Dimensional
Approach by Modern Techniques in Microscopy,
pages 623–628

THREE DIMENSIONAL MICROARCHITECTURE OF ORGANS RECONSTRUCTED ON THE BASIS OF MODERN HISTOLOGICAL OBSERVATION METHODS

R. V. Krstić

Institute of Histology and Embryology
University of Lausanne, Rue du Bugnon 9,
CH-1005 Lausanne, Switzerland

INTRODUCTION

For Malpighi and all other early anatomists self-made drawings were the unique documentation to illustrate their discoveries. Magnificient engravings show their talent. With the development of microphotography the drawing has gradually lost its importance in cytology and microanatomy, and remains actually reduced to more or less schematic sketches. But just at the time when modern observation methods seemed to throw out the drawing from the scientific and didactic literature, it regained importance due to these methods. Indeed modern and often spectacular histological observation techniques - principally transmission and scanning electron microscopy - analyze either the inner structure of cells, tissues, and organs or their spatial arrangement, but without the possibility of depicting both at the same time. Naturally, the integration of these separate information sources into a whole represents a great difficulty to students. However, the solution to this problem is the drawing by hand. It is actually the unique method of obtaining simultaneously precise bi- and three-dimensional microarchitectural image of histological structures in any projection after an analysis and synthesis of data furnished by various modern observation methods. The goal of the present paper is to describe how a three-dimensional artistic reconstruction of a histological structures is prepared for a didactical purpose.

Taking into account that for a successful artistic three-dimensional histological reconstruction the following criteria are required : - scientific accuracy, up-to-dateness, absolute intelligibility, artistic impression, and ease of reproduction, each illustration should be considered as a miniaturized scientific and didactic research study, synthetizing all available data furnished by modern histological observation methods.

MATERIAL AND METHODS

All illustrations are executed with Rotring pens of various sortes using India ink on A4 paper sheets. The preparation of a 3-D reconstruction is as follows : after a first idea, a preliminary sketch is drawn. This is followed by the analysis of my own archive material, and analysis of micrographs published in various atlases (Fujita et al., 1981; Kessel and Kardon, 1979; Motta, 1984; Kurosumi and Fujita, 1975; Rhodin, 1974). In general this is insufficient to prepare a definitive drawing and therefore a study of specialized papers published in periodicals is required. However, experience has shown that this is still insufficient, so it is then frequently necessary to prepare organs for personal light and electron microscopic observations, in order to clarify some imprecisions in the scientific literature. During all these analyses, a sketch is prepared and constantly corrected on the basis of the enumerated criteria. The definitive sketch is then transferred on the new paper sheet and executed in India ink. Through lines the contours are delimited, whereas through stippling (pointillé) with 0,2 or 0,1 points of various densities the shadowing, i.e. the impression of third dimension, is obtained. The technique of stippling has been chosen since it is the closest to a photographic emulsion which gives 3-D impression based on the same principle. Since the stippling technique is a very time-consuming technique, the execution of a drawing lasts at least one week. In case a personal study of some nuclear details is necessary, the executon of a drawing might last several weeks.

RESULTS

In order to honour Malpighi's work on the spleen, the 3D-artistic reconstruction of a splenic sinus (Fig. 1) has been prepared to illustrate the didactic possibilities of such a drawing.

Based on the results of modern observation methods, all components of the sinus have been represented in a cut-away projection: the endothelial cells (EC), reticular cells (RC) with their extensions (RE), ramified reticular fibers (RF) envelopped both in their glycoprotein sheets (G) and extensions of reticular cells (RE). Through openings between endothelial cells pass erythrocytes (E). The internal structure of all enumerated cells has also been depicted. Such a drawing certainly contributes to easier comprehension of histological structures and their spatial arrangement.

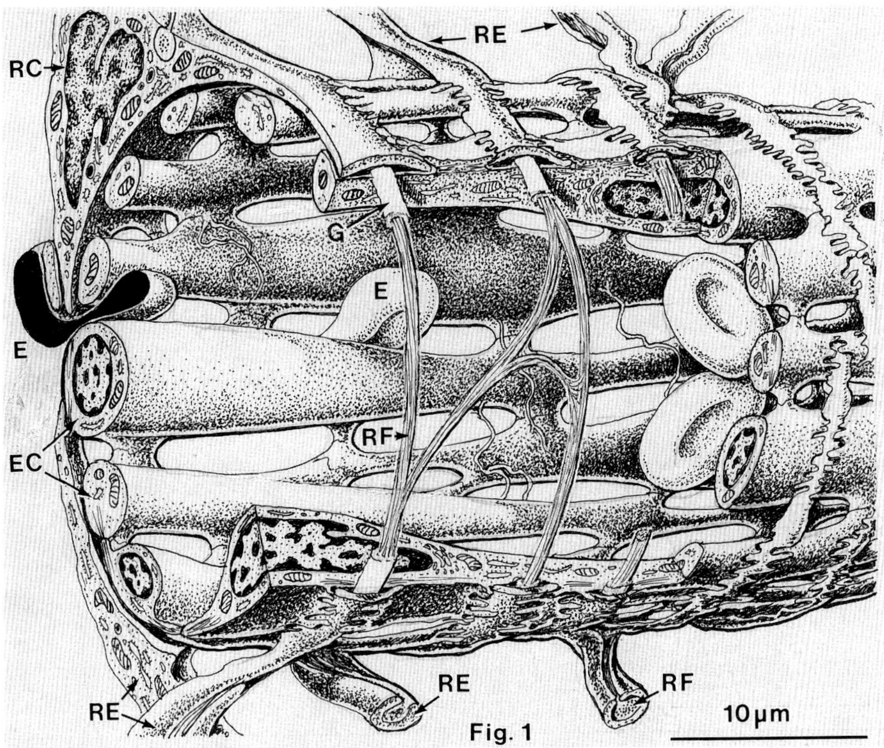

Fig. 1 Splenic sinus. Explanation of a legend in text.

The next two examples illustrate a constant preoccupation of the author to update his drawings.

The Fig. 2 has been taken from an earlier atlas (Krstić, 1985) and represents a segment of cardiac muscle cells, depicted in the cut-away manner, in order to make clear its internal and external configuration. The student can observe the basal lamina (1), the T-tubules, the surrounding collagen and reticular microfibrils (6) and the intercalated disc (5). The sarcoplasmic reticulum (2), dyads (3), Z-lines and glycogen particles (4) are also visible. Taking into account the already-mentioned criteria, this drawing has been updated in function of the recent discovery of Tomita and Ferrans (1987) who observed a channel connecting T-tubules, the so-called tranverse axial tubular system (TAx TS). Thus corrected this drawing is published in the new edition of the same atlas (Krstić, 1988).

For the same edition of this atlas Fig. 3, depicting both the myelinated and unmyelinated nerve fiber, has been completed according to a recent observation of Ushiki and Ide (1986), who described 2 layers of the endoneurial sheet

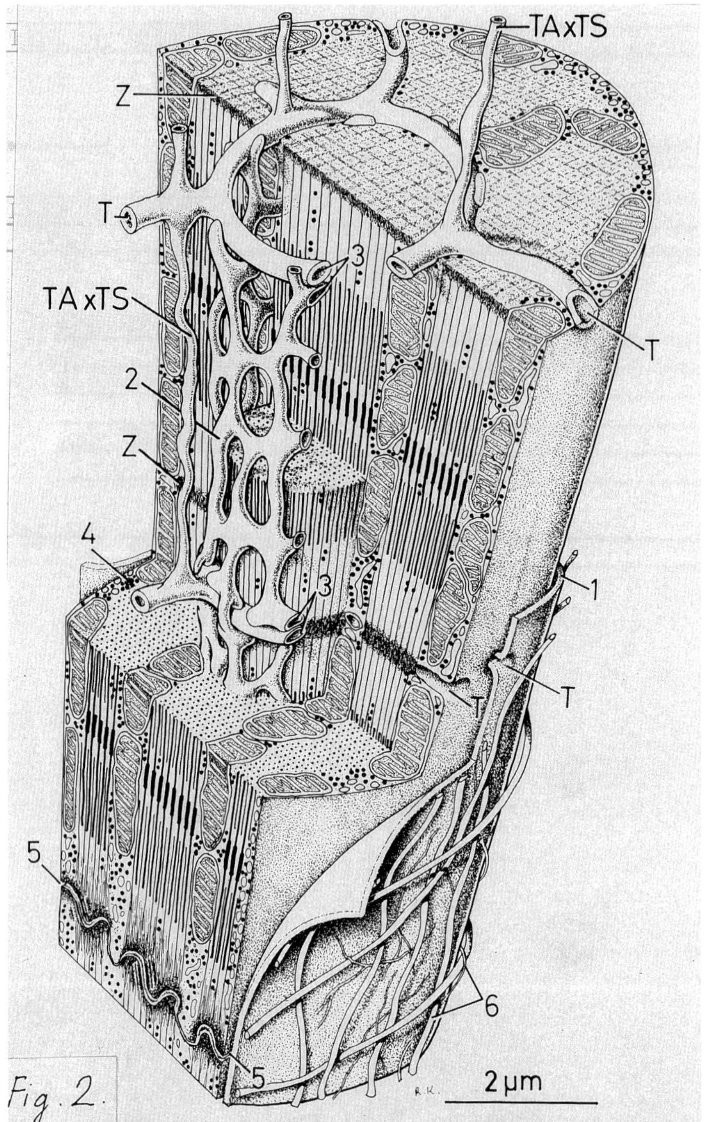

Fig. 2 Cardiac muscle cell. Explanation of a legend in text. From Krstić R (1988), with permission.

(10). Its inner layer consists of a feltwork of collagen microfibrils (10a), whereas the outer layer is composed of longitudinally arranged collagen microfibrils (10b). Otherwise this drawing permits students to analyze the structure of a myelinated nerve fiber (left) with its axon (1), delimited by the axolemma (5) and containing neurotubules (2), smooth endoplasmic reticulum (3) and mitochondria (4),

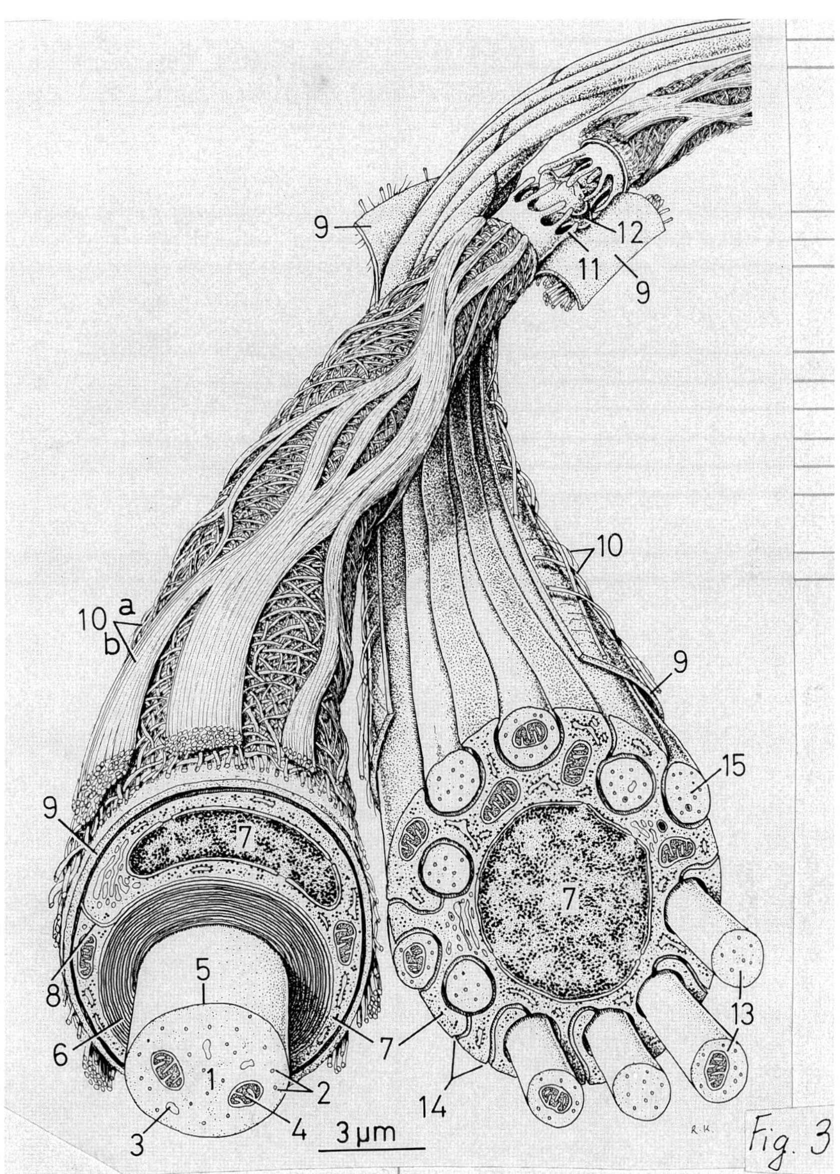

Fig.3 Myelinated and unmyelinated nerve fibers. Explanation of a legend in text. From : Krstić R (1988), with permission.

then the myelin sheet (6), outer mesaxon (8), Schwann cell (7) with its basal lamina (9) and digitiform processes (11). A Ranvier's node is seen between two neighbouring Schwann cells. The structure of an unmyelinated nerve fiber (right) is also visible and can be compared with that of the myelinated fiber. The fact that a Schwann cell(7) in unmyelinated fibers contains several axons (13) - some of them naked (15) - as well as several mesaxons (14), is clearly shown.

CONCLUSION

The given examples have clearly shown that, thanks to modern histological observation methods, a very realistic artistic reconstruction of the inner and outer configuration of cells, tissues and organs is possible. Such a kind of reconstruction facilitates the understanding of histological structures to students, and prepares them to analysise micrographs. Thus, the modern histological observation methods and the drawings are quite compatible and in the future should continue their close relationship.

ACKNOWLEDGMENT

The author wishes to express his gratitude to Mrs D. Nicolas for her excellent technical assistance in preparing specimens for histological analyses.

REFERENCES

Fujita T, Tanaka K, Tokunaga J (1981). "SEM Atlas of Cells and Tissues." Tokyo New York : Igaku-Shoin.
Kessel RG, Kardon RH (1979). "Tissues and Organs : A Text-Atlas of Scanning Electron Microscopy." San Francisco: W.H. Freeman and Company.
Krstić R (1985). "General Histology of the Mammal." Berlin Heidelberg New York Tokyo: Springer-Verlag.
Krstić R (1988). "Die Gewebe des Menschen und der Säugetiere. 2nd ed." Berlin Heidelberg New York Tokyo: Springer-Verlag.
Kurosumi K, Fujita H (1975). "Functional Morphology of Endocrine Glands". Stuttgart: George Thieme Publishers.
Motta, P (1984). "Anatomia microscopica". 3rd ed. Padova: Piccin.
Rhodin JAG (1974). "Histology. A Text and Atlas". New York London Toronto: Oxford University Press.
Tomita Y, Ferrans VJ (1987). Morphological study of the transverse-axial tubular system (TAxTS) in rat heart using ferrocyanide-osmium method and thick sectioning. J Submicrosc Cytol 19: 523-535.
Ushiki T, Ide C (1986). Three-dimensional architecture of the endoneurium with special reference to the collagen fibril arrangement in relation to nerve fibers. Arch Histol Jpn: 553-563.

Cells and Tissues: A Three-Dimensional Approach by Modern Techniques in Microscopy, pages 629–634

COLORED SCANNING ELECTRON MICROSCOPIC PICTURES IN TEACHING ANATOMY

Pietro M. Motta and Stefania A. Nottola

Department of Anatomy, Faculty of Medicine,
University of Rome "La Sapienza",
Via Alfonso Borelli 50, 00161 Rome, Italy

HISTORICAL BACKGROUND. I. THE COOPERATION BETWEEN ANATOMISTS AND ARTISTS AND THE USE OF COLORS IN ANATOMY AND HISTOLOGY

Over the course of ages, the efficacy of a scientific and/or didactic anatomical work was often dependent on a valuable cooperation between anatomists and artists. During the Renaissance, there were numerous examples of such an interchange; just to mention the most famous, Leonardo da Vinci (himself an artist and scientist) decided to produce an anatomy book in collaboration with a celebrated anatomist of his time, Marc'Antonio della Torre. But, unfortunately, Marc'Antonio died and the project could not be completed. Similarly, several years later, Michelangelo Buonarroti, together with Realdo Colombo, a Roman anatomist, also had the idea to write and illustrate a text on human anatomy. But also in this case, the desire became simply unrealizable (Fig. 1). On the contrary, more determined and lucky was Andrea Vesalius, a great anatomist of the Renaissance who, in collaboration with a fantastic artist, Calcar, a celebrated pupil of Tiziano, was able to accomplish the project, a splendidly illustrated book of anatomy entitled "De Humani Corporis Fabrica - Libri Septem". The dazzling anatomical illustrations that followed in the wake of Leonardo and Vesalius grew out of creative efforts in the field of both science and art. The XVI and the XVII centuries saw the creation of artistic anatomical drawings that were as faithful to scientific detail as scientific representations were sensitive to an artistic ideal. Beautiful atlases on human form were published during these years, and scientific illustrators began to take advantage of colors in outlining anatomical parts.

With the advent of light microscopy, artificial coloring of histological samples was introduced, and rapidly became universally used, in order to aid in the recognition and differentiation of tissues, cells and cellular components. Joseph Gerlach (1820-1896) who, in the XIX century, was the first to use a red colorant, carmine, to systematically study stained microscopic sections, can be considered the originator of this "artistic" technique. Indeed, under his monument, in the department of Anatomy at the University of Erlangen, there is an inscription which reads: "Tingendi Arte Innititur Histologia". Gerlach and his predecessors can be considered pioneers of such techniques. For example, Antony van Leeuwenhoeck, in the XVIII century, in the course of his studies on muscles, stained some sections with an alcoholic solution of saffron. Similarly, Alfonso Corti (1822-1876) used carmine to clearly reveal many

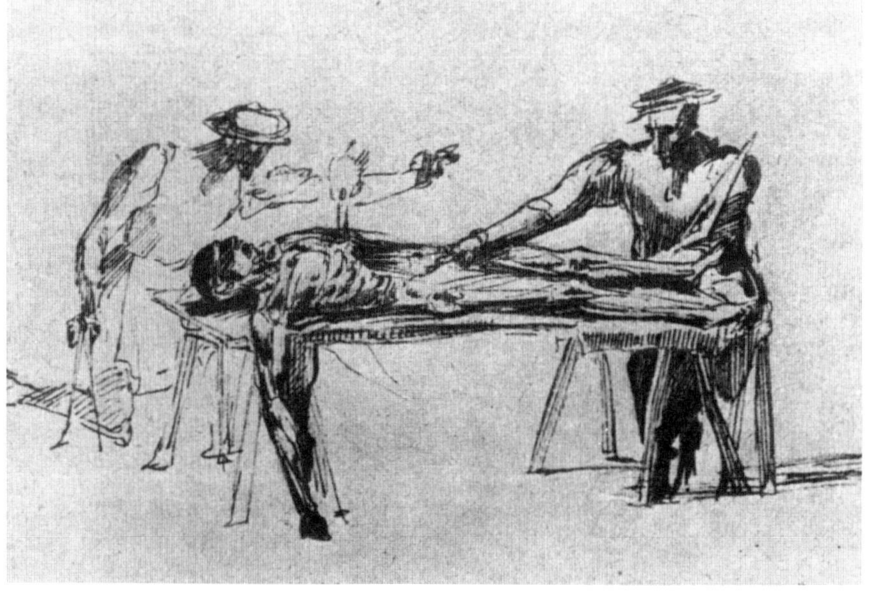

Fig. 1. This pen-sketch drawn by Bartolomeo Passarotti (1529-1592) is from the archives of the University of Oxford, and shows two gentlemen dissecting a cadaver. According to legend, the two dissectors represent Leonardo da Vinci, on the left, and his friend, the anatomist Marc'Antonio della Torre from Verona. According to other opinions, one of them is Michelangelo Buonarroti, and the other is either the Roman anatomist Realdo Colombo, or one of his assistants. Michelangelo and Colombo wanted to create together an illustrated textbook of Anatomy which, unfortunately, was never published.

original histological features of the acoustic organ (Motta, 1986). These techniques rapidly evolved and have become extremely specialized up to the present. This also contributed to the discovery of a fantastic and new microscopic world, revealing essential morphophysiological aspects important to an understanding of the uniqueness of the human body. Consequently, today's students of Anatomy and Histology, as well as biologists and physicians, routinely consider the subjects of their study in terms of colored structures.

HISTORICAL BACKGROUND. II. THE ADVENT AND THE IMPROVEMENT OF ELECTRON MICROSCOPY

The observations of histological sections by light microscopy represented the sole means for studying the structure of biological samples until the 1950s, when the advent of the electron microscope suddenly allowed the observation and description of a great number of details concerning the fine structure of cells and tissues. Nevertheless, neither of these techniques was able to offer a complete three-dimensional view of the tissue architecture itself; so, such spatial reconstructions were based only on the graphic representations attempted by a few ingenious anatomists or on stereological methods and statistical analyses of electron micrographs (Motta, 1984).

More recently, in the late 1960s, the application of scanning electron microscopy to the study of biological materials allowed a direct view of the surfaces of cells and of great portions of tissues, and has permitted us to more rapidly understand and interpret the three-dimensional architecture of tissues, improving on the two-dimensional view originally furnished by light and transmission electron microscopy (Motta et al., 1975). In this vein, in our Institute as well as in many other medical colleges throughout the world, the teaching of scanning electron microscopy has become an integral and fundamental part of the Microanatomy course.

ADVANTAGES OF COLORING SCANNING ELECTRON MICROSCOPIC IMAGES IN TEACHING MICROANATOMY

Having learned from the past, and with a didactic purpose in mind, we have carried out a special artificial coloration of scanning electron microscopic prints (Motta, 1986). Our aim was to render complex microanatomical details and their morphofunctional relationships more comprehensible to the students attending our department of Anatomy. For example, we found this coloration technique useful in clearly demonstrating: 1. the correlation between anatomical and physiological polarity of cells; 2. various plasma membrane structures (cilia, stereocilia, microvilli, blebs etc.) (Fig. 2). Within a given tissue, cells of different morphology, origin and functions can also be more successfully studied (Fig. 2). In addition, the contiguous parts of transitional zones of a particular system (e.g. cardias, pylorus, ileocecal valve) may be easily appreciated if colored in different ways (Motta, 1986; Angela and Motta, 1986).

These colored images, when presented during the microanatomical course in association with traditional light, transmission and scanning electron microscopic pictures, as well as diagrams, are welcomed by the students, improving their understanding and learning capability.

METHODS OF COLORING SCANNING ELECTRON MICROSCOPIC PICTURES

Artificially colored scanning electron microscopic illustrations can be obtained mainly by using photographic filters and other electronic and optical devices, which are also capable of producing spectacular, almost supernatural effects. However, these systems are often very expensive. On the other hand, we believe that a truly artistic result - in which the teacher's task should be to enhance the didactic content while the artist's role should be to successfully render the chromatic effect - can make these images more instructive, attractive and readily interpretable by the students.

Our technique, based on the use of water colors for hand-painting black and white prints, is simple, easily understandable and reproducible. Photo-tinting sheets (Ilford Ltd, London, England; Nicholson's Peerless Transparent Water Colors, Rochester, N.Y., U.S.A) as well as felt-tip pens (Marvy Brush Marker 1500, Japan) were used.

Photo-tinting sheets. These colors are applied in flat washes and are transparent; the detailed highlights and the delicate shadings in the print are thus never obscured, but reflect their own values through the applied color.

In order to prepare the color solution, a piece of tinting sheet should be torn off and immersed in a teaspoonful of cold

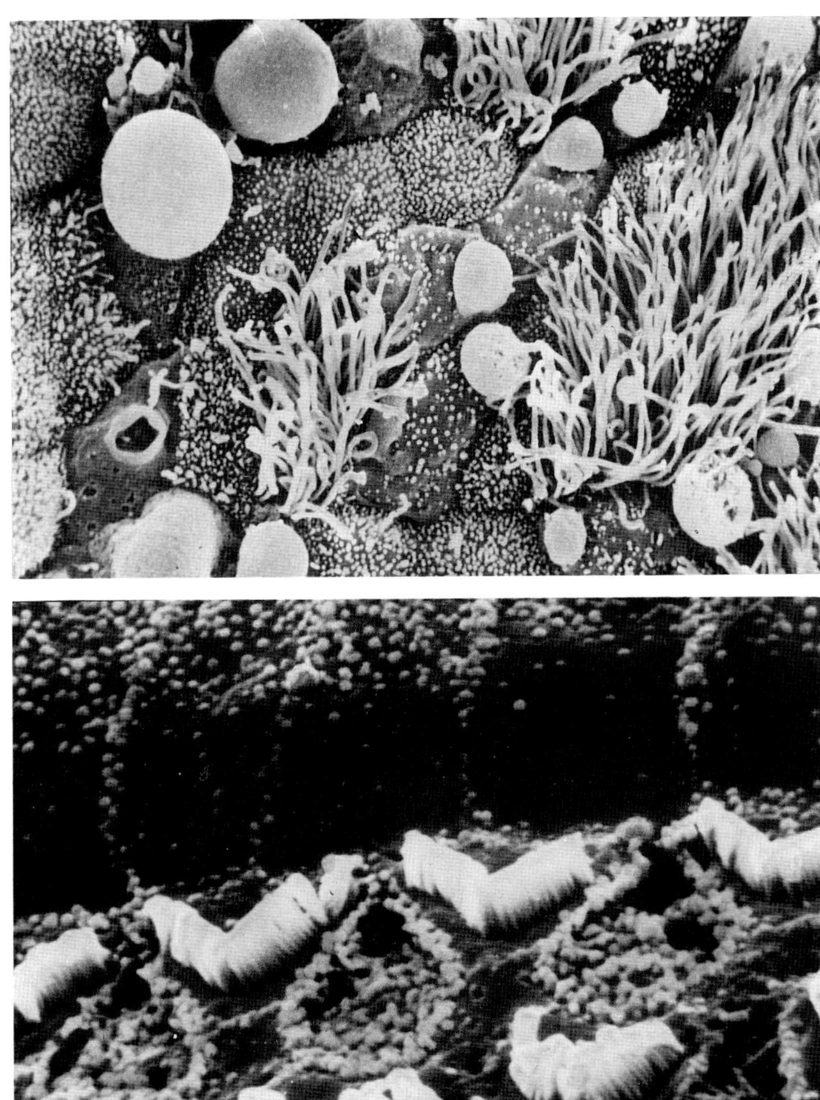

Fig. 2. Top: Endometrial mucosa. Cells with cilia (green) and microvilli (violet) are evident. Numerous secretory granules (yellow) are also noted. (Rabbit. Colored SEM; 2200 X). Bottom: Detailed view of the organ of Corti in the inner ear. Regular rows of outer hair cells (brown-red) possessing typical V-shaped sensory cilia (yellow). These are intermingled with microvillous head plates of inner (blue-green) and outer (blue) pillar cells. (Rat. Colored SEM; 5200 X). SEE NOTE LAST PAGE.

water. The strength of the color solution obtained depends upon the amount of water employed. Two or more colors may be mixed to obtain the desired tint.

The photograph should be printed on resin-coated photographic paper. Before coloring, the print should be washed in cold water. After washing and removing excess water, the print surface should be lightly swabbed with a dilute solution of ammonia; this will better prepare the surface for the application of the colors.

The color should be applied using a soft artist's brush; any excess color can be quickly removed with blotting paper. Color may be totally washed away by immersion in running water and, locally, by the application of dilute acetic acid.

Felt-tip pens. More intense coloring is best obtained either by the application of two or three coats of the above-mentioned diluite solution, or by the use of these pens. In addition, felt-tip pens may be employed for delineating and retouching some particulars on the micrograph.

REFERENCES

Angela P, Motta PM (1986). "Viaggio Nel Corpo Umano". Milano: Garzanti.

Motta PM (1984). The three-dimensional microanatomy of the liver. Arch histol jap 47:1-30.

Motta PM (1986). Teaching scanning electron microscopy in the Anatomical Institute of the University of Rome. Advantages of coloring SEM images. VII International Symposium on Morphological Sciences, Bruxelles. Arch Biol 77:81.

Motta PM, Andrews PM, Porter KR (1975). Scanning electron microscopy of mammalian anatomy. Proceedings of 10th International Congress of Anatomy, Tokyo, p. 364.

NOTE: Original color picture of Fig. 2 will be sent upon reprint request as long as available.

Index

Absorbing peripheral lymphatic vessel, 487–491
Acanthocytosis, 612
Actin, 19
  cytochalasins and, 162–163
  metanephric nephron development, 173–178
  myosin interactions, quick-freeze, deep-etch, rotary-replication technique, 73
  ultra-high resolution SEM, 26
  see also Cytoskeleton
Acute myelogenous leukaemia cells, 377
Acute renal failure, 160–161
Adipose tissue, fetal development of, 281–286
Affinity cytochemistry, lectin, 85–91
Age, bone changes, 121–124
Air vesicles, 9
Alcian blue
  glomerular basement membrane, 167
  proteoglycan quantitation, 115–119
Aldehyde fixation
  microdissection by ultrasonication, 572
  see also specific cell and tissue types
Aldosterone, 197, 411
Alimentary canal
  microdissection by ultrasonication, 573–575
  see also Gastric mucosa
Alpha-atrial natriuretic peptide, 403, 410–411
Alveolar gas exchange, 148–150
Amphibians
  eye, corrosion casts, 527–532
  pinealocytes, 354–356
Anamnia, 353
Annular ligament, 103
ANOVA, 462
Anterior uvea, microcirculation of, 515–518
Antibodies
  quick-freeze, deep-etch, rotary-replication technique, 80
  see also Immunocytochemistry; Monoclonal antibodies
Antigens
  cardiac myocytes, 295–299
  S-antigen-immunoreactive cells, 354, 358
  urothelium, 219
Anurans
  eye, 527–532
  pineal complex, 354, 355, 356
Aorta
  atherosclerosis, 607–608
  cadmium toxicity in pregnant rats, 481–484
  oxytalan, elaunin and elastic fibers, 103–104, 105, 106
Aqueous humor, 506, 511–514
Arachnoid, 390, 391, 392, 393
Arcuate nucleus, tanycytes facing, 364
Artibeus jamacensis, 244
Artificial lung, Kolobow, 598–600
Artistic reconstruction, 623–628
Astrocytic glia, 384
Atherosclerosis, 607–612
ATP, 82
Atrial natriuretic peptide, 403, 410–411
Auerbach plexus, 344, 345
AUM plaques, urothelium, 219
Autoradiography, 214, 264, 266
Axial fibers, alveolar ducts, 148

Backscattered electron imaging (BEI), 214
  atherosclerotic carotid artery, 610
  leukocytes, 612
Bacteriophage, 26, 27
  quick-freeze, deep-etch, rotary-replication technique, 74, 78
  ultra-high resolution SEM, 24, 26, 27
Baglivi, Giorgio, 9, 10
Basement membrane
  glomerular, 167–171
  metanephric nephron development, 173–178
Bats, 244, 246, 247
Bicarbonate secretion, intercalated cells, 201
Biochemistry, cardiac peptides, 410–411
Biomedicine, SEM in, 597–603

Biopsies
  gastric mucosa, 327–331
  see also Clinical application, SEM
Birds, nictitating membrane, 521–525
Blind sacculation model, 181
Blood-brain barrier, 367, 383–386
Blood flow, corrosion cast morphometry, 428
Blood pressure, cardiac hormones and, 414
Blood vessels
  aorta
    atherosclerosis, 607–608
    cadmium toxicity in pregnant rats, 481–484
    oxytalan, elaunin and elastic fibers, 103–104, 105, 106
  atherosclerosis, 607–612
  brain, perivascular glia-endothelial relationships, 383–386
  cardiac hormones and, 402–403
  cochlea, mammalian, 532–534
  endothelium. *See* Endothelium
  eye
    anterior uvea, 515–518
    see also Eye
  and gas exchange
    lung, 147–150
    muscle, 152–154
  kidney
    cardiodilatin and, 412
    glomerular capillary beds, 181–187
  liver and stomach, 317–325
  morphometry, 427
  nasal, 451–456
  pineal, 353, 356
  potassium hydroxide digestion, 469–474
  Purkinje fibers of sheep heart, 423–424
  smooth muscle, 139–144; *see also* Smooth muscle cells
  thyroid, 227–233
  tumor vascularization in nude mice, 475–480
Blood volume, and cardiac hormones, 411, 414
Bone
  age- and radiation-related alterations of, 121–124
  calcification, 109–112
Bone marrow, 494
Bonfiglioli, S., 10
Borelli, Giovanni Alfonso, 8
Brain
  frog, 529
  Malpighi, work of, 4
  microdissection by ultrasonication, 575–576
  pinealocytes, 351–359
  subdural space, 389
Brain natriuretic polypeptide, 403
Butyrate, sodium, 377–382

Calcar, 629
Calcification
  bone changes, age- and radiation-related, 121–124
  matrix vesicles in, 109–112
Calcitonin-gene-related polypeptide, 409–410
Calcium, renal podocyte microfilaments, 162–163
Campylobacter pylori, 327, 329, 331, 615–616
CAPD catheter, 600, 601
Capillaries
  glomerular, 181–187
  lymphatic, dog heart, 435–440
  pulmonary, 148–150
  Purkinje fibers of sheep heart, 423–424
  see also Blood vessels
Capillary loop model, 181, 186
Capillary plexus model, 181, 187
Carbenoxolone, 333, 337, 338, 340
Carbohydrate
  glycoconjugates, glomerular basement membrane, 167
  glycogen synthesis, hepatic, 263–266
  urothelium, 219
  see also Glycoproteins
Cardiac muscle cells, 295–299, 625–626; *see also* Heart
Cardiodilatin, 403
  cleaving enzyme, 404
  immunoreactive material, 405
  immunostaining, 406, 407, 408
  secretory cycle, 410–411
Cardiology, 605–606, 607
Cardionatrin, 403
L-Carnitine, 606, 607
Carotid artery, atherosclerosis, 609–612
Cartilage, proteoglycan quantitation, 115–119
Castelli, Pietro, 8
Cat
  filiform papillae, 305
  fungiform papilla, 305, 306

hypophyseal cleft, 396–398
Catalase, 75
Catecholamines, atrial, 402
Catheters, thrombogenesis, 599, 600, 601
Cationic dyes, 40, 41, 42, 44, 167
Cationic molecules, renal changes, 161–162
Cauchy result, 461
Cavafix catheter, 600, 601
Cell proliferation, gastric mucosa, 309–313
Cell surfaces, nanoanatomy and topochemistry of, 49–56
Central nervous system
cerebellar granule cells, fracture-flip, 55
microdissection by ultrasonication, 575–578
see also Brain; Spinal cord
Cerebrospinal fluid
pinealocytes and, 351
subdural space, 389
Chick embryo aortae, 103
Chief cells. *See* Gastric mucosa
Childhood nephrosis, 159
Cholesterol storage, gallbladder, 269–275
Chondroitin
glomerular basement membrane, 167
proteoglycan quantitation, 115–119
Choriocapillaris, frog, 527, 528, 529
Choroid, 506
Chromatic aberrations, 557–558
Chromatin
freeze fracture, 34
gastric mucosal lesions, 336
Chromosomes
electron microscopy, 59–61
laser microscopy, color, 57–59
optical microscopy, 57
Ciliary body, 506, 507, 515
Ciliary processes
capillaries, 515
vascularization, 516, 517, 518
Ciliary zonule, 102, 105
Cisterna, 19
Clathrin
quick-freeze, deep-etch, rotary-replication technique, 76, 79
rooster comb, 46
Clathrin coated vesicles, 23–24
quick-freeze, deep-etch, rotary-replication technique, 77, 79
tannin-osmium method, 28
Clinical application, SEM, 605–618
atherosclerosis, 607–612
cardiology, 605–606, 607
correlative microscopy, 617–618
gastroenterology, 614–617
hematology, 612–614
ophthalmology, 503–510, 511–514
Clinical conditions
gastric mucosae, 327–331
kidney
acute renal failure, 160–161
nephrotic syndrome, 159–160
Coated pits
limb development, 290–291, 292, 293
quick-freeze, deep-etch, rotary-replication technique, 76
ultra-high resolution SEM, 26
see also Clathrin; Clathrin coated vesicles
Cochlea, rat, 532–534; *see also* Organ of Corti
Coeliac disease, 617–618
Colcemid, 163
Colchicine, 163
Collagen
dura mater, 392
extracellular matrix compounds, 39–48
fibrillar networks
liver, 321
rooster comb, 41
stomach, 325
freeze-etching, 95–99
rat tail tendon, 42
Collagenase, 573
Collecting ducts, 195–201
Colloidal gold, 35
bladder, 214
cardiac myocytes, 295–299
Colloidal iron, proteoglycan quantitation, 115–119
Colombo, Realdo, 629
Colon, 616–617
Color, 557–558
laser microscopy, chromosomes, 57–59
scanning electron microscopic pictures, 629–634
advantages, 631–632
coloring methods, 632–634
history, 629–631
Computer assisted reconstructions
collagen fibril, 97
developing cerebral vessels, 383–386
smooth muscle, 139–144
Concanavalin A
Golgi apparatus, 86, 87

urothelium, 219
Conjunctiva, 505
Connecting tubule, 189–195
Connective tissue, 410
endocrine heart, 409–410
extractable embedding, 39–48
proteoglycan quantitation, 115–119
synovial membrane, 127–130
see also Collagen
Connective tissue papillae, tongue, 303–308
Contractile elements
cardiac myocytes, 295–299
glomerular podocytes, 162–163
see also Muscle; Smooth muscle
Copper binding, proteoglycan quantitation, 115–119
Cornea, 68, 69, 503–505
Correlative studies, clinical applications, 617–618
Corrosion casts, 489
cochlea, 532–534
eye, 517, 518, 527–531
glomerulus, 181–187
microcirculation, 317–325
liver, 318–321
skeletal muscle, 443–449
stomach, 321–325
morphometry, 427–432, 457–460
nasal mucosal blood vessels, 451–456
principles
morphometry, 457–460
projection and thick section stereology, 460–462
statistical experimental design, 462–467
thyroid, 227–233
tumors, in nude mice, 475–480
Cortex, cytoplasmic, 19
Corti, Alfonso, 630
Cortical collecting tubule, 191
Corticosteroids, 199
Critical point drying, 17
Crohn's disease, 616–617
Cross-banding, collagen, 95–98
Cryo-HVEM, 581–586
Cryosubstitution, glomerular basement membrane, 168
Crystalloids, salivary gland protein, 244
Cytochalasins, renal changes, 162–163
Cytochemistry
Golgi apparatus subcompartments, 85–91
see also Histochemistry; Immunocytochemistry
Cytoplasm, 15–19
Cytoplasmic cortex, 19
Cytoprotection, gastric, 333–340
Cytoskeleton
cardiac myocytes, 295–299
limb development, 291, 292
urothelium, 220

Deep-etch replicas, 31–36
cardiac myocytes, 295–299
quick-freeze, deep-etch, rotary-replication technique, 71–83
della Torre, Marc' Antonio, 629, 630
Dense bodies, 15
Dentine, matrix vesicles, 109
De Pulmonibus Observationes Anatomicae, 4, 5, 6
Dermal elastic system, 101
Desmin, 296, 299
Desmodus rotundus, 244
Detergent extraction
extraction procedures, 33, 34, 44, 45
microdissection by ultrasonication, 573
Development, 295–299
adipose tissue, 281–286
brain, third ventricle and hypophyseal cleft, 399
cerebral vessels, perivascular glia-endothelium relationships, 383–386
cultured myocytes, contractile proteins, 295–299
extracellular matrix, 106
kidney, metanephric nephrons, 173–178
limb morphogenesis, 287–293
liver cells, 277–280
De Viscerum Structur, 4
Dexamethasone, 264
Diabetes
and adipogenesis, 285
glycogen breakdown, 264
Dialysis catheters, 600, 602
Dibutyryl-cAMP, 377, 378, 380, 382
Differentiation
gastric mucosa, 303–316
neuroblastoma×glioma hybrids, 377–382
Diffractometry, 559–562
Diffusing capacity, pulmonary, 150
Diffusion studies, matrix structure, 15
Digestion, potassium hydroxide, 469–474
Digitonin, 273, 610
Dimethyl PGE_2, 338
16,16-Dimethyl prostaglandin E_2, 338, 340

Disse, space of, 495
Distal tubules, 198, 208–210, 211–212
Diuresis, actual volume receptors of, 402
DNA-protein interactions
  quick-freeze, deep-etch, rotary-replication technique, 82
  see also Chromatin
Dog
  filiform papillae, 305
  fungiform papilla, 305, 306
  stomach microvasculature, 321–325
Double spiral body, 174
Drawings, 623–628

Ear
  cochlea, corrosion cast, 532–534
  endolymphatic sac, 543–546
  Mauremys caspica, 537–541
Elastic fibers, 45
  atherosclerotic carotid artery, 611
  endocrine heart, 410
  functional significance, 101–106
Elastin layer, aorta, cadmium and, 482, 483
Elaunin, functional significance of, 101–106
Electrolytes
  heart, endocrine function, 401–414
  salivary glands, 243–247
  see also specific ions
Elemental chemical analysis, 214
Elliptocytosis, 612
Embedding media
  polyethylene glycol, 39–48
  see also specific tissues
Embryonic induction, limb morphogenesis, 287–293
Enamel, matrix vesicles, 109
Endocrine heart, 401–444
Endocytosis
  in limb morphogenesis, 287–293
  quick-freeze, deep-etch, rotary-replication technique, 76
  see also Clathrin; Clathrin coated vesicles
Endolymphatic sac, 543–546
Endometrial mucosa, 633
Endoplasmic reticulum, 19, 21
  gastric mucosal lesions, 336
  glycogen metabolism, 264, 266
  thyroid, 228, 229, 230, 231, 232, 233
  ultra-high resolution SEM, 25
  see also specific cells and tissues
Endothelium/endothelial cells
  aorta, cadmium toxicity and, 482
  corneal, 68, 69, 504–505
  fetal liver, 278
  glomerular basement membrane, 169, 170, 171
  liver
    fetal, 280
    Golgi-stained, 258–259
  nasal mucosa, 451, 455, 456
  splenic sinus, 624–625
  thyroid, 228, 229, 230, 231, 232, 233
  trabecular, 505
Endothelium-glia relationships, 383–386
Entactin, 167
Enteric nerve plexuses, 343–347
Enzymatic digestion, microdissection by ultrasonication, 573, 578–579
Epiphyseal cartilage, matrix vesicles, 109, 110
Epithelium/epithelial cells, 328
  bladder. *See* Urothelium
  corneal, 503–504
  gastric mucosa, 309–316, 328
  kidney
    glomerular basement membrane and, 162, 169, 170, 171
    proximal convoluted tubules, 205–206
  limb morphogenesis, 287–293
  tongue, connective tissue papillae and, 303–308
  see also specific cells and tissues
Erythrina cristagalli lectin, 86
Erythrocytes. *See* Red blood cells
Erytoleukaemia, 377
Estrous cycle, and tanycytes, 367
Ethanol, gastric cytoprotection, 333–340
Exercitatio Anatomica De Motu Cordis et Sanguinis In Animalibus, 4, 7
Experimental atherosclerosis, 608
Exposure stacking, 556–557
Extracellular matrix
  atherosclerotic carotid artery, 611
  filamentous structures, 39–48
  oxytalan, elaunin, elastic fibers, 101–106
  renal glomerulus, 167–171
Extractable embedding media, 39–48
Extraction, detergent. *See* Detergent extraction
Eye
  ciliary zonule, 102
  conjunctiva, 505
  cornea, 503–505
  lens, 508, 509

microcirculation, frog, 527–532
nictitating membrane, pigeon, 521–525
retina, 508, 509
sclerocorneal trabecula and glaucoma, 511–514
trabecular meshwork, 505–506
uveal tract, 506, 507
zonular apparatus, 506–508

Fabri, Antonio M., 10
Fascia, muscle, 448
Felt-tip pens, 634
Ferret, 245
Ferritin, 24, 27, 81
Fetal liver cells, 277–280
Fibers, alveolar walls, 148–151
Fibrillar bundles, stomach, 325
Fibrillar sheaths, liver, 321
Fibrils, collagen. *See* Collagen
Fibril structure, collagen, 98–99
Fibrinogen, colloidal gold conjugated, 613
Fibroblasts
dura mater of, 391
rooster comb, 46
3T3 cells, 32–35
ultra-high resolution SEM, 26
Fibronectin, 48, 613
Fibroproliferative zone, atherosclerotic carotid artery, 611
Field emission TEM, 27
Filamentous structures, 19, 39–48
rooster comb, 46
ultra-high resolution SEM, 26
Filaments
types of, 15
urothelium, 220
Filiform papillae, 303–305
Filipin, 610
Filters, 557
Fixation procedures
microdissection by ultrasonication, 572
see also specific cells and tissues
Fluid balance, and cardiac hormones, 411
Foam cells, 272, 273, 274, 275
Foliate papillae, 307
Foot pad, 45
Formaldehyde, 572
Formvar films, 17
Fourier analysis, optical diffractometry, 559–560
Fracture-flip, 49–56
Fracture-permeation, 51
Freeze-cracking apparatus, 23
Freeze-drying, 17, 72, 584–585
Freeze-etching
collagen, 95–99
vs. fracture-flip, 53
Freeze-fracture, 31–36
Freeze-fracture transmission electron-microscopy, 56
Freezing, cryo-HVEM, 581–586
Freezing and thawing, 33, 34, 35
Frogs
eye, 527–532
lung, Malpighi and, 8
pineal complex, 354, 355, 356
Frondose processes, 244
Frontal organ, 354
Frozen-dried cells, 17
Fruit-eating bats, 244
Fungiform papillae, 305–306

Gallbladder, cholesterol storage, 269–275
Ganglia, perikaryal projection, 371–376
Gap junctions, pinealocytes, 357
Gartner, 9
Gas exchange
corrosion cast morphometry, 428
microvasculature and, 151–154
lung, 148–150
muscle, 151–154
Gastric mucosa
cell proliferation, 309–316
cytoprotective agents prior to ethanol exposure, 333–340
gastritis and peptic ulcer, 614–615
human biopsies, 327–331
Gastroduodenal lesions, 329
Gastroenterology, 614–617
Gel, matrix, 17
Gerlach, Joseph, 630
Glaucoma, 511–514, 515–518
Glia-endothelial relationships, 383–386
Glomerulus
cardiodilatin and, 412
metanephric nephron development, 173–178
micromanipulation, 564–566
podocytes, 157–164
vascular architecture, 181–187
GLPS, 339
Glutaraldehyde, 572
Glycerol, and collagen fibrils, 98

Glycoconjugates, glomerular basement membrane, 167
Glycogen synthesis, hepatic, 263–266
Glycoproteins, 613
  glomerular basement membrane, 167
  lectin cytochemistry, 85–91
  salivary glands, 245
  splenic sinus, 624–625
Glycosaminoglycans, 42, 43, 167
Goblet cells, 505
Gold, colloidal. *See* Colloidal gold
Gold-immunolabelling, 613
Golgi apparatus
  lectin cytochemistry, 85–91
  pancreatic acinar cells, 249–255
    continuity of, 251
    identification of, 250
    polarity, 253–254
    position of, 250
    rotation, 254
    structure, 252
    vesicles, 255
    width, 253
  thyroid, 228, 229, 230, 231, 232, 233
Golgi complex, 21
Golgi method, liver sinusoidal wall, 257–262
Granule cells, fracture-flip, 55
Grey level image analysis, corrosion casts, 432
Griffonia simplicifolia, 86, 90
Guinea pig
  filiform papillae, 305
  foliate papillae, 307
  fungiform papilla, 305, 306
Gut
  gastroenterology, 614–617
  microdissection by ultrasonication, 573–575
  see also Gastric mucosa

Habenular ganglion, 357–358
Hair cell, 568–569
Harvey, William, 4, 7, 10
Hearing, phylogenetic scale, 537–541
Heart, 295–299
  capillary length, 153–154
  clinical applications in cardiology, 605–606, 607
  cultured myocytes, contractile proteins, 295–299
  endocrine function, 401–414
    functional role, 412–414
    innervation, microvasculature, and connective tissue, 409–410
    isolation and characterization of hormones, 402–403
    molecular biology, 404–405
    morphology, 405–408
    secretory cycle, 410–411
  lymphatic vessels, dog, 435–440
  muscle cell reconstructions, 625–626
  prosthetic valves, 600, 601, 602
  Purkinje fibers and transitional cells, 419–424
Hedgehog, 246
HeLa cells, 31, 36
Helix pomatia lectin, Golgi apparatus, 86, 90
Hematology, clinical applications of SEM, 612–614
Hematopoietic tissues, 493–498
  bone marrow, 494
  fetal liver, 279, 280
  liver, 493
  lymph nodes and tonsils, 497
  spleen, 494–497
  thymus, 497–498
Henle's loop, 174, 206–208, 210–211
Heparan sulfate, 167
Hepatocytes
  embryos and fetuses, 277–280
  glycogen deposition, 263–266
  mitochondria, 64, 65
  ultra-high resolution SEM, 25
Heterochromatin, gastric mucosal lesions, 336
High-endothelial venules, 497
High-resolution scanning electron microscopy, 31–36
High-voltage electron microscopy (HVEM), 17, 581–586
Histochemistry
  lectin cytochemistry, Golgi apparatus subcompartments, 85–91
  metanephric nephron development, 173–178
  see also Immunocytochemistry; *specific cells and tissues*
Histogenesis
  gastric mucosa, 313–316
  limb morphogenesis, 287–293
  see also Development
Hologram, 561

Hormones
glycogen synthesis, 264
heart, 401–414
function of, 412–414
isolation and characterization, 402–403
molecular biology, 404–405
morphology, 405–408
regulation and transport, 409–410
secretory cycle, 410–411
pinealocytes, 351–359
and renal tubular cells, 192, 197, 199
Humans
filiform papillae, 305
gastric mucosae, 327–331
metanephric nephron development, 173–178
vallate papillae, 306, 307
see also Clinical application, SEM
Hyaloid arteries, 530, 531
Hyaluronan-proteoglycan network, 43
Hyaluronic acid, 167
Hyaluronidase, 573
Hydrogen ion secretion, intercalated cells, 201
Hypercholesterolemia, 608, 613–614
Hypophyseal cleft, 395–399
cat, 396–398
rabbit, 396, 397, 398
rat, 395–396, 397
Hypophyseotrophic area, 365
Hypothalamic releasing hormones, 365

Illustration, 623–628
Image analysis systems, corrosion casts, 430, 431
Immunocytochemistry, 214
cardiac hormones, 405
cardiac myocytes, 295–299
leukemic cells, 613
metanephric nephron development, 173–178
myoendocrine cells, 411
Immunoglobulins, quick-freeze, deep-etch, rotary-replication technique, 80
Index of vascular density, 432
Indoleamines, anuran pineal complex, 355
Inflammatory bowel diseases, 616–617
Infundibular recess, 367
Innervation, endocrine heart, 409–410
Innocent XII, Pope, 10
Intercalated cells, 196, 201
collecting duct, 197, 198, 199, 200
connecting tubules, 190, 192, 193, 194, 195
Intercalated discs, 606
Intercalated ducts, salivary glands, 236–237, 239
Intercellular canaliculus, submandibular gland, 237
Interdigitating cells, 497
Intermediate filaments
cardiac myocytes, 295–299
urothelium, 220
Interphase nucleus, 31–36
Interstitial spaces, origin of cardiac lymphatic vessels in, 436
Interstitial tissue, endocrine heart, 409–410
Intraendothelial channel, lymphatic, 489, 490, 491
Invertebrates, 244
Ion-beam sputter coating, 25, 26, 28
Ion transport, salivary gland, 243, 244
Iris, 506, 507, 515, 516, 517, 518
Iron binding, proteoglycan quantitation, 115–119
Ischemia
heart, 606
renal, 160–161
Isometric contraction, smooth muscle, 133–136
Isotonic contraction, smooth muscle, 136–137
Ito cells, 495

Jamaican fruit bat, 244
Joints, synovia, 127–130
Juxtamedullary layer, glomeruli, 184, 185, 186

Kallikrein, 245
Keratin, proteoglycan quantitation, 115–119
Kerckring, T, 3
Kidney
circulation, cardiodilatin and, 412
collecting duct, 195–201
collecting tubules, organization of, 189–195
glomerular podocytes, 157–164
acute renal failure, 160–161
experimental procedures, 161–164
morphology, 157–159
nephrotin syndrome, 159–160
Malpighi, work of, 4

microdissection, 564–566, 566, 567
microdissection by ultrasonication, 575
structures discovered by Malpighi, 9
vascular architecture, 181–187
Kupffer cells, 277–280

Label-fracture, 51
Lacteal vessel, turtle, 488
Lamina densa, 167, 168
Laminae rarae, fine structure, 167–171
Lamina fibroreticularis, 168
Laminin, glomerular basement membrane, 167
Lancisi, Gian Maria, 10
Laser microscopy, chromosomes, 57–59
Laser trabeculotomy, 512, 513, 514
Lattice
microtrabecular, 19
quick-freeze, deep-etch, rotary-replication technique, 76
Leaf-chinned bat, 247
Lectins
cytochemistry, Golgi apparatus subcompartments, 85–91
intercalated cells, 201
Leeuwenhoeck, Antony van, 630
Leiomyosarcoma, 477, 479
Lens, 506, 509
Lens culinaris lectin, 86, 89
Leonardo da Vinci, 629
Leukaemic cells, 613
Leukocytes, 612–613
Lienal stigmata, 9
Ligand-mediated osmification, 563–570
Ligand-receptor interaction, limb morphogenesis, 287–293
Light cells, 197, 198, 199
Lilium longiflorum, 57, 58
Limb morphogenesis, 287–293
Lipid bilayers, scanning tunneling microscopy, 592–593
Lipoid nephrosis, 159
Little brown bats, 246
Liver
blood vessels
corrosion casts, 318–321
potassium hydroxide digestion, 472, 473
fetal, 277–280
glycogen synthesis, 263–266
hematopoietic tissue, 493
Malpighi, work of, 4
sinusoidal wall in, 257–262
endothelial cells, 258–259
perisinusoidal stellate cells, 259–260
stratification, 261
Liver cells. *See* Hepatocytes
Low-temperature, high-voltage electron microscopy imaging mode (cryo-HVEM), 581–586
Lung, 8
artificial, 598–600
Malpighi, work of, 4, 5, 6, 9
microvasculature, and gas exchange, 147–150
Lymphatic vessels
absorbing vessel, 487–491
liver and stomach, 317–325
liver, 317, 318, 319
stomach, 322–324
myocardial, 435–440
anastomosis, 437, 438
calibre, 436
configuration, 437
lymph nodes, 438–439
origin in interstitium, 437
relations of, 437
vascular walls, 437
structures discovered by Malpighi, 9
Lymph nodes, 9, 497, 498
Lymphocytes, chromosomes, 57–62

Maalox, 340
Macrophages
in atherosclerosis, 607
cholesterol storage, 272, 273, 274, 275
fetal liver, 280
fracture-flip, 54
spleen, 494
supraependymal cells, 399
Malabsorption syndromes, 615–616
Malpighi, Marcello, 3–11
batrachians, 354
foundation of microscopic anatomy, 3–7
life of, 7–11
and lung capillaries, 147–148
Malpighi's glomerular capillary beds, 181–187
Mammalia, 353
Man. *See* Humans
Marginal cells, 395
Massari, Bartolommeo, 7
Massari, Francesca, 7

Matrix, cytoplasmic, 15–19
  rat hepatocyte mitochondria, 65
  see also Cytoskeleton
Matrix, extracellular
  oxytalan, elaunin, elastic fibers, 101–106
  see also Collagen; Connective tissue
Matrix vesicles, calcification, 109–112
Mauremys caspica, hearing, 537–541
Medulloblastomas, 354
Meissner plexus, 344–345, 346
Melanoma, 377, 477
Melatonin, 351, 353
Membranes, cell
  bladder, 216, 217, 218, 219
  clathrin coated structures, 77; *see also* Clathrin coated vesicles
  connecting tubule cell, 197
  fracture-flip, 49–56
  mitochondrial, 66
  organization in limb development, 290–291, 292, 293
  quick-freeze, deep-etch, rotary-replication technique, 76
  renal tubule microvilli, 192
  ultra-high resolution SEM, 26
Membranes, scanning tunneling microscopy, 593–593
Membranes, synovial, 127–130
Membranous impression mounts, dura mater and arachnoid, 390–393
Membranous labyrinth, 192
Mercuric chloride, 161
Mesenteric lymph node, 498
Mesodermal cells, limb morphogenesis, 287–293
Metal coating
  quick-freeze, deep-etch, rotary-replication technique, 71–83
  for ultra-high resolution SEM, 28
Metallic impregnation, bladder, 214
Metanephric nephrons, development of, 173–178
Metaphase chromosomes, 60
Methyl cellulose, 40
N-Methyl-N-nitrosurea (MNU), 221
Mica, 80, 81, 82
Michelangelo Buonarroti, 629, 630
Microcirculation
  cochlea, mammalian, 532–534
  eye
    frog, 527–534
    vascularization of anterior uvea, 515–518
  and gas exchange
    lung, 147–150
    muscle, 151–154
  endocrine heart, 409–410
  potassium hydroxide digestion, 469–474
  skeletal muscle corrosion casts, 443–449
  see also Blood vessels
Microdissection by ultrasonication, 571–579
Microfibrils
  connective tissue, 45
  and elastin, 101
Microfilaments
  cytochalasins and, 162–163
  see also Cytoskeleton
Micromanipulation, 563–570
Micro-PIXE measurements, 116
Microplicae, 200
Microradiography, bone changes, age- and radiation-related, 121–124
Microtrabeculae, 15, 19
Microtubules, 15, 19
  cardiac myocytes, 295–299
  kidney podocytes, 163–164
  tanycytes, 365, 367
  ultra-high resolution SEM, 26
  see also Cytoskeleton
Microvasculature. *See* Blood vessels; Microcirculation
Microvilli, 192
Miles techniques, 460, 461
Mineralocorticoids, 192, 197
Mink, 245
Mitochondria, 19, 21
  oxygen transport, 147–154
  scanning electron microscopy, 63–70
  ultra-high resolution SEM, 25
Mitral valves, prosthetic, 600, 601, 602
Molecular biology of endocrine heart, 404–405
Monoclonal antibodies
  metanephric nephron development, 173–178
  platelet surface receptors, 613
  quick-freeze, deep-etch, rotary-replication technique, 80
Morgagni, Giovanni Battista, 4, 9
Morphogenetic tissue interaction, 287–293
Morphometry, of corrosion casts, 427–432, 457–460
Mouse
  filiform papillae, 303–305
  foliate papillae, 307

fungiform papilla, 305
Mucopolysaccharides, proteoglycan quantitation, 115–119
Mucous neck cells, gastric mucosa, 309–316
Multivariate analysis, 462
Muscle
cardiac myocytes, 295–299
microcirculation
corrosion casts, 443–449
oxygen exchange, 151–154
quick-freeze, deep-etch, rotary-replication technique, 73
smooth, contraction nodes, 133–138
see also Skeletal muscle; Smooth muscle
Mustela vison, 245
Myocytes. *See* Heart
Myoendocrine cells, heart. *See* Heart, endocrine function
Myoepithelial cells, salivary acini, 238–239, 240
Myofibrils, cardiac myocytes, 296–299
Myosin, metanephric nephron development, 173–178
Myotendinous junctions, endocrine heart, 409–410
Myotis lucifugus, 246

Nasal blood vessels, 451–456
Natriuretic peptides, 403, 410–411; *see also* Heart, endocrine
Natural killer (NK) cell, 495
Neoplasia
brain tumors, 354
tumor vascularization, 475–480
urothelium, 220–222
Nephrons
collecting tubules, 189–201
Henle's loop, 206–208, 210–211
micromanipulation, 566, 567
proximal tubule cells, 205–206, 210
Nephrotic syndrome, 159–160
Nervous system
fracture-flip, 55
microdissection by ultrasonication, 575–578
see also Brain; Neurons
Neural tissue, fracture-flip, 55
Neuraminidase, 162
Neuroblastoma×glioma hybrid, 377–382
Neuroendocrine structures, pinealocytes, 351–359
Neurofilament protein, 357
Neurons
enteric plexuses, 343–347
fibers
myelinated and unmyelinated, 625–627
Purkinje fibers of sheep heart, 423, 424
salivary gland function, 247
spinal ganglia, perikaryal projection, 371–376
Neuropeptides, endocrine heart, 409–410
Neurotensin, 409–410
Nictitating membrane, pigeon, 521–525
Nidogen, 167
Norepinephrine model of ischemia, renal, 160–161
North American mink, 245
Nucleus
gastric mucosal lesions, 336
interphase, 31–36
ultra-high resolution SEM, 25
see also specific cells and tissues
Nude mice, human tumors in, 475–480

Ocular hypertension, 515
Open junctions, lymphatic, 490, 491
Ophthalmic artery, 530
Ophthalmology
scanning electron microscopy, 503–510
sclerocorneal trabecula and glaucoma, 511–514
Opsin, 354
Optical correlation, 561
Optical diffraction, 559–562
Optical Fourier transform, 559–560
Optical microscopy, chromosomes, 57
Optic artery, 528
Optic nerve, frog, 528, 530
Organelles, 21
Organic matrix, 109
Organization, cell matrix, 15
Organ of Corti, 566, 568–569, 633
Osmium impregnation, 563–570
Ossification, 109–112
OTOTO-method, 563–570
Oxygen exchange, 147–154
Oxytalan, 101–106

Pancreas
acinar cells, Golgi apparatus, 249–255
blood vessels, potassium hydroxide digestion, 472, 473

Papillary layer, 9
Paraformaldehyde, 48
Paraneurons, 357
Paratendinous tissue, 48
Parenchymal cells, fetal liver, 278
Parietal cells, gastric mucosa, 309-316
Pars convoluta, 208, 209
Pars recta, 208, 209
Passarotti, Bartolomeo, 630
PAS staining
  metanephric nephron development, 173–178
  primitive chief cell, 316
Passive contraction, smooth muscle, 138
Pathophysiology, kidney diseases, 159–164
Peptic ulcer, 614–615
Pericytes, perivascular glia-endothelium relationships, 384
Periendothelial cells, 469
Perikaryon, spinal ganglion neuron, 371–376
Perinuclear matrix, gastric mucosal lesions, 336
Peripheral fibers, alveolar wall, 149
Perisinusoidal stellate cells, 259–260
Perisinusoidal tissue, liver, 472, 493
Peritoneal dialysis catheter, 600, 601
Perivascular cells
  glia-endothelium relationships, in developing cerebral vessels, 383–386
  potassium hydroxide digestion, 469–474
Permeation-chromatography, 81
Photoreceptor
  Mueller cell microvilli, 509
  pineal, 352, 353, 354, 356, 357
Photo-tinting sheets, 632–634
Phylogenesis
  ear morphology, 537–541
  extracellular matrix, 106
Pigeon, nictitating membrane, 521–525
Pigs
  enteric nerve plexuses, 343–347
  dural fibroblasts, 391
  subdural space, 389–393
Pinealocytes
  amphibian, 354–356
  comparative aspects, 351–354
  neuronal features, 356–358
Pineocytomas, 354
Pisum sativum lectin, 86, 89
Pituitary cleft, 395–399
Planimetry, corrosion casts, 430, 431
Plasma membrane. *See* Membranes, cell
Platelets, 613–614
  fracture-flip, 53
  gastric mucosal lesions, 336
Platinum-carbon replica micrograph, freeze-fractured cells, 32
Platinum deposition, quick-freeze, deep-etch, rotary-replication technique, 71–83
Plexus of Meissner, 344–345, 346
Plexus myentericus, 345
Plexus of Schabadasch, 344–345, 346
Podocytes, 157–164, 566–567
Podophyllotoxin, 163
Point counting methods, corrosion casts, 430, 431
Polarity
  intercalated cells, 201
  matrix organization, 15
  pancreatic acinar cells, 250, 253–254
Pollen mother cells, 57, 58
Polycations, and microfilaments, 162–163
Polyethylene glycol, 39–48
Poly-L-lysine, 40, 161–162
Polysomes, 17, 19
Portal vessels, brain, 367, 368
Portasomes, 244
Posttranslational processing, cardiodilatin, 404
Potassium
  fruit bats, 244
  renal tubule microvilli, 192
Potassium hydroxide digestion, 469–474
Pregnancy, cadmium toxicity in, 481–484
Preparation methods
  freeze etching, collagen, 95–99
  glomerular basement membrane, 168
Principal cells, collecting duct, 197, 198, 199, 200
Projection, 460–462
Pronase E, 573
Prostaglandins, 340
Protamine sulfate, renal changes, 161–162
Protein crystals, 75
Protein-rich matrix, 17
Proteins
  contractile
    cardiac myocytes, 295–299
    myosin, metanephric nephron development, 173–178
    see also Actin
  pinealocytes, 357
  quick-freeze, deep-etch, rotary-replication technique, 75, 81, 82

saliva, 244
Protein synthesis, 19
Proteoglycans, 41, 43
atherosclerotic carotid artery, 611
glomerular basement membrane, 167
proton microprobe, 115–119
Proton microprobe, 115–119
Proximal tubules, 205–206
micromanipulation, 566, 567
mitochondria of, 67, 69
Pulmonary capillaries, 148–150
Pulmonary diffusing capacity, 150
Purkinje fibers, myocardium, 419–424
Puromycin aminonucleoside-induced nephrosis (PAN), 159–160
Purple membranes, 592–593

Quantitation, stereology, 457
Quick-freeze, deep-etch, rotary-replication technique, 71–83
bacteriophages, 74, 78
catalase molecules, 75
clathrin, 76, 77
DNA-protein interactions, 82
mica adsorption, 81
muscle, 73
proteins
bacterial Rec A, 82
catalase molecules, 75
permeation chromatograph, 81

Rabbit, subdural space, 389–393
Radiation, bone changes, 121–124
Radioimmunoassay, and cardiac hormones, 411
Rana esculenta, 528
Rana temporaria, 528
Rat
connective tissue
foot pad, 45
tail, 42, 48
dural fibroblasts, 391
filiform papillae, 305
foliate papillae, 307
fungiform papilla, 305
liver, 25
stomach microvasculature, 321–325
subdural space, 389–393
Rayleigh scattering, 557
Rec A protein, 82
Receptor-mediated endocytosis
limb morphogenesis, 287–293
quick-freeze, deep-etch, rotary-replication technique, 76
Red blood cells, 612
fracture-flip, 52
oxygen transport, 147–154
Renal acinus, 9
Renal corpuscle, micromanipulation, 564–566
Renal glomerulus. *See* Glomerulus
Renal ischemia, 160–161
Renal pyramids, 9
Renal tubules, 203–212
micromanipulation, 566, 567
proximal, 205–206, 210
micromanipulation, 566, 567
mitochondria of, 67, 69
three-dimensional structure of cells, 203–212
distal tubules, 208–210, 211–212
Henle's loop, 206–208, 210–211
proximal tubules, 205–206, 210
Resolution, UHS-T1, 27–28
Respiratory gas exchange, corrosion cast morphometry, 428
Rete, Malpighi's, 9
Reticular cells, splenic sinus, 624–625
Reticular fibers, splenic sinus, 624–625
Reticuloendothelial system, 493
Retina, 508, 509
membrana vasculosa retinae, 528, 529
pigment epithelial cells, mitochondria, 64, 66
Retinal S-antigen, 358
Retinoblastomas, 354
Ribosomes, 17, 335
protein synthesis, 19
tannin-osmium method, 28
see also specific cells and tissues
Ricinus communis I lectin, 86, 88
Rod cells, 496
Rooster comb, 41, 43, 44, 45, 46
Rotary replication, 71–83
Rough endoplasmic reticulum, gastric mucosal lesions, 335
Ruthenium hexamine trichloride, 43
Ruthenium red, 167

Safranine O, 167
Salamone, 10
Salivary glands
striated ducts, 243–247
surface microanatomy, 235–240

Sample preparation
quick-freeze, deep-etch, rotary-replication technique, 71–83
see also specific cells and tissues
S-antigen-immunoreactive cells, 354, 358
Sarcoma, 477, 479
Sarcoplasmic reticulum, 625
Sauropsida, 353
S-body, 174
Scanning electron microscopy
aorta, cadmium toxicity in pregnant rats, 481–484
blood vessels
potassium hydroxide digestion, 469–474
vascular corrosion casting method, 181–187; *see also* Corrosion casting
bone changes, 123–124
chromosomes, 61–62
clinical applications. *See* Clinical application, SEM
colored, 629–634
dura mater, 391–393
high-resolution, 31–36
of immunohematopoietic tissues, 493–498
kidney, collecting tubule, 189–195
marginal cells, hypophyseal cleft, 395–396
mitochondria, 63–70
ophthalmology, 503–510
proximal tubule cells, 203–212
spinal ganglion neurons, perikaryal projections, 371–376
synovial membrane, 127–130
thyroid, 227–233
tumor vascularization, 475–480
tunneling, 589–594
ultra-high resolution, 21–29
urinary bladder, 213–222
Scanning tunneling microscopy (STM), 589–594
Schabadasch plexus, 344–345, 346
Schlemm's canal, 505, 511, 515
Sclerocorneal trabecula, 511–514
Secondary electron imaging (SEI), 214
Second Mucus System, 505
Secretory cells
pancreatic acinar cells, 249–255
salivary glands, 235–240
thyroid, 227–233
Sensory ganglia perikarya, 371–376
Septal fibers, alveolar wall, 149
Serotonin, 355
Sialic acid, glomerular epithelium, 162
Signal peptide, cardiodilatin, 404
Silver impregnation, 214
Sinus endothelium, 494
Sinusoids, liver
fetal, 278, 279, 280
perisinusoids, 472
SEM, 493
Skeletal muscle blood vessels, 151–153
corrosion casts, 443–449
potassium hydroxide digestion, 470, 471, 472
Skin, 9
Malpighi, work of, 4
structures discovered by Malpighi, 9
Small intestine, 615–616, 617–618
enteric nerve plexuses, 343–347
microdissection by ultrasonication, 574–575
Smooth endoplasmic reticulum
gastric mucosal lesions, 336
glycogen metabolism, 264, 266
Smooth muscle
computer assisted reconstructions, 139–144
contraction modes, 133–138
eye, 515
glomerulus, 181–187
potassium hydroxide digestion, 470, 471
Smooth muscle cells
aortic media, cadmium toxicity in pregnant rats, 481–484
in atherosclerosis, 607, 610
Sodium
heart, endocrine function, 401–414
renal tubule microvilli, 192
vampire salivary glands, 243, 244
Sodium butyrate, 377–382
Spinal cord
microdissection by ultrasonication, 575, 576, 577
neurons, perikaryal projections, 371–376
Spleen, 4, 9, 494–497, 625
Splenic sinus, 625
Stapes, 103, 105
Statistical experimental design, 462–467
Stearic acid, scanning tunneling microscopy, 592–593
Stellate cells, 259–260, 285
Stereo high-voltage electron microscopy, platelets, 613

Stereology, 457, 460–462
Stereophotogrammetry, corrosion casts, 430, 431
Stereoscopic pairs, stacking of exposures, 556–557
Stomach, 313, 573–575
  ganglionated enteric nerve plexuses, 343–347
  microdissection by ultrasonication, 573–575
  microvascular pattern of, 321–325
  see also Gastric mucosa
Stomatocytosis, 612
Stress fibers, 15, 19
Striated ducts
  cells of, 237, 238
  myoepithelial cells, 239
  protein uptake, 244
  salivary glands, 243–247
Subcommissural organ, 358
Subdural space, 389–393
Substance P, endocrine heart, 409–410
Sucralfate, 337, 340
Sulglycotide, 338, 340
Sulphate groups, proteoglycan quantitation, 115–119
Suncus murinus
  filiform papillae, 303, 304
  fungiform papillae, 305–306
  vallate papillae, 307
Supraependymal cell, 396, 397
Supramarginal cells, hypophyseal cleft, 396, 397
Surfaces
  fracture-flip, 49–56
  salivary glands, 235–240
Synaptophysin, 357
Synovial membrane, SEM and TEM, 127–130
Synthesis
  of cardiodilatin, 405, 406, 407
  of protein, 19

3T3 cells, 32–35
T4 bacteriophage, 74
Tail, rat, 42, 48
Tandem scanning microscope, 551–558
Tannin-osmium method, 28
Tantalum, 612
Tanycytes, 363–368
Tendons, collagen periodic distribution, 96
Thaw-fix technique, 33, 34, 35
Third ventricle
  pineal gland and, 352
  supraependymal cell, 396, 397, 399
  tanycytes, 363–368
Thrombus formation, prosthetic devices and, 601, 602, 603
Thymus, 497–498
Thyroid gland, 227–233
  hyperstimulated, 228, 230
  levothyroxine sodium-treated, 228, 232, 233
  low-iodine treated, 228, 231, 232
  normal, 228, 229
Tight junctions
  perivascular glia-endothelium relationships, 383–386
  tanycytes, 365, 367
Tiziano, 629
Tomkieff equation, 461
Tongue
  connective tissue papillae, 303–308
  Malpighi, work of, 4, 9
Tonsils, 497, 498
Topochemistry, fracture-flip, 49–56
Topography
  bladder, 214, 219, 220
  fracture-flip method, 49–56
Topology, 214
Trabecular meshwork, 19, 505–506, 511–514
Trabecular spaces, 19, 515
Trabeculotomy, 512, 513, 514
Transitional cells, myocardium, 419–424
Transmission electron microscopy
  aorta, cadmium toxicity in pregnant rats, 481–484
  chromosomes, 59
  filamentous structures, 39–48
  marginal cells, hypophyseal cleft, 395–396
  spinal ganglia, 373–374
  synovial membrane, 127–130
  tanycytes, 363–365
  tumor vascularization, 475–480
Triskelia, 79
Triton, 33, 44, 45
Trypsin, 573
T-tubules, 606
Tubular cristae, hepatocyte mitochondria with, 64, 65
Tubular fibrils, glomerular basement membrane, 167
Tubules. *See* Renal tubule

Tubulin, 19, 295–299
Tumors
vascularization, 475–480
see also Neoplasia
Tunneling microscopy, scanning (STM), 589–594
Two phase system, matrix, 17

Ulcerative colitis, 616–617
Ulex europeus I, 86, 90
Ultrafiltration coefficient, glomerular wall, 163
Ultra-high resolution SEM, 21–29
Ultrasound, microdissection by, 571–579
Univariate statistics, 462
Unvertebrate aortae, 104, 106
Urinary bladder
methodology, SEM, 213–215
neoplasia, 220–222
normal urothelium, 215–222
Uvea, 506, 507, 515–518

Vaccinia virus, 24, 26, 27, 26
Vallate papillae, 306, 307
Valsalva, Antonio Maria, 4, 9
Valves, prosthetic, 600, 601, 602
Vampire bats, 243, 244, 246
Variability, statistical analysis, 462
Vascular catheters, 600, 601
Vascular corrosion casting-scanning electron microscopy (SEM). *See* Corrosion casts
Vascular smooth muscle. *See* Smooth muscle; Smooth muscle cells
Vascular walls. *See* Blood vessels; Endothelium/endothelial cells
Vas efferens, 412, 414
Vasoactive intestinal polypeptide, endocrine heart, 409–410
Vasodilatation, cardiodilatin and, 412
Vasopressin, 414
Ventricles, brain
pineal gland and, 352
tanycytes, 363–368
Vesalius, 3, 629
Vesicles
clathrin-coated, 23–24; *see also* Clathrin coated vesicles
Golgi, 255
Vicia villosa lectin, 86
Vimentin, 299
Vinblastine, 163–164
Viruses
quick-freeze, deep-etch, rotary-replication technique, 78
ultra-high resolution SEM, 24, 26, 27
Volume changes, and cardiac hormones, 411

Water-rich matrix, 17, 19
Wheat germ agglutinin-colloidal gold, small intestine mucosa, 618
White blood cells, 612–613
Whole cell preparations, 17

X,Y,Z hypothesis, 504
Xenotransplanted human tumors, on nude mice, 475–480
X-ray microanalysis, 214
atherosclerotic carotid artery, 610
leukocytes, 612

Yag laser, 512, 513, 514

Zonular apparatus, 506–508